Springer Monographs in Mathematics

This series publishes advanced monographs giving well-written presentations of the "state-of-the-art" in fields of mathematical research that have acquired the maturity needed for such a treatment. They are sufficiently self-contained to be accessible to more than just the intimate specialists of the subject, and sufficiently comprehensive to remain valuable references for many years. Besides the current state of knowledge in its field, an SMM volume should ideally describe its relevance to and interaction with neighbouring fields of mathematics, and give pointers to future directions of research.

Mohamed El Kadiri · Bent Fuglede

Classical Fine Potential Theory

Springer

Mohamed El Kadiri
Faculté des Sciences
Mohammed V University in Rabat
Rabat, Morocco

Bent Fuglede
Department of Mathematical Sciences
University of Copenhagen
Copenhagen, Denmark

ISSN 1439-7382 ISSN 2196-9922 (electronic)
Springer Monographs in Mathematics
ISBN 978-981-96-0429-6 ISBN 978-981-96-0430-2 (eBook)
https://doi.org/10.1007/978-981-96-0430-2

Mathematics Subject Classification: 31B05, 31B10, 31C40, 31B25, 32A10, 31B15

This Springer imprint is published by the registered company Springer Nature Singapore Pte Ltd.
The registered company address is: 152 Beach Road, #21-01/04 Gateway East, Singapore 189721, Singapore

To the memory of my parents Ahmed and Fatna,
To my wife Malika,
To my daughters: Asmae, Soumaya, Nadia, Meryem, and Zineb.
To the memory of my co-author Bent Fuglede,
To Dorthea and Einar.
Mohamed El Kadiri

Bent Fuglede (8.10.1925–7.12.2023)

Preface

By classical potential theory, we mean potential theory connected to the Laplace operator for open sets in Euclidean space. By contrast, modern potential theory deals with such things as harmonic spaces (axiomatic theories), potentials on Riemannian manifolds, probabilistic potential theory, potential theory of Markov process, and nonlinear potential theory or complex potential theory (pluripotential theory).

The present book deals with the fine potential theory issued from classical potential theory and founded, in the axiomatic frame, by Fuglede (see [Fu8]) in the beginning of the 1970s. The word "fine" in the title of this book refers to the *fine topology* on $\mathbb{R}^N$, introduced by Henri Cartan in a letter to Marcel Brelot in December 1939. Brelot introduced and studied the concept of a set $E \subset \mathbb{R}^N$ being *thin* at a point $x \in \mathbb{R}^N$.[1] The typical example of thinness of a set E at a point x was $E = \mathbb{R}^N \setminus U$, where U is a domain in $\mathbb{R}^N$ and x is an irregular boundary point for U in the sense of the Dirichlet problem. Cartan noticed that a set $E \subset \mathbb{R}^N$ is thin at point $x \notin E$ if and only if $\mathbb{R}^N \setminus E$ is a neighborhood of x in a certain topology, namely the smallest topology on $\mathbb{R}^N$ in which all subharmonic functions are continuous. This new topology is strictly finer than the Euclidean topology on $\mathbb{R}^N$, and Cartan called it the *fine topology*.

Doob has pointed out in [Do4] that for several years after its appearance, the fine topology was viewed "merely as a tool for phrasing results elegantly". However, it has gradually become a powerful instrument in potential theory, and many results are better understood when formulated in terms of the fine topology.

The fine topology was extended to axiomatic potential theory by Brelot, Doob, and Bauer (harmonic spaces) and later by Boboc, Constantinescu, and Cornea (H-cones). The fine topology, say on $\mathbb{R}^N$, admits a well-known characterization in probabilistic terms (Brownian motion), and many results have been established in the literature by probabilistic arguments. In the present book we only use analytic methods. For the probabilistic setup, see in particular [Do4], [BH3], and [Ng6].

After having shown that the fine topology is locally connected [Fu7], the second named author of the present book felt that perhaps there is a good notion of harmonic and sub/superharmonic functions relative to the fine topology, and this turned out to

[1] The English word "thin" serves as a (somewhat rough) translation of the French "effilé" (thready).

be the case, see [Fu8], and later works by several authors. Again, much of this work was framed in suitable axiomatic settings, especially with $\mathbb{R}^N$ replaced by a harmonic space in the sense of Brelot, often required to satisfy the axiom of domination and sometimes also further axioms. Many results in the beginning of the theory also hold in more general harmonic spaces and without assuming the axiom of domination; see in particular [BBC2], [BC], [LM1], and [LMZ].

The methods of studying the fine topology extend to a certain degree to a topology on $\mathbb{C}^n$ called the *pluri-fine topology*, defined as the smallest topology in which all plurisubharmonic functions on subsets of $\mathbb{C}^n$ are continuous. The pluri-fine topology was first studied thoroughly by Bedford and Taylor [BT3] in 1987, and they applied it to the Monge–Ampère operator on an open subset of $\mathbb{C}^n$. El Marzguioui and Wiegerinck [EMW1] proved the local connectedness of the pluri-fine topology. The trace of the pluri-fine topology on each complex line in $\mathbb{C}^n$ is identical with the Cartan fine topology on that line (identified with $\mathbb{C} \cong \mathbb{R}^2$). This fact led El Kadiri, Fuglede, and Wiegerinck [EKFW] to formulate and establish the main properties of two kinds of plurifinely plurisubharmonic functions (weak and strong) on a plurifinely open subset of $\mathbb{C}^n$. On a Euclidean open subset of $\mathbb{C}^n$ these two kinds of functions are both identical with subharmonicity.

Fine potential theory has led to solutions of classical problems (for example, regarding asymptotic paths for subharmonic functions), or has provided the language to answer natural questions (for example, regarding uniform limits of harmonic functions on compact sets), or has allowed an elegant extension of classical holomorphic functions theory to a wider class of "domains" (namely, finely holomorphic functions), in addition to more recent developments.

The present book does not deal with axiomatic settings, but exclusively with the case of functions (and measures) defined on subsets of $\mathbb{R}^N$, resp. $\mathbb{C}^n$. The book contains five chapters. The first four chapters are about fine potential theory (including a section on finely holomorphic functions), Chap. 1 being preparatory. Chapter 5 deals with pluri-fine pluripotential theory.

An overview on further developments and works on classical, axiomatic, or probabilistic fine potential theory and the use of fine topology in both real and complex analyses are given at the end of the book.

The presentation is largely self-contained. The book will serve as an essential reference for students and researchers alike, and to all with an interest in classical potential theory and its applications. It will be useful both in courses, and for self-study.

In preparing this book, we have benefitted from works of earlier authors. In particular, we acknowledge our indebtedness to Doob [Do4], Armitage and Gardiner [AG], Landkoff [Lan], and Fuglede [Fu8]. We are also grateful to Profs. Christian Berg, Palle E. Jorgensen, and Jan Wiegerinck for reading the book and making helpful suggestions, and for helping to publish it . We would like to extend our thanks to El Kadiri Asmae, El Kadiri Soumaya, El Kadiri Nadia, El Kadiri Meryem, and El Kadiri Zineb for their careful reading and the corrections they provided. We also express our appreciation to the staff of Springer Nature India.

Rabat, Morocco — Mohamed El Kadiri
Copenhagen, Denmark — Bent Fuglede

The Life and Work of Bent Fuglede

A little more than a month after the submission to Springer Nature of the manuscript to the present book, Bent Fuglede passed away.

With the death of Bent Fuglede, Denmark has lost one of its best mathematicians and the Department of Mathematics at the University of Copenhagen has lost a highly valued teacher, researcher, and colleague. He is survived by his son Einar and daughter-in-law Dorthea and two grandsons. His wife Ólafia Einarsdóttir of more than 60 years died in 2017.

Bent Fuglede was born on October 8, 1925, at Frederiksberg, a community within the Copenhagen area. His parents were both working in a bank. He received a high school diploma from Skt. Jørgens Gymnasium in 1943, and he graduated from the University of Copenhagen in 1948. The first two years of study comprised mathematics, physics, astronomy, and chemistry, and in the last part of his studies he specialized in mathematics, but followed also lectures in physics. His advisor in mathematics was Børge Jessen, who proposed him to specialize in the theory of Hilbert spaces and their operators, a subject that was relevant for his interest in quantum mechanics.

Bent Fuglede obtained a grant which made it possible for him to spend two years in the United States, and he arrived by ship to New York at the end of July 1949. He stayed there for a month before traveling West to participate in an AMS conference in Boulder. From there he went on to Stanford, where he spent a year.

During the month in New York, he visited the Institute for Advanced Study in Princeton for some days, curious about the place where he should stay for the academic year 1950–1951. It was during this short stay that he solved a problem, which had been on his mind since his study of Spektraldarstellung linearer Transformationen des Hilbertschen Raumes by Béla v. Sz. Nagy, published in 1942. In the book, there was a claim that it is easy to see that if a bounded operator commutes with a normal operator on a Hilbert space, then it also commutes with the adjoint operator. The claim was not easy to prove and at a meeting between Jessen and Nagy, the latter had admitted that he had overlooked the difficulty, and that he did not know a proof. Fuglede presented his proof at the Boulder meeting, and the participants agreed that it was correct. During the visit at Stanford, Fuglede submitted the proof to John von Neumann, and it was published in the *Proceedings of the National Academy of Sciences* in January 1950. Von Neumann had discussed the problem without solving it in a paper in Portugalia Math. 3 (1942); see Section 30.

During the year at Stanford, he followed lectures by Polya and Szegő, and he was invited to give lectures for a semester about operators on Hilbert space.

During the summer of 1950, he drove across the country to Princeton, where he stayed for the next year. He participated in the International Congress of Mathematicians (ICM) at Harvard, the first ICM after WW2.

Back in Denmark, he became scientific assistant at the Polytechnical Highschool (today Technical University of Denmark, DTU), and later he was appointed associate professor at the Department of Mathematics at the University of Copenhagen. Having defended the dissertation "Extremal Length and Closed Extensions of Partial Differential Operators", he was appointed professor at DTU in 1960. He returned to the University of Copenhagen as professor in 1965. He started by preparing excellent lecture notes for the major course in functional analysis for third year students of mathematics. In the years that followed, he taught at all levels, always with a clear and well-worked presentation and with an easy-to-read blackboard script. His accompanying lecture notes often contained new results and elegant proofs. In 1992, he allowed himself to retire, but continued as an active researcher and professor emeritus at the Department of Mathematics and he was active until shortly before his death.

During the long stay at the Institute for Advanced Study, he started collaboration with the American mathematician Richard V. Kadison on, among other things, determinant theory in infinite dimension. This led to Kadison coming to Copenhagen, and he gained a lifelong connection to Denmark, because he met his wife there. The Fuglede–Kadison determinant continues to be used, e.g. by Lück in his development of L^2-cohomology and in the construction of the Brown measure for non-normal operators.

Subsequently, Fuglede studied partial differential operators, which led to the aforementioned doctoral dissertation, and also to research on potential theory. This theory deals with the solutions to Laplace's equation, i.e. harmonic functions and Newton potentials. In this subject, he quickly became a leading figure worldwide, and he often gave a talk at the potential theory seminar in Paris, Séminaire Brelot–Choquet–Deny named after the three leading French mathematicians in the subject. In the 1960s, Brelot and his students developed an abstract potential theory called harmonic spaces. A harmonic space is a locally compact topological space, where a vector space of continuous real valued functions is specified for each open subset, modeling the harmonic functions in open sets of a Euclidean space. Brelot established a few fundamental axioms that these abstract spaces had to satisfy in order that many of the key results of potential theory could be deduced. The significance of the abstract theory was that one could obtain these key results for solution spaces to PDEs related to the Laplace operator by simply checking the axioms.

In harmonic spaces superharmonic functions can be introduced, but as in the classical theory, these are not always continuous, but only lower semicontinuous. In 1940, Henri Cartan got the idea to replace the classical Euclidean topology by a topology having more open sets, so that superharmonic functions become continuous. Cartan called this new topology the fine topology. The fine topology also appears in the theory of Brownian motion and similar stochastic processes, because a finely open set can be characterized as follows: If the process is in the finely open set at a specific time t_0, then it remains there almost surely for a positive period of time. When Fuglede succeeded proving that the fine topology is locally connected, he was inspired to construct what he called finely harmonic functions in finely open sets, and this was the start of a new era in potential theory and complex analysis. He published a comprehensive presentation of the theory in a Springer Lecture Notes, *Finely Harmonic Functions*, 1972, but already at the ICM in Nice, 1970, he gave an invited lecture on it.

The topic became hot and was further developed in the following years by Fuglede himself and with contributions from many others. He experienced the great honor that Heinz Bauer's plenary lecture on potential theory at the ICM in Vancouver, 1974, was largely about Fuglede's finely harmonic theory.

Just as harmonic functions in the plane are closely related to holomorphic functions, he succeeded in developing a rich theory for finely holomorphic functions in finely open sets of the complex plane. It also turned out that Fuglede's theory was a natural continuation of Borel's theory for monogenic functions, which was published as a monograph as early as in 1917.

Fuglede corresponded with a large number of mathematicians from home and abroad, and he was always generous with advice and improvements to manuscript drafts. He had a rare ability to penetrate to the heart of a mathematical problem, and through this he was often able to provide a surprising solution to the problem.

In 1974, Fuglede wrote a paper in the *Journal of Functional Analysis*: "Commuting Self-Adjoint Partial Differential Operators and a Group Theoretic Problem". It has around 300 citations. The work, which was inspired by a question from Irving Segal, associates a geometric property of an n-dimensional domain G with a property of the Hilbert space $L^2(G)$. The result is later known as Fuglede's conjecture. The Danish mathematician Steen Pedersen from Ohio published a number of papers on the subject, and later Terence Tao published two papers: "Fuglede's Conjecture Holds for Convex Planar Domains" (2001) and "Fuglede's Conjecture Is False in 5 and Higher Dimensions" (2003).

Fuglede decided to retire in the spring of 1992. He felt his health a little failing after a year's stay at the Institute for Advanced Study in Princeton, where he was accompanied by his wife. Fortunately, his health turned out to be fine and he achieved an emeritus period of approximately 30 years, during which he was very active. Out of the 114 of his works mentioned in Math. Sci. Net, roughly 40 are written after he became emeritus. During this period, he made significant contributions to many different areas, e.g. isoperimetric inequalities, Riemannian manifolds, and moment problems. It should also be mentioned that he published a research monograph together with James Eells: *Harmonic Maps Between Riemannian Polyhedra*, Cambridge Tracts in Mathematics Vol. 142, 2001. Together with Natalia Zorii from Kyiv, he has since 2016 written eight papers dealing with energy problems with respect to Riesz kernels, an impressive achievement for a person over 90 years of age.

Fuglede was a member of the Royal Danish Academy of Sciences and Letters, the Finnish Academy of Sciences, and the Bavarian Academy of Sciences. For a number of years, he was a member of the editorial boards of the journals *Expositiones Mathematicae* and *Potential Analysis*.

With his great and broad knowledge in many areas of mathematics, he was an obvious member of assessment committees at home and abroad.

Bent Fuglede passed away on December 7, 2023.

Christian Berg
Copenhagen, Denmark
June 2024

Notations and Terminology

$\mathbb{N}$ is the set of natural integers $1, 2, ...$, and $\mathbb{Z}_+$ is the set of the integers $0, 1, 2, ...$

The space of real valued continuous functions, resp. functions of class $\mathcal{C}^\infty$, with compact support on an open subset $\Omega \subset \mathbb{R}^N$ is denoted by $\mathcal{C}_0(\Omega)$, resp. $\mathcal{C}_0^\infty(\Omega)$ or $\mathcal{D}(\Omega)$.

When speaking of a limit or lim inf or lim sup, say of a function $f(x)$ (with values in $\overline{\mathbb{R}} = [-\infty, +\infty]$) as $x \to x_0$, we follow for example Bourbaki and include $f(x_0)$ in the definition. (The alternative definition, where $f(x_0)$ is not included, is used among others by J. L. Doob, for example in his book [Do4].) A function f on an open subset Ω of a topological space X with values in $\overline{\mathbb{R}}$ is said to be lower semicontinuous (l.s.c.) at x_0, respectively upper semicontinuous (u.s.c.) at $x_0 \in \Omega$ if $\liminf_{x\to x_0} f(x) \geq f(x_0)$, respectively if $-f$ is l.s.c. at x_0, that is, $\limsup_{x\to x_0} f(x) \leq f(x_0)$. The function f is said to be l.s.c., respectively u.s.c. on Ω iff f is l.s.c., respectively u.s.c. at every point $x \in \Omega$.

By a *signed measure* on a locally compact space X we mean a Radon measure on X. A positive Radon measure on X is simply called a measure on X. The σ-algebra of Borel measurable subsets of topological space X be denoted by $\mathcal{B}(X)$. A σ-finite positive measure on $(X, \mathcal{B}(X))$ is called a Borel measure on X. Let $A \subset X$, the *trace* of $\mathcal{B}(X)$ on A, that is, the set of the Borel subsets of A of the form $C \cap A$, $C \in \mathcal{B}(X)$, is denoted by $\mathcal{B}_A(X)$ and its elements be the Borel subsets of the topological space A equipped with the induced topology by X. If $A \in \mathcal{B}(X)$ then $\mathcal{B}_A(X) = \{C \in \mathcal{B}(X) : C \subset A\} \subset \mathcal{B}(X)$. The restriction of a Borel measure μ on X to a μ-measurable set $A \subset X$, also called *the part of* μ *on* A is the measure $\mu_A = 1_A\mu$ on X, also given by $\mu_A(C) = \mu(C \cap A)$ for $C \in \mathcal{B}(X)$. It is generally identified with the Borel measure μ' on the topological space A defined as the restriction of μ to $\mathcal{B}_A(X) = \mathcal{B}(A)$. Likewise, a Borel measure μ' on A is identified with its extension μ to X given by $\mu(C) = \mu'(A \cap C)$ for $C \in \mathcal{B}(X)$ (this identification is compatible with the above because $\mu = \mu_A$). A measure μ is said to be *carried* by a set $A \subset X$ if A is μ-measurable and $\mu(\complement A) = 0$.

The Euclidean norm on $\mathbb{R}^N$ $(N \geq 1)$ is denoted by $|\cdot|$ or $\|\cdot\|$. For $x \in \mathbb{R}^N$ and $r > 0$ we denote by $B(x, r)$, resp. $\overline{B}(x, r)$, the open, resp. the closed, ball of radius r centered at x. In the absence of other indication, Ω denotes a *Greenian domain* (that is, a connected Greenian subset of $\mathbb{R}^N, N \geq 2$), mostly serving as an ambient (locally compact) space; see Sect. 1.1.4. We denote by ℓ_N Lebesgue measure on $\mathbb{R}^N$. The restriction of ℓ_N to a ℓ_N-measurable set $X \subset \mathbb{R}^N$ is also denoted by ℓ_N and $L^p(X)$ denotes the space $L^p(X, \ell_N)$ $(1 \leq p \leq \infty)$.

A map into $\overline{\mathbb{R}} = [-\infty, +\infty]$ is called a *numerical* function. For any function $f : \Omega \to \overline{\mathbb{R}}$ write $f_+ = f \vee 0 (= \max\{f, 0\})$, $f_- = (-f) \vee 0$. Following Bourbaki's terminology, we say that a function is *positive* resp. *negative* if $f_+ = f$, resp. $f_- = -f$.

For the *upper* and *lower* integrals of a numerical function f with respect to a measure μ we follow [Bou6, Chap. 4, § 4, Exercises 5 and 6]. Thus the upper integral $\int^* f\, d\mu$ is defined as the infimum of the integrals $\int g\, d\mu$ where g ranges over all numerical μ-integrable lower l.s.c. functions $\geqslant f$. (If there is no such function g we define $\int^* f\, d\mu$ to be $+\infty$.) The lower integral $\int_* f\, d\mu$ is defined analogously, or equivalently as $-\int^*(-f)\, d\mu$. If $f \geqslant 0$ then $\int^* f\, d\mu$ agrees with the usual upper integral as defined in [Bou6, Chap. 4, § 1, No. 3].

For any increasing sequence of numerical functions f_n such that $\int^* f_n\, d\mu > -\infty$ for some n we have $\int^*(\sup_n f_n)\, d\mu = \sup_n \int^* f_n\, d\mu$.

If $\int^* f\, d\mu = \int_* f\, d\mu$ we may simply write $\int f\, d\mu$ in their place. This occurs in particular if f is *μ-measurable* and if in addition $\int^* f^+\, d\mu < +\infty$ or $\int^* f^-\, d\mu < +\infty$, thus for example if $f \geqslant 0$ is μ-measurable. In order that f be *μ-integrable,* it is necessary and sufficient that $\int^* f\, d\mu$ and $\int_* f\, d\mu$ be equal and finite.

Contents

Chapter 1
Background in Potential Theory

By classical potential theory, we mean the potential theory of the Laplace equation in an open set of Euclidean space $\mathbb{R}^N$ ($N \geq 2$). This chapter is devoted to some background in classical potential theory used in all of the book.

The first section deals with fundamental tools and basic results (without proofs) of this theory. We recall there some fundamental properties of [hyper/super]harmonic functions and potentials. We define Greenian sets, their Greenian kernels and Green potentials, polar sets (sets of singularities of superharmonic functions), and Green capacity and state the fundamental convergence theorem. The Dirichlet problem on balls and on general Greenian open sets, harmonic measures, and regular and irregular points are discussed.

Section 1.2 is devoted to the notion of thin sets, introduced by Brelot in studying the Dirichlet problem, and the Cartan-Brelot fine topology, which is a powerful instrument in potential theory. Deep properties of this topology such as regularity, Baire property, quasi-Lindelöf property, local connectedness (which will be proved in Chap. 2), and Borel measurability of finely continuous functions are also considered. The section is closed by a discussion of thinness at infinity.

Reduction and sweeping of functions, which play a central role in classical potential theory and also in the development of fine potential theory, are thoroughly studied in Sect. 1.3. Polarity and thinness are characterized by means of sweeping of positive functions. Sweeping of measures, studied in Sect. 1.4, allows to study the base of a set or a function and the base operation. It also allows to define the harmonic measure on finely open sets. It is by means of (fine) harmonic measure that the finely [super]harmonic functions on finely open sets are defined in Chap. 2.

In Sect. 1.5, we study "quasi-topological" concepts (quasi-open and quasi-closed sets, quasi-continuous functions) relative to a suitable kind of capacity, not just the Green capacity. Capacity is a suitable countably subadditive set function defined on subsets of a Greenian domain Ω. Inspired by Doob [Do1], we also consider

M. El Kadiri and B. Fuglede, *Classical Fine Potential Theory*, Springer Monographs in Mathematics, https://doi.org/10.1007/978-981-96-0430-2_1

capacity as a suitable sublinear functional on the lattice cone $\mathcal{F}^+$ of all functions $f : \Omega \to [0, +\infty]$. As an application, we obtain the remarkable and very useful Brelot property (see Theorem 1.5.29).

1.1 Basics of Classical Potential Theory

For proofs and further information concerning this section, we refer to the monograph of Armitage and Gardiner [AG] and the monograph of Doob [Do4].

1.1.1 Denote by

$$\Delta = \frac{\partial^2}{\partial x_1^2} + \frac{\partial^2}{\partial x_2^2} + \cdots + \frac{\partial^2}{\partial x_N^2}$$

the Laplacian on $\mathbb{R}^N$, $N \geqslant 2$. Let Ω denote an open subset of $\mathbb{R}^N$. A function $u : \Omega \to \mathbb{R}$ is said to be *harmonic* if $u \in C^2(\Omega)$ and u satisfies the Laplace equation $\Delta u = 0$ identically on Ω. The harmonic functions on Ω form a linear subspace of the vector space of all functions $\Omega \to \mathbb{R}$, closed under locally uniform convergence.

For any point $y \in \mathbb{R}^N$ the function U_y defined on $\mathbb{R}^N \setminus \{y\}$ by

$$U_y(x) = \begin{cases} -\log|x-y|, & (N=2) \\ |x-y|^{2-N}, & (N \geqslant 3) \end{cases}$$

is harmonic on $\mathbb{R}^N \setminus \{y\}$; and U_y is called the *fundamental harmonic function* with pole at y. We extend U_y to y by continuity, so that $U_y(y) = +\infty$.

1.1.2 The *Poisson kernel* of an (open) ball $B = B(x_0, r)$ in $\mathbb{R}^N$ is the function

$$K_B(x, y) = \frac{1}{\sigma_N r} \frac{r^2 - |x - x_0|^2}{|x - y|^N} \quad (y \in \partial B,\ x \in \mathbb{R}^N \setminus \{y\}),$$

where σ_N denotes the surface area of the unit sphere in $\mathbb{R}^N$. Clearly, $K_B(x, y) > 0$ for $x \in B$, $K_B(x, y) < 0$ for $x \in \mathbb{R}^N \setminus \overline{B}$, and $K_B(x, y) = 0$ for $x \in \partial B \setminus \{y\}$. The *Poisson integral* $P_B\mu$ of a signed measure μ on ∂B is well defined by

$$P_B\mu(x) = \int_{\partial B} K_B(x, y)\, d\mu(y) \quad (x \in B)$$

because $K_B(x, \cdot)$ for any $x \in B$ is bounded and continuous on the compact set ∂B carrying μ. Also, $P_B\mu$ is harmonic on B, see, for example, [AG, Theorem 1.3.3], also for (1.1.1) below.

If a function $f : \partial B \to \mathbb{R}$ (or a function defined on some set containing ∂B, for example on $\overline{B}$) is integrable with respect to the surface measure σ on ∂B, we may write $P_B f$ for $P_B(f\sigma)$. For $y \in \partial B$ we have

$$\liminf_{\partial B \ni z \to y} f(z) \leqslant \liminf_{B \ni x \to y} P_B f(x) \leqslant \limsup_{B \ni x \to y} P_B f(x) \leqslant \limsup_{\partial B \ni z \to y} f(z). \tag{1.1.1}$$

If f is continuous (in the extended sense) at $y \in \partial B$, it follows that $P_B f(x) \to f(y)$ for $x \to y$, $x \in B$.

Thus $P_B f$ solves the *Dirichlet problem* for B when f is finite and continuous on ∂B. In other words, $K_B(x, \cdot)\, d\sigma$ is the *harmonic measure* at the point $x \in B$, denoted by μ_x^B, and so

$$P_B f(x) = \int_{\partial B} K_B(x, \cdot) f \, d\sigma = \int f \, d\mu_x^B. \tag{1.1.2}$$

The pointwise supremum of any upper directed family of harmonic functions on a connected open set $\Omega \subset \mathbb{R}^N$ is either harmonic or identically $+\infty$. This is Harnack's *convergence theorem*. It depends on the *Harnack inequalities*. See, for example, [AG, Sections 1.4 and 1.5].

A function $u : \Omega \to \overline{\mathbb{R}} = [-\infty, +\infty]$ (also called a numerical function on Ω) is said to be *hyperharmonic* if (i) u is l.s.c. and $> -\infty$, and (ii)

$$u(x) \geqslant \frac{1}{\sigma_N r^{N-1}} \int^*_{\partial B(x,r)} u \, d\sigma \quad \text{when } \overline{B}(x, r) \subset \Omega, \tag{1.1.3}$$

σ_N denoting the surface area of the unit ball in $\mathbb{R}^N$.

If, in addition to (i) and (ii), we have (iii) $u \not\equiv +\infty$ in each component of Ω, then u is called *superharmonic*. If u satisfies (i) and (iii), one may replace in (ii) the spherical mean-value by the ball mean-value

$$\frac{1}{\ell_N(B(x, r))} \int_{B(x,r)} u \, d\ell_N = (P_{B(x,r)} u)(x),$$

ℓ_N denoting N-dimensional Lebesgue measure. Moreover, one may then replace (ii) by

$$u \geqslant P_B u \quad \text{on } B \text{ for every ball } B \text{ with } \overline{B} \subset \Omega.$$

Beginning after (1.1.3), it suffices to consider balls $B(x, r)$ with sufficiently small radius r depending on x.

It follows from the above that every superharmonic function u on an open set Ω is locally Lebesgue integrable and hence finite Lebesgue almost everywhere, e.g. [AG, Theorem 3.1.3]. In particular, u is finite on a dense subset of Ω. Furthermore, if u_1 and u_2 are superharmonic on Ω and if $u_1 = u_2$ ℓ_N-a.e., then the same holds everywhere, e.g. [AG, Corollary 3.2.7]. Similar assertions hold with ℓ_N replaced by the surface measure σ on a closed ball.

For every $y \in \mathbb{R}^N$ the function U_y is superharmonic on $\mathbb{R}^N$ (and harmonic on $\mathbb{R}^N \setminus \{y\}$).

A function $u : \Omega \to \overline{\mathbb{R}}$ is called *hypoharmonic*, resp. *subharmonic*, if $-u$ is hyperharmonic, resp. superharmonic.

It is known that a numerical function u is harmonic if and only if u is both hyperharmonic and hypoharmonic. Equivalently, u shall be finite and continuous and

$$u(x) = \frac{1}{\sigma_N r^{N-1}} \int_{\partial B(x,r)} u \, d\sigma \quad \text{when } \overline{B}(x,r) \subset \Omega. \tag{1.1.4}$$

Again one may replace the spherical mean-value by the ball mean-value in (1.1.4); and one may replace (1.1.4) by $u = P_B u$ for any ball B with $\overline{B} \subset \Omega$.

A harmonic function on all of $\mathbb{R}^N$ is constant if upper or lower bounded. For this version of Liouville's theorem for holomorphic functions of a complex variable, see, for example, [AG, Theorem 1.2.6].

The hyperharmonic functions on Ω form a convex subcone of $\overline{\mathbb{R}}^\Omega$, stable under pointwise supremum for upper directed families and under pointwise infimum for finite families. The pointwise supremum, resp. infimum, of a family $\mathcal{F}$ of numerical (in particular: of hyperharmonic) functions on Ω is denoted by $\sup \mathcal{F}$, resp. $\inf \mathcal{F}$. We write $u \vee v$ or $\max\{u, v\}$ for $\sup\{u, v\}$ and $u \wedge v$ or $\min\{u, v\}$ for $\inf\{u, v\}$ when u and v are numerical functions, in particular (finite or infinite) constants.

A hyperharmonic function $u \geqslant 0$ on a connected open set $\Omega \subset \mathbb{R}^N$ is either everywhere > 0 or $\equiv 0$. In fact, $\sup_{n\in\mathbb{N}}(nu)$ is hyperharmonic, in particular l.s.c. and (if $\not\equiv +\infty$) locally ℓ_N-integrable, so $\{u > 0\}$ is an open ℓ_N-null set, hence void.

1.1.3 Denote by $\partial^\infty\Omega$ the boundary of an open set $\Omega \subset \mathbb{R}^N$ relative to the one point compactification $\mathbb{R}^N \cup \{\infty\}$ of $\mathbb{R}^N$. (Thus $\partial^\infty\Omega = \partial\Omega$ if Ω is bounded, whereas $\partial^\infty\Omega = (\partial\Omega) \cup \{\infty\}$ if Ω is unbounded.)

Let u be hyperharmonic on an open set $\Omega \subset \mathbb{R}^N$. If

$$\liminf_{x \to y} u(x) \geqslant 0 \text{ for every } y \in \partial^\infty\Omega,$$

then $u \geqslant 0$. This is the classical *boundary minimum principle.*

1.1.4 An open subset Ω of $\mathbb{R}^N$ is called a *Greenian set* if for some, and hence for any, $y \in \Omega$ the fundamental harmonic function U_y has a *greatest harmonic minorant* h_y on Ω. A connected Greenian set is called a Greenian domain. Any open subset of a Greenian set is Greenian.

Suppose that $\Omega \subset \mathbb{R}^N$ is a Greenian set. The function $G_\Omega : \Omega \times \Omega \to \,]0, +\infty]$ defined by

$$G_\Omega(x, y) = U_y(x) - h_y(x)$$

is called the *Green kernel* for Ω. For each $y \in \Omega$, $G_\Omega(\cdot, y)$ is called the *Green function* on Ω with pole at y. The Green kernel is symmetric $G_\Omega(x, y) = G_\Omega(y, x)$, l.s.c., and continuous in the extended sense on $\Omega \times \Omega$; see, for example, [AG, Theorem 4.1.9] or [Do4, 1.VII.4].

For $N \geqslant 3$ (only), the entire space $\mathbb{R}^N$ is a Greenian set, and U_y is then itself the Green function on $\mathbb{R}^N$ with pole at y.

For $N = 2$, an open set $\Omega \subset \mathbb{R}^2$ is a Greenian set if for example $\mathbb{R}^2 \setminus \partial\Omega$ is disconnected, thus in particular if Ω is disconnected or bounded (see further Sect. 1.1.12).

Let $B = B(x_0, r)$ in $\mathbb{R}^N$ $(N \geq 2)$ and

$$\phi(x, y) = \frac{(r^2 - |x - x_0|^2)(r^2 - |y - x_0|^2)}{r^2|x - y|^2} \quad (x, y \in B; x \neq y).$$

Then the Green kernel for the ball B in $\mathbb{R}^N$ is given for $x, y \in B$ with $x \neq y$ by

$$G_B(x, y) = \begin{cases} 2^{-1}\log(1 + \phi(x, y)) & (x \neq y) \\ +\infty & (x = y) \end{cases}$$

if $N = 2$, and

$$G_B(x, y) = \begin{cases} (1 - (1 + \phi(x, y))^{1-\frac{N}{2}})|x - y|^{2-N} & (x \neq y) \\ +\infty & (x = y) \end{cases}$$

if $N \geqslant 3$ (see [AG, p. 91]).[1]

1.1.5 For any (positive) measure μ on a Greenian set $\Omega \subset \mathbb{R}^N$ $(N \geqslant 2)$, the function $G_\Omega\mu$ on Ω defined by

$$G_\Omega\mu(x) = \int_\Omega G_\Omega(x, y)\, d\mu(y)$$

is hyperharmonic. By Fubini's theorem, we have the *reciprocity theorem*: $\int G_\Omega\mu\, d\nu = \int G_\Omega\nu\, d\mu$ for $\mu, \nu \geqslant 0$. If superharmonic, $G_\Omega\mu$ is called the (Green) *potential* of μ. If $f \geqslant 0$ on Ω is Lebesgue integrable, we may write $G_\Omega f$ for $G_\Omega(f\ell_N)$. A measure μ is uniquely determined by its potential $G_\Omega\mu$ (assumed $\not\equiv +\infty$ on each component of Ω); this is contained in Sect. 1.1.6 below.

1.1.6 Let u denote a superharmonic function on a Greenian set $\Omega \subset \mathbb{R}^N$ $(N \geqslant 2)$, and suppose that u has a subharmonic minorant on Ω. Then there exists a unique positive measure μ_u on Ω, called the *Riesz measure* associated with u (relative to Ω), such that u has the decomposition

$$u = G_\Omega\mu_u + h,$$

where h is the greatest harmonic minorant, or equivalently: the greatest subharmonic minorant, of u on Ω. This decomposition is called the *Riesz decomposition* of u, e.g. [Do4, 1.IV.7–8].

---

[1] For the explicit expression for the Green kernel for a ball in $\mathbb{R}^N$ $(N \geqslant 3)$ and a half space in $\mathbb{R}^N$ $(N \geqslant 2)$, see, for example, [AG, p. 92].

On a Greenian set Ω, a superharmonic function $u \geqslant 0$ with associated Riesz measure μ_u is a potential (namely of μ_u) if and only if every harmonic (or equivalently: every subharmonic) minorant of u is $\leqslant 0$ (equivalently: every positive harmonic [or positive subharmonic] minorant of u is the function 0). Clearly, every positive superharmonic minorant of a potential is a potential.

Any finite sum of potentials is a potential. For example, if p_1 and p_2 are potentials and if h is a harmonic minorant of $p_1 + p_2$ then $h - p_1$ is a subharmonic minorant of the potential p_2 and therefore $h - p_1 \leqslant 0$, that is, $h \leqslant p_1$, hence $h \leqslant 0$ since p_1 is a potential. This shows that $p_1 + p_2$ indeed is a potential.

The pointwise sum p of a family $(p_i)_{i\in I}$ of potentials is a potential if p is superharmonic. To see this, let h be a positive harmonic minorant of p. For any finite subset J of I, $h - \sum_{i\in I\setminus J} p_i$ is a hypoharmonic minorant of the potential $\sum_{i\in J} p_i$ and is therefore $\leqslant 0$. It follows that h equals zero on the dense set $\{p < +\infty\}$ (see Sect. 1.1.2) and hence $h \equiv 0$, by continuity.

1.1.7 Let u be hyperharmonic on an open set Ω, and v be hyperharmonic on an open set $\omega \subset \Omega$. If

$$\liminf_{\omega\ni x\to y} v(x) \geqslant u(y) \quad \text{for every } y \in \Omega \cap \partial\omega$$

then the function w defined on Ω by

$$w(x) = \begin{cases} u(x) \wedge v(x) & \text{for } x \in \omega \\ u(x) & \text{for } x \in \Omega \setminus \omega \end{cases}$$

is hyperharmonic on Ω. In fact, w is clearly hyperharmonic on ω and on $\Omega \setminus \overline{\omega}$. And for any point $x \in \Omega \cap \partial\omega$ and any ball $B(x, r)$ small enough so that $\overline{B}(x, r) \subset \Omega$, the hyperharmonic mean-value property holds:

$$w(x) = u(x) \geqslant \frac{1}{\sigma_N r^{N-1}} \int_{\partial B(x,r)}^{*} u \, d\sigma \geqslant \frac{1}{\sigma_N r^{N-1}} \int_{\partial B(x,r)}^{*} w \, d\sigma.$$

From this, we shall obtain the following *relative boundary minimum principle*: Let u be hyperharmonic on an open subset ω of a Greenian set Ω, and suppose that

$$\liminf_{\omega\ni x\to y} u(x) \geqslant 0 \quad \text{for every } y \in \Omega \cap \partial\omega.$$

Suppose in addition that there exists a Green potential p on Ω such that $u \geqslant -p$ on ω. Then $u \geqslant 0$ on ω. In fact, by the above result with u replaced by 0 and v by u, the extension w of $0 \wedge u$ by 0 on $\Omega \setminus \omega$ is hyperharmonic and $\geqslant -p$ on Ω, whence actually $w \geqslant 0$, and in particular $0 \wedge u \geqslant 0$ on ω, that is, $u \geqslant 0$ on ω.

1.1.8 Even if an open set Ω is not Greenian, there is associated with any superharmonic function u on Ω a unique measure μ_u (again called the Riesz measure associated with u) whose restriction to any relatively compact open subset ω of Ω is the Riesz measure associated with the restriction of u to ω; see, for example, [Do4, 1.IV.7]. We have *Poisson's formula*

$$\Delta u = -a_N \mu_u$$

in the distributional sense, where a_N denotes the Poisson constant: $a_N = (N-2)\sigma_N$ for $N > 2$, $a_2 = 2\pi$; see, for example, [AG, Section 4.3].

1.1.9 A superharmonic function u on an open set Ω is *harmonic* on a given open set $\omega \subset \Omega$ if and only if $\mu_u(\omega) = 0$. In particular u is *harmonic* off the (closed) support $\operatorname{Supp}\mu_u$ of μ_u (relative to Ω); see Sect. 1.1.8. At a point $x \in \operatorname{Supp}\mu_u$, u is *continuous* provided that the restriction of u to $\operatorname{Supp}\mu_u$ is continuous at x. This latter assertion is the *continuity principle* of Evans and Vasilescu; see, for example, [AG, Theorem 4.5.1] or [Do4, 1.V.8].

1.1.10 Every hyperharmonic function $\geqslant 0$ on a Greenian set Ω is the pointwise limit of an increasing sequence of finite continuous (even smooth) potentials of measures with compact supports in Ω. See, for example, [Do4, 1.IV.10].

1.1.11 A set E in $\mathbb{R}^N$ is called *polar* if there is a superharmonic function u on some open set $\Omega \subset \mathbb{R}^N$ containing E such that $u = +\infty$ on E. Since u is locally ℓ_N-integrable we have $\ell_N(E) = 0$, similarly with ℓ_N replaced by the surface measure σ on a closed ball in Ω.

An assertion involving a variable point of a given set $A \subset \mathbb{R}^N$ is said to hold *quasi-everywhere* (q.e.) on A if it holds at every point of A off some polar subset. Thus a hyperharmonic function u on an open set $\omega \subset \mathbb{R}^N$ is superharmonic if and only if $u < +\infty$ q.e. on ω. A countable union of polar subsets of $\mathbb{R}^N$ is polar; see, for example, [AG, Corollary 5.1.4]. A single point of $\mathbb{R}^N$ forms a polar set (because $N > 1$). It follows that every countable set is polar.

Let E be a polar subset of a Greenian set Ω and let $x \in \Omega \setminus E$ be given. There exists then a measure μ on Ω with superharmonic potential $G_\Omega\mu$ such that $G_\Omega\mu = +\infty$ on E, $G_\Omega\mu(x) < +\infty$, e.g. [AG, Theorem 5.1.3]. It follows that if E is a relatively closed polar subset of a connected open set $\Omega \subset \mathbb{R}^N$ then $\Omega \setminus E$ is connected, e.g. [AG, Corollary 5.1]. Furthermore, the Riesz measure associated with a *locally bounded* superharmonic function does not charge the polar sets, e.g. [AG, Theorem 5.1.9].

1.1.12 Polar sets are removable singularity sets for [super]harmonic functions: If u is (super)harmonic on $\Omega \setminus E$, where E is a polar and relatively closed subset of Ω, and if u is bounded (from below) near points of E, then u has a unique extension to a (super)harmonic function on Ω; see [AG, p. 128]. For a version of this in which E need not be relatively closed, see [Do4, 1.V.5].

An open set $\Omega \subset \mathbb{R}^2$ is Greenian if and only if $\mathbb{R}^2 \setminus \Omega$ is nonpolar (Myrberg's theorem, e.g. [Do4, 1.V.6], [AG, Theorem 5.3.8]).

1.1.13 For any open set $\Omega \subset \mathbb{R}^N$ the *l.s.c. regularization* $\widehat{f}$ of a numerical function f on Ω is defined as the greatest l.s.c. minorant $\widehat{f}$ of f. Thus

$$\widehat{f}(y) = \liminf_{x \to y} f(x). \tag{1.1.5}$$

Let $\mathcal{F}$ be a family of hyperharmonic functions on an open set Ω in $\mathbb{R}^N$, and let $u = \inf \mathcal{F}$ (pointwise infimum). If $\mathcal{F}$ is locally uniformly bounded from below, that is, if u is locally bounded from below, then $\widehat{u}$ is hyperharmonic,

$$\widehat{u}(y) = \liminf_{x \to y, x \neq y} u(x),$$

and $\widehat{u} = u$ q.e. This result is known as the *fundamental convergence theorem*, due to Brelot and H. Cartan. See, for example, [AG, Theorem 5.7.1 and Notes to it]. Here a topological lemma due to Choquet applies to $\mathcal{F}$. This lemma states that any family $\mathcal{F}$ of numerical functions on a Hausdorff space with a countable basis of open sets has a countable subfamily $\mathcal{F}^*$ such that $\inf \mathcal{F}^*$ and $\inf \mathcal{F}$ have the same l.s.c. regularization. If $\mathcal{F}$ is lower directed, we may arrange that $\mathcal{F}^*$ is formed by the elements of a decreasing sequence. See, for example, [Do4, 1.VI.1 and 1.III.3], [AG, Lemma 3.7.4].

1.1.14 The Green *capacity* $C(K)$ of a compact subset K of a Greenian set Ω is defined by

$$C(K) = \sup\left\{\int d\mu : \mu \in \mathcal{M}_+(K),\ G_\Omega\mu \leqslant 1 \text{ on } \Omega\right\},$$

where $\mathcal{M}_+(K)$ denotes the convex cone of all (positive) measures on Ω supported by K. The *inner capacity* $C_*(A)$ and the *outer capacity* $C^*(A)$ of any subset A of Ω are defined by

$$C_*(A) = \sup\{C(K) : K \text{ compact }, K \subset A\},$$
$$C^*(A) = \inf\{C_*(O) : O \text{ open in } \Omega, O \supset A\}.$$

A set A is called *capacitable* if $C^*(A) = C_*(A)$, and we then define the *capacity* $C(A)$ of A by $C(A) = C^*(A) = C_*(A)$. Clearly, every open set is capacitable. It is well known that so is any compact set K, and the supremum in the above definition of $C(K) = C_*(K)$ is finite and attained uniquely by the so-called *equilibrium measure* on K.

The set function C^* is countably subadditive, and $C^*(A_n) \to C^*(A)$ for any increasing sequence of sets A_n with the union A. As shown by Choquet [Ch1], the latter property implies that every analytic set (in particular every Borel set) is capacitable. As shown by Cartan [Ca1], a set $E \subset \Omega$ is polar if and only if $C^*(E) = 0$; see, for example, [AG, p. 141].

We shall not make much use of the concept of Green capacity until Sect. 1.4, but rather use some analogous set functions defined in terms of *reduction* and *sweeping* of functions; see Sects. 1.3 and 1.5. For the Green capacity, we therefore merely refer to [Do4, 1.XIII], [AG, Sections 5.5–5.9]. For $\Omega = \mathbb{R}^N$ with $N \geqslant 3$, Green capacity is also termed *Newtonian capacity*.

1.1.15 Let u be superharmonic on a Greenian set $\Omega \subset \mathbb{R}^N$ ($N \geqslant 2$), and let B be a ball with $\overline{B} \subset \Omega$. The function u_B, defined to equal $P_B u$ on B and to equal u elsewhere on Ω, is then superharmonic $\leqslant u$ on Ω and harmonic on B; cf. Sect. 1.1.2. This function u_B is called the *Poisson modification* of u relative to B.

A non-empty lower directed family Γ of superharmonic functions on Ω, each of which has a subharmonic minorant, is called a *Perron family* if every Poisson modification of any $u \in \Gamma$ belongs to Γ. For any Perron family Γ it was shown by Perron that, on each component of Ω, the pointwise infimum $\inf \Gamma$ of any Perron family Γ is either harmonic or identically $-\infty$; cf., for example, [AG, Theorem 3.6.2]. The former possibility occurs of course if the functions in Γ are positive. See, for example, [AG, Theorem 3.6.2].

For any superharmonic function u on Ω which has a subharmonic minorant, and for any cover $\mathcal{W}$ of Ω by balls B with $\overline{B} \subset \Omega$, the family of functions of the form $P_{B_n} P_{B_{n-1}} \dots P_{B_1} u$, where $n \in \mathbb{N}$ and $(B_1, \dots, B_n)$ is a finite sequence in $\mathcal{W}$, is termed the *Perron family generated by u and $\mathcal{W}$*. In the absence of other indication, $\mathcal{W}$ is taken as the cover of Ω by all balls B with $\overline{B} \subset \Omega$, and we simply speak of the Perron family generated by u.

For a *positive* superharmonic function u on Ω, the following are equivalent (e.g. [CC3, Proposition 2.2.1]):

(a) u is a potential,

(b) the infimum of some, and hence of any, Perron family containing u is identically 0,

(c) every subharmonic minorant of u is $\leqslant 0$.

It is enough to verify the equivalence of (a), (b), and (c) when Ω is connected.

(a)$\Rightarrow$(b). If u is a potential then so is any positive superharmonic minorant of u, in particular any Poisson modification of u and therefore any Perron family containing u, and likewise the greatest l.s.c. minorant of such a family according to the fundamental convergence theorem (Sect. 1.1.13). As mentioned above, such a minorant is a harmonic potential, hence $\equiv 0$.

(b)$\Rightarrow$(c). Let v be a subharmonic minorant of u. Let Γ be a Perron family generated by u and suppose that $\inf \Gamma \equiv 0$. Then $u - v$ is a positive superharmonic function and $P_B v \geqslant v$ for any ball B such that $\overline{B} \subset \Omega$. Hence

$$(P_{B_n} P_{B_{n-1}} \dots P_{B_1} u) - v \geqslant P_{B_n} P_{B_{n-1}} \dots P_{B_1}(u - v) \geqslant 0,$$

that is, $u' - v \geqslant 0$ for any $u' \in \Gamma$. This implies (c) as follows for any $x \in \Omega$:

$$-v(x) = -v(x) + \inf_{u' \in \Gamma} u'(x) = \inf_{u' \in \Gamma} (u' - v)(x) \geqslant 0.$$

(c)$\Rightarrow$(a) because any positive harmonic minorant of u is subharmonic and hence $\equiv 0$.

1.1.16 The *Dirichlet problem* for a Greenian set $\Omega \subset \mathbb{R}^N$ $(N \geqslant 2)$ is treated by the Perron-Wiener-Brelot (PWB) method. Given an extended real valued function f, defined at least on $\partial^\infty \Omega$, a hyperharmonic and lower bounded function u on Ω is called an *upper function* for f if

$$\liminf_{\Omega \ni x \to y} u(x) \geqslant f(y) \quad \text{for each } y \in \partial^\infty \Omega,$$

and a hypoharmonic upper bounded function u on Ω is called a *lower function* for f if

$$\limsup_{\Omega \ni x \to y} u(x) \leqslant f(y) \quad \text{for each } y \in \partial^\infty \Omega.$$

The class of all upper, resp. lower, functions for f is denoted by $\overline{\Phi}_f^\Omega$, resp. $\underline{\Phi}_f^\Omega$. Clearly $\underline{\Phi}_f^\Omega = -\overline{\Phi}_{-f}^\Omega$.

The *upper* and *lower PWB solutions* to the Dirichlet problem with function f are defined pointwise on Ω by

$$\overline{H}_f^\Omega = \inf \overline{\Phi}_f^\Omega, \qquad \underline{H}_f^\Omega = \sup \underline{\Phi}_f^\Omega.$$

Clearly $\underline{H}_f^\Omega = -\overline{H}_{-f}^\Omega$ because $\underline{\Phi}_f^\Omega = \{-u : u \in \overline{\Phi}_{-f}^\Omega\}$.

On any component of Ω, each of the functions $\overline{H}_f^\Omega$, $\underline{H}_f^\Omega$ is *either harmonic or identically* $+\infty$ *or* $-\infty$. To see this it suffices to consider the case of $\overline{H}_f^\Omega$. Suppose first that Ω is connected. If $\overline{\Phi}_f^\Omega = \{+\infty\}$ then $\overline{H}_f^\Omega \equiv +\infty$. Otherwise $\overline{H}_f^\Omega$ is the pointwise infimum of the family of all *superharmonic* upper functions for f. This family is clearly a Perron family, and its pointwise infimum is therefore either identically $-\infty$ or else harmonic, as mentioned in the preceding subsection. For disconnected Ω; use, for example, [AG, Lemma 6.2.4].

Furthermore, $\overline{H}_f^\Omega \geqslant \underline{H}_f^\Omega$. As above we may suppose that Ω is connected. It is enough to show that if u is a superharmonic upper solution and v a subharmonic lower solution then $u \geqslant v$ on Ω. This inequality will follow from the classical boundary minimum principle (Sect. 1.1.7) if we show that

$$\liminf_{x \to y} (u - v)(x) \geqslant 0$$

for each $y \in \partial^\infty \Omega$. If $f(y)$ is finite then

$$\liminf_{x \to y} (u - v)(x) \geqslant \liminf_{x \to y} u(x) - \limsup_{x \to y} v(x) \geqslant f(y) - f(y) = 0. \qquad (1.1.6)$$

If $f(y) = +\infty$ then $u(x) \to +\infty$ as $x \to y$, while v is upper bounded on Ω and hence $(u - v)(x) \to +\infty$ as $x \to y$. A similar argument yields the same conclusion if $f(y) = -\infty$. Hence (1.1.6) holds at each point $y \in \partial^\infty \Omega$, as required.

A function $f : \partial^\infty\Omega \to \overline{\mathbb{R}}$ is called *resolutive* (relative to Ω) if $\overline{H}_f^\Omega$ and $\underline{H}_f^\Omega$ are identical and finite valued (and hence harmonic) on Ω. In that case we define $H_f^\Omega := \overline{H}_f^\Omega = \underline{H}_f^\Omega$, and we call H_f^Ω the *PWB solution* (or generalized solution) to the Dirichlet problem on Ω with the boundary function f. Every function $f \in \mathcal{C}(\partial^\infty\Omega)$ is resolutive. See, for example, [AG, Chapter 6].

1.1.17 For any point x of a Greenian set Ω, there exists a unique Borel probability measure μ_x^Ω on $\partial^\infty\Omega$ such that

$$H_f^\Omega(x) = \int_{\partial^\infty\Omega} f\, d\mu_x^\Omega \quad \text{for every } f \in \mathcal{C}(\partial^\infty\Omega). \tag{1.1.7}$$

This measure μ_x^Ω is called the *harmonic measure* relative to Ω and x; cf. Sect. 1.1.2. For any function $f : \partial^\infty\Omega \to \overline{\mathbb{R}}$ we have

$$\underline{H}_f^\Omega(x) \leqslant \int_* f\, d\mu_x^\Omega \leqslant \int^* f\, d\mu_x^\Omega \leqslant \overline{H}_f^\Omega(x) \tag{1.1.8}$$

(integrations over $\partial^\infty\Omega$). Therefore, if f is resolutive then f is μ_x^Ω-integrable over $\partial^\infty\Omega$ for every $x \in \Omega$, and equality prevails in each inequality in (1.1.8). On the other hand, if f is μ_x^Ω-measurable and $\int f\, d\mu_x^\Omega$ exists then $\overline{H}_f^\Omega(x) = \underline{H}_f^\Omega(x) = \int f\, d\mu_x^\Omega$. In particular, if f is μ_x^Ω-integrable over $\partial^\infty\Omega$ for some x in each component of Ω then f is resolutive. See, for example, [AG, Theorem 6.4.6 and its corollary]. For any l.s.c. function $f > -\infty$ we have $\overline{H}_f^\Omega = \int^*_{\partial^\infty\Omega} f\, d\mu_x^\Omega$, and f is resolutive if $\overline{H}_f^\Omega < +\infty$ on Ω, [AG, Lemma 6.4.4].

1.1.18 A point z of $\partial^\infty\Omega$ (Ω Greenian) is called *regular* (for Ω) if

$$\lim_{x\to z} H_f^\Omega(x) = f(z) \quad \text{for each } f \in \mathcal{C}(\partial^\infty\Omega),$$

or equivalently if

$$\lim_{\Omega\ni x\to z} G_\Omega(x, y) = 0 \quad \text{for each } y \in \Omega,$$

or just for at least one point y in each component of Ω; see [AG, Theorem 6.8.3]. Otherwise z is called *irregular*. We say that Ω is *regular* if every point of $\partial^\infty\Omega$ is regular. The irregular points of $\partial\Omega$ always form a polar set (Kellogg's theorem); see, for example, [AG, Theorem 6.6.8]. For any Greenian set $\Omega \subset \mathbb{R}^N$ ($N \geqslant 2$), a sufficient condition for regularity of a point $y \in \partial\Omega$ is that $\mathbb{R}^N \setminus \Omega$ contains a truncated cone of revolution with vertex at y; this is the Poincaré-Zaremba exterior cone condition. (For $\Omega \subset \mathbb{R}^2$ the truncated cone may be replaced by a line segment with y as an end-point.) In particular, every ball $B \subset \Omega$ is regular, and the harmonic measure μ_x^B ($x \in B$) has the density $K_B(x, \cdot)$ (the Poisson kernel; see Sect. 1.1.2) with respect to the surface measure on ∂B.

1.2 Thin Sets and Fine Topology

For the classical Dirichlet problem for a Greenian domain $\Omega \subset \mathbb{R}^3$, the Poincaré-Zaremba cone condition (see Sect. 1.1.18) is sufficient for regularity of a boundary point for Ω. On the other hand, the Lebesgue spine (or thorn) in $\mathbb{R}^3$ exhibits a limit to how "sparse" the complement $\mathbb{R}^N \setminus \Omega$ can be near a regular boundary point. A remarkable necessary and sufficient condition for regularity of a boundary point for Ω was obtained by Wiener in terms of the Newtonian capacity of the parts of $\mathbb{R}^N \setminus \Omega$ in small annuli centered at y; see, for example, [AG, Theorem 7.7.2].

A different and often more manageable characterization of regularity for a boundary point y for a Greenian set $\Omega \subset \mathbb{R}^N$ was given by Brelot [Br2] in terms of his notion of "effilement" (in English "thinness") of $\mathbb{R}^N \setminus \Omega$ at y. Although $\mathbb{R}^N \setminus \Omega$ is a closed set, Brelot allowed an arbitrary set E in its place.

Definition 1.2.1 (Brelot [Br2]) A set $E \subset \mathbb{R}^N$ $(N \geqslant 2)$ is said to be thin (effilé) at a point $y \in \mathbb{R}^N$ if either $y \notin \overline{E}$ or else there exists a hyperharmonic function u on an open neighborhood of y such that

$$\liminf_{E\setminus\{y\}\ni x\to y} u(x) > u(y).$$

One may clearly assume here that u is superharmonic and that $u \geqslant 0$ on that neighborhood. In the case $N \geqslant 3$ it suffices to consider u defined on all of $\mathbb{R}^N$. A set which is not thin at a point x is said to be *unthin* at x.

Theorem 1.2.2 (Brelot [Br2]) *A set $E \subset \mathbb{R}^N$ $(N \geqslant 2)$ is thin at a point $y \in \mathbb{R}^N$ if and only if, for some and hence any Greenian set $\omega \subset \mathbb{R}^N$ containing y, there exists a (Green) potential $u \geqslant 0$ on ω such that $u(y) < +\infty$ and*

$$u(x) \to +\infty \quad \textit{as } x \to y \textit{ through points } x \in E \setminus \{y\}.$$

See, for example, [AG, Theorem 7.2.3] or [Do4, Theorem 1.XI.2]. It follows by the Riesz decomposition theorem (Sect. 1.1.6) that one may equivalently replace the potential $u \geqslant 0$ by a superharmonic function on ω in Theorem 1.2.2. Clearly, every finite union of sets which are thin at y is likewise thin at y.

In correspondence with Brelot in 1939, H. Cartan observed that the notion of thinness may be described in terms of a new topology on $\mathbb{R}^N$ which he termed *the fine topology*.

Definition 1.2.3 The fine topology on $\mathbb{R}^N$ $(N \geqslant 2)$ is the coarsest topology in which every superharmonic function on $\mathbb{R}^N$ is continuous in the extended sense.

In other words, the fine topology is the intersection of all topologies in which every superharmonic function on $\mathbb{R}^N$ is continuous (such topologies exist, for example the discrete topology). It is easily shown that every superharmonic function on an open set $\Omega \subset \mathbb{R}^N$ is finely continuous, e.g. [AG, Lemma 7.1.2].

The fine topology is indeed stronger (= finer) than the Euclidean topology, and hence Hausdorff. In fact, every ball $B(y,r) \subset \mathbb{R}^N$ with $N \geqslant 3$ equals $\{x \in \mathbb{R}^N : U_y(x) > r^{2-N}\}$ (recall that U_y denotes the fundamental harmonic function, see Sect. 1.1.1), and so $B(y,r)$ is finely open, U_y being superharmonic on $\mathbb{R}^N$, similarly for $N = 2$.

Actually, the fine topology is *strictly finer* than the Euclidean one because there exist superharmonic (hence finely continuous) functions which are discontinuous in the Euclidean topology. Denoting by E the polar set formed by a sequence of distinct points of $\mathbb{R}^N$ converging to a point $x \in \mathbb{R}^N$, there exists (as mentioned in Sect. 1.1.11) a superharmonic function u on $\mathbb{R}^N$ such that $u = +\infty$ on E, but $u(x) < +\infty$, so u is discontinuous at x.

Topological properties relative to the fine topology will henceforth be distinguished by the qualifier *fine* or *finely* from properties relative to the Euclidean topology. We write f-lim for limit in the fine topology, and similarly f-lim inf and f-lim sup. The fine boundary of a set $A \subset \mathbb{R}^N$ is denoted by $\partial_f A$.

A subbasis for the fine topology on $\mathbb{R}^N$ is formed by the sets $\{u < \alpha\}$ and the sets $\{u > \alpha\}$ with u superharmonic on $\mathbb{R}^N$ and $\alpha \in \mathbb{R}$. The sets $\{u > \alpha\}$ being (Euclidean) open we may replace them by balls, and we obtain another subbasis for the fine topology formed by the sets

$$\{x \in B : u(x) < \alpha\}, \tag{1.2.1}$$

where B is an open ball, u is superharmonic on B, and $\alpha \in \mathbb{R}$. This latter subbasis is actually a *basis* for the fine topology as we will see in Theorem 1.2.6 below.

Theorem 1.2.4 *(H. Cartan) The fine neighborhoods of a point $x \in \mathbb{R}^N$ are the complements of those sets $E \subset \mathbb{R}^N \setminus \{x\}$ which are thin at x.*

Proof For any fine neighborhood V of x there exist a ball $B(x,r) = B$, a superharmonic function u on B, and a number $\alpha > u(x)$, such that $\{y \in B : u(y) < \alpha\} \subset V$. Furthermore, x does not belong to the fine closure of $\complement V$. Since $u \geqslant \alpha$ on $B \setminus V$ it follows that

$$\liminf_{\complement V \ni y \to x} u(y) \geqslant \alpha > u(x),$$

and hence $E := \complement V$ is thin at x without containing x. Conversely, suppose that $E \subset \complement\{x\}$ is thin at x. If $x \notin \overline{E}$, then $V := \complement\overline{E}$ is a neighborhood, in particular a fine neighborhood, of x. If $x \in \overline{E}$, there is a superharmonic function u on a ball $B(x,r) = B$ and a number $\alpha > u(x)$ such that

$$\liminf_{E \ni y \to x} u(y) > \alpha.$$

It follows that $u \geqslant \alpha$ on $E \cap B(x,\delta)$ for some $0 < \delta \leqslant r$. Consequently, $\complement E$ contains $\{y \in B(x,\delta) : u(y) < \alpha\}$, which is a fine neighborhood of x. □

Corollary 1.2.5 *A point x of a set $A \subset \mathbb{R}^N$ is finely isolated in A if and only if A is thin at x. Also, A is finely closed if and only if A is thin at every point not in A.*

Theorem 1.2.6 *The sets of the form $V_{B,u,\alpha} = \{x \in B : u(x) < \alpha\}$, where B is an open ball, u is a superharmonic function on B, and $\alpha \in \mathbb{R}$, form a basis for the fine topology on $\mathbb{R}^N$.*

Proof The sets $V_{B,u,\alpha} := \{x \in B : u(x) < \alpha\}$ are finely open because u is finely continuous on B; see paragraph just after Definition 1.2.3. Let $U \subset \Omega$ be a non-empty finely open set and let $a \in U$. We will prove that U contains a fine neighborhood of a of the form $V_{B,u,\alpha}$ as stated in the theorem. This is trivial if a belongs to the Euclidean interior of U. Consider the case when a is an accumulation point of $\complement U$. By Theorem 1.2.4 there is an open ball B centered at a, a superharmonic function u on B, and a real α such that $\inf_{x \in B \cap \complement U} u(x) > \alpha > u(a)$. This implies that $a \in V_{B,u,\alpha} = \{x \in B : u(x) < \alpha\} \subset U$. □

Remark 1.2.7 We obviously may replace the real α in Theorem 1.2.6 by 0.

For quite some time after its appearance, the fine topology was mostly regarded as an artifice—a means of describing thinness. Gradually, however, the fine topology was recognized as a topology in its own right. In the sequel we state some of its main properties.

Every point $x \in \mathbb{R}^N$ has a fine neighborhood basis consisting of sets that are (Euclidean) *compact*. In fact, any fine neighborhood of x contains a basic fine neighborhood $\{y \in B(x,r) : u(y) < \alpha\}$ of x as in (1.2.1), which in turn contains the compact fine neighborhood $\{y \in \overline{B}(x, r/2) : u(y) \leqslant \beta\}$ of x, where $u(x) < \beta < \alpha$.

It follows, in particular, that the fine topology is *regular*. It is even *completely regular*, or equivalently *uniformizable*, as shown by Brelot [Br17, Theorem I.4 (c)]; or see [LMZ, Theorem 2.3].

The fine topology has the *Baire property*: the intersection of a sequence of finely open finely dense sets is finely dense, e.g. [AG, Theorem 7.1.3].

Every nonempty finely open set V has Lebesgue measure $\ell_N(V) > 0$. We may in fact assume that $V = \{y \in B : u(y) < 0\}$ for some superharmonic function u on a ball B. If $\ell_N(V) = 0$ then $u \geqslant 0$ a.e. on B and hence everywhere on B by the ball mean-value property (see Sect. 1.1.2), contradicting $V \neq \varnothing$. As a consequence there is *no finely isolated point* of $\mathbb{R}^N$. Furthermore, if two finely continuous numerical functions f_1, f_2 on a finely open set U satisfy $f_1 \leqslant f_2$ ℓ_N-a.e. on U then $f_1 \leqslant f_2$ everywhere on U.

Every finely continuous numerical function on Ω is of *Baire class One*; see Theorem 1.4.43 below. That leads to a proof that the fine topology is *not normal*, [Fu5, p. 165], and in particular it is not Lindelöf, and therefore *not metrizable*. The fine topology is *not even first countable*, [Berg]. See also [Br17, p. 7], [LMZ, Theorem 2.1]. However, the fine topology does have the very useful *quasi-Lindelöf property*, due to Doob [Do3]; see [Do4, 1.XI.11] or [AG, Theorem 7.3.11]. It states that every family of finely open subsets of $\mathbb{R}^N$ has a countable subfamily whose union differs at most by a polar set from the union of the whole family. In the case $N \geq 3$, this can be derived from the fundamental convergence theorem (Sect. 1.1.13) together with the existence of a strict potential on Ω (Theorem 1.4.15 below).

On the other hand, the quasi-Lindelöf property implies that if u is *nearly hyperharmonic* (that is, hyperharmonic except that the hypothesis of lower semicontinuity of u is replaced by local boundedness from below) then $\widehat{u}$ is actually hyperharmonic, and $\widehat{u}$ is also the greatest *finely* l.s.c. minorant of u. This is due to Brelot [Br8].

The quasi-Lindelöf property extends to functions as follows [Do3]: *Every family* $(f_\alpha)_{\alpha\in I}$ *of finely l.s.c. functions has a countable subfamily whose pointwise supremum differs at most on a polar set from the supremum of the whole family.* Proof: For any set $J \subset I$ write $f_J = \sup_{\alpha\in J} f_\alpha$, which is finely l.s.c. For any rational $r \in \mathbb{Q}$ we have the family $(\{f_\alpha > r\})_{\alpha\in I}$ of finely open sets with the union $\{f_I > r\}$. By the above quasi-Lindelöf property there is a countable set $J_r \subset I$ and a polar set $E_r \subset \mathbb{R}^N$ such that $\{f_I > r\} = \{f_{J_r} > r\} \cup E_r$, hence $\{f_I > r \geqslant f_{J_r}\} \subset E_r$. Writing $J = \bigcup_{r\in\mathbb{Q}} J_r$ (countable) we infer that indeed $f_I = f_J$ q.e.:

$$\{f_I > f_J\} = \bigcup_{r\in\mathbb{Q}} \{f_I > r \geqslant f_J\} \subset \bigcup_{r\in\mathbb{Q}} \{f_I > r \geqslant f_{J_r}\} \subset \bigcup_{r\in\mathbb{Q}} E_r \quad \text{(polar)}.$$

The fine topology, though not normal, has furthermore two *properties of quasi-normality*, [Fu8, Section 3.10]. See Lemma 1.4.34 and Theorem 1.4.35 below.

Every *polar* set $E \subset \mathbb{R}^N$ is *finely closed* and its points are *finely isolated* in E. Thus E is thin at every point of $\mathbb{R}^N$. It follows that E has an *empty fine interior*. Indeed, as noted in Sect. 1.1.11, for every $x \in E$ there is a superharmonic function $u \geqslant 0$ (even a potential) on some Greenian open neighborhood ω of x which equals $+\infty$ on $\omega \cap E \setminus \{x\}$ whereas $u(x) = 1$. The set $\{u < 2\}$ is then a fine neighborhood of x not meeting $E \setminus \{x\}$. Conversely, as shown in Corollary 1.4.19(b) below, a set E is polar if its points are finely isolated in E, or equivalently if E is thin at every point of E.

A major obstacle for fine potential theory is that *the only finely compact sets are finite sets*. Indeed, let $K \subset \mathbb{R}^N$ be an infinite finely compact set, hence also (Euclidean) compact, and in particular bounded. Then K has a subset E (as above) formed by a point x and a sequence of distinct points x_n converging to x. Being polar and hence finely closed, and contained in K, the sets E and $E \setminus \{x\}$ are likewise finely compact. There exist disjoint balls $B(x_n, r_n)$ not containing x. Together with $\complement(E \setminus \{x\})$ these balls form a cover of E by finely open sets, but this cover clearly has no finite subcover. See, for example, [Br17, p. 7] or [AG, p. 178].

It will be shown (Corollary 2.1.24) that an open set in $\mathbb{R}^N$ is *finely connected* if (and only if) it is connected. The fine topology is *locally connected*, and a finely open set has at most countably many fine components (Corollary 2.1.33). If U is finely open and finely connected (in other words: a *fine domain*) then so is $U \setminus E$ for any polar set E (Corollary 2.1.46). For these properties involving fine connectivity, see also [Fu7], [LMZ, 10.A.7]. Any two points of a fine domain can even be joined by a *polygonal path* in U, [Ly1, Fu17]. See Theorem 4.1.1 below.

A remarkable property of thinness, valid in dimension $N = 2$ only, due to Brelot [Br2, Section 12] and going back to Lebesgue [Leb] and Beurling [Beu, p. 71], is that if a set $E \subset \mathbb{R}^2$ is thin at $x \in \mathbb{R}^2$ then there exist arbitrarily small circles centered at x which do not intersect E. The circles can be replaced more generally by suitable closed curves. For the proof, see, for example, [AG, Theorem 7.3.9].

We close this section with a discussion of thinness at infinity in $\mathbb{R}^N$, $N \geqslant 2$, as introduced by Brelot [Br4]. By adjoining the Alexandrov point ∞ to the locally compact space $\mathbb{R}^N$ we obtain the Riemann sphere $\mathbb{R}^N \cup \{\infty\}$, a compact N-manifold. The neighborhoods of ∞ are the complements of the bounded sets within $\mathbb{R}^N \cup \{\infty\}$. The inverse of a point $x \in \mathbb{R}^N \setminus \{y\}$ with respect to an $(N-1)$-sphere $\partial B(y, \rho)$ in $\mathbb{R}^N$ is the point

$$x^* = \left(\frac{\rho}{|x-y|}\right)^2 (x-y) + y \tag{1.2.2}$$

of $\mathbb{R}^N$. In geometric terms, x^* lies on the ray issuing from y and passing through x, and is determined by $|x-y||x^*-y| = \rho^2$. The inversion map $x \mapsto x^*$ is extended to $\mathbb{R}^N \cup \{\infty\}$ by defining $y^* = \infty$ and $\infty^* = y$. The inversion map is then an involutory self-homeomorphism of $\mathbb{R}^N \cup \{\infty\}$. The image of a set $E \subset \mathbb{R}^N \cup \{\infty\}$ under the (extended) inversion map is denoted by E^*, and we have $E^{**} = E$ for any set E.

If a function u, defined on an open subset ω of $\mathbb{R}^N \setminus \{y\}$, is harmonic/subharmonic/superharmonic then so is the composition u^* of u with the above inversion, defined on ω^*, that is,

$$u^*(x) = \left(\frac{\rho}{|x-y|}\right)^{N-2} u(x^*), \quad x \in \omega^*. \tag{1.2.3}$$

The function u^* from (1.2.3) is called the *Kelvin transform* of u. For example, the constant function 1 and the function $x \mapsto (\rho/|x-y|)^{N-2}$ on $\mathbb{R}^N \cup \{\infty\}$ are the Kelvin transforms of one another (when the latter function is extended to y and to ∞ in the obvious way by continuity). The Kelvin transformation equals its own inverse. And u is a harmonic/subharmonic/superharmonic function on an open set $\omega \subset \mathbb{R}^N \setminus \{y\}$ if and only if u^* has the same property on $\omega^* \subset \mathbb{R}^N$, e.g. [Do4, 1.II.12]. In particular, the function $x \mapsto (\rho/|x-y|)^{N-2}$ is superharmonic off ∞ and subharmonic off y (hence harmonic on $\mathbb{R}^N \setminus \{y\}$) while the function 1 is harmonic on all of $\mathbb{R}^N \cup \{\infty\}$.

A set $E \subset \mathbb{R}^N \cup \{\infty\}$, $N \geq 2$, is defined to be *thin at* ∞, e.g. [AG, Definition 7.6.4], if and only if $(E \setminus \{\infty\})^*$ is thin at ∞^* for some inversion in a sphere, and hence by [Do4, Theorem 1.XI.5 (a)] for any such inversion. For $N = 2$ the point at infinity forms a set which is thin even at that point, and (hence) is defined to be *polar*. For $N > 2$ any set containing the point at ∞ (in particular the singleton $\{\infty\}$) is defined to be *unthin at* ∞ and (hence) to be *nonpolar*. See [Br4, p. 312].

1.3 Reduction and Sweeping of Functions

Throughout the rest of this book (except for Chap. 5) we fix a Greenian domain $\Omega \subset \mathbb{R}^N$ $(N \geqslant 2)$, *that is, a connected Greenian set*, in the absence of other indication. Complements of subsets of Ω are taken relative to Ω, and so are topological concepts, for example Euclidean or fine boundaries. We denote by $\mathcal{U} = \mathcal{U}(\Omega)$ the set of all hyperharmonic functions on Ω, and write $\mathcal{U}^+ = \mathcal{U}^+(\Omega) = \{u \in \mathcal{U} : u \geqslant 0\}$. Unless otherwise mentioned, the restriction of the Lebesgue measure ℓ_N on $\mathbb{R}^N$ to Ω is also denoted by ℓ_N.

For any function $f : \Omega \to [0, +\infty]$ and $x \in \Omega$ define

$$R_f(x) = \inf\{u(x) : u \in \mathcal{U}^+,\ u \geqslant f\}\ (\geqslant f(x)),$$

and denote by $\widehat{R}_f$ the l.s.c. regularization of R_f, that is, the greatest l.s.c. minorant of R_f:

$$\widehat{R}_f = \liminf_{y \to x} R_f(y).$$

The fundamental convergence theorem, see Sect. 1.1.13, where $u, \widehat{u}$ correspond to the present $R_f, \widehat{R}_f$, asserts that $\widehat{R}_f \in \mathcal{U}^+$ and

$$\widehat{R}_f = R_f \text{ q.e.,} \quad \text{hence } \widehat{R}_f \geqslant f \text{ q.e.} \tag{1.3.1}$$

and the following slight sharpening of the above limit relation

$$\widehat{R}_f(x) = \liminf_{y \to x,\, y \neq x} R_f(y). \tag{1.3.2}$$

Clearly, $f_1 \leqslant f_2$ implies $R_{f_1} \leqslant R_{f_2}$ and $\widehat{R}_{f_1} \leqslant \widehat{R}_{f_2}$.

Theorem 1.3.1 (e.g. Doob [Do4, 1.XI.7]) *For any function* $f : \Omega \to [0, +\infty]$, $\widehat{R}_f$ *is also the finely l.s.c. regularization of* R_f, *and*

$$\widehat{R}_f(x) = \text{f} - \lim_{y \to x,\, y \neq x} R_f(y).$$

Proof The polar set $E = \{\widehat{R}_f \neq R_f\}$ being finely discrete (cf. Sect. 1.2), every point $x \in \Omega$ has a fine neighborhood V such that $E \cap (V \setminus \{x\}) = \varnothing$, that is, $\widehat{R}_f = R_f$ on $V \setminus \{x\}$. Since $\widehat{R}_f$ is finely continuous, this leads as follows to the resulting equality in Theorem 1.3.1 in view of (1.3.2):

$$\widehat{R}_f(x) \leqslant \text{f} - \liminf_{y \to x,\, y \in V \setminus \{x\}} R_f(y) = \text{f} - \liminf_{y \to x,\, y \in V \setminus \{x\}} \widehat{R}_f(y) = \widehat{R}_f(x).$$

□

Theorem 1.3.2 *For any function* $f : \Omega \to [0, +\infty]$ *we have* $\widehat{R}_f = \min\{u \in \mathcal{U}^+ : u \geqslant f \text{ q.e. on } \Omega\}$. *In particular, if* $f_1 \leqslant f_2$ *q.e. then* $\widehat{R}_{f_1} \leqslant \widehat{R}_{f_2}$ *everywhere.*

Proof For the former assertion, since $\widehat{R}_f \in \mathcal{U}^+$ it remains in view of (1.3.1) to show that $u \geqslant \widehat{R}_f$ everywhere on Ω for any $u \in \mathcal{U}^+$ such that $u \geqslant f$ q.e. We have $u \geqslant f$ on $\complement E$ for some polar set $E \subset \Omega$. For given $x \in \complement E$ choose $v \in \mathcal{U}^+$ (for example a potential v) such that $v = +\infty$ on E and $v(x) < \varepsilon$ (see Sect. 1.1.11). Then $u + v \geqslant f$ (both on E and on $\complement E$), hence

$$u(x) + \varepsilon \geqslant u(x) + v(x) \geqslant R_f(x) \geqslant \widehat{R}_f(x),$$

and so $u(x) \geqslant \widehat{R}_f(x)$, by letting $\varepsilon \to 0$. By varying $x \in \complement E$ we obtain $u \geqslant \widehat{R}_f$ on $\complement E$, hence q.e. on Ω, and indeed everywhere by fine continuity of u and $\widehat{R}_f$, both in $\mathcal{U}^+$. For the latter assertion of the theorem, consider any $u \in \mathcal{U}^+$ such that $u \geqslant f_2$ q.e. Then $u \geqslant f_1$ q.e. Taking minimum over these u and applying the former assertion of the theorem with f replaced by f_2, we obtain the claimed inequality $\widehat{R}_{f_1} \leqslant \widehat{R}_{f_2}$. □

Theorem 1.3.3 *For any function $f : \Omega \to [0, +\infty]$ we have*

(a) $R_f(x) = f(x) \vee \widehat{R}_f(x)$ *for every* $x \in \Omega$.
(b) *If f is finely l.s.c. on some finely open set $U \subset \Omega$ then $\widehat{R}_f \equiv R_f$ $(\geqslant f)$ on U.*

Proof (a) For a fixed point $x \in \Omega$, we shall prove that $R_f(x) \leqslant f(x) \vee \widehat{R}_f(x)$, the inequality "$\geqslant$" being obvious. Since $R_f \geqslant f$ we may assume that $f(x) < +\infty$. For given $\varepsilon > 0$ there exists $u \in \mathcal{U}^+$ such that $u = +\infty$ on the polar set

$$E := \{y \in \complement\{x\} : \widehat{R}_f(y) < f(y)\},$$

cf. (1.3.1), and that $u(x) < \varepsilon$; see Sect. 1.1.11. It follows that $\widehat{R}_f + u \in \mathcal{U}^+$ and

$$\widehat{R}_f + u \geqslant f \text{ on } \complement\{x\}, \tag{1.3.3}$$

both on $E \cap \complement\{x\} = E$ (where $u = +\infty$) and on $(\complement E) \cap \complement\{x\}$ (where $\widehat{R}_f \geqslant f$ by definition of E).

If $\widehat{R}_f(x) \geqslant f(x)$ then $\widehat{R}_f + u \geqslant f$ on all of Ω in view of (1.3.3), and hence

$$R_f(x) \leqslant \widehat{R}_f(x) + u(x) \leqslant \widehat{R}_f(x) + \varepsilon.$$

If instead $\widehat{R}_f(x) < f(x)$, then $c := f(x) - \widehat{R}_f(x) > 0$ and $c < +\infty$. We then obtain $\widehat{R}_f + u + c \in \mathcal{U}^+$ and $\widehat{R}_f + u + c \geqslant f$ on Ω, whence

$$R_f(x) \leqslant \widehat{R}_f(x) + u(x) + c = f(x) + u(x) \leqslant f(x) + \varepsilon.$$

In any case, $R_f(x) \leqslant (f(x) \vee \widehat{R}_f(x)) + \varepsilon$. By varying first $\varepsilon > 0$ and next $x \in \Omega$ we conclude that $R_f \leqslant f \vee \widehat{R}_f \leqslant R_f$ (since $f, \widehat{R}_f \leqslant R_f$), and so $R_f = f \vee \widehat{R}_f$.

(b) By (a), R_f is finely l.s.c. on U along with f and $\widehat{R}_f$, and hence $R_f = \widehat{R}_f$ on U in view of Theorem 1.3.1. □

Theorem 1.3.4 (Brelot [Br5]) *Let $f : \Omega \to [0, +\infty]$ be a l.s.c. function having a superharmonic majorant. Then $R_f \equiv \widehat{R}_f$ is superharmonic on Ω. Moreover,*

(a) *R_f is harmonic on any open set $\omega \subset \Omega$ such that f is either subharmonic on ω or continuous and $< R_f$ there.*

(b) *R_f is finite and continuous at any point of Ω at which f is finite and continuous.*

Proof We adapt the proof of [CC3, Proposition 2.2.3] (which is in a more general axiomatic setting), where the present requirement of positivity of f is weakened to lower finiteness. Obviously R_f and $\widehat{R}_f$ are identical and superharmonic on Ω.

(a) Let $\omega \subset \Omega$ be open, and let first $f \geqslant 0$ be a subharmonic (hence finite) function on ω. We may clearly assume that ω is connected. Then f is u.s.c. and actually continuous, and R_f is superharmonic on ω. So is therefore $R_f - f$, which is moreover $\geqslant 0$ on ω by Theorem 1.3.3(b) applied to $U = \omega$. According to the last paragraph of Sect. 1.1.2 we may further assume that $R_f - f \not\equiv 0$, and hence $R_f - f > 0$ everywhere on the domain ω.

The proof of (a) thus reduces to the other case where f is continuous and $f < R_f$ on ω. Let $x_0 \in \omega$ be given, and suppose by contradiction that R_f is not harmonic on any ball about x_0. Writing $c = \frac{1}{2}(f(x_0) + R_f(x_0))$ (finite) we have $f(x_0) < c < R_f(x_0)$, and since f is u.s.c. and R_f is l.s.c. on ω there is a ball B about x_0 such that $\overline{B} \subset \omega$ and

$$f < c < R_f$$

on $\overline{B}$. Hence $R_f \leqslant P_B R_f$ on $\overline{B}$, with equality by superharmonicity of R_f, which also shows that R_f is harmonic on B (see Sect. 1.1.2 for the Poisson integral). This contradiction proves assertion (a) for the remaining case where f is finite and continuous on ω and where $R_f > f$ on ω. By varying x_0 we see that R_f is harmonic and hence finite on all of ω, which will be used in the proof of (b) below.

(b) Suppose that f is finite and continuous at some point $x \in \Omega$, and let $\varepsilon > 0$ be given. There is a ball B centered at x and of closure $\overline{B} \subset \Omega$, and small enough so that

$$f \leqslant f(x) + \varepsilon \quad \text{and} \quad R_f \geqslant R_f(x) - \varepsilon \tag{1.3.4}$$

on $\overline{B}$, R_f being l.s.c. and finite valued at x. The function u^* defined on Ω by

$$u^* = \begin{cases} R_f \wedge (P_B R_f + 2\varepsilon) & \text{on } B \\ R_f & \text{on } \complement B \end{cases} \tag{1.3.5}$$

is superharmonic, see Sect. 1.1.7, because the extension of $P_B R_f$ by R_f on $\complement B$ (the Poisson modification of R_f; see the beginning of Sect. 1.1.15) is l.s.c. on Ω. By (1.3.4), $R_f \geqslant R_f(x) - \varepsilon$ on $\partial B \subset \overline{B}$ and hence $P_B R_f \geqslant R_f(x) - \varepsilon$ on B. Thus

$$P_B R_f + 2\varepsilon \geqslant R_f(x) + \varepsilon \geqslant f(x) + \varepsilon \geqslant f$$

on B. It follows from (1.3.5) that $u^* \geqslant R_f \geqslant f$ on B, and trivially on $\complement B$. Thus

$$R_f \leqslant u^* \leqslant P_B R_f + 2\varepsilon$$

on B. In particular, $R_f(x) < \infty$. Since $P_B R_f$ is harmonic and hence continuous on B we obtain

$$\limsup_{y \to x,\, y \in B} R_f(y) \leqslant P_B R_f(x) + 2\varepsilon \leqslant R_f(x) + 2\varepsilon$$

for $x \in B$ because R_f is superharmonic; cf. Sect. 1.1.2. Here ε may be replaced by 0, and so R_f is u.s.c. at x. Being also l.s.c. and finite at x, R_f is indeed finite and continuous at x. □

Theorem 1.3.5 *Let $f : \Omega \to [0, +\infty]$ have a superharmonic majorant s. Then $\widehat{R}_f$ is superharmonic on Ω. Furthermore, $\widehat{R}_f$ is a Greenian potential on Ω provided that f has compact support.*

Proof Clearly $\widehat{R}_f \leqslant R_f \leqslant s$ on Ω since $f \leqslant s$. It follows that $\widehat{R}_f$ indeed is superharmonic on Ω along with s.

Suppose now in addition that f is null outside of a compact set $K \subset \Omega$. Let $s = p + h$ be the Riesz decomposition of s, where p is a (Greenian) potential on Ω and $h \geqslant 0$ a harmonic function on Ω. If $h \equiv 0$ there is nothing to prove because $\widehat{R}_f \leqslant s = p$. For other harmonic $h \geqslant 0$ we have $h > 0$ on Ω. Let $q > 0$ be a potential on Ω. The (well-defined) function q/h is then l.s.c. and > 0 on Ω. There is therefore a finite constant $c > 0$ such that $q/h \geqslant 1/c$ on K, that is, $h \leqslant cq$ on K. Thus $f \leqslant s = p + h \leqslant p + cq$ on K and hence $f \leq p + cq$ on all of Ω because $f \equiv 0$ off K. Consequently, $\widehat{R}_f$ is indeed a potential on Ω along with $p + cq$. □

Theorem 1.3.6 *Let $f_n : \Omega \to [0, +\infty]$ for all $n \in \mathbb{N}$, and let $f : \Omega \to [0, +\infty]$. Then*

(a) *If $f_n \uparrow f$ then $R_{f_n} \uparrow R_f$ and $\widehat{R}_{f_n} \uparrow \widehat{R}_f$.*
(b) *For $f := \sum_n f_n$ we have $R_f \leqslant \sum_n R_{f_n}$ and $\widehat{R}_f \leqslant \sum_n \widehat{R}_{f_n}$.*

Proof (a) Generally, $f_n \uparrow f$ signifies that the sequence (f_n) is increasing and has pointwise limit f. Write $\widehat{R}_{f_n} = u_n$ and $\sup_n u_n = u$. Then $u_n \in \mathcal{U}^+$ and hence $u \in \mathcal{U}^+$. Since $u \geqslant u_n \geqslant f_n$ q.e. by (1.3.1), we have $u \geqslant f$ q.e., and hence $u \geqslant \widehat{R}_f$ by Theorem 1.3.2. Obviously, $\widehat{R}_f \geqslant u$ q.e. and hence everywhere by fine continuity. It follows that $\widehat{R}_f = u$. Finally, we have by Theorem 1.3.3(a)

$$R_f = f \vee \widehat{R}_f = f \vee u = \sup_n (f_n \vee u_n) = \sup_n (f_n \vee \widehat{R}_{f_n}) = \sup_n R_{f_n}.$$

(b) For given $x \in \Omega$ such that $\sum_n \widehat{R}_{f_n}(x) < +\infty$ and for given $\varepsilon > 0$, there exist by Theorem 1.3.2 functions $u_n \in \mathcal{U}^+$ such that $u_n \geqslant f_n$ q.e. and

$$u_n(x) < \widehat{R}_{f_n}(x) + 2^{-n}\varepsilon \ (< +\infty).$$

Defining now $u = \sum_n u_n \in \mathcal{U}^+$ we have $u \geqslant f$ q.e. It follows, again by Theorem 1.3.2, that

$$\widehat{R}_f(x) \leqslant u(x) \leqslant \sum_n \widehat{R}_{f_n}(x) + \varepsilon,$$

and so $\widehat{R}_f \leqslant \sum_n \widehat{R}_{f_n}$, this being obvious at points at which $\sum \widehat{R}_{f_n} = +\infty$. The proof for the case of R in place of $\widehat{R}$ is similar (now omitting "q.e."). □

For a function $f : \Omega \to \overline{\mathbb{R}}$ we denote by $\widetilde{f}$ the least finely u.s.c. majorant of f given by

$$\widetilde{f}(x) = \text{f} - \limsup_{y \to x,\, y \in \Omega} f(y).$$

(This is in harmony with the notation $\widetilde{A}$ for the fine closure of a set $A \subset \Omega$.) Note that

$$R_{\widetilde{f}} = R_f, \quad \widehat{R}_{\widetilde{f}} = \widehat{R}_f \tag{1.3.6}$$

because functions of class $\mathcal{U}^+$ are finely continuous. For the indicator function 1_A of a set $\subset \Omega$, we clearly have

$$\widetilde{1_A} = 1_{\widetilde{A}}$$

where $\widetilde{A}$ denotes the fine closure of A.

We proceed to study in detail the principal case of reduction and sweeping of a function $f = 1_A u$, that is, $f = u$ on a set $A \subset \Omega$ and $f = 0$ on $\complement A$, whereby $u \in \mathcal{U}^+$. Then $R_u^A := R_f$ is called the *reduced function* (réduite) of u on A, and $\widehat{R}_u^A := \widehat{R}_f$ is called the *swept* (function) of u on A, or the sweeping of u on A, the mapping $u \mapsto \widehat{R}_u^A$ being called *sweeping* (balayage) on A. By (1.3.1) and Theorems 1.3.1, 1.3.2, and 1.3.3 we obtain the following equations:

$$\widehat{R}_u^A = \min\{v \in \mathcal{U}^+ : v \geqslant u \text{ q.e. on } A\}, \tag{1.3.7}$$

$$\widehat{R}_u^A = R_u^A \quad \text{q.e., hence } \widehat{R}_u^A = u \text{ q.e. on } A, \tag{1.3.8}$$

$$R_u^A - u \quad \text{on } A \quad \text{(trivial)}, \tag{1.3.9}$$

$$R_u^A = \widehat{R}_u^A \quad \text{on } \complement A, \text{ and on the fine interior } A' \text{ of } A, \tag{1.3.10}$$

$$R_u^A = \widehat{R}_u^A \quad \text{if } A \text{ is finely open}, \tag{1.3.11}$$

$$R_u^A = R_u^{\widetilde{A}}, \quad \widehat{R}_u^A = \widehat{R}_u^{\widetilde{A}}. \tag{1.3.12}$$

Equation (1.3.10) follows because $R_u^A = \widehat{R}_u^A \vee (1_A u) = \widehat{R}_u^A$ on $\complement A$ by Theorem 1.3.3(a) and because $f = 1_A u$ equals $0 \leqslant \widehat{R}_u^A$ on the set $\complement A$. On the other hand $f = 1_A u$ is finely l.s.c. on the finely open set A' by Theorem 1.3.3(b), and hence $\widehat{R}_f \equiv R_f$ on A'. Equation (1.3.11) is a consequence of (1.3.10), whereas (1.3.12) is a particular case of (1.3.6).

Lemma 1.3.7 *For any superharmonic function $u \geqslant 0$ on Ω and any set $A \subset \Omega$, R_u^A is harmonic on $\Omega \setminus \overline{A}$, and $\widehat{R}_u^A = R_u^A$ there.*

Proof We may assume that $A \neq \Omega$. Denote by Γ the class of superharmonic functions $v \geqslant 0$ on Ω such that $v \geqslant u$ on A. Then $R_u^A = \inf \Gamma$ (pointwise infimum). If

$\complement \overline{A}$ is not empty and if B is a ball with closure $\overline{B} \subset \complement \overline{A}$, then the function $P_B v$ is in Γ whenever $v \in \Gamma$ and $P_B v \leqslant v$. Thus R_u^A, considered on B, is the infimum of the lower directed family $\{(P_B v)|_B : v \in \Gamma\}$ of harmonic functions. By the Harnack convergence theorem, R_u^A is then harmonic on each B and therefore on all of $\complement \overline{A}$, and R_u^A equals $\widehat{R}_u^A$ there. □

Corollary 1.3.8 *For any potential p on Ω it holds pointwise on Ω that*

$$\inf\{R_p^{\complement K} : K \text{ compact}\} = 0.$$

Proof For any compact subset K of Ω, $\widehat{R}_p^{\complement K} = R_p^{\complement K}$ by (1.3.11), and this function is harmonic on the interior $\overset{\circ}{K}$ of K, by Lemma 1.3.7. By Harnack's convergence theorem the stated pointwise infimum is harmonic $\geqslant 0$ everywhere on Ω and hence equals 0 there, being majorized by the potential p. □

Lemma 1.3.9 *Let $A \subset \Omega$, $x \in \Omega$, and $u \in \mathcal{U}^+$.*

(a) *If A is unthin at x then $\widehat{R}_u^A(x) = u(x)$, and even $\widehat{R}_u^{A \cap B(x,r)}(x) = u(x)$ for any $r > 0$.*

(b) *If A is thin at x, and if u is bounded on a neighborhood of x, then $\widehat{R}_u^{A \cap B(x,r)}(x) \to 0$ as $r \to 0$.*

Proof (a) Write $v = \widehat{R}_u^{A \cap B(x,r)}$ ($\leqslant u$). By (1.3.1), $v = u$ on $(A \setminus E) \cap B(x,r)$ for some polar set E. If A is unthin at x then so is $(A \setminus E) \cap B(x,r)$ because E is everywhere thin; see Sect. 1.2. Hence x belongs to the fine closure of $(A \setminus E) \cap B(x,r)$, and $u = v$ there because u and v are finely continuous. Thus $u(x) = v(x)$. Of course, $u(x) \geqslant \widehat{R}_u^A(x) \geqslant \widehat{R}_u^{A \cap B(x,r)}(x) = v(x) = u(x)$.

(b) We may assume that $u \leqslant 1$ on $B(x,\rho)$ for some $\rho > 0$. According to Theorem 1.2.2, there exists a superharmonic function $w \geqslant 0$ on $B(x,\rho)$ such that $w(x) = 1$ and

$$\liminf_{y \to x,\, y \in A \setminus \{x\}} w(y) = +\infty.$$

For given $\varepsilon > 0$ choose $r > 0$, $r \leqslant \rho$, so that $w \geqslant 1/\varepsilon$ on $A \cap B(x,r) \setminus \{x\}$. Then $\varepsilon w \geqslant 1 \geqslant u$ q.e. on $A \cap B(x,r)$, and so $\widehat{R}_u^{A \cap B(x,r)}(x) \leqslant \varepsilon w(x) = \varepsilon$ in view of Theorem 1.3.2. □

Lemma 1.3.10 *For any $E \subset \Omega$ the following are equivalent:*

(a) *E is polar.*

(b) *$\widehat{R}_u^E \equiv 0$ for every $u \in \mathcal{U}^+$.*

(c) *There is a superharmonic function $s > 0$ on Ω such that $\widehat{R}_s^E \equiv 0$ on Ω.*

Proof (a)$\Rightarrow$(b). For any $x \in \complement E$ and $\varepsilon > 0$, there is a superharmonic function $s \geqslant 0$ such that $s = +\infty$ on E and $s(x) < \varepsilon$ (see Sect. 1.1.11). It follows that $\widehat{R}_{+\infty}^E(x) \leqslant R_{+\infty}^E(x) < \varepsilon$, and hence $\widehat{R}_{+\infty}^E = 0$ on $\complement E$, which implies (b) because E has empty fine interior (see Sect. 1.2), that is, $\complement E$ is finely dense in Ω, and because $\widehat{R}_u^E$ is finely continuous.

(b)$\Rightarrow$(c) is trivial (take $u \equiv 1$ in (b)).

(c)⇒(a) follows from Theorem 1.3.6(a) applied to $f = +\infty \cdot 1_E = \sup_n(ns \cdot 1_E)$, showing that $\widehat{R}_f = 0$, hence $R_f = f$ by Theorem 1.3.3(a), and this amounts quasi-everywhere to $\widehat{R}_f = f$, by (1.3.1). Thus $f = 0$ q.e., but not on E, which therefore must be polar. □

Lemma 1.3.11 *For $A \subset B \subset \Omega$ we have*

$$\widehat{R}^B_{\widehat{R}^A_u} = \widehat{R}^A_u = \widehat{R}^A_{\widehat{R}^B_u}.$$

Proof From (1.3.8) with A replaced by B and u replaced by $\widehat{R}^A_u$ we obtain $\widehat{R}^B_{\widehat{R}^A_u} = \widehat{R}^A_u$ q.e. on B and hence $\widehat{R}^B_{\widehat{R}^A_u} = u$ q.e. on $A \subset B$, again by (1.3.8). It follows by (1.3.7) (applied with u replaced by $\widehat{R}^A_u$) that $\widehat{R}^B_{\widehat{R}^A_u} \geqslant {}^U\widehat{R}^A_u$, and here the opposite inequality is obvious. And from ${}^U\widehat{R}^B_u = u$ q.e. on $A \subset B$ we find by (1.3.7) that ${}^U\widehat{R}^A_{\widehat{R}^B_u} = \widehat{R}^A_u$. □

Definition 1.3.12 For $u, v \in \mathcal{U}$ write $u \preccurlyeq v$ if there exists $w \in \mathcal{U}^+$ such that $v = u + w$.

This defines an order relation on $\mathcal{U}$ called the *specific order*. Clearly, $u \preccurlyeq v$ implies $u \leqslant v$ (the *natural* or pointwise order on $\mathcal{U}$). The specific infimum (resp. supremum) of a family $\mathcal{F}$ of hyperharmonic functions on Ω is denoted by $\curlywedge\mathcal{F}$ (resp. $\curlyvee\mathcal{F}$). We write $u \curlywedge v$ for $\curlywedge\{u, v\}$, etc. when u and v are hyperharmonic functions.

If u is superharmonic then w in Definition 1.3.12 is uniquely determined by u and v. Indeed, if $u + w_1 = u + w_2$ for distinct functions $w_1, w_2 \in \mathcal{U}^+$ then the non-void finely open set $E := \{w_1 \neq w_2\}$ would be contained in the polar set $\{u = +\infty\}$, which contradicts the fact that a polar set has no finely inner points (see Sect. 1.2).

When applied to (Greenian) *potentials*, the specific order corresponds to the order of the associated Riesz measures:

$$G_\Omega\mu \preccurlyeq G_\Omega\nu \iff \mu \leqslant \nu.$$

For if $w + G_\Omega\mu = G_\Omega\nu$ for some $w \in \mathcal{U}^+$, necessarily a potential $w \leqslant G_\Omega\nu$, hence $w = G_\Omega\tau$ for a certain measure $\tau \geqslant 0$ (see Sect. 1.1.6). It follows that $\mu + \tau = \nu$, and hence $\mu \leqslant \nu$, because $G_\Omega(\mu + \tau) = G_\Omega\mu + G_\Omega\tau = G_\Omega\nu$.

Theorem 1.3.13 (Mokobodzki and Sibony [MS]) *Let $u, v \in \mathcal{U}^+$. Write $f(x) = (u(x) - v(x)) \vee 0$, understood as 0 at points x where $v(x) = +\infty$. Then $R_f \in \mathcal{U}^+$ and $R_f \preccurlyeq u$.*

Proof Since u and v are finely continuous, f is finely l.s.c. on Ω (being finely continuous and > 0 on the finely open set where $u > v$, and $f = 0$ elsewhere). It follows by Theorem 1.3.3(b) that $R_f = \widehat{R}_f \in \mathcal{U}^+$. By the definition of the function f we obviously have

$$u \leqslant v + f \leqslant v + R_f \quad \text{on } \Omega. \tag{1.3.13}$$

Adapting the proof given in [CC3, p. 120] we consider a ball B such that $\overline{B} \subset \Omega$. Let g be a finite and continuous real function on ∂B with $g \leqslant u$. Define

$$u^* = \begin{cases} R_f \wedge (P_B R_f + u - P_B g) & \text{on } B \\ R_f & \text{on } \complement B, \end{cases}$$

where $P_B f$ denotes the Poisson integral of $f|_{\partial B}$ (see Sect. 1.1.2). Since $R_f = \widehat{R}_f$ is l.s.c. and $P_B g \leqslant P_B u \leqslant u$, it follows for every $y \in \partial B$ by (1.1.1) and subsequent lines (applied to the present R_f in place of f there) that

$$\liminf_{x \to y,\, x \in B} (P_B R_f + u - P_B g)(x) \geqslant R_f(y).$$

This implies that $u^* \in \mathcal{U}^+$ (see Sect. 1.1.7).

On ∂B ($\subset \complement B$) we have $g \leqslant u \leqslant v + R_f$ by (1.3.13), and hence $P_B g \leqslant P_B v + P_B R_f \leqslant v + P_B R_f$ on B, which amounts to

$$u = P_B g + (u - P_B g) \leqslant v + (P_B R_f + u - P_B g) \quad \text{on } B.$$

In combination with $u \leqslant v + R_f$ on Ω by (1.3.13) this shows by definition of u^* on B that $u \leqslant v + u^*$ on B, and therefore on all of Ω since $u \leqslant v + R_f = v + u^*$ on $\complement B$. We infer that $f \leqslant u^*$ and hence $R_f \leqslant u^*$ because $u^* \in \mathcal{U}^+$. Again by the definition of u^* we therefore have $R_f \leqslant u^* \leqslant u + P_B R_f - P_B g$ on B. After adding $P_B g$ on both sides we obtain $R_f + P_B g \leqslant u + P_B R_f$ on B, or in view of (1.1.2)

$$R_f(x) + \int_{\partial B} g \, d\mu_x^B \leqslant u(x) + \int_{\partial B} R_f \, d\mu_x^B \quad \text{for } x \in B.$$

By taking supremum over all $g \in \mathcal{C}(\partial B)$ with $g \leqslant u$ it follows that the above inequality holds with g replaced by u, or again by (1.1.2)

$$R_f + P_B u \leqslant P_B R_f + u \quad \text{on } B. \tag{1.3.14}$$

Define $w = u - R_f$ at points of Ω where R_f is finite, and $w = +\infty$ elsewhere on Ω. Then $w \geqslant 0$ because $u \geqslant R_f$ by the definition of f. Furthermore, by (1.3.14),

$$P_B w = P_B u - P_B R_f \leqslant u - R_f = w$$

at points of B where R_f and hence $P_B R_f \leqslant R_f$ are finite. Thus $P_B w \leqslant w$ on all of B. The l.s.c. regularization $\widehat{w}$ of w satisfies $\widehat{w} \geqslant 0$ and $P_B \widehat{w} \leqslant P_B w \leqslant w$ on B. It follows that $P_B \widehat{w} \leqslant \widehat{w}$ on B, $P_B \widehat{w}$ being harmonic and hence continuous on B.

By varying B we conclude that $\widehat{w}$ is hyperharmonic on Ω. Note that $u = R_f + w$ on Ω, both where R_f is finite and where R_f is infinite (for there we have $u \geqslant R_f = +\infty$). By Theorem 1.3.1, $\widehat{w}$ is also the *finely* l.s.c. regularization of w. Furthermore, $w = u - R_f$ is finely continuous and hence equal to $\widehat{w}$ on the finely open set $\{R_f < +\infty\}$. Thus $u = R_f + \widehat{w}$ where R_f is finite, and trivially so at points where R_f and hence u are infinite. We conclude that indeed $R_f \preccurlyeq u$. □

Corollary 1.3.14 (The natural decomposition property [MS]) *Let $u, v, w \in \mathcal{U}^+$. If $u \leqslant v + w$ there exist $v', w' \in \mathcal{U}^+$ such that*

$$v' \leqslant v, \ w' \leqslant w, \ u = v' + w'.$$

Here "natural" refers to the pointwise order. There is a similar property in terms of the specific order, with u, v, w superharmonic $\geqslant 0$; see [Do4, Appendix A.III.6] in the general setting of vector lattices (Riesz spaces).

Proof If $v \equiv +\infty$ (resp. $w \equiv +\infty$) it suffices to take $v' = u$ and $w' = 0$ (resp. $v' = 0$ and $w' = u$). So we may suppose that u and v are superharmonic. With $f = (u - v) \vee 0$ understood as before, Theorem 1.3.13 tells us that $R_f \in \mathcal{U}^+$ and that there exists $v' \in \mathcal{U}^+$ such that

$$u = v' + R_f.$$

Furthermore, (1.3.13) holds, whence by subtraction $v' \leqslant v$, first ℓ_N-a.e., and next everywhere (see Sect. 1.1). It remains to take $w' = R_f$, noting that $R_f \leqslant w$ because $f = (u - v) \vee 0 \leqslant w$ by the hypothesis that $u \leqslant v + w$. □

Theorem 1.3.15 ([CC3, Theorem 4.2.1]) *For any $u, v \in \mathcal{U}^+$ and $A \subset \Omega$,*

$$R^A_{u+v} = R^A_u + R^A_v, \quad \widehat{R}^A_{u+v} = \widehat{R}^A_u + \widehat{R}^A_v.$$

Proof For the latter equality suppose first that $u + v$ is superharmonic. By Theorem 1.3.6(b) we have $\widehat{R}^A_{u+v} \leqslant \widehat{R}^A_u + \widehat{R}^A_v$. According to Corollary 1.3.14 there are $u', v' \in \mathcal{U}^+$ such that

$$u' \leqslant \widehat{R}^A_u, \quad v' \leqslant \widehat{R}^A_v, \quad u' + v' = \widehat{R}^A_{u+v}. \tag{1.3.15}$$

In particular, $u' \leqslant u, v' \leqslant v$. Now $u' + v' = \widehat{R}^A_{u+v} = u + v$ q.e. on A, by (1.3.8), and $u + v$ is superharmonic and hence $\widehat{R}^A_{u+v} < +\infty$ q.e.; cf. Sect. 1.1.11. We therefore obtain

$$u' = u, \quad v' = v \quad \text{q.e. on } A,$$

and hence, by (1.3.7), $u' \geqslant \widehat{R}^A_u$, $v' \geqslant \widehat{R}^A_v$ on Ω. When combined with (1.3.15) this leads to $u' = \widehat{R}^A_u$, $v' = \widehat{R}^A_v$, and so indeed $\widehat{R}^A_{u+v} = u' + v' = \widehat{R}^A_u + \widehat{R}^A_v$.

In the remaining case where $u + v \equiv +\infty$, if moreover A is polar then the latter equality of the theorem becomes $0 = 0 + 0$ since $\widehat{R}^A_w \equiv 0$ for any $w \in \mathcal{U}^+$ by Lemma 1.3.10. And if A is nonpolar then $\widehat{R}^A_{u+v} = u + v = +\infty$ q.e. on A.

The hyperharmonic function $\widehat{R}^A_{u+v}$ thus equals $+\infty$ on a nonpolar set, and so $\widehat{R}^A_{u+v} \equiv +\infty \geqslant \widehat{R}^A_u + \widehat{R}^A_v$, necessarily with equality by Theorem 1.3.6(b).

The remaining former equality of Theorem 1.3.15 reduces at any point of A to $u + v = u + v$ in view of (1.3.9), and reduces at any point of $\complement A$ to the latter equality just established, in view of (1.3.10). □

1.4 Sweeping of Measures. The Base Operation

We denote by[2]

$$\begin{aligned}
\mathcal{P} = \mathcal{P}(\Omega) &= \text{ the cone of all (Green) } \textit{potentials} \text{ on } \Omega,\\
\mathcal{P}^s = \mathcal{P}^s(\Omega) &= \text{ the cone of all } \textit{semibounded potentials} \text{ on } \Omega,\\
\mathcal{P}^b = \mathcal{P}^b(\Omega) &= \text{ the cone of all } \textit{locally bounded potentials} \text{ on } \Omega,\\
\mathcal{P}^c = \mathcal{P}^c(\Omega) &= \text{ the cone of all } \textit{finite continuous potentials} \text{ on } \Omega,\\
\mathcal{P}^c_0 = \mathcal{P}^c_0(\Omega) &= \{p \in \mathcal{P}^c(\Omega) : p \text{ is } \textit{harmonic off some compact subset of } \Omega\}.
\end{aligned}$$

A potential $p = G_\Omega\mu \in \mathcal{P}^c$ is of class $\mathcal{P}^c_0$ if and only if μ has compact support in Ω; cf. Sect. 1.1.9.

We shall often need the following approximation theorem for which we bring the proof given in [Do4, p. 66].

Theorem 1.4.1 (La Vallée-Poussin [VP2], Choquet [Ch2]) *Let μ be a (positive) measure on Ω. Suppose that $G_\Omega\mu$ is superharmonic and that the Borel set $A := \{G_\Omega\mu < +\infty\}$ carries μ. There is then a sequence (μ_n) of measures on Ω for which μ_n is supported by some compact subset of A, $G_\Omega\mu_n$ is bounded and continuous, and $\mu = \sum_{n=1}^\infty \mu_n$, so $G_\Omega\mu = \sum_{n=1}^\infty G_\Omega\mu_n$.*

Proof According to Lusin's theorem (taken by Bourbaki [Bou6, Chapter IV, § 5, Définition 1] as definition of μ-measurability), there exists a sequence of disjoint compact sets $A_n \subset A$ such that $\mu(A \setminus \bigcup_n A_n) = 0$ and that each $G_\Omega\mu|_{A_n}$ is finite and continuous, hence bounded. If μ_n denotes the restriction of μ to A_n then the bounded continuous function $G_\Omega\mu|_{A_n}$ is the sum of the restrictions to A_n of the l.s.c. functions $G_\Omega\mu_n$ and $G_\Omega(\mu - \mu_n)$. These two restrictions are therefore continuous, hence so is $G_\Omega\mu_n$ itself by the continuity principle (Sect. 1.1.9). For the boundedness of $G_\Omega\mu_n$, see Remark 1.4.2.[3] □

Remark 1.4.2 Every finite and continuous (or just locally bounded) potential $G_\Omega\mu$ such that μ has compact support is *bounded*. This is well known and straightforward

[2] For semibounded potentials see Definition 1.5.16 below.

[3] Frostman's maximum principle is not available here. It is the particular case v constant of the domination principle (Theorem 1.4.3), the proof of which, however, uses the present Theorem 1.4.1.

if Ω equals $\mathbb{R}^N$ ($N \geqslant 3$). For the present general Greenian domain $\Omega \subset \mathbb{R}^N$ ($N \geqslant 2$), see [Do4, p. 56 and 1.VII.3 (d)], where it is shown that $G_\Omega\mu$ is bounded off each neighborhood of the compact support of μ, and hence on all of Ω.

Theorem 1.4.3 (Domination principle) *Let $G_\Omega\mu$ be a potential such that $G_\Omega\mu < +\infty$ μ-a.e., and let $v \in \mathcal{U}^+$. If $G_\Omega\mu \leqslant v$ q.e. on some Borel set carrying μ then $G_\Omega\mu \leqslant v$ everywhere on Ω.*

For the proof, based on Sects. 1.1.3 and 1.1.10, we refer to Doob [Do4, p. 67].

We denote by $\Lambda = \Lambda(\Omega)$ the convex cone of all (positive) measures λ on Ω which are *admissible* (this is not a standard terminology) in the sense that $G_\Omega\lambda$ is not identically infinite on Ω [nor on any component of Ω if we were to allow Ω to be a disconnected Greenian set]. Equivalently: $G_\Omega\lambda$ shall be finite q.e. on Ω, that is, $G_\Omega\lambda$ shall be superharmonic on Ω; see Sect. 1.1.11.

Lemma 1.4.4 *A measure λ on Ω is admissible if and only if $\int p\,d\lambda < +\infty$ for some and hence any non-zero $p = G_\Omega\mu \in \mathcal{P}_0^c$. In particular, any bounded measure is admissible.*

Proof Let $\int p\,d\lambda < +\infty$ for some non-zero $p = G_\Omega\mu \in \mathcal{P}_0^c$. Then $\int G_\Omega\lambda\,d\mu = \int G_\Omega\mu\,d\lambda < +\infty$ and hence $G_\Omega\lambda \not\equiv +\infty$. Conversely, suppose that $G_\Omega\lambda$ is superharmonic on Ω, and consider any $p = G_\Omega\mu \in \mathcal{P}_0^c$. Since $G_\Omega\mu$ is finite and continuous and μ has compact support K, there is a constant $a > 0$ such that $G_\Omega\mu \leqslant a$ on K and hence on all of Ω by Theorem 1.4.3. If λ is bounded it follows that indeed $\int p\,d\lambda \leqslant a \int d\lambda < +\infty$. Generally, denote by λ_0 and λ_1 the restriction of λ to ω and $\complement\omega$, respectively, where $\omega \subset \overline{\omega} \subset \Omega$ is an open set containing K. Then λ_0 is bounded and hence admissible, whereas λ_1 is carried by $\complement\omega$, so that $G_\Omega\lambda_1$ is harmonic on ω and therefore $\int G_\Omega\mu\,d\lambda_1 = \int G_\Omega\lambda_1\,d\mu < +\infty$ because μ is carried by $K \subset \omega$. It follows that $\lambda = \lambda_0 + \lambda_1$ is admissible. □

We shall need a further approximation property. For a proof, see, for example, [Do4, p. 95].

Theorem 1.4.5 *Let $f : \Omega \to \mathbb{R}$ be finite and continuous with compact support $K \subset \Omega$, and let L denote a compact neighborhood of K in Ω. Then f can be approximated uniformly by a sequence of differences $p_n - q_n$ of finite continuous potentials p_n, q_n on Ω such that p_n and q_n are harmonic and equal on $\complement L$ (in particular, $p_n, q_n \subset \mathcal{P}_0^c$).*

Corollary 1.4.6 *Two admissible measures λ and μ are equal if $\int p\,d\lambda = \int p\,d\mu$ for every $p \in \mathcal{P}_0^c$.*

In fact, $\int f\,d\lambda = \int f\,d\mu$ holds for $f = p_n - q_n$ with p_n, q_n as stated in Theorem 1.4.5, hence also for any $f \in \mathcal{C}_0(\Omega)$ by the stated uniform approximation.

By application of Theorem 1.4.5, the following extension theorem was established by Hervé [He, Théorème 13.1].

Theorem 1.4.7 *Let U be an open set with compact closure in Ω. For any superharmonic function u on an open neighborhood V of $\overline{U}$ in Ω, there exist two potentials p, q on Ω such that q is finite and continuous on Ω and harmonic on U, and that $p = u + q$ on U. If u is finite and continuous on V then p can be chosen so as to be finite and continuous on Ω.*

We shall consider the following linear subspace $\mathcal{Q}$ of $\mathcal{C}_0(\Omega)$:

$$\mathcal{Q} = \{p - q : p, q \in \mathcal{P}_0^c,\ p = q \text{ off some compact subset of } \Omega\}. \tag{1.4.1}$$

Theorem 1.4.5 shows that $\mathcal{Q}$ is *positively rich* in Ω (in the sense of [Bou6, Chap. III, § 3, Définition 3]).

Theorem 1.4.8 (Brelot [Br5], Hervé [He]) *Let $A \subset \Omega$. For any admissible measure μ on Ω, there is a unique admissible measure μ^A on Ω such that*

$$\int p\, d\mu^A = \int \widehat{R}_p^A\, d\mu \quad \textit{for any } p \in \mathcal{P}_0^c \tag{1.4.2}$$

and even for any function $p \in \mathcal{U}^+$.

The measure μ^A is said to be the *swept* (measure) of μ on A, or the *sweeping* (balayage) of μ on A.

Proof The mapping

$$p - q \mapsto \int (\widehat{R}_p^A - \widehat{R}_q^A)\, d\mu$$

of the above linear subspace $\mathcal{Q}$ of $\mathcal{C}_0(\Omega)$ into $\mathbb{R}$ is well defined and *linear*, by the additivity of sweeping of functions (Theorem 1.3.15), noting also that $\widehat{R}_p^A \leqslant p$, and so $\int \widehat{R}_p^A\, d\mu \leqslant \int p\, d\mu < +\infty$. This linear form is positive because $p \geqslant q$ implies $\widehat{R}_p^A \geqslant \widehat{R}_q^A$. Since $\mathcal{Q}$ is positively rich in Ω the above linear form on $\mathcal{Q}$ extends uniquely to a (positive Radon) measure denoted by μ^A on Ω, that is,

$$\int (p - q)\, d\mu^A = \int (\widehat{R}_p^A - \widehat{R}_q^A)\, d\mu, \quad p - q \in \mathcal{Q}. \tag{1.4.3}$$

To pass from this to (1.4.2), note that there exists an increasing sequence of bounded regular open sets ω_n with the union Ω and such that $\overline{\omega}_n \subset \omega_{n+1}$ (see, for example, [AG, Corollary 6.6.13]). Then $p_n := \widehat{R}_p^{\complement\omega_n}$ is a finite continuous potential on Ω and equal to p off ω_n by Lemma 1.3.9, in particular off $\overline{\omega}_n$ (compact), so $p - p_n \in \mathcal{Q}$, and hence from (1.4.3) with q replaced by p_n

$$\int (p - p_n)\, d\mu^A = \int (\widehat{R}_p^A - \widehat{R}_{p_n}^A)\, d\mu. \tag{1.4.4}$$

Since $\widehat{R}_p^{\complement\omega_{n+1}} \leqslant \widehat{R}_p^{\complement\overline{\omega}_n}$ it follows by Corollary 1.3.8 that $p_n \downarrow 0$ pointwise on Ω as $n \uparrow \infty$, and the same with p_n replaced by the smaller $\widehat{R}_{p_n}^A$. By the monotone convergence theorem, (1.4.4) yields (1.4.2) as $n \to \infty$ because $\int p\,d\mu < +\infty$.

For uniqueness of μ^A, let ν be any admissible measure such that (1.4.2) holds with μ^A replaced by ν. Then (1.4.3) holds with μ^A replaced by ν, and so $\nu = \mu^A$ by the uniqueness of μ^A in (1.4.3).

The extension of (1.4.2) to functions $p \in \mathcal{U}^+$ is justified by use of the approximation property stated in Sect. 1.1.10 combined with Theorem 1.3.6(a) and the monotone convergence theorem. □

Corollary 1.4.9 *For any set $A \subset \Omega$ the self-map $\mu \mapsto \mu^A$ of Λ is linear.*

Note the particular case of Theorem 1.4.8 where $\mu = \varepsilon_x$ (the mass 1 placed at the point $x \in \Omega$):

$$\widehat{R}_u^A(x) = \int u\,d\varepsilon_x^A, \quad u \in \mathcal{U}^+. \tag{1.4.5}$$

Taking $u \equiv 1$ in (1.4.2) we obtain

$$\int d\mu^A = \int \widehat{R}_1^A\,d\mu \leqslant \int d\mu. \tag{1.4.6}$$

For the inequality, note that $\widehat{R}_1^A \leqslant 1$ because the constant 1 is harmonic, and in particular $1 \in \mathcal{U}^+$. Equality prevails here if (but not only if; cf. Theorem 1.4.17 below) μ is carried by the base $\mathfrak{b}(A)$ of A, the set of points of Ω at which A is not thin; see Lemma 1.3.9(a) above and (1.4.11) below.

Corollary 1.4.10 *For $A \subset B \subset \Omega$ and $\mu \in \Lambda$, we have $(\mu^A)^B = \mu^A = (\mu^B)^A$.*

For any $p \in \mathcal{P}^c$ we have in fact

$$\int p\,d(\mu^A)^B = \int \widehat{R}_p^B\,d\mu^A = \int \widehat{R}_{\widehat{R}_p^B}^A\,d\mu = \int \widehat{R}_p^A\,d\mu = \int p\,d\mu^A,$$
$$\int p\,d(\mu^B)^A = \int \widehat{R}_p^A\,d\mu^B = \int \widehat{R}_{\widehat{R}_p^A}^B\,d\mu = \int \widehat{R}_p^A\,d\mu = \int p\,d\mu^A.$$

The third equality in either line follows from Lemma 1.3.11. The remaining three equalities on either line follow from Theorem 1.4.8.

Theorem 1.4.11 *Let $u \geqslant 0$ be superharmonic on Ω and let u_0 denote the extension of u to $\Omega \cup \partial^\infty\Omega$ by $u_0 = 0$ on $\partial^\infty\Omega$. Let ω be an open subset of Ω. Then*

$$R_u^{\complement\omega} = \begin{cases} u & \text{on } \complement\omega \\ \overline{H}_{u_0}^\omega & \text{on } \omega. \end{cases}$$

In particular, $\overline{H}_{u_0}^\omega \leqslant u$ on ω.

Proof Clearly, u_0 is l.s.c. on $\partial^\infty\omega$. We have $u|_\omega \in \overline{\Phi}^\omega_{u_0}$, see Sect. 1.1.16, since $\liminf_{x\to y, x\in\omega} u(x) \geqslant u(y) = u_0(y)$ for $y \in \partial\omega$ whereas $\liminf_{x\to\infty, x\in\omega} u(x) \geqslant 0 = u_0(\infty)$. For any $v \in \mathcal{U}^+$ such that $v \geqslant u$ on $\complement\omega$ we have $v|_\omega \in \overline{\Phi}^\omega_{u_0}$, so $R_u^{\complement\omega} \geqslant \overline{H}^\omega_{u_0}$ on ω.

For the opposite inequality, let $w \in \overline{\Phi}^\omega_u$ and define a function z on Ω by $z = u$ on $\complement\omega$ and $z = u \wedge w$ on ω. Then $z \in \mathcal{U}^+$ (see Sect. 1.1.7), and since $z = u$ on $\complement\omega$ we have $w \geqslant z \geqslant R_u^{\complement\omega}$ on Ω. By varying $w \in \overline{\Phi}^\omega_f$ we obtain $\overline{H}^\omega_{u_0} \geqslant R_u^{\complement\omega}$ on ω. □

The following deep theorem, due to Brelot [Br5, Lemme 4], generalizes the symmetry of the Green kernel (the particular case where $A = \Omega$).

Theorem 1.4.12 *For any set $A \subset \Omega$ we have $\widehat{R}^A_{G_\Omega(x,\cdot)}(y) = \widehat{R}^A_{G_\Omega(\cdot,y)}(x)$, or in other terms*

$$\widehat{R}^A_{G_\Omega\varepsilon_y}(x) = \widehat{R}^A_{G_\Omega\varepsilon_x}(y) \quad \textit{for } x, y \in \Omega. \tag{1.4.7}$$

Proof We bring the version of Brelot's proof given in [AG, Theorem 5.7.6]. Let $x, y \in \Omega$ (with $x \neq y$) and for $n \in \mathbb{N}$ let

$$A_n = A \setminus \big(B(x, \tfrac{1}{n}) \cup B(y, \tfrac{1}{n})\big).$$

If A is relatively closed in Ω then so is each A_n, and hence Theorem 5.7.4 from [AG] applies to the open and therefore Greenian set $\omega := \Omega \setminus A_n$, hence $A_n = \Omega \setminus \omega$. Thus $\widehat{R}^{A_n}_{G_\Omega(\cdot,y)}(x) = \widehat{R}^{\Omega\setminus\omega}_{G_\Omega(\cdot,y)}(x) = G_\Omega(x, y) - G_\omega(x, y)$ is indeed symmetric in (x, y) along with $G_\Omega(x, y)$ and $G_\omega(x, y)$:

$$\widehat{R}^{A_n}_{G_\Omega(y,\cdot)}(x) = \widehat{R}^{A_n}_{G_\Omega(x,\cdot)}(y). \tag{1.4.8}$$

For $n \to \infty$, (A_n) increases to its union $A \setminus \{x, y\}$, and by Theorem 1.3.6(a) we obtain (1.4.7) because $\widehat{R}_u^{A\setminus\{x,y\}} = \widehat{R}_u^A$ for any $u \in \mathcal{U}^+$ by Theorem 1.3.2, $\{x, y\}$ being polar.

If A is open in Ω (instead of relatively closed), consider an increasing sequence (K_n) of compact subsets of A with union A and apply the previous paragraph to obtain

$$\widehat{R}^A_{G_\Omega(\cdot,y)}(x) = \lim_{n\to\infty} \widehat{R}^{K_n}_{G_\Omega(\cdot,y)}(x) = \lim_{n\to\infty} \widehat{R}^{K_n}_{G_\Omega(x,\cdot)}(y) = \widehat{R}^A_{G_\Omega(x,\cdot)}(y).$$

Finally, if A is an arbitrary subset of Ω, then Theorem 5.3.4 (vi) from [AG] together with the conclusion of the last paragraph yields (1.4.8) and (1.4.7) again by letting $n \to \infty$. □

Theorem 1.4.13 ([Br5], Théorème 7) *A set $A \subset \Omega$ is thin at $x \in \Omega$ if and only if $\varepsilon_x^A \neq \varepsilon_x$.*

Proof Suppose first that $\varepsilon_x^A \neq \varepsilon_x$. By Corollary 1.4.6 there exists $p \in \mathcal{P}_0^c$ such that $\int p\, d\varepsilon_x^A \neq \int p\, d\varepsilon_x$, that is, by (1.4.5), $\widehat{R}^A_p(x) \neq p(x)$, and so A is indeed thin at x, by Lemma 1.3.9(a). For the converse, we have for $x, y \in \Omega$ and $A \subset \Omega$

$$G_\Omega \varepsilon_x^A(y) = \int G_\Omega(\cdot, y)\, d\varepsilon_x^A = \widehat{R}^A_{G_\Omega(\cdot,y)}(x) = \widehat{R}^A_{G_\Omega(\cdot,x)}(y) = \widehat{R}^A_{G_\Omega \varepsilon_x}(y)$$

by definition of the Green potential of ε_x^A, next by (1.4.5), and finally by Theorem 1.4.8, noting that $G_\Omega(x, \cdot) = G_\Omega \varepsilon_x$. It follows that if $\varepsilon_x^A = \varepsilon_x$ then $\widehat{R}^A_{G_\Omega(\cdot,x)} = G(x, \cdot)$, and we conclude by Theorem 7.3.2 from [AG] that A is unthin at x. □

Definition 1.4.14 (*Constantinescu* [Co]) A potential $p > 0$ is called *strict* if two measures $\mu, \nu \in \Lambda$ are equal if they satisfy the following two conditions:

$$\int p\, d\mu = \int p\, d\nu < +\infty, \tag{1.4.9}$$

$$\int u\, d\mu \leqslant \int u\, d\nu \quad \text{for all } u \in \mathcal{U}^+. \tag{1.4.10}$$

Compare (1.4.9) with Corollary 1.4.6. It follows from the approximation property in Sect. 1.1.10 that "$u \in \mathcal{U}^+$" may be replaced in (1.4.10) by "$u \in \mathcal{P}_0^c$".

The point in considering strict potentials is that a finite strict potential *determines thinness*; see Theorem 1.4.17 below.

Theorem 1.4.15 (Hervé [He], Bauer [Ba1]) (a) *For any locally Lebesgue integrable function, $f > 0$ ℓ_N-a.e. on Ω, $G_\Omega f$ (see Sect. 1.1.5) is a strict potential if $\not\equiv +\infty$.*

(b) *For any $\lambda \in \Lambda$, there exists a finite continuous function $f > 0$ for which $G_\Omega f > 0$ is a finite continuous (hence strict) λ-integrable potential.*

Proof (a) For $x \in \Omega$ take $u = G_\Omega(x, \cdot)$ in (1.4.10) to obtain $G_\Omega\mu(x) \leqslant G_\Omega\nu(x)$. Since μ is admissible, $G_\Omega\mu < +\infty$ q.e. and hence ℓ_N-a.e., and so $G_\Omega\nu - G_\Omega\mu \geq 0$ ℓ_N-a.e. From (1.4.9) with $p = G_\Omega f$ it then follows by integration with respect to x and applying Fubini's theorem while invoking the symmetry of the Green kernel that

$$\int (G_\Omega\nu - G_\Omega\mu) f\, d\ell_N = 0.$$

Because $f > 0$ ℓ_N-a.e. and $G_\Omega\nu - G_\Omega\mu \geqslant 0$ ℓ_N-a.e., this equation implies that $G_\Omega\mu = G_\Omega\nu$ ℓ_N-a.e., hence everywhere on Ω (see the fifth paragraph after Corollary 1.2.5). It follows that $\mu = \nu$ by uniqueness of the Riesz measure associated with a superharmonic function; see Sect. 1.1.6.

(b) We may obviously suppose that $\lambda \neq 0$. Choose an exhaustion of Ω by a sequence of open sets Ω_n of compact closure $\overline{\Omega}_n \subset \Omega$ so that $\overline{\Omega}_n \subset \Omega_{n+1}$ for all $n \geq 1$ and $\lambda(\Omega_1) > 0$. Now $G_\Omega\lambda$ is superharmonic and hence locally ℓ_N-integrable (see Sect. 1.1.2), so we may define constants

$$c_n = \int_{\overline{\Omega}_n} G_\Omega\lambda\, d\ell_N < +\infty \quad (n \in \mathbb{N}).$$

The sequence (c_n) is increasing, and hence so is $(2^n c_n)$. Define a function f, first on each $\partial\Omega_n$ by $f = 1/(2^{n+1} c_{n+1})$ on $\partial\Omega_n$, noting that the compact sets $\partial\Omega_n$ are disjoint. Next, extend f to a finite continuous function $f > 0$ on each $\overline{\Omega}_n \setminus \Omega_{n-1}$, and thereby on all of Ω, such that

$$f \leqslant \frac{1}{2^n c_n} \quad \text{on } \overline{\Omega}_n \setminus \Omega_{n-1}$$

after defining $\Omega_0 = \varnothing$. Then

$$\int_\Omega G_\Omega f\, d\lambda = \int_\Omega (G_\Omega\lambda) f\, d\ell_N \leqslant \sum_{n\in\mathbb{N}} \int_{\overline{\Omega}_n\setminus\Omega_{n-1}} (G_\Omega\lambda) f\, d\ell_N \leqslant \sum_{n\in\mathbb{N}} 2^{-n} < +\infty.$$

It therefore follows that $G_\Omega f \not\equiv +\infty$, and hence $G_\Omega f$ is a strict potential, by (a). Because f is bounded and continuous, $G_\Omega f = G_\Omega(f 1_{\Omega_n}) + G_\Omega(f 1_{\complement\Omega_n})$ is finite and continuous. In fact, for any n, $G_\Omega(f 1_{\complement\Omega_n})$ is harmonic on Ω_n (being finite q.e.) and $G_\Omega(f 1_{\Omega_n})$ is finite and continuous, being the sum of a harmonic function on Ω and the restriction to Ω of the convolution of the locally ℓ_N-integrable function $x \mapsto |x|^{2-N}$ on $\mathbb{R}^N$ if $N \geq 3$, respectively $x \mapsto -\log|x|$ if $N = 2$, and the bounded Borel measurable function $f 1_{\Omega_n}$ with compact support (e.g. Schwartz [Sc, p. 7]). □

Remark 1.4.16 Although the fine topology is defined on all of $\mathbb{R}^N$, we shall (when nothing else is said) *consider the fine topology relative to a fixed Greenian domain* Ω. However, those concepts and results in what follows which do not involve sweeping (of functions or measures) could equally well be considered relative to the whole of $\mathbb{R}^N$ (even for $N = 2$, where $\mathbb{R}^N$ fails to be Greenian).

The *fine closure* of a set $A \subset \Omega$ (relative to Ω) will be denoted by $\widetilde{A}$, and the *fine interior* of A by A'. The *finely derived set* of A, that is, the set of all fine limit points of A, is also called the *base* of A (relative to Ω) and will be denoted by $\mathfrak{b}(A)$:

$$\mathfrak{b}(A) = \{x \in \Omega : x \in \widetilde{(A \setminus \{x\})}\}, \tag{1.4.11}$$

so $\mathfrak{b}(A)$ is finely closed and of course contained in $\widetilde{A}$. And $\mathfrak{b}(A)$ is the set of points of Ω at which A is not thin. Thus $A \setminus \mathfrak{b}(A)$ is the set of *finely isolated* points of A, in other words the points of A at which A is thin (Corollary 1.2.5). A finely open set has no finely isolated points because $\mathbb{R}^N$ has no such by Corollary 1.2.5 (see also the fifth paragraph after Corollary 1.2.5). By Theorem 1.4.13,

$$\mathfrak{b}(A) = \{x \in \Omega : \varepsilon_x^A = \varepsilon_x\}. \tag{1.4.12}$$

Some purely topological properties are

$$\widetilde{A} = A \cup \mathfrak{b}(A) \supset \mathfrak{b}(A), \quad A' \subset \mathfrak{b}(A') \subset \mathfrak{b}(A), \tag{1.4.13}$$

$$\widetilde{A_1 \cup A_2} = \widetilde{A}_1 \cup \widetilde{A}_2, \quad \mathfrak{b}(A_1 \cup A_2) = \mathfrak{b}(A_1) \cup \mathfrak{b}(A_2) \tag{1.4.14}$$

for A, A_1, $A_2 \subset \Omega$. For (1.4.13), the former part is obvious, and in particular, $\mathfrak{b}(A) \subset A$ if and only if A is finely closed. The latter part holds because A' has no finely isolated points (see before (1.4.12)). And both topological properties (1.4.14) are well known and easily established.

A set $A \subset \Omega$ is said to be subbasic if $A \subset \mathfrak{b}(A)$. It follows from (1.4.13) that every finely open subset of Ω is subbasic.

Theorem 1.4.17 (cf. Hervé [He, Théorème 9.1]) *For any finite strict potential p on Ω and any set $A \subset \Omega$,*

$$\mathfrak{b}(A) = \{x \in \Omega : \widehat{R}_p^A(x) = p(x)\}.$$

Proof Let $x \in \Omega$. If $\widehat{R}_p^A(x) = p(x)$ take $\mu = \varepsilon_x^A$ and $\nu = \varepsilon_x$ in Definition 1.4.14, and invoke (1.4.5):

$$\int p\, d\varepsilon_x^A = \widehat{R}_p^A(x) = p(x) = \int p\, d\varepsilon_x < +\infty,$$
$$\int u\, d\varepsilon_x^A = \widehat{R}_u^A(x) \leqslant u(x) = \int u\, d\varepsilon_x \quad \text{for } u \in \mathcal{U}^+.$$

Since p is strict, it follows that $\varepsilon_x^A = \varepsilon_x$, and hence A is unthin at x, by Theorem 1.4.13. The converse implication follows from Lemma 1.3.9(a). □

The latter assertion in the following corollary generalizes *Kellogg's theorem* according to which the irregular boundary points for an open set form a polar set (a set of capacity 0; see Sect. 1.1.18).

Corollary 1.4.18 *For any set $A \subset \Omega$, the base $\mathfrak{b}(A)$ is a finely closed, Euclidean G_δ-set, and the set $\mathfrak{i}(A) := A \setminus \mathfrak{b}(A) = \widetilde{A} \setminus \mathfrak{b}(A)$ of points of A at which A is thin is polar. Furthermore, $\mathfrak{b}(A)$ is the smallest finely closed set containing A up to a polar set.*

As to $\mathfrak{b}(A)$ being a G_δ, it follows from Theorem 1.4.17 that $\mathfrak{b}(A) = \{\widehat{R}_p^A = p\} = \bigcap_{n\in\mathbb{N}}\{\widehat{R}_p^A > (1 - \frac{1}{n})p\}$ with p a finite continuous strict potential; see Theorem 1.4.15(b). From the equality in (1.4.13) follows $\widetilde{A} \setminus \mathfrak{b}(A) = A \setminus \mathfrak{b}(A)$. For $A \setminus \mathfrak{b}(A)$ being polar, invoke (1.3.8) applied to Theorem 1.4.17. For the last assertion of the corollary, if a finely closed set F contains $A \setminus E$ for some polar set E then $F \cup E \supset A$ and hence by (1.4.14) $F \supset \mathfrak{b}(F) = \mathfrak{b}(F \cup E) \supset \mathfrak{b}(A)$. And $\mathfrak{b}(A) = \mathfrak{b}(\widetilde{A})$ is finely closed and contains $A \setminus E$, where $E := \widetilde{A} \setminus \mathfrak{b}(A)$ is polar.

Corollary 1.4.19 (Brelot [Br3]) *For any set $E \subset \Omega$ the following are equivalent:*

(a) *E is polar.*

(b) *E is thin at each of its points, that is, $E \cap \mathfrak{b}(E) = \varnothing$. Equivalently, the points of E are finely isolated in E.*

(c) *E is everywhere thin, that is, $\mathfrak{b}(E) = \varnothing$. Equivalently, E is finely closed and its points are finely isolated in E.*

For (a)⇒(c), see Sect. 1.2 with reference to [Do4, 1.XI.11] or [AG, Theorem 7.3.11]. Obviously, (c)⇒(b). For (b)⇒(a), note that $E = E \setminus (E \cap \mathfrak{b}(E)) = E \setminus \mathfrak{b}(E)$ is polar, by Corollary 1.4.18.

Corollary 1.4.20 *Two subsets of Ω which differ by at most a polar set have the same base. For any set $A \subset \Omega$ we have*

$$\mathfrak{b}(\widetilde{A}) = \mathfrak{b}(A) = \mathfrak{b}(\mathfrak{b}(A)). \tag{1.4.15}$$

Proof The former assertion follows from Corollary 1.4.19, (a)⇒(c) together with (1.4.14). By Corollary 1.4.18, $\widetilde{A}$ and its subset $\mathfrak{b}(A)$ (cf. (1.4.13)) differ by a polar set and hence have the same base: $\mathfrak{b}(\widetilde{A}) = \mathfrak{b}(\mathfrak{b}(A))$. Since $\mathfrak{b}(A)$ is finely closed and hence contains its own base $\mathfrak{b}(\mathfrak{b}(A))$ (again by (1.4.13)) it follows that

$$\mathfrak{b}(\widetilde{A}) = \mathfrak{b}(\mathfrak{b}(A)) \subset \mathfrak{b}(A) \subset \mathfrak{b}(\widetilde{A}).$$

Thus equality prevails. □

A set $B \subset \Omega$ of the form $B = \mathfrak{b}(A)$ for some $A \subset \Omega$ is simply called a *base*. It is the same as a set B which is its own base: $B = \mathfrak{b}(B)$, the base operator being idempotent, by (1.4.15). Every base B is finely closed.

Theorem 1.4.21 *For any set $A \subset \Omega$, any $u \in \mathcal{U}^+$, and any $\mu \in \Lambda$, we have $R_u^{\widetilde{A}} = R_u^A$ and*

$$\widehat{R}_u^{\widetilde{A}} = \widehat{R}_u^A = \widehat{R}_u^{\mathfrak{b}(A)} = R_u^{\mathfrak{b}(A)}, \quad \mu^{\widetilde{A}} = \mu^A = \mu^{\mathfrak{b}(A)}.$$

Observe the analogy with (1.4.15) (for the base operation).

Proof $R_u^A = R_u^{\widetilde{A}}$ follows from the definition of R_u^A because $v \geqslant u$ on A with $v \in \mathcal{U}^+$ implies $v \geqslant u$ on $\widetilde{A}$, by the fine continuity of u and v ($\in \mathcal{U}^+$). It follows that $\widehat{R}_u^A = \widehat{R}_u^{\widetilde{A}}$. Next, for $\widehat{R}_u^A = \widehat{R}_u^{\mathfrak{b}(A)}$ we prove more:

$$\widehat{R}_u^A = \widehat{R}_u^{A\cup\mathfrak{b}(A)} = \widehat{R}_u^{A\cap\mathfrak{b}(A)} = \widehat{R}_u^{\mathfrak{b}(A)}. \tag{1.4.16}$$

The first equality in (1.4.16), but with "=" replaced by "⩽", is obvious. For the first equality in (1.4.16), now with "=" replaced by "⩾", let $v \in \mathcal{U}^+$ satisfy $v \geqslant u$ on A and hence on $\mathfrak{b}(A) \subset \widetilde{A}$ by fine continuity of u and v. Thus the finely closed set $\{v \geqslant u\}$ contains $\widetilde{A}$ and hence $\mathfrak{b}(A)$, and altogether $\{v \geqslant u\} \supset A \cup \mathfrak{b}(A)$. This means that $v \geqslant R_u^{A\cup\mathfrak{b}(A)}$, and by varying v that $R_u^A \geqslant R_u^{A\cup\mathfrak{b}(A)}$. Consequently, $\widehat{R}_u^A \geqslant \widehat{R}_u^{A\cup\mathfrak{b}(A)}$, actually with equality. The further equality

$$\widehat{R}_u^A = \widehat{R}_u^{A\cap\mathfrak{b}(A)} \tag{1.4.17}$$

(part of (1.4.16)) holds because A and its subset $A \cap \mathfrak{b}(A)$ differ from one another by the set $A \setminus (A \cap \mathfrak{b}(A)) = A \setminus \mathfrak{b}(A)$ which is polar according to Corollary 1.4.18. Finally, it is obvious that

$$A \cap \mathfrak{b}(A) \subset \mathfrak{b}(A) \subset A \cup \mathfrak{b}(A),$$

and we have shown above that $\widehat{R}_u^{A\cap\mathfrak{b}(A)} = \widehat{R}_u^{A\cup\mathfrak{b}(A)}$ (both terms being equal to $\widehat{R}_u^A$). It follows that these two equal terms also equal $\widehat{R}_u^{\mathfrak{b}(A)}$ (the last term in (1.4.16)). The third equality follows from (1.4.2). Indeed, for every $x \in B = \mathfrak{b}(A)$ we have by (1.4.2), (1.4.15), and Theorem 1.4.13

$$\widehat{R}_u^B(x) = \int u d\varepsilon_x^B = \int u d\varepsilon_x = u(x)$$

and hence the required equality.

The three assertions on swept measures follow from the corresponding statements on swept functions, in view of Theorem 1.4.8. □

The *harmonic measure* (see Sects. 1.1.2 and 1.1.17 and Remark 2.1.42) is related to sweeping of ε_x as follows. For a proof of the theorem we refer to [CC3, Theorem 7.1.2], [Do4, Theorem 1.X.2].

Theorem 1.4.22 (La Vallée-Poussin [VP1]) *For any regular open set ω in Ω and any point $x \in \omega$,*

$$\varepsilon_x^{\complement\omega} = \mu_x^\omega,$$

the harmonic measure on $\partial\omega$ relative to $x \in \omega$.

Proof Since ω is regular, $\mathfrak{b}(\complement\omega) = \complement\omega$ is a base, and hence by Theorem 1.4.21 or Theorem 1.3.4(a), $\widehat{R}_u^{\complement\omega} = R_u^{\complement\omega}$ on ω for any hyperharmonic function $u \geqslant 0$ on Ω. With the notations from the proof of Theorem 1.4.8, $p, q \in \mathcal{P}_0^c$ extend by 0 on $\partial^\infty\Omega$ to finite and continuous and hence resolutive functions on $\partial^\infty\omega$ (Sect. 1.1.16). We therefore have from (1.4.5) and Theorem 1.4.11

$$\int (p-q)\, d\varepsilon_x^{\complement\omega} = R_p^{\complement\omega}(x) - R_q^{\complement\omega}(x) = H_p^\omega(x) - H_q^\omega(x) = \int (p-q)\, d\mu_x^\omega,$$

whence the result, $\mathcal{Q}$ from (1.4.1) being positively rich. □

Assertion (b) of the following theorem generalizes Theorem 1.4.13.

Theorem 1.4.23 (Brelot [Br5]) *For $A \subset \Omega$ and $\mu \in \Lambda$ we have*
(a) $\mu^A = \mu^{\mathfrak{b}(A)}$ *is carried by* $\mathfrak{b}(A)$ $(\subset \widetilde{A})$
(b) $\mu^A = \mu$ *holds if and only if μ is carried by* $\mathfrak{b}(A)$.

Proof According to Theorem 1.4.15(b) we may choose a μ-integrable finite continuous strict potential p on Ω. It follows by Theorem 1.4.17 that $\{\widehat{R}_p^A = p\}$ equals $\mathfrak{b}(A)$:

(a) By Theorem 1.4.8 and Lemma 1.3.11,

$$\int (p - \widehat{R}_p^A)\, d\mu^A = \int (\widehat{R}_p^A - \widehat{R}_{\widehat{R}_p^A}^A)\, d\mu = 0.$$

Now, $p - \widehat{R}_p^A > 0$ on $\complement\mathfrak{b}(A)$, and since $\mathfrak{b}(A)$ is a Borel set by Corollary 1.4.18, μ^A is indeed carried by $\mathfrak{b}(A)$. And $\mu^{\mathfrak{b}(A)} = \mu^A$ by Theorem 1.4.21.

(b) If $\mu^A = \mu$ then μ is carried by $\mathfrak{b}(A)$ because μ^A is so, according to (a). Conversely, if μ is carried by $\mathfrak{b}(A)$ then

$$\int p\, d\mu^A = \int \widehat{R}_p^A\, d\mu = \int p\, d\mu \quad \text{for every } p \in \mathcal{P}^c \tag{1.4.18}$$

because $\widehat{R}_p^A = p$ on $\mathfrak{b}(A)$ (see the beginning of the proof). According to Corollary 1.4.6 it follows from (1.4.18) that indeed $\mu^A = \mu$. □

Theorem 1.4.24 (Brelot [Br5]) *For $A \subset \Omega$ and $\mu \in \Lambda$ we have $G_\Omega\mu^A = \widehat{R}^A_{G_\Omega\mu}$.*

Proof By Theorems 1.4.8 and 1.4.12 we have for any $y \in \Omega$

$$G_\Omega\mu^A(y) = \int G_\Omega\varepsilon_y\, d\mu^A = \int \widehat{R}^A_{G_\Omega\varepsilon_y}(x)\, d\mu(x) = \int \widehat{R}^A_{G_\Omega\varepsilon_x}(y)\, d\mu(x).$$

For $y \in \mathfrak{b}(A)$ we have $\widehat{R}^A_{G_\Omega\varepsilon_x}(y) = G_\Omega\varepsilon_x(y)$ by Lemma 1.3.9(a), and so we altogether obtain

$$G_\Omega\mu^A(y) = \int \widehat{R}^A_{G_\Omega\varepsilon_x}(y)\, d\mu(x) = \int G_\Omega\varepsilon_x(y)\, d\mu(x) = G_\Omega\mu(y) \quad \text{for } y \in \mathfrak{b}(A),$$

whence by Theorem 1.4.21

$$G_\Omega\mu^A \geqslant \widehat{R}^{\mathfrak{b}(A)}_{G_\Omega\mu} = \widehat{R}^A_{G_\Omega\mu}.$$

By integration with respect to any $\nu \in \Lambda$ we therefore obtain from the reciprocity theorem (see Sect. 1.1.5) and (1.4.2)

$$\int \widehat{R}^A_{G_\Omega\mu}\, d\nu \leqslant \int G_\Omega\mu^A\, d\nu = \int G_\Omega\nu\, d\mu^A = \int \widehat{R}^A_{G_\Omega\nu}\, d\mu,$$

and so, by symmetry, generalizing Theorem 1.4.12 (and thereby the symmetry of G_Ω),

$$\int \widehat{R}^A_{G_\Omega\mu}\, d\nu = \int \widehat{R}^A_{G_\Omega\nu}\, d\mu. \tag{1.4.19}$$

Taking in particular $\nu = \varepsilon_x$ leads (again in view of Theorem 1.4.8) to the desired result

$$\widehat{R}^A_{G_\Omega\mu}(x) = \int \widehat{R}^A_{G_\Omega\mu}\, d\varepsilon_x = \int \widehat{R}^A_{G_\Omega\varepsilon_x}\, d\mu = \int G_\Omega\varepsilon_x\, d\mu^A = G_\Omega\mu^A(x).$$

□

We formulate (1.4.19) as follows.

Corollary 1.4.25 *For any $A \subset \Omega$ and $\mu, \nu \in \Lambda$,*

$$\int_\Omega \widehat{R}^A_{G_\Omega \mu}\, d\nu = \int_\Omega \widehat{R}^A_{G_\Omega \nu}\, d\mu.$$

From Theorems 1.4.23(b) and 1.4.24 we derive the following.

Corollary 1.4.26 *For any $A \subset \Omega$, a measure $\mu \in \Lambda$ is carried by $\mathfrak{b}(A)$ if and only if $\widehat{R}^A_{G_\Omega \mu} = G_\Omega \mu$.*

Lemma 1.4.27 ([Do4, 1.XI.16]) *For any potential $p = G_\Omega \mu$ on Ω and any constant $c \in [0, +\infty]$ we have*

$$\widehat{R}_p^{\{p \leqslant c\}} = \widehat{R}_p^{\{p < c\}} = p \wedge c.$$

Proof Recall that $u \wedge v = \min\{u, v\}$ (pointwise infimum) for numerical functions u, v. The case $c = 0$ is trivial, and so is the case $c = +\infty$, where Theorem 1.3.2 applies to $f_1 := 1_{\{p<+\infty\}} p$ and $f_2 = 1_{\{p \leqslant +\infty\}} p$, for which $\widehat{R}_{f_1} = \widehat{R}_{f_2}$ since $f_1 = f_2$ q.e., $\{p = +\infty\}$ being polar. Thus we may assume that $0 < c < +\infty$. Clearly,

$$\widehat{R}_p^{\{p<c\}} \leqslant \widehat{R}_p^{\{p \leqslant c\}} = \widehat{R}_{p \wedge c}^{\{p \leqslant c\}} \leqslant p \wedge c.$$

Here the latter inequality is an equality. In fact, $p \wedge c = (G_\Omega \mu) \wedge c$ is a finite potential $G_\Omega \nu$, and $\operatorname{Supp} \nu \subset \{p \leqslant c\}$ because $G_\Omega \nu = p \wedge c$ is constant ($= c$) and hence harmonic off the closed set $\{p \leqslant c\}$; see Sect. 1.1.9. For $v \in \mathcal{U}^+$ with $v \geqslant p (= p \wedge c)$ q.e. on $\operatorname{Supp} \nu$ we find by the domination principle (Theorem 1.4.3) that $v \geqslant p \wedge c$ on Ω, so indeed $\widehat{R}_p^{\{p \leq c\}} = p \wedge c$ in view of (1.3.7), and hence $\widehat{R}_p^{\{p<c\}} = p \wedge c$ by (1.3.12). □

In analogy with the concept of regular points for a Euclidean open set, a point $x \in \partial_f U$ (U finely open in Ω) is termed (finely) *regular* if $\complement U$ is unthin at x, that is, $x \in \mathfrak{b}(\complement U)$. For a (Euclidean) open set, the regular points in this sense are precisely those usual regular points which lie on the fine boundary $\partial_f U$ ($\subset \partial U$). A finely open set U is said to be (finely) regular if every point of $\partial_f U$ is regular, that is, if $\partial_f U \subset \mathfrak{b}(\complement U)$. Every finely open set $U \subset \Omega$ *can be regularized*, that is, be replaced by the regular finely open set ($\subset \widetilde{U}$), obtained by adjoining to U the polar set of all finely isolated points of $\complement U$. More generally, for an arbitrary set $A \subset \Omega$, $\mathfrak{r}(A)$ is defined by duality as follows:

$$\complement \mathfrak{r}(A) = \mathfrak{b}(\complement A) \quad \text{(complements relative to } \Omega), \tag{1.4.20}$$

and $\mathfrak{r}(A)$ will be termed the*regularizer* of A (Brelot called it the nucleus or kernel of A). Note the analogy with $\complement A' = \widetilde{\complement A}$ (recall that A' denotes the fine interior of A).

By duality from Corollary 1.4.18, $\mathfrak{r}(U)$ is the largest finely open set contained in U up to a polar set. Furthermore, $\mathfrak{r}(A)$ is a regular finely open F_σ-set and

$$A' \subset \mathfrak{r}(A) \subset \mathfrak{b}(A) \subset \widetilde{A}. \tag{1.4.21}$$

The second inclusion amounts to $\mathfrak{b}(\complement A) \supset \complement\mathfrak{b}(A)$. And this holds because A and $\complement A$ are not thin at the same point, for then $A \cup \complement A = \Omega$ would be thin at that point, contradicting the absence of finely isolated points in $\mathbb{R}^N$; see Sect. 1.2 (fifth paragraph after Corollary 1.2.5).

If U is a *fine domain* (that is, a finely connected finely open set) then so is $\mathfrak{r}(U)$ (and moreover regular). It follows from the regularity (in the topological sense) of the fine topology (see Sect. 1.2) that *the regular finely open sets form a basis for the fine topology*. The fine interior A' of a finely closed set $A \subset \Omega$ is *regular* in view of (1.4.20) because the fine closure $\widetilde{B}$ of any finely open set $B := \complement A$ equals its base $\mathfrak{b}(B)$; see text after Remark 1.4.16.

A (Euclidean) open set U is finely regular if and only if U is regular in the usual sense. For if U is finely regular then also regular in the usual sense because $\partial U = \partial_f U$. In fact, if $x \in (\partial U) \setminus (\partial_f U)$ then $\varepsilon_x^{\complement U} = \varepsilon_x$ (since $\complement U$ is unthin at x) and $\varepsilon_x^{\complement U} \neq \varepsilon_x$ (since $\varepsilon_x^{\complement U}$ is carried by $\partial_f U$ in view of Corollary 1.4.26, but ε_x is not). Furthermore $\varepsilon_x^{\complement U} = \mu_x^U$ (the usual harmonic measure), according to Theorem 1.4.22. Conversely, a usual regular open set U is finely open, and regular as such because $\partial_f U \subset \partial U$ (and actually $\partial_f U = \partial U$, as just shown).

Definition 1.4.28 The (fine) harmonic measure relative to a regular fine domain $V \subset \Omega$ and a point $x \in V$ equals $\varepsilon_x^{\complement V} = {}^\Omega\varepsilon_x{}^{\complement V}$.

Inspired by (1.4.5) we propose to extend the concept of sweeping of a positive hyperharmonic function onto a set $A \subset \Omega$. Recall that ε_x^A is carried by the base $\mathfrak{b}(A)$ (Theorem 1.4.23(a)).

Definition 1.4.29 For any set $A \subset \Omega$, any numerical function f defined at least on $\mathfrak{b}(A)$, and any point $x \in \Omega$, define

$$f^{*A}(x) = \int^* f \, d\varepsilon_x^A, \quad f_*^A(x) = \int_* f \, d\varepsilon_x^A.$$

For the notions of upper and lower integrals of an arbitrary numerical function with respect to a measure, in this case ε_x^A, see "Notation and Terminology". If $f^{*A}(x) = f_*^A(x)$ we may write simply $f^A(x)$ instead. In particular, $f^{*A}(x) = f_*^A(x) \in \mathbb{R}$ if and only if f is ε_x^A-integrable, thus in particular if f is bounded and ε_x^A-measurable (noting that $\int d\varepsilon_x^A = \widehat{R}_1^A(x) \leqslant 1$ by Theorem 1.4.8 for $p = 1$).

Let $\mathcal{B} = \mathcal{B}(\Omega)$ denote the class of all *Borel measurable* functions $f : \Omega \to \overline{\mathbb{R}}$, and $\mathcal{B}_+ = \mathcal{B}_+(\Omega)$ the positive cone in $\mathcal{B}$. For any $f \in \mathcal{B}_+$ (and $A \subset \Omega$) one has $f^{*A} = f_*^A$ $(= f^A)$. Furthermore, $f^A \in \mathcal{B}$, a non-trivial result for which we refer to [Do4, 1.X.4].[4] It means that we have a self-map $f \mapsto f^A$ of $\mathcal{B}_+$, in other words a *kernel*.

---

[4] Doob [Do4] takes Theorem 1.4.24 as an equivalent definition of the swept measure μ^A; cf. Theorem 1.4.24.

In the particular case of a positive hyperharmonic function $u : \Omega \to [0, +\infty]$ (hence l.s.c. and therefore of class $\mathcal{B}_+$), we have in view of (1.4.5)

$$u^A = \widehat{R}_u^A \quad \text{for } u \in \mathcal{U}^+. \tag{1.4.22}$$

Theorem 1.4.30 *For any Borel measurable function $f : \Omega \to [0, +\infty]$ and any (positive) measure μ on Ω, we have*

$$\int f^A \, d\mu = \int f \, d\mu^A. \tag{1.4.23}$$

This reduces to $\mu^A = \int \varepsilon_x^A \, d\mu(x)$ in the sense of integration of measures[5] as follows:

$$\int f \, d\mu^A = \int\Big(\int f \, d\varepsilon_x^A\Big) d\mu(x) = \int f^A(x) \, d\mu(x).$$

Proof Suppose first that μ is carried by a compact of Ω. Consider first the case where $f \in \mathcal{C}_0^+(\Omega)$. As noted before in Theorem 1.4.8, the linear subspace $\mathcal{Q}$ of $\mathcal{C}_0(\Omega)$ defined in (1.4.1) is positively rich. Hence there exist two sequences of potentials $p_n, q_n \in \mathcal{P}_0^c$ such that $p_n - q_n \in \mathcal{C}_0(\Omega)$ and

$$|f - (p_n - q_n)| \leqslant 1/n.$$

For any $x \in \Omega$ it follows by (1.4.3) and (1.4.6) that

$$\Big|\int f \, d\varepsilon_x^A - (\widehat{R}_{p_n}^A(x) - \widehat{R}_{q_n}^A(x))\Big| = \Big|\int \big(f - (p_n - q_n)\big) \, d\varepsilon_x^A\Big| \leqslant \frac{1}{n} \int d\varepsilon_x^A \leqslant \frac{1}{n}.$$

This shows that $\widehat{R}_{p_n}^A(x) - \widehat{R}_{q_n}^A(x) \to \int f \, d\varepsilon_x^A$ uniformly for $x \in \Omega$ as $n \to \infty$. The functions $\widehat{R}_{p_n}^A - \widehat{R}_{q_n}^A$ are finely continuous and Borel measurable on Ω, and harmonic off $\overline{A}$ by Theorem 1.3.4(a). We conclude that indeed

$$\begin{aligned}\int f^A \, d\mu &= \int\Big(\int f \, d\varepsilon_x^A\Big) d\mu(x) = \lim_{n\to\infty} \int (\widehat{R}_{p_n}^A - \widehat{R}_{q_n}^A) \, d\mu \\ &= \lim_{n\to\infty} \int (p_n - q_n) \, d\mu^A = \int f \, d\mu^A.\end{aligned}$$

[5] In terms of Radon measures, as in [Bou2, § 3, Proposition 4], this requires that the family $x \mapsto \varepsilon_x^A$ is vaguely μ-adequate in the sense of loc. cit. Definition 1—a fact mostly overlooked in the literature. For a proof of this μ-adequacy in the present situation, see [FuZ, Lemma 3.16]. However, the reader who prefers to use the abstract concept of measure as a countably additive set function may instead follow Doob [Do4, Part 1, Chapters X and XI], where no such extra verification is needed.

Next, for $f : \Omega \to [0, +\infty]$ l.s.c. and μ-integrable there is an increasing sequence of functions $f_n \in \mathcal{C}_0(\Omega)$ converging pointwise to f. It follows that

$$\int f_n^A \, d\mu = \int f_n \, d\mu^A \uparrow \int f \, d\mu^A,$$

$$\int f \, d\mu^A = \sup_n \int f_n \, d\mu^A = \sup_n \int f_n^A \, d\mu = \int \sup_n f_n^A \, d\mu = \int f^A \, d\mu.$$

For the general case $f \in \mathcal{B}_+$, fix a potential $q \in \mathcal{P}_0^c$, $q > 0$, and a measure $\mu \in \Lambda$. Let $\mathcal{A}$ denote (in this proof) the set of all functions $f \in \mathcal{B}_+$ such that

$$\int (f \wedge (kq)) \, d\mu^A = \int (f \wedge (kq))^A \, d\mu \quad \text{for every } k \in \mathbb{N},$$

which implies (1.4.23) by letting $k \to \infty$. By the dominated convergence theorem, $\mathcal{A}$ is closed under limits of pointwise convergent sequences in $\mathcal{B}_+$ in view of Lemma 1.4.4. As shown above, $\mathcal{C}_0^+(\Omega) \subset \mathcal{A}$, and we conclude that $\mathcal{A}$ contains all Baire functions $\Omega \to [0, +\infty]$, and so $\mathcal{A} \supset \mathcal{B}_+$. This establishes (1.4.23) for every $f \in \mathcal{B}_+$.

For general μ write $\mu = \sum_j \mu_j$, where μ_j is carried by a compact subset of Ω. Then we have $\mu^A = \sum_j \mu_j^A$, an easy consequence of (1.4.2), and hence

$$\int f^A d\mu = \sum_j \int f^A d\mu_j = \sum_j \int f d\mu_j^A = \int f d\mu^A.$$

□

The following lemma is a strengthening of the fact that the Euclidean interior of the set $\{u = c\}$ has μ_u-measure 0 because u is constant and hence harmonic on this interior; see Sect. 1.1.9. Later we shall see that this simple reasoning for the case of the Euclidean interior carries over to the generalization of Lemma 1.4.31 to a *finely* superharmonic function u on a fine domain U (replacing Ω) and its associated Riesz measure μ; see Theorem 2.6.16 below.

Lemma 1.4.31 ([Br17, Lemma VIII.7], [Do4, 1.XI.17]) *Let u be superharmonic on Ω, and let μ be the Riesz measure associated with u. For any constant c the fine interior $\{u = c\}'$ of $\{u = c\}$ has μ-measure* 0.

As noted shortly before this lemma, the fine interior $\{u = c\}'$ of the finely closed set $\{u = c\}$ is regular in view of (1.4.20) (see the third paragraph after the proof of Lemma 1.4.27).

Proof It suffices to prove that for any given point $x \in \Omega$, there exists a ball $B(x, r) \subset \overline{B}(x, r) \subset \Omega$ such that $\mu\big(B(x, r) \cap \{u = c\}'\big) = 0$. For any ball B with $\overline{B} \subset \Omega$, the Riesz measure associated with $u|_B$ relative to B equals the restriction $\mu|_B$ of the Riesz measure μ (relative to Ω); see Sect. 1.1.8. This reduces the assertion of the lemma to the case where Ω is replaced by $B(x,r')$, $r<r'$ and $\overline{B}(x,r')\subset\Omega$, and where u is lower bounded, or even $u\geqslant 0$ after adding a constant (which is permissible by the arbitrariness of $c\in\mathbb{R}$). Finally, we may assume that u is a potential on $B(x,r')$, otherwise replace u by $\widehat{R}_u^{B(x,r)}=R_u^{B(x,r)}$ (cf. (1.3.11)) which is a potential (cf. Theorem 1.3.5), equal to u on $B(x,r)$ and hence having the same Riesz measure after restriction to $B(x,r)$, again by Sect. 1.1.8.

With the lemma in this context, apply Lemma 1.4.27 with p replaced by the potential u on $\Omega=B(x,r')$. Thus

$$\widehat{R}_{u\wedge c}^{\{u<c\}}=\widehat{R}_u^{\{u<c\}}=u\wedge c,$$

the former equality being straightforward. Here $u\wedge c$ is a potential along with u. Writing $u\wedge c=G_\Omega\mu_c$ we have by Theorem 1.4.24 $\widehat{R}_{u\wedge c}^{\{u<c\}}=G_\Omega((\mu_c)^{\{u<c\}})$, and this measure $(\mu_c)^{\{u<c\}}$ is carried by $\mathfrak{b}(\{u<c\})$ according to Theorem 1.4.23(a). And $\{u=c\}'$ does not meet $\mathfrak{b}(\{u<c\})$. (In fact, if the finely open set $\{u=c\}'$ meets $\mathfrak{b}(\{u<c\})$ then it meets $\{u<c\}$ itself, in contradiction with $\{u=c\}'\subset\{u=c\}$.) It follows that $\mu_c(\{u=c\}')=0$.

By Lemma 1.4.27 applied to the potentials $u=G_\Omega\mu$ and $u_c=G_\Omega\mu_c$ we obtain

$$\widehat{R}_u^{\{u\leqslant c\}}=u\wedge c,\quad \mu^{\{u\leqslant c\}}=\mu_c.$$

Writing for brevity $A=\{u=c\}'$ we finally obtain by Theorem 1.4.30 applied to the Borel measurable indicator function $f=1_A$

$$0=\int 1_A\,d\mu_c=\int 1_A\,d\mu^{\{u\leqslant c\}}=\int (1_A)^{\{u\leqslant c\}}\,d\mu=\int\Big(\int 1_A\,d\varepsilon_x^{\{u\leqslant c\}}\Big)\,d\mu(x).$$

Because $\varepsilon_x^{\{u\leqslant c\}}=\varepsilon_x$ for $x\in\mathfrak{b}(\{u\leqslant c\})$ (Theorem 1.4.23(b)), in particular for

$$x\in A=\{u=c\}'\subset\mathfrak{b}(\{u=c\})\subset\mathfrak{b}(\{u\leqslant c\})$$

(see (1.4.13)), we conclude that

$$0=\int\Big(\int 1_A\,d\varepsilon_x\Big)\,d\mu(x)=\int 1_A(x)\,d\mu(x)=\mu(A).$$

□

Theorem 1.4.32 ([Br17, VIII.8], [He, Proposition 28.4]) *Let A be a subset of Ω and μ an admissible measure on Ω.*

(a) *The implication*

$$\mu(E)=0\ \Rightarrow\ \mu^A(E)=0 \tag{1.4.24}$$

holds for any polar set $E\subset\mathfrak{b}(A)$, and also for any subset E of the fine interior $\mathfrak{b}(A)'$ of $\mathfrak{b}(A)$.

(b) *If* $\mu(\mathfrak{b}(A)) = 0$ *then* μ^A *is carried by* $\mathfrak{b}(A) \cap \mathfrak{b}(\complement A)$ *and hence by the base of the fine boundary* $\partial_f A$ *of* A *(complement and fine boundary being taken relative to* Ω*). Furthermore,* μ^A *does not charge any polar subset of* Ω.

Since $(\partial_f A) \setminus \mathfrak{b}(\partial_f A)$ is polar, by Corollary 1.4.18, it follows that μ^A is carried even by $\mathfrak{b}(\partial_f A)$ (when $\mu(\mathfrak{b}(A)) = 0$). Actually, it will be shown later that $\mathfrak{b}(A) \cap \mathfrak{b}(\complement A) = \mathfrak{b}(\partial_f A) = \partial_f(\mathfrak{b}A))$ for any set $A \subset \Omega$.

Proof (a) We have by Definition 1.4.29 and Theorem 1.4.30 applied to $f = 1_E$

$$\mu^A(E) = \int \varepsilon_x^A(E)\, d\mu(x). \tag{1.4.25}$$

Consider first the case of a polar set $E \subset \mathfrak{b}(A)$ with $\mu(E) = 0$. For that we give a proof simpler than those in [Br17] and [He] with $\mu(E) = 0$. As noted in Sect. 1.1.11 there exists for each $x \in \Omega \setminus E$ a potential p on Ω such that $p = +\infty$ on E and $p(x) < +\infty$, hence

$$(+\infty)\varepsilon_x^A(E) = \int_E p\, d\varepsilon_x^A \leqslant \int_\Omega p\, d\varepsilon_x^A = \widehat{R}_p^A(x) \leqslant p(x) < +\infty$$

and therefore $\varepsilon_x^A(E) = 0$. It then follows from (1.4.25) that

$$\mu^A(E) = \int_E \varepsilon_x^A(E)\, d\mu(x) = 0.$$

Let μ_1 be the restriction of μ to the set of points of Ω where A is thin; by Theorem 1.4.23, it suffices to show that the set of points of Ω where $\complement A$ is thin, which is μ_1-negligible, is also μ_1^A-negligible. Let p be a finite strict potential on Ω, then $\mathfrak{b}(\complement A) = \{x \in \Omega : \widehat{R}_p^{\complement A}(x) = p(x)\}$ according to Theorem 1.4.17, and the set $\Omega \setminus \mathfrak{b}(\complement A)$ is contained in A up to polar set. It then follows that $\widehat{R}^A_{\widehat{R}_p^{\complement A}} = \widehat{R}_p^A$ in $\mathfrak{b}(\complement A)$ by Theorem 1.3.2, in particular at each point of Ω where A is thin. Consequently, $\int \widehat{R}^A_{\widehat{R}_p^{\complement A}} d\mu_1 = \int \widehat{R}_p^A d\mu_1$, that is, $\Omega \setminus \mathfrak{b}(\complement A)$ is μ_1^A-negligible.

(b) When $x \notin \mathfrak{b}(A)$, hence $x \notin \mathfrak{r}(A)$, we have just found that ε_x^A is carried by

$$\complement(\mathfrak{r}(\mathfrak{b}(A)) = \mathfrak{b}(\complement\mathfrak{b}(A)) = \mathfrak{b}(\mathfrak{r}(\complement A)) \subset \mathfrak{b}(\mathfrak{b}(\complement A)) = \mathfrak{b}(\complement A),$$

cf. (1.4.15), and so altogether by

$$\mathfrak{b}(A) \cap \mathfrak{b}(\complement A) \subset \widetilde{A} \cap (\complement A)^{\sim} = \partial_f A.$$

Finally, for any polar set $E \subset \mathfrak{b}(A)$ we have $\varepsilon_x^A(E) = 0$ because $\varepsilon_x^A(\{x\}) = 0$ by Theorem 1.4.23(a) and because $\varepsilon_x^A(E \setminus \{x\}) = 0$ by the former assertion of (a) of the present theorem (applied to the polar set $E \setminus \{x\}$ not charged by ε_x). $\square$

Theorem 1.4.32 will chiefly be used later for $A = \complement U$ with U finely open. In the following corollary it is used that $\partial_f U \setminus \mathfrak{b}(\partial_f U)$ is a polar set; see Corollary 1.4.18.

Corollary 1.4.33 *Let $U \subset \Omega$ be finely open, and let $x \in U$. The measure $\varepsilon_x^{\complement U}$ does not charge the polar sets, and it is carried by $\mathfrak{b}(U) \cap \mathfrak{b}(\complement U) \subset \partial_f U$, and hence by $\mathfrak{b}(\partial_f U) \subset \partial_f U$.*

It will be shown in Lemma 2.1.47 that actually $\mathfrak{b}(U) \cap \mathfrak{b}(\complement U) = \mathfrak{b}(\partial_f U)$.

Here are two fine topology results of Tietze-Urysohn type. Recall that the fine topology, though completely regular, is not normal (as mentioned in Sect. 1.2).

Lemma 1.4.34 (First property of quasi-normality) *Let B_0 and B_1 be two disjoint bases in Ω. There exists a finely continuous function f on Ω such that $f = 0$ on B_0, $f = 1$ on B_1, and $0 < f < 1$ elsewhere on Ω.*

In view of Theorem 1.4.17, this analog of the definition of normality of a topological space is proved simply by choosing a finite continuous strict potential p (see Theorem 1.4.15) and putting $f = f_0/(f_0 + f_1)$, where

$$f_i := p - \widehat{R}_p^{B_i}, \quad i = 0, 1.$$

Note the strict inequalities here and in the following theorem and its proof.

Theorem 1.4.35 (Second property of quasi-normality) *Let f denote a finely continuous numerical function defined on a base $B \subset \Omega$. There exists a finely continuous numerical function g on Ω such that $g = f$ on B and (unless f is constant on B) that*

$$\inf_{x \in B} f(x) < g < \sup_{x \in B} f(x) \quad \text{on } \complement B.$$

Proof Like in the proof of the Tietze-Urysohn theorem for a normal space (see, for example, [Bou5, §4, Théorème 2]), it can be verified by induction that it suffices for the proof of Theorem 1.4.35 to show that, for any base $B \subset \Omega$ and any finely continuous function $u : B \to [-1, 1]$, there exists a finely continuous function $v : \Omega \to [-\frac{1}{3}, \frac{1}{3}]$ such that

$$|u - v| \leqslant \tfrac{2}{3} \quad \text{on } B, \quad |v| < \tfrac{1}{3} \quad \text{on } \complement B.$$

To prove that such v exists, let $H_0 := B \cap \{-1 \leqslant u \leqslant -\frac{1}{3}\}$ and $H_1 := B \cap \{\frac{1}{3} \leqslant u \leqslant 1\}$. These two sets are finely closed and disjoint, and by the above lemma (slightly transformed) there exists a finely continuous function $v : \Omega \to [-\frac{1}{3}, \frac{1}{3}]$ such that $v = -\frac{1}{3}$ on $\mathfrak{b}(H_0) \subset H_0$, $v = \frac{1}{3}$ on $\mathfrak{b}(H_1) \subset H_1$, and $|v| < \frac{1}{3}$ elsewhere on Ω, in particular on $\complement B$. The inequality $|u - v| \leqslant \frac{2}{3}$ holds on $\mathfrak{b}(H_0) \cup \mathfrak{b}(H_1)$, and by the triangle inequality also on the rest of B. □

Theorem 1.4.36 *Every finely continuous numerical function $f : \Omega \to \overline{\mathbb{R}}$ is G_δ-measurable in the sense that, for any constant $a \in \mathbb{R}$, the sets $\{f \leqslant a\}$ and $\{f \geqslant a\}$ are (Euclidean) G_δ-sets. Thus f is in particular Borel measurable.*

Proof For the former set $\{f \leqslant a\}$ we have in fact

$$\{f \leqslant a\} = \bigcap_n \{f < a + \tfrac{1}{n}\} \subset \bigcap_n \mathfrak{b}(\{f < a + \tfrac{1}{n}\})$$
$$\subset \bigcap_n \mathfrak{b}(\{f \leqslant a + \tfrac{1}{n}\}) \subset \bigcap_n \{f \leqslant a + \tfrac{1}{n}\} = \{f \leqslant a\},$$

the first inequality because $\{f < a + \frac{1}{n}\}$ is finely open and hence contained in its base in view of (1.4.13). The third inequality holds because $\{f \leqslant a + \frac{1}{n}\}$ is finely closed and hence contains its base, again in view of (1.4.13). Thus equality prevails throughout this display, and hence $\{f \leqslant a\}$ is a G_δ along with the third (and the fourth) set in the display, the base of any set being a G_δ (see Corollary 1.4.18), and the intersection of a sequence of G_δ-sets being likewise a G_δ.

The latter set $\{f \geqslant a\}$ equals $\{-f \leqslant -a\}$ and is therefore a G_δ because $-f$ is likewise finely continuous. □

Actually, as a sharpening of the above theorem, every G_δ-measurable function f is of Baire class One, see, for example, [Kur], and so is therefore in particular every finely continuous numerical function on Ω, but this will not be needed in the present book. For a different proof that every G_δ-measurable function is of Baire class One, see [LMZ, Corollary 3.8].[6]

Combined with Theorem 1.4.35 the above theorem yields the following.

Corollary 1.4.37 *Every finely continuous numerical function f defined on a base $B \subset \Omega$ is Borel measurable relative to B.*

We close this section by discussing the base operation $\mathfrak{b}$ on *functions* and the regularizer $\mathfrak{r}$ of functions, as introduced and studied by Doob [Do3, Do4]. This will lead us showing that every finely continuous numerical function on a finely open set $U \subset \Omega$ is a Borel function relative to U (Theorem 1.4.44), a property to be used in Sects. 2.3 (see Theorem 2.3.11) and 2.6 (see Corollary 2.6.6).

Denote by $\mathcal{F}$ the family of all functions $\Omega \to [-\infty, +\infty]$, and by $\mathcal{G}$, resp. $\mathcal{H}$, the cone of all l.s.c., resp. u.s.c., functions of class $\mathcal{F}$. A function $f \in \mathcal{F}$ is said to be of class $\mathcal{H}_\sigma$, resp. $\mathcal{G}_\delta$, if f is the pointwise supremum, resp. infimum, of sequence of functions of class $\mathcal{H}$, resp. $\mathcal{G}$. We shall also consider the class $\mathcal{H}_{\sigma\delta}$ and $\mathcal{G}_{\delta\sigma}$ in the obvious sense. The notations $\widetilde{f}$, resp. $\widehat{f}$, will be extended so as to designate the least finely u.s.c. majorant, resp. the greatest finely l.s.c. minorant of a function $f \in \mathcal{F}$.

Definition 1.4.38 The *base* $\mathfrak{b}(f)$ and the *regularizer* $\mathfrak{r}(f)$ of $f \in \mathcal{F}$ are defined by

$$\mathfrak{b}(f) = \mathrm{f-}\limsup_{y\to x, y\neq x} f(y), \quad \mathfrak{r}(f) = \mathrm{f-}\liminf_{y\to x, y\neq x} f(y)$$

for $x \in \Omega$.

---

[6] Regrettably, an alternative proof that f is of Baire class One which Fuglede attempted at in [Fu9] is faulty.

Clearly (1.4.21) extends to

$$\widehat{f} \leq \mathfrak{r}(f) \leq \mathfrak{b}(f) \leq \widetilde{f} \text{ for } f \in \mathcal{F}. \tag{1.4.26}$$

The duality (1.4.20) between $\mathfrak{b}$ and $\mathfrak{r}$ as set functions obviously extends to $\mathfrak{b}$ and $\mathfrak{r}$ as follows:

$$\mathfrak{r}(-f) = -\mathfrak{b}(f). \tag{1.4.27}$$

In view of this duality we henceforth state results mostly for $\mathfrak{b}$.

Lemma 1.4.39 *$\mathfrak{b}(f)$ is determined by a Dedekind cut in $[-\infty, +\infty]$ as follows, cf. [Do3, p. 590]:*

$$\mathfrak{b}(f)(x) = \sup\{t : x \in \mathfrak{b}(\{f \geq t\})\} \tag{1.4.28}$$

$$= \inf\{s : x \in \complement\mathfrak{b}(\{f \geq s\})\}. \tag{1.4.29}$$

Proof For the equivalence of (1.4.28) and (1.4.29), we merely use that the base operation on the sets is *increasing*. For $x \in \Omega$ and $t \in [-\infty, +\infty]$ we have by Definition 1.4.38, denoting by $\mathcal{V}_x$ a deleted fine neighborhood base of x,

$$\begin{aligned} \mathfrak{b}(f)(x) \geq t &\iff \forall V \in \mathcal{V}_x, \forall s < t, \exists y \in V : f(y) > s \\ &\iff \forall s < t : x \in \mathfrak{b}(\{f \geq s\}) \\ &\iff \inf\{s : x \in \complement\mathfrak{b}(\{f \geq s\}) \geq t. \end{aligned}$$

This establishes (1.4.28), two members having the same minorants t being equal. □

The relationship between $\mathfrak{b}$ and $\mathfrak{r}$ as set functions and as functionals is

$$\mathfrak{b}(1_A) = 1_{\mathfrak{b}(A)}, \ \mathfrak{r}(1_A) = 1_{\mathfrak{r}(A)},$$

for any set $A \subset \Omega$. For $\mathfrak{b}$ this follows for example from (1.4.28). For $\mathfrak{r}$, use the duality (1.4.27).

Lemma 1.4.40 *For any function $f \in \mathcal{F}$ we have $\mathfrak{b}(f) \in \mathcal{H}_{\sigma\delta}$ and $\mathfrak{r}(f) \in \mathcal{G}_{\delta\sigma}$. If moreover $f \geq 0$ then $\mathfrak{b}(f) \in \mathcal{G}_\delta$.*

Proof Choose a finite continuous *strict* potential p on Ω (cf. Definition 1.4.14 and Theorem 1.4.15). For any set $A \subset \Omega$ and $n \in \mathbb{N}$ write

$$b_n(A) = \{\widehat{R}_p^A > (1 - \frac{1}{n})p\}.$$

The sets $b_n(A)$ are open in Ω and decrease with intersection

$$\bigcap_n b_n(A) = \{\widehat{R}_p \geq p\} = \{\widehat{R}_p = p\} = \mathfrak{b}(A) \tag{1.4.30}$$

by Theorem 1.4.17. (This is the usual proof that $\mathfrak{b}(A)$ is a G_δ-set; cf. Corollary 1.4.18.) For each $n \in \mathbb{N}$ the functions b_n are increasing. In analogy with (1.4.28) we therefore put

$$g_n(x) = \sup\{t : x \in \complement b_n(\{f \geq t\})\} = \sup_t t.1_{\complement b_n(\{f \geq t\})}(x). \tag{1.4.31}$$

Then $g_n \in \mathcal{H}_\sigma$, because $\complement b_n(\{f \geq t\})$ is closed. And $g_n \geq \mathfrak{b}(f)$ because $b_n(A) \supset \mathfrak{b}(A)$. For the proof that $\mathfrak{b}(f) \in \mathcal{H}_{\sigma\delta}$ it therefore remains to show that $\inf_n g_n \leq \mathfrak{b}(f)$. For given $x \in \Omega$ let $s > \mathfrak{b}(f)(x)$, that is, by (1.4.29), there exists $\sigma \in [0, s[$ such that $x \in \complement\mathfrak{b}(\{f \geq \sigma\})$. In view of (1.4.30) it follows that $x \in \complement b_n(\{f \geq \sigma\})$ for some n, and hence $g_n(x) \leq \sigma < s$ because

$$g_n(x) = \inf\{s : x \in \complement b_n(\{f \geq s\})\}$$

by (1.4.31) in view of the analogy of the identity of (1.4.28) and (1.4.29). By varying s with $s > \mathfrak{b}(f)(x)$ we conclude that indeed $\inf_n g_n(x) \leq \mathfrak{b}(f)(x)$ for every $x \in \Omega$. Having thus shown that $\mathfrak{b}(f) \in \mathcal{H}_{\sigma\delta}$ it follows by duality, see (1.4.27), that $\mathfrak{r}(f) \in \mathcal{G}_{\delta\sigma}$.

If $f \geq 0$ then $g_n(x) = \sup_{t\geq 0} t.1_{\complement b_n(\{f\geq t\})}(x)$, hence g_n is finely u.s.c. (the sets $b_n(\{f \geq t\})$ being finely closed) and therefore $\mathfrak{b}(f) \in \mathcal{G}_\delta$. □

Lemma 1.4.41 (a) *Every finely l.s.c. function $f \in \mathcal{F}$ is subbasic: $f \leq \mathfrak{b}(f)$.*

(b) *A function $f \in \mathcal{F}$ is finely u.s.c. if and only if f is superbasic: $f \geq \mathfrak{b}(f)$.*

Proof (a) For given $x \in \Omega$ let $t < f(x)$. The $\{f > t\}$ is finely open set, hence subbasic (see (1.4.13), and so

$$x \in \{f > t\} \subset \mathfrak{b}(\{f > t\} \subset \mathfrak{b}(\{f \geq t\}),$$

from which it follows by (1.4.28) that $\mathfrak{b}(f)(x) \geq t$.

(b) The "if" part follows from Definition 1.4.38. For "only if" part, let $x \in \Omega$ and $s > f(x)$. Then $\{f \geq s\}$ is finely closed and hence

$$x \in \complement\{f \geq s\} \subset \mathfrak{b}(\complement\{f \geq s\}),$$

from which it follows by (1.4.29) that $\mathfrak{b}(f)(x) \leq s$, and hence $\mathfrak{b}(f)(x) \leq f(x)$. □

Corollary 1.4.42 *For any finely continuous function $f \in \mathcal{F}$ we have $\mathfrak{b}(f) = f$, and by duality $\mathfrak{r}(f) = f$. In particular $f \in \mathcal{G}_\delta \cap \mathcal{H}_\sigma$.*

Theorem 1.4.43 *A function $f \in \mathcal{F}$ is of Baire class One on Ω if and only if $f \in \mathcal{G}_\delta \cap \mathcal{H}_\sigma$, thus in particular if f is finely continuous.*

Proof Suppose first that $f \in \mathcal{G}_\delta \cap \mathcal{H}_\sigma$. Thus there are sequences (g_n) in $\mathcal{G}$ and (h_n) in $\mathcal{H}$ such that $g_n \downarrow f$ and $h_n \uparrow f$ pointwise. According to the "Between Theorem" of Hahn, see Hausdorff [Hau] for a simple proof, there exist continuous functions f_n such that $h_n \leq f_n \leq g_n$. Clearly $f_n \to f$ pointwise, so f is of Baire class One.

Conversely, suppose that a sequence (f_n) of continuous functions converges pointwise to a function $f \in \mathcal{F}$. Then $f \in \mathcal{G}_\delta \cap \mathcal{H}_\sigma$ because

$$f = \limsup_n f_n = \inf_n \sup_{p>n} f_p, \quad \sup_{p>n} f_p \in \mathcal{G},$$

$$f = \liminf_n f_n = \sup_n \inf_{p>n} f_p, \quad \inf_{p>n} f_p \in \mathcal{H}.$$

□

Theorem 1.4.44 *Let $U \subset \Omega$ be finely open. Every finely continuous function $f : U \to [-\infty, +\infty]$ is a Borel measurable function in the relative Euclidean topology.*

Proof Extend f to functions $f_{+\infty}, f_{-\infty} : \Omega \to [-\infty, +\infty]$ by defining $f_{\pm\infty} = \pm\infty$ on $\Omega \backslash U$. Then $f_{-\infty} = f = f_{+\infty}$ on U. Furthermore, $f_{-\infty}$ is finely l.s.c. on Ω and $f_{+\infty}$ is finely u.s.c. on Ω. Since

$$f_{-\infty} \leq \mathfrak{r}(f_{-\infty}) \leq \left\{ \begin{matrix} \mathfrak{r}(f_{+\infty}) \\ \mathfrak{b}(f_{-\infty}) \end{matrix} \right\} \leq \mathfrak{r}(f_{+\infty}) \leq f_{+\infty}$$

with equality on U, by Lemma 1.4.40, $\mathfrak{b}(f_{+\infty}) \in \mathcal{H}_{\sigma\delta}$ and $\mathfrak{r}(f_\infty) \in \mathcal{G}_{\delta\sigma}$. Thus $f = \mathfrak{b}(f_{+\infty}) = \mathfrak{r}(f_{-\infty})$ on U and consequently f is of class $\mathcal{H}_{\sigma\delta} \cap \mathcal{G}_{\delta\sigma}$. □

Proposition 1.4.45 *The base $\mathfrak{b}(f)$ of any function $f \in \mathcal{F}$ is finely u.s.c.*

Proof In fact, $\{\mathfrak{b}(f) \geq t\}$ is finely closed for any $t \in [0, +\infty]$ because

$$\{\mathfrak{b}(f) \geq t\} = \bigcap_{s<t} \mathfrak{b}(\{f \geq s\})$$

according to Lemma 1.4.39. □

Corollary 1.4.46 *For every finely l.s.c. function $f \in \mathcal{F}$, the base $\mathfrak{b}(f)$ of f is the least finely u.s.c. majorant of f.*

Proof Let $\widetilde{f}$ be the least finely u.s.c. majorant of f. Then we have $\widetilde{f} \geq \mathfrak{b}(\widetilde{f}) \geq \mathfrak{b}(f) \geq f$ according to Definition 1.4.38 and Lemma 1.4.41. Hence $\widetilde{f} = \mathfrak{b}(f)$ by Proposition 1.4.45. □

1.5 Quasi-Topology and Fine Topology

The results in the present section form the foundation for "fine potential theory", to be introduced in the next section.

We begin by studying "quasi-topological" concepts relative to a suitable kind of *capacity*, not just the Green capacity (Sect. 1.1.14). Capacity is still a suitable countably subadditive set function defined on subsets A of Ω. But inspired by Doob [Do1] we also consider capacity as a suitable *sublinear functional* on the lattice cone $\mathcal{F}^+$ of all functions $f : \Omega \to [0, +\infty]$; see also [Fu6]. The special case of an indicator function $f = 1_A$ then leads to the corresponding set function.

Following Brelot [Br17] and earlier work of his, we shall consider two concrete capacities (besides the Green capacity). First we define the "capacity" of a function $f \in \mathcal{F}^+$ simply to be $\widehat{R}_f$. This "capacity" is not a number, but an element of the lattice cone $\mathcal{U}^+$ of all hyperharmonic functions $\Omega \to [0, +\infty]$.

Next, a related capacity of $f \in \mathcal{F}^+$, now with numerical values, is

$$C_m(f) = \int \widehat{R}_f \, dm, \tag{1.5.1}$$

where m denotes a given non-zero bounded (positive Radon) measure on Ω which does not charge any polar set, and for which $\int u \, dm < +\infty$ for every $u \in \mathcal{S}^+$. We have simply $C_m(f) = \int f \, dm$ when $f \in \mathcal{U}^+$ since then $\widehat{R}_f = f$.

A particular case of this latter capacity C_m arises when m is the *harmonic measure* $\rho = \rho_{x,r} = \mu_x^B$ on (the boundary of) a closed ball $\overline{B}(x, r) \subset \Omega$; see Sect. 1.1.2. The third capacity to be considered is defined as the functional $f \mapsto R_f(x_0)$ (not $\widehat{R}_f(x_0)$) on $\mathcal{F}^+$, where x_0 is a given point of Ω. The associated set function is $A \mapsto R_1^A(x_0)$.

An intimate relationship between *quasi-topological* notions and fine topology notions will be established relative to either of the above two notions of capacity in Theorems 1.5.12 and 1.5.13 below. For convenience we mostly formulate the results relative to the capacity C_m. In the absence of other indication it will then be tacitly understood that corresponding results hold for the "capacity" $f \mapsto \widehat{R}_f$ with the corresponding proof, usually replacing inf with $\widehat{\inf}$; see, for example, Lemma 1.5.1 and Definition 1.5.2.

As before, Ω denotes a fixed Greenian domain in $\mathbb{R}^N$ $(N \geqslant 2)$. Complements and topological notions are understood relative to Ω. For any family Φ of functions $f : \Omega \to \overline{\mathbb{R}}$ we denote by $\widehat{\inf}\, \Phi$ the greatest l.s.c. minorant of Φ, or what amounts to the same: of the pointwise infimum of Φ. Recall the notations $\mathcal{P}$, $\mathcal{P}^s$, $\mathcal{P}^b$, $\mathcal{P}^c$, $\mathcal{P}_0^c$ from the beginning of Sect. 1.4, and the class Λ of all admissible (positive) measures on Ω.

Obviously C_m is *increasing*: $C_m(f_1) \leqslant C_m(f_2)$ if $f_1 \leqslant f_2$; *positive homogeneous* $C_m(af) = a\, C_m(f)$ for $a \in [0, +\infty[$; and *countably sublinear*:

$$C_m\Big(\sum_{n\in\mathbb{N}} f_n\Big) \leqslant \sum_{n\in\mathbb{N}} C_m(f_n).$$

Such functionals are studied in [Fu6].

The capacity functional C_m from (1.5.1) has the further property of being an *upper* capacity (now on the topological space Ω).

Lemma 1.5.1 *For any* $f \in \mathcal{F}^+$ *we have* $C_m(f) = \inf\{C_m(g) : g \in \mathcal{G},\ g \geqslant f\}$, *where* $\mathcal{G}$ *denotes the class of all l.s.c. functions* $\Omega \to [0, +\infty]$.

Proof If $C_m(f) = +\infty$ take $g \equiv +\infty$. Otherwise the set $E := \{\widehat{R}_f < R_f\}$ is polar, and hence there is a superharmonic function $u \geqslant 0$ on Ω with $u \equiv +\infty$ on E. For any $\varepsilon > 0$ write $g := \widehat{R}_f + \varepsilon u \geqslant R_f \geqslant f$. Then g is a positive superharmonic majorant of f, and

$$C_m(g) = \int \widehat{R}_g \, dm = \int g \, dm = \int \widehat{R}_f \, dm + \varepsilon \int u \, dm \downarrow \int \widehat{R}_f \, dm = C_m(f)$$

as $\varepsilon \downarrow 0$, $\int u \, dm$ being finite by definition of C_m. Thus C_m is upper. □

The capacity C_m determines a kind of metric (the C_m-"metric") on $\mathcal{F}$, whereby the "distance" between $f_1, f_2 \in \mathcal{F}$ is defined as $C_m(|f_1 - f_2|)$ ($\leqslant +\infty$).[7] This "metric" determines in the usual way a topology on $\mathcal{F}$, called the *C_m-metric topology*.

With the functional C_m we associate (as already mentioned) a *set function*, likewise denoted by C_m and termed a capacity. It is defined for all sets $A \subset \Omega$ by

$$C_m(A) := C_m(1_A) = \int \widehat{R}_1^A \, dm.$$

Clearly this set function C_m is positive, increasing, and *countably subadditive*. Such set functions are studied in [Fu5].

By Lemma 1.3.10, a set $E \subset \Omega$ is *polar* if and only if $\widehat{R}_1^E \equiv 0$, or equivalently if and only if $C_m(E) = \int \widehat{R}_1^E \, dm = 0$, noting that $\int \widehat{R}_1^E \, dm = 0$ if and only if $\widehat{R}_1^E = 0$ m-a.e., which by the requirements on m amounts to $\widehat{R}_1^E \equiv 0$, $\widehat{R}_1^E$ being superharmonic. Also, E is polar if and only if $\widehat{R}_1^E(x_0) = 0$ for some and hence any point $x_0 \in \Omega$.

Definition 1.5.2 A function $f : \Omega \to \overline{\mathbb{R}}$ is said to be *quasi-continuous* if

$$\widehat{\inf}\{\widehat{R}_1^\omega : \omega \text{ open in } \Omega, \ f|_{\complement\omega} \text{ continuous}\} = 0. \tag{1.5.2}$$

Equivalently,

$$\inf\{C_m(\omega) : \omega \text{ open in } \Omega, \ f|_{\complement\omega} \text{ continuous}\} = 0 \tag{1.5.3}$$

for some and hence any measure m as above; see (1.5.1) and ensuing text.

For the equivalence of (1.5.2) and (1.5.3) apply the fundamental convergence theorem (including Choquet's topological lemma) to obtain a decreasing sequence of open sets ω_n such that $\inf_n \widehat{R}_1^{\omega_n} = u$ pointwise q.e., where we define $u = \widehat{\inf}\{\widehat{R}_1^\omega : \omega \text{ open in } \Omega, \ f|_{\complement\omega} \text{ continuous}\} \in \mathcal{U}^+$. Then

$$\int \inf_n \widehat{R}_1^{\omega_n} \, dm = \inf_n \int \widehat{R}_1^{\omega_n} \, dm = \int u \, dm.$$

[7] $|f_1 - f_2|$ is defined to be $+\infty$ at points where $f_1 = f_2 = +\infty$.

In fact, $\int \widehat{R}_1^{\omega_n}\, dm \leqslant \int dm < +\infty$. The equivalence of (1.5.2) and (1.5.3) follows, corresponding to the case $u \equiv 0$.

Similarly we have two equivalent definitions of f to be *quasi-u.s.c.*, resp. *quasi-l.s.c.*, obtained by replacing "continuous" with "u.s.c.", resp. "l.s.c.". Clearly, f is quasi-u.s.c. if and only if $-f$ is quasi-l.s.c. And f is quasi-continuous if and only if f is both quasi-u.s.c. and quasi-l.s.c.

Remark 1.5.3 In Definition 1.5.2 one may obviously drop the requirement that ω be open because the set function C_m is *outer* in the sense that

$$C_m(E) = \inf\{C_m(\omega) : \omega \text{ open}, \omega \supset E\},$$

as a consequence of the functional C_m being upper. In fact, for given $0 < \varepsilon < 1$ there is $g \in \mathcal{G}$ with $g \geqslant 1_E$ and

$$C_m(g) \leqslant C_m(1_E) + \varepsilon = C_m(E) + \varepsilon.$$

From $g \geqslant 1 > 1 - \varepsilon$ on E follows $g > 1 - \varepsilon$ on some open set $G \supset E$, hence

$$C_m(G) \leqslant C_m\Big(\frac{g}{1-\varepsilon}\Big) = \frac{1}{1-\varepsilon} C_m(g) < \frac{1}{1-\varepsilon}(C_m(E) + \varepsilon),$$

which tends to $C_m(E)$ as $\varepsilon \to 0$.

Definition 1.5.4 A set $A \subset \Omega$ is said to be *quasi-closed* (resp. *quasi-open*) if

$$\widehat{\inf_{\omega}\{\widehat{R}_1^{\omega}} : \omega \text{ open}, A \text{ closed (resp. open) relative to } \complement\omega\} = 0.$$

Equivalently,

$$\inf\{C_m(\omega) \colon \omega \text{ open}, A \text{ closed (resp. open) relative to } \complement\omega\} = 0 \tag{1.5.4}$$

for some and hence any measure m as above; see (1.5.1).

This equivalence follows easily by the fundamental convergence theorem in the same way as in Definition 1.5.2. The words "relative to $\complement\omega$" may clearly be omitted (just diminish ω suitably). Thus A is quasi-closed (resp. quasi-open) if and only if

$$\inf\{C_m(A \triangle E) : E \text{ closed (resp. open)}\} = 0, \tag{1.5.5}$$

whereby $A \triangle E = (A \backslash E) \cup (E \backslash A)$ denotes the symmetric difference of two sets A and E.

As in Remark 1.5.3 the requirement in Definition 1.5.4 that ω be open can be omitted.

Clearly, A is quasi-closed (resp. quasi-open) if and only if the indicator function 1_A is quasi-u.s.c. (resp. quasi-l.s.c.), and also if and only if $\complement A$ is quasi-open (resp. quasi-closed).

Lemma 1.5.5 *A set $A \subset \Omega$ is quasi-open, resp. quasi-closed, if (and only if)*

$$\inf\{C_m(E\backslash A) : E \text{ open in } \Omega, A \subset E\} = 0,$$

resp.

$$\inf\{C_m(A\backslash E) : E \text{ closed in } \Omega, E \subset A\} = 0.$$

Proof By[8] duality it suffices to consider the case where A is quasi-closed. For given $\varepsilon > 0$ let $C_m(A\Delta E) < \varepsilon$ and hence $C_m(E\backslash A) < \varepsilon$. By openness of C_m (Remark 1.5.3) there is an open set $G \subset \Omega$ containing $E\backslash A$ so that $C_m(G) < \varepsilon$. Then $E\backslash G$ is closed and contained in A, and $A\backslash(E\backslash G) \subset (A\backslash E) \cup G$, whence indeed $C_m(A\backslash(E\backslash G)) \leqslant \varepsilon + \varepsilon = 2\varepsilon$ is as small as we please. □

Lemma 1.5.6 *Any countable infimum or finite supremum of quasi-u.s.c. functions $\Omega \to [-\infty, +\infty]$ is quasi-u.s.c. Any countable supremum or finite infimum of quasi-l.s.c. functions $\Omega \to [-\infty, +\infty]$ is quasi-l.s.c.*

Proof Let for example (f_n) be a sequence of quasi-l.s.c. functions, and write $f := \sup_n f_n$. Choose a sequence (ω_n) of open subsets of Ω so that each $C_m(\omega_n) < \varepsilon/2^n$ and that each $f_n|_{\complement\omega_n}$ is l.s.c. Then $\omega := \bigcup_n \omega_n$ is open with $C_m(\omega) < \varepsilon$ and each $f_n|_{\complement\omega}$ is l.s.c., hence $f|_{\complement\omega}$ is l.s.c., and consequently f is indeed quasi-l.s.c. The remaining three assertions are established similarly. □

Corollary 1.5.7 *Any countable intersection or finite union of quasi-closed sets is quasi-closed. Any countable union or finite intersection of quasi-open sets is quasi-open.*

Lemma 1.5.8 *A function $f : \Omega \to \overline{\mathbb{R}}$ is quasi-u.s.c. (resp. quasi-l.s.c.) if and only if $\{f \geqslant t\}$ (resp. $\{f \leqslant t\}$) is quasi-closed for every $t \in \mathbb{R}$ (or just for every t from some dense subset of $\mathbb{R}$).*

Proof The "only if part" being obvious, suppose (say in the case of a quasi-u.s.c. function f) that $\{f \geqslant t\}$ is quasi-closed for every t from some dense subset of $\mathbb{R}$. That subset has a countable dense subset $T = \{t_n\}$. For a given $\varepsilon > 0$ there is for every $n \in \mathbb{N}$ an open set $\omega_n \subset \Omega$ with $C_m(\omega_n) < \varepsilon/2^n$ such that $\{f \geqslant t_n\}\backslash\omega_n$ is closed in Ω. In terms of the open set $\omega := \bigcup_n \omega_n$ it follows *a fortiori* that $\{f \geqslant t_n\}\backslash\omega$ is closed in Ω for every $n \in \mathbb{N}$, and hence $f|_{\complement\omega}$ is u.s.c. Since such an open set ω with $C(\omega) < \varepsilon$ has been found for every $\varepsilon > 0$, we conclude that f indeed is quasi-u.s.c. □

As a preparation for the next theorem, we bring the following lemma and corollary from [Br17, Lemmas VIII, 13 and 14].

Lemma 1.5.9 ([Br17, p. 69]) *For any set $A \subset \Omega$ and $x_0 \in \Omega$, there exists a potential p on Ω such that $p(x_0) < +\infty$ and for every $x \in \overline{A}\backslash\mathfrak{b}(A)$*

$$p(y) \to +\infty \text{ as } y \to x,\ y \in A.$$

---

[8] This proof replaces an incomplete proof of Lemma 2.2 in [Fu5].

Proof To avoid triviality assume that A is a nonvoid proper subset of Ω. Choose a finite continuous *strict* potential $q = G_\Omega\mu$ on Ω (Definition 1.4.14 and Theorem 1.4.15). Suppose first that A is a *base*: $A = \mathfrak{b}(A)$. Decompose μ into $\mu_1 + \mu_2$, where

$$\mu_1 = 1_A\mu,\ \mu_2 = 1_{\complement A}\mu.$$

The measures μ_1 and μ_2 have finite continuous potentials $G_\Omega\mu_1$ and $G_\Omega\mu_2$ along with $q = G_\Omega\mu$. Furthermore, $\mu_1^A = \mu_1$ and $\mu_2^A(\complement A) = 0$, by Theorem 1.4.23. Choose a sequence of open sets $U_n \supset \complement A$ so that $\mu_2^A(U_n) \to 0$, and write

$$\nu_n = 1_{U_n}\mu_2^A.$$

Then ν_n is carried by $A \cap U_n$, and $\nu_n \to 0$ vaguely (that is, in the weak* topology on the space of Radon measures on Ω). It follows that

$$G_\Omega\nu_n(x_0) = \int_{U_n} G_\Omega\varepsilon_{x_0}\,d\mu_2^A \to 0$$

as $n \to \infty$, by the dominated convergence theorem, noting that by Theorem 1.4.8

$$\int_\Omega G_\Omega\varepsilon_{x_0}\,d\mu_2^A = \int_\Omega \widehat{R}^A_{G_\Omega\varepsilon_{x_0}}\,d\mu_2 \leqslant \int_\Omega G_\Omega\varepsilon_{x_0}\,d\mu = G_\Omega\mu(x_0) = q(x_0) < +\infty.$$

After passing to a subsequence we may therefore assume that

$$\sum_n G_\Omega\nu_n(x_0) < +\infty.$$

Then $p := \sum_n G_\Omega\nu_n$ is a potential, finite at x_0 (see Sect. 1.1.6). Since q is strict it follows from Theorem 1.4.17 that

$$R_q^A = q \text{ on } A,\ R_q^A < q \text{ on } \complement A,$$

and hence for $x \in \overline{A}\setminus\mathfrak{b}(A)$

$$\liminf_{y\to x,\,y\in A} \widehat{R}_q^A(y) = \liminf_{y\to x,\,y\in A} q(y) > \widehat{R}_q^A(x), \tag{1.5.6}$$

q being finite and continuous, and $A = \mathfrak{b}(A)$. Next,

$$\widehat{R}_q^A - G_\Omega\nu_n = G_\Omega(\mu^A) - G_\Omega\nu_n = G_\Omega(\mu_1^A) + G_\Omega(\mu_2^A - \nu_n) = G_\Omega\mu_1 + G_\Omega(1_{\complement U_n}\mu_2^A),$$

which is continuous at x because $G_\Omega\mu_1$ is continuous on Ω along with $G_\Omega\mu$, while $G_\Omega(1_{\complement U_n}\mu_2^A)$ is harmonic on U_n. From (1.5.6) we therefore obtain

$$\liminf_{y\to x, y\in A} G_\Omega\nu_n(y) - G_\Omega\nu_n(x) = q(x) - \widehat{R}_q^A(x) > 0,$$

and since $p = \sum_n G_\Omega\nu_n$ we conclude that, in case A is a base,

$$\liminf_{y\to x, y\in A} p(y) \geqslant \sum_n \liminf_{y\to x, y\in A} G_\Omega\nu_n(y) \geqslant (+\infty)(q(x) - \widehat{R}_q^A(x)) = +\infty. \quad (1.5.7)$$

For an arbitrary set $A \subset \Omega$ we have $A \subset \widetilde{A} = \mathfrak{b}(A) \cup E$, where $E = \widetilde{A}\setminus\mathfrak{b}(A)$ is polar, by Corollary 1.4.18. Choose a potential $r > 0$ on Ω with $r = +\infty$ on $E\setminus\{x_0\}$ and $r(x_0) < +\infty$. Since r is l.s.c. it follows in view of (1.5.7) with A replaced by $\mathfrak{b}(A)$ (which does not affect $\widehat{R}_q^A$) that $p(y) + r(y) \to +\infty$ as $y \to x$, $y \in A \subset \mathfrak{b}(A) \cup E$. □

Corollary 1.5.10 *For any set $A \subset \Omega$ and point $x_0 \in \Omega$,*

$$\inf_\omega\{R_1^{A\cap\omega}(x_0) : \omega \text{ open, } \omega \supset \overline{A}\setminus\widetilde{A}\} = 0.$$

In fact, for any $\varepsilon > 0$, every point $x \in \overline{A}\setminus\widetilde{A}$ has an open neighborhood σ_x such that the potential v from the above lemma satisfies $v > 1/\varepsilon$ on $A \cap \sigma_x$. The union of these σ_x over $x \in \overline{A}\setminus\widetilde{A}$ is an open set ω for which $\varepsilon v > 1$ on $A \cap \omega$ and hence $R_1^{A\cap\omega}(x_0) \leqslant \varepsilon v(x_0)$. It remains to let $\varepsilon \to 0$.

Remark 1.5.11 In [Ch3] from 1959 Choquet proved that the set $\complement\mathfrak{b}(A)$ of points at which a given set $A \subset \Omega$ is thin can be enclosed in an open set which meets A in a set of Green capacity (see Sect. 1.1.14) as small as we please. In view of its many applications Brelot has termed this the *Choquet property*, and he used the same term for the somehow equivalent property that *every finely closed set is quasi-closed* (with respect to the Green capacity). Brelot has established that also the present capacity C_m; see (1.5.1) (now as a set function) as well as the underlying $\mathcal{U}^+$-valued set function $A \mapsto \widehat{R}_1^A$ both have the Choquet property, and furthermore that so has a third capacity, namely $A \mapsto R_1^A(x_0)$ for a given point $x_0 \in \Omega$. See the following theorem.

Theorem 1.5.12 *Every finely closed set $A \subset \Omega$ is quasi-closed relative to each of the three capacities $A \mapsto \widehat{R}_1^A$, $A \mapsto \int \widehat{R}_1^A\,dm$, and $A \mapsto R_1^A(x_0)$. Thus*

$$\inf_\omega \widehat{R}_1^{A\cap\omega} = 0, \quad \inf_\omega \int \widehat{R}_1^{A\cap\omega}\,dm = 0, \quad \inf_\omega R_1^{A\cap\omega}(x_0) = 0, \quad (1.5.8)$$

where ω ranges over all open subsets of Ω containing $\complement A$, and where x_0 is an arbitrary point of Ω and m is an arbitrary measure as described in connection with (1.5.1).

Proof For given $\varepsilon > 0$ there exists by the above corollary an open set $\omega \supset \overline{A}\setminus\widetilde{A}$ with $R_1^{A\cap\omega}(x_0) < \varepsilon$. The open set $\omega' := \omega \cup \complement\overline{A}$ contains $\complement\widetilde{A}$, and $A \cap \omega' = A \cap \omega$, so $R_1^{A\cap\omega'}(x_0) < \varepsilon$, thus establishing the last and hence the first equation (1.5.8). For the first equation (1.5.8) write $u := \widehat{\inf_\omega} \widehat{R}_1^{A\cap\omega}$ (with ω ranging as described). By the fundamental convergence theorem, there is a decreasing sequence of open sets $\omega_n \supset \complement A$ such that

$$u = \widehat{\inf_n} \widehat{R}_1^{A\cap\omega_n} \leqslant \widehat{\inf_n} R_1^{A\cap\omega_n} = 0$$

according to the last equation (1.5.8). For the remaining second equation

$$\inf_n \int \widehat{R}_1^{A\cap\omega_n}\, dm = \int \inf_n \widehat{R}_1^{A\cap\omega_n}\, dm = 0,$$

noting that $\int dm < +\infty$ and that m does not charge any polar set. □

Theorem 1.5.12 can now be inversed and thus extended as follows.

Theorem 1.5.13 *A set $A \subset \Omega$ is quasi-closed (resp. quasi-open) if and only if A differs at most by a polar set from some finely closed (resp. finely open) set, for example from the fine closure $\widetilde{A}$ or from the base $\mathfrak{b}(A)$ of A (resp. from the fine interior A' or the regularizer $\mathfrak{r}(A)$ of A).*

Proof It suffices to consider the "closed" case; cf. (1.4.20). The "if part" then follows from the preceding theorem because if A is quasi-closed then so is any set which differs from A at most by a polar set in view of Theorem 1.3.2 and Eq. (1.3.1). For the "only if part" consider first the "capacity" $\widehat{R}_1^{\bullet}$. Suppose that A is quasi-closed, and consider any open set $\omega \subset \Omega$ such that $A|_{\complement\omega}$ is closed, or equivalently that $F := A\backslash\omega$ is closed. Then $A \subset F \cup \omega$ and hence

$$\widetilde{A} \subset F \cup \widetilde{\omega} = A \cup \widetilde{\omega}.$$

It follows that $\widetilde{A}\backslash A \subset \widetilde{\omega}$ and hence $\widehat{R}_1^{\widetilde{A}\backslash A} \leqslant \widehat{R}_1^{\widetilde{\omega}}$. By varying ω we obtain from (1.3.12)

$$\widehat{R}_1^{\widetilde{A}\backslash A} \leqslant \widehat{\inf_\omega}\ \widehat{R}_1^{\widetilde{\omega}} = \widehat{\inf_\omega}\ \widehat{R}_1^{\omega} = 0,$$

obviously with equality. This implies by Lemma 1.3.10 that $\widetilde{A}\backslash A$ indeed is polar, and since $\widetilde{A}\backslash\mathfrak{b}(A)$ is polar by Corollary 1.4.18, A also differs by a polar set from $\mathfrak{b}(A)$. □

Corollary 1.5.14 *A function $f \in \mathcal{F}$ is quasi-u.s.c. if and only if f at most differs on a polar set from some finely u.s.c. function (for example from $\widetilde{f}$).*

This corollary (and its counterpart with u.s.c. replaced throughout by l.s.c.) is actually equivalent with Theorem 1.5.13 (the case of indicator functions in the corollary).

To verify the corollary (in the u.s.c. case), let first f be finely u.s.c. Then $\{f \geqslant t\}$ is finely closed for every $t \in \mathbb{R}$. By the theorem, $\{f \geq t\}$ is quasi-closed for every such t, and so f is indeed quasi-u.s.c., by Lemma 1.5.8.

In particular, the corollary implies that *a function* $f \in \mathcal{F}$ *is quasi-continuous if and only if* f *is finely continuous q.e.* (But there need not be any finely continuous function equal q.e. to f.)

We shall not use the following generalization of this, and refer to [Fu5, Theorem 4.3] for the easy proof. See also [Br15, p. 39].

Theorem 1.5.15 *A map f of Ω into a topological space Y with a countable basis of open sets is quasi-continuous (in the obvious sense) if and only if f is finely continuous q.e.*

We proceed to study semibounded potentials, introduced by Brelot [Br15].

Definition 1.5.16 A potential $p = G_\Omega\mu$ (where $\mu \in \Lambda$, that is, $G_\Omega\mu \not\equiv +\infty$) is said to be *semibounded*, and we write $p \in \mathcal{P}^s = \mathcal{P}^s(\Omega)$, if p has the following equivalent properties, where for example $\{p > c\}$ (with $c \in [0, +\infty[$) shall stand for $\{x \in \Omega : p(x) > c\}$,

(i) $\widehat{\inf}_{c>0}\widehat{R}_p^{\{p>c\}} = 0$ on Ω, or equivalently $\widehat{\inf}_{c>0}\widehat{R}_{(p-c)^+} = 0$ on Ω,
(ii) $p = G_\Omega\mu$ is representable as the pointwise sum of a sequence of potentials $p_n = G_\Omega\mu_n$ of class $\mathcal{P}^b$, or equivalently (see Sect. 1.1.10) of class $\mathcal{P}^c$ or $\mathcal{P}_0^c$,
(iii) p is the specific supremum of its specific minorants of class $\mathcal{P}^b$, or equivalently of class $\mathcal{P}^c$ or $\mathcal{P}_0^c$,
(iv) $G_\Omega\mu < +\infty$ μ-a.e.,
(v) μ does not charge any polar set.

The name "semibounded" comes from (i), according to which *every locally bounded potential p is semibounded.* For any compact set $K \subset \Omega$ we have in fact $\{p > c\} \subset \complement K$ for large c, and so by Corollary 1.3.8

$$\inf_c R_p^{\{p>c\}} \leqslant \inf_K R_p^{\complement K} = 0. \tag{1.5.9}$$

Also note that by Theorem 1.3.3(b) $\widehat{R}_p^{\{p>c\}} \equiv R_p^{\{p>c\}}$. For the specific order, see Definition 1.3.12.

Proof of equivalence of (i) *through* (v). The equivalence between the two formulations of (i) follows from the inequalities $(p - c)^+ \leqslant 1_{\{p>c\}}p \leqslant (2p - c)^+$.

(i)$\Rightarrow$ (ii). Suppose that $\widehat{\inf}_{c>0}\widehat{R}_p^{\{p>c\}} = 0$. It suffices to consider integers $c = n \in \mathbb{N}$. Let (K_n) denote an increasing sequence of compact sets whose interiors cover Ω. Write

$$E_n = \{p > n\} \cup \complement K_n.$$

Then

$$p = \widehat{R}_p^{\Omega} \leqslant \widehat{R}_p^{\complement E_n} + \widehat{R}_p^{E_n},$$

and by the natural decomposition property (Corollary 1.3.14) there are $q_n, r_n \in \mathcal{U}^+$ such that $p = q_n + r_n$, and

$$q_n \leqslant \widehat{R}_p^{\complement E_n} \leqslant n \widehat{R}_1^{K_n},$$

$$r_n \leqslant \widehat{R}_p^{E_n} \leqslant \widehat{R}_p^{\{p>n\}} + \widehat{R}_p^{\complement K_n}.$$

It follows that $q_n \in \mathcal{P}^b$ because $\widehat{R}_1^{K_n} \in \mathcal{P}^b$ by Theorem 1.3.5. Furthermore,

$$\curlywedge_n r_n \leqslant \widehat{\inf_n r_n} = 0$$

by (1.5.9) and Corollary 1.3.8, and here equality prevails because $r_n \geqslant 0$. Since $p = q_n + r_n$ we obtain $\curlyvee_n q_n = p$. Writing $q_n' = q_1 \curlyvee \ldots \curlyvee q_n$, we have

$$p = \curlyvee_n q_n' = \sup_n q_n'.$$

Thus (iii) holds with $p_1 = q_1'$, $p_n = q_n' - q_{n-1}'$ for $n \geqslant 2$ because $q_n' \in \mathcal{P}^b$ (indeed, $q_n' \leqslant q_1 + \ldots + q_n$), and because $q_n' \succcurlyeq q_{n-1}'$. We conclude that $p_n \in \mathcal{P}^b$. According to Theorem 1.4.1 each p_n is the pointwise sum of a sequence of potentials of class $\mathcal{P}_0^c$, and so is therefore $p = \sum_n p_n$ (by the "diagonal argument").

(ii)⇒(iii). Obvious.

(iii)⇒(i). Let $q \in \mathcal{P}^b$, $q \preccurlyeq p$, and $c > 0$. Write $p = q + s$, whereby $s \in \mathcal{P}$. Consider any $u \in \mathcal{U}^+$ such that $u \geqslant q$ on $\{q > c\}$. We show that

$$p \leqslant u + 2s \quad \text{on } \{p > 2c\}. \tag{1.5.10}$$

It suffices to verify this at points x where $s(x) < +\infty$ and hence $q(x) = p(x) - s(x)$. On $\{q \leqslant c\} \cap \{p > 2c\}$ we have $p > 2q = 2p - 2s$, and hence $p < +\infty$ and $p < 2s \leqslant u + 2s$, as desired. On $\{q > c\}$ we have $u \geqslant q$, and hence $p = q + s \leqslant u + s \leqslant u + 2s$, again as desired. Altogether this establishes (1.5.10). It follows that $R_p^{\{p>2c\}} \leqslant u + 2s$ for $u \in \mathcal{U}^+$ such that $u \geqslant q$ on $\{q > c\}$. Taking pointwise infimum over these u we obtain after inserting $s = p - q$

$$R_p^{\{p>2c\}} \leqslant R_q^{\{q>c\}} + 2(p - q),$$

and consequently,

$$\inf_{c>0} R_p^{\{p>2c\}} \leqslant \inf_{c>0} R_q^{\{q>c\}} + 2(p - q), \tag{1.5.11}$$

valid for any $q \in \mathcal{P}^b$, $q \preccurlyeq p$. Since $q \in \mathcal{P}^b$ we may for any compact $K \subset \Omega$ choose $c > 0$ so that $q \leqslant c$ on K and hence $\{q > c\} \subset \complement K$. This implies that

$$\inf_{c>0} R_q^{\{q>c\}} \leqslant \inf_K \{R_q^{\complement K} : K \text{ compact}\} = 0$$

by Corollary 1.3.8. With q ranging over $\{q \in \mathcal{P}^b : q \preccurlyeq p\}$ we infer from (1.5.10) and (1.5.11) that

$$\inf_{c>0} R_p^{\{p>2c\}} \leqslant 2(p-q) \tag{1.5.12}$$

which approaches 0 at any point of $\{p < +\infty\}$ as q varies, hence quasi-everywhere on Ω. Since $\widehat{\inf_{c>0} R_p^{\{p>c\}}} \in \mathcal{U}^+$ is finely continuous, it follows that (i) holds true.

(ii)⇒(v). We may assume that the polar set E in question is compact. From (ii) we have $\mu = \sum_n \mu_n$ with $G_\Omega\mu_n$ of class $\mathcal{P}_0^c$. There is a measure $\lambda \in \Lambda$ such that $G_\Omega\lambda = +\infty$ on E; see Sect. 1.1.11. Since $G_\Omega\mu_n \in \mathcal{P}_0^c$ we have $\int G_\Omega\lambda\, d\mu_n = \int G_\Omega\mu_n\, d\lambda < +\infty$ by Lemma 1.4.4, and hence $\mu_n(E) = 0$ for every n. We conclude that indeed $\mu(E) = \sum_n \mu_n(E) = 0$.

(v)⇒(iv). Obvious since $\{G_\Omega\mu = +\infty\}$ is polar and hence has μ-measure 0.

(iv)⇒(ii). According to Theorem 1.4.1 there is an increasing sequence (ν_n) of measures ν_n with $G_\Omega\nu_n \in \mathcal{P}_0^c$ such that

$$G_\Omega\mu = \sup_n G_\Omega\nu_n = \sum_n G_\Omega(\nu_n - \nu_{n-1})$$

(writing $\nu_0 = 0$), whereby each $G_\Omega(\nu_n - \nu_{n-1}) \in \mathcal{P}_0^c$. □

We denote by $\mathcal{G}$, resp. $\mathcal{H}$, the class of all l.s.c., resp. u.s.c. and finite valued, functions $\Omega \to [0, +\infty]$, and by $\mathcal{H}_0$ the compactly supported functions of class $\mathcal{H}$. The closures $\mathcal{G}^*$ of $\mathcal{G}$, $\mathcal{H}^*$ of $\mathcal{H}$, and $\mathcal{H}_0^*$ of $\mathcal{H}_0$ in the C_m-metric topology are given respectively by

$$f \in \mathcal{G}^* \iff \inf\{C_m(|f-g|) : g \in \mathcal{G}\} = 0, \tag{1.5.13}$$
$$f \in \mathcal{H}^* \iff \inf\{C_m(|f-h|) : h \in \mathcal{H}\} = 0, \tag{1.5.14}$$
$$f \in \mathcal{H}_0^* \iff \inf\{C_m(|f-h|) : h \in \mathcal{H}_0\} = 0. \tag{1.5.15}$$

In analogy with Lemma 1.5.5 we have the followng.

Lemma 1.5.17 *In the above three bi-equivalences it suffices to take infimum over $g \in \mathcal{G}$, $g \geqslant f$, resp. over $h \in \mathcal{H}$, $h \leqslant f$, resp. over $h \in \mathcal{H}_0$, $h \leqslant f$.*

Proof First[9] suppose that the right-hand statement in (1.5.13) holds for some given $f \in \mathcal{G}^*$. For given $\varepsilon > 0$ choose $g \in \mathcal{G}$ so that

$$C_m(|f-g|) < \varepsilon. \tag{1.5.16}$$

By Lemma 1.5.1 there exists $\varphi \in \mathcal{G}$ such that $\varphi \geqslant |f-g| \geqslant f-g$ and by (1.5.1)

$$C_m(\varphi) < \varepsilon. \tag{1.5.17}$$

Then $g+\varphi \in \mathcal{G}$, and $g+\varphi \geqslant f$ (see above). Finally,

$$C_m((g+\varphi)-f) \leqslant C_m(\varphi) + C_m(|f-g|) < 2\varepsilon$$

[9] The present proof replaces an incomplete proof in [Fu6, Lemma 3.2].

by (1.5.16), (1.5.17). Letting $\varepsilon \to 0$ we see for $\psi := g + \varphi$ that indeed $\inf\{C_m(f - \psi) : \psi \in \mathcal{G}, \psi \geqslant f\} = 0$.

Suppose instead that the right-hand statement in (1.5.14) holds for some given $f \in \mathcal{H}^*$. For given $\varepsilon > 0$ choose $h \in \mathcal{H}$ so that $C_m(|f - h|) < \varepsilon$. By Lemma 1.5.1 there exists $\varphi \in \mathcal{G}$ such that $\varphi \geqslant |f - h| \geqslant h - f$ and that $C_m(\varphi) < \varepsilon$. Then $(h - \varphi)^+ \in \mathcal{H}$, and $(h - \varphi)^+ \leqslant f$ because $h - \varphi \leqslant f$ (see above) and $0 \leqslant f$. Finally,

$$C_m(f - (h - \varphi)^+) \leqslant C_m(h - (h - \varphi)^+) + C_m(|f - h|) < C_m(\varphi) + \varepsilon < 2\varepsilon$$

because $0 \leqslant h - (h - \varphi)^+ = \varphi \wedge h \leqslant \varphi$. Letting $\varepsilon \to 0$ we see for $\psi := (h - \varphi)^+$ that $\inf\{C_m(\psi) : \psi \in \mathcal{H}, \psi \leqslant f\} = 0$.

The proof for the case of $\mathcal{H}_0^*$ is obtained by replacing throughout $\mathcal{H}$ by $\mathcal{H}_0$. □

Lemma 1.5.18 *$\mathcal{G}^*$ is stable under countable supremum and finite infimum. $\mathcal{H}^*$ and $\mathcal{H}_0^*$ are stable under countable infimum and finite supremum.*

Proof Follows easily from the corresponding properties of $\mathcal{G}$, $\mathcal{H}$, and $\mathcal{H}_0$ in view of the inequality (for countable families)

$$|\inf_n f_n - \inf_n h_n| \leqslant \sup_n |f_n - h_n| \leq \sum_n |f_n - h_n|$$

and the same with "inf" replaced by "sup" (twice). □

Lemma 1.5.19 *Every semibounded potential p is of class $\mathcal{H}_0^*$.*

Proof Let $p \in \mathcal{P}^s$ and $\varepsilon > 0$. By Definition 1.5.16(iii) there is a sequence of potentials $p_n = G_\Omega \mu_n \in \mathcal{P}_0^c$ such that $p_n \uparrow p$ specifically. Since $p - p_n$ is then a potential we have

$$C_m(p - p_n) = \int \widehat{R}_{p-p_n}\, dm = \int (p - p_n)\, dm \downarrow 0$$

as $n \uparrow \infty$ because $\int p\, dm < \infty$ in the definition of C_m; see (1.5.1). It follows that there is an integer n such that $C_m(p - p_n) < \frac{\varepsilon}{2}$. Let (Ω_j) be an increasing sequence of open subsets of Ω such that $K_j = \overline{\Omega}_j$ is a compact subset of Ω for each j and $\bigcup \Omega_j = \Omega$. It follows from Corollary 1.3.8 that $\widehat{R}_{p_n}^{\complement K_j} \downarrow 0$ pointwise on Ω as $j \uparrow \infty$ and hence $\int \widehat{R}_{p_n}^{\complement K_j}\, dm \downarrow 0$ as $j \uparrow \infty$ by Lebesgue's integration theorem because $\int p_n\, dm \leqslant \int p\, dm$, which is finite, as again required in the definition of C_m. Thus we may choose j large enough that $\int_\Omega R_{p_n}^{\complement K_j}\, dm < \frac{\varepsilon}{2}$ so that

$$C_m(p_n - 1_{K_j} p_n) = \int_\Omega \widehat{R}_{p_n - 1_{K_j} p_n} dm = \int_\Omega R_{p_n}^{\complement K_j} dm < \frac{\varepsilon}{2},$$

and hence

$$C_m(p - 1_{K_j} p_n) \leq C_m(p - p_n) + C_m(p_n - 1_{K_j} p_n) < \varepsilon.$$

Since p_n is finite and continuous, K_n is compact and 1_{K_n} is u.s.c., $1_{K_n} p_n$ is finite and u.s.c. and supported by K_n, it follows that $1_{K_n} p_n \in \mathcal{H}_0$. For $\varepsilon \to 0$ this implies that indeed $p \in \mathcal{H}_0^*$. □

Lemma 1.5.20 *Let $f \in \mathcal{F}^+$. Then*

(a) $f \in \mathcal{G}^*$ if and only if f is quasi-l.s.c.
(b) The following are equivalent:

(i) $f \in \mathcal{H}_0^*$,
(ii) f is quasi-u.s.c. and has a majorant of class $\mathcal{H}_0^*$,
(iii) f is quasi-u.s.c. and $\inf\{C_m((f-h)^+) : h \in \mathcal{H}_0\} = 0$.

Proof (a) For the "only if part" (not to be used) we refer to [Fu6, Theorem 2.5].[10] For the "if part", suppose that f is quasi-l.s.c. For any $n \in \mathbb{N}$ the function $f_n := f \wedge n$ is likewise quasi-l.s.c. We propose to show that $f_n \in \mathcal{G}^*$. Consider an open set $\omega \subset \Omega$ such that f_n is l.s.c. relative to $\complement\omega$. Replacing f_n by n on ω gives a function $\varphi_n \geqslant f_n$ of class $\mathcal{G}$. Then $C_m(|\varphi_n - f_n|) \leqslant nC_m(\omega)$, which becomes as small as we please by a suitable choice of ω. Hence each f_n is of class $\mathcal{G}^*$, and so is therefore $f = \sup_n f_n$ according to Lemma 1.5.18.

(b) We begin by proving that $f \in \mathcal{H}_0^*$ holds provided that f is quasi-u.s.c. and has a majorant $h \in \mathcal{H}_0$. Since C_m is "upper", see Remark 1.5.3, there are by Definition 1.5.2 open sets ω with $C_m(\omega)$ as small as we please such that $f|_{\complement\omega}$ is u.s.c. Hence $\varphi := 1_{\complement\omega} f$ is u.s.c. and therefore of class $\mathcal{H}_0$ along with h. Since f is bounded along with h, say $f \leqslant c$, we find that

$$C_m(|f - \varphi|) = C_m(1_\omega f) \leqslant c\, C_m(1_\omega) = c\, C_m(\omega)$$

is as small as we please, so $f \in \mathcal{H}_0^*$.

(i)⇒(ii). There is a sequence $(h_n) \subset \mathcal{H}_0$ with $C_m(|f - h_n|) < 4^{-n}$. Writing

$$M_n := \{|f - h_n| > 2^{-n}\}, \quad N_n := \bigcup_{j>n} M_j,$$

we have $1_{M_n} \leqslant 2^n |f - h_n|$ and

$$C_m(M_n) \leqslant 2^n C_m(|f - h_n|) \leqslant 2^{-n},$$
$$C_m(N_n) \leqslant \sum_{j>n} C_m(M_j) \leqslant \sum_{j>n} 2^{-j} = 2^{-n}.$$

For each j, $h_n(x) \to f(x)$ uniformly for $x \in \complement N_j$. Hence f is u.s.c. relative to $\complement N_j$ (along with each h_n), and since $C_m(N_j) \leqslant 2^{-j}$ (small) this shows that indeed f is quasi-u.s.c. (and of course majorized by $f \in \mathcal{H}_0^*$).

[10] In [Fu6, Theorem 2.5] the hypothesis that C be "upper" is missing.

(ii)⇒(iii). Suppose that f is quasi-u.s.c. and has a majorant $F \in \mathcal{H}_0^*$. Then

$$(f-h)^+ \leqslant (F-h)^+ \leqslant |F-h|$$

for any $h \in \mathcal{H}_0^*$, and here $C_m(|F-h|)$ and hence $C_m((f-h)^+)$ can be made as small as we please by a suitable choice of $h \in \mathcal{H}_0^*$.

(iii)⇒(i). For given $\varepsilon > 0$ choose $h \in \mathcal{H}_0$ so that $C_m((f-h)^+) < \varepsilon$. Then $0 \leqslant f - f \wedge h \leqslant (f-h)^+$ and hence

$$C_m(|f - f \wedge h|) \leqslant C_m((f-h)^+) < \varepsilon. \tag{1.5.18}$$

Since $f \wedge h$ is quasi-u.s.c. along with f, and since $f \wedge h \leqslant h \in \mathcal{H}_0$, we see from the beginning of the proof of (b) that $f \wedge h \in \mathcal{H}_0^*$, and we conclude from (1.5.18) that indeed $f \in \mathcal{H}_0^*$, $\mathcal{H}_0^*$ being closed in the C_m-metric topology. □

The following theorem is essentially contained in [Fu6, Theorem 3.6]. It uses the fact that the present upper capacity C_m from (1.5.1) is *locally finite* in the sense that $C_m(K) < +\infty$ for any compact set K (actually for any set $K \subset \Omega$ because $\int_\Omega dm < +\infty$).

Theorem 1.5.21 (a) *For any family* $(h_\alpha) \subset \mathcal{H}_0$ *we have* $\inf_\alpha h_\alpha \in \mathcal{H}_0$. *If* (h_α) *is lower directed then*

$$C_m(\inf_\alpha h_\alpha) = \inf_\alpha C_m(h_\alpha).$$

(b) *For any sequence* $(f_n) \subset \mathcal{H}_0^*$ *we have* $\inf_n f_n \in \mathcal{H}_0^*$. *If* (f_n) *is decreasing then*

$$C_m(\inf_n f_n) = \inf_n C_m(f_n).$$

Proof (a) Let $h := \inf_\alpha h_\alpha$. Clearly, $h \in \mathcal{H}_0$ because $\operatorname{Supp} h \subset \operatorname{Supp} h_\alpha$ (compact) for any α. Now let (h_α) be lower directed. Obviously, $C_m(h) \leq \inf_\alpha C_m(h_\alpha)$. For the opposite inequality, for any number $t > C_m(h)$ there exists by Lemma 1.5.1 a function $g \in \mathcal{G}$ such that $g \geqslant h$ and $C_m(g) < t$. For any such function g the lower directed family of functions $((h_\alpha - g)^+)_\alpha \subset \mathcal{H}_0$ converges pointwise to 0. By Dini's theorem the convergence is uniform. For any $\varepsilon > 0$ there is thus an index α such that $(h_\alpha - g)^+ < \varepsilon$ everywhere on Ω. We may take $\alpha \geqslant \beta$, where β denotes a fixed index. Denoting by K the compact support of h_β, we obtain $h_\alpha \leqslant g + \varepsilon 1_K$, and hence

$$C_m(h_\alpha) \leqslant C_m(g + \varepsilon 1_K) \leqslant C_m(g) + \varepsilon C_m(K) < t$$

by a suitable choice, first of ε and next of $\alpha \geqslant \beta$. This implies that $\inf_\alpha C_m(h_\alpha) < t$, and consequently, by varying t, that $\inf_\alpha C_m(h_\alpha) \leqslant C_m(h)$.

(b) Given $\varepsilon > 0$, for each n choose $\varphi_n \in \mathcal{H}_0$ so that $\varphi_n \leqslant f_n$ and $C_m(f_n - \varphi_n) < \varepsilon/2^n$, cf. Lemma 1.5.17, and put

$$h_n := \varphi_1 \wedge \ldots \wedge \varphi_n \leqslant f_n, \quad h := \inf_n \varphi_n = \inf_n h_n \leqslant \inf_n f_n.$$

Then $(h_n)_{n\in\mathbb{N}}$ is a decreasing sequence of functions of class $\mathcal{H}_0$. According to (a) above,

$$C_m(h_n) \leqslant C_m(h) + \varepsilon \leqslant C_m(\inf_n f_n) + \varepsilon$$

for all sufficiently large n. If (f_n) is decreasing we have for every n

$$0 \leqslant f_n - h_n \leqslant \sup_{j\leqslant n}(f_j - \varphi_j) \leqslant \sum_{j\leqslant n}(f_j - \varphi_j).$$

Consequently,

$$C_m(f_n) \leqslant C_m(h_n) + C_m(f_n - h_n) \leqslant C_m(h_n) + \sum_{j\leqslant n} \varepsilon/2^j \leqslant C_m(\inf_n f_n) + \varepsilon$$

for n sufficiently large, and here we let $\varepsilon \to 0$. □

The following two approximation theorems are instrumental for our study of fine harmonicity and superharmonicity. Recall that $\mathcal{H}_0$ denotes the class of all positive, finite, compactly supported, u.s.c. functions on Ω, and that $\mathcal{G}$ is the class of all positive l.s.c. functions on Ω. The "only if" parts of Theorem 1.5.22 will not be used in the sequel.

Theorem 1.5.22 *Let $p \in \mathcal{P}^s$, let $f : \Omega \to [0, +\infty]$, and suppose that $f \leqslant p$.*
(a) *The function f is quasi-u.s.c. (equivalently: finely u.s.c. q.e.) if and only if*

$$\widehat{\inf}\{\widehat{R}_{f-\varphi} : \varphi \in \mathcal{H}_0,\ \varphi \leqslant f\} = 0. \tag{1.5.19}$$

The function f is quasi-continuous (equivalently: finely continuous q.e.) if and only if

$$\widehat{\inf}\{\widehat{R}_{|f-\varphi|} : \varphi \in C_0^+(\Omega),\ \varphi \leqslant p\} = 0. \tag{1.5.20}$$

(b) *Same conclusions with $\widehat{R}_{\bullet}$ replaced by $\int \widehat{R}_{\bullet}\,dm$ (and $\widehat{\inf}$ by $\inf$).*

Proof Being semibounded, p is of class $\mathcal{H}_0^*$ by Lemma 1.5.19.
(b) When f is quasi u.s.c. (and $\leqslant p \in \mathcal{H}_0^*$) we have $f \in \mathcal{H}_0^*$ by Lemma 1.5.20(b). By Lemma 1.5.17 this means that

$$\inf\{C_m(f - \varphi) : \varphi \in \mathcal{H}_0,\ \varphi \leqslant f\} = 0, \tag{1.5.21}$$

which amounts to (1.5.19). For the converse, (1.5.19) in the form (1.5.21) shows that $f \in \mathcal{H}_0^*$ and hence f is quasi-u.s.c. by Lemma 1.5.20(b).

Next, let f be quasi-continuous. Then f is both quasi u.s.c. (hence of class $\mathcal{H}_0^*$ as above and (1.5.21) holds), and quasi-l.s.c. (that is, $f \in \mathcal{G}^*$, by Lemma 1.5.20(a)). To obtain (1.5.20) in the form

$$\inf\{C_m(|f - \varphi|) : \varphi \in C_0^+(\Omega),\ \varphi \leqslant p\} = 0, \tag{1.5.22}$$

note that there is for given $\varepsilon > 0$ by Lemma 1.5.20(a) a function $h \in \mathcal{H}_0$ such that $h \leqslant f\,(\leqslant p)$ and $C_m(f-h) < \varepsilon$, and a function $g \in \mathcal{G}^*$ such that $g \geqslant f$ and $C_m(g-f) < \varepsilon$. Replacing g by $g \wedge p$ we arrange that $g \leqslant p$. Since h is u.s.c. and g is l.s.c., and since $h \leqslant g$, there exists by the "Between Theorem" of Hahn (see Hausdorff [Hau] for a simple proof) a continuous function φ on Ω such that $h \leqslant \varphi \leqslant g$. In the present case, φ may be taken with compact support, like h, so altogether $\varphi \in \mathcal{C}_0^+(\Omega)$. Since $|f-\varphi| \leqslant g-h = (g-f)+(f-h)$ we obtain in view of Lemma 1.5.17

$$\begin{aligned}&\inf C_m(|f-\varphi|) : \varphi \in \mathcal{C}_0^+(\Omega),\ \varphi \leqslant p\}\\&\quad \leqslant \inf\{C_m(g-f) : g \in \mathcal{G},\ g \geqslant f\} + \inf\{C_m(f-h) : h \in \mathcal{H}_0,\ h \leqslant f\} = 0.\end{aligned}$$

In fact, $\inf_h C_m(f-h) = 0$ by (1.5.19) (with φ replaced by h). And the term $\inf_g C_m(g-f)$ equals 0 because $f \in \mathcal{G}^*$. This establishes (1.5.22) and hence (1.5.20). Conversely, (1.5.22) implies that f is quasi-continuous, being again both of class $\mathcal{G}^*$ and $\mathcal{H}_0^*$, hence both quasi-l.s.c. and quasi-u.s.c. according to Lemma 1.5.20, and so indeed quasi-continuous.

(a) Suppose again that f is quasi-u.s.c. By Choquet's topological lemma (cf. Sect. 1.1.13) applied to the lower directed family $\{\widehat{R}_{f-\varphi} : \varphi \in \mathcal{H}_0,\ \varphi \leqslant f\} \subset \mathcal{U}^+$ has a decreasing subsequence $(\widehat{R}_{f-\varphi_n})_{n\in\mathbb{N}}$ such that

$$u := \widehat{\inf}\{\widehat{R}_{f-\varphi} : \varphi \in \mathcal{H}_0,\ \varphi \leqslant f\} = \widehat{\inf_n}\,\widehat{R}_{f-\varphi_n},$$

$$\begin{aligned}\int u\,dm &= \int (\inf_n \widehat{R}_{f-\varphi_n})\,dm = \inf_n \int \widehat{R}_{f-\varphi_n}\,dm\\&\geqslant \inf\Big\{\int \widehat{R}_{f-\varphi}\,dm : \varphi \in \mathcal{H}_0, \varphi \leqslant f\Big\} = 0\end{aligned}$$

by (1.5.19). The opposite inequality is trivial, and so $\int u\,dm = 0$. This implies that $u \equiv 0$, and thus (1.5.20) holds. Conversely, if (1.5.19) or equivalently (1.5.21) holds then f is clearly quasi-continuous. □

Theorem 1.5.22 has the following fine topology extension; cf. [Br13, Theorem 13]. We bring an alternative proof, much as in [Fu8, Theorem 6.6].

Theorem 1.5.23 *Let Φ be a lower directed family of finely u.s.c. functions $\Omega \to [0,+\infty]$ majorized by a semibounded potential p. Then* $\inf \Phi$ *is finely u.s.c., and*

$$\widehat{R}_{\inf\Phi} = \widehat{\inf_{f\in\Phi}}\,\widehat{R}_f, \qquad \int \widehat{R}_{\inf\Phi}\,dm = \inf_{f\in\Phi} \int \widehat{R}_f\,dm.$$

Proof The latter equality in the theorem means that $C_m(\inf\Phi) = \inf_{f\in\Phi} C_m(f)$, which provides a fine topology extension of Theorem 1.5.22(a). By Choquet's topological lemma (Sect. 1.1.13) applied to the lower directed families Φ and $(\widehat{R}_f : f \in \Phi)$, there is a decreasing sequence $(f_n) \subset \Phi$ such that

$$\widehat{\inf_{f\in\Phi}} f = \widehat{\inf_n} f_n, \quad \widehat{\inf_{f\in\Phi}} \widehat{R}_f = \widehat{\inf_n} \widehat{R}_{f_n}.$$

(In the first place there are two decreasing sequences, one for each case, but we let (f_n) denote a common decreasing subsequence of the two.) The finely u.s.c. functions $f_n \in \Phi$ are quasi-u.s.c. by Corollary 1.5.14, and majorized by $p \in \mathcal{H}_0^*$, see Lemma 1.5.19, so $f_n \in \mathcal{H}_0^*$ by Lemma 1.5.20. Clearly, $\inf_n R_{f_n} \geqslant R_{\inf_n f_n}$ and hence

$$\widehat{\inf_n} \widehat{R}_{f_n} \geqslant \widehat{R}_{\inf_n f_n}. \tag{1.5.23}$$

By integration

$$\begin{aligned}\int \widehat{\inf_n} \widehat{R}_{f_n}\, dm &= \int \inf_n \widehat{R}_{f_n}\, dm = \inf_n \int \widehat{R}_{f_n}\, dm \\ &= \inf_n C_m(f_n) = C_m(\inf_n f_n) = \int \widehat{R}_{\inf_n f_n}\, dm,\end{aligned}$$

the first equality because m does not charge any polar set, the second equality because $\int f_n\, dm \leqslant \int p\, dm < +\infty$, and the fourth equality by Theorem 1.5.21(b). Combining the conclusion in this display with (1.5.23) leads to equality m-a.e. in (1.5.23). To obtain the asserted equality everywhere we choose m as Lebesgue measure on a ball B with $\overline{B} \subset \Omega$. This m has all the properties required in the text following (1.5.1). Both members of the inequality (1.5.23) being hyperharmonic and hence finely continuous, there is equality in (1.5.23) on B and hence on all of Ω because any finely open Lebesgue null set is void; see Sect. 1.2. □

Lemma 1.5.24 *A finely u.s.c. function $f : \Omega \to \overline{\mathbb{R}}$ differs only on a polar set from some Borel measurable function. In particular, f is measurable with respect to any measure on Ω which does not charge the polar sets.*

Proof For any $n \in \mathbb{N}$ there is by the Choquet property, cf. Theorem 1.5.12, an open set $\omega_n \subset \Omega$ with $C_m(\omega_n) < 1/n$ such that f coincides off ω_n with the u.s.c. and hence Borel measurable function $1_{\complement\omega_n} f$ on Ω. Hence f coincides off the polar set $\bigcap_n \omega_n$ with the Borel measurable function $1_{\complement\bigcap\omega_n} f$ on Ω. □

Remark 1.5.25 It follows from Lemma 1.5.24 that every numerical function f on Ω which differs only on a polar set from some finely semicontinuous function on Ω is ε_x^A-measurable for any set $A \subset \Omega$ and any $x \in \Omega$. Indeed, $\varepsilon_x^A = \varepsilon_x$ if $x \in \mathfrak{b}(A)$ (Theorem 1.4.23), whereas ε_x^A does not charge any polar set if $x \notin \mathfrak{b}(A)$ (Corollary 1.4.33).

We proceed to study sweeping of a semibounded potential on the intersection of a sequence of finely closed sets. (The easy extension of this to the intersection of an arbitrary family of such sets will be omitted.)

Lemma 1.5.26 *Let $p \in \mathcal{P}^s$, and let $(A_n)_{n\in\mathbb{N}}$ be a sequence of finely closed sets with the intersection A. If $\widehat{R}_p^{A_n} = p$ for every $n \in \mathbb{N}$ then $\widehat{R}_p^A = p$.*

Proof Writing $p = G_\Omega\mu$ we have, by Theorem 1.4.24, $\widehat{R}_p^E = G_\Omega\mu^E$ for any set $E \subset \Omega$. The lemma then asserts that, if $\mu^{A_n} = \mu$ for every $n \in \mathbb{N}$, then $\mu^A = \mu$. Equivalently (by Theorem 1.4.23(b)): if μ is carried by each $\mathfrak{b}(A_n)$ then also it is by $\mathfrak{b}(A)$. But μ does not charge the polar sets $A_n \setminus \mathfrak{b}(A_n)$, by Theorem 1.4.32(b) and Definition 1.5.16(v). According to Corollary 1.4.18 $\mathfrak{b}(A)$ and $\mathfrak{b}(A_n)$ $(n \in \mathbb{N})$ are G_δ-sets, it then follows by Lemma 1.5.24 applied to the finely u.s.c. functions 1_A and 1_{A_n} that A and each A_n are therefore μ-measurable, and the assertion of the lemma thus reduces to the obvious statement that if μ is carried by each A_n then also by their intersection A. □

For any finite family of subsets $A_1, \ldots, A_n$ of Ω and any Borel measurable function $f : \Omega \to [0, +\infty]$ we define by recursion, cf. Definition 1.4.29 and subsequent paragraphs,

$$f^{A_1,A_2,\ldots,A_n} = (\ldots((f^{A_1})^{A_2})\ldots)^{A_n}. \tag{1.5.24}$$

Lemma 1.5.27 *Let $p \in \mathcal{P}^s$, and let $(A_n)_{n\in\mathbb{N}}$ be a sequence of finely closed sets with the intersection A. Consider a sequence $(B_m)_{m\in\mathbb{N}}$ such that each B_m equals $A_{n(m)}$ for some $n(m) \in \mathbb{N}$, and that each A_n occurs infinitely often among the B_m. Then*

$$\widehat{R}_p^A = p^A = \widehat{\inf_{m\in\mathbb{N}}} \, p^{B_1,\ldots,B_m}.$$

We may, for example, take for (B_m) the sequence $A_1, A_1, A_2, A_1, A_2, A_3, \ldots$.
Proof of Lemma 1.5.27. Note that the sequence $(p^{B_1,\ldots,B_m})_{m\in\mathbb{N}}$ is decreasing in view of (1.4.22). Denote by q the right-hand member of the latter equality in the above display; then $q \leqslant p$. By repeated application of Lemma 1.3.11 with $B = A$,

$$p^A = p^{A,\ldots,A} \leqslant p^{B_1,\ldots,B_m},$$

the inequality because $A \subset B_i$ for every $i \in \mathbb{N}$. For $m \to +\infty$ this shows that $p^A \leqslant q$.

To establish the opposite inequality $q \leqslant p^A$ we show that $q^{A_i} = q$ for every $i \in \mathbb{N}$; it will then follow from Lemma 1.5.26 that indeed $q = q^A \leqslant p^A$ because $q \leqslant p$. The decreasing sequence $(p^{B_1,\ldots,B_m})_{m\in\mathbb{N}}$ converges pointwise q.e. to q according to the fundamental convergence theorem; cf. Sect. 1.1.13. For fixed $i \in \mathbb{N}$ the likewise decreasing sequence $(p^{B_1,\ldots,B_m,A_i})_{m\in\mathbb{N}}$ therefore converges pointwise q.e. as $m \to +\infty$ to q^{A_i}, by Theorem 1.5.23. Thus

$$q^{A_i} = \widehat{\inf_{m\in\mathbb{N}}} \, p^{B_1,\ldots,B_m,A_i}. \tag{1.5.25}$$

For any $m, i \in \mathbb{N}$ we have

$$p^{B_1,\ldots,B_m,A_i} \geqslant p^{B_1,\ldots,B_m,\ldots,B_n} \quad \text{for } n \geqslant N(m),$$

where $N(m)$ is chosen large enough so that A_i occurs among $B_{m+1}, \dots, B_{N(m)}$. It follows that

$$p^{B_1,\dots,B_m,A_i} \geqslant \widehat{\inf_{n \geqslant N(m)}} p^{B_1,\dots,B_n} = q,$$

and hence $q^{A_i} \geqslant q$, by (1.5.25). □

The following theorem extends Lemma 1.5.26 by replacing the potential $p \in \mathcal{P}^s$ with a more general function $f \geqslant 0$. The proof is based on Lemma 1.5.26, and the theorem will serve in the next chapter as a key to fine hyperharmonicity, in establishing a fine boundary minimum principle (Theorem 2.1.10).

Theorem 1.5.28 *Let $(A_n)_{n \in \mathbb{N}}$ be a sequence of finely closed subsets of Ω with the intersection A. Let $f : \Omega \to [0, +\infty]$ be finely u.s.c. quasi-everywhere, $\varepsilon_x^{A_n}$-measurable for each n, and majorized q.e. by some semibounded potential $p \in \mathcal{P}^s$. If $f \leqslant f^{A_n}$ q.e. on Ω for every $n \in \mathbb{N}$ then $f \leqslant f^A$ q.e. on Ω.*

Proof Fix $h \in \mathcal{H}_0$ (that is, $h : \Omega \to [0, +\infty[$ u.s.c. with compact support) such that $h \leqslant f$. Consider a function $\varphi \in \mathcal{C}_0^+(\Omega)$ such that $\varphi \geqslant h$. For given $\varepsilon > 0$ there exist by Theorem 1.4.5 potentials $p, q \in \mathcal{P}_0^c$ such that $p - q \in \mathcal{C}_0^+$ and that

$$|p - q - \varphi| \leqslant \varepsilon. \tag{1.5.26}$$

It follows that $h \leqslant \varphi \leqslant (p - q) + \varepsilon$, and hence

$$f \leqslant h + R_{f-h} \leqslant (p - q) + \varepsilon + \widehat{R}_{f-h} \quad \text{q.e. on } \Omega. \tag{1.5.27}$$

As in Lemma 1.5.27, consider a sequence (B_m) such that each B_m equals some A_n and that each A_n equals B_m for infinitely many $m \in \mathbb{N}$. The notation from (1.5.24) then carries over, and we write

$$f_n := f^{B_1,\dots,B_n}, \quad p_n := p^{B_1,\dots,B_n}, \quad q_n := q^{B_1,\dots,B_n}.$$

Since $p - q \geqslant 0$ we have in view of (1.5.24)

$$p_n - q_n = (p - q)^{B_1,\dots,B_n} \geqslant 0.$$

By repeated application of the hypothesis $f \leqslant f^{A_n}$ q.e. it follows by (1.5.24), (1.5.27), and Lemma 1.5.26 that

$$f \leqslant f_n \leqslant (p_n - q_n) + \widehat{R}_{f-h} + \varepsilon \quad \text{q.e.} \tag{1.5.28}$$

From Lemma 1.5.27 combined with the fundamental convergence theorem (see Sect. 1.1.13) we obtain

$$p^A = \widehat{\inf_{n \in \mathbb{N}}} p_n = \inf_{n \in \mathbb{N}} p_n \quad \text{q.e.},$$

and similarly for q^A. The sequences (p_n) and (q_n) being pointwise decreasing, it follows that

$$p^A - q^A = \lim_{n\to+\infty} (p_n - q_n) \quad \text{pointwise q.e.} \tag{1.5.29}$$

From (1.5.26) we have

$$p^A - q^A = (p-q)^A \leqslant \varphi^A + \varepsilon. \tag{1.5.30}$$

In the limit as $n \to +\infty$, (1.5.28) leads in combination with (1.5.29) and (1.5.30) to

$$f \leqslant \varphi^A + \widehat{R}_{f-h} + 2\varepsilon \quad \text{q.e.} \tag{1.5.31}$$

The function h (see the beginning of the proof) is the pointwise limit of a decreasing sequence of functions $\varphi_j \in \mathcal{C}_0^+(\Omega)$ $(j \geq 1)$. Replacing in the above analysis φ by φ_j, and ε by ε/j, we obtain from (1.5.31) for $j \to +\infty$

$$f \leqslant h^A + \widehat{R}_{f-h} \leqslant f^A + \widehat{R}_{f-h} \quad \text{q.e.}$$

because $h \leqslant f$. It follows by (1.5.19) in Theorem 1.5.22 that indeed $f \leqslant f^A$ q.e. □

The Choquet property relative to the capacity $R_1^\bullet(x_0)$, given in the third equation in (1.5.8), but now in the equivalent formulation corresponding to the original formulation by Choquet (who considered the Green capacity), cf. Remark 1.5.11, reads as follows:

$$\inf\{R_1^{A\cap\omega}(x_0) : \omega \text{ open}, \ \omega \supset \complement b(A)\} = 0. \tag{1.5.32}$$

From (1.5.32) we shall derive the following remarkable and very useful result which goes back to Brelot in [Br14, p. 43, lines 1-6]; cf. Remark 1.5.30 below.

Theorem 1.5.29 (The Brelot property, cf. [Fu13]) *Let f be a finely continuous map from a finely open set $U \subset \Omega$ into a separable metrizable topological space Y. Every point $x_0 \in U$ then has a compact fine neighborhood $K \subset U$ such that $f|_K$ is continuous on K (with the Euclidean topology).*

Proof Y can be embedded as a subspace of the cube $[0, 1]^{\mathbb{N}}$, e.g. [Bou5, §2, Proposition 12], and it therefore suffices to consider the case $Y = [0, 1]^{\mathbb{N}}$. For any $x \in U$ we may then write

$$f(x) = (f_n(x))_{n\in\mathbb{N}},$$

where each f_n is a finely continuous function $U \to [0, 1]$. Let g_n and h_n denote the extensions of f_n from U to Ω obtained by putting $g_n = 0$ and $h_n = 1$ on $\complement U$. Since U is finely open, it follows that g_n is finely l.s.c. and h_n is finely u.s.c. on Ω. For any $n \in \mathbb{N}$ there exists by Theorem 1.5.12 relative to the capacity $A \mapsto R_1^A(x_0)$ an open set $\omega_n \subset \Omega$ such that $g_n|_{\complement\omega_n}$ is l.s.c. and $h_n|_{\complement\omega_n}$ is u.s.c. and that moreover

$$R_1^{\omega_n}(x_0) < 2^{-n},$$

in particular $x_0 \in \complement\omega_n$. Choose a compact fine neighborhood V of x_0 so that $V \subset U$, and write

$$\omega = \bigcup_{n \in \mathbb{N}} \omega_n, \quad K = V \setminus \omega.$$

Then

$$R_1^{\omega}(x_0) \leqslant \sum_{n \in \mathbb{N}} R_1^{\omega_n}(x_0) < 1,$$

and ω is therefore thin at x_0, by Lemma 1.3.9(a); and furthermore $x_0 \in \complement\omega$. It follows that K is a compact fine neighborhood of x_0 in U. Since $f_n = g_n = h_n$ on U and $K \subset U$ we deduce that each $f_n|_K$ is continuous, and so is $f|_K$. □

Remark 1.5.30 Theorem 1.5.29 obviously extends to the case of a *countable family* of finely continuous maps $f_n : U \to Y_n$ into spaces like Y above. (It suffices to consider the product mapping $x \mapsto (f_n(x))_n$ of U into the separable metric product space $\Pi_{n \in \mathbb{N}} Y_n$.) In the particular case where $Y = \overline{\mathbb{R}}$ and where $f : U \to \overline{\mathbb{R}}$ is finely l.s.c., resp. u.s.c., the proof of Theorem 1.5.29 shows that there is a compact fine neighborhood K of x_0 in U such that $f|_K$ is l.s.c., resp. u.s.c. When applied to an indicator function of a finely open, resp. finely closed set $A \subset \Omega$, this shows that $A \cap K$ is open relative to K, resp. compact. It is essentially this case of Theorem 1.5.29 which was included in [Br14, p. 43] (but regrettably was overlooked in [Fu13]). Again, one may choose K so as to apply simultaneously to a countable family of finely semicontinuous functions f_n, or of finely open, resp. finely closed sets $A_n \subset U$.

Notes: The material of Sects. 1.1, 1.2, 1.3, and 1.4 is classical and may be found in the monographs of Armitage and Gardiner [AG] and the monograph of Doob [Do4], and also in the works of Brelot and other authors cited in the bibliography. We refer the reader to these references for more details. Some of the original results were slightly improved in this chapter. The content of Sect. 1.5 is essentially due to Fuglede [Fu5, Fu9].

Chapter 2
Fundamentals of Fine Potential Theory

This chapter is devoted to the fundamentals of fine potential theory, that is, the theory of finely [hyper/super]harmonic functions on a finely open set and related problems. In Sect. 2.1, we prove an extension result (Theorem 2.1.25) and show that every finely hyperharmonic function is finely continuous. We extend the fundamental convergence theorem to finely hyperharmonic functions (Theorem 2.1.34) and prove a convergence property for sequences of finely harmonic functions on a finely open set and that the fine topology is locally connected (Corollary 2.1.33). We also prove removable singularities theorems for finely [hyper]harmonic functions. Reduction and balayage relative to a finely open set are defined and their properties are studied in Sect. 2.2. The generalized fine Dirichlet problem is solved in Sect. 2.3 by the PWB method extended to the present setting.

In Sect. 2.4, we define and study the finely superharmonic functions and the fine potentials. We prove a relative fine boundary minimum principle and a domination property and that, on Euclidean open sets, locally lower bounded finely superharmonic functions and fine potentials are the same as superharmonic functions and potentials, respectively. We introduce the notion of quasi-support and the notion of specific multiplication of suitable positive functions by semibounded fine potentials.

In Sect. 2.5, we study the localization in fine potential theory and define the Green function for a fine domain and study its properties. At the end of this section we prove that the cone $\mathcal{S}_+(U)$ of all positive finely superharmonic functions on a fine domain $U \subset \mathbb{R}^N$ is a standard H-cone of functions, see [BBC2]. In Sect. 2.6 we study the integral representation of fine potentials on a fine domain $U \subset \Omega$ (Ω Greenian domain $\subset \mathbb{R}^N$) by the means of the Green function G_U and Borel measures on U (as a topological subspace of Ω) and we prove that every fine potential p on U is the Green potential $G_U\mu$ of a unique (σ-finite) Borel measure μ on U.

Section 2.7 deals with the particular case $N = 2$. We prove in this case some variants of the fine boundary minimum principle (Theorems 2.7.3 and 2.7.4) and

M. El Kadiri and B. Fuglede, *Classical Fine Potential Theory*, Springer Monographs in Mathematics, https://doi.org/10.1007/978-981-96-0430-2_2

the identity of finely [hyper]harmonic and [hyper]harmonic functions on Euclidean open sets. This is no longer the case for $N \geq 3$ (see, for example, [Fu11]).

In Sect. 2.8, we study the problem of approximation of continuous functions on a closed set $A \subset \mathbb{R}^N$ by harmonic and superharmonic functions on open neighborhoods of A.

Section 2.9 is devoted to the study of the invariant functions, that is, the finely superharmonic functions without non-zero fine potential minorant in the specific order.

2.1 Finely Hyperharmonic and Finely Harmonic Functions

Recall from Sect. 1.3 that Ω is a fixed Greenian domain in $\mathbb{R}^N$, $N \geqslant 2$, and that complements of subsets of Ω are taken relative to Ω, and likewise topological concepts (e.g. closure and fine closure of sets).

Definition 2.1.1 Let U denote a finely open subset of the Greenian domain $\Omega \subset \mathbb{R}^N$. A function $f : U \to \overline{\mathbb{R}}$ is said to be *finely hyperharmonic* if f is finely l.s.c.$> -\infty$ on U, and if the induced fine topology on U has a basis $\mathfrak{V}$ consisting of finely open sets V of fine closure $\widetilde{V} \subset U$ such that

$$f(x) \geqslant \int^* f \, d\varepsilon_x^{\complement V} \quad \text{for } x \in V. \tag{2.1.1}$$

Recall that the swept measure $\varepsilon_x^{\complement V}$ on Ω is carried by the fine boundary $\partial_f V \subset U$ and does not charge the polar sets (Corollary 1.4.33). Every finely l.s.c. function f on U differs at most on some polar (hence $\varepsilon_x^{\complement V}$-null) set from some Borel measurable function (Lemma 1.5.24), and so f is $\varepsilon_x^{\complement V}$-measurable on Ω. Since moreover those *relatively compact* sets $V \in \mathfrak{V}$ on which f is *lower bounded* likewise form a basis $\mathfrak{V}'$ of finely open sets V with $\widetilde{V} \subset U$, the upper integral in (2.1.1) equals the corresponding lower integral $\int_* f \, d\varepsilon_x^{\complement V}$ when $V \in \mathfrak{V}'$ (see the second paragraph after Definition 1.4.29), and we may then write simply $\int f \, d\varepsilon_x^{\complement V}$ for $V \in \mathfrak{V}'$. Also note that it suffices in Definition 2.1.1 (and likewise in Definitions 2.1.3 and 2.1.4 below) to consider *regular*, hence Borel measurable, finely open sets V (with $\widetilde{V} \subset U$) because the regular finely open sets form a basis for the fine topology (see the third paragraph after the proof of Lemma 1.4.27). For such V, $\complement V$ is a base and hence $\varepsilon_x^{\complement V} = \varepsilon_x$ for every $x \in \complement V$ (Theorem 1.4.23(b)).

Remark 2.1.2 By use of Choquet theory and the Brelot property (Theorem 1.5.29) it was shown by Lyons [Ly3] that a finely l.s.c. function $f > -\infty$ on a finely open set U is finely hyperharmonic if (and only if) every point $x \in U$ has a fine neighborhood basis consisting of finely open sets V with $\widetilde{V} \subset U$ such that $f(x) \geqslant \int^* f \, d\varepsilon_x^{\complement V}$. This "pointwise" characterization of fine hyperharmonicity, akin to the definition of usual hyperharmonicity (see Sect. 1.1.2), will only be used in the sequel in the few instances where it allows for a simpler proof.

Definition 2.1.3 With U as above, a function f on U is said to be *finely hypoharmonic* if $-f$ is finely hyperharmonic, that is, if f is finely u.s.c. and $< +\infty$ and if the induced fine topology on U has a basis $\mathfrak{V}$ consisting of finely open sets V of fine closure $\widetilde{V} \subset U$ such that

$$f(x) \leqslant \int_* f \, d\varepsilon_x^{\complement V} \quad \text{for every } x \in V.$$

Definition 2.1.4 With U as above, a function $f : U \to \mathbb{R}$ is said to be *finely harmonic* if f is finely continuous and if there is a basis of finely open sets V with $\widetilde{V} \subset U$ such that

$$f(x) = \int f \, d\varepsilon_x^{\complement V} \quad \text{for every } x \in V,$$

this being understood to imply that f is $\varepsilon_x^{\complement V}$-integrable on $\partial_f V$ (which carries $\varepsilon_x^{\complement V}$ when $x \in V$).

Clearly, every finely harmonic function is both finely hyperharmonic and finely hypoharmonic. The converse likewise holds (see Corollary 2.1.19 below), but is by no means obvious because the finely open subsets V of U need not be the same in Definitions 2.1.1 and 2.1.3.

We recall the following difference between (i) harmonic functions on the one hand and (ii) finely harmonic functions on finely open sets, on the other. Indeed, many properties of the more familiar class of harmonic functions on open sets in Euclidean space $\mathbb{R}^N$ are not shared by finely harmonic functions on finely open sets. Example: An application of the Poisson integral representation formula to harmonic functions shows that these functions (i) are quasi-analytic. However, this property may fail for (ii) finely harmonic functions, as was shown by Lyons (see [Ly4]).

Remark 2.1.5 The finely hyperharmonic functions on finely open subsets of Ω form a *sheaf* (strictly speaking: a complete pre-sheaf), that is: (i) if u is finely hyperharmonic on a finely open set $U \subset \Omega$ then the restriction of u to any finely open subset of U is finely hyperharmonic, and (ii) a function u on U is finely hyperharmonic on a finely open set $U \subset \Omega$ provided that u has finely hyperharmonic restrictions to each member of a family $\mathfrak{V}$ of finely open sets having the union U. Here (i) is obviously fulfilled. Concerning (ii), u is finely l.s.c. on each $V \in \mathfrak{V}$ and hence on U. For each $V \in \mathfrak{V}$ there exists by Definition 2.1.1 a basis $\mathcal{B}_V$ for the fine topology on V such that $u(x) \geqslant \int^* u \, d\varepsilon_x^{\complement W}$ for any $W \in \mathcal{B}_V$ and any point $x \in W$. To obtain a similar basis for $U = \bigcup\{V : V \in \mathfrak{V}\}$, take the collection $\mathcal{B}_U$ of all the sets in $\mathcal{B}_V$ as V ranges over $\mathfrak{V}$. This collection $\mathcal{B}_U$ is a basis for the fine topology on U, for if Y is any finely open subset of U and if $y \in Y$, then by regularity of the fine topology there exists $Z \in \mathcal{B}_V$ containing y and with fine closure $\widetilde{Z} \subset Y$. It follows that $u(z) \geqslant \int^* u \, d\varepsilon_z^{\complement Z}$ for any $Z \in \mathcal{B}_V$ and any point $z \in Z$. This shows that u indeed is finely hyperharmonic on

U. Similarly, the finely hypoharmonic functions on U form a sheaf, and so do the finely harmonic functions on U.

The key example of finely hyperharmonic functions on a finely open set U is given by

Theorem 2.1.6 *Let μ be a signed measure on Ω such that $G_\Omega\mu_- < +\infty$ on U and $(\mu_-)_*(U) = 0$. Then $G_\Omega\mu = G_\Omega\mu_+ - G_\Omega\mu_-$ is well defined and finely hyperharmonic on U.*

For any measure μ on Ω, μ_* denotes the corresponding inner measure. Later we shall see (Theorem 2.6.4) that the two conditions on the signed measure μ in the above theorem are essentially necessary for $G_\Omega\mu$ to be finely hyperharmonic.

Proof of Theorem 2.1.6. Consider first a measure $\mu \geqslant 0$ on Ω such that $G_\Omega\mu < +\infty$ on U and $\mu_*(U) = 0$. We show that $G_\Omega\mu$ is finely harmonic on U. For any regular finely open set V such that $\widetilde{V} \subset U$ we have $\mu(V) = 0$ because V is a Borel by Corollary 1.4.18 applied to $A = \complement V = b(\complement V)$. We obtain

$$\int G_\Omega\mu \, d\varepsilon_x^{\complement V} \underset{1}{=} \int G\varepsilon_x^{\complement V} \, d\mu \underset{2}{=} \int \widehat{R}^{\complement V}_{G_\Omega\varepsilon_x} \, d\mu \underset{3}{=} \int G_\Omega\varepsilon_x \, d\mu^{\complement V}$$
$$\underset{4}{=} \int G_\Omega\varepsilon_x \, d\mu \underset{5}{=} \int G_\Omega\mu \, d\varepsilon_x = G_\Omega\mu(x).$$

Here the equalities marked 1 and 5 follow from the reciprocity theorem (see Sect. 1.1.5), equality 2 from (1.4.7) together with (1.4.2) (or the more general Theorem 1.4.24), equality 3 from (1.4.2), and equality 4 follows from Theorem 1.4.23(b) according to which $\mu^{\complement V} = \mu$ since μ is carried by $\complement V = b(\complement V)$. The above sets V form a basis for the fine topology on U, and since $G_\Omega\mu$ is finite and finely continuous on U, $G_\Omega\mu$ is indeed finely harmonic there, according to Definition 2.1.4.

For a signed measure μ on Ω such that $G_\Omega\mu_- < +\infty$ on U and $(\mu_-)_*(U) = 0$, the above result $\int G_\Omega\mu \, d\varepsilon_x^{\complement V} = G\mu(x)$ holds with μ replaced by μ_-, again for any regular finely open set V with $\widetilde{V} \subset U$. For such V we therefore have by Theorem 1.4.24

$$\int G_\Omega\mu_+ \, d\varepsilon_x^{\complement V} = \widehat{R}^{\complement V}_{G_\Omega\mu_+}(x) \leqslant G_\Omega\mu_+(x) \quad \text{for } x \in V.$$

By subtraction this leads to (2.1.1) for $f = G_\Omega\mu_-$ because $\mu_+ - \mu_- = \mu$ and $G_\Omega\mu_- < +\infty$ on V ($\subset U$). Since $G_\Omega\mu = G_\Omega\mu_+ - G_\Omega\mu_-$ on U with $G_\Omega\mu_-$ finite there, $G_\Omega\mu$ is finely continuous, in particular finely l.s.c., on U, and so $G_\Omega\mu$ is indeed finely hyperharmonic on u, according to Definition 2.1.1. □

Corollary 2.1.7 *Suppose instead that $G_\Omega|\mu| < +\infty$ on U and that $|\mu|_*(U) = 0$. Then $G_\Omega\mu$ is well defined and finely harmonic on U.*

Example 2.1.8 The hypothesis in Corollary 2.1.7 that $G_\Omega|\mu| < +\infty$ on U cannot be omitted. We give an example of an admissible measure $\mu \geqslant 0$ on $\Omega = \mathbb{R}^3$ and a finely open Borel set $U \subset \mathbb{R}^3$ such that $\mu(U) = 0$ and that $G_\Omega\mu(x) = +\infty$ for

$x = 0 \in U$.[1] Consider a sequence $(x_n) \subset \mathbb{R}^3$ such that $|x_n| = 2^{-n}$, and the measure $\mu = \sum_n 2^{-n}\varepsilon_{x_n}$. The set $\{x_n\}$ is a polar Borel set, hence finely closed, and its finely open complement U is μ-null. But the potential $G\mu$ equals $\sum_n 2^{-n}|x_n|^{-1} = \sum_n 1 = +\infty$ at $0 \in U$.

Remark 2.1.9 For *finely local properties* the requirement that the given finely open set U be a subset of a Greenian domain in $\mathbb{R}^N$ is not needed in the non-trivial case $N = 2$. We may define the notion of a finely [hyper or hypo]harmonic function f on an arbitrary finely open set $U \subset \mathbb{R}^2$ by requiring that U can be covered by Greenian domains $\Omega \subset \mathbb{R}^2$ such that $f|_\Omega$ is finely [hyper or hypo]harmonic on $U \cap \Omega$ (in the sense considered up to now) for each Ω. In this way, these concepts acquire the usual sheaf properties relative to the fine topology. Finely local properties of fine harmonicity, etc., thus make sense even for functions defined on a finely open subset of a *Riemann surface*.

Theorem 2.1.10 (A first fine boundary minimum principle) *Let f be finely hyperharmonic on a finely open set $U \subset \Omega$, and suppose that*

$$\text{f-}\liminf_{x\to y,\, x\in U} f(x) \geqslant 0 \tag{2.1.2}$$

for quasi-every $y \in \partial_f U$. If moreover there exists a semibounded potential p on Ω such that $f \geqslant -p$ on U, then $f \geqslant 0$ on U.

Proof As noted after Definition 2.1.1, f is $\varepsilon_x^{\complement V}$-measurable for each finely open V with $\widetilde{V} \subset U$. Furthermore, there is a basis $\mathfrak{V}$ of *regular* (hence Borel measurable) finely open sets $V \subset U$ of fine closure $\widetilde{V} \subset U$ with f *lower bounded* on V and such that (2.1.1) holds for each of these V, even with the lower integral replaced by an integral. By the quasi-Lindelöf property (see paragraph 6 after Corollary 1.2.5) there is a countable subfamily (V_n) of $\mathfrak{V}$ and a polar set $E \subset \Omega$ such that $\bigcup_n V_n = U \setminus E$.

The positive function h defined on Ω by $h = f_-$ on U and $h = 0$ on $\complement U$ is finely u.s.c. quasi-everywhere on Ω, namely everywhere except precisely at those points $y \in \partial_f U$ at which (2.1.2) fails. Clearly $h \leqslant p$ on all of Ω. Since $h = f_-$ on U, h is $\varepsilon_x^{\complement V_n}$-measurable for each n.

With reference to Definition 1.4.29 we have

$$h(x) \leqslant h^{\complement V_n}(x) = \int h\, d\varepsilon_x^{\complement V_n},$$

not only for $x \in V_n$ but also for $x \in \complement V_n$. Indeed, $\varepsilon_x^{\complement V_n} = \varepsilon_x$ by Theorem 1.4.23(b) because V_n is regular and hence $b(\complement V_n) = \complement V_n$. Since $-f \leqslant f^- = h$ on U we thus obtain from (2.1.1) if $h(x) > 0$

$$h(x) = -f(x) \leqslant -\int^* f\, d\varepsilon_x^{\complement V_n} \leqslant \int h\, d\varepsilon_x^{\complement V_n}$$

[1] This is simpler than the Example [Fu8, p. 74].

because $h \geqslant 0$ is $\varepsilon_x^{\complement V_n}$-measurable. The resulting inequality above holds trivially if $h(x) = 0$. The function h is finely u.s.c. quasi-everywhere and hence quasi-u.s.c. by Corollary 1.5.14, and since $h \leqslant p$ on Ω it follows from Theorem 1.5.28 that

$$0 \leqslant h \leqslant h^{\bigcap_n \complement V_n} = h^{\complement U} = 0 \quad \text{q.e. on } U.$$

The former equality holds everywhere on U, for if $x \in U$ then $\varepsilon_x^{\complement U}$ does not charge the polar set $U \setminus \bigcup_n V_n$ by Corollary 1.4.33. The latter equality holds on U because $h = 0$ on the finely closed set $\complement U \supset b(\complement U)$ which carries $\varepsilon_x^{\complement U}$ by Theorem 1.4.23(a). Thus $f_- = h = 0$ q.e. on U.

For given $x \in U$ choose $V \in \mathfrak{V}$ so that $x \in V$. By Definition 2.1.1 we have

$$f(x) \geqslant \int^* f \, d\varepsilon_x^{\complement V} \geqslant \int (-h) \, d\varepsilon_x^{\complement V} \geqslant 0$$

because $f \geqslant -h = 0$ q.e. on U and because $\varepsilon_x^{\complement V}$ is carried by $\partial_f V \subset \widetilde{V} \subset U$ and $\varepsilon_x^{\complement V}$ does not charge any polar set. □

Remark 2.1.11 The semiboundedness of p in Theorem 2.1.10 cannot be omitted because we only require (2.1.2) to hold quasi-everywhere on $\partial_f U$. Example: $U = \complement\{x\}$ for some point $x \in \Omega$. If (2.1.2) is required to hold at *every* point $x \in \partial_f U$ and if $f \geqslant -p$ for some potential p (not necessarily semibounded), then the conclusion $f \geqslant 0$ subsists, see Theorem 2.4.14 below.

Remark 2.1.12 The case of Theorem 2.1.10 where U is *open* and relatively compact, and f is hyperharmonic (hence finely hyperharmonic, see below) and bounded from below is contained in Brelot [Br6].

Lemma 2.1.13 *Let $p \in \mathcal{P}^s$, let $B \subset \Omega$ be a base, and let $f : B \to \overline{\mathbb{R}}$ be a finely continuous function such that $|f| \leqslant p$ on B. Then f^B is defined, finite and finely continuous on the finely open set $\{p < +\infty\}$, and finely harmonic on the finely open set $\{p < +\infty\} \setminus B$. Furthermore $|f^B| \leqslant p$, and*

$$(f^B)^A = f^B \text{ on } \{p < +\infty\}$$

for any set $A \subset \Omega$ containing B.

Proof Being finely continuous on the base B, f extends to a finely continuous function on Ω, by Theorem 1.4.35, and this extension is Borel measurable (Theorem 1.4.36). Clearly, $|f|^B \leqslant p^B = \widehat{R}_p^B \leqslant p$ by Definition 1.4.29 and (1.4.22). The fine harmonicity of f^B (see Definition 1.4.29) on $U := \{p < +\infty\} \setminus B$ will follow from the other assertions when we take $A = \complement V$ in the display in the lemma, V denoting a finely open subset of Ω of fine closure $\widetilde{V} \subset U$. (Note that U is finely open because p is finely continuous and B is finely closed relative to Ω; furthermore the sets V form a basis for the fine topology on U.)

For the proof of the remaining assertions of the lemma it suffices to consider the case where $f \geqslant 0$. By Theorem 1.4.35, f admits a finely continuous extension to all of Ω. Since p is a finely continuous on Ω we may assume that this extension, likewise denoted by f, satisfies $0 \leqslant f \leqslant p$ on Ω (otherwise replace the extended f by $f \wedge p$).

For any given function $\varphi \in \mathcal{C}_0^+(\Omega)$ and any $\varepsilon > 0$ there exist by Theorem 1.4.5 potentials $p_1, p_2 \in \mathcal{P}^c$ such that $p_1 - p_2 \in \mathcal{C}_0^+(\Omega)$ and

$$|\varphi - (p_1 - p_2)| \leqslant \varepsilon.$$

Since $f - \varphi$ is finely continuous, $|f - \varphi| \leqslant \widehat{R}_{|f-\varphi|}$ everywhere, by Theorem 1.3.3(b), and hence

$$-v \leqslant f - (p_1 - p_2) \leqslant v, \tag{2.1.3}$$

where we define $v \in \mathcal{U}^+$ by

$$v := \widehat{R}_{|f-\varphi|} + \varepsilon \leqslant p + \widehat{R}_{\varphi} + \varepsilon.$$

Writing $u := f^B \leqslant p$ we obtain from (2.1.3) by integration with respect to ε_x^B

$$-v \leqslant -v^B \leqslant u - (p_1^B - p_2^B) \leqslant v^B \leqslant v \tag{2.1.4}$$

because $v^B = \widehat{R}_v^B \leqslant v$ by (1.4.22).

Since p_1^B, p_2^B, and v are finely continuous the inequalities (2.1.4) remain valid when u is replaced by its least finely u.s.c. majorant $\widetilde{u}$ or its greatest finely l.s.c. minorant $\widehat{u}$. It follows that

$$\widetilde{u} - \widehat{u} \leqslant 2v = 2\widehat{R}_{|f-\varphi|} + 2\varepsilon \quad \text{on } \{p < +\infty\}$$

because $\{p < +\infty\} \subset \{v < +\infty\} \subset \{\widetilde{u} < +\infty\}$ (noting that $\widetilde{u} \leqslant p_1^B + v$). For $\varepsilon \to 0$ this implies that

$$\widetilde{u}(x) - \widehat{u}(x) \leqslant 2\widehat{R}_{|f-\varphi|}(x) \quad \text{for } x \in \{p < +\infty\}.$$

By varying φ we obtain $\widetilde{u}(x) - \widehat{u}(x) = 0$ on $\{p < +\infty\}$ according to (1.5.20) in Theorem 1.5.22. Consequently, $u = f^B$ *is finely continuous on* $\{p < +\infty\}$.

For any set A such that $B \subset A \subset \Omega$ it follows in the same way from (2.1.4), now by integration with respect to ε_x^A, that

$$-v \leqslant u^A - (p_1 - p_2)^B \leqslant v. \tag{2.1.5}$$

In fact, for $i = 1, 2$,

$$(p_i^B)^A = \widehat{R}^A_{p_i^B} = \widehat{R}^A_{\widehat{R}^B_{p_i}} = \widehat{R}^B_{p_i} = p_i^B$$

by (1.4.22) and Lemma 1.3.11 since $B \subset A$. Combining (2.1.4) and (2.1.5), we obtain

$$|u^A - u| \leqslant 2v \quad \text{on } \{p < +\infty\}$$

whence $u^A = u$ on $\{p < +\infty\}$, by varying ε and φ as above. □

The following theorem contains the fine boundary minimum principle (Theorem 2.1.10) which however is used in the proof.

Theorem 2.1.14 (The mean-value inequality for finely hyperharmonic functions) *Let U be a finely open subset of Ω. Let $f : \widetilde{U} \to \overline{\mathbb{R}}$ be finely l.s.c. on $\widetilde{U}$, and suppose that $f|_U$ is finely hyperharmonic. If moreover $f \geqslant -p$ on U for some $p \in \mathcal{P}^s$ then*

$$f(x) \geqslant \int^* f \, d\varepsilon_x^{\complement U}$$

for every $x \in U \cap \{p < +\infty\}$ (thus in particular q.e. on U).

Proof Write $\mathfrak{b}(\complement U) = B$, hence $\varepsilon_x^{\complement U} = \varepsilon_x^B$ by Theorem 1.4.21. Let $g : \Omega \to \overline{\mathbb{R}}$ denote the greatest finely l.s.c. minorant of the function equal to f on U and $+\infty$ on $\complement U$. Then $g \geqslant -p$ on Ω, $g \geqslant f$ q.e. on $\partial_f U$, and $g = f$ on U.

The fine topology being completely regular (see Sect. 1.2), hence uniformizable, g is the pointwise supremum of the family $(g_\alpha)_{\alpha \in I}$ of all finely continuous minorants $g_\alpha : \Omega \to \overline{\mathbb{R}}$ of g, see, for example, [Bou5, § 1, no. 6]. By the quasi-Lindelöf property for functions, see Sect. 1.2, there exists a sequence $(f_n)_{n \in \mathbb{N}}$ of finely continuous minorants $f_n \geqslant -\infty$ of g on Ω, converging q.e. to g. Replacing f_n by $f_1 \vee \ldots \vee f_n$ ($\leqslant g$) we further achieve that the sequence (f_n) becomes increasing.

Let q be a fixed non-zero potential of class $\mathcal{P}^c$. Replacing f_n by $-p$ on $\{f_n < -p\}$ and by nq on $\{f_n > nq\}$ we may assume that $-p \leqslant f_n \leqslant nq$, and hence

$$-p \leqslant f_n^B \leqslant nq.$$

For $m > n$ the function $f_{n,m}$ defined on Ω by

$$f_{n,m} = f_n \vee (-mq)$$

is finite and finely continuous, and $|f_{n,m}| \leqslant mq$ since $f_n \leqslant nq \leqslant mq$. According to Lemma 2.1.13 (with p replaced by $mq < +\infty$), $f_{n,m}^B$ is finite and finely continuous, hence Borel measurable (Theorem 1.4.36), and

$$(f_{n,m}^B)^A = f_{n,m}^B \quad \text{for any set } A \text{ with } B \subset A \subset \Omega. \tag{2.1.6}$$

For $m \uparrow +\infty$ we obtain $f_{n,m} \downarrow f_n$, and hence $f_{n,m}^B \downarrow f_n^B$ because $f_{n,m}^B \leqslant nq^B \leqslant nq < +\infty$. It follows that f_n^B is finely u.s.c., Borel measurable, $-\infty \leqslant f_n^B \leqslant nq$, and from (2.1.6) that

$$(f_n^B)^A = f_n^B \quad \text{for } A \supset B \tag{2.1.7}$$

because $nq^A \leqslant nq < +\infty$. Now g is finely hyperharmonic on U since $g = f$ there, and hence there is by Definition 2.1.1 a basis $\mathfrak{V}$ for the fine topology on U consisting of finely open sets V with $\widetilde{V} \subset U$ such that

$$\int^* g \, d\varepsilon_x^{\complement V} \leqslant g(x) \quad \text{for } x \in V \in \mathfrak{V}. \tag{2.1.8}$$

Applying (2.1.7) to $A = \complement V$ ($\supset \complement U \supset b(\complement U) = B$) we obtain

$$(f_n^B)^{\complement V} = f_n^B \tag{2.1.9}$$

on Ω. It follows that $g - f_n^B$ is defined, finely l.s.c. and $\geqslant -(p + nq)$ on $U \cap \{p < +\infty\}$. Moreover, $g - f_n^B$ is finely hyperharmonic on U since $g - f_n^B = f - f_n^B > -\infty$ there, and because

$$\int^* (g - f_n^B) \, d\varepsilon_x^{\complement V} \leqslant g(x) - f_n^B(x) \quad \text{for } x \in V$$

for every $V \in \mathfrak{V}$ by (2.1.8), (2.1.9). For $x \in (\partial_f U) \cap \{p < +\infty\}$ we have

$$\text{f-}\liminf_{y \to x,\, y \in U} (g(y) - f_n^B(y)) \geqslant g(x) - f_n^B(x)$$

because g and $-f_n^B$ are finely l.s.c. For quasi-every $x \in \partial_f U$ we have $p(x) < +\infty$, $g(x) \geqslant f(x)$, and $x \in b(\complement U) = B$, hence $f_n^B(x) = \int f_n \, d\varepsilon_x^B = f_n(x) \leqslant g(x)$, and consequently $g(x) - f_n^B(x) \geqslant 0$. It therefore follows by the fine boundary principle (Theorem 2.1.10) that $g - f_n^B \geqslant 0$ on U. That is,

$$g(x) \geqslant \int f_n \, d\varepsilon_x^B \quad \text{for every } x \in U. \tag{2.1.10}$$

Finally, as $n \uparrow +\infty$, f_n increases and $f_n \uparrow g$ q.e. Since $\varepsilon_x^B = \varepsilon_x^{\complement U}$ does not charge the polar sets when $x \in U$ and since $g \geqslant f$ q.e. on $\partial_f U$ which carries $\varepsilon_x^{\complement U}$, we conclude from (2.1.10) that indeed

$$f(x) = g(x) \geqslant \sup_n \int f_n \, d\varepsilon_x^B = \int g \, d\varepsilon_x^{\complement U} \geqslant \int^* f \, d\varepsilon_x^{\complement U}$$

at any point $x \in U$ such that $p(x) < +\infty$. □

Corollary 2.1.15 (The mean-value equality for finely harmonic functions) *Let U be a finely open subset of Ω. Let $f : \widetilde{U} \to \overline{\mathbb{R}}$ be finely continuous on $\widetilde{U}$, and suppose that $f|_U$ is finely harmonic. If moreover $|f| \leqslant p$ on U for some $p \in \mathcal{P}^s$ then*

$$f(x) = \int f \, d\varepsilon_x^{\complement U}$$

for every $x \in U \cap \{p < +\infty\}$.

Here is a finely local consequence of Theorem 2.1.14 applied to V in place of U.

Corollary 2.1.16 *Let f be a finely hyperharmonic function on a finely open set $U \subset \Omega$. Then*

$$f(x) \geqslant \int f \, d\varepsilon_x^{\complement V} \quad \textit{for every } x \in V \cap \{p < +\infty\}$$

holds for every finely open set V with $\widetilde{V} \subset U$ such that $f \geqslant -p$ on V for some semibounded potential p on Ω.

Applying Corollary 2.1.16 to the finely harmonic functions $f = \pm 1$ leads to the following extension of part of Theorem 1.4.22.

Corollary 2.1.17 *Let V be a finely open subset of Ω such that there exists a finite potential p on Ω with $p \geqslant 1$ on V. Then $\varepsilon_x^{\complement V}$ is a probability measure: $\int d\varepsilon_x^{\complement V} = 1$*

The hypothesis on the finely open set V is fulfilled in particular if V has compact (Euclidean) closure $\overline{V} \subset \Omega$, in other words if V is of class $\mathfrak{V}_U(1)$, cf. the following remark.

Remark 2.1.18 For any function $f : U \to \overline{\mathbb{R}}$ (U a given finely open subset of Ω) we denote by $\mathfrak{V}_U(f)$ the class of all finely open sets V of compact closure $\overline{V}$ (in the Euclidean topology) contained in U, and such that f is *bounded* on $\overline{V}$. We shall often use the fact that there exists for any $V \in \mathfrak{V}_U(f_-)$ a potential $p \in \mathcal{P}^c$ such that $f \geqslant -p$ on $\widetilde{V}$ ($\subset \overline{V}$). (Choose $q \in \mathcal{P}^c$, $q > 0$, and note that f_-/q is bounded on $\overline{V}$, say $f_-/q \leqslant k$, and so $f_- \leqslant p := kq$ on $\overline{V}$.)

If f is finely locally bounded, say from below, on U then $\mathfrak{V}_U(f_-)$ is a basis for the fine topology on U. (This is because every fine neighborhood of a point of U contains a compact one.)

According to Corollary 2.1.16 one may assume in Definition 2.1.1 that $\mathfrak{V} = \mathfrak{V}_U(f_-)$. Similarly one may assume in Definition 2.1.3 that $\mathfrak{V} = \mathfrak{V}_U(f_+)$, and in Definition 2.1.4 that $\mathfrak{V} = \mathfrak{V}_U(f)$. Since $\mathfrak{V}_U(f) = \mathfrak{V}_U(f_+) \cap \mathfrak{V}_U(f_-)$ we therefore have

Corollary 2.1.19 *A function $f : U \to \mathbb{R}$ (U finely open in Ω) is finely harmonic if (and only if) f is both finely hyperharmonic and finely hypoharmonic.*

For then f is finite and finely continuous on U, and hence $f(x) = \int f \, d\varepsilon_x^{\complement V}$ for every $x \in V \in \mathfrak{V}_U(f_+) \cap \mathfrak{V}_U(f_-) = \mathfrak{V}_U(f)$, and $\mathfrak{V}_U(f)$ is a basis for the fine topology on U.

Remark 2.1.20 The particular basis $\mathfrak{V}_U(f_-)$ for the fine topology on U (f being finely locally bounded from below) depends on the order of magnitude of f_-, unlike the situation for ordinary hyperharmonic functions on a Euclidean open set U (such functions being bounded from below on every compact set).

This observation applies, in particular, to any finite family of finely hyperharmonic functions. There are of course similar observations regarding finely harmonic functions. As a final corollary to Theorem 2.1.14 we have

Corollary 2.1.21 *The family $\mathcal{U}(U)$ of all finely hyperharmonic functions on a finely open set $U \subset \Omega$ is a convex cone (that is, $\mathcal{U}(U)$ is stable under addition and multiplication by positive constants) stable under pointwise supremum for upper directed families, and under pointwise infimum for finite families. The finite valued finely hyperharmonic functions form a convex cone in the vector space of all finite valued functions, and this cone is closed under finely locally uniform convergence. The finely harmonic functions on U form a linear subspace of the vector space of all finely continuous functions $U \to \mathbb{R}$, and this subspace is closed under finely locally uniform convergence.*

We will see later that the finely hyperharmonic functions are finely continuous (see Theorem 2.1.26 below). It therefore follows that the finite valued finely hyperharmonic functions form a convex cone in the vector space of all finite valued finely continuous functions.

Theorem 2.1.22 *Let U be a (Euclidean) open subset of Ω.*

(a) *A function $u : U \to \,]-\infty, +\infty]$ is hyperharmonic if and only if u is finely hyperharmonic and furthermore (Euclidean) locally bounded from below.*

(b) *A function $u : U \to \mathbb{R}$ is harmonic if and only if u is finely harmonic and furthermore (Euclidean) locally bounded.*

Proof (a) Suppose first that u is hyperharmonic on U, hence l.s.c. $> -\infty$, and in particular locally bounded from below. We may suppose that u is superharmonic (otherwise write $u = \sup_{n\in\mathbb{N}}(u \wedge n)$ and use Corollary 2.1.21). For the proof that u is finely hyperharmonic, let $x_0 \in U$ and let U_0 be an open neighborhood of x_0 with compact closure $\overline{U}_0 \subset U$. By Theorem 1.4.7 there are potentials p, q on Ω such that q is harmonic on U_0 and that $u = p - q$ on U_0. Consider a ball ω about x_0 so that $\overline{\omega} \subset U_0$. For any finely open set V with $\widetilde{V} \subset \omega$ and any $x \in V$ we have

$$p(x) \geqslant \widehat{R}_p^{\complement V}(x) = \int p \, d\varepsilon_x^{\complement V},$$

$$q(x) = \int q \, d\varepsilon_x^{\complement \omega} = \widehat{R}_q^{\complement \omega}(x) \leqslant \widehat{R}_q^{\complement V}(x) = \int q \, d\varepsilon_x^{\complement V},$$

and so $q(x) \leq \int q \, d\varepsilon_x^{\complement V}$. Consequently, $u(x) = p(x) - q(x) \geqslant \int u \, d\varepsilon_x^{\complement V}$ for any $x \in V$. By Definition 2.1.1, u is finely hyperharmonic on ω and hence on U_0 and indeed on U, by the sheaf property (Remark 2.1.5).

Conversely, let u be finely hyperharmonic on U. For any ball ω with $\overline{\omega} \subset U$ it follows from Corollary 2.1.16 that $u \geqslant u^{\complement \omega}$ on ω because u is supposed to be bounded from below on $\overline{\omega}$ (compact), and so $\omega \in \mathfrak{V}_U(u_-)$, cf. Remark 2.1.18. Having thus shown that u is *nearly hyperharmonic* on U we conclude that u actually is hyperharmonic there, being also finely l.s.c. (see Sect. 1.2, 7th paragraph after Corollary 1.2.5).

(b) If u is harmonic then $\pm u$ is hyperharmonic and hence finely hyperharmonic, and it follows by Corollary 2.1.19 that u is finely harmonic (and of course locally bounded). Conversely, if u is finely harmonic and locally bounded then $\pm u$ is finely hyperharmonic and locally bounded from below, hence hyperharmonic, by (a), and so u is indeed harmonic. □

Remark 2.1.23 In dimension $N = 2$ the local boundedness assumption in (a) or (b) can be omitted, as shown in Theorem 2.7.2. For $N \geqslant 3$ this is no longer the case, see [Fu11, Théorème 3.2], but it will be shown in Theorem 2.4.25 below that it suffices to assume in (b) that u be locally bounded *from one side*.

Corollary 2.1.24 *A (Euclidean) open set $U \subset \Omega$ is finely connected if and only if U is (Euclidean) connected.*

Proof For any decomposition $U = V \cup W$ of U into disjoint finely open sets V and W, the function u on U defined by $u = +\infty$ on V and $u = 0$ on W is finely hyperharmonic on U, hence hyperharmonic there, in particular l.s.c., and so V must be open. Similarly W is open, and consequently either V or W must be void since U is connected. □

For a more direct proof of the corollary see [Fu4, Theorem 2]. The corollary has the following equivalent formulation:

The fine components of a (Euclidean) open set U are same as the (Euclidean) components.

The following theorem is a "fine" version of the local extension property for superharmonic functions (Theorem 1.4.7), but with the important difference that we need not assume (fine) continuity of the given finely hyperharmonic function. On the contrary, this fine extension property *implies* the fine continuity of finely hyperharmonic functions (see Theorem 2.1.26 below), which is indispensable for the rest of this book.

Theorem 2.1.25 (Finely local extension property) *Let $U \subset \Omega$ be a finely open set. Let $a \in \mathbb{R}$, $a > 0$, and let $x_0 \in U$. Then there exists a finely open fine neighborhood V of x_0 in U and a locally bounded potential v on Ω, finely harmonic on V, such that, for any finely hyperharmonic function f on U with $|f| \leqslant a$, there is a locally bounded potential u on Ω such that $f = u - v$ on $\widetilde{V}$.*

Proof Choose a relatively compact (in Ω) finely open fine neighborhood U_1 of x_0 in U and a strict potential $p \in \mathcal{P}^c$ on Ω (Theorem 1.4.15(b)), normalized so that $p \geqslant a$ on U_1. Let f be finely hyperharmonic on U with $|f| \leqslant a$, and hence $|f| \leqslant p$ on U_1. Choose a relatively compact regular finely open set U_0 with $x_0 \in U_0 \subset \overline{U}_0 \subset U_1$ (see Sect. 1.2 and the paragraph following (1.4.21)). Writing for brevity $q = \widehat{R}_p^{\complement U_0}$ we have by Theorem 1.4.17

$$p > q \text{ on } U_0, \quad p = q \text{ on } \complement U_0. \tag{2.1.11}$$

Since $f \geqslant -p$ on U_1 it follows by Corollary 2.1.16 with U replaced by U_1 that

$$f(x) \geqslant \int f \, d\varepsilon_x^{\complement U_0} \geqslant -\int p \, d\varepsilon_x^{\complement U_0} = -\widehat{R}_p^{\complement U_0}(x) = -q(x) \qquad (2.1.12)$$

for $x \in U_0$, and trivially for $x \in U_1 \backslash U_0$, hence altogether for $x \in U_1$. Thus

$$-q \leqslant f \leqslant p \quad \text{on } U_1.$$

Since $p(x_0) > q(x_0)$, by (2.1.11), there is a constant $c > 1$ (independent of f) such that

$$cp(x_0) - cq(x_0) > p(x_0) + 2q(x_0). \qquad (2.1.13)$$

The set

$$V := \{x \in U_0 : cp(x) - cq(x) > p(x) + 2q(x)\}$$

is finely open along with U_0 by the fine continuity of p and q. Consider the following function u on Ω:

$$u = \begin{cases} cp \wedge (f + (c+2)q) & \text{on } U_1 \\ cp & \text{on } \complement \widetilde{U}_0. \end{cases}$$

This definition makes sense because $U_1 \cup \complement \widetilde{U}_0 = \Omega$ and because $cp \leqslant f + (c+2)q$ on $U_1 \backslash \widetilde{U}_0$ since $f \geqslant -p$ on U_1 and $p = q$ on $\complement U_0$ by (2.1.11). Clearly, u is finely hyperharmonic on U_1, cf. Corollary 2.1.21 and Theorem 2.1.22(a), and likewise on $\complement \widetilde{U}_0$, hence on all of Ω. In view of (2.1.12), $u \geqslant 0$ on U_1 and hence on Ω. Again by Theorem 2.1.22(a), u is even hyperharmonic on Ω, and $u \in \mathcal{P}^b$ because $u \leqslant cp \in \mathcal{P}^b$. Defining $v = (c+2)q$ we have $v \in \mathcal{P}^b$ and v is finely harmonic on $V \subset U_0$ along with $q := \widehat{R}_p^{\complement U_0} = p^{\complement U_0}$ by (1.4.22) and Lemma 2.1.13. Finally $u = v + f$ on $\widetilde{V} \subset U_0$ because

$$c(p - q) \geqslant p + 2q \geqslant f + 2q \quad \text{on } \widetilde{V} \subset \widetilde{U}_0 \subset U_1$$

by definition of V and by fine continuity of p and q. Thus $cp \geqslant f + (c+2)q$ on $\widetilde{V}$, and so indeed $u = f + (c+2)q = f + v$ on $\widetilde{V}$. Note that $f = u - v$ is finely continuous on $\widetilde{V}$ along with $u, v \in \mathcal{P}^b$. □

Theorem 2.1.26 *Every finely hyperharmonic function is finely continuous.*

Proof Let $f : U \to]-\infty, +\infty]$ be finely hyperharmonic (U finely open in Ω). Then f is finely l.s.c., hence bounded from below on some fine neighborhood W of any given point $x_0 \in U$. It remains to prove that f is finely u.s.c. at x_0. We may clearly assume that $f(x_0) < +\infty$. For any constant $c > f(x_0)$, $f \wedge c$ is finely hyperharmonic on W, by Corollary 2.1.21, and bounded there, say $|f \wedge c| \leqslant a$. The above theorem therefore shows that $f \wedge c$ is finely continuous at x_0. The set $\{f < c\} = \{f \wedge c < c\}$ is therefore a fine neighborhood of x_0. Since this holds for every $c > f(x_0)$, f is indeed finely u.s.c. at x_0. This completes the proof of Theorem 2.1.26. □

Theorem 2.1.26 means that there is so to speak no (new) "fine-fine" topology.

Corollary 2.1.27 *Let a numerical function $u \geqslant 0$ be finely hyperharmonic on a fine domain U. Then either $u > 0$ or $u \equiv 0$ on U.*

The set $E := \{x \in U : u(x) > 0\}$ is finely open. The pointwise supremum $\sup_{n\in\mathbb{N}}(nu) = +\infty 1_E$ is finely hyperharmonic on U by Corollary 2.1.21, and hence finely continuous by the above theorem. Thus E is also finely closed, and since U is finely connected it follows that either $E = U$ or $E = \varnothing$.

Theorem 2.1.28 *Let u be a finely [hyper]harmonic function on a (Euclidean) open set $U \subset \mathbb{R}^N$. The largest (Euclidean) open set $U_0 \subset U$ on which u is [hyper]harmonic is non-void and even dense everywhere in U.*

Proof Let first the hypothesis be that u be finely hyperharmonic function on U. The function u is finely continuous by Theorem 2.1.26 and hence of the Baire class One (see 6th paragraph of the text following Corollary 1.2.5). It follows that u has a dense set $D \subset U$ of points of continuity. As $u > -\infty$, each point of D has an open neighborhood $\omega(\subset U)$ on which u is lower bounded and hence hyperharmonic according to Theorem 2.1.22(a). This implies that $D \subset U_0$, and so U_0 is dense in U. In the remaining case the hypothesis is that u is finely harmonic on U, hence u and $-u$ are finely hyperharmonic there by Corollary 2.1.19, and we have just found that the largest open U_1 and U_2 on which u and $-u$, respectively, are hyperharmonic are dense in U. It then follows that the open set $U_1 \cap U_2$, on which u is harmonic, is everywhere dense in U, and so is U_0 because $U_1 \cap U_2 \subset U_0$. □

We shall see later (cf. Theorem 2.7.2) that $U_0 = U$ for $N = 2$. For $N \geq 3$, Gardiner has sharpened in [Ga3] the result of Theorem 2.1.28 by proving that given an open set U in $\mathbb{R}^N$ and a dense open subset V of U, there is a finely harmonic function u on U such that V is the largest open subset of U on which u is harmonic.

We bring a fine topology version of Theorem 1.3.4. (There is no such "fine" version of Theorem 1.3.5.) See Theorem 2.4.19 below for a still more general fine topology version of (a) and (b).

Theorem 2.1.29 *Let $f : \Omega \to [0, +\infty]$ have a positive superharmonic majorant. Then*

(a) *$\widehat{R}_f$ is superharmonic. $\widehat{R}_f$ is finite and continuous at any point of Ω at which f is finite and continuous,*

(b) *$\widehat{R}_f$ is finely harmonic on any finely open set $U \subset \Omega$ on which $\widehat{R}_f$ is finite and f is either finely hypoharmonic or finely continuous and $< \widehat{R}_f$.*

Proof Part (a) was established in the proof of Theorem 1.3.4. For (b), let $U \subset \Omega$ be finely open, and suppose first that f is finely hypoharmonic on U. We may clearly assume that U is finely connected. By Theorem 2.1.26, f is finely continuous (and finite) on U. By Theorem 2.1.22, $\widehat{R}_f$ is in particular finely hyperharmonic on Ω and hence on U, and so is therefore $\widehat{R}_f - f$. We may clearly assume that $\widehat{R}_f - f \not\equiv 0$ and hence $\widehat{R}_f - f > 0$ everywhere on the fine domain U according to Corollary 2.1.27. The proof of (b) thus reduces to the case where f is finely continuous and $< \widehat{R}_f$ on U. Let V be a regular finely open set in $\mathfrak{V}_U(f|_U)$, then we have by (1.4.22) and Theorem 1.4.21 (applied with $A = \complement V = b(\complement V)$ because V is regular) $(\widehat{R}_f)^{\complement V} = \widehat{R}_f$ on $\complement V$ and, for every $x \in V$,

$$(\widehat{R}_f)^{\complement V}(x) = \int \widehat{R}_f d\varepsilon_x^{\complement V} \geq \int f d\varepsilon_x^{\complement V} \geq f(x),$$

the former inequality because $\widehat{R}_f \geq f$ and the latter by Corollary 2.1.16 and Remark 2.1.18, f being finely hypoharmonic on U. Hence $(\widehat{R}_f)^{\complement V} = \widehat{R}_f$ on Ω. Since the function $\widehat{R}_f$ is finely continuous and finite on U, it then follows that it is finely harmonic there, see Definition 2.1.4 (the regular finely open sets form a basis for the fine topology, cf. the third paragraph after the proof of Lemma 1.4.27). Now suppose that f, not necessarily finely hypoharmonic, is finely continuous on U and satisfies the inequality $f < \widehat{R}_f$ on U, and let $v \in \mathcal{U}^+(\Omega)$ be such that $v \geq f$ on Ω and let x be a given point in U. Then we have $v(x) \geq R_f(x) > f(x)$. By fine continuity of f and $\widehat{R}_f$ at x there exists a constant c and a finely open neighborhood Z of x with $Z \subset U$ and $f < c < \widehat{R}_f$ on Z. Let g be the function defined by $g = f \vee c1_Z$. Then $v \geq g \geqslant f$ on Ω and $g = c$ on Z. Hence $R_f \leqslant R_g \leqslant v$. It then follows that $R_f = R_g$ and therefore $\widehat{R}_f = \widehat{R}_g$. Since g is finely hypoharmonic on Z, it follows from the first case above applied to g, that $\widehat{R}_f$ is finely harmonic on Z. By the sheaf property of finely harmonic functions, cf. Remark 2.1.5, it follows that indeed f is finely harmonic on U. □

The most frequent application of this theorem is the following.

Corollary 2.1.30 *For any function $f : \Omega \to [0, +\infty]$ the function $\widehat{R}_f$ is finely harmonic on any finely open set $U \subset \{f = 0\} \cap \{\widehat{R}_f < \infty\}$.*

Theorem 2.1.31 (Directed nets of finely harmonic functions) *The pointwise limit f of an upper or lower directed net of finely harmonic functions on a finely open set $U \subset \Omega$ is finely harmonic on the finely open set $\{|f| < +\infty\}$.*

Proof Consider, for example, an *upper* directed net $(f_i)_{i\in I}$. By Corollary 2.1.21, $f = \sup_i f_i$ is finely hyperharmonic on U, along with each f_i. By Theorem 2.1.26 f is therefore finely continuous, and hence the set $U_1 := \{|f| < +\infty\} = \{f < +\infty\}$ is finely open.

By the quasi-Lindelöf property for functions (see text following Corollary 1.2.5) there is a countable subfamily $(f_{i_n})_{n\in\mathbb{N}}$ of $(f_i)_{i\in I}$ such that $\sup_n f_{i_n} = \sup f_i$ quasi-everywhere, and indeed everywhere because the two suprema are finely hyperharmonic on U by Corollary 2.1.21, and hence finely continuous by Theorem 2.1.26, and finally because a polar set has empty fine interior (see Sect. 1.2). Clearly, we may achieve that the sequence (f_{i_n}) becomes increasing, by replacing i_n by $j_n \in I$ defined recursively so that $f_{j_{n+1}} \geqslant f_{i_1}, \dots, f_{i_n}, f_{j_n}$, the index set I being upper directed.

Every point $x_0 \in U_1$ has a finely open fine neighborhood V of compact closure $\overline{V} \subset U_1$ on which f and f_{i_1} are bounded. It follows by Corollary 2.1.16 and Remark 2.1.18 applied to the finely harmonic and hence finely continuous (and in particular Borel measurable according to Theorem 1.4.36) functions $\pm f_{i_n}$, that $f_{i_n} = f_{i_n}^{\complement V}$ on V and hence $f = f^{\complement V}$ by the dominated convergence theorem. Consequently, the finely continuous function f is finely harmonic on U_1. □

Remark 2.1.32 The above proof shows, moreover, that every upper directed net of finely *hyperharmonic* functions on U admits a countable upper directed subnet with the same pointwise supremum (which is finely hyperharmonic).

From Theorem 2.1.31 we deduce the former assertion in the following corollary in the same way as in [CC3, Theorem 1.1.1] for the Euclidean topology:

Corollary 2.1.33 *The fine topology on Ω is locally connected, and the set of fine components of any finely open set $U \subset \Omega$ is countable.*

Proof Given a finely open set $U \subset \Omega$ and a point $x_0 \in U$, denote by $\mathfrak{V}$ the family of all sets $V \subset U$ containing x_0 such that V is both finely open and finely closed relative to U. Then $U \in \mathfrak{V}$, and $\mathfrak{V}$ is stable under finite unions; in particular, $\mathfrak{V}$ is upper directed. For each $V \in \mathfrak{V}$ the function h_V which equals 0 on V and 1 on $U \setminus V$ is finely harmonic on U by the sheaf property of finely harmonic functions, see Remark 2.1.5. By Theorem 2.1.31 applied to the net $(h_V)_{V \in \mathfrak{V}}$, the limit function $h_0 = \sup_{V \in \mathfrak{V}} h_V : U \to \{0, 1\}$ is finely harmonic, in particular finely continuous by Theorem 2.1.26. The subset $U_0 := \{h_0 < 1\} = \{h_0 = 0\}$ is therefore both finely open and finely closed, and contains x_0 since $h_0(x_0) = h_V(x_0) = 0$. Clearly, $\{h_0 = 0\} = \bigcap_{V \in \mathfrak{V}} V$, the smallest set of class $\mathfrak{V}$. It follows that U_0 is finely connected, for otherwise there would exist a proper subset W of U_0 containing x_0 and both finely open and finely closed, that is, $W \in \mathfrak{V}$, contradicting the minimality of $\{h_0 = 0\}$.

For the latter assertion of the corollary, there is by the quasi-Lindelöf property a countable family of fine components of U such that the union of the remaining fine components of U is polar and finely open, hence void. □

We proceed to extend the fundamental convergence theorem (see Sect. 1.1.13) to the present setting.

Theorem 2.1.34 (The "fine" convergence theorem) *Let $\mathcal{G}$ be a family of finely hyperharmonic functions $U \to \overline{\mathbb{R}}$ (U finely open in Ω), and let $u = \inf \mathcal{G}$ (pointwise infimum). If u is finely locally bounded from below (equivalently: if $\widehat{u} > -\infty$) then $\widehat{u}$ is finely hyperharmonic and $\widehat{u} = u$ q.e. on U, whence*

$$\widehat{u}(x) = \underset{y \to x,\, y \neq x}{\text{f-lim}}\, u(y).$$

Proof [2]Let $x \in U$. The family $\mathcal{G}$ being finely locally bounded, thus there is a finely open fine neighborhood U_x of x such that $U_x \subset U$ and $u \geqslant \alpha$ on U_x for some

[2] This theorem is [Fu8, Theorem 11.8], where however the brief indication of proof fails. The present proof extends right away to the axiomatic setting in [Fu8] because we may assume there that the constant functions are superharmonic.

$\alpha \in \mathbb{R}, \alpha < 0$. For each $n \in \mathbb{N}$ we have $u \wedge n = \inf_{w\in\mathcal{G}}(w \wedge n)$ and $\widehat{u \wedge n} = \widehat{u} \wedge n$. We propose to apply Theorem 2.1.25 with the finely open set U_x, $a = n \vee (-\alpha)$ and the given point x of U_x. Thus there exists a finely open fine neighborhood V_x of x in U_x and a locally bounded potential q_n on Ω, finely harmonic on V_x, such that for every $w \in \mathcal{G}$, there is a locally bounded potential p_w on Ω such that $w \wedge n = p_w - q_n$ in V_x, and hence $\inf_{w\in\mathcal{G}}(w \wedge n) = \inf_{w\in\mathcal{G}} p_w - q_n$ on V_x. By fine continuity of q_n it follows that

$$\widehat{u} \wedge n = \widehat{\inf_{w\in\mathcal{G}}}(w \wedge n) = \widehat{\inf_{w\in\mathcal{G}}} p_w - q_n \text{ on } V_x. \tag{2.1.14}$$

Denoting (just here) by $\widetilde{g}$ the Euclidean l.s.c. regularized of a function $g : \Omega \to \overline{\mathbb{R}}$, we have $\widetilde{g} \le \widehat{g} \le g$ because the fine topology is finer than the Euclidean topology on Ω. By the classical convergence theorem, see Sect. 1.1.3, the function $\widetilde{\inf}_{w\in\mathcal{G}} p_w$ is hyperharmonic and hence finely hyperharmonic on Ω according to Theorem 2.1.22(a). It follows that $\widetilde{\inf}_{w\in\mathcal{G}} p_w$ is finely hyperharmonic on V_x, and so is $\widetilde{\inf}_{w\in\mathcal{G}} p_w - q_n$ because q_n is finely harmonic on V_x. Again by the classical convergence theorem in Ω we have $\widetilde{\inf}_{w\in\mathcal{G}} p_w = \inf_{w\in\mathcal{G}} p_w$ q.e. in Ω, that is, in $\Omega\setminus F$ for some polar set $F \subset \Omega$, and hence by fine continuity of $\widetilde{\inf}_{w\in\mathcal{G}} p_w$ and because F has an empty fine interior (see Sect. 1.2, 10th paragraph after Corollary 1.2.5)

$$\begin{aligned}\widetilde{\inf}_{w\in\mathcal{G}} p_w(z) &= \text{f-}\lim_{y\to z} \widetilde{\inf}_{w\in\mathcal{G}} p_w(y) = \text{f-}\lim_{y\to z, y\notin F} \widetilde{\inf}_{w\in\mathcal{G}} p_w(y)\\ &= \text{f-}\lim_{y\to z, y\notin F} \inf_{w\in\mathcal{G}} p_w(y) \ge \widehat{\inf}_{w\in\mathcal{G}} p_w(z)\end{aligned}$$

for every $z \in \Omega$. Since $\widehat{\inf} p_w \ge \widetilde{\inf} p_w$ it then follows that $\widetilde{\inf}_{w\in\mathcal{G}} p_w = \widehat{\inf}_{w\in\mathcal{G}} p_w$ and thus by (2.1.14)

$$\widehat{u} \wedge n = \widehat{\inf}_{w\in\mathcal{G}} p_w - q_n = \widetilde{\inf}_{w\in\mathcal{G}} p_w - q_n$$

on V_x, so that $\widehat{u} \wedge n$ is finely hyperharmonic on V_x because $\widetilde{\inf}_{w\in\mathcal{G}} p_w - q_n$ is finely hyperharmonic on V_x as stated above. By varying x in U and using the sheaf property of finely hyperharmonic functions (see Remark 2.1.5) we infer that $\widehat{u} \wedge n$ is finely hyperharmonic on U and so is $\widehat{u}$ by letting $n \to +\infty$ while invoking Theorem 2.1.21. On the other hand, we have $\widetilde{\inf}_{w\in\mathcal{G}} p_w = \inf_{w\in\mathcal{G}} p_w$ q.e. in Ω as seen above, and hence

$$\widehat{u} \wedge n = \widehat{\inf_{w\in\mathcal{G}}} p_w - q_n = \widetilde{\inf_{w\in\mathcal{G}}} p_w - q_n = \inf_{w\in\mathcal{G}} p_w - q_n = \inf_{w\in\mathcal{G}} w \wedge n = u \wedge n \quad \text{q.e. on } V_x.$$

This means that there is a polar set $e_{n,x} \subset V_x$ such that $\widehat{u} \wedge n = u \wedge n$ in $V_x\setminus e_{n,x}$. By the quasi-Lindelöf property of the fine topology (see Sect. 1.2) there is a sequence $(x_j) \subset U$ and a polar set $e_n \subset U$ such that $U = \bigcup_j V_{x_j} \cup e_n$. We then have $\widehat{u} \wedge n = u \wedge n$ in $U\setminus E_n$, where E_n is the polar set $e_n \cup \bigcup_j e_{n,x_j}$. By taking supremum over all $n \in \mathbb{N}$, we have $\widehat{u} = u$ in $U\setminus \bigcup_n E_n$, whence $\widehat{u} = u$ q.e. on U, the set $\bigcup_n E_n$ being polar. Finally, the stated limit expression for $\widehat{u}(x)$ in the theorem follows in the same way as in Theorem 1.3.1 because $\widehat{u}$ is finely continuous and the polar exceptional set E is finely discrete. □

Theorem 2.1.35 *The pointwise limit u of a pointwise convergent sequence of finely harmonic functions u_n on U is finely harmonic provided that* $\sup_n |u_n|$ *is finely locally bounded on U.*

Proof In the relations

$$u = \lim_n u_n = \sup_n \inf_{p>n} u_p \geqslant v := \sup_n \widehat{\inf_{p>n}} u_p$$

the inequality sign may be replaced by equality quasi-everywhere on account of the preceding theorem. Note that v is finely hyperharmonic according to Theorem 2.1.34 and Corollary 2.1.21. Replacing u_n by $-u_n$, we similarly find for some finely hyperharmonic function w that $-u \geqslant w$ with equality q.e. Altogether, $v \leqslant u \leqslant -w$, with equality quasi-everywhere, and hence everywhere by fine continuity of v and w. Consequently, $u = v$ and $u = -w$ are both finely hyperharmonic on U, and u is therefore finely harmonic by Corollary 2.1.19. □

Remark 2.1.36 According to Remark 2.1.32, the proof of the above applies and the theorem is still true if we replace the sequence (u_n) by a net $(u_\alpha)_{\alpha\in A}$ of harmonic functions such that $\sup_{\alpha\in A} |u_\alpha|$ is locally bounded and $\lim u_n$ by $\lim u_\alpha$ along the filter of sections of A.

Remark 2.1.37 Unlike the situation for harmonic functions, there exists a finely open set $U \subset \Omega$ and an increasing sequence of finely harmonic functions $f_n \geqslant 0$ on U such that the pointwise limit function $f : U \to]0, +\infty]$ is finite (and hence finely harmonic) quasi-everywhere on U, but not everywhere on U. In Example 2.1.8 let $B_n = \{x \in \mathbb{R}^3 : |x| < 1/n\}$, and denote by μ_n the measure $1_{\complement B_n}\mu$. Then $\mu_n(U) = 0$ because $\mu(U) = 0$. Take $f_n = G_\Omega\mu_n$. Then f_n is finely harmonic on U, by Corollary 2.1.7. But $\sup_n f_n = G_\Omega\mu = f$ is infinite at $x = 0$. Thus $\sup_n f_n = f$ is not finely harmonic on U, but only *invariant*, in the sense of Definition 2.9.1 below.

We shall need an extension of Lemma 2.1.13 to a finely l.s.c. function f, now with *finite* and hence semibounded potential p (see Definition 1.5.16).

Lemma 2.1.38 *Let p denote a finite potential on* Ω, *let* $B \subset \Omega$ *be a base, and let* $f : B \to]-\infty, +\infty]$ *be a finely l.s.c. function on B such that* $f \geqslant -p$ *on B. The function* $u = f^B$ *exists, defined for* $x \in \Omega$ *by*

$$u(x) = \int f\, d\varepsilon_x^B,$$

and is finely l.s.c. and $\geqslant -p$ *on* Ω, *finely hyperharmonic on* $\complement B$, *and finely harmonic on the finely open set* $\{u < +\infty\}\setminus B$. *Furthermore,* $u = u^A$ *for any set A such that* $B \subset A \subset \Omega$.

Proof According to Remark 1.5.25, f is ε_x^B-measurable for every $x \in B$. Since $\int p\, d\varepsilon_x^B = \widehat{R}_p^B(x) \leqslant p(x) < \infty$, $u(x)$ is well defined and $\geqslant -p$ by Definition 1.4.29

and subsequent text. The extension of f to Ω by $f = +\infty$ on $\complement B$ remains finely l.s.c. and $\geqslant -p$, and has unchanged f^B because ε_x^B is carried by B.

Fix $q \in \mathcal{P}^c$, $q > 0$. The fine topology being completely regular and hence uniformizable (Sect. 1.2), the finely l.s.c. function f is (see [Bou5, § 1, no. 6]) the pointwise supremum of the family Γ of all finely continuous (hence Borel measurable by Theorem 1.4.36) functions φ on Ω such that $-p \leqslant \varphi \leqslant f$, and that φ/q is bounded from above (not necessary here, but used in the next paragraph). Since ε_x^B does not charge the polar sets when $x \in \complement B$ (Corollary 1.4.33), it follows easily by the quasi-Lindelöf property for functions (see Sect. 1.2) and by Remark 1.5.25 that

$$u = f^B = \sup\{\varphi^B : \varphi \in \Gamma\}.$$

(Note that this holds trivially on B because $f^B = f$ and $\varphi^B = \varphi$ there.)

Now Lemma 2.1.13 applies to each $\varphi \in \Gamma$, and the asserted properties of $u = f^B$ follow on account of Corollary 2.1.21 and Theorem 2.1.31. In fact, as above (now with f, B replaced by u, A),

$$u^A = \sup\{(\varphi^B)^A : \varphi \in \Gamma\} = \sup\{\varphi^B : \varphi \in \Gamma\} = u$$

for any set A with $B \subset A \subset \Omega$. □

Using Lemma 2.1.38 (but avoiding the hypothesis $f \geqslant -p$) we shall establish the following theorem which in turn is crucial for treating the generalized fine Dirichlet problem (Sect. 2.3).

Theorem 2.1.39 *Let $B \subset \Omega$ be a base, and consider a numerical function $f : B \to \overline{\mathbb{R}}$. The function $u = f^{*B}$, resp. $u = f_*^B$, defined for $x \in \Omega$ by*

$$u(x) = \int^* f \, d\varepsilon_x^B, \quad \textit{resp.} \quad u(x) = \int_* f \, d\varepsilon_x^B,$$

is finely hyperharmonic on the finely open set $U_- := \{u > -\infty\}\setminus B$, finely hypoharmonic on the finely open set $U_+ := \{u < +\infty\}\setminus B$, and hence finely harmonic on the finely open set $U_+ \cap U_-$ of all points of $\complement B$ at which u is finite. Furthermore, $u = u^{*\complement V}$, resp. $u = u_*^{\complement V}$, holds everywhere on $\complement B$ for any finely open set V such that $\widetilde{V} \subset \complement B$.

Proof Again it suffices to consider the case involving the upper integral, replacing otherwise f by $-f$. Clearly u is well defined on Ω because ε_x^B is carried by the base B, by Theorem 1.4.23(a).

Fix a finite potential $p > 0$ on Ω. Consider first the case $0 \leqslant f \leqslant p$, whereby

$$u(x) = \int^* f \, d\varepsilon_x^B = \inf\{g^B(x) : g \text{ l.s.c.}, (0 \leqslant) f \leqslant g \leqslant p\}$$

for every $x \in \Omega$ (see Notations and Terminology for the latter equality, noting that g is ε_x^B-integrable along with p). Then $0 \leqslant u \leqslant g^B \leqslant p^B \leqslant p < +\infty$ on Ω, and it follows by Lemma 2.1.38, applied with f replaced by g, that each g^B is finely harmonic on $\complement B$. And so is u according to Theorem 2.1.31, the family $\{g^B : g \text{ l.s.c.}, f \leqslant g \leqslant p\}$ being finitely inf-stable and hence lower directed.

For general function $f \geqslant 0$ on B, choose $p \in \mathcal{P}^c$, $p > 0$, and define

$$f_n(x) = f(x) \wedge (np(x)), \quad u_n(x) = \int^* f_n \, d\varepsilon_x^B$$

for $x \in \Omega$ and $n \in \mathbb{N}$. As just shown (with p replaced by np) each u_n is finely harmonic on $\complement B$. We have $u = \sup_n u_n$ by the monotone convergence theorem. By Corollary 2.1.21 and Theorem 2.1.31, u therefore is finely hyperharmonic (in particular finely continuous) on $\complement B$, and finely harmonic on the finely open set $\{u < +\infty\} \backslash B$. From $0 \leqslant f_n \leqslant np$ follows $0 \leqslant u_n \leqslant np$, and we may therefore apply Corollary 2.1.15 with U replaced by any finely open set V such that $\widetilde{V} \subset \complement B$ and with f replaced by u_n. Thus $u_n = u_n^{*\complement V}$ on V, and hence $u = u^{*\complement V}$ on V for every V as stated, by letting $n \to \infty$.

Next consider the case $f \leq 0$. Here (with $\mathcal{H}_0$ denoting the class of u.s.c., finite valued and compactly supported functions $\Omega \to [0, +\infty]$)

$$-u(x) = \int_* (-f) d\varepsilon_x^B = \sup\{h^B(x) : h \in \mathcal{H}_0, \ h \leq -f\} \tag{2.1.15}$$

for every $x \in \Omega$. Any $h \in \mathcal{H}_0$ is majorized by a potential $q \in \mathcal{P}^b (\subset \mathcal{P}^s)$, see Definition 1.5.16(ii), and hence each h^B above is $\leq q$ in Ω, and finely harmonic in $\complement B$ by the preceding lemma applied to $-h$. Consequently u is finely hypoharmonic in $\complement B$ and finely harmonic in $(\complement B) \cap \{u > -\infty\}$ according to Theorem 2.1.31. The relation $u = u^{\complement V}$ in V for V as stated reduces to similar relation for h^B (valid by Corollary 2.1.15) by application of the quasi-Lindelöf principle to (2.1.15), noting that $-u$ is l.s.c. in $\complement B$, and that each h^B is even finely continuous there.

Finally, the case of quite arbitrary function f reduces to the cases just considered because

$$u(x) = \int^* f d\varepsilon_x^B = \int^* f^+ d\varepsilon_x^B + \int^* (-f^-) d\varepsilon_x^B$$

in $U^+ \cup U^-$, viz. for $x \in \complement B$ where this latter difference is not of the form $(+\infty) - (\infty)$. With $\leq$ in place of equality, this follows from [Bou6, Chap. IV, § 4, exerc. 5 b)]. As to the opposite inequality, we may assume that there exist l.s.c. and ε_x^B-integrable functions $g \geq f$. For any such g,

$$\int g^+ d\varepsilon_x^B + \int (-g^-) d\varepsilon_x^B \geq \int^* f^+ d\varepsilon_x^B + \int^* (-f^-) d\varepsilon_x^B,$$

whence the result by taking the infimum with respect to g. □

Corollary 2.1.40 *Let again $B \subset \Omega$ be a base and $f : B \to \overline{\mathbb{R}}$. If $f \geqslant -p$, resp. $f \leqslant p$, for some finite potential p on Ω then, for any $y \in B$,*

$$\text{f-}\liminf_{x\to y,\, x\in\Omega} \int_* f\, d\varepsilon_x^B = \text{f-}\liminf_{x\to y,\, x\in B} f(x), \tag{2.1.16}$$

resp.

$$\text{f-}\limsup_{x\to y,\, x\in\Omega} \int^* f\, d\varepsilon_x^B = \text{f-}\limsup_{x\to y,\, x\in B} f(x).$$

In particular, if $|f| \leqslant p$ then each of the extensions $x \mapsto \int_* f\, d\varepsilon_x^B$ and $x \mapsto \int^* f\, d\varepsilon_x^B$ of f from B to Ω is finely continuous at any point $y \in B$ at which f is finely continuous (relative to B).

For (2.1.16) the inequality $\leqslant$ is obvious. To establish the inequality $\geqslant$, let $\widehat{f} : B \to \overline{\mathbb{R}}$ denote the greatest finely l.s.c. minorant of f on B. Then $-p \leqslant \widehat{f} \leqslant f$ (on B), and

$$\int \widehat{f}\, d\varepsilon_x^B \leqslant \int_* f\, d\varepsilon_x^B, \text{ for every } x \in \Omega,$$

cf. Remark 1.5.25, whence the result by Lemma 2.1.38, according to which $\int \widehat{f}\, d\varepsilon_x^B$ is finely l.s.c.$\geqslant -p$ as a function of $x \in \Omega$, for it takes at every $y \in B$ the value

$$\widehat{f}(y) = \text{f-}\liminf_{x\to y,\, x\in B} f(x).$$

Proposition 2.1.41 *Let Ω_1 and Ω_2 be intersecting Greenian domains in $\mathbb{R}^N$, $N \geqslant 2$, and let V be a regular finely open set with fine closure contained in $\Omega_1 \cap \Omega_2$. For any $x \in V$ we then have ${}^{\Omega_1}\varepsilon_x^{\Omega_1\setminus V} = {}^{\Omega_2}\varepsilon_x^{\Omega_2\setminus V}$.*

Here the left superscript $\Omega_i, i = 1, 2$, indicates that Ω_i takes the place of Ω when defining the sweeping of ε_x on $\Omega_i \setminus V$.

Proof Let $x \in V$, let W be the fine component of V containing x and let p be a finite and continuous potential on Ω_1. Then we have ${}^{\Omega_1}\widehat{R}_p^{\Omega_1\setminus V}(x) = {}^{\Omega_1}\widehat{R}_p^{\Omega_1\setminus W}(x)$ because the functions ${}^{\Omega_1}\widehat{R}_p^{\Omega_1\setminus V}$ and ${}^{\Omega_1}\widehat{R}_p^{\Omega_1\setminus W}$ are equal on W. Indeed, we obviously have ${}^{\Omega_1}\widehat{R}_p^{\Omega_1\setminus V} \leq {}^{\Omega_1}\widehat{R}_p^{\Omega_1\setminus W}$. For the inverse inequality let s be a positive hyperharmonic on Ω_1 such that $s \geq p$ on $\Omega_1\setminus V$, then $s \geq p$ on $\partial_f W \subset \Omega_1\setminus V$ so that $s(x) \geq \int s d^{\Omega_1}\varepsilon_x^{\Omega_1\setminus W} \geq \int p d^{\Omega_1}\varepsilon_x^{\Omega_1\setminus W} = {}^{\Omega_1}\widehat{R}_p^{\Omega_1\setminus W}(x)$ for $x \in W$ by (1.4.22) and because ${}^{\Omega_1}\varepsilon_x^{\Omega_1\setminus W}$ is carried by $\partial_f W$ according to Corollary 1.4.33. Hence ${}^{\Omega_1}\widehat{R}_p^{\Omega_1\setminus V} \geq {}^{\Omega_1}\widehat{R}_p^{\Omega_1\setminus W}$ on W. It then follows by Corollary 1.4.6 that ${}^{\Omega_1}\varepsilon_x^{\Omega_1\setminus V} = {}^{\Omega_1}\varepsilon_x^{\Omega_1\setminus W}$. This allows us to suppose that V is finely connected. Let ω be the (Euclidean) component of $\Omega_1 \cap \Omega_2$ containing V (or, equivalently, $\widetilde{V}$). We have for every $x \in V$

$$\int p d^{\Omega_1}\varepsilon_x^{\Omega_1\setminus V} = {}^{\Omega_1}\widehat{R}_p^{\Omega_1\setminus V}(x)$$

and

$$\int p d^{\omega}\varepsilon_x^{\omega\setminus V} = {}^{\omega}\widehat{R}_p^{\omega\setminus V}(x),$$

and hence $\int pd^{\Omega_1}\varepsilon_x^{\Omega_1\setminus V} = \int pd^{\omega}\varepsilon_x^{\omega\setminus V}$. Indeed, since the functions ${}^{\Omega_1}\widehat{R}_p^{\Omega_1\setminus V}$ and ${}^{\omega}\widehat{R}_p^{\omega\setminus V}$ are finely harmonic on V by Lemma 2.1.38, and equal on $\partial_f V$, then they are equal on V according to Theorem 2.1.10. It therefore follows by Corollary 1.4.6 that ${}^{\Omega_1}\varepsilon_x^{\Omega_1\setminus V} = {}^{\omega}\varepsilon_x^{\omega\setminus V}$. The same holds with Ω_2 instead of Ω_1, so that ${}^{\Omega_1}\varepsilon_x^{\Omega_1\setminus V} = {}^{\Omega_2}\varepsilon_x^{\Omega_2\setminus V}$ as stated. □

Remark 2.1.42 According to Proposition 2.1.41 and Remark 2.1.18 the concepts of finely hyperharmonic, finely hypoharmonic, and finely harmonic functions on U are *independent of the choice of the Greenian domain* Ω containing U. Also recall Definition 1.4.28 according to which $\varepsilon_x^{\complement V}$ serves as the *(fine) harmonic measure* relative to V and x, again independently of Ω.

Theorem 2.1.43 (Removable singularities for finely hyperharmonic functions) *Let $U \subset \Omega$ be a finely open set and let $E \subset U$ be polar, hence finely closed. Then a finely hyperharmonic function $u : U\setminus E \to \,]-\infty, +\infty]$ admits a finely hyperharmonic extension to U if (and only if) u is bounded from below on some deleted fine neighborhood of each point of the finely discrete set E. The extension is then unique and given by*

$$u(x) = \operatorname*{f\text{-}lim\,inf}_{y\to x,\; y\in U\setminus E} u(y), \quad x \in E. \tag{2.1.17}$$

Proof Being finely discrete, a polar set E has an empty fine interior (since the points of E are finely isolated, see Sect. 1.2). Thus $U\setminus E$ is finely dense in U. The extension is therefore unique in view of Theorem 2.1.26 and determined as stated in the theorem. Suppose that the stated (obviously necessary) condition for the existence is fulfilled. Then the extension, given by (2.1.17), is $> -\infty$ on U.

Let V be a regular finely open set such that $\widetilde{V} \subset U$ and that $u \geqslant -p$ on V for some finite continuous potential p on Ω (p may depend on V), for example, $V \in \mathfrak{V}_U(u_-)$ as in Remark 2.1.18 ($\mathfrak{V}_U(u_-)$ is a basis for the fine topology on U, see Remark 2.1.18). Since u is finely hyperharmonic and $u \geqslant -p$ on $V\setminus E$, and since u is finely l.s.c. on $\widetilde{(V\setminus E)} = \widetilde{V} \subset U$ it follows from Theorem 2.1.14 that

$$u(x) \geqslant \int^* u\, d\varepsilon_x^{\complement(V\setminus E)} = \int^* u\, d\varepsilon_x^{\complement V} \quad \text{for } x \in V\setminus E \tag{2.1.18}$$

because $\varepsilon_x^{\complement(V\setminus E)} = \varepsilon_x^{\complement V}$ by polarity of E. Let $u : \Omega \to \,]-\infty, +\infty]$ denote the extension of u to Ω by $+\infty$ off $\partial_f V$. Then u is finely l.s.c. and $\geqslant -p$ and hence $\varepsilon_x^{\complement V}$-measurable (Remark 1.5.25). By Lemma 2.1.38 applied to the base $B = \complement V$ the function $u^{\complement V}$ is therefore well defined (Definition 1.4.29), finely l.s.c., and $\geqslant -p$ on Ω; and $u^{\complement V}$ is finely hyperharmonic on V.

Having thus established that $u^{\complement V}$ is finely l.s.c. on V, in particular at every point of $V \cap E$, we find that the inequality $u \geqslant u^{\complement V}$ on $V\setminus E$, see (2.1.18), extends to all of V. For any $x \in V \cap E$ we in fact have by (2.1.17)

$$u(x) = \text{f-}\liminf_{y\to x,\, y\in U\setminus E} u(y) = \text{f-}\liminf_{y\to x,\, y\in V\setminus E} u(y)$$
$$\geqslant \text{f-}\liminf_{y\to x,\, y\in V\setminus E} u^{\complement V}(y) \geqslant u^{\complement V}(x),$$

noting for the second equality that $V\setminus E$ is a deleted finely open fine neighborhood of x, and that $V\setminus E \subset U\setminus E$. The inequality $u \geqslant u^{\complement V}$ on V thus obtained for every finely open subset V with $\widetilde{V} \subset U$ as described above shows that u indeed is finely hyperharmonic on U (Definition 2.1.1). □

Remark 2.1.44 In view of Theorem 2.1.22(a), the above theorem contains as a particular case the following well-known result: If u is hyperharmonic and lower bounded on a Euclidean open set ω then u has a fine limit at every irregular boundary point x for ω. (Just apply Theorem 2.1.43 to the finely open set $U = \omega \cup \{x\}$.)

Theorem 2.1.45 (Removable singularities for finely harmonic functions) *Let $U \subset \Omega$ be finely open, and let $E \subset U$ be polar. Then a finely harmonic function $u : U\setminus E \to \mathbb{R}$ admits a finely harmonic extension to U if (and only if) u is bounded on some deleted fine neighborhood of each point of E. The extension is then unique, and given by*

$$u(x) = \text{f-}\liminf_{y\to x,\, y\in U\setminus E} u(y), \quad x \in E.$$

Proof As in the proof of Theorem 2.1.43, the extension is unique and determined as stated in the present theorem. Suppose that the stated (obviously necessary) condition for the existence of the extension is fulfilled. By Theorem 2.1.43, u extends to a finely hyperharmonic function u_1 on U, and also to a finely hypoharmonic function u_2 on U. These two extensions are finely continuous, by Theorem 2.1.26, and agree with u on the finely dense subset $U\setminus E$ of U, and so $u_1 = u_2$ on U, thus yielding the desired finely harmonic extension. □

The following corollary to Theorem 2.1.43 was established more directly in [Fu7].

Corollary 2.1.46 *Let again $U \subset \Omega$ be finely open and let $E \subset U$ be polar. The finely open set $U\setminus E$ is then finely connected if and only if U is finely connected.*

Proof If $U\setminus E$ is finely connected then so is U because $U\setminus E$ is finely dense in U (as noted in the beginning of the proof of Theorem 2.1.43). Conversely, suppose that U is finely connected. Let V_0 and V_1 be two disjoint finely open sets with the union $U\setminus E$. The function u defined on $U\setminus E$ by $u = 0$ on V_0 and $u = 1$ on V_1 is finely harmonic and bounded. By Theorem 2.1.45 u admits a finely harmonic and hence finely continuous extension to all of U. Since U is finely connected, it follows that $u \equiv 0$ or $u \equiv 1$ on U, whence either V_0 or V_1 is void. □

Lemma 2.1.47 *For any set $A \subset \Omega$ we have*

$$\mathfrak{b}(\partial_f A) = \mathfrak{b}(A) \cap \mathfrak{b}(\complement A). \tag{2.1.19}$$

Proof Because $\partial_f A \subset \widetilde{A}$ it follows from (1.4.15) that $\mathfrak{b}(\partial_f A) \subset \mathfrak{b}(\widetilde{A}) = \mathfrak{b}(A)$. Since $\partial_f A = \partial_f(\complement A)$ this yields $\mathfrak{b}(\partial_f A) \subset \mathfrak{b}(\complement A)$, and hence

$$\mathfrak{b}(\partial_f A) \subset \mathfrak{b}(A) \cap \mathfrak{b}(\complement A).$$

For the opposite inclusion, let $x \in \complement\mathfrak{b}(\partial_f A)$, that is, $\partial_f A$ is thin at x. Thus x is a finely isolated point of $\partial_f A$. The fine topology on Ω being locally connected (Corollary 2.1.33), there exists a fine domain $W \subset \Omega$ such that $W \cap \partial_f A = \{x\}$ and hence $(W \backslash \{x\}) \cap \partial_f A = \varnothing$. Since $\{x\}$ is polar, $W \backslash \{x\}$ is finely connected by Corollary 2.1.46, and hence either $W \backslash \{x\} \subset A$ or $W \backslash \{x\} \subset \complement A$. Suppose for a moment that $x \in A$. Then $W \backslash \{x\} \subset \complement A$, the alternative $W \backslash \{x\} \subset A$ being ruled out since it would imply that $W \subset A$, hence $W \cap \complement A = \varnothing$, in contradiction with $x \in \partial_f A$. But $W \backslash \{x\} \subset \complement A$ shows that $\complement A$ is a deleted fine neighborhood of x, which by Theorem 1.2.4 implies that A is thin at x, that is, $x \notin \mathfrak{b}(A)$. In the present case $x \in A$ this proves the stated opposite inclusion, and hence the whole lemma. If instead $x \in \complement A$, then we have therefore established (2.1.19) with A replaced by $\complement A$, and this is clearly equivalent to (2.1.19) as it stands in the lemma. □

Corollary 2.1.48 *A finely open set U is regular, that is, $\complement U$ is a base, if and only if $\partial_f U$ is a base.*

Proof In fact, the finely open set U is not thin at any of its points, so $U \subset \mathfrak{b}(U)$. By (1.4.13) we therefore have $\widetilde{U} = U \cup \mathfrak{b}(U) = \mathfrak{b}(U)$. Thus $\mathfrak{b}(\partial_f U) = \mathfrak{b}(U) \cap \mathfrak{b}(\complement U) = \widetilde{U} \backslash \mathfrak{r}(U)$, which equals $\partial_f U = \widetilde{U} \backslash U$ if and only if $\mathfrak{r}(U) = U$. □

Definition 2.1.49 The regular part $\partial_{f,r} U$ and the irregular part $\partial_{f,i} U$ of the fine boundary $\partial_f U$ of a finely open set $U \subset \Omega$ are defined by

$$\partial_{f,r} U = (\partial_f U) \cap \mathfrak{b}(\complement U) = (\partial_f U) \backslash \mathfrak{r}(U), \tag{2.1.20}$$

$$\partial_{f,i} U = (\partial_f U) \backslash \mathfrak{b}(\complement U) = (\partial_f U) \cap \mathfrak{r}(U), \tag{2.1.21}$$

where $\mathfrak{r}(U) = \complement\mathfrak{b}(\complement U)$ is the regularization of U (the least regular finely open set containing U).

Corollary 2.1.50 *Let a numerical function u be finely hyperharmonic on a regular finely open set $U \subset \Omega$. Then the finely open set $V := \{x \in U : u(x) > 0\}$ is regular.*

For if $x \in \Omega$ is a finely isolated point of $\complement V$ then $x \in U$ since otherwise x would be finely isolated in $\complement U$, contradicting the regularity of U. Thus there exists according to Corollary 2.1.33 a fine domain $W \subset U$ such that $x \in W$ and $W \backslash \{x\} \subset V$, hence $u > 0$ on $W \backslash \{x\}$, and so by fine continuity $u(x) \geqslant 0$, and in fact $u(x) = 0$ because $x \notin V$. This would contradict Corollary 2.1.27.

The following lemma (a "fine" version of Subsection 1.1.7 of Sect. 1.1) is often useful.

Lemma 2.1.51 *Let u, resp. v, be finely hyperharmonic functions on a finely open set U, resp. V, where $V \subset U$. Suppose that*

$$\text{f-}\liminf_{x \to y,\, x \in V} v(x) \geqslant u(y) \quad \textit{for } y \in U \cap \partial_f V. \tag{2.1.22}$$

Then the following function w on U is finely hyperharmonic:

$$w(x) = \begin{cases} u(x) \wedge v(x) & \textit{for } x \in V \\ u(x) & \textit{for } x \in U \setminus V. \end{cases}$$

Proof For every integer $n \geq 1$ let w_n be the function w defined as in the lemma with u replaced by the finely hyperharmonic function $u_n := u \wedge n - \frac{1}{n}$ on U. It is clear that w_n is finely hyperharmonic on V in view of Corollary 2.1.21. Furthermore, we have

$$\text{f-}\liminf_{V \ni x \to y} v(x) \geq u(y) > u_n(y)$$

for every point $y \in U \cap \partial_f V$, so that w_n equals u_n in a finely open fine neighborhood of $U \setminus V$ and hence it is finely hyperharmonic there. It then follows by the sheaf property (see Remark 2.1.5) that w_n indeed is finely hyperharmonic on all of U. Since the sequence (w_n) is increasing and $\sup_n w_n = w$, we deduce that w is finely hyperharmonic on U according to Corollary 2.1.21. □

Theorem 2.1.52 *Let u be a finely hyperharmonic function on a finely open set $U \subset \Omega$. Let V denote a regular finely open set such that $\widetilde{V} \subset U$. Suppose that $u \geqslant -p$ on V for some finite potential p on Ω. Then $u^{\complement V}$ (considered on U) is the smallest function which is finely hyperharmonic on U, majorizes u on $U \setminus V$, and majorizes $-p$ on V. Furthermore, $u^{\complement V}$ is finely harmonic on $V \cap \{u^{\complement V} < +\infty\}$ (and $u^{\complement V} \leqslant u$ on U, $u^{\complement V} = u$ on $U \setminus V$).*

Proof From Lemma 2.1.38 applied to u we see that $u^{\complement V}$ exists and is $> -\infty$ by finiteness of p. According to Theorem 2.1.10 we have $u^{\complement V} \leqslant u$ on V. By Definition 1.4.29 and (1.4.12) we have $u^{\complement V} = u$ on $U \cap \mathfrak{b}(\complement V) = U \setminus V$. Next apply Lemma 2.1.38 instead to the base $B = \complement V$ and to the finely l.s.c. function f defined on Ω by

$$f = \begin{cases} u & \text{on } \widetilde{V} \\ +\infty & \text{on } \complement\widetilde{V}. \end{cases}$$

Because $u \geq -p$ on V (hence on $\widetilde{V}$), we have $f \geq -p$ on Ω, and consequently $f^{\complement V}$ exists and is finely l.s.c. and $\geq -p$ there, finely hyperharmonic on V, and finely harmonic on $V \cap \{f^{\complement V} < +\infty\}$.

Now $u^{\complement V} = f^{\complement V}$ on $\widetilde{V}$. For $x \in V$ this is because $\varepsilon_x^{\complement V}$ then is carried by $\partial_f V \subset \widetilde{V}$ (Corollary 1.4.33), and for $x \in \partial_f V \subset \complement V$ we have $\varepsilon_x^{\complement V} = \varepsilon_x$. Thus $u^{\complement V}$ is finely l.s.c. relative to $\widetilde{V}$ and $\geqslant -p > -\infty$ on $\widetilde{V}$, finely hyperharmonic on V, and finely harmonic on $V \cap \{u^{\complement V} < +\infty\}$. Since $\varepsilon_x^{U \setminus V} = \varepsilon_x$ for $x \in U \setminus V$ we have $u^{\complement V} = u$ on

$U \backslash V$, and we conclude that $u^{\complement V}$ indeed is finely hyperharmonic on the whole of U, by the above lemma applied to the functions u (on U) and $v := u^{\complement V} = u \wedge v$ (on V).

For any finely hyperharmonic function w on U such that $w \geqslant u$ $(= u^{\complement V})$ on $U \backslash V$ and $w \geq -p$ on V, we similarly find from Theorem 2.1.14 that $w \geqslant w^{\complement V} \geqslant u^{\complement V}$ on V and hence $w \geqslant u^{\complement V}$ on all of U. This proves the stated minimality of $u^{\complement V}$. □

Theorem 2.1.53 *Let u be finely hyperharmonic on a finely open set $U \subset \Omega$, and suppose that u admits a finely hypoharmonic minorant on U. The pointwise supremum (within $\overline{\mathbb{R}}$) of all finely hypoharmonic minorants of u on U is then finely harmonic on the finely open set $\{x \in U : u(x) < +\infty\}$. Under the further hypothesis that $u < +\infty$ on U it follows that u has a greatest finely hypoharmonic minorant h on U and that h is finely harmonic on U.*

Proof Fix a regular finely open set V of class $\mathfrak{V}_U(u)$ relative to U (see Remark 2.1.18). For any finite valued finely hypoharmonic function $v \leqslant u$ (on U) it follows from Theorem 2.1.14 that (since $V \in \mathfrak{V}_U(v_+)$)

$$v \leqslant v^{\complement V} \leqslant u^{\complement V} \leqslant u \quad \text{on } V.$$

The inequality $v^{\complement V} \leqslant u$ holds on all of U because $v^{\complement V} = v$ on $U \backslash V$. According to Theorem 2.1.52 (applied to $-v$ in place of u) $v^{\complement V}$ is finely hypoharmonic (and $\geqslant v > -\infty$) on all of U, and finely harmonic on V because $u^{\complement V}$ is finite on V.

When v ranges over the class $\mathcal{M}$ of all finite valued finely hypoharmonic minorants of u on U, it therefore follows for every $x \in U$ with $u(x) < +\infty$ (hence $u < +\infty$ on some regular finely open fine neighborhood V of x with $\widetilde{V} \subset U$) that

$$h(x) := \sup_{v \in \mathcal{M}} v(x) = \sup_{v \in \mathcal{M}} v^{\complement V}(x)$$

since $v^{\complement V} \in \mathcal{M}$ when $v \in \mathcal{M}$. Clearly, $-\infty < h \leqslant u < +\infty$ on V because $V \in \mathfrak{V}_U(u)$. Invoking Theorem 2.1.31 we conclude that h is finely harmonic on V.

Since u is finite and finely continuous (hence finely locally bounded) on the finely open subset $\{u < +\infty\}$ of U, the sets V in question form a basis for the fine topology on $\{u < +\infty\}$, and consequently h is indeed finely harmonic on $\{u < +\infty\}$. □

2.2 Reduction and Sweeping of Functions Relative to a Finely Open Set

We continue the study of the convex cone $\mathcal{U}(U)$ of all finely hyperharmonic functions on a fixed finely open subset U of the Greenian domain Ω in $\mathbb{R}^N$ $(N \geqslant 2)$. It follows easily from Theorem 2.1.26 that the (induced) fine topology on U is the coarsest topology on U in which every function of class $\mathcal{U}(U)$ is continuous. Generalizing Definition 1.2.1 this implies that a subset A of U is thin at a point $y \in U$ if and only if either $y \notin \widetilde{A}$ or else there exists a finely hyperharmonic function u on a finely open set containing y such that

$$\operatorname{f-lim\,inf}_{A\setminus\{y\}\ni x\to y} u(x) > u(y).$$

Since Ω is Greenian it suffices here to consider functions $u \in \mathcal{U}^+(U)$.

For a given numerical function f on U, recall from Remark 2.1.18 the notation $\mathfrak{V}_U(f)$ for the family of all finely open sets V with compact Euclidean closure $\overline{V} \subset U$ such that f is bounded on $\overline{V}$. When f is locally bounded, $\mathfrak{V}_U(f)$ is a basis for the fine topology on U.

For any function $f : U \to \overline{\mathbb{R}}$ we denote by $\widetilde{f}$, resp. $\widehat{f}$, the least finely u.s.c. majorant, resp. the greatest finely l.s.c. minorant of f:

$$\widetilde{f}(x) = \operatorname{f-lim\,sup}_{y\to x} f(y), \quad \widehat{f}(x) = \operatorname{f-lim\,inf}_{y\to x} f(y).$$

For any non-void family $\mathcal{F}$ of numerical functions on U we accordingly denote by $\widehat{\inf}\mathcal{F}$ the greatest finely l.s.c. minorant of $\mathcal{F}$, that is, of the pointwise infimum $\inf \mathcal{F}$ of $\mathcal{F}$. We continue to understand complements of subsets of Ω (for example, subsets of U) and topological notions for subsets of Ω relative to Ω (not relative to U).

As a tool for our study of reduction and sweeping of finely hyperharmonic functions we first give a "fine" version of Brelot's concept of nearly hyperharmonic functions, [Br11], see also Bauer [Ba3].

Definition 2.2.1 A function $v : U \to \overline{\mathbb{R}}$ is called *nearly finely hyperharmonic* if v is finely locally bounded from below, and if

$$v(x) \geqslant \int^* v \, d\varepsilon_x^{\complement V} \quad \text{for every } x \in V \in \mathfrak{V}_U(v_-).$$

In view of Theorem 1.4.23(a) (applied to $A = \complement V$), and Equation (1.4.20) with subsequent text, it suffices to verify the conditions in Definition 2.2.1 for *regular* sets $V \in \mathfrak{V}_U(v_-)$.

Every finely hyperharmonic function on U is also nearly finely hyperharmonic according to Corollary 2.1.16.

Lemma 2.2.2 *The pointwise infimum of any finely locally uniformly lower bounded family $\mathcal{F}$ of nearly finely hyperharmonic functions on U is nearly finely hyperharmonic on U.*

Proof Clearly $u := \inf \mathcal{F}$ (pointwise) is finely locally lower bounded. For any $v \in \mathcal{F}$, any finely open set $V \in \mathfrak{V}_U(u_-)$, and any point $x \in V$ we have

$$u(x) = \inf_{v\in\mathcal{F}} v(x) \geqslant \inf_{v\in\mathcal{F}} \int^* v \, d\varepsilon_x^{\complement V} \geqslant \int^* \Big(\inf_{v\in\mathcal{F}} v\Big) \, d\varepsilon_x^{\complement V} = \int^* u \, d\varepsilon_x^{\complement V}.$$

This completes the proof that u is nearly finely hyperharmonic. $\square$

The following lemma is a "fine" version of a result by Brelot [Br8].

Lemma 2.2.3 *The greatest finely l.s.c. minorant $\widehat{v}$ of a nearly finely hyperharmonic function v on U is finely hyperharmonic on U. For every $x \in U$,*

$$\widehat{v}(x) = \sup\Big\{\int^* v\, d\varepsilon_x^{\complement V} : V \in \mathfrak{V}_U(v_-), x \in V\Big\}. \tag{2.2.23}$$

Proof Since v is nearly finely hyperharmonic we have $\widehat{v}(x) > -\infty$ for every $x \in U$, and from $\widehat{v} \leqslant v$ follows

$$\int^* \widehat{v}\, d\varepsilon_x^{\complement V} \leqslant \int^* v\, d\varepsilon_x^{\complement V} \leqslant v(x) \quad \text{for } x \in V \in \mathfrak{V}_U(v_-).$$

In the latter inequality, $v(x)$ may be replaced by $\widehat{v}(x)$ because the second integral is finely hyperharmonic, in particular finely l.s.c., as a function of $x \in V \in \mathfrak{V}_U(v_-)$ according to Theorem 2.1.39. From the resulting inequality concerning $\widehat{v}$ it follows by Definition 2.1.1 and Remark 2.1.18 that the finely l.s.c. function $\widehat{v}$ indeed is finely hyperharmonic.

By varying $V \in \mathfrak{V}_U(v_-)$ we have so far established (2.2.23) with the equality replaced by $\geqslant$. For given $x \in U$ let $\alpha < \widehat{v}(x)$. As noted above $\mathfrak{V}_U(v_-)$ is a basis for the fine topology on U. Since $\widehat{v}$ is finely l.s.c. there hence exists $V \in \mathfrak{V}_U(v_-)$ such that $x \in V$ and $\alpha < \widehat{v}(y)$ for every $y \in \overline{V}(\subset U)$. In view of Corollary 2.1.17

$$\alpha = \alpha \int d\varepsilon_x^{\complement V} \leqslant \int \widehat{v}\, d\varepsilon_x^{\complement V} \leqslant \int^* v\, d\varepsilon_x^{\complement V},$$

from which the remaining inequality $\leqslant$ in (2.2.23) follows. □

Lemma 2.2.4 *Let (v_n) denote a sequence of nearly finely hyperharmonic functions $\geqslant 0$ on U. Then*

(a) *$\sum_n v_n$ is a nearly finely hyperharmonic function on U and*

$$\widehat{\sum_n v_n} = \sum_n \widehat{v_n}.$$

(b) *If (v_n) is pointwise increasing then the pointwise supremum $\sup_n v_n$ is nearly finely hyperharmonic on U and*

$$\widehat{\sup_n v_n} = \sup_n \widehat{v_n}.$$

Proof (b) Clearly, $v := \sup_n v_n$ is nearly finely hyperharmonic. The stated formula for $\widehat{v}$ follows from (2.2.23) in the preceding lemma by interchanging the two suprema.

(a) In view of (b) it suffices to consider the case of a finite sequence, or just a pair, of nearly finely hyperharmonic functions v_1, v_2. Clearly, $v_1 + v_2$ is nearly finely hyperharmonic. The stated formula for $\widehat{v_1 + v_2}$ follows again by use of (2.2.23), the inequality $\widehat{v_1 + v_2} \geqslant \widehat{v_1} + \widehat{v_2}$ being obvious. □

Corollary 2.2.5 *For any sequence of finely hyperharmonic (or just nearly finely hyperharmonic) functions $u_n \geqslant 0$ on U, the function*

$$u := \liminf_n u_n$$

is nearly finely hyperharmonic on U, and its greatest finely l.s.c. minorant is

$$\widehat{u} = \sup_n \widehat{\inf_{p>n} u_p}.$$

The former assertion follows immediately from Lemma 2.2.2 and Lemma 2.2.4(b). The latter assertion follows by application of Lemma 2.2.4(b).

We denote by $\mathcal{U}^+(U) = \{u \in \mathcal{U}(U) : u \geqslant 0\}$ the positive cone in the cone $\mathcal{U}(U)$ of all finely hyperharmonic functions on a given finely open set $U \subset \Omega$.

If U is (Euclidean) open in $\mathbb{R}^N$ then it follows from Theorem 2.1.22(a) that $\mathcal{U}^+(U)$ coincides with the cone of all hyperharmonic functions $\geqslant 0$ on U. For $N = 2, \mathcal{U}(U)$ (with U open in $\mathbb{R}^N$) coincides with the cone of all hyperharmonic functions on U, see Theorem 2.7.2 below, but that does not extend to dimensions $N \geqslant 3$.

Definition 2.2.6 Let $A \subset U$, and let a function f with values in $[0, +\infty]$ be defined (at least) on U. The (fine) *reduced function* ${}^U R_f$ of f relative to U is defined for $x \in U$ by

$${}^U R_f(x) = \inf\{u(x) : u \in \mathcal{U}^+(U), u \geqslant f\}.$$

The (fine) *swept function* ${}^U \widehat{R}_f$ of f relative to U is defined on U as the greatest finely l.s.c. minorant of ${}^U R_f$:

$${}^U \widehat{R}_f(x) = \text{f-}\liminf_{y \to x,\, y \in U} {}^U R_f(y), \quad x \in U.$$

For $u \in \mathcal{U}^+(U)$ and $A \subset U$, define ${}^U R_u^A = {}^U R_{1_A u}$, ${}^U \widehat{R}_u^A = {}^U \widehat{R}_{1_A u}$.

Note that ${}^U R_f$ is nearly finely hyperharmonic on U, and hence ${}^U \widehat{R}_f$ is finely hyperharmonic, by Lemma 2.2.3. Clearly ${}^U R_{\widetilde{f}} = {}^U R_f$ and hence ${}^U \widehat{R}_{\widetilde{f}} = {}^U \widehat{R}_f$.

In the case $U = \Omega$, Definition 2.2.6 agrees with the usual definition of R_f and $\widehat{R}_f$ (see Sect. 1.3) on account of Theorems 1.3.1 and 2.1.22(a).

Lemma 2.2.7 *For any polar set $E \subset U$ we have ${}^U R_{+\infty}^E = 0$ on $U \setminus E$ and hence ${}^U \widehat{R}_{+\infty}^E = 0$.*

Proof For given $x \in U \setminus E$ there exists a hyperharmonic function $u \geqslant 0$ (even a potential) on Ω with $u = +\infty$ on E, but $u(x) < +\infty$ (see Sect. 1.1.11). By Theorem 2.1.22(a), u is finely hyperharmonic on Ω and hence on U. We have $\varepsilon u \geq +\infty 1_E$ and hence $\varepsilon u(x) \geq {}^U R_{+\infty}^E(x)$. By letting $\varepsilon \to 0$ we get ${}^U R_{+\infty}^E(x) = 0$, and by varying

x in $U\backslash E$ this yields ${}^U\widehat{R}^E_{+\infty} = {}^U R^E_{+\infty} = 0$ q.e. on U, and hence everywhere on U by fine continuity. □

Corollary 2.2.8 *Let $f_1, f_2 \in \mathcal{U}^+(U)$. If $f_1 \leqslant f_2$ q.e. on U then ${}^U\widehat{R}_{f_1} \leqslant {}^U\widehat{R}_{f_2}$ everywhere on U.*

In fact, with the polar set $E := \{f_1 > f_2\}$ it is easily seen that ${}^U R_{f_1} \leqslant {}^U R_{f_2} + {}^U R^E_{+\infty}$ on U. It follows by Lemma 2.2.4(a) that ${}^U\widehat{R}_{f_1} \leqslant {}^U\widehat{R}_{f_2} + {}^U\widehat{R}^E_{+\infty} = {}^U\widehat{R}_{f_2}$ according to Definition 2.2.6 and the above lemma.

Corollary 2.2.9 *For any function $f : U \to [0, +\infty]$,*

$$ {}^U\widehat{R}_f = {}^U R_f \quad \textit{q.e. on } U, \tag{2.2.24}$$

$$ {}^U\widehat{R}_f = \min\{u \in \mathcal{U}^+(U) : u \geqslant f \textit{ q.e. on } U\}. \tag{2.2.25}$$

The former equality follows from Theorem 2.1.34 applied to the family $\mathcal{F}$ of all $u \in \mathcal{U}^+(U)$ with $u \geqslant f$. For the latter equality note that by (2.2.24) ${}^U\widehat{R}_f$ competes for the minimum in (2.2.25), and for any competing u we have by Corollary 2.2.8

$$u = {}^U\widehat{R}_u \geqslant {}^U\widehat{R}_f.$$

The following Corollary is an easy consequence of Theorem 2.1.52 and (2.2.25).

Corollary 2.2.10 *Let $u \in \mathcal{U}^+(U)$. Then*

$$u^{\complement V} = {}^U\widehat{R}_u^{U\backslash V}$$

for any regular finely open set $V \subset \widetilde{V} \subset U$.

Theorem 2.2.11 *A set $A \subset U$ is thin at a point $x \in U\backslash A$ if and only if ${}^U R_1^{A\cap V}(x) < 1$ for some finely open set $V \subset U$ such that $x \in V$.*

Proof Suppose first that $x \in U\backslash A$ and that the stated condition is fulfilled. Thus there exists a finely open set $V \subset U$ containing x and a function $u \in \mathcal{U}^+(U)$ such that $u \geqslant 1$ on $A \cap V$ and $u(x) < 1$. It follows that $x \in \{u < 1\} \cap V \subset U\backslash A$ and hence A is thin at x because $\{u < 1\} \cap V$ is finely open by Theorem 2.1.26. Conversely, let A be thin at $x \in U\backslash A$. We may assume that $x \in \widetilde{A}$, for otherwise there is a finely open set $V \subset U$ containing x and not meeting A, whence ${}^U R_1^{A\cap V}(x) = 0$. Then there exists $u \in \mathcal{U}^+(U)$ with

$$u(x) < \operatorname{f-}\liminf_{y\to x,\, y\in A} u(y) = \sup_V \inf u(A \cap V),$$

where V ranges over all finely open subsets of U containing x. By replacing u by a suitable constant multiple of u one achieves that

$$u(x) < 1 < \sup_V \inf u(A \cap V).$$

Thus there exists V as stated such that $1 < \inf u(A \cap V)$ and hence ${}^U R_1^{A\cap V} \leqslant u$ and in particular ${}^U R_1^{A\cap V}(x) \leqslant u(x) < 1$. □

As a generalization of the identity $\widehat{R}_u^B = R_u^B$ when B is a base in Ω (see Theorem 1.4.21) we have in the present setting relative to U:

Lemma 2.2.12 *For any set $A \subset U$ and any $u \in \mathcal{U}^+(U)$ we have ${}^U R_u^{\widetilde{A}\cap U} = {}^U R_u^A$ and*

$$ {}^U\widehat{R}_u^A = {}^U\widehat{R}_u^{U\cap \mathfrak{b}(A)} = {}^U R_u^{U\cap \mathfrak{b}(A)} \quad on\ U. \tag{2.2.26}$$

Proof Copy word by word the proof of Theorem 1.4.21 while replacing throughout Ω (implicitly) by U, and hence $\mathcal{U}^+$ by $\mathcal{U}^+(U)$, and $\widetilde{A}$ by $\widetilde{A} \cap U$ (the fine closure of A relative to U), and finally $\mathfrak{b}(A)$ by $\mathfrak{b}(A) \cap U$ (the base of A relative to U), and by using Corollary 2.2.10 to $V = \complement\mathfrak{b}(A)$. □

Here is a “fine” version of Lemma 1.3.9(a):

Lemma 2.2.13 *Let $A \subset U$, $x \in U$, and $u \in \mathcal{U}^+(U)$ with $u \leqslant p$ for some finite (and hence semibounded) potential p on Ω. If A is unthin at x then ${}^U\widehat{R}_u^A(x) = u(x)$, and even ${}^U\widehat{R}_u^{A\cap V}(x) = u(x)$ for any fine neighborhood V of x in U.*

Proof Let W be a finely open fine neighborhood of x such that $\overline{W} \subset V$. When A is unthin at x, so is $A \cap W$ according to Theorem 2.2.11, and hence $x \in B := \mathfrak{b}(A \cap W)$, a base in Ω contained in $\widetilde{W} \subset U$. By Corollary 1.4.18, $(A \cap W)\backslash B$ is polar, and it follows by Lemma 2.2.12 and by Corollary 2.2.8, that

$$u(x) \leqslant {}^U R_u^B(x) = {}^U\widehat{R}_u^B(x) \leqslant {}^U\widehat{R}_u^{A\cap V}(x) \leqslant u(x),$$

and so indeed ${}^U\widehat{R}_u^{A\cap V}(x) = u(x)$. □

Generalizing Theorem 1.3.3 to the “fine” setting we obtain

Theorem 2.2.14 *For any function $f : U \to [0, +\infty]$ we have*
(a) ${}^U R_f(x) = f(x) \vee {}^U\widehat{R}_f(x)$ *for every* $x \in U$.
(b) *If f is finely l.s.c. on U then* ${}^U\widehat{R}_f \equiv {}^U R_f$ $(\geqslant f)$.

Proof (a) Since ${}^U R_f \geqslant f$ on U we may assume that $f(x) < +\infty$. For given $\varepsilon > 0$ there exists a function $u \in \mathcal{U}^+$ on Ω such that $u \equiv +\infty$ on the polar set

$$E = \{y \in \Omega\backslash\{x\} : {}^U\widehat{R}_f(y) < f(y)\},$$

and that $u(x) < \varepsilon$ (see Sect. 1.1.11). It follows that ${}^U\widehat{R}_f + u|_U \in \mathcal{U}^+(U)$ and

$${}^U\widehat{R}_f + u|_U \geqslant f \quad \text{on } U\backslash\{x\}.$$

The rest of the proof is quite analogous to the corresponding part of that of Theorem 1.3.3: If ${}^U\widehat{R}_f(x) \geqslant f(x)$ then ${}^U\widehat{R}_f + u|_U \geqslant f$ on all of U, and hence

$$ {}^U R_f(x) \leqslant {}^U\widehat{R}_f(x) + u(x) \leqslant {}^U\widehat{R}_f(x) + \varepsilon. $$

If instead ${}^U\widehat{R}_f(x) < f(x)$, then $c := f(x) - {}^U\widehat{R}_f(x) > 0$ and $c < +\infty$. We then obtain ${}^U\widehat{R}_f + u|_U + c \in \mathcal{U}^+(U)$ and ${}^U\widehat{R}_f + u|_U + c \geqslant f$ on U, whence

$$ {}^U R_f(x) \leqslant {}^U\widehat{R}_f(x) + u(x) + c = f(x) + u(x) \leqslant f(x) + \varepsilon. $$

In any case, ${}^U R_f(x) \leqslant (f(x) \vee {}^U\widehat{R}_f(x)) + \varepsilon$. By varying first $\varepsilon > 0$ and next $x \in U$ we conclude that ${}^U R_f \leqslant f \vee {}^U\widehat{R}_f \leqslant {}^U R_f$ (since $f, {}^U\widehat{R}_f \leqslant {}^U R_f$), and so indeed ${}^U R_f = f \vee {}^U\widehat{R}_f$.

(b) By (a), ${}^U R_f$ is finely l.s.c. along with f and ${}^U\widehat{R}_f$, and hence ${}^U\widehat{R}_f = {}^U R_f$ by definition of ${}^U\widehat{R}_f$. □

Theorem 2.2.15 *Let $f_n : U \to [0, +\infty]$ for $n \in \mathbb{N}$, and let $f : U \to [0, +\infty]$. Then*

(a) *If $f_n \uparrow f$ then ${}^U R_{f_n} \uparrow {}^U R_f$ and ${}^U\widehat{R}_{f_n} \uparrow {}^U\widehat{R}_f$.*

(b) *If $f := \sum_n f_n$ then we have ${}^U R_f \leqslant \sum_n {}^U R_{f_n}$ and ${}^U\widehat{R}_f \leqslant \sum_n {}^U\widehat{R}_{f_n}$.*

Proof (a) Generally, $f_n \uparrow f$ signifies that the sequence (f_n) is increasing and has pointwise limit f. Write ${}^U\widehat{R}_{f_n} = u_n$ and $\sup_n u_n = u$. Then $u_n \in \mathcal{U}^+(U)$ and (u_n) is increasing by Corollary 2.2.8 and hence $u \in \mathcal{U}^+(U)$ according to Corollary 2.1.21. Since $u \geqslant u_n \geqslant f_n$ q.e. by (2.2.24), we have $u \geqslant f$ q.e., and hence $u \geqslant {}^U\widehat{R}_f$ by Corollary 2.2.8. Obviously, ${}^U\widehat{R}_f \geqslant u$, and so ${}^U\widehat{R}_f = u$, that is, ${}^U\widehat{R}_{f_n} \uparrow {}^U\widehat{R}_f$. Finally, we have by Theorem 2.2.14(a)

$$ {}^U R_f = f \vee {}^U\widehat{R}_f = f \vee u = \sup_n (f_n \vee u_n) = \sup_n (f_n \vee {}^U\widehat{R}_{f_n}) = \sup_n {}^U R_{f_n}. $$

(b) For given $x \in U$ such that $\sum_n {}^U\widehat{R}_{f_n}(x) < +\infty$ and for given $\varepsilon > 0$ there exist by (2.2.25) functions $u_n \in \mathcal{U}^+(U)$ such that $u_n \geqslant f_n$ q.e. and

$$ u_n(x) \leqslant {}^U\widehat{R}_{f_n}(x) + 2^{-n}\varepsilon \ (< +\infty). $$

Defining now $u = \sum_n u_n \in \mathcal{U}^+(U)$ we have $u \geqslant f$ q.e. It follows, again by (2.2.25) that

$$ {}^U\widehat{R}_f(x) \leqslant u(x) \leqslant \sum_n {}^U\widehat{R}_{f_n}(x) + \varepsilon, $$

and so ${}^U\widehat{R}_f \leqslant \sum_n {}^U\widehat{R}_{f_n}$, this being obvious at points at which $\sum_n {}^U\widehat{R}_{f_n} = +\infty$. The proof for the case of ${}^U R$ in place of ${}^U\widehat{R}$ is similar (now omitting "q.e."). □

Theorem 2.2.16 *For $A \subset B \subset U$ and $u \in \mathcal{U}^+(U)$, we have ${}^U\widehat{R}^B_{({}^U\widehat{R}^A_u)} = {}^U\widehat{R}^A_u = {}^U\widehat{R}^A_{({}^U\widehat{R}^B_u)}$.*

In particular for $A = B$, the sweeping $u \mapsto {}^U\widehat{R}_u^A$ is an *idempotent* selfmap of $\mathcal{U}^+(U)$.

Proof From (2.2.24) with f replaced by $1_B({}^U\widehat{R}_u^A)$ we obtain ${}^U\widehat{R}^B_{({}^U\widehat{R}_u^A)} = {}^U\widehat{R}_u^A$ q.e. on B and hence ${}^U\widehat{R}^B_{({}^U\widehat{R}_u^A)} = u$ q.e. on $A \subset B$, again by (2.2.24). It follows by (2.2.25) (applied with f replaced by $1_A({}^U\widehat{R}_u^A)$) that ${}^U\widehat{R}^B_{({}^U\widehat{R}_u^A)} \geqslant {}^U\widehat{R}_u^A$, and here the opposite inequality is obvious. And from ${}^U\widehat{R}_u^B = u$ q.e. on $A \subset B$ we find by (2.2.25) that ${}^U\widehat{R}^A_{({}^U\widehat{R}_u^B)} = {}^U\widehat{R}_u^A$. □

Above we have studied sweeping of a positive finely hyperharmonic *function*, defined on U, onto a subset of U and relative to U. But we do not define sweeping of a *measure* on U onto a subset of U and relative to U. For any point $x \in U$ and any $A \subset U$, ε_x^A is defined only relative to Ω (see Theorem 1.4.8).

The specific order "$\preccurlyeq$" on the cone $\mathcal{U}(U)$ is defined by

$$u \preccurlyeq v \iff \exists w \in \mathcal{U}_+(U) : v = u + w$$

for $u, v \in \mathcal{U}(U)$. We proceed to extend the theorem of Mokobodzki and Sibony (Theorem 1.3.13) to the present "fine" setting.

Theorem 2.2.17 *Let $u, v \in \mathcal{U}^+(U)$. Write $f = (u - v)_+$, understood as 0 at points $x \in U$ where $v(x) = +\infty$. Then ${}^U R_f \in \mathcal{U}^+(U)$ and ${}^U R_f \preccurlyeq u$.*

Proof Choose a bounded potential $q > 0$ on Ω and define $u_n = u \wedge (nq)$ on U. Each of the functions $(u_n - v)_+$ is finely continuous on U and hence $(u - v)_+ = \sup_n (u_n - v)_+$ is finely l.s.c. It follows by Theorem 2.2.14(b) that ${}^U R_f = {}^U\widehat{R}_f \in \mathcal{U}^+(U)$. Now fix a regular finely open set V with $\widetilde{V} \subset U$. The function w_n defined on U by $w_n = {}^U R_f$ on $U \backslash V$ and

$$w_n = {}^U R_f \wedge (({}^U R_f)^{\complement V} + u - u_n^{\complement V}) \quad \text{on } V$$

is finely hyperharmonic on V because $u_n^{\complement V} = {}^U\widehat{R}_{u_n}^{U\backslash V}$ (see Corollary 2.2.10) is finely harmonic on V by Theorem 2.1.52 which also shows that w_n is finely continuous on U, and since $w_n = {}^U R_f$ on $U \backslash V$ because $u - u_n^{\complement V} \geqslant 0$ we conclude by Lemma 2.1.51 that $w_n \in \mathcal{U}^+(U)$. From here on, the proof of Theorem 1.3.13 carries over when g is replaced by u_n, balls B with $\overline{B} \subset U$ by finely sets W with $\widetilde{W} \subset U$, and whence P_B by integration with respect to $\varepsilon_x^{\complement W}$ for $x \in W$. Note that, by Theorem 2.1.26, $\int_U u_n \, d\varepsilon_{\cdot}^{\complement V} = {}^U\widehat{R}_{u_n}^{U\backslash V}$ is finely continuous on U. Of course we shall often add the word "fine(ly)". □

The cone $\mathcal{U}(U)$ is a lattice, both in the natural (pointwise) order "$\leqslant$" and in the specific order "$\preccurlyeq$". The supremum and the infimum of a subfamily $\mathcal{F}$ of $\mathcal{U}(U)$ in the natural order are denoted by $\bigvee$ and $\bigwedge$, and in the specific order by $\curlyvee$ and $\curlywedge$. Recall the fine convergence theorem (Theorem 2.1.34) according to which $\mathcal{U}(U)$ has the following property of existence and value of $\bigwedge \mathcal{F}$, $\mathcal{F} \subset \mathcal{U}(U)$.

Property of finely l.s.c. regularization. $\mathcal{U}(U)$ is lower directed in the natural order, and for any non-void family $\mathcal{F} \subset \mathcal{U}(U)$ having a natural minorant, the natural infimum of $\mathcal{F}$ is

$$\bigwedge \mathcal{F} = \widehat{\inf} \mathcal{F} \in \mathcal{U}(U),$$

the greatest finely l.s.c. minorant of $\inf \mathcal{F}$.

Property of upper directed families: For any upper directed family $\mathcal{F} \subset \mathcal{U}(U)$ the natural supremum exists and equals

$$\bigvee \mathcal{F} = \sup \mathcal{F}$$

(the pointwise supremum of $\mathcal{F}$ as subfamily of $\overline{\mathbb{R}}$). This was established in Corollary 2.1.21 to Theorem 2.1.14 (the mean-value property of finely hyperharmonic functions).

Property of natural decomposition. Let $u, u_1, u_2 \in \mathcal{U}^+(U)$. If $u \leqslant u_1 + u_2$, there exist $v_1, v_2 \in \mathcal{U}^+(U)$ such that

$$v_1 \leqslant u_1, \; v_2 \leqslant u_2, \; u = u_1 + u_2.$$

This is obtained from Theorem 2.2.17 with v replaced by the present u_1; taking $f = (u - u_1)_+ \leqslant u_2$, whereby ${}^U R_f \leqslant u_2$ and ${}^U R_f \in \mathcal{U}^+(U)$. Since ${}^U R_f \preccurlyeq u$ there exists $w \in \mathcal{U}^+(U)$ such that $u = w + {}^U R_f$. Since $u \leqslant u_1 + f \leqslant u_1 + {}^U R_f$ we obtain

$$u = (w + {}^U R_f) \wedge (u_1 + {}^U R_f) = (w \wedge u_1) + {}^U R_f,$$

and we may take $v_1 = w \wedge u_1 \leqslant u_1$ and $v_2 = {}^U R_f \leqslant u_2$.

In [CC3, Chapter 4], Constantinescu and Cornea considered a *Baire topological space* X and a convex cone $\mathcal{W}$ of continuous functions $X \to]-\infty, +\infty]$ among which the function identical to $+\infty$. They set up three axioms, and derived from them various lattice properties of sweeping of functions in $\mathcal{W}$ on subsets of X.

According to Theorem 2.2.17 and the three properties stated thereafter, we may take for X the given finely open subset U of Ω endowed with the fine topology on U, and for $\mathcal{W}$ the cone $\mathcal{U} = \mathcal{U}(U)$ of all finely hyperharmonic functions on U. Recall that U (endowed with the fine topology) is a Baire topological space, cf. paragraph 4 after Corollary 1.2.5. The properties stated in the following two corollaries are therefore consequences of the axiomatic theory developed in [CC3, Chapter 4].

Corollary 2.2.18 *Let U be a finely open subset of Ω.*

(a) *$\mathcal{U}(U)$ is upper complete in the specific order "$\preccurlyeq$". Thus any non-void family $\mathcal{F} \subset \mathcal{U}^+(U)$ has a specific supremum*

$$\curlyvee \mathcal{F} \succcurlyeq \bigvee \mathcal{F}$$

with equality if $\mathcal{F}$ is specifically upper directed, in which case it therefore follows that

$$\curlyvee\mathcal{F} = \bigvee \mathcal{F} = \sup \mathcal{F}.$$

(b) *For any specifically lower directed and naturally minorized non-void family $\mathcal{F} \subseteq \mathcal{U}(U)$ we have*

$$\curlywedge\mathcal{F} = \bigwedge \mathcal{F} \leqslant \inf \mathcal{F},$$

with equality at points where $\inf \mathcal{F} < +\infty$.

See [CC3, Theorem 4.1.1, Corollary 4.1.3, and Proposition 4.1.4]. Concerning (a), note that the constant function $+\infty$ is a specific majorant of any function in $\mathcal{U}^+(U)$. It follows from (a) that $\mathcal{U}(U)$ is conditionally lower complete. In particular, *$\mathcal{U}^+(U)$ is a complete lattice in the specific order.*

Corollary 2.2.19 *Let U be a finely open subset of Ω.*

(a) *For any $u \in \mathcal{U}^+(U)$ and $A \subseteq U$ such that $u < +\infty$ on A we have*

$${}^U R_u^A = \inf\{{}^U R_u^V : V \textit{ finely open, } A \subseteq V \subseteq \{u < +\infty\}\}.$$

(b) *For any $u, v \in \mathcal{U}^+(U)$ and any set $A \subseteq U$,*

$${}^U R_{u+v}^A = {}^U R_u^A + {}^U R_v^A, \quad {}^U \widehat{R}_{u+v}^A = {}^U \widehat{R}_u^A + {}^U \widehat{R}_v^A.$$

(c) *For any $u \in \mathcal{U}^+(U)$ and $A, B \subseteq U$,*

$$\begin{aligned}
{}^U R_u^{A\cup B} + {}^U R_u^{A\cap B} &\leqslant {}^U R_u^A + {}^U R_u^A,\\
{}^U \widehat{R}_u^{A\cup B} + {}^U \widehat{R}_u^{A\cap B} &\leqslant {}^U \widehat{R}_u^A + {}^U \widehat{R}_u^A,\\
{}^U \widehat{R}_u^{A\cup B} &\preccurlyeq {}^U \widehat{R}_u^A + {}^U \widehat{R}_u^B.
\end{aligned}$$

See [CC3, Theorem 4.2.1, Theorem 4.2.2, and exerc. 4.2.8].

2.3 The Fine Dirichlet Problem

Theorem 2.3.1 (The proper fine Dirichlet problem) *Let U be a regular finely open subset of Ω, and let $f : \widetilde{U} \to \mathbb{R}$ be a finely continuous function. There exists then at most one finely continuous extension u of $f|_{\partial_f U}$ to $\widetilde{U}$ such that u is finely harmonic on U and majorized there in absolute value by some semibounded potential on Ω (which may depend on u). If U is regular and if $|f| \leqslant p$ on $\partial_f U$ for some finite (hence semibounded) potential p on Ω then the said extension of $f|_{\partial_f U}$ to $\widetilde{U}$ exists and is given by $u = f^{\complement U}$, that is,*

$$u(x) = \int f \, d\varepsilon_x^{\complement U} \quad \textit{for } x \in U.$$

Recall from Theorem 1.4.36 that every finely continuous function $f : U \to \overline{\mathbb{R}}$ is Borel measurable, from Definition 1.5.16(iv) that every finite potential in Ω is semibounded, and from Corollary 2.1.48 that a finely open set U is regular if and only if $\partial_f U$ is a base. As usual, the cone of all finely hyperharmonic functions on U is denoted by $\mathcal{U}(U)$ and its positive subcone by $\mathcal{U}^+(U)$.

Proof The uniqueness of u follows from the fine boundary minimum principle (Theorem 2.1.10) applied to $u_1 - u_2$ and $u_2 - u_1$, where u_1 and u_2 denote two extensions of $f|_{\partial_f U}$ of stated kind. In fact let p_1, p_2 be semibounded potentials on Ω such that $|u_j| \leqslant p_j$ on U for $j = 1, 2$. Then $u_1 - u_2 \geq -(p_1 + p_2)$ and

$$\text{f-}\liminf_{x \to y,\, x \in U} (u_1(x) - u_2(x)) \geqslant 0$$

for $p_1(y) + p_2(y) < +\infty$ (whereby $f_1(y)$ and $f_2(y)$ are finite), thus q.e. for $y \in \partial_f U$, and so $u_1 - u_2 \geqslant 0$ and similarly $u_2 - u_1 \geqslant 0$. Consequently $u_1 = u_2$ on U. The remaining assertions of the theorem follow immediately by Lemma 2.1.13 applied to the restriction of f to the base $\partial_f U$. In addition, it is found in this way that $|u| \leqslant p$ on $\widetilde{U}$. □

Even when the proper fine Dirichlet problem has no solution, there may be a generalized solution. We proceed to show that the Perron-Wiener-Brelot method extends to the present fine setting. In the rest of the present section, U denotes an arbitrary finely open subset of Ω, and f denotes an extended real-valued function defined at least on $\mathfrak{b}(\partial_f U)$ (hence q.e. on $\partial_f U$). Recall from Lemma 2.1.47 that $\mathfrak{b}(\partial_f U) = \mathfrak{b}(\complement U) \cap \partial_f U = \partial_{f,r} U$ (the regular part of $\partial_f U$).

Definition 2.3.2 The upper PWB class $\overline{\Phi}_f^U$ is the class of all $u \in \mathcal{U}(U)$ such that

$$\text{f-}\liminf_{x \to y,\, x \in U} u(x) \geqslant f(y) \quad \text{for every } y \in \partial_f U \tag{2.3.1}$$

and that $u \geqslant -p$ for some finite potential p on Ω (which may depend on u). The lower PWB class $\underline{\Phi}_f^U$ is defined by $\underline{\Phi}_f^U = -\overline{\Phi}_{-f}^U$. The upper and lower PWB solutions to the fine Dirichlet problem (with fine boundary function f) are defined by

$$\overline{H}_f^U = \inf \overline{\Phi}_f^U, \quad \underline{H}_f^U = \sup \underline{\Phi}_f^U.$$

Thus $\underline{H}_f^U = -\overline{H}_{-f}^U$. For $u_1, u_2 \in \overline{\Phi}_f^U$ we have $u_1 \wedge u_2 \in \overline{\Phi}_f^U$, and for $u_1, u_2 \in \underline{\Phi}_f^U$ we have $u_1 \vee u_2 \in \underline{\Phi}_f^U$. Furthermore, $\overline{H}_f^U \geqslant \underline{H}_f^U$, for if $u \in \overline{\Phi}_f^U$ and $v \in \underline{\Phi}_f^U$ then $u - v \geqslant 0$, again by Theorem 2.1.10. We may omit U from notations and simply write $\overline{\Phi}_f$, $\underline{\Phi}_f$, $\overline{H}_f$, $\underline{H}_f$ when there is no risk of ambiguity. A function in $\overline{\Phi}_f^U$, respectively $\underline{\Phi}_f^U$, is called an *upper function* (*superfunction*), respectively a *lower function* (*subfunction*), for f.

Definition 2.3.3 A function $f : \partial_f U \to \overline{\mathbb{R}}$ is called (finely) resolutive (relative to U) if

$$-\infty < \underline{H}_f^U(x) = \overline{H}_f^U(x) < +\infty \quad \text{for every } x \in U.$$

If f is resolutive we may write H_f^U for $\overline{H}_f^U$ and $\underline{H}_f^U$.

Lemma 2.3.4 *For every fine component V of U and every function $f : \partial_f U \to \overline{\mathbb{R}}$ we have $\overline{H}_f^U = \overline{H}_f^V$ and $\underline{H}_f^U = \underline{H}_f^V$ on V. Hence f is resolutive if and only if its restriction to each fine component of $\partial_f U$ is resolutive.*

Proof Since $\partial_f V \subset \partial_f U$, f is defined on $\partial_f V$. If $u \in \overline{\Phi}_f^U$ then $u|_V \in \overline{\Phi}_f^V$, and so $\overline{H}_f^V \leqslant \overline{H}_f^U$ on V. Conversely; if $v \in \overline{\Phi}_f^V$ the function equal to v on V and $+\infty$ on $U \backslash V$ belongs to $\overline{\Phi}_f^U$, so $\overline{H}_f^V \geqslant \overline{H}_f^U$ on V. □

Remark 2.3.5 A function f on $\partial_f U$ is resolutive if each fine component V of U contains a point x at which $\underline{H}_f^U(x) = \overline{H}_f^U(x) \in \mathbb{R}$. For then $\overline{H}_f^U - \underline{H}_f^U$ is finely hyperharmonic and $\geqslant 0$ on V, with equality at x, and so identically 0 on V, by Corollary 2.1.27, and hence identically 0 on U.

Remark 2.3.6 $\overline{H}_f^U$ and $\underline{H}_f^U$ do not change if, in (2.3.1) in the definition of $\overline{\Phi}_f^U$, $\partial_f U$ is replaced by $\mathfrak{b}(\partial_f U)$, or even if "for every $y \in \partial_f U$" is replaced by "q.e. $y \in \partial_f U$". To see this, let $u \in \overline{\Phi}_f^U$ in the modified sense. Then $u + \varepsilon w \in \overline{\Phi}_f^U$ in the original sense when $w \in \mathcal{U}^+ = \mathcal{U}^+(U)$ equals $+\infty$ on the polar set of points of $\partial_f U$ at which the inequality in (2.3.1) fails to hold, whereas $w(x) < +\infty$ for a given point $x \in U$. It follows that $u + \varepsilon w \geqslant \overline{H}_f$ in the original sense. By varying ε we conclude that $u \geqslant \overline{H}_f^U$ (original), and varying u that $\overline{H}_f^U$ (new) $\geqslant \overline{H}_f^U$ (original). The opposite inequality is obvious. The concept of resolutivity therefore likewise remains unchanged.

Theorem 2.3.7 (The generalized fine Dirichlet problem) *For any extended real valued function f defined (at least) on $\mathfrak{b}(\partial_f U) = \mathfrak{b}(\complement U) \cap \partial_f U = \partial_{f,r} U$ we have*

$$\overline{H}_f^U(x) = \int^* f \, d\varepsilon_x^{\complement U}, \quad \underline{H}_f^U(x) = \int_* f \, d\varepsilon_x^{\complement U} \tag{2.3.2}$$

for every $x \in U$. Each of the functions $\overline{H}_f^U$ and $\underline{H}_f^U$ is therefore finely harmonic on the finely open set of points of U at which it is finite. In particular, f is finely resolutive if and only if f is integrable with respect to $\varepsilon_x^{\complement U}$ for every $x \in U$. In the affirmative case,

$$H_f^U(x) = \int f \, d\varepsilon_x^{\complement U} \quad \textit{for every } x \in U.$$

Proof For any $u \in \overline{\Phi}_f^U$ the extension $\widetilde{u}$ of u to $\widetilde{U}$, defined by putting

$$\widetilde{u}(y) = \text{f-}\liminf_{x \to y,\, x \in U} u(x) \quad \text{for every } y \in \partial_f U,$$

is finely l.s.c. on $\widetilde{U}$, and $\geqslant f$ on $\partial_f U$. If p denotes a finite potential on Ω such that $u \geqslant -p$ on U, then $\widetilde{u} \geqslant -p$ on $\widetilde{U}$, by fine continuity. For any $x \in U$ it therefore follows by Theorem 2.1.14 that

$$u(x) \geqslant \int \widetilde{u}\, d\varepsilon_x^{\complement U} \geqslant \int^* f\, d\varepsilon_x^{\complement U},$$

and hence, by varying $u \in \overline{\Phi}_f^U$, that $\overline{H}_f^U(x) \geqslant \int^* f\, d\varepsilon_x^{\complement U}$.

In proving the opposite inequality for any $x \in U$ we may suppose that $\int^* f\, d\varepsilon_x^{\complement U} < +\infty$. For any real $\delta > 0$ there exists a l.s.c. function g on Ω, integrable with respect to $\varepsilon_x^{\complement U}$, such that $g \geqslant f$ on $\mathfrak{b}(\partial_f U)$ (which carries $\varepsilon_x^{\complement U}$) and that

$$\int g\, d\varepsilon_x^{\complement U} < \int^* f\, d\varepsilon_x^{\complement U} + \delta < +\infty.$$

Choose a finite and continuous potential $q > 0$ on Ω and define $g_n = g \vee (-nq)$ on Ω, $n \in \mathbb{N}$. Each g_n is then l.s.c. $\varepsilon_x^{\complement U}$-integrable. Furthermore

$$-nq \leqslant g_n \leqslant g_+,$$

and $g_n \geqslant g \geqslant f$ on $\mathfrak{b}(\partial_f U)$. Finally, $g_n^{\complement U} \geqslant (-nq)^{\complement U} \geqslant -nq$, and

$$g_n^{\complement U}(x) = \int g_n\, d\varepsilon_x^{\complement U} \leqslant \int g_+\, d\varepsilon_x^{\complement U} < +\infty$$

for every $x \in U$ whereas $g_n \geqslant f$ on $\mathfrak{b}(\partial_f U)$. According to Lemma 2.1.38 with B, f, p replaced by $\mathfrak{b}(\partial_f U)$, g_n, nq, we have $g_n^{\complement U} \in \overline{\Phi}_{g_n}^U \subset \overline{\Phi}_f^U$. From $g_n \downarrow g$ pointwise on Ω, it follows that $g_n^{\complement U} \downarrow g^{\complement U}$ on U, and hence for $x \in U$

$$\overline{H}_f^U(x) \leqslant \inf_n g_n^{\complement U}(x) = g^{\complement U}(x) < \int^* f\, d\varepsilon_x^{\complement U} + \delta.$$

By varying δ we obtain $\overline{H}_f^U(x) \leqslant \int^* f\, d\varepsilon_x^{\complement U}$ and similarly $\underline{H}_f^U(x) \geqslant \int_* f\, d\varepsilon_x^{\complement U}$ for $x \in U$. This completes the verification of (2.3.2). The remaining assertions of the theorem follow from Theorem 2.1.39. □

Corollary 2.3.8 *For any function $f : \mathfrak{b}(\partial_f U) \to \overline{\mathbb{R}}$ such that $f \geqslant -p$, resp. $f \leqslant p$, for some finite potential on Ω, we have for any $y \in \mathfrak{b}(\partial_f U)$*

$$\text{f-}\liminf_{x \to y,\, x \in U} \underline{H}_f^U(x) \geqslant \text{f-}\liminf_{x \to y,\, x \in \mathfrak{b}(\partial_f U)} f(x),$$

resp.

$$\text{f-}\limsup_{x \to y,\, x \in U} \overline{H}_f^U(x) \leqslant \text{f-}\limsup_{x \to y,\, x \in \mathfrak{b}(\partial_f U)} f(x).$$

In particular, if $|f| \leqslant p$ for some finite potential p on Ω, then

$$\text{f-}\lim_{x \to y,\, x \in U} \overline{H}_f^U(x) = \text{f-}\lim_{x \to y,\, x \in U} \underline{H}_f^U(x) = f(y) \tag{2.3.3}$$

holds for any *regular* fine boundary point $y \in \mathfrak{b}(\partial_f U) = \mathfrak{b}(\complement U) \cap \partial_f U$ at which f is finely continuous (relative to $\mathfrak{b}(\partial_f U)$). Thus $\overline{H}_f^U = \underline{H}_f^U = H_f^U$ solves the proper fine Dirichlet problem with finely continuous fine boundary function f such that $|f| \leq p$ on $\mathfrak{b}(\partial_f U)$ for some finite potential p on Ω. On account of Theorem 2.3.7, all this follows immediately from Corollary 2.1.40 applied to the base $B = \mathfrak{b}(\partial_f U)$.

The PWB solution is connected with the reduced functions as shown by the following theorem (which reduces to Theorem 2.1.52 in case $\widetilde{V} \subset U$ and V is regular).

Theorem 2.3.9 *Let U and V be finely open in Ω and suppose that $V \subset U$. Let $u \geqslant 0$ be a finely hyperharmonic function on U, and define $u_0 = u$ on $(\partial_f V) \cap U$ and $u_0 = 0$ on $\partial_f U$. Then*

$${}^U R_u^{U \setminus V} = \begin{cases} \overline{H}_{u_0}^V = \int_U u \, d\varepsilon_{\cdot}^{\complement V} & \text{on } V \\ u & \text{on } U \setminus V. \end{cases}$$

In particular, $\overline{H}_{u_0}^V \leqslant u$ on V. If $\overline{H}_{u_0}^V < +\infty$ on V then u_0 is resolutive relative to V.

Proof Obviously, ${}^U R_u^{U\setminus V} = u$ on $U \setminus V$. The function u_0 is finely l.s.c. on U and hence $u|_V \in \overline{\Phi}_{u_0}^V$. Let $v \in \mathcal{U}^+(U)$ and suppose that $v \geqslant u$ on $U \setminus V$. Then $v|_V \in \overline{\Phi}_{u_0}^V$, and so ${}^U R_u^{U\setminus V} \geqslant \overline{H}_{u_0}^V$ on V. Conversely, let $w \in \overline{\Phi}_{u_0}^V$, and define $w^* = u \wedge w$ on V and $w^* = u$ on $U \setminus V$. Then $w^* \in \mathcal{U}^+(U)$ by Lemma 2.1.51, and hence $w^* \geqslant {}^U R_u^{U\setminus V}$ on U. In particular, $w \geqslant w^* \geqslant {}^U R_u^{U\setminus V}$ on V, and altogether $\overline{H}_{u_0}^V = {}^U R_u^{U\setminus V}$ on V.

It remains to show that u_0 is $\varepsilon_x^{\complement V}$-measurable for every $x \in V$. Let $x \in V$ be given. Since every polar set is $\varepsilon_x^{\complement V}$-null it suffices to prove that u_0 can be changed on some polar set so as to become Borel measurable on the Borel set $\mathfrak{b}(U)$ (which differs from $\widetilde{U}$ by a polar set). And this is true because u is finely continuous on U and hence Borel measurable (in the relative Euclidean topology on U), by Theorem 1.4.36, and because $u_0 = 0$ on $\partial_f U$. □

Remark 2.3.10 In the classical setting, where $U = \Omega$, $u \not\equiv +\infty$ (or, equivalently, u superharmonic), cf. Sect. 1.1.2, and V is (Euclidean) open in Ω, the condition $\overline{H}_{u_0}^V < +\infty$ on V is automatically fulfilled because $\overline{H}_{u_0}^V = R_u^{U\setminus V}|_V$ then is harmonic on V by Lemma 1.3.7. In our setting it can occur that $\overline{H}_{u_0}^V = {}^U R_u^{U\setminus V}|_V$ takes value ∞ on V, even if $U = \Omega$, as shown by Example 2.1.8.

Theorem 2.3.11 *If $u \in \mathcal{U}_+(U)$ and $A \subset U$ then ${}^U \widehat{R}_u^A(x) = \int_U u \, d\varepsilon_x^{A \cup \complement U}$ for $x \in U$.*

Proof The integral exists because $u \geq 0$ is Borel measurable in the relative Euclidean topology on $U \subset \Omega$, by Theorem 1.4.44. Since ${}^U\widehat{R}_u^A$ and $\int_U u\, d\varepsilon_x^{A\cup\complement U}$ remain unchanged if A is replaced by its base $\mathfrak{b}(A) \cap U$ relative to U and since the base operation is idempotent we may assume that $A = \mathfrak{b}(A) \cap U$. Consequently, ${}^U\widehat{R}_u^A = {}^U R_u^A$ (see Lemma 1.3.7) and the set $V := U\setminus A$ is finely open (and regular). Let u_0 denote the extension of u to $\widetilde{U} = U \cup \partial_f U$ by the value 0 on $\partial_f U$. Every function $s \in \mathcal{U}_+(U)$ with $s \geqslant u$ on A is an upper function (superfunction) for u_0 relative to V (see Theorem 2.3.9). It follows that

$$
{}^U\widehat{R}_u^A(x) \geq \int_\Omega^* u_0\, d\varepsilon_x^{\complement V} = \int_U^* u\, d\varepsilon_x^{A\cup\complement U}
$$

for $x \in V$. For the opposite inequality, consider any upper function v for u_0 relative to V. In particular, $v \geq -p$ on V for some finite and hence semibounded potential p on Ω. Since $u_0 \geq 0$ it then follows that $v \geq 0$ according to Theorem 2.1.10. Define $w = u \wedge v$ on V and $w = u$ on $A = U\setminus V$. Then

$$
\operatorname{f-}\liminf_{y\to x,\, y\in V} w(y) \geqslant u(x) \wedge \operatorname{f-}\liminf_{y\to x,\, y\in V} v(y) = u(x)
$$

for every $x \in U \cap \partial_f V$. It therefore follows by Lemma 2.1.51 that w is finely hyperharmonic on U. Moreover, w is an upper function for u relative to V because $w = u \wedge v \geqslant 0$ on V. Since $w = u$ on A we have $w \geqslant \widehat{R}_u^A$ on U, and in particular $v \geqslant w \geqslant \widehat{R}_u^A$ on V. By varying v we obtain $\overline{H}_{u_0}^V \geq \widehat{R}_u^A$ on V. Altogether we have established the asserted equality for $x \in V$. It also holds for $x \in U\setminus V = A = \mathfrak{b}(A) \cap U$ because $\widehat{R}_u^A(x) = R_u^A(x) = u(x)$ and because $\varepsilon_x^{A\cup\complement U} = \varepsilon_x$, noting that $x \in \mathfrak{b}(A) \subset \mathfrak{b}(A \cup \complement U)$. □

The following corollary extends (1.4.5) (in which $U = \Omega$).

Corollary 2.3.12 *For any finely hyperharmonic function $u \geqslant 0$ on U and any finely open set V of fine closure $\widetilde{V}$ contained in $\mathfrak{r}(U)$ we have*

$$
{}^U\widehat{R}_u^{U\setminus V}(x) = \int u\, d\varepsilon_x^{\complement V} \text{ for } x \in U.
$$

Proof It is easily verified that

$$
{}^U\widehat{R}_u^{U\setminus V} = {}^{\mathfrak{r}(U)}\widehat{R}_u^{\mathfrak{r}(U)\setminus V} \quad \text{on } U,
$$

when u on the right means the extension of the given finely hyperharmonic function $\geqslant 0$ on U by the fine continuity from U to $\mathfrak{r}(U)$, cf. Theorem 2.1.43, noting that $\mathfrak{r}(U)\setminus U$ is polar. From the inclusion $\widetilde{V} \subset \mathfrak{r}(U)$ it follows that $V \subset \mathfrak{r}(U)$. By applying Theorem 2.3.11 with $A = \mathfrak{r}(U)\setminus V$ we then get ${}^U\widehat{R}_u^{U\setminus V}(x) = {}^{\mathfrak{r}(U)}\widehat{R}_u^{\mathfrak{r}(U)\setminus V}(x) = \int u d\varepsilon_x^{\complement V}$ for all $x \in U$. □

2.4 Finely Superharmonic Functions and Fine Potentials

Recall from Corollary 2.1.33 that the fine topology on Ω is locally connected and that a finely open subset U of Ω has only countably many fine components. Furthermore from Corollary 2.1.24 that a (Euclidean) open subset of Ω is finely connected if and only if it is connected. In particular, our Greenian domain Ω is also a fine domain.

Definition 2.4.1 A *finely superharmonic* function on a finely open set $U \subset \Omega$ is a finely hyperharmonic function u having one of the following equivalent properties:
(i) u is not identically $+\infty$ on any fine component of U,
(ii) u is finite on some finely dense subset of U,
(iii) u is finite quasi-everywhere on U.

The implication (iii)⇒(ii) is obvious because a polar set has no finely interior point, and (ii)⇒(i) follows from the local connectedness of the fine topology. For the remaining implication (i)⇒(iii) we may assume that U is finely connected because a countable union of polar sets is polar. We shall prove that the set $E := \{x \in U : u(x) = +\infty\}$ is polar. Consider a non-void fine domain $V \subset U$ of class $\mathfrak{V}_U(u_-)$, see Remark 2.1.18. Then $u \geqslant c$ on V for some constant $c \in \mathbb{R}$. We have $V \backslash E \neq \varnothing$ since otherwise $V \subset E$, which is absurd by the hypothesis (i). For any $\varepsilon > 0$ we have $\varepsilon(u - c) = +\infty$ on $V \cap E$, and hence

$$\varepsilon(u - c) \geqslant {}^V R_{+\infty}^{V \cap E} \geqslant {}^V \widehat{R}_{+\infty}^{V \cap E},$$

cf. Definition 2.2.6. For $\varepsilon \to 0$ this implies that ${}^V \widehat{R}_{+\infty}^{V \cap E} = 0$ on $V \backslash E$, and hence by Corollary 2.1.27 that ${}^V \widehat{R}_{+\infty}^{V \cap E} = 0$ on all of V. By the fine convergence theorem (Theorem 2.1.34) it follows that

$$+\infty 1_{V \cap E} \leqslant {}^V \widehat{R}_{+\infty}^{V \cap E} \quad \text{q.e.},$$

and so $V \cap E$ is polar. By the quasi-Lindelöf property, there is a sequence $(V_n) \subset \mathfrak{V}_U(u_-)$ and a polar set F such that $U = \bigcup_n V_n \cup F$ and therefore $E = \bigcup_n (V_n \cap E) \cup (F \cap E)$ is polar.

A function u on U is said to be *finely subharmonic* if $-u$ is finely superharmonic.

We denote by $\mathcal{S}(U)$ the family of all finely superharmonic functions on U (Definition 2.4.1) and by $\mathcal{S}_+(U)$ the positive cone of $\mathcal{S}(U)$, consisting of all finely superharmonic functions $\geqslant 0$ on U. We proceed to show that $\mathcal{S}_+(U)$ has many properties in common with the positive cone $\mathcal{S}_+$ in the family $\mathcal{S}$ of all superharmonic functions on the whole of Ω. Recall from Theorem 2.1.22 (with $U = \Omega$), we actually have $\mathcal{S}_+(\Omega) = \mathcal{S}_+$. More precisely we show in Sect. 2.5 that $\mathcal{S}_+$ and $\mathcal{S}_+(U)$ are H-cones in the sense of [BBC2, p. 36]. It is known that they are even standard *H-cones* of functions. For $\mathcal{S}_+$ see [BBC2, p. 113], for $\mathcal{S}_+(U)$ see the paragraph following the proof of Theorem 2.5.15 below.

Definition 2.4.2 A *fine potential* on U is a finely superharmonic function $p \geqslant 0$ such that every finely subharmonic minorant of p is $\leqslant 0$, or equivalently: 0 is the only finely subharmonic minorant $\geqslant 0$ of p.

For the equivalence, let v be any finely subharmonic minorant of p. Then $v_+ := v \vee 0$ is a finely subharmonic $\geqslant 0$ minorant of p, hence $v_+ = 0$, that is, $v \leqslant 0$.

As shown by Gardiner and Hansen [GH] there exists a finely superharmonic function $s \geqslant 0$ on a finely open subset of $\mathbb{R}^N$, $N \geqslant 3$, such that 0 is the only finely *harmonic* minorant $\geqslant 0$ of s, and yet s is not a fine potential. This answers a question posed in [Fu8]. The case $N = 2$ remains unsettled.

However, a *finite valued* finely superharmonic function $s \geqslant 0$ without finely harmonic minorants $\geqslant 0$ is indeed a fine potential. This appears from the following fine version of the classical Riesz decomposition theorem (for finite valued functions).

Theorem 2.4.3 *Let s be a finite finely superharmonic function on U (finely open in Ω), and suppose that s has a finite finely subharmonic minorant v. Then s admits a unique decomposition $s = p + h$, where p is a fine potential and h is finely harmonic on U. Furthermore, h is the greatest finely subharmonic minorant of s, so in particular $s \geq 0$ implies $h \geq 0$.*

Proof Fix for a while a regular finely open set $V \subset U$ of class $\mathfrak{V} := \mathfrak{V}_U(s_+ \vee v_-)$ relative to U (see Remark 2.1.18), that is, s and v are bounded from above and below on V, and V has a compact closure $\overline{V} \subset U$. There exists then a bounded potential q_V on Ω such that

$$-q_V \leqslant v \leqslant s \leqslant q_V \quad \text{on } V,$$

and hence on $\widetilde{V} \subset U$, by fine continuity.

Consider any finely superharmonic function u on U such that $v \leqslant u \leqslant s$ on U (s is a such function), and hence $|u| \leqslant q_V$ on V. Since $U \supset \widetilde{V} \supset \mathfrak{b}(V)$, $u^{\complement V}$ is defined in all of U, cf. Definition 1.4.29, and $|u^{\complement V}| \leqslant q_V$. We have

$$u^{\complement V} \leqslant u \text{ on } U, \; u^{\complement V} = u \text{ on } U \setminus V, \tag{2.4.1}$$

the inequality by Corollary 2.1.16 because $u \geqslant -q_V$ on U. Finally

$$u^{\complement V} \text{ is finely harmonic on } V \tag{2.4.2}$$

according to Theorem 2.1.39 applied to $B = \mathfrak{b}(\complement V)$ and any extension f of u from U to Ω. (Alternatively apply Lemma 2.1.38 to the extension of u by q_V in $\complement\widetilde{V}$, this extension being finely l.s.c. on Ω because $u \leqslant q_V$ on $\widetilde{V}$.)

If instead u is finely subharmonic, again with $v \leqslant u \leqslant s$ and $|u| \leqslant q_V$ on V, then it follows from the above that u replaced by $-u$ and s by $-v$ and v by $-s$ that (2.4.1) and (2.4.2) hold except that now $u^{\complement V} \geqslant u$ on U.

We are now prepared to apply Perron's method. The finely open sets V of class $\mathfrak{V} := \mathfrak{V}_U(s^+ \vee v^-)$ cover U.

For any finite sequence $(V_1, \ldots, V_n)$ of sets $V_i \in \mathfrak{V}$ and any Borel measurable function u on U between s and v write

$$u_{V_1,\ldots,V_n} = (\ldots((u^{\complement V_1})^{\complement V_2})\ldots)^{\complement V_n}.$$

If such u is finely superharmonic (resp. subharmonic) on U then so is $u_{V_1,\ldots,V_n}$. For any finely subharmonic minorant $w \geqslant v$ of s we have

$$w \leqslant w_{V_1,\ldots,V_n} \leqslant s_{V_1,\ldots,V_n} \leqslant s \text{ on } U.$$

Denote by $\mathcal{V}$ the family of all functions $s_{V_1,\ldots,V_n}$ as $n \in \mathbb{N}$ and $(V_1, \ldots, V_n)$ vary. This family $\mathcal{V}$, termed the *Perron family generated by s and* $\mathfrak{V}$, is pointwise lower directed. In fact, for any other finite sequence, say $(V_{n+1}, \ldots, V_m)$ of sets $V_j \in \mathfrak{V}$, and in view of (2.4.1)

$$s_{V_1,\ldots,V_m} \leqslant s_{V_1,\ldots,V_n} \wedge s_{V_{n+1},\ldots,V_m}. \tag{2.4.3}$$

The pointwise infimum

$$h := \inf_{V_1,\ldots,V_n \in \mathfrak{V}} s_{V_1,\ldots,V_n}$$

satisfies

$$v \leqslant w \leqslant h \leqslant s \tag{2.4.4}$$

for any finely subharmonic minorant $w \geqslant v$ of s.

For fixed $V \in \mathfrak{V}$ it follows from (2.4.2) that any function $s_{V_1,\ldots,V_n,V}$ with $V_1, \ldots, V_n \in \mathfrak{V}$ is finely harmonic on V. The family $s_{V_1,\ldots,V_n}$ with variable $(V_1, \ldots, V_n) \in \mathfrak{V}^n$ $(n \in \mathbb{N})$ is lower directed because (2.4.3) implies

$$s_{V_1,\ldots,V_m,V} \leqslant s_{V_1,\ldots,V_n,V} \wedge s_{V_{n+1},\ldots,V_m,V}.$$

And $s_{V_1,\ldots,V_n,V}$ is finely harmonic in V by (2.4.2) applied to $u = s_{V_1,\ldots,V_n} \in \mathcal{V}$. As $V_1, \ldots, V_n$ range over $\mathfrak{V}$ it therefore follows by Theorem 2.1.31 that

$$h = \inf_{V_1,\ldots,V_n,V \in \mathfrak{V}, n \in \mathbb{N}} s_{V_1,\ldots,V_n,V}$$

likewise is finely harmonic on V. Since the sets $V \in \mathfrak{V}$ cover U we conclude that h is finely harmonic on U.

Now consider any finely subharmonic minorant w of s on U. Such is $v \vee w$, and hence by (2.4.4) with w replaced by $v \vee w$

$$v \leqslant v \vee w \leqslant h.$$

Thus h is indeed the greatest finely subharmonic (or finely harmonic) minorant of s. Finally, write $s - h = p$. Then $p \geqslant 0$ is finely superharmonic on U along with s. For any finely subharmonic minorant z of p, and hence of s, we have $z + h \leqslant s$, and since $z + h$ is finely subharmonic along with z, we conclude that $z + h \leqslant h$, that is, $z \leqslant 0$, and so p is indeed a fine potential on U. □

Remark 2.4.4 The restriction $s < +\infty$ is irrelevant in the uniqueness of the decomposition but cannot be omitted for the existence thereof even in the case $s \geqslant 0$. This appears from Example 2.1.8. In fact, let μ_n denote the measure $\sum_{j=1}^{n} 2^{-j}\varepsilon_{x_j}$ $(n \in \mathbb{N})$. Since $\mu(\{0\}) = 0$ we then have $G_\Omega\mu = \sup_n G_\Omega\mu_n$ with $G_\Omega\mu_n$ harmonic and hence finely harmonic on U, cf. Corollary 2.1.7 and Theorem 2.1.22(b). The restriction s of $G_\Omega\mu$ to U thus is finely superharmonic but has no greatest finely harmonic minorant because we would have $h \geqslant G_\Omega\mu_n|_U$ and hence $h(0) \geqslant \sup_n G_\Omega\mu_n(0) = G_\Omega\mu(0) = +\infty$.

Theorem 2.4.5 *The pointwise sum p of a countable family of fine potentials p_n on U is a fine potential on U if (and only if) p is finely superharmonic on U.*

Note that p is always finely hyperharmonic on U, by Corollary 2.1.21, so Definition 2.4.1 applies.

Proof of Theorem 2.4.5. Let v be a finely hypoharmonic minorant of p. For any $n \in \mathbb{N}$;

$$v - \sum_{j>n} p_j \leqslant p_1 + \ldots + p_n$$

and the left hand member is finely hypoharmonic because $\sum_{j>n} p_j$ is finely hyperharmonic. Hence $v \leqslant \sum_{j>n} p_j$ and so $v \leqslant 0$ at any point of $\{p < +\infty\}$. The latter set is finely dense in U because p is finely superharmonic, by Definition 2.4.1(ii). Since v is finely continuous, $v \leqslant 0$ on all of U, that is, p is indeed a fine potential. □

We continue denoting by $\mathcal{S}(U)$ the family of all superharmonic functions on U, and by $\mathcal{S}_+(U)$ its positive cone. Furthermore, $\mathcal{P}(U)$ will denote the convex subcone (cf. Theorem 2.4.5) of $\mathcal{S}_+(U)$ consisting of all fine potentials on U. Clearly, $\mathcal{S}_+(U)$ and $\mathcal{P}(U)$ are *solid*, or *hereditary*, in $\mathcal{U}^+(U)$ (the finely hyperharmonic functions $\geqslant 0$ on U), in the sense that, for any $u \in \mathcal{U}^+(U)$ and $s \in \mathcal{S}_+(U)$,

$$u \leqslant s \Rightarrow u \in \mathcal{S}_+(U),$$

and similarly with $\mathcal{P}(U)$ in place of $\mathcal{S}_+(U)$.

The fine convergence theorem (Theorem 2.1.34) evidently remains valid when "finely hyperharmonic" is replaced by "finely superharmonic" or by "fine potential". Hence $\mathcal{S}(U)$ is conditionally lower complete in the sense of lattice theory, in the natural order.

The specific order on $\mathcal{U}^+(U)$ induces the specific order "$\preccurlyeq$" on $\mathcal{S}_+(U)$, defined for $s, t \in \mathcal{S}_+(U)$ by

$$s \preccurlyeq t \iff \exists u \in \mathcal{S}_+(U) : s + u = t.$$

In the affirmative case u is uniquely determined, first q.e. on U and next everywhere on U, by fine continuity. We may then write $t - s = u$.

According to Corollary 2.2.18, $\mathcal{S}_+(U)$ is a *lower complete* and *conditionally upper complete lattice* in the specific order. Moreover $(\mathcal{S}_+(U), \preccurlyeq)$ is a *prevector lattice* in the sense of [CC3, Section 8.1], that is, for any $s, t, u \in \mathcal{S}_+(U)$ we have

$$s + u = t + u \Rightarrow s = t,$$

$$s + t = 0 \Rightarrow s = t = 0.$$

In other words, $(\mathcal{S}_+(U), \preccurlyeq)$ can be imbedded as the positive cone in a canonical way vector lattice (Riesz space) $[\mathcal{S}_+(U)]$. See [CC3, Proposition 11.2.1], [Do4, Appendix III]. See also Sect. 2.5, where we show that $\mathcal{S}_+(U)$ even is an H-cone, and hence [BBC2, Theorem 2.1.5] applies.

For the following and further lattice properties of $\mathcal{S}_+(U)$ with the specific order, see [CC3, Proposition 8.1.1].

Distributivity. For any family $(u_i)_{i\in I}$ in $\mathcal{S}_+(U)$ and any $u \in \mathcal{S}_+(U)$ we have

$$u \curlyvee (\curlywedge_{i\in I} u_i) = \curlywedge_{i\in I}(u \curlyvee u_i), \;\; u + \curlywedge_{i\in I} u_i = \curlywedge_{i\in I}(u + u_i).$$

For any specifically majorized family $(u_i)_{i\in I}$ in $\mathcal{S}_+(U)$ and any $u \in \mathcal{S}_+(U)$ we have

$$u \curlywedge (\curlyvee_{i\in I} u_i) = \curlyvee_{i\in I}(u \curlywedge u_i), \;\; u + \curlyvee_{i\in I} u_i = \curlyvee_{i\in I}(u + u_i).$$

Decomposition property. If $(u_i)_{i\in I}$ and $(v_j)_{j\in J}$ are two finite families in $\mathcal{S}_+(U)$ such that

$$\sum_{i\in I} u_i = \sum_{j\in J} v_j$$

then there exists a family $(w_{ij})_{(i,j)\in I\times J}$ in $\mathcal{S}_+(U)$ such that for any $i \in I$ and $j \in J$ we have

$$u_i = \sum_{k\in J} w_{ik}, \;\; v_j = \sum_{k\in I} w_{kj}.$$

Property of decomposition in the specific order. Let $u, u_1, u_2 \in \mathcal{S}_+(U)$. If $u \preccurlyeq u_1 + u_2$ there exist $v_1, v_2 \in \mathcal{S}_+(U)$ such that

$$v_1 \preccurlyeq u_1, \;\; v_2 \preccurlyeq u_2, \;\; u = v_1 + v_2.$$

In fact, there exists $v \in \mathcal{S}_+(U)$ with $u + v = u_1 + u_2$. By the Decomposition property there exist w_{ij}, $i, j \in \{1, 2\}$, such that

$$u = w_{11} + w_{12},\ v = w_{21} + w_{22},\ u_1 = w_{11} + w_{21},\ u_2 = w_{12} + w_{22}.$$

The functions $v_1 = w_{11}$ and $v_2 = w_{12}$ satisfy the required conditions.

Theorem 2.4.6 *The convex subcone $\mathcal{P}(U)$ of $\mathcal{S}_+(U)$ is a band in $\mathcal{S}_+(U)$ in the sense that*

(a) *$\mathcal{P}(U)$ is specifically solid in $\mathcal{S}_+(U)$, that is, for any $s \in \mathcal{S}_+(U)$ and $p \in \mathcal{P}(U)$,*

$$s \preccurlyeq p \Rightarrow s \in \mathcal{P}(U).$$

(b) *If a non-void specifically upper directed subfamily $\mathcal{F}$ of $\mathcal{P}(U)$ has a specific majorant of class $\mathcal{S}_+(U)$ then $\curlyvee \mathcal{F} \in \mathcal{P}(U)$.*

Proof (a) $\mathcal{P}(U)$ is clearly solid in $\mathcal{S}_+(U)$, even in the natural order "$\leqslant$". And (b) is an extension of Theorem 2.4.5 with a similar proof: Write $s = \curlyvee \mathcal{F} = \bigvee \mathcal{F}$, cf. Corollary 2.2.18(a). For any finely hypoharmonic pointwise minorant v of s and any $p \in \mathcal{F}$ we have

$$v - (s - p) \leqslant p,$$

where the left hand member is finely hypoharmonic because $p \preccurlyeq s$, that is, $s - p \in \mathcal{S}_+(U)$. Since $p \in \mathcal{P}(U)$ it follows that $v - (s - p) \leqslant 0$, that is, $p \leqslant s - v$. This holds for any $p \in \mathcal{F}$, and so

$$s = \bigvee \mathcal{F} \leqslant s - v,$$

whence $v \leqslant 0$ q.e. (namely on $\{s < +\infty\}$). Since v is finely continuous we conclude that indeed $v \leqslant 0$. □

The orthogonal band $\mathcal{P}(U)^\perp$ within $\mathcal{S}_+(U)$ is properly contained in the set of all $s \in \mathcal{S}_+(U)$ which are finely harmonic q.e. on U. This will follow from Theorem 2.4.11 below because $\mathcal{P}^s(U) \subset \mathcal{P}(U)$ with strict inclusion (assuming that $U \neq \emptyset$). The elements of $\mathcal{P}(U)^\perp$ are termed *invariant* functions. They will be studied in Sect. 2.9.

Polar sets are removable for fine potentials:

Theorem 2.4.7 *Let E be a polar subset of finely open set $U \subset \Omega$. Every fine potential p on $U \setminus E$ extends by fine continuity to a unique fine potential on U.*

Proof By the corresponding result for finely superharmonic functions (cf. Theorem 2.1.43), p extends by fine continuity to a (unique) finely superharmonic function $s \geq 0$ on U. For any subharmonic minorant v of s on U, the restriction $v|_{U \setminus E}$ is a finely subharmonic minorant of p on $U \setminus E$, and hence $v \leqslant 0$ on $U \setminus E$. Since v is finely continuous on U, and E has no finely inner points, it follows that $v \leqslant 0$ on U, and so s is indeed a fine potential on U. □

The situation is different when we consider the *restriction* of a fine potential on U to $U \setminus E$:

Definition 2.4.8 A function $p \in \mathcal{S}_+(U)$ is called a *semibounded (or stable) fine potential* if it has one of the following equivalent properties:

(i) for any polar set $E \subset U$, the restriction $p|_{U\setminus E}$ is a fine potential on the finely open set $U\setminus E$, [3]

(ii) $p|_{\{p<+\infty\}}$ is a fine potential.

Note that (i) implies that p even is a fine potential on U (take $E = \varnothing$). Obviously (i)⇒(ii) (take $E = \{p = +\infty\}$ in (i)).

For the remaining implication (ii)⇒(i) let v denote a finely subharmonic minorant of p on $U\setminus E$. Since p is finite and finely continuous on $\{p < +\infty\}$ it follows by Theorem 2.1.43 on removable singularities that v extends by fine continuity to a finely hypoharmonic, and indeed finely subharmonic function v' on $\{p < +\infty\}$, the difference set $\{p < +\infty\}\setminus(U\setminus E)$ being polar (a subset of E). By hypothesis, $p|_{\{p<+\infty\}}$ is a fine potential, and hence $v' \leqslant 0$, in particular $v \leqslant 0$.

The following theorem is an easy consequence of Definition 2.4.8 and Theorem 2.4.5.

Theorem 2.4.9 *Every finite valued fine potential is semibounded. The pointwise sum of a countable family of semibounded fine potentials is a semibounded fine potential, if finely superharmonic.*

We denote by $\mathcal{P}^s(U)$ the convex cone of all semibounded fine potentials on U. Clearly, $\mathcal{P}^s(U)$ is *solid* (or hereditary) in $\mathcal{S}_+(U)$ in the sense that if $p \in \mathcal{P}^s(U)$ and if $q \in \mathcal{S}_+(U)$, and if $q \leqslant p$, then $q \in \mathcal{P}^s(U)$. The latter statement of Theorem 2.4.9 extends as follows:

Theorem 2.4.10 *$\mathcal{P}^s(U)$ is a band in $\mathcal{P}(U)$ and hence in $\mathcal{S}_+(U)$.*

Proof Given a polar set $E \subset U$, consider the restriction map $\varrho : \mathcal{S}_+(U) \to \mathcal{S}_+(U\setminus E)$, where $\varrho(s) := s|_{U\setminus E}$ for $s \in \mathcal{S}_+(U)$. In view of Theorem 2.1.43 on removable singularities ϱ is a bijection and an order isomorphism in the specific orders on U and on $U\setminus E$. For any non-void specifically upper directed family $\Phi \subset \mathcal{P}^s(U)$ having a specific majorant $s \in \mathcal{S}_+(U)$ we have

$$\varrho(\curlyvee \Phi) = \curlyvee \varrho(\Phi) \in \mathcal{P}(U\setminus E)$$

because $\varrho(\Phi) \subset \mathcal{P}(U\setminus E)$ and $\mathcal{P}(U\setminus E)$ is a band in $\mathcal{S}_+(U\setminus E)$, by Theorem 2.4.6. For $p = \curlyvee \Phi$ we thus have $p|_{U\setminus E} = \varrho(p) \in \mathcal{P}(U\setminus E)$, and we conclude by varying E that indeed p is semibounded. □

Theorem 2.4.11 *Every function $s \in \mathcal{S}_+(U)$ has a unique decomposition $s = p + h$, where $p \in \mathcal{P}^s(U)$, and $h \in \mathcal{S}_+(U)$, h finely harmonic on $\{h < +\infty\}$.*

---

[3] The condition (ii) in [Fu8, Lemma 10.9] should have been omitted since it obviously does not follow from p being a stable fine potential. (The said condition is, however, nowhere referred to in [Fu8], nor in other work of the authors.)

Proof The restriction s' of s to $\{s < +\infty\}$ is a finite finely superharmonic function $\geqslant 0$ and hence representable uniquely, by Theorem 2.4.3, as

$$s' = p' + h'$$

where p' is a fine potential and h' a finely harmonic function $\geqslant 0$ on $\{s < +\infty\}$. By Theorem 2.1.43, p' and h' extend by fine continuity to functions p and h of class $\mathcal{S}_+(U)$. By Definition 2.4.8(ii), $p \in \mathcal{P}^s(U)$ whereas h is finely harmonic on $\{s < +\infty\}$, hence q.e. on U, and indeed everywhere on $\{h < +\infty\}$ according to Theorem 2.1.45. The uniqueness of the decomposition is straightforward. □

We shall write

$$\begin{aligned}\mathcal{P}^s(U)^\perp &= \{h \in \mathcal{S}_+(U) : h|_{\{h<+\infty\}} \text{ is finely harmonic}\} \\ &= \{h \in \mathcal{S}_+(U) : h \text{ is finely harmonic q.e. on } U\}.\end{aligned}$$

For the equivalence of these two definitions of $\mathcal{P}^s(U)^\perp$ note that, if $h|_{\{h<+\infty\}}$ is finely harmonic then h is finely harmonic q.e.. Conversely, if h is finely harmonic on $U \setminus E$ for some polar set E then h is finely harmonic on $\{h < +\infty\}$ in view of Theorem 2.1.45, by fine continuity of $h \in \mathcal{S}_+(U)$.

According to Theorem 2.1.31, the pointwise sum h of a countable family of functions of class $\mathcal{P}^s(U)^\perp$ belongs to $\mathcal{P}^s(U)^\perp$ if (and only if) $h \in \mathcal{S}_+(U)$. We therefore have the following

Theorem 2.4.12 *$\mathcal{P}^s(U)^\perp$ is a band, and indeed the orthogonal complement to $\mathcal{P}^s(U)$ in $\mathcal{S}_+(U)$.*

The uniqueness in Theorem 2.4.11 amounts in fact to the orthogonal bands $\mathcal{P}^s(U)$ and $\mathcal{P}^s(U)^\perp$ being the orthogonal complements of each other (relative to the prevector lattice $\mathcal{S}_+(U)$).

The semibounded fine potentials can be characterized similarly to the semibounded potentials. In particular (i), (ii), and (iii) in Definition 1.5.16 carry over as follows (recall from Definition 2.2.6 and subsequent text the notations ${}^U R_u^A$ and ${}^U \widehat{R}_u^A$).

Theorem 2.4.13 *A fine potential p on U is semibounded if and only if it has the following equivalent properties*
(i) $\widehat{\inf}_{c>0} {}^U R_p^{\{p>c\}} = 0$, *or equivalently* $\widehat{\inf}_{c>0} {}^U \widehat{R}_p^{\{p>c\}} = 0$.
(ii) *p is representable as the pointwise sum of finite fine potentials on U.*
(iii) *p is the specific supremum of those specific minorants of p which are finite fine potentials.*

Since every finite fine potential is semibounded (Theorem 2.4.9); Assertion (iii) in the above theorem amounts to *the band $\mathcal{P}^s(U)$* being *generated by the finite fine potentials on U*. We show later that the remaining properties (iv) and (v) in Definition 1.5.16 carry over so as to cover the semibounded potentials.

Proof of Theorem 2.4.13. Let p be a fine potential on U. If p is semibounded then (i) holds. In fact each ${}^U\widehat{R}_p^{\{p\geq n\}}$ with $n \in \mathbb{N}$ is finely harmonic on $\{p < n\}$ wherever finite, by Theorem 2.1.52 applied with $u = p$ and $V = \{p < n\}$. Thus ${}^U\widehat{R}_p^{\{p\geq n\}}$ is finely harmonic q.e. on $\{p < n\}$, and decreases as n increases. By Theorem 2.1.31 it follows that $\inf_n {}^U\widehat{R}_p^{\{p\geq n\}}$ is finely harmonic (and $\leqslant p$) q.e. on $\{p < +\infty\}$, and hence q.e. on U. So is therefore the semibounded potential $\widehat{\inf_n} {}^U\widehat{R}_p^{\{p\geq n\}}$, by the fine fundamental convergence theorem (Theorem 2.1.34). Consequently, $\widehat{\inf_n} {}^U R_p^{\{p>n\}} = \widehat{\inf_n} {}^U R_p^{\{p\geq n\}} = 0$ in view of Theorem 2.4.11. Thus semiboundedness of p implies (i).

(i)⇒(ii) is established as the same implication in Definition 1.5.16, now with $R_p^{\{p>n\}}$ replaced by ${}^U R_p^{\{p>n\}}$ and $\widehat{R}_p^{\{p>n\}}$ by ${}^U\widehat{R}_p^{\{p>n\}}$, using the fine property of natural decomposition (see the fourth paragraph following the proof of Theorem 2.2.17). Also replace $\mathcal{P}^b$ and $\mathcal{P}_0^c$ by the finite fine potentials on U.

(ii)⇒(iii) is obvious, and (iii)⇒(ii) holds in view of the quasi-Lindelöf property for functions (see Sect. 1.2). Finally, (ii) implies $p \in \mathcal{P}^s$ according to Theorem 2.4.9. □

Theorem 2.4.14 (The relative fine boundary minimum principle) *Let u be finely hyperharmonic on a finely open subset V of U where U is finely open in Ω. Let $p \in \mathcal{P}(U)$, and suppose that $u \geqslant -p$ on V. If*

$$\text{f-}\liminf_{x\to y,\, x\in V} u(x) \geqslant 0 \quad \textit{for every } y \in U \cap \partial_f V \tag{2.4.5}$$

then $u \geqslant 0$ on V. If $p \in \mathcal{P}^s(U)$ and if instead (2.4.5) holds q.e. for $y \in U \cap \partial_f V$, then again $u \geqslant 0$ on V.

Proof In view of Lemma 2.1.51 the inequality (2.4.5) implies that the extension of $u \wedge 0$ by 0 on $U \setminus V$ is a finely hyperharmonic majorant of $-p$, hence $u \geqslant 0$ on V. If $p \in \mathcal{P}^s(U)$ and if (2.4.5) holds on $U \cap \partial_f V \setminus E$ for some polar set $E \subset U \cap \partial_f V$, then the above applies with U, p replaced by $U \setminus E$, $p|_{U\setminus E}$ (while keeping V), noting that $p|_{U\setminus E} \in \mathcal{P}(U \setminus E)$ according to Definition 2.4.8(i). □

Corollary 2.4.15 (The strengthened relative fine boundary minimum principle) *Let u be finely hyperharmonic on a finely open subset V of U where U is finely open in Ω. Let $p \in \mathcal{P}(U)$, and suppose that $u \geqslant -p$ on V. If*

$$\text{f-}\liminf_{x\to y,\, x\in V} u(x) \geqslant 0 \quad \textit{for every } y \in U \cap \partial_f V \setminus e \tag{2.4.6}$$

and

$$\text{f-}\liminf_{x\to y,\, x\in V} u(x) > -\infty \quad \textit{for every } y \in U \cap \partial_f V \cap e \tag{2.4.7}$$

for some polar set $e \subset \Omega$, then $u \geqslant 0$ on V.

Proof Let s be a positive finely superharmonic function on Ω such that $s = +\infty$ on e. By applying the above theorem to the function $u + \varepsilon s$, $\varepsilon > 0$, we have $u + \varepsilon s \geq 0$ on V. By letting ε to 0 this leads to $u \geq 0$ q.e. and hence everywhere on V by fine continuity. □

Corollary 2.4.16 *Let u be finely hyperharmonic on a finely open subset V of U where U is finely open in Ω. Let $p \in \mathcal{P}(U)$, and suppose that $u \geqslant -p$ on V. If*

$$\text{f-}\liminf_{x\to y,\, x\in V} u(x) \geqslant 0 \quad \textit{for every } y \in U \cap \partial_{f,r} V \tag{2.4.8}$$

and

$$\text{f-}\liminf_{x\to y,\, x\in V} u(x) > -\infty \quad \textit{for every } y \in U \cap \partial_{f,i} V \tag{2.4.9}$$

then $u \geqslant 0$ on V.

Indeed, it suffices to apply the above corollary with the polar set $e = \partial_{f,i} V$.

Remark 2.4.17 For any given finely open set $U \subset \Omega$, both assertions of Theorem 2.4.14 would break down if p were replaced by any function $s \in \mathcal{S}_+(U)$ which is not a fine, resp. semibounded potential on U. To see this in the case of former assertion, take $V = U$ and $u = -v$, where v denotes a non-zero finely subharmonic minorant $\geqslant 0$ of s, cf. Definition 2.4.2. Then $u \geqslant -s$ and (2.4.5) holds trivially because $U \cap \partial_f V = \varnothing$, but $u = -v$ takes values < 0 on V. For the latter assertion of Theorem 2.4.14, let $s = p + h$ be the decomposition of s from Theorem 2.4.11, now with h taking values > 0 on $V := \{x \in U : h(x) < +\infty\}$, and with $u = -h$. Then $u \geqslant -s$ on V, and (2.4.5) now holds trivially q.e. on the polar set $U \cap \partial_f V = U \setminus V = \{x \in U : h(x) = +\infty\}$, noting that $h \in \mathcal{S}_+(U)$. Nevertheless, $u = -h$ takes values < 0 on V.

Theorem 2.4.18 (The fine domination property) *Let $p \in \mathcal{P}^s(U)$, and suppose that p is finely harmonic q.e. on $U \setminus S$ for some $S \subset U$ which is finely closed relative to U. Let $u \in \mathcal{U}^+(U)$ and suppose that $u \geqslant p$ q.e. on S. Then $u \geqslant p$ everywhere on U. Thus (Corollary 2.2.9) ${}^U\widehat{R}_p^S = p$.*

Proof The set $e := \{x \in U : p(x) = +\infty\}$ is polar, by Definition 2.4.1(iii), and hence finely closed (see Sect. 1.2). Now, p is finite and finely continuous on the finely open set $V := U \setminus (S \cup e)$, and hence finely locally bounded there. Being moreover finely harmonic q.e. on V, p is finely harmonic on all of V according to Theorem 2.1.45. Now apply the latter case of Theorem 2.4.14 to the finely hyperharmonic function $u - p$ ($\geqslant -p$) on V, observing that by the fine continuity of u and p, the boundary inequality (2.4.5) in that theorem is fulfilled q.e. on $U \cap \partial_f V$, namely at any point $x \in (U \cap \partial_f V) \setminus e (\subset S)$ such that $u(x) \leqslant p(x)$. Consequently, $u \geqslant p$ on V and hence q.e. on U, namely q.e. on $U \setminus e = S \cup V$. Finally, $u \geqslant p$ on all of U because $\{x \in U : u(x) < p(x)\}$ is polar and finely open, hence void, the fine topology on $\mathbb{R}^N$ having no isolated points, see Sect. 1.2. □

We bring a "fine" extension of an important result of Brelot [Br5].

Theorem 2.4.19 *Let U be a finely open subset of Ω, and let $f : U \to [0, +\infty]$ be a finely l.s.c. function having a finely superharmonic majorant on U. Then*

(a) ${}^U R_f \equiv {}^U \widehat{R}_f$ *is such a majorant on U. In particular ${}^U R_f$ is finely continuous on U.*

(b) ${}^U R_f$ *is finely harmonic on any finely open set $V \subset U$ such that either (i) ${}^U R_f$ is finite on V and $f|_V$ is finely subharmonic or (ii) $f|_V$ is finely continuous and $f|_V < ({}^U R_f)|_V < +\infty$.*

Proof (a) Since f is supposed to be finely l.s.c. on U we have ${}^U R_f = {}^U \widehat{R}_f$ on U, by Theorem 2.2.14(b). As noted after Definition 2.2.6, the function $v := {}^U \widehat{R}_f$ is finely hyperharmonic on U and hence finely continuous there, thus establishing the fine continuity of ${}^U R_f$ on U. Moreover, $f \leq v$ and v is finely superharmonic because it is majorized by a finely superharmonic function along with f.

(b) Put $u = {}^U R_f$. Consider first case (i) where f is finely subharmonic on V and hence finite there. Let W be a finely open set with compact closure $\overline{W} \subset V$ on which f is upper bounded. According to Theorem 2.3.9, we have ${}^U R_u^{U \setminus W} = \overline{H}_u^W$ on W. Hence ${}^U R_u^{U \setminus W} \geqslant \underline{H}_u^W \geqslant \underline{H}_f^W \geqslant f$ on W and trivially on $U \setminus W$ where ${}^U R_u^{U \setminus W} = u \geqslant f$. It then follows that $u \geqslant {}^U R_u^{U \setminus W} \geqslant f$ on all of U. Hence ${}^U \widehat{R}_u^{U \setminus W} \geq f$ q.e. on U by the fine convergence theorem (Theorem 2.1.34), and by fine lower semicontinuity of f the polar set $\{{}^U \widehat{R}_u^{U \setminus W} < f\} \cap U$ is finely open and hence void. Thus ${}^U \widehat{R}_u^{U \setminus W} \geqslant f$ everywhere on U, and hence $u \geqslant {}^U R_u^{U \setminus W} \geqslant {}^U \widehat{R}_u^{U \setminus W} \geqslant {}^U R_f = u$ on U so that $u = {}^U R_u^{U \setminus W}$ and therefore $u = \overline{H}_u^W$ on W. The function $u = {}^U R_f$ on U being finite on W it follows from Theorem 2.3.7 that indeed u is finely harmonic on W and hence on all of V by the sheaf property of finely harmonic functions, see Remark 2.1.5.

Next consider case (ii) where f is finely continuous on a finely open set $V \subset U$ and $f < u < +\infty$ on V. For given $x \in V$ there exists a constant c and a finely open set $Z \subset V$ such that $x \in Z$ and $f < c < u$ on Z. Let g be the function defined by $g = f \vee c 1_Z$. Then $u \geq g \geqslant f$ on U and $g = c$ on Z. Moreover, g is finely l.s.c. on U along with f and $c 1_Z$. Hence $u = {}^U R_f \leqslant {}^U R_g \leqslant u$ and therefore ${}^U R_g = u$. Since g is finely subharmonic on Z, it follows from case (i) above, applied with V replaced by Z, that u is finely harmonic on Z. We conclude, again by the sheaf property of finely harmonic functions, that u is finely harmonic on V. □

Corollary 2.4.20 *For any finely superharmonic function $u \geqslant 0$ on U and any set $A \subset U$, ${}^U \widehat{R}_u^A$ is finely harmonic on $\{{}^U \widehat{R}_u^A < +\infty\} \setminus \mathfrak{b}(A)$ and hence finely harmonic q.e. on $U \setminus \mathfrak{b}(A)$.*

Since ${}^U \widehat{R}_u^A = {}^U \widehat{R}_u^{U \cap \mathfrak{b}(A)}$ by Lemma 2.2.12, this is obtained by applying Theorem 2.4.19(b) to $f = 1_{\mathfrak{b}(A)} u$ and $V = U \setminus \mathfrak{b}(A)$.

Corollary 2.4.20 will be further extended in Lemma 2.9.7 in terms of invariant functions.

We proceed to compare notions of potentials and fine potentials on a (Euclidean) open set U. In view of Remark 2.1.42, it suffices to consider the particular case where $U = \Omega$.

In the first place, $\mathcal{S}_+(\Omega) = \mathcal{S}_+$, that is, the positive finely superharmonic functions on Ω are the same as the positive superharmonic functions, by Theorem 2.1.22(a) and in view of Definition 2.4.1(iii). Next, every fine potential p on Ω is a potential. Indeed, p is finely superharmonic and $\geqslant 0$, hence superharmonic, as just noted, and every subharmonic minorant of p is finely subharmonic, hence $\leqslant 0$, and so p is indeed a potential. The converse will be established in Theorem 2.4.22 below, after first considering the semibounded potentials in the next Theorem.

Theorem 2.4.21 *A semibounded fine potential on Ω is the same as a semibounded potential.*

Proof As noted just before Theorem 2.4.21, a fine potential p on Ω (Definition 2.4.2) is a potential on Ω. And Theorem 2.4.13(i) (applied to $U = \Omega$) shows that Definition 1.5.16(i) is satisfied.

Conversely, every semibounded potential p on Ω is finely superharmonic on Ω, and p must be a semibounded fine potential relative to Ω in view of the fine boundary minimum principle (Theorem 2.1.10) combined with Remark 2.4.17 to the relative fine boundary minimum principle (Theorem 2.4.14), now with $U = \Omega$ and with V replaced by U (hence with $U \cap \partial_f V$ replaced by $\partial_f U$). □

Theorem 2.4.22 *A fine potential on Ω is the same as a potential on Ω.*

Proof As noted in the first part of the proof of the preceding theorem, every fine potential on Ω is a potential. Conversely, let p be a potential on Ω. In particular, $p \geqslant 0$ is superharmonic and hence finely hyperperharmonic, by Theorem 2.1.22(a). Since p is finite q.e. it follows from Definition 2.4.1(iii) that p is finely superharmonic. By Theorem 2.4.11, now with $U = \Omega$, we therefore have

$$p = p' + h,$$

where p' is a semibounded fine potential on Ω and h is finely harmonic on $\{h < +\infty\}$. Besides $h \geqslant 0$ is finely superharmonic on Ω, hence superharmonic, again by Theorem 2.1.22(a), and $h \leqslant p$ is therefore a potential on Ω. This reduces the remaining part of the proof of Theorem 2.4.22 to the case where the given potential p is finely harmonic on $\{p < +\infty\}$.

In that case, let v denote a finely subharmonic minorant $\geqslant 0$ of p. Then $q := p - v$ is finely superharmonic and $\geqslant 0$, and hence superharmonic.

Since $q \leqslant p$, q is a potential. For fixed $c > 1$ the set

$$A = \{p > cv\}$$

is finely open because p and v are finely continuous. Hence $cq = cp - cv \geqslant (c - 1)p$ on A, and

$$A \subset \mathfrak{b}(A) \subset \widetilde{A} \subset \{p \geqslant cv\},$$

$$q \geqslant \frac{c-1}{c} p \quad \text{on } A,$$

$$q \geqslant \frac{c-1}{c}\widehat{R}_p^A \quad \text{on } \Omega.$$

To complete the proof we show that $\widehat{R}_p^A = p$ on Ω for every choice of $c > 1$, and hence $q \geqslant p$, that is, by letting $c \to +\infty$ in the last inequality in the above display.

For that purpose we apply Corollary 2.1.16 to the finely open set $U = \{p < +\infty\}$ (where the given potential p is finely harmonic) and $V := V_n = \{p < n\}$ with fine closure $\widetilde{V}_n = \{p \leqslant n\} \subset U$, the finely (hyper)harmonic functions $f = \pm p$ on U, and the (semi)bounded potential $p \wedge n$ on Ω. Thus $p = p^{\complement V_n} = \widehat{R}_p^{\complement V_n}$ on V_n, and of course on $\mathfrak{b}(\complement V_n)$ (cf. Definition 1.4.29 and Eq. (1.4.22)), hence altogether q.e. on $V_n \cup \complement V_n = \Omega$. Since p and $p^{\complement V_n} = \widehat{R}_p^{\complement V_n}$ are finely continuous it follows that $p = p^{\complement V_n}$ everywhere on Ω.

We may write $p = G_\Omega\lambda$, $q = G_\Omega\mu$ with admissible measures $\lambda, \mu \in \Lambda$. The relation

$$G_\Omega\lambda = (G_\Omega\lambda)^{\complement V_n} = G_\Omega(\lambda^{\complement V_n}),$$

cf. Theorem 1.4.24, shows that $\lambda = \lambda^{\complement V_n}$, that is, λ is carried by $\mathfrak{b}(\complement V_n) \subset \complement V_n$, see Theorem 1.4.23(b), and so $\lambda(V_n) = 0$. Since the sequence (V_n) covers $\{p < +\infty\}$, it follows that λ is carried by $\{p = +\infty\}$. In particular, λ is carried by the above finely open set $A = \{p > cv\}$ and hence by $\mathfrak{b}(A) \supset A$. We conclude, again by Theorem 1.4.23(b), that $\lambda^A = \lambda$, and hence

$$\widehat{R}_p^A = (G_\Omega)^A = G_\Omega(\lambda^A) = G_\Omega\lambda = p,$$

as claimed. This completes the proof of Theorem 2.4.22. □

Corollary 2.4.23 *The fine boundary minimum principle given in Theorem 2.1.10 remains valid with p replaced by an arbitrary potential, provided that the boundary inequality is required to hold everywhere on* $\partial_f U$.

This follows from the former assertion of Theorem 2.4.14 in view of Theorem 2.4.22. The next corollary is similar to the fine domination principle (Theorem 2.4.18).

Corollary 2.4.24 *Let p be a potential on* Ω*, and let S be a finely closed subset of* Ω*. If p is finely harmonic on* $\complement S$ *and finite on* $\partial_f S$*, then* $R_p^S = p$.

To see this, let $u \geqslant 0$ be hyperharmonic on Ω, and suppose that $u \geqslant p$ on S. Then $u - p$ is finely hyperharmonic and $u - p \geqslant -p$ on $U := \complement S$, and finely continuous on $\widetilde{U}$ (because u and p are finely continuous on $\widetilde{U}$). It therefore follows from the version of Theorem 2.1.10 given in Corollary 2.4.23 that $u - p \geqslant 0$ on $U = \complement S$. Hence $u \geqslant p$ on all of Ω, and so indeed $R_p^S = p$.

We are now prepared to improve the relation between harmonic and finely harmonic functions on Ω (or any Euclidean open set), given in Theorem 2.1.22(b).

Theorem 2.4.25 *A function $u : U \to \mathbb{R}$ (U Euclidean open) is harmonic if and only if u is finely harmonic and (Euclidean) locally bounded either from above or from below.*

Proof For the remaining "if" part, suppose that u is finely harmonic and locally bounded, say from below, on U. Then U can be covered by Greenian domains V on each of which u is bounded from below, say $u \geqslant c_V$ on V, u is hyperharmonic there by Theorem 2.1.22(a). Being also finite, u is even superharmonic $\geqslant c_V$ on V, and hence $u = p + h$ there, with p a finite potential and $h \geqslant c_V$ harmonic, in particular finely harmonic there. Thus $p = u - h$ is finely harmonic on V, and since p is also a fine potential on V, by Theorems 2.4.21 or 2.4.22, we conclude that $p = 0$, that is, $u = h$, and so u is indeed harmonic on each V, and therefore on U. □

Remark 2.4.26 The requirement in Theorem 2.4.25 that u be locally bounded from one side cannot be removed when $N > 2$, as shown by an example in [Fu11] (cf. Remark 2.1.23). For $N = 2$, this boundedness assumption can, however, be omitted, see Theorem 2.7.2 below.

Definition 2.4.27 For any numerical function f on a finely open set $U \subset \Omega$ we denote by $S(f)$ the *quasiharmonic support* of f relative to U, that is, the complement within U of the largest finely open subset of U on which f is finely harmonic quasi-everywhere.

Note that $S(f)$ exists in view of the quasi-Lindelöf principle (see Sect. 1.2). Indeed, f is finely harmonic q.e. on the union of any countable family of finely open sets on each of which f is finely harmonic q.e. Furthermore, $S(f)$ is finely closed relative to U, and even a *base* relative to U, in the sense that

$$U \cap \mathfrak{b}(S(f)) = S(f)$$

because $S(f) \backslash \mathfrak{b}(S(f))$ is polar. Clearly, $f_1 = f_2$ q.e. implies $S(f_1) = S(f_2)$. In particular

$$S(f) = S(\mathfrak{b}(f)). \tag{2.4.10}$$

Recall that the face of $\mathcal{S}_+(U)$ formed by all *finite* valued fine potentials relative to U is contained in $\mathcal{P}^s(U)$. For any finite fine potential p on U, $S(p)$ equals the fine harmonic support of p (in the obvious sense) relative to U on account of Theorem 2.1.43. Note that there is no good concept of fine harmonic support for a general fine potential on U.

The following proposition asserts that the quasiharmonic support S of semi-bounded fine potentials on a finely open set $U \subset \Omega$ is an *abstract carrier* in the sense of [CC3, p. 186]:

Theorem 2.4.28 (a) *For any $p \in \mathcal{P}^s(U)$, $S(p) = \varnothing$ if and only if $p = 0$,*
(b) *for $p_1, p_2 \in \mathcal{P}^s(U)$ such that $p_1 \preccurlyeq p_2$ we have $S(p_1) \subset S(p_2)$,*

(c) *for any $p \in \mathcal{P}^s(U)$ and any two sets $A_1, A_2 \subset U$ which together cover U and are finely closed relative to U, there exist $p_1, p_2 \in \mathcal{P}^s(U)$ such that $p_1 + p_2 = p$ and $S(p_i) \subset A_i$ for $i = 1, 2$.*

Proof (a) $S(p) = \varnothing$ means that p is finely harmonic q.e. on U, that is, $p \in \mathcal{P}^s(U)^\perp$, which amounts to $p = 0$.

(b) When $p_1 \preccurlyeq p_2$, p_1 is of course finely harmonic on any finely open set $V \subset U$ on which p_2 is finely harmonic, thus on $U \setminus S(p_2)$ off some polar set. It follows that $S(p_1) \subset S(p_2)$.

(c) Using the last assertion of Corollary 2.2.19(c) we obtain

$$p = {}^U\widehat{R}_p^U = {}^U\widehat{R}_p^{A_1 \cup A_2} \preccurlyeq {}^U\widehat{R}_p^{A_1} + {}^U\widehat{R}_p^{A_2}$$

Here ${}^U\widehat{R}_p^{A_i} \leqslant p$, and so ${}^U\widehat{R}_p^{A_i} \in \mathcal{P}^s(U)$ for $i = 1, 2$. By the property of decomposition in specific order property (last two paragraphs before Theorem 2.4.6) there exist $p_1, p_2 \in \mathcal{P}^s(U)$ such that $p_1 + p_2 = p$ and $p_i \preccurlyeq {}^U\widehat{R}_p^{A_i}$ for $i = 1, 2$. It follows by Corollary 2.4.20 that ${}^U\widehat{R}_p^{A_i}$ is finely harmonic q.e. in $U \setminus A_i$ and hence p_i is finely harmonic q.e. on $U \setminus A_i$ in view of (b), that is, $S(p_i) \subset A_i$ for $i = 1, 2$. □

Definition 2.4.29 Let $p \in \mathcal{P}^s(U)$. The specific restriction $1_A \cdot p$ of p relative to a relatively finely closed set $A \subset U$ is defined by

$$1_A \cdot p = \curlyvee\{h \in \mathcal{P}^s(U) : h \preccurlyeq p \quad \text{and } S(h) \subset A\}.$$

In other words, $1_A \cdot p$ is the specific supremum within $\mathcal{P}^s(U)$ of all specific minorants h of p which are finely harmonic q.e. on $U \setminus A$. It follows that $1_A \cdot p \preccurlyeq p$.

More generally, there is a good notion of specific multiplication $f \cdot p \in \mathcal{P}^s(U)$ for suitable (for example, bounded Borel measurable) function $f \geqslant 0$ on U (and $p \in \mathcal{P}^s(U)$), see [Fu8, Sections 11.19–11.21]. Indeed, any $p \in \mathcal{P}^s(U)$ has an integral representation $p = G_U\mu = \int_U G_U(\cdot, y)\, d\mu(y)$ in terms of the Green function G_U on U and a unique measure $\mu \geqslant 0$ on U, see Sects. 2.5 and 2.6, and then we simply have

$$f \cdot p = f \cdot G_U\mu = G_U(f\mu) = \int_U G(\cdot, y) f(y)\, d\mu(y).$$

The properties of this specific multiplication obtained in [Fu8] become immediate consequences of the corresponding well-known properties of the measures $f\mu$ with density f with respect to μ. The above notion of specific restriction of $p \in \mathcal{P}^s(U)$ to a relatively finely closed set $A \subset U$ will, however, be used in the proof of the integral representation of $p \in \mathcal{P}^s(U)$, see Theorem 2.6.11.

Consider two relatively finely closed subsets A, B of U. If $A \subset B$ (or just if $A \setminus B$ is polar) then clearly $1_A \cdot p \preccurlyeq 1_B \cdot p$. Hence, if A and B differ by a polar set then $1_A \cdot p = 1_B \cdot p$. The last observation applies in particular to A and $\mathfrak{b}(A)$, so

$$1_A \cdot p = 1_{\mathfrak{b}(A)} \cdot p.$$

Theorem 2.4.30 *For any $p \in \mathcal{P}^s(U)$ and any relatively finely closed subset A of U we have*

$$1_A \cdot p = p \curlywedge {}^U\widehat{R}_p^A.$$

Proof The set $V := U \setminus A$ is finely open, and $V \subset \mathfrak{r}(V)$ and $\mathfrak{r}(V) \setminus V$ is polar, see the beginning of the paragraph containing (1.4.21). If the theorem holds when V is regular we have

$$1_A \cdot p = 1_{\mathfrak{b}(A)} \cdot p = p \curlywedge {}^U\widehat{R}_p^{\mathfrak{b}(A)} = p \curlywedge {}^U\widehat{R}_p^A.$$

We may therefore assume that A is a base, that is, V is regular.

For any $t \in \mathcal{P}^s(U)$ with $t \preccurlyeq p$ and $S(t) \subset A$ we show that $t \preccurlyeq {}^U\widehat{R}_p^A$. There is a unique $u \in \mathcal{P}^s(U)$ such that $t + u = p$. Since $S(t) \subset A$, t is finely harmonic q.e. on V, and so is ${}^U\widehat{R}_p^A$ by Corollary 2.4.20. It follows by Theorem 2.4.18 that $t = \widehat{R}_t^A \leqslant \widehat{R}_p^A$, and hence there is a unique finely continuous function $v \geqslant 0$ on U such that $t + v = {}^U\widehat{R}_p^A$. Clearly, $v \leqslant p$ because ${}^U\widehat{R}_p^A \leqslant p$. Thus v is finely harmonic q.e. on V (along with ${}^U\widehat{R}_p^A$ and t) and hence finely hyperharmonic on V by Theorem 2.1.43 since it is positive and finely continuous on V. The set $E := (A \cap \{{}^U\widehat{R}_p^A \neq p\}) \cup \{p = +\infty\}$ is polar and $v = u + {}^U\widehat{R}_p^A - p$ on $U \setminus E$. Hence

$$\text{f-}\liminf_{x \to y,\, x \in V} v(x) = u(y) + {}^U\widehat{R}_p^A(y) - p(y) = u(y)$$

for $y \in U \cap \partial_f V \setminus E$. According to Lemma 2.1.51 applied with U, V replaced by $U \setminus E, V \setminus E$ (with V regular), $w = v|_{U \setminus E}$ is finely hyperharmonic. Since $v > -\infty$ on U, $v|_{U \setminus E}$ extends by fine continuity to a finely hyperharmonic function $v' \geqslant 0$ on U, by Theorem 2.1.43. Hence $t + v' = {}^U\widehat{R}_p^A$ on U. This shows that $t \preccurlyeq {}^U\widehat{R}_p^A$. Since $t + u = p$ it follows that $t \preccurlyeq p \curlywedge {}^U\widehat{R}_p^A$, and by varying t that $1_A \cdot p \preccurlyeq p \curlywedge {}^U\widehat{R}_p^A$.

On the other hand, the function $\tau := p \curlywedge {}^U\widehat{R}_p^A$ is of class $\mathcal{P}^s$ along with p. Clearly $\tau \preccurlyeq p$, and τ is finely harmonic q.e. on $U \setminus A$ along with ${}^U\widehat{R}_p^A$. Thus $\tau \preccurlyeq 1_A \cdot p$, and we conclude that indeed $1_A \cdot p = p \curlywedge {}^U\widehat{R}_p^A$. □

2.5 Localization in Fine Potential Theory. The Fine Green Kernel

The idea of capturing a local potential theoretic structure from a global one by a process of localization is classical, for example, when expressing the Green function on a domain in terms of that on the entire space. In more recent time, this process enters in fundamental work by Bliedtner and Hansen [BH1, BH2] on simplicial cones and by Boboc, Bucur, and Cornea [BBC1, BBC2] on H-cones.

In the present section and the next we shall similarly express the H-cone $\mathcal{S}_+(U)$ of positive finely superharmonic functions on a given finely open subset $U \subset \Omega$ by localization of the H-cone $\mathcal{S}_+(\Omega)$ of positive superharmonic functions on Ω as in [Fu21].

For brevity, write $B = \mathfrak{b}(\complement U)$, the base of $\complement U$ relative to the Greenian set $\Omega \subset \mathbb{R}^N$, $N \geqslant 2$. For $s \in \mathcal{S}_+ = \mathcal{S}_+(\Omega)$ we keep the notation

$$s^B = \widehat{R}_s^B = \widehat{R}_s^{\complement U}, \tag{2.5.1}$$

see Definition 1.4.29 and (1.4.22). Recall that $s^B = s$ on B and

$$s^B = \overline{H}_s^U = \int s \, d\varepsilon_{\cdot}^{\complement U} \tag{2.5.2}$$

on U, see Theorem 2.3.9 (with U, V replaced by Ω, U). Note that $s \in \mathcal{S}_+$ is l.s.c. and in particular Borel measurable. Recall from Sect. 2.4 the prevector lattice $\mathcal{S}_+(U)$ of all finely superharmonic functions $u \geqslant 0$ on U (Definition 2.4.1). In addition to the pointwise order "$\leqslant$" we have on $\mathcal{S}_+(U)$ the specific order "$\preccurlyeq$" on $\mathcal{S}_+(U)$ defined by

$$u \preccurlyeq v \iff \text{there exists } w \in \mathcal{S}_+(U) \text{ such that } u + w = v.$$

In the affirmative case, w is uniquely determined, first q.e. on U and next everywhere on U, by fine continuity. We may write $v - u = w$.

Recall furthermore the following convex subcones of $\mathcal{S}_+(U)$:

$\mathcal{P}(U) =$ the fine potentials on U (Definition 2.4.2). $\mathcal{P}(U)$ is a band in $\mathcal{S}_+(U)$ (Theorem 2.4.6), and $\mathcal{P}(U)$ is solid (=hereditary) in $\mathcal{S}_+(U)$.

$\mathcal{P}^s(U)$= the semibounded (or stable) fine potentials on U (Definition 2.4.8). $\mathcal{P}^s(U)$ is a band in $\mathcal{P}(U)$ and hence in $\mathcal{S}_+(U)$, and $\mathcal{P}^s(U)$ is pointwise solid.

In particular, $\mathcal{S}_+(\Omega) = \mathcal{S}_+$, $\mathcal{P}(\Omega) = \mathcal{P}$ (the potentials on Ω, Theorem 2.4.22), and $\mathcal{P}^s(\Omega) = \mathcal{P}^s$ (the semibounded potentials on Ω, Theorem 2.4.21).

The orthogonal band $\mathcal{P}^s(U)^\perp$ of $\mathcal{P}^s(U)$ in $\mathcal{S}_+(U)$ consists of all functions $h \in \mathcal{S}_+(U)$ which are finely harmonic quasi-everywhere on U, or equivalently: finely harmonic on $\{h < +\infty\}$ (Theorems 2.4.11 and 2.4.12).

For any $h \in \mathcal{P}^s(U)^\perp$ and any $s \in \mathcal{S}_+(U)$, $h \leqslant s$ is equivalent to $h \preccurlyeq s$, and in the affirmative case we may write $s - h$ for the unique function $t \in \mathcal{S}_+(U)$ for which $s = h + t$.

In particular, $\widehat{R}_s^{\complement U}|_U \in \mathcal{P}^s(U)^\perp$ for any $s \in \mathcal{S}_+$ (Corollary 2.1.30), and hence the following definition produces a function $s_U \in \mathcal{S}_+(U)$.

Definition 2.5.1 The *localization operator* associated with U is the map $s \mapsto s_U$ of $\mathcal{S}_+$ into $\mathcal{S}_+(U)$ defined by

$$s_U = s|_U - \widehat{R}_s^{\complement U}|_U = s|_U - \overline{H}_s^U.$$

The last expression for $s|_U$ follows from (2.5.2). It shows that the localization operator is *countably additive* (and positive homogeneous): If $(s_n)_{n\in\mathbb{N}}$ is a sequence on $\mathcal{S}_+$ and if $s \in \mathcal{S}_+$ satisfies $s = \sum_{n\in\mathbb{N}} s_n$ then

$$s_U = \sum_{n\in\mathbb{N}} (s_n)_U. \tag{2.5.3}$$

In fact, $s_U + \overline{H}_s^U = s|_U$ q.e. on U, hence everywhere on U, by fine continuity. When applied to s_n in place of s this leads after summation to $s_U + H_s^U = \sum_n (s_n)_U + H_s^U$, whence the assertion.

For any pointwise upper directed family $\mathcal{F} \subset \mathcal{S}_+$ with pointwise supremum $\bigvee \mathcal{F} = \sup \mathcal{F} = s \in \mathcal{S}_+$ we have

$$s_U = \bigvee \{t_U : t \in \mathcal{F}\}.$$

In view of the quasi-Lindelöf principle for finely l.s.c. functions $\geqslant 0$ it suffices to consider the case of an increasing sequence in $\mathcal{F}$, and hence the assertion follows from Definition 2.5.1 much as in the Proof of (2.5.3).

If $V = \mathfrak{r}(U) = \complement B$ (the regularized of U) then $V = U \cup \partial_{f,i} U$ and s_V is the extension of s_U by fine continuity, in view of Theorem 2.1.43, noting that $\mathfrak{b}(\complement V) = \mathfrak{b}(\complement U) = B$. When U is regular we have $U = \complement B$, and $s - s^B$ is the extension of s_U by 0 in $\complement U = B$ because $s^B = s$ on B.

The following three lemmas are particular cases of [BBC2, Proposition 5.1.1 and Proposition 5.1.2] for H-cones with a given sweeping of $s \in \mathcal{S}_+ = \mathcal{S}_+(\Omega)$ on $B = \complement U$ for the given finely open set $U \subset \Omega$, cf. (2.5.1).

Lemma 2.5.2 *Let $s \in \mathcal{S}_+$ and write $t = s - s \curlywedge s^B$. Then $s_U = t_U$ and $t \curlywedge t^B = 0$.*

Proof Write $u = s \curlywedge s^B$. Then $u \preccurlyeq s^B$, that is, $s^B = u + v$ for some $v \in \mathcal{S}_+$. Since sweeping of functions of class $\mathcal{S}_+$ is additive, contractive, and idempotent (see Theorem 1.3.15 and Lemma 1.3.11) it follows

$$s^B = (s^B)^B = u^B + v^B \leqslant u + v = s^B.$$

Thus equality prevails, and hence $u^B = u$ because $v^B \leqslant v$. We conclude that, in fact, $t_U = s - s^B$, and so $t_U = s_U$. Furthermore $t \curlywedge t^B = (s - u) \curlywedge (s^B - u) = 0$. □

The next lemma shows that the localization operator is *specifically increasing*:

Lemma 2.5.3 *Let $s, t \in \mathcal{S}_+$. If $s \preccurlyeq t$ (in $\mathcal{S}_+$) then $s_U \preccurlyeq t_U$ (in $\mathcal{S}_+(U)$).*

Proof There exists $r \in \mathcal{S}_+$ such that $r + s = t$ and hence $r_U + s_U = t_U$ with $r_U, s_U, t_U \in \mathcal{S}_+(U)$, so $s_U \preccurlyeq t_U$. □

In the opposite direction we have

Lemma 2.5.4 *Let $s, t \in \mathcal{S}_+$. Then*
(a) *If $s_U \leqslant t_U$ then $s - s^B + t^B \in \mathcal{S}_+$.*
(b) *If $s_U \leqslant t_U$ and $s \curlywedge s^B = 0$ then $s \leqslant t$ and $s^B \leqslant t^B$.*

Proof We may assume that U is regular, that is, $U = \complement B$. From $s_U \leqslant t_U$ follows $s - s^B \leqslant t - t^B$ on $\complement B$, and trivially on B, where $s^B = s$ and $t^B = t$. Thus $s + t^B \leqslant s^B + t$. By the natural decomposition property (Corollary 1.3.14) there exist functions $s', t' \in \mathcal{S}_+$ with $s' \leqslant s^B$ and $t' \leqslant t$ such that

$$s + t^B = s' + t', \tag{2.5.4}$$

and hence by additivity and idempotency of sweeping

$$s^B + t^B = (s')^B + (t')^B.$$

But $(s')^B \leqslant (s^B)^B = s^B$ and $(t')^B \leqslant t^B$, so equality prevails. It follows that $s' \leqslant s^B = (s')^B \leqslant s'$, that is, $s' = s^B$, and $(t')^B = t^B$. We therefore obtain

$$s - s^B + t^B = s - s' + t^B = t' \in \mathcal{S}_+, \tag{2.5.5}$$

then establishing (a). For (b) we have from (2.5.4)

$$s \preccurlyeq s + t^B = s^B + t'.$$

When $s \curlywedge s^B = 0$ we conclude that

$$s = s \curlywedge s \preccurlyeq s \curlywedge s^B + s \curlywedge t' = s \curlywedge t',$$

and so $s \preccurlyeq t' \leq t$. And from $s \preccurlyeq t'$ follows indeed $s^B \preccurlyeq (t')^B = t^B$ because we also showed before (2.5.5) that $(t')^B = t^B$. □

Theorem 2.5.5 *Let $p \in \mathcal{P}$. Then*

$$\text{f-}\lim_{y \to x,\, y \in U} p_U(y) = 0 \text{ for } x \in \partial_{f,r} U \cap \{p < +\infty\}. \tag{2.5.6}$$

If $p < +\infty$ on $\partial_f U$ then $p_U \in \mathcal{P}(U)$. If $p \in \mathcal{P}^s$ then $p_U \subset \mathcal{P}^s(U)$.

Proof Write $B = \mathrm{b}(\complement U)$. As noted before Definition 2.5.1, $p_U \in \mathcal{S}_+(U)$ and $p - p^B$ is finely continuous. Furthermore $p = p^B$ on B, in particular on $\partial_{f,r} U = B \cap \partial_f U$, the regular part of $\partial_f U$ (see Definition 2.1.49). Since $p - p^B = 0$ on $\partial_{f,r} U \cap \{p < +\infty\}$, (2.5.6) follows.

Suppose that $p < +\infty$ on $\partial_f U$. Consider a finely subharmonic minorant v of $p_U \leqslant p$ on U. Then $v \leqslant p$ on U, and

$$\text{f-}\limsup_{y \to x,\, y \in U} v(y) \begin{cases} \leqslant 0 & \text{for } x \in \partial_{f,r} U \\ \leqslant p(x) < +\infty & \text{for } x \in \partial_{f,i} U \end{cases}$$

in view of 2.5.6. By the strengthened fine boundary minimum principle (Corollary 2.4.16) it follows that $v \leqslant 0$ on U. Since $p_U \in \mathcal{S}_+(U)$ we see from Definition 2.4.2 that indeed $p_U \in \mathcal{P}(U)$.

Suppose instead that $p \in \mathcal{P}^s$. A finely subharmonic minorant v of p_U on U now satisfies

$$\text{f-}\limsup_{y \to x,\, y \in U} v(y) \le 0 \text{ q.e. on } \partial_f U$$

according to (2.5.6) because $\partial_{f,i} U$ and $\{p = +\infty\}$ are polar. Since $v \leqslant p_U \leqslant p$ on U and $p \in \mathcal{P}^s$, it follows by the fine boundary minimum principle (Theorem 2.1.10) that $v \leqslant 0$, and so $p_U \in \mathcal{P}(U)$. Writing $p = \sum_n p_n$ with $p_n \in \mathcal{P}^b(U)$, see Definition 1.5.16(ii), we obtain $p_U = \sum_n (p_n)_U$, a sum of finite fine potentials, hence semibounded according to Theorem 2.4.9. □

Theorem 2.5.6 *Let $U \subset \Omega$ be finely open and write $B = \mathfrak{b}(\complement U)$. The convex cone $\{s_U : s \in \mathcal{S}_+\}$ in $\mathcal{S}_+(U)$ equals its subcone $\{t_U : t \in \mathcal{S}_+$ and $t \curlywedge t^B = 0\}$, and is solid (=hereditary) even in the pointwise order.*

Proof The former assertion follows from Lemma 2.5.2 in view of (2.5.1). For solidity assertion, suppose to begin with that U is *regular*, that is, $B = \complement U$. Let $s \in \mathcal{S}_+$ and $u \in \mathcal{S}_+(U)$ satisfy $u \leqslant s_U$ (on U), that is,

$$v := u + \widehat{R}_s^{\complement U} \leqslant s.$$

Then $v \in \mathcal{S}_+(U)$ along with u and $\widehat{R}_s^{\complement U}$. Since u and $\widehat{R}_s^{\complement U}$ are finely continuous and $u \geqslant 0$ on Ω we have

$$\text{f-}\liminf_{y \to x,\, y \in U} v(y) \geqslant \widehat{R}_s^{\complement U}(x) = s(x)$$

for $x \in \partial_f U \subset \complement U = B$, a base. By Lemma 2.1.51 (applied with U, V and u replaced with Ω, U and s) the function

$$t := \begin{cases} s \wedge v & \text{on } U \\ s & \text{on } \complement U \end{cases}$$

is of class $\mathcal{S}_+(\Omega) = \mathcal{S}_+$. Clearly $t_U = t - \widehat{R}_t^{\complement U} = u$ on U because $t = v$ on U and $t = s$ on $B = \complement U$.

Next, if U is not regular, we pass to the larger regular finely open set $V = \mathfrak{r}(U) = \complement B \subset \widetilde{U}$, whereby $V \setminus U$ is polar, (see Sect. 1.4, paragraph containing (1.4.21)). Let again $s \in \mathcal{S}_+$ and $u \in \mathcal{S}_+(U)$, and suppose that $u \leqslant s_U$. Since s_V is the extension of s_U by fine continuity and since u extends by fine continuity to a function $v \in \mathcal{S}_+(V)$ according to Theorem 2.1.43, it follows since V is regular that $v = r_V$ for some $r \in \mathcal{S}_+$. By restriction to U we conclude that $u = r_U$ because $\complement V = \mathfrak{b}(\complement U)$ and hence $\widehat{R}_r^{\complement V} = \widehat{R}_r^{\complement U}$. □

Remark 2.5.7 Theorem 2.5.6 remains valid if the cone $\mathcal{S}_+$ is replaced by throughout by $\mathcal{P}$ or $\mathcal{P}^s$, or indeed by any other convex subcone $\mathcal{Q}$ of $\mathcal{S}_+$ which is pointwise solid, for the former assertion of the theorem this follows again from Lemma 2.5.2 because $q^B \leqslant q$ for $q \in \mathcal{Q}$. For the solidity assertion, suppose that $u \in \mathcal{S}_+(U)$ satisfies $u \leqslant t_U$ for some $t \in \mathcal{Q}$. Since $t \in \mathcal{S}_+$ there exists (by the theorem as it stands) $s \in \mathcal{S}_+$ such that $s \curlywedge s^B = 0$ and $u = s_U$, hence $s_U \leqslant t_U$, and consequently $s \leqslant t$, by Lemma 2.5.4. Since $t \in \mathcal{Q}$ and $\mathcal{Q}$ is solid it follows that $s \in \mathcal{Q}$ and hence $u = s_U$ with $s \in \mathcal{Q}$, as claimed.

Theorem 2.5.8 *The band $\mathcal{P}^s(U)$ of stable fine potentials on a finely open set $U \subset \Omega$ is generated by the cone of all p_U, where p ranges over $\mathcal{P}^s$, or $\mathcal{P}^c$.*

Proof We shall prove that every $p \in \mathcal{P}^s(U)$ is representable as the specific supremum of a specifically increasing sequence $((p_n)_U)$ with $p_n \in \mathcal{P}^c$, cf. Corollary 2.2.18. Since the band $\mathcal{P}^s$ is generated by $\mathcal{P}^c$, it suffices to establish this assertion with $\mathcal{P}^s$ in place of $\mathcal{P}^c$. In view of Theorem 2.4.13(iii) we may similarly assume that p is finite valued.

Choose a finite strict potential $q \in \mathcal{P}^b$, cf. Definition 1.4.14, and note that $q_U > 0$ on U. In fact, $\widehat{R}_q^{\complement U} < q$ on $\mathfrak{r}(U) \supset U$ according to Theorem 1.4.17 applied to $A = \complement U$, hence $\mathfrak{b}(A) = \complement\mathfrak{r}(U)$ by (1.4.20) (with A replaced by U). Let q_n denote the specific restriction $1_{A_n} \cdot p$ of p to the relatively finely closed subset

$$A_n := \{x \in U : p(x) \leqslant n q_U(x)\}$$

of U, cf. Definition 2.4.29. Then q_n is finely harmonic q.e. on $U \setminus A_n$ (in fact everywhere on $U \setminus A_n$, by Theorem 2.1.45). It follows by Theorem 2.4.18 (the fine domination property) that $q_n \leqslant n q_U$ on all of U. Thus Theorem 2.5.6 applies to show that $q_n = (p_n)_U$ for some $p_n \in \mathcal{P}^s$. By Definition 2.4.29, the sequence $(1_{A_n} . p)$ is specifically increasing because (A_n) is increasing. Since the sets A_n cover U we have

$$\curlyvee_n 1_{A_n} \cdot p = p$$

by [Fu8, Theorem 11.21] applied to $f_n = 1_{A_n}$. □

Theorem 2.5.9 *Let $U \subset \Omega$ be finely open. The following are equivalent for any function $u : U \to [0; +\infty]$:*

(i) *u is finely hyperharmonic on U,*

(ii) *u is the pointwise supremum of a pointwise increasing sequence (u_n) of functions $u_n = (s_n)_U$, $s_n \in \mathcal{S}_+$.*

(iii) *For any $s \in \mathcal{S}_+$ there exists $t \in \mathcal{S}_+$ such that $u \wedge s_U = t_U$.*

In assertions (ii) *and* (iii), *$\mathcal{S}_+$ may be replaced by any pointwise solid convex subcone of $\mathcal{S}_+$.*

Proof Choose a strict potential $q \in \mathcal{P}^b$ (cf. Theorem 1.4.15(a)), and note that $q_U \in \mathcal{P}^s(U)$ by Theorem 2.5.5, and $q_U > 0$, by Theorem 1.4.17.

(i)$\Rightarrow$(ii). Writing $u_n = u \wedge (n q_U)$ for $n \in \mathbb{N}$ we have $u_n \leqslant (nq)_U$ and hence $u_n \in \mathcal{P}^s(U)$ along with q_U. By Theorem 2.5.6 there exists $s_n \in \mathcal{S}_+$ such that $u_n = (s_n)_U$. Clearly, $u_n \uparrow u$ pointwise. Conversely, (ii)$\Rightarrow$(i) by Corollary 2.1.21.

(ii)⇒(iii). Let $s \in \mathcal{S}_+$ be given. Without changing s_U we may arrange by Lemma 2.5.2 that $s \curlywedge s^B = 0$. Let $u_n = (s_n)_U$ with $s_n \in \mathcal{S}_+$ and suppose that $u_n \uparrow u$ pointwise. Again we may arrange without changing $(s_n)_U$ that $s_n \curlywedge s_n^B = 0$. Since $u_n \wedge s_U \leqslant u_n$ there exists by Theorem 2.5.6 some $t_n \in \mathcal{S}_+$ such that $u_n \wedge s_U = (t_n)_U$ and $t_n \curlywedge t_n^B = 0$. It follows that $(t_n)_U \leqslant (t_{n+1})_U \leqslant s_U$, and hence $t_n \leqslant t_{n+1} \leqslant s$ according to Lemma 2.5.4. The pointwise limit $t = \lim_n t_n$ is hyperharmonic on Ω, and $t \leqslant s$, so $t \in \mathcal{S}_+$. It follows that t and t^B are finite q.e. on Ω. We thus obtain

$$(t_n)_U = t_n - (t_n)^B \to t - t^B = t_U \text{ q.e. on } U,$$

and indeed everywhere on U, by fine continuity of $\lim_n (t_n)_U = \sup_n (t_n)_U \in \mathcal{S}_+(U)$ and t_U. On the other hand

$$(t_n)_U = u_n \wedge s_U \uparrow u \wedge s_U,$$

and we conclude that $u \wedge s_U = t_U$. – Conversely (iii)⇒(ii), for if u satisfies (iii) and if we fix $s \in \mathcal{S}_+$ with $s_U > 0$ then there exists, by (iii), for each $n \in \mathbb{N}$ some $t_n \in \mathcal{S}_+$ such that $u \wedge (ns_U) = (t_n)_U$, and it follows that $(t_n)_U \uparrow u$ as $n \to +\infty$.

The final assertion of the theorem follows easily in view of Remark 2.5.7. □

Recall from (1.4.20) that the regularizer $\mathfrak{r}(U)$ of a set $U \subset \Omega$ is defined by $\complement\mathfrak{r}(U) = \mathfrak{b}(\complement U)$. In the present situation, U is finely open in Ω, and

$$\mathfrak{r}(U) = U \cup \partial_{f,i} U$$

is therefore the least regular finely open subset of Ω containing U. Clearly $\mathfrak{r}(U)\setminus U = \partial_{f,i} U$ is polar.

By localization of the Green function $G_\Omega(\cdot, y)$ with pole at $y \in U$ we shall obtain the Green function $G_U(\cdot, y) = G_\Omega(\cdot, y)_U$ on U with that pole. For simplicity we assume that U is *finely connected* and thus a fine domain.

Definition 2.5.10 Let $U \subset \Omega$ be a *fine domain*. For any $y \in U$ the *Green function for U* with pole at y is defined on Ω by

$$G_U(\cdot, y) = G_\Omega(\cdot, y) - \widehat{R}^{\complement U}_{G_\Omega(\cdot, y)}, \tag{2.5.7}$$

extended from $\complement\{y\}$ to Ω by fine continuity.

Note that $G_U = G_{\mathfrak{r}(U)}$ in view of Theorem 1.4.21. (It would therefore be sufficient to consider the case where U is regular.) For $y \in \complement\mathfrak{r}(U) = \mathfrak{b}(\complement U)$ we have $G_U(\cdot, y) := G_\Omega(\cdot, y) - \widehat{R}^{\complement U}_{G_\Omega(\cdot, y)} \equiv 0$; this holds in particular for $y \in \partial_{f,r} U = \partial_f \mathfrak{r}(U)$. By Corollary 2.1.30 or Theorem 2.4.19(b) the right hand member of (2.5.7) is finely harmonic and $\geqslant 0$ on $U\setminus\{y\}$, and the extension by fine continuity of that member to $\mathfrak{r}(U)$, and in fact to Ω, exists on U and is finely superharmonic there according to Theorem 2.1.43.

Theorem 2.5.11 *Let $U \subset \Omega$ be a fine domain and let y be a given point of $\mathfrak{r}(U)$.*

(a) *G_U is symmetric: $G_U(x, y) = G_U(y, x)$ for $x, y \in \mathfrak{r}(U)$.*

(b) *$G_U(\cdot, y)$ is a fine potential > 0 relative to $\mathfrak{r}(U)$ (or to U) and finely harmonic on $\mathfrak{r}(U)\setminus\{y\}$: these two properties characterize $G_U(\cdot, y)$ up to constant factor > 0.*

(c)

$$G_U(\cdot, y) = G_\Omega(\cdot, y) - \int G_\Omega\varepsilon_y \, d\varepsilon_{\cdot}^{\complement U} = G_\Omega\varepsilon_y - G_\Omega\varepsilon_y^{\complement U} \tag{2.5.8}$$

on $\Omega\setminus\{y\}$.

(d) *$G_U(y, y) = +\infty$.*

(e) *$G_U(\cdot, y)$ is subharmonic on $\complement\{y\}$ and equals 0 on $\complement\mathfrak{r}(U)$, in particular on $\partial_{f,r}\mathfrak{r}(U)$.*

Proof For brevity, write $G_U(\cdot, y) = p$.

(a) follows from the symmetry of G_Ω (see Sect. 1.1.4) together with Theorem 1.4.12.

(b) It follows from the second assertion of Theorem 2.5.5 that $p \in \mathcal{P}(U)$ because $G_\Omega(\cdot, y) \in \mathcal{P}$ and $y \notin \partial_{f,i}U$ (since $y \in U$). Since $G_\Omega(\cdot, y)$ is harmonic (hence finely harmonic) on $\complement\{y\}$ we see from (2.5.7) that p is finely harmonic on $U\setminus\{y\}$ (and hence $\mathfrak{r}(U)\setminus\{y\}$) because $-\widehat{R}_{G_\Omega(\cdot,y)}^{\complement U}$ is so.

We proceed to establish the asserted unique characterization of $p = G_U(\cdot, y)$. We may assume that $y \in U$. Consider any non-zero $q \in \mathcal{P}(U)$ which is finely harmonic on $U\setminus\{y\}$ and fix a point $x_0 \in U\setminus\{y\}$. Since U is finely connected it follows from Corollary 2.1.27 that $q(x_0) > 0$. After multiplying q by a suitable constant > 0 we therefore achieve that $q(x_0) = p(x_0)$. Consider the finely open set

$$V := \{x \in U\setminus\{y\} : q(x) < p(x)\},$$

and suppose to begin with that $V \neq \varnothing$. The function $u := q - p$ defined on $U\setminus\{y\}$ is finely harmonic and < 0 on V. For a given $z \in \partial_{f,i}V$ we proceed to determine

$$\lambda := \text{f-}\liminf_{x\to z, x\in V} u(x).$$

Being finely harmonic, u is finely continuous on $U\setminus\{y\}$, and hence $\lambda = 0$ if $z \in U\setminus\{y\}$.

Suppose first that $z \neq y$. If $z \in \partial_{f,r}V$ then either $z \in U\setminus\{y\}$ or $z \in \partial_{f,r}U$, so anyway $\lambda \geqslant 0$. If $z \in \partial_{f,i}V$ then there exists a deleted fine neighborhood $Z \subset V \subset U\setminus\{y\}$ of $z \neq y$, and so $p(z) < +\infty$, whence $\lambda > -\infty$. According to Corollary 2.4.16 it follows that $u \geqslant 0$ on V, which is a contradiction.

Thus actually $z = y$, and $V \supset Z$ is a deleted fine neighborhood of y in $U\setminus\{y\}$. By hypothesis, U is finely connected, and hence so is $U\setminus\{y\}$ by Corollary 2.1.46, $\{y\}$ being polar. Like V, the set

$$W := \{x \in U\setminus\{y\} : q(x) > p(x)\}$$

is finely open, and since $V \cup W = U\backslash\{y\}$ and $V \neq \varnothing$, we have $W = \varnothing$, that is, $q \leqslant p$ on $U\backslash\{y\}$, where p, q and hence $p - q$ are finely harmonic. Since $p - q \geqslant 0$ on $U\backslash\{y\}$ with equality at x_0, it follows that $p - q = 0$ on $U\backslash\{y\}$, that is, $p = q$ on $U\backslash\{y\}$, and indeed on all of U, by fine continuity.

This was under the hypothesis that $V \neq \varnothing$. If $V = \varnothing$ then $q \geqslant p$ on $U\backslash\{y\}$, with equality at x_0, so that $q = p$, as above. This completes the proof of (b).

(c) Follows from (1.4.2) in view of symmetry of G_Ω.

(d) If $p(y)$ were finite then p would be finely harmonic on all of U, by Theorem 2.1.45, and hence $p \equiv 0$, which is false.

(e) Since $G_\Omega(\cdot, y)$ is harmonic on $\complement\{y\}$ it follows from (2.5.7) that $G_\Omega(\cdot, y)$ is subharmonic on $\complement\{y\}$. Clearly, $G_U(\cdot, y) = 0$ on $\complement\mathfrak{r}(U) = \mathfrak{b}(U) \supset \partial_{f,r}U = \partial_f\mathfrak{r}(U)$. □

Remark 2.5.12 It follows from (2.5.8) that, according to Lemma 2.1.41, G_U is independent of the open Greenian domain Ω containing U.

Lemma 2.5.13 *For any fine domain U the extended Green kernel* $G_U : \Omega \times \Omega \to [0, +\infty]$ *is Borel measurable.*

Proof We may assume that U is regular. It suffices to establish Borel measurability of the compensation term

$$\widehat{R}^{\complement U}_{G_\Omega(\cdot,y)}(x) = \int_\Omega G_\Omega(x, \cdot)\, d\varepsilon_y^{\complement U}$$

as function of $(x, y) \in \Omega \times \Omega$, cf. Theorem 1.4.12 and (1.4.2), the latter applied to $p = G_\Omega(x, \cdot) = G_\Omega\varepsilon_x$. In fact, (2.5.8) holds as it stands except perhaps for $x = y$, in which $G_U(x, y) = +\infty$ or 0 according as $x \in U$ or $x \in \complement U$, noting that U is a Borel set. Being l.s.c., G_Ω is representable as the pointwise supremum of an increasing sequence of continuous functions $\geqslant 0$ of compact support in $\Omega \times \Omega$. We may therefore replace G_Ω by such a function $G \in \mathcal{C}_0(\Omega \times \Omega)$. Next, we approximate such G by a finite sum of the form $\varphi(x)f(y)$ with $\varphi, f \in \mathcal{C}_0^+(\Omega)$, see [Bou6, Chap. III, § 5]. This reduces the claim to the fact that the function $f^{\complement U} = \int f\, d\varepsilon_\cdot^{\complement U}$, see Definition 1.4.29, is finely continuous and hence of Baire class One, (see Theorem 1.4.43). For fine continuity of $f^{\complement U}$ see Lemma 2.1.38 applied to $\pm f$, noting that $\widehat{R}_1^K$ is a bounded potential on Ω for any compact $K \subset \Omega$, by Theorem 1.3.5. □

The following Lemma will enter in the proof of the integral representation of semibounded fine potentials.

Lemma 2.5.14 *Let* $U \subset \Omega$ *be finely open. Let* $f : \Omega \to [0; +\infty]$ *equal* 0 *on* $\complement U$, *and let* $f|_U \in \mathcal{P}(U)$. *Suppose that f is majorized on U by some semibounded potential on* Ω, *that is,*

$$p := \widehat{R}_f^U \in \mathcal{P}^s.$$

Then $p_U \preccurlyeq f|_U$. *If in addition f is finely harmonic on a fine boundary strip* $U\backslash F$ *(F finely closed,* $F \subset U$*) and if moreover p is finite valued, then*

$$\text{f-}\liminf_{y\to x,\, y\in U} f(y) = 0 \text{ at every } x \in \partial_{f,r}U, \tag{2.5.9}$$

and, moreover, $p_U \equiv 0$ *implies* $f \equiv 0$.

Proof If $f \leqslant q$ on U (and hence on Ω) for some $q \in \mathcal{P}^s$ then $f \leqslant R_f^U = \widehat{R}_f^U \leqslant q$ by Theorem 1.3.3(b) (because $f 1_U$ is finely l.s.c.), and so $f \leqslant p = \widehat{R}_f^U \in \mathcal{P}^s$. The converse is obvious.

In view of Theorems 2.1.43 and 2.1.45 on removable singularities for finely [hyper]harmonic functions we may suppose that F and $\complement U$ are basic sets. Otherwise replace F by $\mathfrak{b}(F)$ and U by the regular finely open set $\mathfrak{r}(U) = \complement\mathfrak{b}(\complement U)$ (see the first two paragraphs following the proof of Lemma 1.4.27), and recall from Lemma 2.1.47 and Definition 2.1.49 that $\partial_{f,r}U = \partial_f(\mathfrak{r}(U))$. For the proof that $p_U \preccurlyeq f|_U$ under the hypothesis $f|_U \in \mathcal{P}(U)$ and $f \leqslant p \in \mathcal{P}^s$ (and $f|_{\complement U} = 0$), note that $p_U \in \mathcal{P}^s(U)$ by Theorem 2.5.5, and that p is finely harmonic on the finely open set

$$V := \{x \in U : f(x) < p(x) < +\infty\}$$

by Theorem 2.1.29(b) because $\widetilde{f} = f$ on U (where $f|_U \in \mathcal{P}(U)$ is finely continuous) and $p = \widehat{R}_f^U$. Recall that $f \leqslant p$ on Ω. It follows that $f - p$ is finely superharmonic on V, and even on the finely open set $U_0 = U \setminus \{p = +\infty\} \supset V$ on account of Lemma 2.1.51. In fact, f and p are finely continuous on U, and hence $f - p$ is finely continuous on U_0, and furthermore $f - p = 0$ on $U_0 \cap \partial_f U \subset U_0 \setminus V$, whereas $f - p \leqslant 0$ on all U_0. It follows by the definition of p_U that the function

$$f - p_U = (f - p) + \widehat{R}_p^{\complement U} \text{ on } U_0$$

is finely superharmonic on U_0. It remains to prove that $f - p_U \geqslant 0$ on U_0, for that will show that $f - p_U$ extends by fine continuity to a function of class $\mathcal{S}_+(U)$ according to Theorem 2.1.43.

For the proof that $f - p_U \geqslant 0$ on U_0, apply the fine boundary minimum principle (Theorem 2.1.10) to the finely superharmonic function $(f - p_U)|_V$, noting that $f - p_U \geqslant -p_U$ with $p_U \in \mathcal{P}^s$; and that

$$\text{f-}\liminf_{y\to x,\, y\in V} (f(y) - p_U(y)) \geqslant 0 \text{ q.e. on } \partial_f V,$$

namely q.e. for $x \subset \partial_f U$ by Theorem 2.5.5 since $f \geqslant 0$, and everywhere on $U_0 \cap \partial_f V (\subset U_0 \setminus V)$ because f and p_U are finely continuous on $U_0 = U \setminus \{p = +\infty\}$. This shows that $f - p_U \geqslant 0$ on V, whereas $f = p \geqslant p_U$ on $U_0 \setminus V$ by definition of V and U_0. Consequently, $f \geqslant p_U$ on U_0, and hence indeed on all of U by fine continuity of f and p_U on U. This completes the proof that $p_U \preccurlyeq f|_U$.

Now pass to the Proof of (2.5.9) and the implication $p_U \equiv 0 \Rightarrow f \equiv 0$, under the extra hypothesis that $p < +\infty$ and that f is finely harmonic on $U \setminus F$ for some finely closed subset F of Ω, $F \subset U$. As noted in the beginning of the proof, we may assume that F and $\complement U$ are bases. The restriction $f|_{U\setminus F}$ is a fine superfunction for the restriction of f to $\partial_f(U\backslash F) = (\partial_f U) \cup (\partial_f F)$, cf. Sect. 2.3. Writing

$$u := \overline{H}_f^{U\backslash F} \text{ on } U\backslash F,$$

we therefore have $f \geqslant u \geqslant 0$ on $U\backslash F$ because $f \geqslant 0$. In view of (2.3.3) it follows that

$$\text{f-}\lim_{y\to x,\, y\in U\backslash F} u(y) = \begin{cases} 0 & \text{for } x \in \partial_f U \\ f(x) & \text{for } x \in \partial_f F. \end{cases} \tag{2.5.10}$$

From (2.5.10) we arrive at (2.5.9) (where now $\partial_{f,r}U = \partial_f U$) once we have shown that $f = u$ on $U\backslash F$. Now, $f - u$ is finely harmonic and $\geqslant 0$ on $U\backslash F$ by hypothesis, and $u = \overline{H}_f^{U\backslash F}$ is so by Theorem 2.3.7 because $0 \leqslant u \leqslant p < +\infty$ on $U\backslash F$. In view of (2.5.10) the extension w of $f - u$ from $U\backslash F$ to U by 0 in F is therefore finely subharmonic and $\geqslant 0$ on U, by Lemma 2.1.51. But $w \leqslant f|_U \in \mathcal{P}(U)$ on U, and we conclude that $w = 0$ on U and hence that indeed $f = u$ on $U\backslash F$.

From (2.5.10) we therefore infer, again by Lemma 2.1.51, that f is finely subharmonic on $\complement F$. It follows by Theorem 2.1.29(b) that $p = \widehat{R}_f$ is finely harmonic on $\complement F$.

Now suppose that $p_U = 0$, that is, $p = \widehat{R}_p^{\complement U}$. Then p is finely harmonic on U, according to Corollary 2.1.30, and so altogether on $U \cup \complement F = \Omega$. Consequently, $p \equiv 0$, and hence indeed $f \equiv 0$ for $0 \leqslant f \leq p$. □

We close this section by showing that the cone $\mathcal{S}_+(U)$ is a standard H-cone of functions. This result will be useful in Sect. 3.1. We will use in the sequel the notions of kernels, resolvent family of kernels and excessive functions of resolvent family of kernels on a measurable space. For these notions and their properties we refer the reader to the monographs [BBC2, BeB].

Theorem 2.5.15 *There exists a resolvent family (W_λ) of kernels on U which are absolutely continuous with respect to a (σ-finite) measure σ on U such that $\mathcal{S}_+(U)$ is the cone of excessive functions which are finite σ-a.e.*

For the definition of absolutely continuous resolvent see [BBC2, p. 20].

Sketch of proof. We may assume without loss of generality that U is regular, and hence a Borel set. Let $p = G_\Omega\tau$ be a strict bounded continuous potential on the Greenian domain Ω in $\mathbb{R}^N$. Then the measure τ does not charge the polar sets and we have $\tau(\omega) > 0$ for any finely open subset ω of Ω. Denote by V the Borel measurable kernel on Ω defined by

$$Vf(x) = \int G_\Omega(x, y) f(y) d\tau(y)$$

for any Borel measurable function $f \geqslant 0$ in Ω and for $x \in \Omega$, and by (V_λ) the resolvent family of kernels whose kernel potential is V. According to [CC3, Proposition 10.2.2], the cone of excessive functions of the resolvent (V_λ) is the cone

$\mathcal{S}_+(\Omega) \cup \{+\infty\}$ (and hence $\mathcal{S}_+(\Omega)$ is the cone of excessive functions of (V_λ) which are finite τ-a.e.). Define a kernel W on U by

$$Wf = V\bar{f} - \widehat{R}_{V\bar{f}}^{\complement U}$$

(restricted to U) for any bounded Borel measurable function $f \geqslant 0$ in U, where $\bar{f}$ denotes the extension of f to Ω by 0 in $\complement U$. Here the sweeping $\widehat{R}_{V\bar{f}}^{\complement U}$ is taken relative to Ω. Then by [BeB, Proposition 3.6.2] there exists a resolvent family (W_λ) of Borel measurable kernels on U having the potential kernel W. The resolvent (W_λ) is absolutely continuous with respect to the restriction σ to U of the measure τ. Moreover, the kernel W_λ is given by

$$W_\lambda g = (V_\lambda \bar{g} - \widehat{R}_{V_\lambda \bar{g}}^{\complement U})|_U$$

for every bounded Borel measurable function g on U.

We proceed to determine the excessive functions for the resolvent (W_λ). Let u be excessive for (W_λ). It follows by [DM, Théorème 17, p. 11], that $u = \sup_j Wf_j$ for some increasing sequence of bounded potentials Wf_j of bounded Borel functions $f_j \geq 0$ on U. Since the functions $Wf_j = V\bar{f}_j - \widehat{R}_{V\bar{f}_j}^{\complement U}$ are finely superharmonic on U, it follows that u is finely hyperharmonic on U, that is, $u \in \mathcal{S}_+(U) \cup \{+\infty\}$ by property (i) in Definition 2.4.1. Let us now prove that every finely hyperharmonic function $v \geq 0$ is excessive for the resolvent (W_λ). By Theorem 2.5.9(ii) this reduces to the case $v = s_U$, where $s \geq 0$ is superharmonic on Ω. Since s is excessive for the resolvent (V_λ), there exists again by [DM, Théorème 17, p. 11] an increasing sequence (Vf_j) of bounded potentials of bounded Borel functions $f_j \geq 0$ on Ω, such that $s = \sup Vf_j$ and hence $s_U = s|_U - \widehat{R}_s^{\complement U}|_U = \liminf W(f_j|_U)$ quasi-everywhere. Since by Fatou's lemma $\liminf W(f_j|_U)$ is supermedian for the resolvent (W_λ), it follows that s_U is σ-a.e. equal to an excessive function u for (W_λ). But u is finely hyperharmonic as seen above, and hence $s_U = u$ everywhere by fine continuity because a σ-null set has an empty fine interior. This shows that s_U is excessive for (W_λ). It then follows that any $u \in \mathcal{S}_+(U) \cup \{+\infty\}$ is excessive for (W_λ), according to Theorem 2.5.9(ii). The proof is complete. □

It follows from Theorem 2.5.15 by [BBC2, Theorem 4.4.6] that $\mathcal{S}_+(U)$ is a *standard H-cone of functions* on U.

2.6 Integral Representation of Fine Potentials

In the absence of other indications, U shall denote a *fine domain* in Ω, that is, a finely open finely connected subset of Ω. The least regular finely open set $\mathfrak{r}(U) = \complement\mathfrak{b}(\complement U)$ containing U is likewise a fine domain. For if $\mathfrak{r}(U) = V_1 \cup V_2$ with V_1, V_2 two disjoint finely open sets, then $E := \mathfrak{r}(U) \setminus U$ is polar (see paragraph containing (1.4.21), so are $E_1 := E \cap V_1$ and $E_2 := E \cap V_2$. We obtain

$$U = \mathfrak{r}(U) \setminus E = (V_1 \setminus E_1) \cup (V_2 \setminus E_2),$$

a union of disjoint finely open subsets of U. For example, $V_2 \setminus E_2 = \varnothing$, hence $V_2 \subset E_2$ and so $V_2 = \varnothing$.

We mostly identify a (σ-finite Borel) measure μ on U (as a topological subspace of Ω) with the measure μ' on Ω defined by $\mu'(E) = \mu(E \cap U)$ for $E \in \mathcal{B}(\Omega)$ (the Borel subsets of Ω), hence $E \cap U \in \mathcal{B}(U)$. If U is regular, hence a K_σ, then of course $U \in \mathcal{B}(\Omega)$.

For any measure μ on Ω the inner μ-measure μ_* and the outer μ-measure μ^* are defined as usual for any set $A \subset \Omega$ by

$$\mu_*(A) = \sup\{\mu(E) : E \in \mathcal{B}(\Omega),\ E \subset A\},$$

$$\mu^*(A) = \inf\{\mu(E) : E \in \mathcal{B}(\Omega),\ E \supset A\}.$$

Recall Definition 2.5.10 of the extended Green kernel $G_U = G_{\mathfrak{r}(U)}$ and the fact that $G_U(x, \cdot)$ is finely continuous and hence Borel measurable relative to $\mathfrak{r}(U)$ (Theorem 1.4.36) we write for any measure μ on Ω

$$G_U\mu(x) = \int_\Omega G_U(x, y)\, d\mu(y), \ x \in \Omega. \tag{2.6.1}$$

We have $G_U\mu = G_{\mathfrak{r}(U)}\mu$ because $G_U = G_{\mathfrak{r}(U)}$. Also note that $G_U\mu$ depends on the restriction of μ to $\mathfrak{r}(U)$ only. Clearly,

$$G_U\mu = 0 \text{ on } \complement\mathfrak{r}(U),$$

and if $G_U\mu(x) = 0$ for some $x \in \mathfrak{r}(U)$ then $G_U\mu = 0$.

By $\Lambda(U)$ we denote the set of positive σ-finite Borel measures μ on U (or sometimes on $\mathfrak{r}(U)$) with the Euclidean relative topology, such that

$$G_U\mu \not\equiv +\infty \text{ on } U.$$

(This notation conforms with that of Sect. 1.4 in the particular case $U = \Omega$.) For every $\mu \in \Lambda(U)$ the restriction of $G_U\mu$ to U is also denoted by $G_U\mu$.

Lemma 2.6.1 *Every $\mu \in \Lambda(U)$ such that $\mu_*(\complement U) = 0$ can be decomposed into the sum $\mu = \sum_{n \geq 1} \mu_n$ of a sequence of measures μ_n on Ω of compact support* Supp $\mu_n \subset \mathfrak{r}(U)$.

Proof Fix $x_0 \in U$ with $G_U\mu(x_0) < +\infty$. By Theorem 2.5.11(e), $G_U(x_0, \cdot)$ is u.s.c. relative to the $\complement\{x_0\}$, hence $\{y \in \Omega : G_U(x_0, y) \geqslant \frac{1}{n}\}$ ($\subset \mathfrak{r}(U)$ because $G_U(x_0, y) > 0$ when $y \in \mathfrak{r}(U)$) is closed relative to $\complement\{x_0\}$, and even in Ω because $G_U(x_0, x_0) = +\infty$ by Theorem 2.5.11(d). The union of these finely closed sets is $\mathfrak{r}(U)$. Choose an increasing sequence of compact sets K_n with the union Ω, and for every integer $n \geq 1$ write

$$A_n = K_n \cap \{G_U(x_0, \cdot) \geqslant \frac{1}{n}\}. \tag{2.6.2}$$

The sets A_n are compact, and they form an increasing sequence with the union U, furthermore

$$\mu(A_n) \leqslant n \int G_U(x_0, \cdot)\, d\mu = nG_U\mu(x_0) < +\infty$$

and the desired decomposition $\mu = \sum_n \mu_n$ is therefore obtained by taking for μ_n the restriction of μ on $A_n \setminus A_{n-1}$ (with $A_0 = \varnothing$). □

Theorem 2.6.2 *For any measure $\mu \in \Lambda(U)$, $G_U\mu$ is finely superharmonic on U. If $\mu^*(\partial_{f,i}U) = 0$ (for example, if U is regular) then $G_U\mu$ is a fine potential relative to U.*

Proof In view of Theorem 2.4.5 and Definition 2.4.2 for fine potentials on U (and similarly for finely superharmonic functions on U), we may assume by Lemma 2.6.1 that μ is a finite measure of compact support $\operatorname{Supp}\mu \subset \mathfrak{r}(U)$ (and that $\mu^*(\partial_{f,i}U) = 0$). In that case we have $G_\Omega\mu \not\equiv +\infty$ and hence $G_\Omega\mu$ is superharmonic $\geqslant 0$ on Ω, hence finely superharmonic on U. The compensating term $\widehat{R}^{\complement U}_{G_\Omega\mu}$, cf. Definition 2.5.1 (with $s = G_\Omega\mu$), is finely harmonic on the finely open set

$$V := \{x \in \mathfrak{r}(U) : \widehat{R}^{\complement U}_{G_\Omega\mu}(x) < +\infty\}$$

according to Corollary 2.1.30 because $\mathfrak{b}(\complement U) = \complement\mathfrak{r}(U)$. It therefore follows that the difference $G_U\mu$ is finely superharmonic $\geqslant 0$ on V. Moreover, $G_U\mu$ is even finely harmonic on $U \cap \complement \operatorname{Supp}\mu$. To prove that $G_U\mu$ is finely superharmonic on some fine neighborhood of a prescribed point $x_0 \in U \setminus V$ we decompose μ as follows:

$$\mu = \mu_0 + \mu_1 + \mu_2 + \ldots$$

where $\mu_0 = a\varepsilon_{x_0}$ for $a = \mu(\{x_0\})$ while $\operatorname{Supp}\mu_n$ does not contain x_0 for $n \geqslant 1$. Then $G_U\mu_0 = aG_U(\cdot, x_0) \in \mathcal{P}(U) \subset \mathcal{S}_+(U)$, whereas (for $n \geqslant 1$) $G_U\mu_n$ is finely superharmonic $\geqslant 0$ on $V \cup \{x_0\}$, which is finely open because $U \setminus V$ is polar. Indeed, $G_U\mu_n$ is finely superharmonic on V, as shown above, and likewise on $U \setminus \operatorname{Supp}\mu_n$, a fine neighborhood of x_0. Invoking Corollary 2.1.21, we conclude that

$$G_U\mu = \sum_n G_U\mu_n$$

is at least finely hyperharmonic in the fine neighborhood $V \cup \{x_0\}$ of x_0, hence actually on all of U, by varying x_0. Since $G_U\mu \leqslant G_\Omega\mu < +\infty$ q.e., we have shown that $G_U\mu \in \mathcal{S}_+(U)$.

To prove finally that $G_U\mu \in \mathcal{P}(U)$ we choose a decreasing sequence of open sets $\omega_n \subset \Omega$ containing $\partial_{f,i}U$ so that $\mu(\omega_n) \to 0$. Denoting by ν_n the restriction of μ to $\omega_{n-1}\backslash\omega_n$ (whereby $\omega_0 := \Omega$) we have $\mu = \sum_n \nu_n$, and it therefore suffices to prove that each $G_U\nu_n \in \mathcal{P}(U)$. We shall verify that every finely subharmonic minorant v of $G_U\nu_n$ in U is $\leqslant 0$, cf. Definition 2.4.2. As shown above, $G_U\nu_n$ is finely superharmonic on U, finite and finely continuous on the complement of the compact support of ν_n and hence on $\partial_{f,i}U \subset \omega_n$ in view of Definition 2.5.1 because $G_\Omega\nu_n < +\infty$ there; and $G_U\nu_n = 0$ on $\partial_{f,r}U$. Consequently,

$$\text{f-}\limsup_{x\to z, x\in U} v(x) \leqslant G_U\nu_n(z) \text{ for } z \in \partial_f U.$$

Since $v \leqslant G_U\nu_n \leqslant G_\Omega\nu_n$ on U we conclude from Corollary 2.4.16 applied to $-u$ (with v replaced by $-u$) that indeed $v \leqslant 0$ on U. □

Replacing Ω, U in Definition 2.5.10 by U, V the following Lemma is to be expected.

Lemma 2.6.3 *Let U and V be fine domains in Ω such that $\widetilde{V} \subset \mathfrak{r}(U)$. Then*

$$G_U(\cdot, y) = G_V(\cdot, y) + {}^U\widehat{R}^{U\backslash V}_{G_U(\cdot,y)} \text{ on } U, \tag{2.6.3}$$

$$G_U(x, y) = G_V(x, y) + \int G_U(\cdot, y)\, d\varepsilon_x^{\complement V}, \tag{2.6.4}$$

for $x, y \in U$. Consequently,

$$G_U\mu = G_V\mu + {}^U\widehat{R}^{U\backslash V}_{G_U\mu} \text{ on } U, \tag{2.6.5}$$

$$G_U\mu(x) = G_V\mu(x) + \int G_U\mu\, d\varepsilon_x^{\complement V}, \tag{2.6.6}$$

for $x \in U$, $\mu \in \Lambda(U)$.

Proof By Theorem 2.6.2, $G_U\varepsilon_x^{\complement V}$ and $G_U\mu$ are finely superharmonic on U, so Corollary 2.3.12 applies. Hence (2.6.4) implies (2.6.3), and (2.6.6) implies (2.6.5). By definition 2.5.10

$$G_\Omega(\cdot, y) = G_U(\cdot, y) + \widehat{R}^{\complement U}_{G_\Omega(\cdot,y)} \text{ on } \Omega, \tag{2.6.7}$$

whence by integration with respect to $\varepsilon_x^{\complement V}$ for $x \in U$, and application of (1.4.5) to $A = \complement V$ and $u = G_\Omega(\cdot, y)$ or $\widehat{R}^{\complement U}_{G_\Omega(\cdot,y)}$,

$$\widehat{R}^{\complement V}_{G_\Omega(\cdot,y)}(x) = \int G_U(\cdot, y)\, d\varepsilon_x^{\complement V} + \widehat{R}^{\complement V}_{\widehat{R}^{\complement U}_{G_\Omega(\cdot,y)}}(x) \tag{2.6.8}$$

on U. Here the latter term on the right simplifies to $\widehat{R}^{\complement U}_{G_\Omega(\cdot,y)}$ according to Lemma 1.3.11, applicable because $\complement U \subset \complement V$. Furthermore, the integral in (2.6.8) equals

${}^{U}\widehat{R}^{U\setminus V}_{G_U(\cdot,y)}$ by Corollary 2.3.12 applied to $u = G_U(\cdot, y)$. Inserting the point x in (2.6.7) and subtracting the simplified version of (2.6.8) we obtain (2.6.3) from Definition 2.5.10 (with U replaced by V), first q.e. on U and next everywhere on U, by fine continuity of three terms. And (2.6.4) implies (2.6.6) by integration over U with respect to $d\mu(y)$, using Fubini's theorem. □

By application of Lemma 2.6.3 we obtain an extension and a converse of Theorem 2.1.6. For a more general result in terms of invariant functions, see Theorem 2.9.11.

Theorem 2.6.4 *For any $\mu \in \Lambda(U)$ and any finely open set $V \subset U$ on which $G_U\mu$ is finite we have*

$$G_U\mu \; is\; finely\; harmonic\; on\; V \iff \mu^*(V) = 0 \iff \mu_*(V) = 0.$$

Proof It is understood that μ is viewed as a measure on Ω taking the value $\mu(E \cap U)$ for any $E \in \mathcal{B}(\Omega)$. Suppose first that $\mu_*(V) = 0$. For any fine domain W of fine closure $\widetilde{W} \subset V$ the restriction of μ to the K_σ-set $\mathfrak{r}(W)(\subset \widetilde{W})$ then is 0, and so $G_W\mu = 0$. By the Lemma 2.6.3 applied with V replaced by W,

$$G_U\mu(x) = \int G_U\mu \, d\varepsilon_x^{\complement W} \text{ for } x \in W.$$

This shows that the finely continuous function $G_U\mu$ (finite on V) is finely harmonic on V according to Definition 2.1.4 and in view of the local connectedness of the fine topology (Corollary 2.1.33).

Conversely, suppose that $G_U\mu$ is finely harmonic (in particular finite) on V. Replacing U by $\mathfrak{r}(U)$ we may suppose that U is regular (since $G_{\mathfrak{r}(U)} = G_U$). For the proof that $\mu^*(V) = 0$ choose a finite continuous potential $q > 0$ on Ω and for $n \in \mathbb{N}$ write $V_n = \{x \in V : G_U\mu(x) < nq(x)\}$. Since the Borel sets V_n cover V it suffices to prove that each $\mu^*(V_n) = 0$. Clearly, V_n is finely open and by fine continuity we have $G_U\mu \leq nq$ on $\mathfrak{r}(V_n)$ ($\subset U \cap \widetilde{V}$). According to Theorem 2.1.45, $G_U\mu$ is finely harmonic in $\mathfrak{r}(V_n)$. Replacing V by a fine component of $\mathfrak{r}(V_n)$ we may assume that V is a regular fine domain and that $G_U\mu \leq q$ in V. Each fine component W of some V_n is likewise regular, and their number is countable, by the quasi-Lindelöf principle. It therefore suffices to prove that each $\mu(W) = 0$. Since $\widetilde{W} \subset \widetilde{V}_n \subset V \subset U$ it follows from the mean-value property of finely harmonic functions (Corollary 2.1.15) that

$$G_U\mu(x) = \int G_U\mu \, d\varepsilon_x^{\complement W} \text{ for every } x \in W.$$

Inserting this in (2.6.6) with V replaced by W we obtain $G_W\mu = 0$ on W, and hence $\mu(W) = 0$ because $G_W > 0$ on $W \times W$ by Theorem 2.5.11(b). □

There is a similar, more general result for finely hyperharmonic functions on U, to be obtained after we have established in Theorem 2.6.16 the integral representation of fine potentials.

By application of the above Theorem 2.6.4 we obtain

Theorem 2.6.5 *Let $\mu \in \Lambda(U)$ and let E be a polar subset of U. The restriction of $G_U\mu$ to the fine domain $U\backslash E$ is a fine potential on $U\backslash E$ if and only if $\mu_*(E) = 0$ relative to U, that is, $\mu(e) = 0$ for any $e \in \mathcal{B}(U)$ with $e \subset E$. (When either $U \in \mathcal{B}(\Omega)$ or $E \in \mathcal{B}(\Omega)$ the stated condition is equivalent to $\mu_*(E) = 0$, whereby μ is regarded as a measure on Ω.)*

Proof Since E is polar, $U\backslash E$ is a fine domain along with U, by Corollary 2.1.46. When $\mu(e) = 0$ for any $e \in \mathcal{B}(U)$ with $e \subset E$ (in particular $\mu_*(E) = 0$), then μ may be regarded as a measure on $U\backslash E$, and hence $G_U\mu|_{U\backslash E} = G_{U\backslash E}\mu$ is a fine potential on $U\backslash E$, by Theorem 2.6.2.

For the converse, suppose that $\mu_*(E) > 0$. The measure μ on U is identified with the measure μ' on Ω defined by

$$\mu'(B) = \mu(B \cap U) \text{ for } B \in \mathcal{B}(\Omega).$$

By Lemma 2.6.1 there is a non-zero measure ν on Ω with compact support K contained in E such that $\nu \leq \mu'$ and $\nu_*(E) > 0$. It follows that

$$0 < G_U\nu \leqslant G_\Omega\nu < +\infty \text{ on } U \cap \complement K \supset U\backslash E.$$

Since $\nu(U\backslash K) = 0$ and $K \subset E$ it follows by Theorem 2.6.4 that $G_U\nu|_{U\backslash E}$ is finely harmonic. But $G_U\nu|_{U\backslash E} \leqslant G_U\mu|_{U\backslash E}$, and consequently $G_U\mu|_{U\backslash E}$ is not a fine potential.

The final assertion of the theorem is easily verified. □

Corollary 2.6.6 *Let $\mu \in \Lambda(U)$.*

(a) $G_U\mu \in \mathcal{P}^s(U)$ if and only if $\mu_*(E) = 0$ for every polar set $E \subset U$.

(b) $G_U\mu \in \mathcal{P}^s(U)^\perp$ if and only if $\mu^*(\complement E) = 0$ for every polar set $E \subset U$.

In either assertion it suffices to take $E = \{x \in U : G_U\mu(x) = +\infty\}$, whereby μ_*, resp. μ^* may be replaced by μ.

(a) follows from Theorem 2.6.5 and (b) from Theorem 2.6.4. The remaining assertion holds because the polar set $\{x \in U : G_U\mu(x) = +\infty\}$ is a relative Borel subset of U, the finely continuous function $G_U\mu$ being a Borel function relative to U, by Theorem 1.4.44.

Theorem 2.6.7 *Let $\mu, \nu \in \Lambda(U)$. If $G_U\mu = G_U\nu$ q.e. on U then $\mu = \nu$.*

Proof Replacing, if necessary, U by $\mathfrak{r}(U)$ we may assume that U is regular, because $\mathfrak{r}(U)\backslash U$ is polar anyway, and $G_{\mathfrak{r}(U)} = G_U$. Since $G_U\mu$ and $G_U\nu$ are finely continuous on U we actually have $G_U\mu = G_U\nu$ everywhere on U (and trivially on $\complement U$, where $G_U\mu = 0 = G_U\nu$).

Consider any compact set $A \subset U$ such that $\mu(A) < \infty$, $\nu(A) < \infty$. Then $G\mu_A$ and $G\nu_A$ are harmonic in $\complement A$ (writing μ_A for the measure $1_A\mu$). We proceed to show that $\mu_A = \nu_A$.

Let likewise B be compact, $B \subset U$, $\nu(B) < \infty$, and suppose that $A \cap B = \varnothing$. From $G_U\mu = G_U\nu$ therefore follows that

$$G_U\mu_A + G_U\mu_{\complement A} = G_U\nu_B + G_U\nu_{\complement B}$$

and hence by specific decomposition property, see Sect. 2.4, that

$$G_U\mu_A = p + q \text{ in } U,$$

where $p \preceq G_U\nu_B$ and $q \preceq G_U\nu_{\complement B}$ are fine potentials on U (because $G_U\mu_A$ is one by Theorem 2.6.2). Since $G_U\mu_A \leq G_\Omega\mu_A < \infty$ in $\complement A$, it follows from Theorem 2.6.4 that $G_U\mu_A$ is finely harmonic in $U \backslash A$, and similarly $G_U\nu_B$ is finely harmonic in $U \backslash B$. But $p \preccurlyeq G_U\mu_A$ and $p \preccurlyeq G_U\nu_B$, cf. above. The potential p is therefore finely harmonic in $(U\backslash A) \cup (U\backslash B) = U$ and hence $p = 0$. Consequently

$$G_U\mu_A = q \preccurlyeq G_U\nu_{\complement B},$$

in particular $G_U\mu_A \leq G_U\nu_{\complement B}$ in U (and trivially in $\complement U$). By a construction similar to that in the proof of Lemma 2.6.1 (now with $\bigcup_n K_n = \complement A$ in place of Ω, and ν in place of μ), we can let B range over an increasing sequence of compacts B_n such that $\nu(B_n) < \infty$ and $\bigcup_n B_n = U \backslash A$. Then $G_U\nu_{\complement B_n}$ decreases to $G_U\nu_A$ q.e. in Ω, and we conclude that $G_U\mu_A \leq G_U\nu_A$ (first q.e., next everywhere, by fine continuity). By symmetry,

$$G_U\mu_A = G_U\nu_A \text{ in } \Omega,$$

or explicitly, by integrating (2.6.7) with respect to μ_A, resp. ν_A, and using (1.4.2),

$$G_\Omega(\mu_A + (\nu_A)^{\complement U}) = G_\Omega(\nu_A + (\mu_A)^{\complement U}) \text{ in } \Omega$$

whence by the unique representation of a potential on Ω as a G_Ω-potential

$$\mu_A + (\nu_A)^{\complement U} = \nu_A + (\mu_A)^{\complement U}.$$

Since μ_A and ν_A are carried by U, whereas $(\mu_A)^{\complement U}$ and $(\nu_A)^{\complement U}$ are carried by $\complement U (= b(\complement U))$, it follows that indeed $\mu_A = \nu_A$. In order to conclude finally that $\mu = \nu$, it remains to cover U by an increasing sequence of compacts A_n such that $\mu(A_n) < \infty$, $\nu(A_n) < \infty$ as in the proof of Lemma 2.6.1 (choosing now x_0 so that $G_U\mu(x_0) < \infty$ and $G_U\nu(x_0) < \infty$). Then $\mu_{A_n} = \nu_{A_n}$, as shown above, and consequently $\mu = \nu$. □

The next two Lemmas are keys to the integral representation of fine potentials.

Lemma 2.6.8 *Let $p \in \mathcal{P}(U)$, $\mu \in \Lambda(U)$, and suppose that $p \preccurlyeq G_U\mu$. Then there exists $\lambda \in \Lambda(U)$ such that $p = G_U\lambda$ and (necessarily) $\lambda \leqslant \mu$.*

Proof It is easy to reduce to the case where U is regular. The hypothesis $p \preccurlyeq G_U\mu$ means that $p + q = G_U\mu$ for some $q \in \mathcal{P}(U)$. Once it has been shown that there exists $\lambda \in \Lambda(U)$ such that $G_U\lambda = p$, and therefore likewise $\nu \in \Lambda(U)$ such that $G_U\nu = q$ then $G_U\lambda + G_U\nu = G_U\mu$ and hence $\lambda + \nu = \mu$ by Theorem 2.6.7, where indeed $\lambda \leqslant \mu$.

Consider first the case where $\mu \in \Lambda = \Lambda(\Omega)$. Since $p \leqslant G_U\mu$ we have by (2.6.5) with U, V replaced by Ω, U

$$v := p + \widehat{R}^{\complement U}_{G_\Omega\mu} \leq G_\Omega\mu \text{ on } U. \tag{2.6.9}$$

Here v is finely hyperharmonic on U, and

$$\text{f-}\liminf_{y\to x,\, y\in U} v(y) \geqslant \widehat{R}^{\complement U}_{G_\Omega\mu}(x) = G_\Omega\mu(x)$$

for every $x \in \partial_f U$ because $p \geqslant 0$ and $\partial_f U \subset \complement U$, which is a base. According to Lemma 2.1.51 it follows from (2.6.9) that the function

$$w := \begin{cases} v(\leqslant G_\Omega\mu) & \text{on } U \\ G_\Omega\mu & \text{on } \complement U \end{cases}$$

is finely hyperharmonic (and $\geqslant 0$) on Ω, hence also hyperharmonic by Theorem 2.1.22(a). From $w \leqslant G_\Omega\mu$ on U we therefore conclude that w is a potential $G_\Omega\lambda$, $\lambda \in \Lambda(\Omega)$. Since $G_\Omega\lambda = w = G_\Omega\mu$ on $\complement U$ it follows that $\widehat{R}^{\complement U}_{G_\Omega\lambda} = \widehat{R}^{\complement U}_{G_\Omega\mu}$, and hence by (2.6.5)

$$p = v - \widehat{R}^{\complement U}_{G_\Omega\mu} = w - \widehat{R}^{\complement U}_{G_\Omega\mu} = G_\Omega\lambda - \widehat{R}^{\complement U}_{G_\Omega\lambda} = G_U\lambda$$

q.e. on U, and consequently $p = G_U\lambda$ everywhere on U, by fine continuity. To achieve that $\lambda(\complement U) = 0$ it only remains to replace λ by its restriction λ_U to U, noting that $G_U\lambda_{\complement U} = 0$.

In the general case where $\mu \in \Lambda(U)$, let μ_n denote the restriction of μ to the compact set A_n defined in (2.6.2). Then $G_U\mu_n$ increases specifically to $G_U\mu$:

$$\curlyvee_n G_U\mu_n = \sup_n G_U\mu_n = G_U\mu,$$

cf. Corollary 2.2.18(a). It follows that

$$p_n := p \curlywedge G_U\mu_n$$

increases specifically to $p \curlywedge G_U\mu = p$, cf. Distributivity of specific order, paragraph 6 after the proof of Theorem 2.4.5. Now $p_n \preccurlyeq G_U\mu_n$, and since $G_U\mu_n \not\equiv +\infty$, U being finely connected, we have shown above that there exists $\lambda_n \in \Lambda(U)$ such that

$$p_n = G_U\lambda_n,\ \lambda_n \leqslant \mu_n,$$

and consequently $G_\Omega\lambda_n \leqslant G_\Omega\mu_n \not\equiv +\infty$. Because $G_U\lambda_n \preccurlyeq G_U\lambda_{n+1}$ and $G_\Omega\lambda_n \not\equiv +\infty$, we likewise conclude from the former part of the proof that $\lambda_n \leqslant \lambda_{n+1}$. For there exists $q_n \in \mathcal{P}(U)$ with $G_U\lambda_n + q_n = G_U\lambda_{n+1}$, and hence a measure ϱ_n on U such that $q_n = G_U\varrho_n$, and so $G_U(\lambda_n + \varrho_n) = G_U\lambda_{n+1}$. It follows by Theorem 2.6.7

that $\lambda_n + \varrho_n = \lambda_{n+1}$, and consequently $\lambda_n \leqslant \lambda_{n+1}$. Defining

$$\lambda(E) = \sup_n \lambda_n(E), \quad E \in \mathcal{B}(U),$$

we infer from $\lambda_n \leqslant \mu_n$ that $\lambda \leqslant \mu$, showing that λ is a Borel measure on U, like μ. In particular, $\lambda \in \Lambda(U)$. Finally,

$$G_U\lambda = \sup_n G_U\lambda_n = \sup_n p_n = p.$$

Recall from Theorem 2.4.6 that $\mathcal{P}(U)$ is a *band* in the lower complete prevector lattice $\mathcal{S}_+(U)$ of all finely superharmonic functions $\geqslant 0$ on U.

Theorem 2.6.9 *The convex cone* $G_U\Lambda(U) = \{G_U\mu : \mu \in \Lambda(U)\}$ *is a band in* $\mathcal{P}(U)$, *and hence in* $\mathcal{S}_+(U)$. *The affine σ-additive mapping* $\mu \mapsto G_U\mu$ *is an order isomorphism between* $\Lambda(U)$ *and* $G_U\Lambda(U)$, *the latter being equipped with the specific order.*

Proof By Lemma 2.6.8, $G_U\Lambda(U)$ is specifically solid in $\mathcal{P}(U)$. The map $\mu \mapsto G_U\mu$ of $\Lambda(U)$ onto $G_U\Lambda(U)$ is an order isomorphism:

$$\lambda \leqslant \mu \iff G_U\lambda \preccurlyeq G_U\mu,$$

being injective according to Theorem 2.6.7. To complete the proof that $G_U\Lambda(U)$ is a band in $\mathcal{P}(U)$ it remains to show that if a specifically increasing net $(p_\alpha) = (G_U\mu_\alpha)$ in $G_U\Lambda(U)$ is specifically majorized in $\mathcal{P}(U)$ and hence possesses a specific supremum p in $\mathcal{P}(U)$, then $p \in G_U\Lambda(U)$. Defining

$$\mu(E) = \sup_\alpha \mu_\alpha(E), \quad E \in \mathcal{B}(U),$$

we show that μ is σ-finite, hence a measure on U which clearly satisfies $G_U\mu = p$. Choosing $x_0 \in U$ with $p(x_0) < +\infty$ and writing $U = \bigcup_n V_n$ with $V_n = \{G_U(x_0, \cdot) > \frac{1}{n}\}$ we have $\mu_\alpha(V_n) \leqslant nG_U\mu_\alpha \leqslant np(x_0)$ and hence $\mu(V_n) \leqslant np(x_0) < +\infty$. $\square$

Lemma 2.6.10 *Let p denote a fine potential on U, majorized by some finite potential on Ω, and suppose that p is finely harmonic on a fine boundary strip $U \setminus F$ (F finely closed in Ω, $F \subset U$). If $p \not\equiv 0$ there exists a non-zero measure $\mu \in \Lambda(U)$ such that $G_U\mu \preccurlyeq p$.*

Proof Clearly $s := \widehat{R}_p^U$ is a finite potential on Ω, hence semibounded by Definition 1.5.16(iv). According to Lemma 2.5.14, $s_U \preccurlyeq p$ and $s_U \not\equiv 0$ (because $p \not\equiv 0$). Writing $s = G_\Omega\mu$ with $\mu \in \Lambda(\Omega)$ we have on U by (2.6.5) (with U, V replaced by Ω, U)

$$G_U\mu = G_\Omega\mu - \widehat{R}_{G_\Omega\mu}^{\complement U} = s - \widehat{R}_s^{\complement U} = s_U.$$

It remains to replace μ by its restriction to U, a Borel measure on U. This does not change $G_U\mu$ because $\mu^*(\mathfrak{r}(U)\backslash U) = 0$ by Corollary 2.6.6(a) applied to the polar set $\mathfrak{r}(U)\backslash U$. □

We are now prepared to show that every semibounded fine potential p is representable as $p = G_U\mu$ for some measure $\mu \in \Lambda(U)$.

Theorem 2.6.11 *Every semibounded fine potential p on U equals $G_U\mu$ for a unique measure $\mu \in \Lambda(U)$.*

Proof In view of Theorem 2.4.13 and Theorem 2.6.9 we may assume that p is finite valued. We may further suppose that p is orthogonal to $G_U\Lambda(U)$, and we shall then deduce that $p = 0$. Fix a point $x_0 \in U$. The sets

$$A_n = \{y \in U : G_U(x_0, y) \geqslant \frac{1}{n} \text{ and } p(y) \leqslant n\}$$

($n \geq 1$) are finely closed relative to U and form an increasing sequence with the union U. Let $p_n = 1_{A_n}.p$ denote the specific restriction of p to A_n (see Definitions 2.4.27 and 2.4.29):

$$\begin{aligned} p_n &= 1_{A_n}.p = \curlyvee\{h \in \mathcal{P}^s(U) : h \preccurlyeq p, S(h) \subset A_n\} \\ &= p \curlywedge \widehat{R}_p^{A_n} \end{aligned}$$

according to Theorem 2.4.30. The sequence (p_n) is specifically increasing because $\{h \in \mathcal{P}^s(U) : h \preccurlyeq p,\ S(h) \subset A_n\}$ increases with A_n. We have $\bigvee_n {}^U\widehat{R}_p^{A_n} = \sup_n {}^U\widehat{R}_p^{A_n} = p$ by Theorem 2.2.15(a) because (A_n) is increasing to $\bigcup_n A_n = U$. It follows by Corollary 2.2.18(a) and the distributivity of specific order, cf. paragraph 6 after the proof Theorem 2.4.5, that

$$\curlyvee_n(p \curlywedge {}^U\widehat{R}_p^{A_n}) = p \curlywedge \curlyvee_n {}^U\widehat{R}_p^{A_n} \succcurlyeq p \curlywedge p = p$$

and so $\curlyvee_n p_n = p$. It therefore remains to show that each $p_n = 0$.

Now p_n is finely continuous on U, and finely harmonic on $U\backslash A_n$ because $S(p_n) \subset A_n$ and because p_n is finite valued, cf. Theorem 2.1.45. The function $n - p_n$ thus is finely hyperharmonic on $U\backslash A_n$, and finely continuous and $\geqslant -p_n$ on U. Moreover,

$$n - p_n \geqslant n - p \geqslant 0 \text{ on } A_n.$$

It therefore follows from the relative fine boundary minimum principle (Theorem 2.4.14) that $n - p_n \geqslant 0$ on $U\backslash A_n$, and so on all of U. Consequently, Lemma 2.6.10 applies, if $p_n \neq 0$, to produce a non-zero measure $\mu_n \in \Lambda(U)$ such that $G_U\mu_n \preccurlyeq p_n$ and thus $G_U\mu_n \preccurlyeq p$. But p was supposed to be orthogonal to $G_U\Lambda(U)$, which is a contradiction. □

Corollary 2.6.12 *The class $G_U\Lambda(U)$ of all fine potentials of the form $G_U\mu$, $\mu \in \Lambda(U)$, is solid even in the pointwise order. In other words, if $s \in \mathcal{S}_+(U)$ and if $s \leqslant G_U\mu$ for some $\mu \in \Lambda(U)$ then $s \in G_U\Lambda(U)$.*

In fact, by Theorem 2.4.11, $s = p + h$ with $p \in \mathcal{P}^s(U)$ and $h \in \mathcal{S}_+(U)$, h finely harmonic q.e. on U. By Theorem 2.6.11, $p \in G_U\Lambda(U)$. Since $h \leqslant s \leqslant G_U\mu$ we even have $h \preccurlyeq G_U\mu$ (Theorem 2.1.43 on removable singularities) and hence $h \in G_U\Lambda(U)$, by Lemma 2.6.8. Altogether, $s = p + h \in G_U\Lambda(U)$.

Lemma 2.6.13 *If μ is a weak* cluster point of a sequence of uniformly bounded measures μ_n on Ω supported on a fixed compact set $A \subset \Omega$ then*

$$G_U\mu \geqslant \liminf_n G_U\mu_n \text{ q.e. on } \Omega.$$

Proof Recall the definition (2.6.1) of $G_U\mu$. It is not required that $A \subset U$. It is well known that (μ_n) has a subsequence (μ'_n) that converges weakly* to μ. If the lemma holds for (μ'_n) then

$$G_U\mu \geqslant \liminf_n G_U\mu'_n \geqslant \liminf_n G_U\mu_n$$

holds q.e. on Ω. We may therefore assume that the given sequence (μ_n) converges weakly* to μ. We begin by proving that

$$G_U\mu \geqslant \liminf_n G_U\mu_n \text{ q.e. on } X \tag{2.6.10}$$

for any compact X in Ω such that $A \subset X$. This will be derived from the Brelot-Cartan convergence theorem, cf. [Br9, p. 36] or [Fu3, Lemme 2.1], which asserts that

$$G_\Omega\mu = \liminf_n G_\Omega\mu_n \text{ q.e. on } X$$

when $\mu_n \to \mu$ weakly*. And that in turn is related to the fundamental convergence theorem, see Sect. 1.1.13. Since $G_\Omega\varepsilon_x^{\complement U}$ is l.s.c. we have

$$\int G_\Omega\varepsilon_x^{\complement U}\,d\mu \leqslant \liminf_n \int G_\Omega\varepsilon_x^{\complement U}\,d\mu_n.$$

From (2.6.6) with U, V replaced by Ω, U we therefore obtain q.e. for $x \in X$

$$\begin{aligned}G_U\mu(x) &= G_\Omega\mu(x) - \int G_\Omega\varepsilon_x^{\complement U}\,d\mu \\ &\geqslant \liminf_n G_\Omega\mu_n(x) - \liminf_n \int G_\Omega\varepsilon_x^{\complement U}\,d\mu_n \\ &\geqslant \liminf_n \Big(G_\Omega\mu_n(x) - \int G_\Omega\varepsilon_x^{\complement U}\,d\mu_n\Big) \\ &= \liminf_n G_U\mu_n(x),\end{aligned}$$

again by (2.6.6), now with μ replaced by μ_n. By recovering Ω by a sequence of compact sets as above we infer that

$$G_U\mu \geqslant \liminf_n G_U\mu_n \text{ q.e. on } \Omega.$$ □

Lemma 2.6.14 *Let $p \in \mathcal{P}(U)$ and let $A \subset U$ be relatively compact. Suppose there exists a point $x_0 \in U$ such that $p(x_0) < +\infty$ and $\inf\{G_U(x_0, y) : y \in A\} > 0$. Then ${}^U\widehat{R}_p^A \in G_U\Lambda(U)$.*

Proof For $n \in \mathbb{N}$ the finite fine potential

$$p_n := {}^U\widehat{R}_{p\wedge n}^A = {}^U\widehat{R}_{p\wedge n}^{b(A)}$$

is semibounded and therefore representable as $p_n = G_U\mu_n$ for some $\mu_n \in \Lambda(U)$, by Theorem 2.6.11. By Corollary 2.4.20, p_n is finely harmonic on $U\setminus\widetilde{A}$, and hence μ_n is carried by $\widetilde{A} \subset \overline{A}$ in view of Theorem 2.6.4. By hypothesis there is a constant $c > 0$ such that $G_U(x_0, y) \geqslant c$ for $y \in A$, and even for $y \in \widetilde{A}$ by fine continuity. It follows that

$$c\mu_n(\widetilde{A}) \leqslant \int G_U(x_0, y)\,d\mu_n(y) = p_n(x_0) \leq p(x_0) < +\infty,$$

and so the measures μ_n are uniformly bounded. It is well known that there exists a weak* cluster point μ of (μ_n), and by Lemma 2.6.13 such a measure μ satisfies

$$G_U\mu \geqslant \liminf_n G_U\mu_n = \lim_n p_n = {}^U\widehat{R}_p^A$$

q.e. on U by definition of p_n and by Theorem 2.2.15. This extends by fine continuity to $G_U\mu \geqslant {}^U\widehat{R}_p^A$ everywhere on U. According to Corollary 2.6.12 this implies that indeed ${}^U\widehat{R}_p^A \in G_U\Lambda(U)$. □

Lemma 2.6.15 *Let $p \in \mathcal{P}(U)$ be orthogonal to $G_U\Lambda(U)$, and let $e \subset U$ be relatively compact such that $p = +\infty$ on e. Suppose there exists $x_0 \in U$ such that $p(x_0) < +\infty$ and $\inf\{G_U(x_0, y) : y \in e\} > 0$. Then*

$$\inf\{{}^U\widehat{R}_p^V(x_0) : V \text{ finely open},\ e \subset V \subset U\} = 0.$$

Proof Choose a relatively compact open set $\omega \subset \Omega$ containing $\overline{e}$, and a constant $c > 0$ such that $G_U(x_0, \cdot) > c$ on e. The finely open set

$$A := \omega \cap \{y \in U : G_U(x_0, y) > c\}$$

satisfies $e \subset A \subset U$ and ${}^U\widehat{R}_p^A \in G_U\Lambda(U)$ according to the preceding lemma. Let $\mathcal{V}$ denote the family of finely open sets V such that $e \subset V \subset A$, and write

$$q(x) = \inf\{{}^U\widehat{R}_p^V(x) : V \in \mathcal{V}\}, \ x \in U, \tag{2.6.11}$$

and $E = \{x \in U : p(x) = +\infty\}$. For any $V \in \mathcal{V}$, ${}^U\widehat{R}_p^V$ is finely harmonic on $U \setminus (\widetilde{V} \cup E)$, by Corollary 2.4.20 applied to $A = \widetilde{V}$ (since $\mathfrak{b}(V) \subset \widetilde{V}$), and so is therefore q, by Theorem 2.1.31. Since the fine topology is regular and e is polar, hence finely closed, we have $\bigcap\{\widetilde{V} : V \in \mathcal{V}\} = e \subset E$. It follows that q is finely harmonic on the finely open set $U \setminus E$ which contains x_0. The greatest finely l.s.c. minorant $\widehat{q}$ of q on U equals q on $U \setminus E$, and in particular at x_0.

By the fine convergence theorem (Theorem 2.1.34), $\widehat{q}$ is finely hyperharmonic on U. Furthermore

$$\widehat{q} \leqslant q \leqslant {}^U\widehat{R}_p^V \leqslant {}^U\widehat{R}_p^A \in G_U\Lambda(U)$$

by (2.6.11), Corollary 2.2.8 and Lemma 2.6.14. Since E is polar it follows from $\widehat{q} \leqslant q$ and $\widehat{q} \leqslant {}^U\widehat{R}_p^A$ that $\widehat{q} \preccurlyeq p$ and $\widehat{q} \preccurlyeq {}^U\widehat{R}_p^A$. By hypothesis on p this shows that $\widehat{q} = 0$ and hence $q(x_0) = \widehat{q}(x_0) = 0$. In view of (2.6.11) this proves lemma. □

We are now prepared to establish the main theorem of the present section.

Theorem 2.6.16 *Every fine potential p on a fine domain U has a unique integral representation $p = G_U\mu$, $\mu \in \Lambda(U)$.*

Proof Uniqueness was obtained in Theorem 2.6.7. According to Theorem 2.6.9, the convex cone $G_U\Lambda(U)$ of all representable fine potentials on U is a band in the lower complete prevector lattice $\mathcal{P}(U)$ of all fine potentials on U. The given fine potential $p \in \mathcal{P}(U)$ therefore decomposes uniquely into the sum of a representable fine potential $G_U\mu$ and a fine potential orthogonal to $G_U\Lambda(U)$. We may therefore assume from the beginning that $p \perp G_U\Lambda(U)$, that is, p has no non-zero specific minorant of class $G_U\Lambda(U)$. The proof reduces then to that $p = 0$.

As in the proof of Lemma 2.6.15 we put $E = \{p = +\infty\}$ and fix a point $x_0 \in \complement E$. We now cover Ω by a sequence of relatively compact open sets ω_n $(n \geq 1)$, and define

$$A_n = \omega_n \cap \{G_U(x_0, \cdot) > \frac{1}{n}\}, \quad e_n = E \cap A_n.$$

According to Lemma 2.6.15, there exists a finely open set V_n such that $e_n \subset V_n \subset A_n$ and

$${}^U\widehat{R}_p^{V_n}(x_0) < 2^{-n}. \tag{2.6.12}$$

According to Lemma 2.6.14, there exists a measure μ_n on U such that

$$ {}^U\widehat{R}_p^{V_n} = G_U\mu_n. \tag{2.6.13}$$

Writing $\mu = \sum_n \mu_n$ we therefore have

$$G_U\mu = \sum_n G_U\mu_n \not\equiv +\infty,$$

that is, $\mu \in \Lambda(U)$, because $G_U\mu(x_0) < +\infty$ in view of (2.6.12) and (2.6.13). By Theorem 2.2.14(b) applied to $f = 1_{V_n}p$ we have ${}^U\widehat{R}_p^{V_n} = p$ on V_n, and hence $\sum_n {}^U\widehat{R}_p^{V_n} \geq p$ on V_n, that is, by (2.6.13), $G_U\mu \geqslant p$ on $V := \bigcup_n V_n$. This extends by fine continuity to

$$G_U\mu \geqslant p \text{ on } \widetilde{V} \cap U. \tag{2.6.14}$$

According to Theorem 2.6.11, $\mathcal{P}^s(U) \subset G_U\Lambda(U)$. Since $p \perp G_U\Lambda(U)$ we therefore have $p \perp \mathcal{P}^s(U)$, and hence p is finely harmonic on $\{p < +\infty\}$ in view of Theorem 2.4.11. It follows that $G_U\mu - p$ is finely hyperharmonic on $U\setminus E$, in particular on the finely open set $W := U\setminus\widetilde{V}$ because

$$E = \bigcup_n e_n \subset \bigcup_n V_n = V \subset \widetilde{V}.$$

Furthermore, $G_U\mu - p$ is finely continuous on $U \cap \widetilde{W} \subset U\setminus V$, and $\geqslant 0$ on $U \cap \partial_f W$, by (2.6.4). Finally, $G_U\mu - p \geqslant -p$ on U, and we conclude by the relative fine boundary minimum principle (Theorem 2.4.14), applied with u and V replaced by $G_U\mu - p$ and W, that $G_U\mu \geqslant p$ on $U\setminus\widetilde{V}$. When combined with (2.6.14) this shows that $G_U\mu \geqslant p$ on U. According to Corollary 2.6.12 (or Lemma 2.6.8 since we even have $G_U\mu \succcurlyeq p$), this implies that $p \in G_U\Lambda(U)$ and hence indeed that $p = 0$. □

Definition 2.6.17 A function $s \in \mathcal{S}_+(U)$ is termed extremal (in $\mathcal{S}_+(U)$) if every specific minorant of s in $\mathcal{S}_+(U)$ is proportional to s.

In other words, s is extremal if and only if for any $s_1, s_2 \in \mathcal{S}_+(U)$ with $s_1 + s_2 = s$, s_1 and s_2 are constant multiples of s.

For any point $y \in V$, the fine potential $G_U\varepsilon_y = G_U(\cdot, y)$ is extremal in $\mathcal{S}_+(U)$ (Definition 2.6.17) and hence in $\mathcal{P}(U)$. Indeed, let $p_1, p_2 \in \mathcal{P}(U)$, and suppose that $p_1 + p_2 = G_U\varepsilon_y$. Then for $p_i \preccurlyeq G_U\varepsilon_y$ for $i = 1, 2$, and hence p_i must be finely harmonic on $U\setminus\{y\}$ along with $G_U\varepsilon_y$, whence Theorem 2.5.11(b) applies. Actually, we have by Theorem 2.6.16 the following

Corollary 2.6.18 *The extremal fine potentials on U are precisely the functions $cG_U(\cdot, y)$ with $y \in U$ and $c \in [0; +\infty[$.*

In view of Theorem 2.6.16 this reduces to the extremal measures on U being $c\varepsilon_y$, with $y \in U$ and $c \in [0; +\infty[$.

The integral representation of fine potentials allows us furthermore to strengthen the fine boundary minimum principles obtained earlier. Recall from Definition 2.1.49 the regular part $\partial_{f,r}U$ and the irregular part $\partial_{f,i}U$ of a finely open set $U \subset \Omega$:

$$\partial_{f,r}U = (\partial_f U) \cap \mathfrak{b}(\complement U) = (\partial_f U) \setminus \mathfrak{r}(U),$$

$$\partial_{f,i}U = (\partial_f U) \setminus \mathfrak{b}(\complement U) = (\partial_f U) \cap \mathfrak{r}(U) = \mathfrak{r}(U) \setminus U.$$

The case $U = \Omega$ of the following theorem was obtained in [Fu10, Section 2], where it was shown to be best possible when it is required that $u \geqslant -p$ for some potential p on Ω see [also] Remark 2.4.17).

Theorem 2.6.19 *Let $V \subset U$ be a finely open set. Let u be finely hyperharmonic on V and suppose there exists a fine potential $p = G_U\mu$ on U such that $u \geqslant -p$ on V. Suppose in addition that*

$$\text{f-} \liminf_{y \to x,\, y \in V} u(y) \geqslant 0$$

for every $x \in U \cap \partial_{f,r}V$ not belonging to some subset of zero inner (fine harmonic) measure $(\varepsilon_z^{\complement V})_*$ for some (and hence any) point z of each fine component of V. Suppose furthermore that

$$\text{f-} \liminf_{y \to x,\, y \in V} u(y) > -\infty$$

for every $x \in U \cap \partial_{f,i}V$ not belonging to some subset of $\partial_{f,i}V$ of zero inner measure μ_*. Then $u \geqslant 0$.

Proof We may of course suppose that $U \cap \partial_f V \neq \varnothing$. We define two subsets E and e of $\partial_f V$ by

$$E = \{x \in U \cap \partial_f V : \text{f-} \liminf_{y \to x,\, y \in V} u(y) < 0\},$$

$$e = \{x \in U \cap \partial_f V : \text{f-} \liminf_{y \to x,\, y \in V} u(y) = -\infty\}.$$

The assertion of the theorem is (with as in the above enunciation of the theorem) that if

$$(\varepsilon_z^{\complement V})_*(E \cap \partial_{f,r}V) = 0 \text{ and} \tag{2.6.15}$$

$$\mu_*(e \cap \partial_{f,i}V) = 0 \tag{2.6.16}$$

then $u \geqslant 0$ on V.

Denote by v the extension of $u_- = (-u) \vee 0$ from V to Ω obtained by putting $v = 0$ on $\complement V$. Since u and hence u_- are finely continuous on V, $v \geqslant 0$ is finely l.s.c. on Ω, and v is therefore subbasic: $v \leqslant \mathfrak{b}(v)$, by Lemma 1.4.41(a). The least finely u.s.c. majorant $\widetilde{v} = \mathfrak{b}(v)$ of v, cf. Corollary 1.4.46, therefore is of class $\mathcal{G}_\delta$ according to Lemma 1.4.40, that is, $\widetilde{v}$ is the pointwise infimum of a sequence of l.s.c. functions $\geqslant 0$ on Ω. Clearly, $\widetilde{v} = v = u_-$ on V. Furthermore,

$$\widetilde{v}(x) = \text{f-}\limsup_{y \to x,\, y \in V} u_-(y) \leqslant p(x) = G_U\mu(x) \tag{2.6.17}$$

for every $x \in U \cap \partial_f V$. Since $\widetilde{v} = 0$ on $\complement\widetilde{V}$ we conclude that

$$E = \{x \in U : \widetilde{v}(x) > 0\} \cap \partial_f V, \tag{2.6.18}$$

$$e = \{x \in U : \widetilde{v}(x) = +\infty\} \subset \partial_f V, \tag{2.6.19}$$

and since $\widetilde{v} = \mathfrak{b}(v) \in \mathcal{G}_\delta$, e is a G_δ relative to U. Note that $e \subset \{x \in U : p(x) = +\infty\}$, so e is polar. Furthermore, $\{x \in U : \widetilde{v}(x) > 0\}$ is a $G_{\delta\sigma}$ relative to U, and so is therefore, by (2.6.18), the set

$$E \cap \partial_{f,r} U = \{x \in U : \widetilde{v}(x) > 0\} \cap \partial_{f,r} V \tag{2.6.20}$$

because $\partial_{f,r} U = \mathfrak{b}(U) \cap \mathfrak{b}(\complement U)$ is a G_δ in Ω. We shall now consider three particular cases which gradually lead to the theorem in general.

Case 1. $E = \varnothing$ and hence $e = \varnothing$. Here the theorem follows from Theorem 2.4.14 (the relative boundary minimum principle, with $p \in \mathcal{P}(U)$).

Case 2. E is polar and $e = \varnothing$. We fix a point $x_0 \in V$ and choose a superharmonic function $s \geqslant 0$ on Ω so that $s = +\infty$ on E and $s(x_0) = \delta$ for some given $\delta > 0$, cf. Sect. 1.1.11. The function $u + s|_U$ is finely hyperharmonic and $\geqslant -p$ on V. Furthermore,

$$\text{f-}\liminf_{y \to x,\, y \in V} (u(y) + s(y)) \geqslant 0$$

for every $x \in U \cap \partial_f V$ because $e = \varnothing$, $s(y) \to +\infty$ as $y \to x$ finely when $x \in E$, and the displayed fine lim inf is $\geqslant 0$ for $x \in U \cap (\partial_f V)\backslash E$. Thus the first particular case considered in the proof prevails when u is replaced by $u + s|_U$, and so $u + s \geqslant 0$ on V, in particular $u(x_0) + \delta \geqslant 0$, and consequently $u(x_0) \geqslant 0$, by varying δ.

Case 3. E is polar and $\mu_*(e) = 0$. According to Theorem 2.4.11 we have, by integral representation, the decomposition

$$p = G_U\mu = G_U\lambda + G_U\nu, \;\; G_U\lambda \in \mathcal{P}^s(U), \;\; G_U\nu \in \mathcal{P}^s(U)^\perp.$$

Thus $\mu = \lambda + \nu$, and $G_U\nu$ is finely harmonic on the finely open set

$$W := \{x \in U : G_U\nu(x) < +\infty\}.$$

It follows by Theorem 2.6.4 that $\nu^*(W) = 0$. Since $\nu^*(U\backslash e) = 0$ by Corollary 2.6.6(b), $G_U\nu$ being in $\mathcal{P}^s(U)^\perp$, we have

$$\nu_*(U\backslash W) = \nu_*(\{x \in U : G_U\nu(x) = +\infty\}) \leqslant \nu_*(e) \leqslant \mu_*(e) = 0$$

by (2.6.19) and by hypothesis, and hence $\nu(U) \leqslant \nu_*(U\backslash W) + \nu^*(W) = 0$, hence $\nu = 0$ and $G_U\nu = 0$. From $u \geqslant -G_U\mu$ on V therefore follows $u \geqslant -G_U\lambda$ on V, and since $G_U\lambda \in \mathcal{P}^s(U)$ we conclude by Theorem 2.4.14 (the relative fine boundary minimum principle, with $p \in \mathcal{P}^s(U)$) that indeed $u \geqslant 0$ on V.

The general case. Recall that e is a Borel set relative to U. We have the decomposition $p = p_e + p_{\complement e}$, where

$$p_e = G_U(1_e\mu), \ \ p_{\complement e} = G_U(1_{\complement e}\mu).$$

Consider an arbitrary *fine lower function* w on V for the function $f : \partial_f V \to \overline{\mathbb{R}}$ defined by

$$f(x) = \text{f-}\liminf_{y\to x,\ y\in V}(u(y) + p_e(y)),\ \ x \in \partial_f V, \tag{2.6.21}$$

cf. Definition 2.3.2. Explicitly, w shall be finely hypoharmonic on V and satisfy $w \leqslant q$ on V for some finite continuous potential $q = G_\Omega\lambda$, $\lambda \in \Lambda(\Omega)$, and finally we shall have

$$\text{f-}\limsup_{y\to x,\ y\in V} w(y) \leqslant f(x) \quad \text{for every } x \in \partial_f V. \tag{2.6.22}$$

It follows that $u + p_e - w$ is finely hyperharmonic and $\geqslant -(p_{\complement e} + q)$ on V, and hence in view of (2.6.21) and (2.6.22)

$$\text{f-}\liminf_{y\to x,\ y\in V}(u(y) + p_e(y) - w(y)) \geqslant f(x) - f(x) = 0 \quad \text{for } x \in U \cap \partial_{f,r} V. \tag{2.6.23}$$

And if $f(x) = +\infty$ then since $w \leqslant q$ on V

$$\text{f-}\liminf_{y\to x,\ y\in V}(u(y) + p_e(y) - w(y)) \geqslant f(x) - q(x) = +\infty \geqslant 0.$$

Since e is a polar G_δ relative to U, cf. text after (2.6.19), and since λ does not charge the Borel polar sets, cf. Corollary 2.6.6(a), we have $\lambda(e) = 0$. Furthermore, $p_{\complement e} = G_U(1_{\complement U}\mu)$ with $(1_{\complement e}\mu)(e) = 0$, and so altogether $(1_{\complement e}\mu + \lambda)(e) = 0$. We are thus in Case 3 with u and E replaced by the present $u + p_e - w$ and e while keeping e. This shows that

$$u + p_e \geqslant w \quad \text{on } V.$$

Finally, when w ranges over all fine lower functions for f on $\partial_f V$, defined in (2.6.21), we obtain for every $x \in V$, by Theorem 2.3.7,

$$\begin{aligned} u(x) + p_e(x) \geqslant w(x) \geqslant \underline{H}_f^V(x) &= \int_* f\, d\varepsilon_x^{\complement V} \\ &\geqslant \int p_e\, d\varepsilon_x^{\complement V} = \widehat{R}_{p_e}^{\complement V}(x) = G_\Omega(1_e\mu)^{\complement V}(x) \\ &= G_\Omega(1_e\mu)(x) = p_e(x). \end{aligned}$$

For the third inequality note that $\varepsilon_x^{\complement V}$ is carried by $\partial_{f,r}V (= (\partial_f V) \cap \mathfrak{b}(\complement V))$, and that $f \geqslant p_e$ on $(\partial_{f,r}V)\setminus E$ by hypothesis, cf. (2.6.21), and $(\varepsilon_x^{\complement V})_*(E) = 0$. For any Borel measurable lower bounded function $\varphi \geqslant f$ on the Borel set $\partial_{f,r}V$ we have $\int \varphi \, d\varepsilon_x^{\complement V} \geqslant \int p_e \, d\varepsilon_x^{\complement V}$ (p_e being finely continuous and hence Borel measurable). By varying φ we obtain $\int_* f \, d\varepsilon_x^{\complement V} \geqslant \int p_e \, d\varepsilon_x^{\complement V}$. Having thus shown that $u + p_e \geqslant p_e$ on V we conclude that $u \geqslant 0$ q.e. and hence everywhere on V. □

2.7 The Two-Dimensional Case $N = 2$

For $N = 2$ some of the results obtained in Sects. 2.1 and 2.4 can be strengthened, in view of the property particular for the planar case mentioned toward the end of Sect. 1.2. It states that every fine neighborhood of a point $x_0 \in \mathbb{R}^2$ contains arbitrarily small circles centered at x_0. We begin by considering finely [hyper]harmonic functions on a (Euclidean) open subset of $\mathbb{R}^2$.

The following lemma is a fine boundary minimum principle for usual hyperharmonic functions on a Euclidean open subset of $\mathbb{R}^2$. In Theorem 2.7.3 the lemma will be extended to the fine setting.

Lemma 2.7.1 *Let u be hyperharmonic on a (Euclidean) open set $U \subset \mathbb{R}^2$, and suppose*

$$\text{f-}\liminf_{y \to x,\, y \in U} u(y) \geqslant 0 \quad \textit{for every } x \in \partial_f^\infty U. \tag{2.7.1}$$

Then $u \geqslant 0$ on U.

We denote by $\partial_f^\infty U$ the fine boundary of U relative to $\mathbb{R}^2 \cup \{\infty\}$. This means that $\partial_f^\infty U = \partial_f U$ if U is thin at ∞, in the sense that the inversion of U with respect to the unit circle (or any other circle in $\mathbb{R}^2$) is thin at 0. Otherwise we have $\partial_f^\infty U = \partial_f U \cup \{\infty\}$. In particular, every bounded subset of $\mathbb{R}^2$ is thin at ∞.

Proof of Lemma 2.7.1. For a given point $x_0 \in U$ and a given member $\varepsilon > 0$ we shall prove that $u(x_0) \geqslant -\varepsilon$. Every point $y \in \partial_f^\infty U$ differs from x_0 and has therefore a fine neighborhood W_y in $\mathbb{R}^2 \cup \{\infty\}$ such that $x_0 \notin W_y$ and

$$u \geqslant -\varepsilon \quad \text{on } U \cap W_y. \tag{2.7.2}$$

(If $y \in \partial^\infty U \setminus \partial_f^\infty U$, take $W_y = \complement U$ in $\mathbb{R}^2 \cup \{\infty\}$. For $y \in \partial_f^\infty U$ the existence of W_y follows from (2.7.1).)

Denote by D_y an open disk centered at y such that $\partial^\infty D_y \subset W_y$ and $\overline{D}_y \subset \complement\{x_0\}$. (If $y = \infty$, a fine neighborhood W_∞ of ∞ in $\mathbb{R}^2 \cup \{\infty\}$ is the image of a fine neighborhood of 0 under the inversion i in the unit circle in $\mathbb{R}^2$, and $u \circ \mathrm{i}$ is then finely hyperharmonic, hence finely l.s.c., on $\mathrm{i}^{-1}(U)$. Furthermore, D_∞ shall be the complement in $\mathbb{R}^2 \cup \{\infty\}$ of a closed disk centered at 0 such that $\partial D_\infty \subset W_\infty$.)

In the sequel, closures and complement are taken relative to $\mathbb{R}^2 \cup \{\infty\}$. Since $\partial^\infty U$ is compact there exists a finite subset $\{y_1, \dots, y_m\}$ of $\partial^\infty U$ such that the open set

$$\omega = \bigcup_{i=1}^{m} D_{y_i} \subset \mathbb{R}^2 \cup \{\infty\}$$

covers $\partial^\infty U$. Note that

$$\overline{\omega} \subset \bigcup_{i=1}^{m} \overline{D}_{y_i} \subset \complement\{x_0\}$$

and

$$\partial^\infty \omega \subset \bigcup_{i=1}^{m} \partial^\infty D_{y_i} \subset \bigcup_{i=1}^{m} W_{y_i}.$$

The open subset $V := U \backslash \overline{\omega}$ of $\mathbb{R}^2$ contains x_0 and u is hyperharmonic on V, and l.s.c. on $\overline{V}$ because $\partial^\infty U \subset \omega$ and hence

$$\overline{V} \subset \overline{U} \backslash \omega = U \backslash \omega.$$

Consequently,

$$\partial^\infty V = \overline{V} \backslash V \subset U \cap \partial^\infty \omega \subset \bigcup_{i=1}^{m} (U \cap W_{y_i}).$$

It follows by (2.7.2) that $u + \varepsilon \geqslant 0$ on $\partial^\infty V$, and hence by the classical boundary minimum principle (Sect. 1.1.3) that $u + \varepsilon \geq 0$ on V, in particular at x_0. □

Theorem 2.7.2 *On an open set $U \subset \mathbb{R}^2$, [hyper]harmonic functions are the same as finely [hyper]harmonic functions.*

Proof According to Corollary 2.1.19, the proof can be reduced to the case of finely hyperharmonic functions. Let U_0 be the largest (Euclidean) open set $\subset U$ on which u is hyperharmonic, cf. Theorem 2.1.28. Suppose by contradiction that $U \backslash U_0 \neq \varnothing$. By Theorem 2.1.26 u is finely continuous on U and hence of the Baire class One in view of Theorem 1.4.43. It then follows that the restriction of u to $U \backslash U_0$ is of the Baire class One and then has a point of continuity $x_0 \in U \backslash U_0$. Fix a real $\lambda < u(x_0)$ and choose a (Euclidean) open set $\omega \subset U$ containing x_0 such that

$$u > \lambda \text{ on } \omega \backslash U_0. \tag{2.7.3}$$

Since u is finely continuous at x_0 there exists a fine neighborhood $V \subset \omega$ of x_0 such that

$$u > \lambda \text{ on } V. \tag{2.7.4}$$

As mentioned above, there exists an open disk centered at x_0 such that $\overline{D} \subset \omega$ and $\partial^\infty D \subset V$. From 2.7.3 and 2.7.4 it follows that $u - \lambda \geq 0$ on $\partial_f(D \cap U_0)$ or even more on $\partial(D \cap U_0)$ because

$$\partial(D \cap U_0) \subset (\partial D) \cup (\overline{D} \cap \partial U_0) \subset V \cup (\omega \backslash U_0).$$

We can now apply Lemma 2.7.1 to the hyperharmonic function $u - \lambda$ on the bounded (Euclidean) open set $D \cap U_0$, for $u - \lambda$ is finely continuous on the fine closure of $D \cap U_0$ (or even more the (Euclidean) closure $\overline{D \cap U_0} \subset \overline{D} \subset \omega \subset U$). This prove that $u \geq \lambda$ on $D \cap U_0$ and hence on D according to (2.7.3). Consequently, u is hyperharmonic on D by Theorem 2.1.22(a), hence $x_0 \in U_0$, which is a contradiction since $x_0 \notin U_0$. It follows that $U_0 = U$, so that u is hyperharmonic in U. □

Theorem 2.7.3 (A variant fine boundary minimum principle for $N = 2$) *Let u be finely hyperharmonic on a finely open set $U \subset \mathbb{R}^2$, and suppose that*

$$\text{f-}\liminf_{y \to x,\, y \in U} u(y) \geqslant 0 \quad \text{for every } x \in \partial_f^\infty U. \tag{2.7.5}$$

Then $u \geqslant 0$ on U.

In comparison with Corollary 2.4.15 we have thus (for $N = 2$) omitted the hypothesis $u \geqslant -p$ (for some potential p on a Greenian domain $\Omega \subset \mathbb{R}^N$ containing U). Instead, we have included ∞ as a possible fine boundary of U relative to $\mathbb{R}^2 \cup \{\infty\}$ when $N = 2$.

Proof of Theorem 2.7.3. Let $x \in \partial_f^\infty U$. After replacing u by $u \wedge 0$ we may suppose that $u \leqslant 0$ on U. The extension of u from U to $\mathbb{R}^2$ obtained by putting $u = 0$ on $\mathbb{R}^2 \backslash U$ is then finely hyperharmonic on $\mathbb{R}^2$ according to Lemma 2.1.51 in view of (2.7.5). This extension u is therefore hyperharmonic on $\mathbb{R}^2$, by Theorem 2.7.2. Furthermore, by (2.7.5)

$$\text{f-}\liminf_{y \to x,\, y \in \mathbb{R}^2} u(y) = \text{f-}\liminf_{y \to x,\, y \in U} u(y) \geqslant 0 \quad \text{if } x \in \mathbb{R}^2 \cap \partial_f^\infty U.$$

In any case,

$$\text{f-}\liminf_{y \to x,\, y \in \mathbb{R}^2} u(y) \geqslant 0$$

because this holds with equality if $x \in \partial_f^\infty \mathbb{R}^2 \backslash \partial_f^\infty U$, that is, if $x = \infty$ and U is thin at ∞. For then we have $u = 0$ in the deleted fine neighborhood $\mathbb{R}^2 \backslash U$ of ∞. Consequently, it follows from Lemma 2.7.1 that $u \geqslant 0$. □

Combining Theorem 2.7.3 with the proof of Theorem 2.6.19 leads to the following

Theorem 2.7.4 (A strengthening of Theorem 2.7.3) *Let again u be finely hyperharmonic on a finely open set $U \subset \mathbb{R}^2$. Suppose that*

$$\text{f-}\liminf_{y\to x,\, y\in U} u(y) \geqslant 0$$

for every $x \in \partial_{f,r}U$ *not belonging to some subset of* $\partial_{f,r}U$ *of* 0 *inner harmonic measure* $(\varepsilon_x^{\complement U})_*$ *for some point* x *of each fine component of* U. *Furthermore suppose that*

$$\text{f-}\liminf_{y\to x,\, y\in U} u(y) > -\infty$$

for every $x \in \partial_f U$. *Then* $u \geqslant 0$.

Proof Without loss of generality we may assume that U is a fine domain. Proceed as in the proof of Theorem 2.6.19, now with $e = \varnothing$ and

$$E = \{x \in \partial_f U : \text{f-}\liminf_{y\to x, \in U} u(y) < 0\}.$$

Then (2.6.17) reads $\widetilde{v}(x) < +\infty$ for every $x \in \partial_f U$, and (2.6.18 is therefore redundant, whereas (2.6.20) remains and shows that $E \cap \partial_{f,r}U$ is a $G_{\delta\sigma}$. Case 1 ($E = \varnothing$) is now covered by Theorem 2.7.3. Case 2 (E polar) carries over since $e = \varnothing$ (deleting the inequalities $s \geq 0$ and $u + s \geqslant -p$). Case 3 is now the same as Case 2 because there is no potential $p = G_\Omega\mu$ to be considered. Finally the proof in the general case carries over with $e = \varnothing$, omitting $p = G_\Omega\mu$, and replacing p_e by 0. Thus the definition (2.6.21) of f becomes

$$f(x) = \text{f-}\liminf_{y\to x,\, y\in U} u(y),\ x \in \partial_f U,$$

and (2.6.23) reads

$$\text{f-}\liminf_{y\to x,\, y\in U} (u(y) - w(y)) \geqslant 0 \text{ for } x \in \partial_f U,$$

where w denotes any fine lower function for f from the lower PWB class $\underline{\Phi}_f^U$ (see Definition 2.3.2). We conclude by Case 1 that $u \geqslant w$ on U. The rest of the proof simplifies as follows:

$$u(y) \geqslant \underline{H}_f^U(y) = \int_* f\, d\varepsilon_y^{\complement U} \geqslant 0 \text{ for } y \in U. \qquad \square$$

We also bring a variant of mean-value property from Theorem 2.1.10 for $N = 2$. We first consider the case where $U \subset \mathbb{R}^2$ is not thin at ∞ (that is, $\partial_f^\infty U = \partial_f U$). For the general case see Remark 2.7.6.

Theorem 2.7.5 (A variant of mean-value property for $N = 2$) *Let* U *be a finely open subset of* $\mathbb{R}^2$ *with fine closure* $\widetilde{U}$ *relative to* $\mathbb{R}^2 \cup \{\infty\}$ *contained in* $\mathbb{R}^2$. *Let* $u : \widetilde{U} \to \overline{\mathbb{R}}$ *be finely l.s.c. everywhere on* $\widetilde{U}$ *and finely hyperharmonic on* U. *Then*

$$u(x) \geqslant \int_* u\, d\varepsilon_x^{\complement U} \text{ for every } x \in U.$$

For any finely open set $V \subset U$ we have

$$\int_* u\, d\varepsilon_x^{\complement V} \geqslant \int_* u\, d\varepsilon_x^{\complement U} \text{ for every } x \in U.$$

Proof Since $\mathbb{R}^2 \setminus \widetilde{U}$ is a deleted neighborhood of ∞ in $\mathbb{R}^2 \cup \{\infty\}$ there exists a compact nonpolar subset K of $\mathbb{R}^2 \setminus \widetilde{U}$. According to the Szegő-Myrberg theorem, $\Omega := \mathbb{R}^2 \setminus K$ is a Green subset of $\mathbb{R}^2$ containing $\widetilde{U}$. Complement and topological notions are henceforth taken relative to Ω. For $x \in U$ the swept measure $\varepsilon_x^{\complement U}$ serves as the (fine) harmonic measure $\varrho(U, x)$ relative to U and x.

Denoting by B the base of $\complement U$ relative to Ω (that is, B is the set of points of Ω at which $\complement U$ is not thin) we have

$$U \subset \complement B \subset \widetilde{U} \subset \Omega$$

and $\varepsilon_x^{\complement U} = \varepsilon_x^B$ (see Theorem 1.4.21). Extend u from $\widetilde{U}$ to Ω as a finely l.s.c. function, e.g. by putting $u = +\infty$ on $\complement \widetilde{U}$. Choose a finite continuous potential $q > 0$ on Ω. For $n \in \mathbb{N}$ and $x \in U$ write for brevity $u_n(x) = u(x) \wedge nq(x)$, and as usual

$$u_n^B(x) = \int u_n\, d\varepsilon_x^B$$

(see Definition 1.4.29 and Remark 1.5.25). We may assume $\int_* u d\varepsilon_x^{\complement U} > -\infty$. Note that q is $\varepsilon_x^{\complement U}$-integrable; hence $\int u_n\, d\varepsilon_x^B = \int u_n\, d\varepsilon_x^{\complement U}$ exists and is $> -\infty$. For fixed n let Φ denote the upper directed family of all usual u.s.c. functions φ on Ω such that $-\infty \leqslant \varphi \leqslant u_n$; hence $\varphi \leqslant nq$. For any $\varphi \in \Phi$ it follows from Lemma 2.1.38 applied to $-\varphi$ that φ^B is finely u.s.c. and $\leqslant nq < +\infty$ on Ω, and finely hypoharmonic on $\complement B$. The function $u_n - \varphi^B$ is therefore well defined and finely l.s.c. on Ω, finely hyperharmonic on $\complement B$, and moreover equal to $u_n - \varphi \geqslant 0$ on B because $\varepsilon_x^B = \varepsilon_x$ for $x \in B$. It therefore follows from Theorem 2.7.3 that $u_n - \varphi^B \geqslant 0$ on $\complement B$ and hence on all of Ω. For every $x \in \Omega$ we therefore have

$$\begin{aligned} u_n(x) \geqslant \sup_{\varphi \in \Phi} \varphi^B &= \sup_{\varphi \in \Phi} \int \varphi\, d\varepsilon_x^{\complement U} \\ &= \int u_n\, d\varepsilon_x^{\complement U} = u_n^B(x), \end{aligned} \tag{2.7.6}$$

whence the former assertion of the theorem, by letting $n \to +\infty$ while observing that $u_n(y) = u(y)$ for any $y \in U$ such that $u(y) \leqslant 0$.

Finally, let $V \subset U$ be finely open and denote by A the base of $\complement V$ relative to Ω. Then $A \supset B$, and we obtain from (2.7.6)

$$u_n^A \geqslant (u_n^B)^A = u_n^B$$

according to the last assertion of Lemma 2.1.38. In other terms,

$$\int u_n \, d\varepsilon_x^{\complement V} \geqslant \int u_n \, d\varepsilon_x^{\complement U} \quad \text{for every } x \in \Omega,$$

whence the latter assertion of the theorem, by letting $n \to +\infty$. □

Remark 2.7.6 Theorems 2.7.3 and 2.7.5 extend to the case where U is allowed to be any finely open subset of a hyperbolic Riemann surface Ω, see [Fu8, Fu11, Fu10, Fu28].

2.8 Uniform Approximation by Harmonic or Subharmonic Functions

In this section, A denotes a closed subset of $\mathbb{R}^N$, $N \geqslant 2$, and A' denotes the fine interior of A. We denote by $\mathcal{C}(A)$ the space of continuous real-valued functions on A. Uniform closure of a subset of $\mathcal{C}(A)$ is indicated by a bar. When A is a compact subset K of $\mathbb{R}^N$, the uniform topology on $\mathcal{C}(A) = \mathcal{C}(K)$ coincides with the topology defined by the norm of uniform convergence and $\mathcal{C}(K)$ is a Banach space for this norm. By $H(A)$, resp. $S(A)$, we denote the linear subspace, resp. the convex subcone of all restrictions to A of harmonic, resp. finite continuous superharmonic, functions defined on open subsets of $\mathbb{R}^N$ containing A.

Theorem 2.8.1 *The uniform closure of $H(A)$, resp. of $S(A)$, is the set of all functions in $\mathcal{C}(A)$ which are finely harmonic, resp. finely superharmonic, on the fine interior A' of A.*

Proof The statement being obvious if $A' = \varnothing$, we may suppose in the proof that $A' \neq \varnothing$.

Proof in the case where A is a compact set $K \subset \mathbb{R}^N$. According to Corollary 2.1.21 every function in $\overline{H}(K)$, resp. $\overline{S}(K)$ is finely harmonic, resp. finely superharmonic, on K'. Conversely, suppose that $f \in \mathcal{C}(K)$ is finely harmonic, resp. finely superharmonic, on K'. With appeal to the Hahn-Banach theorem we shall consider a continuous linear form on $\mathcal{C}(K)$, that is, a Radon measure μ on K, and a constant $c \in \mathbb{R}$ such that $\mu \geqslant c$ on $H(K)$, resp. $S(K)$, and we must then prove that $\mu(f) \geqslant c$, cf. [Bou4, chap. II, p. 173, cor. 4]. Since $H(K)$ is a vector space the requirement $\mu \geqslant c$ on $H(K)$ amounts to $\mu = 0$ on $H(K)$, and since $S(K)$ is a cone the requirement $\mu \geqslant c$ on $S(K)$ amounts to $\mu \geqslant 0$ on $S(K)$. And we shall prove that $\mu(f) = 0$, resp. $\mu(f) \geqslant 0$.

Choose a Greenian domain $\Omega \subset \mathbb{R}^N$ containing K. Consider first the case of $H(K)$. For any open set ω such that $K \subset \omega \subset \Omega$, and any locally bounded potential s on Ω, $\widehat{R}_s^{\complement\omega}$ is harmonic on ω (complements and sweeping relative to Ω) according to Theorem 1.3.4(a). Its restriction to K thus satisfies $\widehat{R}_s^{\complement\omega}|_K \in H(K)$ and hence, by hypothesis on μ, $\int \widehat{R}_s^{\complement\omega} d\mu = 0$. Applying this to a decreasing sequence of open sets (ω_n) with the intersection K we obtain $\widehat{R}_s^{\complement\omega_n} \uparrow \widehat{R}_s^{\complement K}$ by Theorem 1.3.6(a), and hence

$$\begin{aligned}\int s d(\mu_+)^{\complement K} &= \int \widehat{R}_s^{\complement K} d\mu_+ = \lim_n \int \widehat{R}_s^{\complement\omega_n} d\mu_+ = \lim_n \int \widehat{R}_s^{\complement\omega_n} d\mu_- \\ &= \int \widehat{R}_s^{\complement K} d\mu_- = \int s d(\mu_-)^{\complement K}.\end{aligned} \tag{2.8.1}$$

By varying s it follows by Corollary 1.4.6 that $(\mu_+)^{\complement K} = (\mu_-)^{\complement K}$, to be written shortly as $\mu^{\complement K} = 0$.

As noted in the paragraph following (1.4.21), K' is regular, that is, $\complement K'$ is a base. Since $|f| \leqslant s$ for some finite potential s on Ω (proceed as in Remark 2.1.18) it follows from Theorem 2.3.1 applied to $U = K'$ that the bounded function h on K defined by

$$h(x) = \varepsilon_x^{\complement K}(f) = \varepsilon_x^{\complement K'}(f), \quad x \in K,$$

is the unique function on K which is finely harmonic on K', finely continuous on K, and equal to f on $K \setminus K'$. We obtain

$$\int h\, d\mu_\pm = \int \varepsilon_x^{\complement K}(f)\, d\mu_\pm = (\mu_\pm)^{\complement K}(f).$$

Here $\int \varepsilon_x^{\complement K}(f)\, d\mu_\pm$ denotes the integral with respect to μ_+ or μ_- of the family $(\varepsilon_x^{\complement K})_{x\in K}$ of positive Radon measures on K, cf. e.g. [Bou2]. The latter equality above follows from Theorem 1.4.30 (with μ replaced by μ_+ or μ_-). Since $(\mu_+)^{\complement K} = (\mu_-)^{\complement K}$ we conclude that

$$\int h\, d\mu = \int h\, d\mu_+ - \int h\, d\mu_- = (\mu_+)^{\complement K}(f) - (\mu_-)^{\complement K}(f) = 0. \tag{2.8.2}$$

This settles the harmonic case because $h = f$ in that case and hence $\int f\, d\mu = \int h\, d\mu = 0$. In the superharmonic case the function $g := f - h$ on K is finely superharmonic on K' along with f because h is finely harmonic on K'. Since $g = 0$ on $K \setminus K'$ it follows easily by the fine boundary minimum principle (Theorem 2.1.10) that $g|_{K'}$ is a fine potential. According to Theorem 2.5.8 there exists a sequence of finite continuous potentials p_n on Ω such that

$$g = \sum_n (p_n)_{K'} = \sum_n (p_n - \widehat{R}_{p_n}^{\complement K'}) \quad \text{on } K'.$$

Here $\int p_n\, d\mu \geqslant 0$ because $p_n|_K \in S(K)$. Furthermore,

$$\int \widehat{R}_{p_n}^{\complement K'} d\mu = \int \widehat{R}_{p_n}^{\complement K} d\mu = 0$$

by (2.8.1) applied to $s = p_n$. Because $p_n = \widehat{R}_{p_n}^{\complement K}$ in $\complement K$ by (1.3.11), it follows that

$$\int_{K'} (p_n - \widehat{R}_{p_n}^{\complement K'})\, d\mu \geqslant 0,$$

and hence by summation $\int_K g\, d\mu = \int_{K'} g\, d\mu \geqslant 0$ since $g = 0$ on $K \setminus K'$. By adding (2.8.2) to this we conclude that

$$\int f\, d\mu = \int (g + h)\, d\mu \geqslant \int h\, d\mu = 0.$$

Proof in the case of an arbitrary closed subset of $\mathbb{R}^N$. The case where $A = \mathbb{R}^N$ being trivial, we may suppose that A is proper subset of $\mathbb{R}^N$. Let f be a real valued function on A. Suppose first that $f \in \overline{H}(A)$. Let K be any closed ball in $\mathbb{R}^N$, then $f|_{A \cap K} \in \overline{H}(A \cap K)$. Thus by the above case f is continuous on $A \cap K$ and finely harmonic in the fine interior of $A \cap K$. Since this is true for any closed ball K, it follows that f is continuous on A and finely harmonic in the fine interior of A. Conversely, suppose that f is continuous on A and finely harmonic on the fine interior of A. Again, let K be a closed ball. Since no point of the boundary of K lies in the fine interior of $A \cap K$, it follows that f is continuous on $A \cap K$ and finely harmonic on the fine interior of $A \cap K$. According to the above case, we have $f|_{A \cap K} \in \overline{H}(A \cap K)$. Since this is so for every closed ball K, it follows by the localization theorem for harmonic approximation sets, see [GaH, Theorem 2.3.2 and Corollary 2.3.8], that $f \in \overline{H}(A)$.

For the approximation by superharmonic functions in the general case of a closed set, using similar idea as above, we refer to [BG]. □

From Theorem 2.8.1 we derive the following local versions.

Corollary 2.8.2 *A numerical function u on a finely open set $U \subset \Omega$ is finely harmonic, resp. finite valued and finely superharmonic, if and only if every point $x \in U$ has a compact fine neighborhood $V \subset U$ such that $u|_V \in \overline{H}(V)$, resp. $\overline{S}(V)$, that is, $u|_V$ is the uniform limit of a sequence of restrictions to V of harmonic, resp. finite continuous superharmonic functions, u_n on open sets $\omega_n \supset V$.*

Proof The "if part" follows in either case from Corollary 2.1.21 because each $u_n|_{V'}$ is finely harmonic, resp. finite and finely superharmonic, by Theorem 2.8.1. For the "only if part" it follows by the Brelot property (Theorem 1.5.29) that every point $x \in U$ has a compact fine neighborhood $V \subset U$ such that $u|_V \in \mathcal{C}(V)$. Since $u|_{V'}$ is finely harmonic, resp. finite and finely superharmonic, it follows from Theorem 2.8.1 that $u|_V$ belongs to $\overline{H}(V)$, resp. $\overline{S}(V)$. □

2.9 Invariant Functions

Definition 2.9.1 Let $U \subset \Omega$ be a finely open set. A function $h \in \mathcal{S}_+(U)$ is called *invariant* if the only fine potential which minorizes h in the specific order $\preccurlyeq$ is 0.

This means that the essential supremum of all $p \in \mathcal{P}(U)$ such that $p \preccurlyeq h$ equals 0, in other words that the invariant functions on U form the orthogonal band $\mathcal{P}(U)^\perp$ in $\mathcal{S}_+(U)$ to the band $\mathcal{P}(U)$ of all fine potentials on U. Together with Theorem 2.6.16, this shows (if U is a fine domain) that every function $u \in \mathcal{S}_+(U)$ has a unique integral representation of the form

$$u(x) = \int G_U(x, y)d\mu(y) + h(x), \quad x \in U, \tag{2.9.1}$$

where μ is a (positive) Borel measure on U, G_U is the (fine) Green kernel for U, and h is an invariant function on U. This is the closest we get at a Riesz decomposition of (positive) finely superharmonic functions (when U is only Euclidean open, cf. Sect. 1.1.6, see [GH]). For *finite* valued finely superharmonic functions we have the expected Riesz decomposition Theorem 2.4.3.

It follows easily from Definition 2.9.1 that $h \in \mathcal{S}_+(U)$ is invariant if and only if its restriction to every fine component of U is invariant.

Lemma 2.9.2 *A finite valued invariant function is the same as a finite finely harmonic function* ≥ 0.

Proof If $h \geq 0$ is finite and invariant then by Theorem 2.4.3 we have $h = p + k$ for some $p \in \mathcal{P}(U)$ and some finely harmonic $k \geq 0$. Thus $p \preccurlyeq h$ and hence $p = 0$, and $h = k$ is finely harmonic. Conversely, if $h \geq 0$ is finite and finely harmonic then again $h = p + k$ with $p \in \mathcal{P}(U)$ and k finite and invariant, hence finely harmonic, as just shown, and here $p = 0$ by the uniqueness in Theorem 2.4.3. □

Lemma 2.9.3 *Every invariant function h on U is finely harmonic q.e. on U, more precisely: finely harmonic everywhere on* $\{h < +\infty\}$.

Proof In fact, $\mathcal{P}(U) \supset \mathcal{P}^s(U)$ (the semibounded fine potentials). Hence $\mathcal{P}(U)^\perp \subset \mathcal{P}^s(U)^\perp$, whence the former assertion by Theorem 2.4.11. The latter assertion then follows by the removable singularities theorem for finely harmonic functions (Theorem 2.1.43) applied with U replaced by $\{h < +\infty\}$. □

The term "invariant" reflects a property established in Theorem 2.9.9 below and generalizing fine harmonicity (for positive functions).

For a Euclidean open set $U \subset \Omega$, $\mathcal{S}_+(U)$ equals the cone of positive superharmonic functions (cf. Theorem 2.1.22), $\mathcal{P}(U)$ the cone of potentials (Theorem 2.4.22), and hence the invariant functions $\mathcal{P}(U)^\perp$ become the positive harmonic functions on U.

Lemma 2.9.4 *If h is invariant on U, if $u \in \mathcal{S}_+(U)$, and if $h \leqslant u$, then $h \preccurlyeq u$.*

Proof By the above lemma there is a polar set $E \subset U$ such that h is finely harmonic on $U \setminus E$, and hence $u - h$ is a positive finely superharmonic on $U \setminus E$. According to Theorem 2.1.43, $u - h$ extends by fine continuity to a function $s \in \mathcal{S}_+(U)$ such that $h + s = u$ on $U \setminus E$ and hence on all of U, whence $h \preccurlyeq u$. □

Theorem 2.9.5 (a) *If $h_1, h_2 \in \mathcal{S}_+(U)$ are invariant, $p \in \mathcal{P}(U)$, and if $h_1 \leqslant h_2 + p$, then $h_1 \preccurlyeq h_2$ and $h_2 - h_1$ is invariant.*

(b) *If (h_j) is an increasing sequence of invariant functions on U majorized by some $u \in \mathcal{S}_+(U)$, then $\sup_j h_j$ is invariant.*

Proof (a) By Lemma 2.9.4 there exists $s \in \mathcal{S}_+(U)$ such that $h_1 + s = h_2 + p$. We have $s = h + q$ with h invariant and $q \in \mathcal{P}(U)$. It follows that $(h_1 + h) + q = h_2 + p$ with $h_1 + h$ invariant, and hence $h_1 + h = h_2$ (and $q = p$).

(b) According to Corollary 2.1.21 $\sup_j h_j \in \mathcal{S}_+(U)$ along with u and, by the Riesz decomposition property, we have $\sup_j h_j = h + p$ with h invariant and p a fine potential on U. It follows from (a) that $h_j \leq h$ for every j and hence $\sup_j h_j = h$ (and $p = 0$), so that $\sup_j h_j$ is invariant. □

It follows from Theorem 2.9.5 that in the Riesz decomposition $s = p + h$ of $s \in \mathcal{S}_+(U)$, p fine potential and h invariant, h is the greatest invariant minorant of s.

Remark 2.9.6 In view of the fourth paragraph after the proof of Theorem 2.4.5 and Theorem 2.9.5 the set $\mathcal{H}_i(U)$ of all invariant functions on U is clearly a lower complete and conditionally upper complete sublattice of $\mathcal{S}_+(U)$ in the specific order. According to Lemma 2.9.4 the specific order on $\mathcal{H}_i(U)$ coincides with the pointwise order.

Henceforth we mostly write simply R for ${}^U R$ and $\widehat{R}$ for ${}^U\widehat{R}$, thus taking for granted that reduction and sweeping of functions are taken relative to U (in the absence of other indication).

Lemma 2.9.7 *Let $u \in \mathcal{S}_+(U)$ and let A be a subset of U. The restriction of $\widehat{R}_u^A$ to any finely open subset V of $U \setminus \mathfrak{b}(A)$ is invariant.*

Proof We have $\widehat{R}_u^A = \sup_{k \in \mathbb{N}} \widehat{R}_{u \wedge k}^A$ by Theorem 2.2.15(a). Each of the functions $\widehat{R}_{u \wedge k}^A$ is finely harmonic on V by Corollary 2.4.20, and hence $\widehat{R}_u^A$ is invariant on V in view of Lemma 2.9.4 because the invariant functions on V form a band. □

As shown by the following example, the invariant functions on finely open (not Euclidean open) subsets of U need not form a sheaf.

Example 2.9.8 Let μ be the one-dimensional Lebesgue measure on a line segment E (of length $|E| > 0$) in $\Omega := U := \mathbb{R}^3$. Then E is polar and hence everywhere thin in $\mathbb{R}^3$. Every point $x \in \mathbb{R}^3$ therefore has a fine neighborhood V_x with $\mu(V_x) = 0$. (For $x \in E$ take $V_x = \{x\} \cup \complement L$, where L denotes the whole line extending E.) Thus μ does not have a (minimal) fine support (unlike measures which do not charge any polar set). The Green potential $G_\Omega \mu$ is invariant on each V_x (but of course not on Ω). For if p denotes a non-zero fine potential on V_x, $x \in E$, such that $p \preccurlyeq G_\Omega \mu$ on V_x then p is finely harmonic on $V_x \setminus \{x\} = \complement L$ along with $G_\Omega \mu$. Hence p behaves on V_x

near x like a constant times $G_{V_x}\varepsilon_x$ and is therefore of the order of magnitude $1/r$, where r denotes the distance from x, cf. Definition 2.5.10 and Theorem 2.5.11. But on the plane through x orthogonal to L, $G_\Omega\mu$ behaves like a constant times $\log(1/r)$, in contradiction with $p \leqslant G_\Omega\mu$.

Theorem 2.9.9 *Let $U \subset \Omega$ be a fine domain. A function $s \in \mathcal{S}_+(U)$ is invariant if and only if*

$$\widehat{R}_s^{U\setminus V} = s \tag{2.9.2}$$

for some countable covering of U by finely open sets $V \subset \mathfrak{r}(U)$ such that $\widetilde{V} \subset \mathfrak{r}(U)$. In the affirmative case, (2.9.2) holds for any set V as stated and such that moreover V has compact closure in Ω and

$$\inf_{y\in V} G_U(x,y) > 0 \quad \textit{for some } x \in U \textit{ with } s(x) < +\infty. \tag{2.9.3}$$

Proof Suppose first that (2.9.2) holds for some countable cover $\mathcal{V}$ of U by finely open sets V such that $\widetilde{V} \subset \mathfrak{r}(U)$. We may suppose that each $V \in \mathcal{V}$ is finely connected because a finely open set has only countably many fine components, by the quasi-Lindelöf principle (see Sect. 1.2). Let $p = G_U\mu$ denote the fine potential in the Riesz decomposition of s, cf. (2.9.1). We shall prove that $p = 0$. From $s = p + h$ we have

$$s = \widehat{R}_s^{U\setminus V} = \widehat{R}_p^{U\setminus V} + \widehat{R}_h^{U\setminus V} \leqslant p + h = s,$$

and hence (2.9.2) holds with s replaced by p, first q.e. and next everywhere on U, by fine continuity. It follows by Lemma 2.6.3 that $G_V\mu = 0$ on $\mathfrak{r}(U)$, in particular on V. Hence $\mu(\mathfrak{r}(V)) = 0$, and so $\mu = 0$ and finally $p = 0$, showing that indeed s is invariant.

Conversely, suppose that s is invariant. Let V be a finely open set of fine closure $\widetilde{V} \subset \mathfrak{r}(U)$ and compact (Euclidean) closure $\overline{V}$ in Ω, and suppose that (2.9.3) holds. To prove (2.9.2) we fix a finite fine potential $q > 0$ on U and write $p = \widehat{R}_{s-\widehat{R}_s^{U\setminus V}}$ in U where $s - \widehat{R}_s^{U\setminus V}$ is understood as 0 at points where $\widehat{R}_s^{U\setminus V}$ is infinite, and $p_n = \widehat{R}_{\min(s-\widehat{R}_s^{U\setminus V}, nq)}$ for each integer n. According to the inequality of Mokobodzki and Sibony, cf. Theorem 2.2.17, we have $p \preccurlyeq s$. We proceed to show that p is a fine potential and hence $p = 0$ by Definition 2.9.1, so (2.9.2) holds q.e., and indeed everywhere on U by fine continuity.

Since $p_n \leqslant nq$ we have $p_n \in \mathcal{P}^s(U)$, and hence $p_n = G_U\mu_n$ for a unique measure μ_n on U by Theorem 2.6.11. Since $s - \widehat{R}_s^{U\setminus V} = 0$ on $U \cap \mathfrak{b}(\complement V)$, hence on the finely open set $U\setminus\widetilde{V}$, we infer from Theorem 2.4.19(b) that p_n, by its definition, is finely harmonic on $U\setminus\widetilde{V}$. It follows by Theorem 2.6.4 that $\mu_n(U\setminus\widetilde{V}) = 0$. Thus μ_n is carried by $\widetilde{V}_n$ and may therefore be considered as a measure on $\overline{V}_n$ ($\supset \widetilde{V}_n$). Since

$$\int G_U(x,y)\,d\mu_n(y) = p_n(x) \leqslant p(x) \leqslant s(x) < +\infty,$$

it follows easily from (2.9.3) (in which V may be replaced by $\widetilde{V}$, by fine continuity) that the measures μ_n on $\overline{V}$ are uniformly bounded. If μ denotes a vague cluster point for the sequence (μ_n) the following holds q.e. on U according to Lemma 2.6.13:

$$G_U\mu \geqslant \liminf_n G_U\mu_n = \lim_n p_n = p,$$

the last equation stemming from Theorem 2.2.15. Hence $G_U\mu \geqslant p$ on all of U, by fine continuity, and consequently p is a fine potential, necessarily 0. □

Corollary 2.9.10 *Let $U \subset \Omega$ be a fine domain. A function $s \in \mathcal{S}_+(U)$ is a fine potential if and only if*

$$\inf_n \widehat{R}_s^{U\setminus V_n} = 0 \ \ q.e. \ in \ U \tag{2.9.4}$$

for some countable covering of U by an increasing sequence finely open sets $V_n \subset \mathfrak{r}(U)$ such that $\widetilde{V}_n \subset \mathfrak{r}(U)$. It suffices to verify (2.9.4) for a single increasing sequence (V_n) as stated and such that moreover V_n has compact closure in Ω and

$$\inf_{y\in V_n} G_U(x, y) > 0 \ \ for \ some \ x \in U \ with \ s(x) < +\infty. \tag{2.9.5}$$

Proof Suppose that s is a potential. Let (V_n) be a covering of U by finely open sets as stated in the Corollary and put $h = \widehat{\inf R_s^{U\setminus V_n}}$. Then $h \preccurlyeq s$ and $\widehat{R}_h^{U\setminus V_n} = h$ for each V_n. Indeed, for a fixed k, the $\widehat{R}_h^{U\setminus V_n}$ are finely harmonic q.e. on V_k for every $n \geq k$ according to Theorem 2.4.19 and hence, by varying k, finely harmonic q.e. in U in view of Corollary 2.1.21. Since $h \leq s$, it follows that $h \preccurlyeq s$ by Theorem 2.1.43. Furthermore, by the monotone convergence theorem we have for every $x \in U\setminus e$ for some polar set $e \subset U$ (including the set where s is infinite)

$$\begin{aligned}\widehat{R}_h^{U\setminus V_k}(x) &= \int h d\varepsilon_x^{\complement V_k} = \int \inf_{n\geq k} \widehat{R}_s^{U\setminus V_n} d\varepsilon_x^{\complement V_k} \\ &= \inf_{n\geq k} \int \widehat{R}_s^{U\setminus V_n} d\varepsilon_x^{\complement V_k} = \inf_{n\geq k} \widehat{R}_s^{U\setminus V_n} = h(x)\end{aligned}$$

by Theorem 2.3.9. It follows from Theorem 2.9.9 that h is invariant and hence $h = 0$. Conversely, suppose that $\inf_n \widehat{R}_s^{U\setminus V_n} = 0$ q.e. on U and let $s = p + h$ the Riesz decomposition of s into a fine potential p and an invariant function h. According to Theorem 2.9.9 we have $h = \inf_n \widehat{R}_h^{U\setminus V_n} \preceq \inf_n \widehat{R}_s^{U\setminus V_n} = 0$ q.e. and hence $h = 0$ so that $s = p$ is a fine potential. □

For a (positive) Borel measure μ on U we denote by μ_* and μ^* the inner and the outer μ-measure, respectively.

Theorem 2.9.11 *Let $U \subset \Omega$ be a fine domain. Let $p = G_U\mu$ be a fine potential on U and let V be a finely open subset of U. Then we have the bi-implications*

$$p|_V \ invariant \iff \mu_*(V) = 0 \iff \mu^*(V) = 0.$$

Proof Replacing U, if necessary, by $\mathfrak{r}(U)$ we may without loss of generality suppose that U is regular. Indeed, we have $G_{\mathfrak{r}(U)}\mu = G_U\mu$ (see text in the beginning of Sect. 2.6). Suppose first that $\mu_*(V) = 0$ and let V_0 be a fine component of V. Then we have $\mu_*(V_0) = 0$. For any regular fine domain W with $\widetilde{W} \subset V_0$ the part of μ on the K_σ set W equals 0, and hence $G_W\mu \equiv 0$. It follows by Lemma 2.6.3 that

$$p = \int p\, d\varepsilon_x^{\complement W}$$

on U, in particular on V_0. Fix a point $x_0 \in V_0$. The finely open sets $W_j := \{x \in V_0 : G_{V_0}(x, x_0) > \frac{1}{j}\}$ $(j \in \mathbb{N})$ form a countable cover of V_0 such that $\widetilde{W}_j \subset V_0$. It therefore follows by Theorem 2.9.9 (with U replaced by V_0 and s by $p|_{V_0}$) that $p|_{V_0}$ indeed is invariant. Hence $p|_V$ is invariant, see paragraph just before Lemma 2.9.2.

Conversely, suppose that $p|_V$ is invariant, and let us prove that $\mu^*(V) = 0$. Consider again a fine component V_0 of V. Under the extra hypothesis that $\widetilde{V}_0 \subset U$ it now follows by Lemma 2.6.3 that

$$p = G_{V_0}\mu + \widehat{R}_p^{U\setminus V_0}$$

on U. By Lemma 2.9.7 the latter term on the right is invariant on V_0, and so is therefore the difference $q = G_{V_0}\mu$, which however is a fine potential on V_0, and so $G_{V_0}\mu = 0$ on V_0, that is, $\mu(\mathfrak{r}(V_0)) = 0$, whence $\mu^*(V_0) = 0$. Without the above extra hypothesis that $\widetilde{V}_0 \subset U$ we cover V_0 by a sequence of finely open sets

$$W_j := \{x \in V_0 : G_{\mathfrak{r}(V_0)}(x, x_0) > \frac{1}{j}\} \subset \{x \in \mathfrak{r}(V_0) : G_{\mathfrak{r}(V_0)}(x, x_0) \geqslant \frac{1}{j}\}.$$

The last set is finely closed subset of U, and so $\widetilde{W}_j \subset U$. As shown above, it follows that $\mu^*(W_j) = 0$, and hence indeed $\mu^*(V_0) = 0$. Since the set of fine component is countable, it then follows that $\mu^*(V) = 0$. □

The invariant functions on finely open subsets of U do, however, have a kind of countable sheaf property, as shown by (a) and (b) in the following theorem:

Theorem 2.9.12 (a) *Let $u \in \mathcal{S}_+(U)$ be invariant and let V be a finely open subset of U. Then $u|_V$ is invariant.*

(b) *Let $u \in \mathcal{S}_+(U)$, and let (U_j) be a countable cover of U by regular finely open subsets of U. If each $u|_{U_j}$ is invariant then u is invariant.*

(c) *Let (u_α) be a decreasing net of invariant functions in $\mathcal{S}_+(U)$. Then $\widehat{\inf}_\alpha u_\alpha$ is invariant. Moreover, the set $V = \{\inf_\alpha u_\alpha < +\infty\}$ is finely open, and $\widehat{\inf}_\alpha u_\alpha = \inf_\alpha u_\alpha$ on V.*

Proof (a) Let $p \in \mathcal{P}(V)$ satisfy $p \preccurlyeq u$ on V. We may assume without loss of generality that V is a fine domain. Write $p = G_V\mu$ (considered on V) in terms of the associated Borel measure μ on Ω such that the inner measure $\mu_*(\complement V) = 0$. We shall prove that $p = 0$. An extension of μ from V to a larger subspace of Ω by 0 off V

will also be denoted by μ. Thus $p = G_V\mu$ extends to the fine potential $G_{U\cap\mathfrak{r}(V)}\mu$ on $U \cap \mathfrak{r}(V)$ and further extends to a finely continuous function $f : U \to [0, +\infty]$ which equals 0 off $U \cap \mathfrak{r}(V)$.

Suppose to begin with that μ is finite and (after extension to Ω) carried by a Euclidean compact subset K of Ω contained in $\mathfrak{r}(V)$. Let $q := \widehat{R}_f$ (sweeping relative to U). Then $f \leqslant G_U\mu$ and hence $q \leqslant G_U\mu < +\infty$ q.e. on U, and so q is a fine potential on U along with $G_U\mu$. On the other hand, $q \leqslant u$ because $f \leqslant u$ on U. In the first place, $f = p \leqslant u$ on V and hence $f \leqslant u$ on $U \cap \mathfrak{r}(V) \subset U \cap \widetilde{V}$ by fine continuity of f and u. Secondly, $f = 0$ on $U \backslash \mathfrak{r}(V)$. Since $p \preccurlyeq u$ on V we have $u = p + s$ on V for a certain $s \in \mathcal{S}_+(V)$. Since $U \cap \mathfrak{r}(V) \backslash V$ is polar, s extends by fine continuity to a similarly denoted $s \in \mathcal{S}_+(U \cap \mathfrak{r}(V))$, and we have $u = G_{U\cap\mathfrak{r}(V)}\mu + s$ on $U \cap \mathfrak{r}(V)$. Since f-lim $G_{U\cap\mathfrak{r}(V)}\mu = f = 0$ at $U \cap \partial_f\mathfrak{r}(V) \subset U\backslash\mathfrak{r}(V)$ we have f-lim $s = u$ at $U \cap \partial_f\mathfrak{r}(V)$. It follows by Lemma 2.1.51 that the extension of s to U by u on $U\backslash\mathfrak{r}(V)$ is of class $\mathcal{S}_+(U)$. Denoting also this extension by s we have $u = s + f$ on U. By Mokobodzki-Sibony's inequality in our setting, see Theorem 2.2.17, we infer that $q = \widehat{R}_f \preccurlyeq u$. Since u is invariant on U and $q \in \mathcal{P}(U)$ it follows that $q = 0$ and hence $p \leqslant q = 0$ on V, showing that indeed $u|_V$ is invariant.

Dropping the above temporary hypothesis that μ be finite and carried by a Euclidean compact subset of $\mathfrak{r}(V)$ we decompose μ in accordance with Lemma 2.6.1 into the sum of a sequence of finite measures μ_j with Euclidean compact supports $K_j \subset \mathfrak{r}(V)$. Since $G_V\mu_j \preccurlyeq G_V\mu = p \preccurlyeq u|_V$, the result of the above paragraph applies with μ replaced by μ_j. It follows that $G_V\mu_j = 0$ and hence $p = G_V\mu = \sum_j G_V\mu_j = 0$. Thus $u|_V$ is indeed invariant on V.

(b) For each index j there is by Theorem 2.9.9 a countable finely open cover $(V_{j,k})_k$ of U_j such that $\widetilde{V}_{j,k} \subset \mathfrak{r}(U_j) = U_j$ and (with sweeping relative to U_j)

$$^{U_j}\widehat{R}_{u|_{U_j}}^{U_j\backslash V_{j,k}} = u|_{U_j}$$

for each k. It follows that (with sweeping relative to U, resp. U_j)

$$u \geq \widehat{R}_u^{U\backslash V_{j,k}} \geq {}^{U_j}\widehat{R}_{u|_{U_j}}^{U_j\backslash V_{j,k}} = u$$

on each U_j, so equality prevails here. In particular, $u = \widehat{R}_u^{U\backslash V_{j,k}}$ on $V_{j,k}$ for each j, k. Consequently, u is invariant according to the quoted theorem applied to the countable cover $(V_{j,k})_{j,k}$ of U.

(c) For indices α, β with $\alpha < \beta$ we have $u_\beta \leqslant u_\alpha$ and hence $u_\beta \preccurlyeq u_\alpha$ by Lemma 2.9.4. The claim therefore follows from Corollary 2.2.18(b). □

Notes: The concepts of finely [hyper/super]harmonic functions on a finely open set are due to Fuglede. They appeared for the first time in the book by Fuglede [Fu1] in 1972, in the general setting of harmonic spaces in the Brelot axiomatic theory. The theory of finely harmonic functions had led to better comprehension and new applications of classical potential theory. The results in the Sects. 2.1–2.4 of the present chapter are adaptations of those of [Fu1] to the classical potential theory in

$\mathbb{R}^N$. To these are added in Sects. 2.5, 2.6 and 2.7, other new results by Fuglede in other themes such as the localization in fine potential theory and the fine Green kernel, the integral representation of fine potentials, and particular results in the case $N = 2$. The results of Sect. 2.8 on uniform approximation of continuous functions by harmonic functions are due to Debiard and Gaveau [DG1] for compact sets, and to Gauthier and Ladouceur [GaL] for closed sets. The results of approximation by continuous superharmonic functions are due to Blietdner and Hansen [BH2] for compact sets, and to Bensouda and Gauthier [BG] for closed sets. Results on invariant functions (introduced by Fuglede) in Sect. 2.9 are due to El Kadiri and Fuglede.

Chapter 3
The Martin Boundary of a Fine Domain

This chapter is devoted to the Martin boundary of a fine domain U in the Greenian domain $\Omega \subset \mathbb{R}^N$ and the integral representation of invariant functions on U. In Sect. 3.1, we define a topology on the cone $\mathcal{S}_+(U)$ of positive finely superharmonic functions on U for which it has a compact base **B**. We establish the integral representation in $\mathcal{S}_+(U)$ by means of extreme points of **B** and prove that any point of the set $\text{Ext}(\mathbf{B})$ of extreme points of **B** is of the form $\alpha G_U(\cdot, y)$ for some $\alpha > 0$ and some $y \in U$, or an invariant function (called a minimal invariant function). We prove that the sets $\text{Ext}_p(\mathbf{B})$ of extreme fine potentials in **B** and $\text{Ext}_i(\mathbf{B})$ of minimal invariant functions in **B** are Borel subsets of **B** and that every fine potential, resp. invariant function, is uniquely represented by a measure μ on **B** supported by $\text{Ext}_p(\mathbf{B})$, resp. $\text{Ext}_i(\mathbf{B})$.

In Sect. 3.2 we define the Martin compactification of U as the closure $\overline{U}$ in **B** of U, U being identified with a subset of **B**. The set $\Delta(U) = \overline{U} \setminus U$ is called the Martin boundary of U. The main difference from the classical case is that the Martin boundary in our setting is not necessarily compact, but only a G_δ-set. We define the (fine) Riesz-Martin kernel K for U and study its regularity. Then $\Delta(U)$ and the set $\Delta_1(U)$ of minimal invariant functions in $\Delta(U)$ are Borel subsets of $\overline{U}$ and every invariant function is uniquely represented by means of K and a unique measure μ on $\overline{U}$ supported by the set $\Delta_1(U)$.

In Sect. 3.3 we prove a Fatou-Naïm-Doob theorem for finely superharmonic functions and in Sect. 3.4 we define and study the sweeping on subsets of the Martin space $\overline{U}$ of U. In Sect. 3.5 we introduce, as in the classical case, the minimal-fine topology on $\overline{U}$ associated with the minimal thinness of a subset of U at a point of $\Delta_1(U)$ and we prove the integral representation of the sweeping of a function $u \in \mathcal{S}_+(U)$ on a set $A \subset \overline{U}$. This result seems to be new even in the case where U is a Euclidean open set. Section 3.6 is devoted to the Dirichlet problem at the Martin boundary of U.

M. El Kadiri and B. Fuglede, *Classical Fine Potential Theory*, Springer Monographs in Mathematics, https://doi.org/10.1007/978-981-96-0430-2_3

3.1 Integral Representation in the Cone of Positive Finely Superharmonic Functions

Throughout this chapter U is a regular fine domain in our Greenian domain $\Omega \subset \mathbb{R}^N$ $(N \geq 2)$. In particular, U is a Euclidean K_σ subset of Ω.

In this section we define a topology on the cone $\mathcal{S}_+(U)$ of positive finely superharmonic functions on U, for which $\mathcal{S}_+(U)$ has a compact base, and we get the integral representation of positive finely superharmonic by the Choquet's method.

Recall that the convex cone of fine potentials on U is denoted by $\mathcal{P}(U)$. The cone of *invariant* functions on U is denoted by $\mathcal{H}_i(U)$; it is the orthogonal band to $\mathcal{P}(U)$ relative to $\mathcal{S}_+(U)$, see Definition 2.9.1. By G_U we denote the (fine) Green kernel for U, see Sect. 2.5.

We continue to take for granted that reduction and sweeping of functions are taken relative to U (in the absence of other indications). This applies until Definition 3.4.3 of the more general sweeping relative to the Martin compactification of U.

Recall from Eq. (2.9.1) the unique integral representation $u = G_U\mu + h$ of $u \in \mathcal{S}_+(U)$, where $G_U\mu$ is the potential of a (positive) Borel measure on U [or Ω] with respect to the (fine) Green kernel G_U on U, and h is an invariant function on U. In order to utilize Choquet's theorem on integral representation [Ch5] (or see Phelps' booklet [Ph]), Brelot [Br8] and Hervé [He, no. 19–21] introduced a topology on the cone of all positive superharmonic functions on the domain Ω (or in a suitable harmonic space). In this section, we shall generalize this topology to the present cone $\mathcal{S}_+(U)$. By identifying that topology with the natural topology on the H-cone $\mathcal{S}_+(U)$, cf. [BBC2], we shall show that $\mathcal{S}_+(U)$ has a compact base $\mathbf{B}$ and by Choquet's theorem that every function $u \in \mathcal{S}_+(U)$ admits a unique integral representation of the form

$$u(x) = \int_{\mathbf{B}} s(x)d\mu(s), \quad x \in U,$$

where μ is a finite measure on $\mathbf{B}$ carried by the (Borel) set of all extreme finely superharmonic functions belonging to $\mathbf{B}$.

Following [BBC2, Section 4.3] we give the standard H-cone $\mathcal{S}_+(U)$ of functions (see last paragraph of Sect. 2.5) the natural topology. This topology on $\mathcal{S}_+(U)$ is metrizable and induced by the weak topology on a locally convex topological vector space in which $\mathcal{S}_+(U)$ is embedded as a proper convex cone. This cone is well capped with compact caps, but we show that the cone $\mathcal{S}_+(U)$ even has a compact base, and that is crucial for our investigation.

Lemma 3.1.1 *There exists a sequence (s_n) of functions in $\mathcal{S}_+(U)$ such that every function $s \in \mathcal{S}_+(U)$ is the upper envelope of a subsequence (s_{n_k}) of (s_n).*

The lemma is a direct consequence of Theorem 2.5.15, [BBC2, Condition S2) of Definition, p. 104], [BBC2, Definition, p. 113] and [BBC2, Theorem 4.4.6].

Lemma 3.1.2 *There exists a sequence (H_j) of Euclidean compact subsets of U, each unthin at any of its points, and a polar set P such that*

(1) $U = \bigcup_j H_j \cup P$.

(2) For any j the restriction of any function from $\mathcal{S}_+(U)$ to H_j is l.s.c. in the Euclidean topology.

Proof Let (s_n) be the sequence from Lemma 3.1.1. It follows from the Brelot property, see Theorem 1.5.29, that every $x \in U$ has a regular finely open fine neighborhood U_x such that $\overline{U}_x \subset U$ and each function $s_n|_{U_x}$ is continuous on U_x (with the Euclidean topology). By the quasi-Lindelöf property, see Sect. 1.2, there is a sequence (x_j) and a polar set $P \subset U$ such that $U = \bigcup U_j \cup P$, where $U_j = U_{x_j}$. For any j, every function $s \in \mathcal{S}_+(U)$ is l.s.c. on U_j, s being the upper envelope of a subsequence (s_{n_k}) of (s_n) according to Lemma 3.1.1. Let $(U_{j,k})$ be the sequence of the fine components of U_j and, for every integer $l > 0$, put $H_{j,k,l} = \{x \in U_{j,k} : G_{U_{j,k}}(x, y_{j,k}) \geq \frac{1}{l}\}$ where $y_{j,k}$ is a fixed point in $U_{j,k}$. Since the function $G_{U_{j,k}}(\cdot, y_{j,k})$ (defined on all of Ω, see Definition 2.5.10) is subharmonic on $\Omega \setminus \{y_{j,k}\}$ and null outside of $U_{j,k}$ (which is necessarily regular), and since $y_{j,k} \in H_{j,k,l}$, it follows that $H_{j,k,l}$ is (Euclidean) compact. It is clear that $U = \bigcup_{j,k,l} H_{j,k,l} \cup P$. Moreover, $H_{j,k,l}$ is unthin at any of its points. Indeed, if $H_{j,k,l}$ were thin at $z \in H_{j,k,l}$, then $G_{U_{j,k}}(z, y_{j,k}) = \frac{1}{l}$ (because a finely open set is unthin at any of its points) and there would exist a finely connected finely open set $W \subset U_{j,k} \setminus \{y_{j,k}\}$ containing z such that $(W \setminus \{z\}) \cap H_{j,k,l} = \varnothing$ and hence $W \cap H_{j,k,l} = \{z\}$. The function $\frac{1}{l} - G_{U_{j,k}}(\cdot, y_{j,k})$ being finely harmonic and ≥ 0 on W and null at z, it then follows that $G_{U_{j,k}}(\cdot, y_{j,k}) - \frac{1}{l} = 0$ on W according to Corollary 2.1.27 and hence $W \subset H_{j,k,l}$, which is absurd. □

We shall now use the sequence (H_j) from this lemma to define in analogy with [Mo] a locally compact topology on the cone $\mathcal{S}_+(U)$. For each j let $\mathcal{C}_l(H_j)$ denote the space of l.s.c. functions on H_j with values in $\overline{\mathbb{R}}_+$, and provide this space with the topology of convergence in graph (cf. [Mo]). It is known that $\mathcal{C}_l(H_j)$ is a compact metrizable space in this topology. Let d_j denote a distance compatible with this topology. We define a pseudo-distance d on $\mathcal{S}_+(U) \cup \{+\infty\}$ by

$$d(u, v) = \sum_j \frac{1}{2^j \delta(\mathcal{C}_l(H_j))} d_j(u|_{H_j}, v|_{H_j})$$

for each couple (u, v) of functions from $\mathcal{S}_+(U) \cup \{+\infty\}$, where $\delta(\mathcal{C}_l(H_j))$ denotes the diameter of $\mathcal{C}_l(H_j)$. Since two finely hyperharmonic functions are identical if they coincide quasi-everywhere it follows that d is a true distance on $\mathcal{S}_+(U) \cup \{+\infty\}$. We denote by $\mathcal{T}$ the topology on $\mathcal{S}_+(U) \cup \{+\infty\}$ defined by the distance d.

For any filter $\mathcal{F}$ on $\mathcal{S}_+(U) \cup \{+\infty\}$ we write

$$\lim \widehat{\inf}_{\mathcal{F}} = \sup_{M \in \mathcal{F}} \widehat{\inf_{u \in M}} u,$$

where the l.s.c. regularized $\widehat{\inf}_{u \in M} u$ is taken with respect to the fine topology.

Theorem 3.1.3 *The cone $\mathcal{S}_+(U) \cup \{+\infty\}$ is compact in the topology $\mathcal{T}$. For any convergent filter $\mathcal{F}$ on $\mathcal{S}_+(U) \cup \{+\infty\}$ we have*

$$\lim_{\mathcal{F}} = \lim \widehat{\inf}_{\mathcal{F}}.$$

Proof Let $\mathcal{U}$ be an ultrafilter on $\mathcal{S}_+(U) \cup \{+\infty\}$. For any $M \in \mathcal{U}$ put $u_M = \inf_{u \in M} u$. For each j the ultrafilter base $\mathcal{U}_j$ obtained from $\mathcal{U}$ by taking restrictions to the Euclidean compact H_j from Lemma 3.1.2, converges in the compact space $\mathcal{C}_l(H_j)$ to the function $u_j := \sup_{M \in \mathcal{U}} \widehat{u_M}^j$ (where $\widehat{v}^j$ for $v \in \mathcal{S}_+(U) \cup \{+\infty\}$ denotes the (Euclidean) l.s.c. regularized of the restriction of v to H_j), cf. [Mo, Proposition 1]. The finely l.s.c. regularized $\widehat{u_M}$ of u_M in U is l.s.c. in H_j by Lemma 3.1.2 and minorizes u_M, whence $\widehat{u_M} \leqslant \widehat{u_M}^j$. On the other hand there exists by Theorem 2.1.34 a polar set $A \subset \Omega$ such that $u_M = \widehat{u_M}$ in $U \setminus A$, and so $\widehat{u_M}^j \leqslant \widehat{u_M}$ in $H_j \setminus A$. But for $x \in A$ we have $\widehat{u_M}^j(x) \leqslant \widehat{u_M}(x)$ because $\widehat{u_M}$ is finely continuous on U and x is in the fine closure of $H_j \setminus A$ since H_j is unthin at x. We conclude that $u_j = \lim \widehat{\inf}_{\mathcal{U}}$ in H_j for each j. Since the function $u := \lim \widehat{\inf}_{\mathcal{U}}$ belongs to $\mathcal{S}_+(U) \cup \{+\infty\}$ according to Theorem 2.1.34, Corollary 2.1.21 and Definition 2.4.1 it follows that the filter $\mathcal{U}$ converges to u in the topology $\mathcal{T}$. This proves that $\mathcal{S}_+(U) \cup \{+\infty\}$ is compact in the topology $\mathcal{T}$. The remaining assertion follows easily from [Mo, Proposition 1] and the definition of the topology $\mathcal{T}$. □

Corollary 3.1.4 *The topology of convergence in graph coincides with the natural topology on $\mathcal{S}_+(U)$.*

This follows immediately from Theorem 3.1.3 and [BBC2, Theorem 4.5.8] and the property of finely l.s.c. regularization, cf. second paragraph after the proof of Theorem 2.2.17.

Corollary 3.1.5 *The cone $\mathcal{S}_+(U)$ endowed with the natural topology has a compact base.*

From Theorem 3.1.3 and Corollary 3.1.4 it follows that the natural topology on $\mathcal{S}_+(U)$ is locally compact, and we infer by a theorem of Klee, see [Al, Theorem II.2.6], that indeed $\mathcal{S}_+(U)$ has a compact base.

Corollary 3.1.6 *For given $x \in U$ the affine forms $u \mapsto u(x)$ and $u \mapsto \widehat{R}_u^A(x)$ ($A \subset U$) are l.s.c. in the natural topology on $\mathcal{S}_+(U)$.*

Proof Clearly, the map $u \mapsto \widehat{R}_u^A(x)$ is affine for fixed $x \in U$ (see Corollary 2.2.19(b)). Let (u_j) be a sequence in $\mathcal{S}_+(U)$ converging naturally to $u \in \mathcal{S}_+(U)$. For any index k we have

$$\widehat{\inf_{j \geqslant k}} \widehat{R}_{u_j}^A(x) \geqslant \widehat{R}^A_{\widehat{\inf}_{j \geqslant k} u_j}(x).$$

Either member of this inequality increases with k, and, by Theorem 2.2.15(a), we get for $k \to \infty$

$$\liminf_k \widehat{R}_{u_k}^A(x) \geqslant \lim_k \widehat{\inf} \widehat{R}_{u_k}^A(x) \geqslant \widehat{R}^A_{\lim \widehat{\inf}_k u_k}(x) = \widehat{R}_u^A(x),$$

and so the map $u \mapsto \widehat{R}_u^A(x)$ is indeed l.s.c. on $\mathcal{S}_+(U)$. For the map $u \mapsto u(x)$ take $A = U$. □

Lemma 3.1.7 *Let* **B** *a compact base of* $\mathcal{S}_+(U)$. *There exists a unique continuous function* $\Phi : \mathcal{S}_+(U) \to [0, +\infty[$ *with the following properties*

(1) $\Phi(0) = 0$,

(2) $\Phi(\alpha u + \beta v) = \alpha\Phi(u) + \beta\Phi(v)$ *for all* $\alpha, \beta \geq 0$ *and* $u, v \in \mathcal{S}_+(U)$,

(3) $\Phi(s) > 0$ *for every* $s \in \mathcal{S}_+(U) \setminus \{0\}$,

(4) $\mathbf{B} = \{s \in \mathcal{S}_+(U) : \Phi(s) = 1\}$.

Proof By definition of a base of the cone $\mathcal{S}_+(U)$, see [Be, p. xii (Notations and definitions)] and [Ph, p. 59], there is a function $\Phi : \mathcal{S}_+(U) \to [0, +\infty[$ satisfying (1), (2) and (4). The continuity of Φ and the property (3) follow easily from the compacity of **B**. The uniqueness of Φ follows from (2) and (4). □

The function Φ will be called the *form defining* the base **B**.

Let **B** a compact base of $\mathcal{S}_+(U)$. As shown by Choquet (cf. [Al, Corollary I.4.4]) the set $\text{Ext}(\mathbf{B})$ of extreme elements of **B** is a G_δ subset of **B**. On the other hand it follows by the fine Riesz decomposition theorem, see (2.9.1), that every element of $\text{Ext}(\mathbf{B})$ is either a fine potential or an invariant function. We denote by $\text{Ext}_p(\mathbf{B})$ (resp. $\text{Ext}_i(\mathbf{B})$) the cone of all extreme fine potentials (resp. allextreme invariant functions) in **B**. According to Corollary 2.6.18, any element of $\text{Ext}_p(\mathbf{B})$ has the form $\alpha G_U(\cdot, y)$, where α is a constant > 0 and $y \in U$.

Theorem 3.1.8 $\text{Ext}_p(\mathbf{B})$ *and* $\text{Ext}_i(\mathbf{B})$ *are Borel subsets of* **B**.

Proof Let Φ the form defining **B** (see Lemma 3.1.7 and the line following its proof). By Corollary 2.6.18 we have

$$\text{Ext}_p(\mathbf{B}) = \left\{\frac{G_U(\cdot, y)}{\Phi(G(\cdot, y))} : y \in U\right\}.$$

Let (B_l) be a sequence of open balls in $\mathbb{R}^N$ with Euclidean closures $\overline{B}_l \subset \Omega$ with the union Ω. For integers $k, l > 0$ put

$$A_{k,l} = \{y \in U : \Phi(G_U(\cdot, y)) \geqslant 1/k\} \cap \overline{B}_l$$

and

$$\mathbf{C}_{k,l} = \left\{\frac{G_U(\cdot, y)}{\Phi(G(\cdot, y))} : y \in A_{k,l}\right\}.$$

The sets $A_{k,l}$ cover U because $G_U(\cdot, y) > 0$ and hence $\Phi(G_U(\cdot, y)) > 0$, whence $\text{Ext}_p(\mathbf{B}) = \bigcup_{k,l} \mathbf{C}_{k,l}$. We first show that each $\mathbf{C}_{k,l}$ is compact in **B** with the natural topology. Let (y_j) be a sequence of points of $A_{k,l}$. After passing to a subsequence we may suppose that (y_j) converges Euclidean to a point $y \in \overline{B}_l$, and that the sequences $(G_U(\cdot, y_j))$ and $(\widehat{R}^{\complement U}_{G_\Omega(\cdot, y_j)}|_U)$ converge in $\mathcal{S}_+(U) \cup \{+\infty\}$ and hence necessarily in $\mathcal{S}_+(U)$. It follows that

$$\Phi(G_\Omega(\cdot, y_j)|_U) = \Phi(G_U(\cdot, y_j)) + \Phi(\widehat{R}^{\complement U}_{G_\Omega(\cdot,y_j)}|_U),$$

and hence by passing to the limit in $\mathcal{S}_+(U)$ as $j \to +\infty$

$$\Phi(G_\Omega(\cdot, y)|_U) = \lim_j \Phi(G_U(\cdot, y_j)) + \lim_j \Phi(\widehat{R}^{\complement U}_{G_\Omega(\cdot,y_j)}|_U). \tag{3.1.1}$$

On the other hand,

$$\widehat{R}^{\complement U}_{G_\Omega(\cdot,y)} = \widehat{R}^{\complement U}_{\lim \widehat{\inf}_j G_\Omega(\cdot,y_j)} \leq \lim_j \widehat{\inf}\, \widehat{R}^{\complement U}_{G_\Omega(\cdot,y_j)}.$$

The function $\widehat{R}^{\complement U}_{G_\Omega(\cdot,y)}|_U$ being invariant according to Lemma 2.9.7, it follows by Lemma 2.9.4 that $\widehat{R}^{\complement U}_{G_\Omega(\cdot,y)}|_U \preccurlyeq \lim \widehat{\inf}_j \widehat{R}^{\complement U}_{G_\Omega(\cdot,y_j)}|_U$ and hence

$$\Phi(\widehat{R}^{\complement U}_{G_\Omega(\cdot,y)}|_U) \leq \Phi(\lim \widehat{\inf}_j \widehat{R}^{\complement U}_{G_\Omega(\cdot,y_j)}|_U) = \lim_j \Phi(\widehat{R}^{\complement U}_{G_\Omega(\cdot,y_j)}|_U).$$

We infer from (3.1.1) by the definition of $A_{k,l}$ that

$$\Phi(G_\Omega(\cdot, y)|_U) \geqslant \frac{1}{k} + \Phi(\widehat{R}^{\complement U}_{G_\Omega(\cdot,y)}|_U),$$

and consequently $y \in U$ and $\Phi(G_U(\cdot, y)) \geqslant 1/k$, whence $y \in A_{k,l}$. Now put $s = \lim_j G_U(\cdot, y_j)$. We have $s > 0$ because $y_j \in A_{k,l} \subset U$ and hence

$$\Phi(s) = \lim_j \Phi(G_U(\cdot, y_j)) \geq 1/k.$$

It then follows that

$$\frac{G_U(\cdot, y_j)}{\Phi(G_U(\cdot, y_j))} \to \frac{s}{\Phi(s)} \in \mathcal{S}_+(U).$$

By passing to the limit as $j \to +\infty$ in the equality

$$G_\Omega(\cdot, y_j)|_U = G_U(\cdot, y_j) + \widehat{R}^{\complement U}_{G_{(\cdot,y_j)}}|_U$$

we obtain

$$G_\Omega(\cdot, y)|_U = s + \lim \widehat{\inf}_j \widehat{R}^{\complement U}_{G_\Omega(\cdot,y_j)}|_U \geqslant s + \widehat{R}^{\complement U}_{G_\Omega(\cdot,y)}|_U.$$

It follows from the equality that s is finely harmonic on $U \setminus \{y\}$, and from the inequality that $s \leqslant G_U(\cdot, y)$ and hence s is a fine potential > 0 on U, so that s is of the form $\alpha G_U(\cdot, y)$ for some $\alpha > 0$ according to Theorem 2.5.11(b). Hence

$$\frac{G_U(\cdot, y_j)}{\Phi(G_U(\cdot, y_j))} \to \frac{\alpha G_U(\cdot, y)}{\Phi(\alpha G_U(\cdot, y))} = \frac{G_U(\cdot, y)}{\Phi(G_U(\cdot, y))} \in \mathbf{C}_{k,l}$$

(convergence in $\mathcal{S}_+(U)$). This shows that $\mathbf{C}_{k,l}$ is compact in the natural topology on $\mathbf{B}$, and consequently $\mathrm{Ext}_p(\mathbf{B}) = \bigcup_{k,l} \mathbf{C}_{k,l}$ is indeed a K_σ of $\mathbf{B}$. From the equality $\mathrm{Ext}_i(\mathbf{B}) = \mathrm{Ext}(\mathbf{B}) \setminus \mathrm{Ext}_p(\mathbf{B})$ and the well-known fact that $\mathrm{Ext}(\mathbf{B})$ is a G_δ of $\mathbf{B}$ it then follows that $\mathrm{Ext}_i(\mathbf{B})$ is a Borel subset of $\mathbf{B}$, □

When μ is a Radon measure on $\mathbf{B}$ there exists a unique element s of $\mathcal{S}_+(U)$ such that

$$l(s) = \int_{\mathbf{B}} l(u) d\mu(u)$$

for every continuous affine form $l : \mathbf{B} \to [0, +\infty[$ on $\mathbf{B}$. If $\mu = 0$ this is obvious and $s = 0$. If $\mu \neq 0$, then $s = \mu(\mathbf{B}) b(\mu)$, where $b(\mu) \in \mathbf{B}$ denotes the barycenter (or resultant) of the probability measure $\frac{1}{\mu(\mathbf{B})}\mu$, cf., e.g. [Ph, p. 1]. In particular if μ is a probability measure, then $s = b(\mu)$. For any l.s.c. affine form $\varphi : \mathbf{B} \to [0, +\infty]$ there exists by [Al, Corollary I.1.4] an increasing sequence of continuous affine forms on $\mathbf{B}$ with values in $[0, +\infty[$, which converges to φ, and hence

$$\varphi(s) = \int_{\mathbf{B}} \varphi(u) d\mu(u).$$

In particular, for fixed $x \in U$ and $A \subset U$, the affine forms $u \mapsto u(x)$ and $u \mapsto \widehat{R}_u^A(x)$ are l.s.c. according to Corollary 3.1.6, and hence

$$s(x) = \int_{\mathbf{B}} u(x) d\mu(u) \quad \text{and} \quad \widehat{R}_s^A(x) = \int_{\mathbf{B}} \widehat{R}_u^A(x) d\mu(u).$$

The function s is denoted by

$$\int u d\mu(u).$$

The following theorem is a particular case of Choquet's theorem.

Theorem 3.1.9 *For any $s \in \mathcal{S}_+(U)$ there exists a unique Radon measure on* $\mathbf{B}$ *carried by* $\mathrm{Ext}(\mathbf{B})$ *such that*

$$s(x) = \int_{\mathbf{B}} u(x) d\mu(u), \quad x \in U.$$

A measure μ on $\mathbf{B}$ such that

$$s(x) = \int_{\mathbf{B}} u(x) d\mu(x), \ x \in U,$$

($s \in \mathcal{S}_+(U)$) is called a *measure representing s* or a *representing measure for s*. The measure μ given by Theorem 3.1.9 is called the *measure representing s carried by* $\mathrm{Ext}(\mathbf{B})$ and is denoted by μ_s.

Theorem 3.1.10 *Let $s \in \mathcal{S}_+(U)$. Then*
(a) *s is invariant if and only if μ_s is carried by* $\mathrm{Ext}_i(\mathbf{B})$,
(b) *s is a fine potential if and only if μ_s is carried by* $\mathrm{Ext}_p(\mathbf{B})$.

Proof (a) Suppose that s is invariant and let $x \in U$ be such that $s(x) < +\infty$. Since the function $u \mapsto u(x)$ is s.c.i. on $\mathbf{B}$, see Corollary 3.1.6, it follows that the set $F(x) = \{q \in \mathrm{Ext}(\mathbf{B}) : q(x) < +\infty\} = \mathrm{Ext}(\mathbf{B}) \cap \{q \in \mathbf{B} : q(x) < +\infty\}$ is a Borel subset of $\mathbf{B}$ and μ_s is carried by $F(x)$ because $\int q(x)d\mu_s(q) = s(x) < +\infty$. Let $x_0 \in U$ such that $s(x_0) < +\infty$ be fixed. For each integer $n > 0$, let V_n be the finely open set

$$V_n = \{y \in U : G_U(x_0, y) > \frac{1}{n}\}.$$

By Corollary 2.1.33 V_n has at most countably many fine components. Choose in each fine component V_n^j of V_n a point y_j^n such that $s(y_j^n) < +\infty$. Then, according to Lemmas 2.9.7, 2.9.4 and Corollary 2.1.27, for every $q \in \mathcal{S}_+(U)$ finite at each y_j^n, we have ${}^U\widehat{R}_q^{U\setminus V_n} = q$ if only if ${}^U\widehat{R}_q^{U\setminus V_n}(y_j^n) = q(y_j^n)$ for every j. By Theorem 2.9.9 we have $s = \widehat{R}_s^{U\setminus V_n}$. It therefore follows that

$$\int q(y_j^n)d\mu_s(q) = s(y_j^n) = \widehat{R}_s^{U\setminus V_n}(y_j^n) = \int \widehat{R}_q^{U\setminus V_n}(y_j^n)d\mu_s(q)$$

as seen above (see text after the proof of Theorem 3.1.8) and hence $\int_{F(y_j^n)}(q(y_j^n) - \widehat{R}_q^{U\setminus V_n}(y_j^n))d\mu_s(q) = 0$, so that μ_s is carried by the set

$$\bigcap_n(\bigcap_j(F(y_j^n) \cap \{\widehat{R}_q^{U\setminus V_n}(y_j^n) = q(y_j^n)\}) \subset \mathrm{Ext}(\mathbf{B}) \cap (\bigcap_n\{\widehat{R}_q^{U\setminus V_n} = q\}) \subset \mathrm{Ext}_i(\mathbf{B})$$

while invoking Theorem 2.9.9. Conversely, suppose that μ_s is carried by $\mathrm{Ext}_i(\mathbf{B})$ and let $n \in \mathbb{N}$, then, as seen above,

$$\widehat{R}_s^{U\setminus V_n} = \int \widehat{R}_q^{U\setminus V_n}d\mu_s(q) = \int q d\mu_s(q) = s,$$

and hence s is invariant in view of Theorem 2.9.9.

(b) Suppose that μ_s is carried by $\mathrm{Ext}_p(\mathbf{B})$ and let $s = p + h$ be the Riesz decomposition of s, where p is a fine potential and h invariant. It then follows that $\mu_s = \mu_p + \mu_h$ and hence $\mu_h = 0$ by (a), so that $s = p$ is a fine potential. Conversely, suppose that s is a fine potential. Let ν be the restriction of μ_s to $\mathrm{Ext}_i(\mathbf{B})$ (that is, $\nu = 1_{\mathrm{Ext}_i(\mathbf{B})}\mu_s$), then $h = \int q d\nu(q)$ is invariant by (a) and we have $h \preccurlyeq s$. Hence $\nu = 0$ so that μ_s is carried by $\mathrm{Ext}_p(\mathbf{B})$. □

The following three theorems are easy consequences of Theorem 3.1.9 and Theorem 3.1.10:

Theorem 3.1.11 *For any fine potential p on U there exists a unique Radon measure μ on $\mathbf{B}$ carried by* $\mathrm{Ext}_p(\mathbf{B})$ *such that*

$$p(x) = \int_{\mathbf{B}} q(x)d\mu(q), \quad x \in U.$$

Theorem 3.1.12 *For any invariant function $h \in \mathcal{S}_+(U)$ there exists a unique Radon measure μ on $\mathbf{B}$ carried by* $\mathrm{Ext}_i(\mathbf{B})$ *such that*

$$h(x) = \int_{\mathbf{B}} k(x)d\mu(k), \quad x \in U.$$

Theorem 3.1.13 *Let $A \subset \mathrm{Ext}_p(\mathbf{B})$, resp. $A \subset \mathrm{Ext}_i(\mathbf{B})$, be a Borel set, and let μ be a Radon measure on $\mathbf{B}$. Then $\int_A u d\mu(u)$ is a fine potential, resp. an invariant function.*

Remark 3.1.14 On a Euclidean domain U the invariant functions are the same as the harmonic functions ≥ 0. It is well known in view of Harnack's principle that the set of these functions is closed in $\mathcal{S}_+(U)$ with the natural topology (which coincides with the topology of R.-M. Hervé [He]). However, when U is just a regular fine domain in a Green space Ω, the set $\mathcal{H}_i(U)$ of invariant functions on U need not be closed in the induced natural topology on U, as shown by the following example: Let $y \in U$ be a Euclidean non-inner point of U, and let (y_k) be a sequence of points of $\complement U$ which converges Euclidean to y. The sequence $(G_\Omega(\cdot, y_k)|_U)$ then converges naturally in $\mathcal{S}_+(U)$ to $G_\Omega(\cdot, y)|_U$, which does not belong to $\mathcal{H}_i(U)$ because its fine potential part $G_U(\cdot, y)$ is non-zero, see Theorem 2.5.11(b).

3.2 Martin Compactification of a Fine Domain, the Martin Kernel, and Martin Representation of Invariant Functions

In this section, we introduce and study the Martin compactification $\overline{U}$ and in particular the Martin boundary $\Delta(U)$ of the (regular) fine domain U. We define the Martin kernel for U and establish the Martin representation of invariant functions. The results are very close to those in the classical case, where U is a usual domain in Ω, see [Do4, 1.XII], [AG, Section 8]. The main difference is that the Martin boundary in our setting is not necessarily compact, but only a G_δ-set, see Examples 3.2.8 and 3.2.9 below.

Let $\mathbf{B}$ be a compact base of the cone $\mathcal{S}_+(U)$, and let Φ the form defining $\mathbf{B}$ (see Lemma 3.1.7 and the line following its proof). Consider the mapping $\varphi : U \to \mathbf{B}$ defined by

$$\varphi(y) = P_y := \frac{G_U(\cdot, y)}{\Phi(G_U(\cdot, y))}. \tag{3.2.1}$$

Note that φ is injective because $\{y\} = \{x \in U : G_U(x, y) = +\infty\} = \{x \in U : \varphi(y)(x) = +\infty\}$. We may therefore identify $y \in U$ with $\varphi(y) = P_y \in \mathbf{B}$ and hence U with $\varphi(U) = \mathrm{Ext}_p(\mathbf{B})$, the latter equality by Corollary 2.6.18.

We denote by $\overline{U}$ the closure of U in $\mathbf{B}$ (with the natural topology), and write $\Delta(U) = \overline{U} \setminus U$. Then $\overline{U}$ is compact in $\mathbf{B}$ and is called the *Martin compactification* of U, and $\Delta(U)$ is called the *Martin boundary* of U.

If $\mathbf{B}$ and $\mathbf{B}^*$ are two bases of $\mathcal{S}_+(U)$ the Martin compactifications of U relative to $\mathbf{B}$ and to $\mathbf{B}^*$ are clearly homeomorphic.

Throughout the rest of this chapter we fix the compact base $\mathbf{B}$ of the cone $\mathcal{S}_+(U)$ and the continuous affine form $\Phi : \mathcal{S}_+(U) \to]0, +\infty[$ defining this base.

For any $Y \in \overline{U}$ consider the function $K(\cdot, Y) \in \mathbf{B} \subset \mathcal{S}_+(U) \setminus \{0\}$ defined on U by

$$K(x, Y) = \begin{cases} \varphi(Y)(x) & \text{if } Y \in U \\ Y(x) & \text{if } Y \in \Delta(U) \end{cases} \tag{3.2.2}$$

for $x \in U$. Clearly the map $Y \mapsto K(\cdot, Y)$ is a bijection of $\overline{U}$ on its image in $\mathbf{B}$.

Definition 3.2.1 The function $K : U \times \overline{U} \to]0, +\infty]$ defined by $K(x, Y) = K(\cdot, Y)(x)$ is called the (fine) Riesz-Martin kernel for U, and its restriction to $U \times \Delta(U)$ is called the (fine) Martin kernel for U.

Theorem 3.2.2 *The Riesz-Martin kernel* $K : U \times \overline{U} \to]0, +\infty]$ *has the following properties,* $\overline{U}$ *being given the natural topology:*

(i) *For any* $x \in U$, $K(x, \cdot)$ *is l.s.c. on* $\overline{U}$.

(ii) *For any* $Y \in \overline{U}$, $K(\cdot, Y) \in \mathcal{S}_+(U)$ *is finely continuous on* U.

(iii) K *is l.s.c. on* $U \times \overline{U}$ *when* U *is given the fine topology and* $\overline{U}$ *the natural topology.*

Proof (i) follows from Corollary 3.1.6 applied to $u = K(\cdot, Y)$ while identifying $K(\cdot, Y)$ with Y.

(ii) is obvious.

(iii) Let $x_0 \in U$, $Z \in \overline{U}$, and let (V_j) be a fundamental system of open neighborhoods of Z in $\overline{U}$ such that $V_{j+1} \subset V_j$ for any j. For a given constant $c > 0$ consider the increasing sequence of functions

$$k_j := \inf_{Y \in V_j} K(\cdot, Y) \wedge c$$

and their finely l.s.c. regularizations $\widehat{k}_j \in \mathcal{S}_+(U)$. By the Brelot property, cf. Theorem 1.5.29, there exists a fine neighborhood H of x_0 in U such that H is compact (in the Euclidean topology) and that the restrictions of the functions $\widehat{k}_j \in \mathcal{S}_+(U)$ and of $K(\cdot, Z) \wedge c \in \mathcal{S}_+(U)$ to H are continuous on H (again with the induced Euclidean topology). By (i) we have pointwise on U

$$K(\cdot, Z) \wedge c = \liminf_{Y \to Z} K(\cdot, Y) \wedge c = \sup_j \inf_{Y \in V_j} K(\cdot, Y) \wedge c,$$

which quasi-everywhere and hence everywhere on U equals $\sup_j \widehat{\inf_{Y \in V_j}} K(\cdot, Y) \wedge c \in \mathcal{S}_+(U)$. By Theorem 3.1.3 and Dini's theorem there exists for given $\varepsilon > 0$ an integer $j_0 > 0$ and fine neighborhood W of x_0 with $W \subset H$ such that

$$K(\cdot, Z) \wedge c = \sup_j \widehat{\inf_{Y \in V_j}} K(\cdot, Y) \wedge c = \sup_j \widehat{k}_j < \widehat{k}_j + \varepsilon$$

on H for any $j \geqslant j_0$ and

$$K(\cdot, Z) \wedge c \geq K(x_0, Z) \wedge c - \varepsilon$$

on W, so that

$$\begin{aligned}\liminf_{(x,Y)\to(x_0,Z)} K(x, Y) &\geq \inf_{x \in W, Y \in V_j} K(x, Y) \wedge c = \inf_{x \in W} k_j(x) \geqslant \inf_{x \in W} \widehat{k}_j(x) \\ &\geq \inf_{x \in W} K(x, Z) \wedge c - \varepsilon \geqslant K(x_0, Z) \wedge c - 2\varepsilon\end{aligned}$$

for $j \geqslant j_0$. The assertion (iii) follows by letting $\varepsilon \to 0$ and next $c \to +\infty$. □

Remark 3.2.3 The Riesz-Martin kernel K is in general not l.s.c. in the product of the induced natural topology on U and the natural topology on $\overline{U}$. Not even the function $K(\cdot, y) = G_U(\cdot, y)/\Phi(G_U(\cdot, y))$, or equivalently $G_U(\cdot, y)$ itself, is l.s.c. on U with the induced natural topology for fixed $y \in U$. For if the set $V := \{x \in U : G_U(x, y) > 1\}$ were (natural) open for every $y \in U$ then U would be a natural neighborhood of y for every $y \in U$, that is, U would be naturally open in $\overline{U}$. But that is in general not the case because $\overline{U} \setminus U = \Delta(U)$ need not be naturally compact, see Examples 3.2.8 and 3.2.9 below.

Remark 3.2.4 A set $A \subset U$ is termed a Euclidean nearly Borel set if it differs by a polar set from a Euclidean Borel set. We denote by $\mathcal{B}(U)$, resp. $\mathcal{B}^*(U)$, the σ-algebra of all Euclidean Borel, resp. nearly Borel subsets of U. Every finely open set $V \subset U$ is Euclidean nearly Borel because its regularized $\mathfrak{r}(V)$ is a Euclidean F_σ-set and $\mathfrak{r}(V) \setminus V$ is polar. It follows that every open subset W of $U \times \overline{U}$, now with the fine topology on U (and of course the natural topology on $\overline{U}$), belongs to the σ-algebra $\mathcal{B}^*(U) \times \mathcal{B}(\overline{U})$ generated by all sets $A_1 \times A_2$ where $A_1 \in \mathcal{B}^*(U)$ and where $A_2 \in \mathcal{B}(\overline{U})$, that is, A_2 is a Borel subset of $\overline{U}$. In view of Theorem 3.2.2(iii) every set $\{(x, Y) \in U \times \overline{U} : K(x, Y) > \alpha\}$ $(\alpha \in \mathbb{R})$ is such an open set W and therefore belongs to $\mathcal{B}^*(U) \times \mathcal{B}(\overline{U})$. This means that the Riesz-Martin kernel K is measurable with respect to $\mathcal{B}^*(U) \times \mathcal{B}(\overline{U})$.

Definition 3.2.5 An invariant function $h \in \mathcal{S}_+(U)$ is termed *minimal* if it belongs to an extreme generator of the cone $\mathcal{S}_+(U)$.

An interesting problem is whether every minimal invariant function is finite valued and hence finely harmonic on U. A negative answer to this question was obtained in dimension $n > 2$ by Gardiner and Hansen, see [GH]. For $N = 2$ the problem remains open. As a consequence of the result of Gardiner and Hansen, there is for every $N \geq 3$ an invariant function which is not the sum of a sequence of finely positive harmonic functions.

Recall that Ext(**B**) denotes the set of extreme points of **B**, and $\mathrm{Ext}_i(\mathbf{B})$ the subset of minimal invariant functions in **B**.

Theorem 3.2.6 *For any point $Y \in \Delta(U)$, the function $K(\cdot, Y)$ is an extreme point of* **B** *if and only if $K(\cdot, Y)$ is a minimal invariant function.*

Proof Suppose that $K(\cdot, Y)$ is an extreme point of **B**. By the Riesz decomposition of functions from $\mathcal{S}_+(U)$, $K(\cdot, Y)$ is either a fine potential on U or else a minimal invariant function on U. In the former case it follows by Corollary 2.6.18 that $K(\cdot, Y)$ must be equal to P_y for some $y \in U$, which contradicts $Y \in \Delta(U)$. Thus $K(\cdot, Y)$ is indeed a minimal invariant function. The converse is obvious. □

Definition 3.2.7 A point $Y \in \Delta(U)$ is termed minimal if the function $K(\cdot, Y)$ is minimal, that is, it belongs to an extreme generator of the cone $\mathcal{S}_+(U)$.

We denote by $\Delta_1(U)$ the set of all minimal points of $\Delta(U)$. Contrary to the case where U is Euclidean open in Ω, $\Delta(U)$ is in general not compact (in the natural topology), as shown by the following examples.

Example 3.2.8 Let ω be a bounded non-regular Hölder domain in $\mathbb{R}^N$ $(N \geqslant 2)$ and take $U = \omega \cup \partial_i\omega$ ($\partial_i\omega$ is the set of the irregular boundary points of ω). Suppose that $\partial_i\omega$ is not a relatively (Euclidean) open subset of $\partial\omega$. The map $\psi : \mathcal{S}_+(U) \to \mathcal{S}_+(\omega)$ defined by $\psi(u) = u|_\omega$ is an affine homeomorphism by Theorems 2.1.22, 2.1.43 and 3.1.3 and satisfies $\psi(0) = 0$. The image $\mathbf{B}' = \psi(\mathbf{B})$ of the base **B** of $\mathcal{S}_+(U)$ is a base of $\mathcal{S}_+(\omega)$. We have $\overline{U} = \overline{\psi^{-1}(\omega \cup \partial_i\omega)} = \overline{\psi^{-1}(\omega)} \cup \overline{\partial_i\omega}$ and hence $\overline{U} = \overline{\omega} \cup \overline{\partial_i\omega}$ (the closures are taken relative to natural topology) by identifying each subset of $\mathbf{B}'$ with its image in **B** under ψ^{-1}. For every $z \in \partial_i\omega$ we have $P_z = \text{f-}\lim_{\omega \ni y \to z} P_y$, so that $z \in \overline{\omega}$, where z is identified with P_z. It follows that $\partial_i\omega \subset \overline{\omega}$ and hence $\overline{\partial_i\omega} \subset \overline{\omega}$, so that $\Delta(U) = \overline{U} \setminus U = \Delta(\omega) \setminus \partial_i\omega$. By [Ai, Theorem 3.1 and Theorem 1.5 (iv)] we have $\partial\omega \subset \Delta(\omega)$ (where $\partial\omega$ is the Euclidean boundary of ω), and since $\partial\omega$ is compact, we infer that $\Delta(U) = \Delta(\omega) \setminus \partial_i\omega$ is not compact because $\partial_i\omega$ is not a relatively open subset of $\partial\omega$.

Example 3.2.9 In $\Omega = \mathbb{R}^3$, with the Green kernel $G(x, y) = |x - y|^{-1}$, let $\omega = B(0, 1) \setminus F$, where

$$F = \{0\} \cup \{(x', x_3) : x_3 > 0, \|x'\| \geq \exp(-x_3^{-1})\},$$

is the Lebesgue spine. By [AG, Remark 6.6.17] 0 is an irregular point of $\partial\omega$, and all the other points of $\partial\omega$ are regular by the Poincaré-Zaremba criterion, cf., e.g. [AG, theorem 6.6.15]. It follows that $U = \omega \cup \{0\}$ is a fine domain of $\mathbb{R}^3$. By considering as in the above example the affine homeomorphism $\psi : \mathcal{S}_+(U) \to \mathcal{S}_+(\omega)$ defined by $\psi(u) = u|_\omega$ we have $\overline{U} = \overline{\omega} \cup \{0\}$ (the closures are taken relative to natural topology) by identifying each subset of $\mathbf{B}'$ with its image in **B** under ψ^{-1}. As in the precedent example $0 \in \overline{\omega}$ (0 being identified with P_0), and hence $\overline{U} = \overline{\omega}$, so that $\Delta(U) = \overline{U} \setminus U = \Delta(\omega) \setminus \{0\}$. Since $\Delta(\omega)$ is compact and 0 is not (naturally) isolated, we infer that $\Delta(U)$ is not compact.

However, we have the following

Theorem 3.2.10 *$\Delta(U)$ is a G_δ of $\overline{U}$.*

Proof With the notations from the proof of Theorem 3.1.8 we have $\Delta(U) = \overline{U} \setminus U = \bigcap_{k,l}(\overline{U} \setminus A_{k,l})$, and this prove the theorem because the sets $A_{k,l}$ (identified with the $\mathbf{C}_{k,l}$) are compact. □

Corollary 3.2.11 *$\Delta_1(U)$ is a G_δ of $\overline{U}$.*

Because $\Delta_1(U) = \Delta(U) \cap \mathrm{Ext}_i(\mathbf{B}) = \Delta(U) \cap \mathrm{Ext}(\mathbf{B})$ (the former equality by Theorem 3.2.6), it suffices according to Theorem 3.2.10 to use the well-known fact that $\mathrm{Ext}(\mathbf{B})$ is a G_δ of $\mathbf{B}$.

The proof of Theorem 3.2.10 also establishes

Corollary 3.2.12 *For any integers $k, l > 0$ the set $\mathbf{C}_{k,l} := \{P_y : y \in A_{k,l}\}$ is compact in $\mathbf{B}$, and the mapping $\varphi : y \mapsto P_y$ is a homeomorphism of $A_{k,l}$ onto $\mathbf{C}_{k,l}$.*

Corollary 3.2.13 *Let $K \subset U$ be compact in the Euclidean topology. Then the set $C_K := \{P_y : y \in K\}$ is a (natural) K_σ in $\mathbf{B}$, and so is therefore $\{P_y : y \in U\}$ (identified with U).*

Since $C_K = \bigcup_{k,l} \varphi(K \cap A_{k,l})$ the assertion follows by the preceding corollary.

Cover Ω by a sequence of Euclidean open balls B_k $(k \in \mathbb{N})$ with closures $\overline{B}_k$ contained in Ω and such that each ball B_k occurs infinitely often in the sequence, and consider the compact sets $A_{k,l} \subset U$ in the proof of Theorem 3.1.8.

Lemma 3.2.14 (a) *The mapping $U \ni y \mapsto G_U(\cdot, y) \in \mathcal{S}_+(U)$ is continuous from U with the fine topology into $\mathcal{S}_+(U)$ with the natural topology.*

(b) *The function $U \ni y \mapsto \Phi(G_U(\cdot, y)) \in]0, +\infty[$ is finely continuous.*

(c) *The sets*

$$V_k = \{y \in U : \Phi(G_U(\cdot, y)) > 1/k\} \cap B_k \ (k \in \mathbb{N})$$

form a countable cover of U by finely open sets which are relatively naturally compact in U.

Proof (a) Consider a net $(y_\alpha)_{\alpha \in I}$ on U converging finely to some $y \in U$, in other words $s(y_\alpha) \to s(y)$ for every $s \in \mathcal{S}_+(U)$. Taking $s = G_U(x, \cdot)$ with $x \in U$ we have in particular $G_U(x, y_\alpha) \to G_U(x, y)$ for every $x \in U$. Since $G_U(\cdot, y_\alpha) \in \mathcal{S}_+(U) \cup \{+\infty\}$, which is naturally compact by Theorem 3.1.3, we may assume that $G_U(\cdot, y_\alpha) \to z$ naturally for some $z \in \mathcal{S}_+(U) \cup \{+\infty\}$. Furthermore, we then have

$$z = \liminf_\alpha{}^{\widehat{}}\, G_U(\cdot, y_\alpha) = G_U(\cdot, y),$$

where the latter equality holds first q.e., and next everywhere on U by fine continuity of z and $G_U(\cdot, y)$. Thus $G_U(\cdot, y_\alpha) \to G_U(\cdot, y)$ (in natural topology), which proves assertion (a).

(b) It follows from (a) that this function is finely continuous on U because the mapping $\Phi : \mathcal{S}_+(U) \to [0, +\infty[$ is (naturally) continuous, as recalled above.

(c) According to (b), each V_k is finely open along with $U \cap B_k$. By the proof of Theorem 3.1.8 we have $V_k \subset A_{k,k} \subset U$ with $A_{k,k}$ naturally compact. Clearly, $\bigcup_k V_k = U$. □

Corollary 3.2.15 *The mapping $\varphi : U \ni y \mapsto P_y = K(\cdot, y) \in \mathcal{S}_+(U)$ is continuous from U with the fine topology to $\mathcal{S}_+(U)$ with the natural topology. In other words, the fine topology on U is finer than the topology on U induced by the natural topology on $\overline{U}$.*

Proof Recall that $P_y = G_U(\cdot, y)/\Phi(G_U(\cdot, y))$ for $y \in U$. The corollary follows immediately from (a) and (b) in the above lemma. □

Lemma 3.2.16 *Every Euclidean Borel subset of U is a natural Borel subset of U.*

Proof By Corollary 3.2.13 U (being regular) is a natural K_σ. Let A be a Euclidean compact subset of Ω. By replacing B_l in the proof of Theorem 3.1.8 by A we see that the set $A_k = A \cap \{y \in U : \Phi(G_U(\cdot, y)) \geq \frac{1}{k}\}$ is naturally compact for any integer $k > 0$ and hence $A \cap U = \bigcup A_k$ is a natural Borel subset of U. Since any Euclidean closed subset of U is the countable union of sets of the form $A \cap U$, A (Euclidean) compact subset of Ω, it then follows that any Euclidean Borel subset of U is a natural Borel subset of U. □

It follows from Lemma 3.2.16 that the σ-algebra $\mathcal{B}(U)$ (see Remark 3.2.4) is contained in the σ-algebra of natural Borel subsets of U, and hence the restriction of a finite measure μ on $U = \mathrm{Ext}_p(\mathbf{B})$ to $\mathcal{B}(U)$ is a Borel measure on U, again denoted by μ, and so is the measure ν defined by $d\nu(y) = \frac{1}{\Phi(G_U(\cdot,y))} d\mu(y)$, the function $y \mapsto \Phi(G_U(\cdot, y))$ being finely continuous in view of Lemma 3.2.14 and hence Borel measurable relative to Euclidean topology by Theorem 1.4.44. Furthermore, ν belongs to $\Lambda(U)$ for we have $G_U\nu = \int_U q\,d\mu(q) \in \mathcal{P}(U)$.

Theorem 3.2.17 *Let $u \in \mathcal{S}_+(U)$ and let $A \subset \overline{A} \subset U$, where $\overline{A}$ denotes the closure of A in $\overline{U}$. The measure μ on $\mathbf{B}$ carried by the extreme points of $\mathbf{B}$ and representing $\widehat{R}_u^A$ is then carried by $\overline{A}$.*

Proof Let p be a finite fine potential > 0 on U. For any integer $j > 0$ the function $\widehat{R}^A_{u\wedge jp}$ is a fine potential on U and finely harmonic on $U \setminus \overline{A}$ by Corooary 2.4.20. By Theorem 3.1.11, there exists a measure μ_j on $\mathbf{B}$ carried by $\mathrm{Ext}_p(\mathbf{B}) = U$ such that

$$\widehat{R}^A_{u\wedge jp} = \int q\,d\mu_j(q) = \int_U P_y\,d\mu_j(y) = \int_U G_U(\cdot, y)\,d\nu_j(y),$$

where ν_j is the measure defined by $\nu_j(dy) = \frac{1}{\Phi(G_U(\cdot,y))}\mu_j(dy)$. Since $\widehat{R}^A_{u\wedge jp}$ is finely harmonic on $U \setminus \widetilde{A}$, the measure ν_j is carried by $\widetilde{A}$ by Theorem 2.6.4, and so is

μ_j. By Corollary 3.2.15, $\widetilde{A} \subset \overline{A}$ and hence μ_j is carried by $\overline{A}$. The sequence of probability measures $\tau_j = \frac{1}{\Phi(R^A_{u \wedge jp})}\mu_j$ on $\mathbf{B}$ has a subsequence (τ_{j_k}) which converges to a probability measure τ on $\mathbf{B}$ carried by $\overline{A}$. We thus have

$$\widehat{R}^A_u = \lim_{k\to\infty} \widehat{R}^A_{u\wedge j_k p} = \Phi(\widehat{R}^A_u) \lim_{k\to\infty} \int q\, d\tau_{j_k}(q) = \Phi(\widehat{R}^A_u) \int q\, d\tau(q).$$

The assertion now follows by the fact that $\overline{A} \subset U \subset \mathrm{Ext}(\mathbf{B})$ and from the uniqueness of the integral representation in Choquet's theorem. □

Theorem 3.2.18 *Every extreme element of the base* $\mathbf{B}$ *of the cone* $\mathcal{S}_+(U)$ *belongs to* $\overline{U}$. *In particular, any minimal invariant function* h *in* $\mathbf{B}$ *has the form* $h = K(\cdot, Y)$ *where* $Y \in \Delta_1(U)$.

Proof Let p be an extreme element of $\mathbf{B}$. It follows from the proof of Theorem 3.1.8 that there exists an increasing sequence of compact subsets (K_j) of $\overline{U}$ (of the form $A_{k,l}$) such that $\bigcup_j K_j = U$. For each j, by Theorem 3.2.17, the measure μ_j on $\mathbf{B}$ carried by $\mathrm{Ext}(\mathbf{B})$ and representing $\widehat{R}^{K_j}_p$ is carried by K_j, and hence $\frac{\widehat{R}^{K_j}_p}{\Phi(\widehat{R}^{K_j}_p)}$ belongs to the closed convex hull $\mathrm{cch}(\overline{U})$. It then follows that $p = \lim_{j\to+\infty} \frac{\widehat{R}^{K_j}_p}{\Phi(\widehat{R}^{K_j}_p)} \in \mathrm{cch}(\overline{U})$, and by Millman's theorem, cf. [Ph, Proposition 1.5], $p \in \overline{U}$. The remaining follows from Theorem 3.2.6, the identification of U with the subset $\{P_y : y \in U\} = \mathrm{Ext}_p(\mathbf{B})$ of $\mathbf{B}$ and the definition of $\Delta(U)$ and $\Delta_1(U)$. □

Corollary 3.2.19 $\mathrm{Ext}_p(\mathbf{B}) = U$, $\mathrm{Ext}_i(\mathbf{B}) = \Delta_1(U)$ *and* $\mathrm{Ext}(\mathbf{B}) = U \cup \Delta_1(U)$.

Proof Every extreme element of $\mathbf{B}$ is either the fine potential of a measure supported by a single point $y \in U$, hence of the form P_y, or else a minimal invariant function, hence of the form $K(\cdot, Y)$ with $Y \in \Delta_1(U)$, according to Theorem 3.2.18. This establishes the inclusions $\mathrm{Ext}_p(\mathbf{B}) \subset U$ and $\mathrm{Ext}_i(\mathbf{B}) \subset \Delta_1(U)$. The opposite inclusions and the latter equality in theorem are evident. □

Theorem 3.2.20 *For any invariant function* h *on* U *there exists a unique Radon measure* μ *on* $\overline{U}$ *carried by* $\Delta_1(U)$ *such that*

$$h(x) = \int_{\Delta(U)} K(x, Y)\, d\mu(Y), \quad x \in U.$$

Proof The theorem follows immediately from Theorem 3.1.12 and Corollary 3.2.19. □

Theorem 3.2.21 *Let* $u \in \mathcal{S}_+(U)$ *and let* V *be a finely open Borel subset of* U. *Let* μ *be the measure on* $\mathbf{B}$ *carried by* $\mathrm{Ext}(\mathbf{B})$ *and representing* u. *Then* $\mu(V) = 0$ *if and only if the restriction* $u|_V$ *is invariant.*

Proof Without loss of generality we may suppose that V is a fine domain. Write $u = p + h$ with $p \in \mathcal{P}(U)$ and h invariant on U. Then we have $\mu = \mu_p + \mu_h$, $\mu_p(\Delta(U)) = 0$ and $\mu_h(U) = 0$. Let λ denote the restriction of μ_p to U and let τ be the Borel measure on U defined by $\tau(dy) = \frac{1}{\Phi(G_U(\cdot,y))}\lambda(dy)$, see text preceding Theorem 3.2.17. By Lemma 2.6.3, we have

$$p = G_U\tau = G_V(\tau|_V) + \widehat{R}^{U\setminus V}_{G_U\tau} \preccurlyeq u \quad \text{on} V,$$

and hence $G_V(\tau|_V) \preccurlyeq u$ on V. If $u|_V$ is invariant the fine potential $G_V(\tau|_V)$ must therefore be 0, whence $\lambda(V) = 0$ and finally $\mu(V) = \lambda(V) + \mu_h(V) = 0$ because $0 \leqslant \mu_h(V) \leqslant \mu_h(U) = 0$. Conversely, suppose that $\mu(V) = 0$ and hence $\tau(V) = 0$. In the above display $\widehat{R}^{U\setminus V}_{G_U\tau}$ is invariant on V by Lemma 2.9.7, and so is therefore p. It follows that $u = p + h$ is invariant on V, h being invariant on U and hence on V by Theorem 2.9.12(a). □

For any (positive) Borel measure μ on $\overline{U}$ define a function $K\mu : U \to [0, +\infty]$ by

$$K\mu = \int K(\cdot, Y)d\mu(Y), \quad x \in U.$$

This integral exists in view of Theorem 3.2.2(i). If $\mu(\overline{U}) = +\infty$, then for any $x \in U$ we have $K\mu(x) = +\infty$. Indeed, the function $K(x, \cdot)$ is > 0 and l.s.c. on $\overline{U}$ (see Theorem 3.2.2(i)) and therefore $K(x, \cdot) \geq c$ for some constant $c > 0$, so that $K\mu(x) \geq c\mu(\overline{U}) = +\infty$.

Theorem 3.2.22 (a) *For any Radon measure μ on $\overline{U}$, $K\mu$ is finely superharmonic on U, that is, $K\mu \in \mathcal{S}_+(U)$.*

(b) *Every function $u \in \mathcal{S}_+(U)$ has a unique integral representation $u = K\mu$ in terms of a finite Borel measure μ on $U \cup \Delta_1(U)$.*

Proof (a) We have by the definition of the kernel K

$$K\mu = \int_{\overline{U}} K(\cdot, Y)d\mu(Y) = \int_{\overline{U}} Y d\mu(Y)$$

and hence $K\mu \in \mathcal{S}_+(U)$, see text following the proof of Theorem 3.1.8.

(b) This follows immediately from Choquet's integral representation theorem applied to the cone $\mathcal{S}_+(U)$ with the base **B** and Corollary 3.2.19. □

Corollary 3.2.23 *For any finite measure μ on U, resp. $\Delta_1(U)$, the function $h = \int K(\cdot, Y)d\mu(Y)$ is a fine potential, resp. an invariant function.*

Proof We give the proof in case of invariant functions, the case of fine potentials is similar. Let ν be the measure on **B** defined by $\nu(A) = \mu(\Delta_1(U) \cap A)$ for any Borel set $A \subset \mathbf{B}$. Then ν is a finite measure on **B** carried by $\Delta_1(U) = \mathrm{Ext}_i(\mathbf{B})$. Then $h = \int_{\Delta_1(U)} K(\cdot, Y)d\mu(Y) = \int_{\mathrm{Ext}_i(\mathbf{B})} q d\mu(q)$, and hence h is invariant according to Corollary 3.2.19 and Theorem 3.1.13. □

The following theorem, which is just another formulation of Theorem 3.1.9, is an easy consequence of the precedent corollary.

Theorem 3.2.24 *Let μ be a finite Borel measure on $U \cup \Delta_1(U)$. The function $u = K\mu$ is a fine potential, resp. an invariant function, if and only μ is carried by U, resp. by $\Delta_1(U)$.*

Corollary 3.2.25 *Let $u \in \mathcal{S}_+(U)$ and $A \subset \overline{A} \subset U$, where $\overline{A}$ denotes the closure of A in $\overline{U}$. Then $\widehat{R}_u^A$ is a fine potential on U.*

Proof According to Theorem 3.2.17 the measure μ on $U \cup \Delta_1(U)$ representing $\widehat{R}_u^A$ is carried by $\overline{A}$. It follows by the preceding theorem that $\widehat{R}_u^A$ is a fine potential. □

Lemma 3.2.26 *For any set $A \subset U$ and any Radon measure μ on $\overline{U}$ we have $\widehat{R}_{K\mu}^A = \int \widehat{R}_{K(\cdot,Y)}^A d\mu(Y)$.*

Proof The integral exists according to the text following the proof of Theorem 3.1.8 and the definition of the kernel K. As in the proof of Theorem 2.3.11, $\widehat{R}_{K\mu}^A$ and $\widehat{R}_{K(\cdot,Y)}^A$ remain unchanged when A is replaced by $\mathfrak{b}(A) \cap U$, and we may therefore assume that $A = \mathfrak{b}(A) \cap U$, whence $\widehat{R}_{K\mu}^A = R_{K\mu}^A$ and $\widehat{R}_{K(\cdot,Y)}^A = R_{K(\cdot,Y)}^A$, see Lemma 2.2.12. The kernel K on $U \times \overline{U}$ is measurable with respect to the product σ-algebra $\mathcal{B}^*(U) \times \mathcal{B}(\overline{U})$ in view of the conclusion in Remark 3.2.4. Furthermore, $K\mu$ is finely superharmonic on U, and we have by Theorem 2.3.11 and Fubini's theorem for any $x \in U$

$$\begin{aligned}\widehat{R}_{K\mu}^A(x) &= \int_U K\mu \, d\varepsilon_x^{A\cup(\complement U)} = \int_{\overline{U}} d\mu(Y) \int_U K(\cdot, Y) d\varepsilon_x^{A\cup(\complement U)} \\ &= \int_{\overline{U}} \widehat{R}_{K(\cdot,Y)}^A(x) d\mu(Y).\end{aligned}$$

□

The following characterizations of the invariant functions and the fine potentials on U are analogous (but only partly comparable) to Theorem 2.9.9, where the present condition $\overline{V} \subset U$ was replaced by the weaker condition $\widetilde{V} \subset U$.

Theorem 3.2.27 *Let $u \in \mathcal{S}_+(U)$. Then u is invariant if and only if $\widehat{R}_u^{U\setminus V} = u$ for any regular finely open set $V \subset U$ such that the closure $\overline{V}$ of V in the natural topology on $\mathbf{B}$ is contained in U.*

Proof Suppose that u is invariant. For any V as stated we have $u \leqslant \widehat{R}_u^V + \widehat{R}_u^{U\setminus V}$. By the propriety of natural decomposition (cf. Sect. 2.2) there are functions $u_1, u_2 \in \mathcal{S}_+(U)$ such that $u = u_1 + u_2$ with $u_1 \leqslant \widehat{R}_u^V$ and $u_2 \leq \widehat{R}_u^{U\setminus V}$. But $\widehat{R}_u^V$ is a fine potential by Corollary 3.2.25, and so is therefore u_1. It follows that $u_1 = 0$ because $u_1 \preccurlyeq u$ and u is invariant. Consequently, $u \leqslant \widehat{R}_u^{U\setminus V}$, and so indeed $\widehat{R}_u^{U\setminus V} = u$. Conversely, suppose that $\widehat{R}_u^{U\setminus V} = u$ for any regular finely open (and finely connected, if we like) set $V \subset U$ with $\overline{V} \subset U$. Let μ be the (finite) measure on $U \cup \Delta_1(U)$ which represents u. Then

$$u = \int_U P_y d\mu(y) + \int_{\Delta_1(U)} K(\cdot, Y) d\mu(Y).$$

For any regular fine domain $V \subset U$ with $\overline{V} \subset U$ we have by hypothesis and by Lemma 3.2.26

$$u = \widehat{R}_u^{U \setminus V} = \int_U \widehat{R}_{P_y}^{U \setminus V} d\mu(y) + \int_{\Delta_1(U)} \widehat{R}_{K(\cdot,Y)}^{U \setminus V} d\mu(Y),$$

and hence

$$\int_U (P_y - \widehat{R}_{P_y}^{U \setminus V}) d\mu(y) = 0.$$

Since V is regular and since $P_y - \widehat{R}_{P_y}^{U \setminus V} = G_V(\cdot, y)/\Phi(G_U(\cdot, y)) > 0$ on V in view of Lemma 2.6.3 and Theorem 2.5.11(b), it follows that $\mu(V) = 0$. This implies by Lemma 3.2.14(c) (together with the fact that a finely open set has only countably many fine components) that μ is carried by $\Delta_1(U)$. Consequently, u is invariant according to Theorem 3.2.24. □

Corollary 3.2.28 *Let $u \in \mathcal{S}_+(U)$. Then u is a fine potential if and only if $\widehat{\inf}_j \widehat{R}_u^{U \setminus V_j} = 0$ for some, and hence any, cover of U by an increasing sequence (V_j) of regular finely open sets such that $\overline{V}_j \subset V_{j+1} \subset U$ for every j, where $\overline{V}_j$ is the closure of V_j in the natural topology on* **B**.

Proof Suppose first that u is a fine potential, and consider any cover (V_j) of U as stated. Denote $v = \widehat{\inf}_j \widehat{R}_u^{U \setminus V_j}$, which is likewise a fine potential. For each index k, $\widehat{R}_u^{U \setminus V_j}$ is invariant on V_k for any $j \geqslant k$ according to Lemma 2.9.7. It follows by Theorem 2.9.12(c) that v likewise is invariant on V_k. By varying k we see from Theorem 2.9.12(b) applied to the regular finely open sets V_k that v is invariant on $\bigcup_k V_k = U$. Consequently $v = 0$. Conversely, suppose that $\widehat{\inf}_j \widehat{R}_u^{U \setminus V_j} = 0$ for some cover (V_j) as stated in the corollary. We have $u = p + h$, where p is a fine potential and h is an invariant function. By the preceding theorem, $h = \widehat{R}_h^{U \setminus V_j} \leqslant \widehat{R}_u^{U \setminus V_j}$ for every j, and hence $h = 0$, showing that u is a fine potential. □

3.3 The Fatou-Naïm-Doob Theorem for Finely Superharmonic Functions

This section is inspired by the axiomatic approach to the Fatou-Naïm-Doob theorem given in [T]. These axioms are, however, only partially fulfilled in our setting. In particular, our invariant functions, which play the role of positive harmonic functions, may take the value $+\infty$. We therefore choose to give the proof of the Fatou-Naïm-Doob theorem without reference to the proofs in [T].

In the present section we shall not consider the whole Riesz-Martin space $\overline{U}$ and the full Riesz-Martin kernel $K : U \times \overline{U} \to]0, +\infty]$ (Definition 3.2.1), but only the Martin boundary $\Delta(U)$ and the Martin kernel, the restriction of the Riesz-Martin kernel to $U \times \Delta(U)$, and K will henceforth stand for this restriction. It is understood that U and $\Delta(U)$ are given the natural topology (induced by the natural topology on the Riesz-Martin space $\overline{U}$). In particular, by Lemma 3.2.14(c), U is a K_σ subset of $\overline{U}$, and we know that $\Delta(U)$ and the minimal Martin boundary $\Delta_1(U)$ are G_δ subsets of $\overline{U}$ (Theorem 3.2.10 and Corollary 3.2.11).

For any set $A \subset U$, R_u^A and $\widehat{R}_u^A$ are understood as reduction and sweeping of a function u on A relative to U, whereas ε_x^A stands for sweeping of ε_x on A relative to all of Ω.

Lemma 3.3.1 *For any set $E \subset U$ and any $Y \in \Delta_1(U)$ we have $\widehat{R}^E_{K(\cdot,Y)} \neq K(\cdot, Y)$ if and only if $\widehat{R}^E_{K(\cdot,Y)} \in \mathcal{P}(U)$.*

Proof Note that, for any $u \in \mathcal{S}_+(U)$ and $E \subset U$, we have $R_u^E \neq u \iff \widehat{R}_u^E \neq u$. Proceeding much as in [T, proof of G)], see also [Gow], we first suppose that $\widehat{R}^E_{K(\cdot,Y)} \neq K(\cdot, Y)$, and hence necessarily $R^E_{K(\cdot,Y)} \neq K(\cdot, Y)$. Since $R^E_{K(\cdot,Y)} \leqslant K(\cdot, Y)$ there exists $x_0 \in U$ with $R^E_{K(\cdot,Y)}(x_0) < K(x_0, Y)$. Thus there exists $u \in \mathcal{S}_+(U)$ such that $u \geqslant K(\cdot, Y)$ on E and $u(x_0) < K(x_0, Y)$. Replacing u by $u \wedge K(\cdot, Y) \in \mathcal{S}_+(U)$ we arrange that $u \leqslant K(\cdot, Y)$ on all of U. Writing $u = q + h$ with $q \in \mathcal{P}(U)$ and $h \in \mathcal{H}_i(U)$ we have $h \leqslant u \leqslant K(\cdot, Y)$ on U. Because $K(\cdot, Y) \in \mathcal{H}_i(U)$ is minimal invariant and $h \preccurlyeq K(\cdot, Y)$ by Lemma 2.9.5 (with $p = 0$), there exists $\alpha \in [0, 1]$ such that $h = \alpha K(\cdot, Y)$, and here $\alpha < 1$ since $h(x_0) \leq u(x_0) < K(x_0, Y)$. On E we have $q = u - h = K(\cdot, Y) - h = (1 - \alpha)K(\cdot, Y)$, and hence

$$K(\cdot, Y) = p := \frac{1}{1-\alpha} q \quad \text{on } E.$$

Thus $p \in \mathcal{P}(U)$ and $\widehat{R}^E_{K(\cdot,Y)} = \widehat{R}^E_p \leqslant p$ on U, so indeed $\widehat{R}^E_{K(\cdot,Y)} \in \mathcal{P}(U)$. Conversely, suppose that $\widehat{R}^E_{K(\cdot,Y)} = K(\cdot, Y)$ and (by contradiction) that $\widehat{R}^E_{K(\cdot,Y)} \in \mathcal{P}(U)$. Being thus both a fine potential and invariant, $K(\cdot, Y)$ must equal 0, which is false. □

Definition 3.3.2 A set $E \subset U$ is said to be minimal-thin at a point $Y \in \Delta_1(U)$ if $\widehat{R}^E_{K(\cdot,Y)} \neq K(\cdot, Y)$, or equivalently (by the preceding lemma) if $\widehat{R}^E_{K(\cdot,Y)}$ is a fine potential on U.

Corollary 3.3.3 *For any $Y \in \Delta_1(U)$ the sets $W \subset U$ for which $U \setminus W$ is minimal-thin at Y form a filter on U.*

This follows from Lemma 3.3.1 which easily implies that for any $W_1, W_2 \subset U$ such that $\widehat{R}^{U\setminus W_i}_{K(\cdot,Y)} \neq K(\cdot, Y)$ for $i = 1, 2$, we have $\widehat{R}^{U\setminus(W_1\cup W_2)}_{K(\cdot,Y)} \neq K(\cdot, Y)$.

The filter from Corollary 3.3.3 is called the minimal-fine filter at Y and will be denoted by $\mathcal{F}(Y)$. Limit, resp. upper limit along that filter will be called minimal-fine limit, resp. minimal-fine upper limit, and will be denoted by $\lim_{\mathcal{F}(Y)}$, resp. $\limsup_{\mathcal{F}(\mathcal{Y})}$.

For any two functions $u, v \in \mathcal{S}_+(U)$ with $v \neq 0$ the quotient u/v is assigned some arbitrary value, say 0, on the polar set of points at which both functions take the value $+\infty$. The choice of such a value does not affect a possible minimal-fine limit of u/v at a point $Y \in \Delta_1(U)$ because every polar set E clearly is minimal-thin at any point of $\Delta_1(U)$.

For any two measures μ, ν on a measurable space we denote by $d\nu/d\mu$ the Radon-Nikodým derivative of the absolutely continuous component of ν relative to that of μ, cf., e.g. [Do4, p. 773], [AG, p. 305].

For any function $u \in \mathcal{S}_+(U)$ with Riesz decomposition $u = p + h$, where $p \in \mathcal{P}(U)$ and $h \in \mathcal{H}_i(U)$, we denote by μ_u the unique measure on $\Delta_1(U)$ which represents the invariant part h of u, that is, $h = \int K(\cdot, Y)d\mu_u(Y)$.

We may now formulate the Fatou-Naïm-Doob theorem in the present setting of finely superharmonic functions on a regular fine domain U. It clearly contains the classical Fatou-Naïm-Doob theorem for which we refer to [Do4, 1.XII.19], [AG, Section 9.4].

Theorem 3.3.4 *Let $u, v \in \mathcal{S}_+(U)$, where $v \neq 0$. Then u/v has minimal-fine limit $d\mu_u/d\mu_v$ at μ_v-a.e. point Y of $\Delta_1(U)$.*

For the proof of Theorem 3.3.4 we begin by establishing the following important particular case, cf. [T, Theorem 1.2].

Theorem 3.3.5 ([T]) *Let $u \in \mathcal{S}_+(U)$ and $h \in \mathcal{H}_i(U) \setminus \{0\}$, and suppose that $u \wedge h \in \mathcal{P}(U)$. Then u/h has minimal-fine limit $d\mu_u/d\mu_h = 0$ at μ_h-a.e. point Y of $\Delta_1(U)$.*

Proof Write $u \wedge h = p$. Given $\alpha \in \,]0, 1[$, consider any point $Y \in \Delta_1(U)$ such that $\limsup_{\mathcal{F}(Y)} \frac{u}{h} > \alpha$. Then $\{u \leqslant \alpha h\} \notin \mathcal{F}(Y)$, that is, $\{u > \alpha h\}$ is not minimal-thin at Y:

$$\widehat{R}_{K(\cdot,Y)}^{\{u>\alpha h\}} = K(\cdot, Y). \tag{3.3.1}$$

It follows that (always with Y ranging over $\Delta_1(U)$)

$$\{Y : \limsup_{\mathcal{F}(Y)} \frac{u}{h} > \alpha\} \subset A_\alpha := \{Y : \widehat{R}_{K(\cdot,Y)}^{\{u>\alpha h\}} = K(\cdot, Y)\}. \tag{3.3.2}$$

We show that $\mu_h(A_\alpha) = 0$. Consider the measure $\nu = 1_{A_\alpha}\mu_h$ on $\Delta_1(U)$ and the corresponding function

$$v = \int K(\cdot, Y)d\nu(Y) = \int \widehat{R}_{K(\cdot,Y)}^{\{u>\alpha h\}} d\nu(Y) = \widehat{R}_{K\nu}^{\{u>\alpha h\}} = \widehat{R}_{v}^{\{u>\alpha h\}},$$

the second equality by (3.3.1) and the third equality by Lemma 3.2.26. Since $v \leqslant \int K(\cdot, Y) d\mu_h = h$ and $0 < \alpha < 1$ it follows that

$$v = \widehat{R}_v^{\{u > \alpha h\}} \leq \widehat{R}_h^{\{u > \alpha h\}} \leqslant \frac{u}{\alpha} \wedge h \leqslant \frac{u}{\alpha} \wedge \frac{h}{\alpha} = \frac{p}{\alpha}.$$

Because $v = \int K(\cdot, Y) d\nu(Y) \in \mathcal{H}_i(U)$ and $p/\alpha \in \mathcal{P}(U)$ we find by Lemma 2.9.5(a) (applied to $h_1 = v$, $h_2 = 0$) that $v = 0$, that is,

$$v = \int K(\cdot, Y) d\nu(Y) = \int_{A_\alpha} K(\cdot, Y) d\mu_h(Y) = 0,$$

and since $K(\cdot, Y) > 0$ it follows that $\mu_h(A_\alpha) = 0$. By varying α through a decreasing sequence tending to 0 we conclude from (3.3.2) that indeed $\mu_h(\{Y : \limsup_{\mathcal{F}(Y)} u/h > 0\}) = 0$. □

The rest of the proof of Theorem 3.3.4 proceeds much as in the classical case. For the convenience of the reader we bring most of the details, following in part [AG, Section 9.4].

Corollary 3.3.6 *Let $u, h \in \mathcal{H}_i(U)$, where $h \neq 0$ and μ_u, μ_h are mutually singular. Then $u \wedge h \in \mathcal{P}(U)$, and u/h has minimal-fine limit 0 μ_h-a.e. on $\Delta_1(U)$.*

Proof There are Borel subsets A_1, A_2 of U such that $A_1 \cup A_2 = \Delta_1(U)$ and $\mu_u(A_1) = \mu_h(A_2) = 0$. Write $u \wedge h = p + k$ with $p \in \mathcal{P}(U), k \in \mathcal{H}_i(U)$. Then $k \leqslant u$, hence $k \preccurlyeq u$ by Lemma 2.9.4, and so $\mu_k \leqslant \mu_u$. Similarly, $\mu_k \leqslant \mu_h$. It follows that $\mu_k(\Delta_1(U)) \leqslant \mu_u(A_1) + \mu_h(A_2) = 0$ and hence $k = 0$, and so $u \wedge h = p \in \mathcal{P}(U)$. The remaining assertion now follows from Theorem 3.3.5. □

Corollary 3.3.7 *Let $h \in \mathcal{H}_i(U) \setminus \{0\}$, let A be a Borel subset of $\Delta(U)$, and let $h_A = K(1_A \mu_h) = \int_A K(\cdot, Y) d\mu_h(Y)$. Then h_A/h has minimal-fine limit $1_A(Y)$ at Y μ_h-a.e. for $Y \in \Delta_1(U)$.*

Proof We may assume that $\mu_h(A)\mu_h(\Delta(U) \setminus A) > 0$. Write $u = h - h_A \in \mathcal{H}_i(U)$, which is invariant because $h_A \preccurlyeq h$. Since $h_A = K(1_A \mu_h)$ and because $1_A \mu_h$ is carried by $\Delta_1(U)$ along with μ_h, we have $\mu_{h_A} = 1_A \mu_h$, which is carried by $A \cap \Delta_1(U)$. Similarly, $\mu_u = \mu_h - \mu_{h_A} = 1_{U \setminus A} \mu_h$ is carried by $\Delta_1(U) \setminus A$. In particular, μ_u and μ_h are mutually singular. It follows by Corollary 3.3.6 that $h_A/h \leq h_A/u$ and $u/h = 1 - h_A/h \leq u/h_A$ have minimal fine limit 0 μ_h-a.e. on A and on $\Delta_1(U) \setminus A$, respectively, whence the assertion. □

Definition 3.3.8 For any function $h \in \mathcal{H}_i(U) \setminus \{0\}$ and any μ_h-integrable function f on $\Delta_1(U)$ we define

$$u_{f,h}(x) = \int K(x, Y) f(Y) d\mu_h(Y) \quad \text{for } x \in U.$$

The function $u_{f,h}$ is invariant according to Theorem 3.2.24.

Theorem 3.3.9 *Let $h \in \mathcal{H}_i(U) \setminus \{0\}$ and let f be a μ_h-integrable function on $\Delta(U)$. Then $u_{f,h}/h$ has minimal-fine limit $f(Y)$ at μ_h-a.e. point Y of $\Delta_1(U)$.*

Proof We may assume that $f \geq 0$. The case where f is a (Borel) step function follows easily from Corollary 3.3.7. For the general case we refer to the proof of [AG, Theorem 9.4.5], which carries over entirely. □

We are now prepared to prove Theorem 3.3.4, the Fatou-Naïm-Doob theorem in our setting.

(*Proof of Theorem* 3.3.4). Write $v = p + h$ with $p \in \mathcal{P}(U)$ and $h \in \mathcal{H}_i(U)$. By our definition of μ_v we then have $\mu_v = \mu_h$. We may assume that $h \neq 0$, for otherwise $\mu_v = 0$ and the assertion becomes trivial. Let ν be the singular component of μ_u with respect to $\mu_v = \mu_h$. Write $u = q + k$ with $q \in \mathcal{P}(U)$ and $k \in \mathcal{H}_i(U)$. Then $\mu_u = \mu_k = f\mu_h + \nu$, and hence in view of Definition 3.3.8

$$u = q + u_{f,h} + \int K(\cdot, Y)d\nu(Y).$$

By applying Theorem 3.3.5 with u replaced by q (hence μ_u by $\mu_q = 0$), next Corollary 3.3.6 with u replaced by $K\nu$ (hence μ_h replaced by ν and $d\mu_u/d\mu_h$ by $d\nu/d\mu_h = 0$), and finally by applying Theorem 3.3.9 to the present $u_{f,h}$, we see that u/h has minimal-fine limit $f(Y)$ at μ_v-a.e. point Y of $\Delta_1(U)$. Since u/v is defined quasi-everywhere in U and

$$\frac{u}{v} = \frac{u/h}{1 + p/h},$$

the theorem now follows by applying Theorem 3.3.5 with u there replaced by p (and hence μ_u by the present $\mu_p = 0$). □

3.4 Sweeping on Subsets of the Martin Space of a Fine Domain

In the present section, we extend the notions of reduction R_u^A and sweeping $\widehat{R}_u^A$ of a function $u \in \mathcal{S}_+(U)$ onto a set A to the case where A is any subset of the Riesz-Martin space $\overline{U}$. In addition to the previously considered particular case $A \subset U$ there is the other important case where A is a subset of the Martin boundary $\Delta(U)$ of U. Most properties of sweeping onto subsets of U are preserved by this extension.

We shall need an ad hoc the following concept of a (fine) Perron family.

Definition 3.4.1 A non-void lower directed family $\mathcal{F} \subset \mathcal{S}_+(U)$ is called a (fine) Perron family if $\widehat{R}_u^{U \setminus V_k} \in \mathcal{F}$ for every k and every $u \in \mathcal{F}$, where the V_k are the finely open sets given by Lemma 3.2.14.

Theorem 3.4.2 *If $\mathcal{F} \subset \mathcal{S}_+(U)$ is a Perron family then $\widehat{\inf}\,\mathcal{F}$ is an invariant function, and $\widehat{\inf}\,\mathcal{F} = \inf\,\mathcal{F}$ on $\{\inf\,\mathcal{F} < +\infty\}$.*

Proof Fix k. Clearly

$$\inf\,\mathcal{F} = \inf\{\widehat{R}_u^{U \setminus V_k} : u \in \mathcal{F}\},$$

and the family $\{\widehat{R}_u^{U \setminus V_k} : u \in \mathcal{F}\}$ is lower directed in U. For fixed k, there are infinitely many $k_j > k$ such that $V_k \subset V_{k_j}$. By Lemma 2.9.7 and Theorem 2.9.12(a) each $\widehat{R}_u^{U \setminus V_{k_j}}$ is invariant in V_k, and so is therefore $\widehat{\inf}\,\mathcal{F}|_{V_k}$ according to Theorem 2.9.12(c), and we have $\widehat{\inf}\mathcal{F} = \inf\mathcal{F}$ on $\{\inf\mathcal{F} < +\infty\} \cap V_k$ again by Theorem 2.9.12(c) applied to the family $\{\widehat{R}_u^{U \setminus V_{k_j}}|_{V_k} : u \in \mathcal{F}\}$ on V_k. Consequently, $\widehat{\inf}\,\mathcal{F}$ is likewise invariant, by Theorem 2.9.12(b) and $\widehat{\inf}\mathcal{F} = \inf\mathcal{F}$ on $\{\inf\mathcal{F} < +\infty\}$. □

We are now prepared to study sweeping on $\overline{U}$, following in part the classical procedure, cf. [Do4], [AG, Section 8.2], the main deviations being caused by the possible non-compactness of $\Delta(U)$.

Definition 3.4.3 Let $A \subset \overline{U}$. For any function $u \in \mathcal{S}_+(U)$ the reduction of u on A is defined by

$$R_u^A = \inf\{v \in \mathcal{S}_+(U) : v \geqslant u \text{ on } A \cap U \text{ and on } W \cap U \text{ for some } W \in \mathcal{W}(A)\},$$

where $\mathcal{W}(A)$ denotes the family of all open sets $W \subset \overline{U}$ with the natural topology such that $W \supset A \cap \Delta(U)$. The sweeping $\widehat{R}_u^A$ of u on A is defined as the regularization of R_u^A, that is, the greatest finely l.s.c. minorant of R_u^A.

Thus $\widehat{R}_u^A$ is of class $\mathcal{S}_+(U)$. It is convenient to express R_u^A and $\widehat{R}_u^A$ in terms of reduction and sweeping on subsets of U, cf. [Do4, 1.III.5]:

$$R_u^A = \inf\{R_u^{(A \cup W) \cap U} : W \in \mathcal{W}(A)\},$$

$$\widehat{R}_u^A = \widehat{\inf}\,\{\widehat{R}_u^{(A \cup W) \cap U} : W \in \mathcal{W}(A)\}.$$

In particular, for any subset A of U, the present reduction R_u^A and sweeping $\widehat{R}_u^A$ relative to $\overline{U}$ reduce to the similarly denoted usual reduction and sweeping on A relative to U. Note that if $A \subset \Delta(U)$ we may replace $A \cup W$ by W in the above expressions for R_u^A and $\widehat{R}_u^A$.

By the "fine" convergence theorem (Theorem 2.1.34) and the quasi-Lindelöf property for finely l.s.c. functions (cf. Section 2, paragraph 8 of the text following Corollary 1.2.5 or [Do3] for finely l.s.c. functions), there is a decreasing sequence (W_j) of sets $W_j \in \mathcal{W}(A)$ (depending on u) such that it suffices to take for W the sets W_j, in the above definitions and alternative expressions.

Remark 3.4.4 If $A \subset \Delta(U)$ then $\mathcal{W}(A)$ is the family of all open sets $W \subset \overline{U}$ containing A, and it then suffices to take for W a decreasing sequence of open sets $W_j \supset A$ (depending on u) such that $\bigcap_j \overline{W}_j \subset \overline{A}$. In fact, $\overline{A}$ is the intersection of a decreasing sequence of open sets $V_j \subset U$ such that $\overline{V}_j \subset V_{j+1}$ for every j, and we merely have to replace the above (W_j) by the decreasing sequence of open sets $W_j \cap V_j \in \mathcal{W}(A)$ whose intersection clearly is contained in $\overline{A}$. If A is a compact subset of $\Delta(U)$ we may therefore take $W_j = V_j$ (independently of u).

Theorem 3.4.5 *Let A and B be two subsets of $\overline{U}$ and let $u, v \in \mathcal{S}_+(U)$ and $0 \leq \alpha < +\infty$. Then*

(1) $\widehat{R}_u^{A\cup B} \leq \widehat{R}_u^A + \widehat{R}_u^B$.

(2) *If $A \subset B$ then $\widehat{R}_u^A \leqslant \widehat{R}_u^B$.*

(3) *If 0 times $+\infty$ is defined to be 0 then $\widehat{R}_{\alpha u}^A = \alpha \widehat{R}_u^A$.*

(4) $\widehat{R}_{u+v}^A = \widehat{R}_u^A + \widehat{R}_v^A$.

(5) *For any decreasing sequence of functions $u_j \in \mathcal{S}_+(U)$ we have $\widehat{\inf}_j \widehat{R}_{u_j}^A = \widehat{R}_{\widehat{\inf}_j u_j}^A$.*

(6) *If $A \subset B$ then $\widehat{R}_{\widehat{R}_u^A}^B = \widehat{R}_{\widehat{R}_u^B}^A = \widehat{R}_u^A$.*

Proof Property (1) is established just as in [Do4, (4.1), p. 39] (with $v = 0$): For $u_A \in \mathcal{S}_+(U)$ with $u_A \geqslant u$ on $(A \cup W_A) \cap U$ for some $W_A \in \mathcal{W}(A)$ and analogous u_B, W_B we have $u_A + u_B \in \mathcal{S}_+(U)$ and $u_A + u_B \geqslant u$ on $(A \cup B \cup W_A \cup W_B) \cap U$ with $W_A \cup W_B \in \mathcal{W}(A \cup B)$. This implies $R_u^{A\cup B} \leqslant R_u^A + R_u^B$. The asserted inequality therefore holds quasi-everywhere and hence everywhere on U, by fine continuity. Property (2) follows from $\mathcal{W}(A) \supset \mathcal{W}(B)$. Property (3) follows from $\widehat{R}_{\alpha u}^{(A\cup W)\cap U} = \alpha \widehat{R}_u^{(A\cup W)\cap U}$ by taking $\widehat{\inf}$ over $W \in \mathcal{W}(A)$. As to (4) we have $\widehat{R}_{u+v}^{(A\cup W)\cap U} = \widehat{R}_u^{(A\cup W)\cap U} + \widehat{R}_v^{(A\cup W)\cap U}$ for any $W \in \mathcal{W}(A)$, whence the asserted equation by taking the natural limits of the decreasing nets on $\mathcal{S}_+(U)$ in question as the index W ranges over the lower directed family $\mathcal{W}(A)$, cf. Theorem 3.1.3.

Concerning (5), according to Theorems 1.4.44 and 2.3.11, u_j is Euclidean Borel measurable and

$$\widehat{R}_{u_j}^{(A\cup W)\cap U}(x) = \int_U u_j \, d\varepsilon_x^{\complement(U\setminus(A\cup W))} \leqslant u_j(x) \leqslant u_1(x)$$

(complements relative to Ω). For given $x \in U$ with $u_1(x) < +\infty$ and $W \in \mathcal{W}(A)$ consider the equalities

$$\begin{aligned} \inf_j \widehat{R}_{u_j}^{(A\cup W)\cap U}(x) &= \inf_j \int_U u_j \, d\varepsilon_x^{\complement(U\setminus(A\cup W))} = \int_U \inf_j u_j \, d\varepsilon_x^{\complement(U\setminus(A\cup W))} \\ &= \int_U \widehat{\inf_j} u_j \, d\varepsilon_x^{\complement(U\setminus(A\cup W))} = \widehat{R}_{\widehat{\inf}_j u_j}^{(A\cup W)\cap U}(x). \end{aligned}$$

The first and the last equalities hold by Theorem 2.3.11. The second equality is obvious (Lebesgue), the integrals being finite by hypothesis. The third equality holds if $\widehat{\inf}_j u_j(x) = \inf_j u_j(x)$, for either $x \in U \cap \mathfrak{b}((A \cup W) \cap U)$, and then

$\varepsilon_x^{\complement(U\setminus(A\cup W))} = \varepsilon_x$; or else $x \in U \setminus b((A \cup W) \cap U)$, and then $\varepsilon_x^{\complement(U\setminus(A\cup W))}$ does not charge the polar set $\{\inf_j u_j \neq \widehat{\inf}_j u_j\}$ (cf. Corollary 1.4.33). The resulting equality in the above display thus holds q.e. for $x \in U$, and hence also everywhere on U after finely l.s.c. regularization of both members.

Property (6) is known for $A, B \subset U$, see Theorem 2.2.16. For general $A, B \subset \overline{U}$, say with $A \subset B$, the first and the second member of the equalities in (6) lie between $\widehat{R}^A_{\widehat{R}^A_u}$ and $\widehat{R}^A_u$ in view of (2), and it therefore suffices to consider the case where $A = B$. For given $W \in \mathcal{W}(A)$ consider a decreasing sequence $(W_j) \subset \mathcal{W}(A)$ such that $\widehat{R}^A_u = \widehat{\inf}_j \widehat{R}_u^{(A\cup W_j)\cap U}$. Replacing W_j by $W_j \cap W$ we achieve that $W_j \subset W$, and hence

$$\widehat{R}^{(A\cup W)\cap U}_{\widehat{R}_u^{(A\cup W_j)\cap U}} = \widehat{R}_u^{(A\cup W_j)\cap U}$$

by Theorem 2.2.16. According to (5) this implies (6) by taking $\widehat{\inf}_j$ and next taking $\widehat{\inf}_{W\in\mathcal{W}(A)}$. □

Theorem 3.4.6 *Let $u \in \mathcal{S}_+(U)$. For any subset A of $\Delta(U)$ the function $\widehat{R}^A_u$ is invariant, and we have $\widehat{R}^A_u \preccurlyeq u$.*

Proof Consider the family

$$\mathcal{F} := \{\widehat{R}_u^{W\cap U} : W \in \mathcal{W}(A)\}.$$

Clearly, $\mathcal{F}$ is lower directed. Consider the compact sets $A_{k,l} \subset U$ in the proof of Theorem 3.1.8. Fix k and V_k from Lemma 3.2.14. In view of that lemma and the text preceding it, $\overline{U} \setminus A_{k,k}$ is open in $\overline{U}$ and contains $\Delta(U)$. In Definition 3.4.3 of $\widehat{R}^A_u$ it therefore suffices to consider open sets $W \supset A$ such that $W \subset \overline{U} \setminus A_{k,k}$, whereby $W \cap U \subset U \setminus A_{k,k} \subset U \setminus V_k$. By (6) in Theorem 3.4.5 we then have $\widehat{R}^{U\setminus V_k}_{\widehat{R}_u^{W\cap U}} = \widehat{R}_u^{W\cap U}$. The lower directed family $\mathcal{F} := \{\widehat{R}_u^{W\cap U} : W \in \mathcal{W}(A)\}$ is therefore a Perron family in the sense of Definition 3.4.1. By Definition 3.4.3 we have $\widehat{R}^A_u = \widehat{\inf}\,\mathcal{F}$, and it therefore follows by Theorem 3.4.2 that $\widehat{R}^A_u$ indeed is invariant. Consequently, $\widehat{R}^A_u \preccurlyeq u$ in view of Lemma 2.9.4. □

Theorem 3.4.7 *Let $u \in \mathcal{S}_+(U)$.* (a) *For any increasing sequence (A_j) of subsets of $\overline{U}$ we have $\widehat{R}_u^{\bigcup_j A_j} = \sup_j \widehat{R}_u^{A_j}$.*

(b) *For any sequence (A_j) of subsets of $\overline{U}$ we have $\widehat{R}_u^{\bigcup_j A_j} \leqslant \sum_j \widehat{R}_u^{A_j}$.*

Proof For (a) we proceed much as in [Do4, p. 74, Proof of (e)] (where U is a Euclidean Green domain). Writing $A = \bigcup_j A_j$ and $v = \sup_j \widehat{R}_u^{A_j}$ the inequality $v \leqslant \widehat{R}^A_u$ is obvious. For the opposite inequality we shall also consider $\widehat{R}_u^{A_j\cap\Delta(U)}$. Consider a point $x \in U$ for which $u(x) < +\infty$ and $\widehat{R}_u^{A_j\cap\Delta(U)}(x) = R_u^{A_j\cap\Delta(U)}(x)$. For any integer $j > 0$ there exists $W_j \in \mathcal{W}(A_j \cap \Delta(U)) = \mathcal{W}(A_j)$ and $v_j \in \mathcal{S}_+(U)$ such that $v_j \geqslant u$ on $W_j \cap U$ and

$$v_j(x) \leqslant R_u^{A_j\cap\Delta(U)}(x) + 2^{-j} = \widehat{R}_u^{A_j\cap\Delta(U)}(x) + 2^{-j}.$$

The swept function $\widehat{R}_u^{A_j \cap \Delta(U)}$ is invariant by Theorem 3.4.6, and $\widehat{R}_u^{A_j \cap \Delta(U)} \leqslant \widehat{R}_u^{W_j \cap U} \leqslant R_u^{W_j \cap U} \leqslant v_j$. Hence $\widehat{R}_u^{A_j \cap \Delta(U)} \preccurlyeq v_j$. For any integer $k > 0$ the function

$$u'_k := v + \sum_{j \geqslant k} (v_j - \widehat{R}_u^{A_j \cap \Delta(U)}) \tag{3.4.1}$$

is finely hyperharmonic on U by Corollary 2.1.21 and we have

$$u'_k(x) = v(x) + \sum_{j \geq k} (v_j(x) - R_u^{A_j \cap \Delta(U)}(x)) \leq v(x) + 2^{1-k} < +\infty,$$

so that u'_k is of class $\mathcal{S}_+(U)$. For any $W \in \mathcal{W}(A_j)$ we have $\widehat{R}_u^{(A_j \cup W) \cap U} = u$ q.e. on $(A_j \cup W) \cap U$, in particular q.e. on $A_j \cap U$. By Definition 3.4.3 we have $\widehat{R}_u^{A_j} = u$ q.e. on $A_j \cap U$ (because it suffices to consider a suitable sequence of sets W). It follows that $v = u$ q.e. on each $A_j \cap U$ and hence also q.e. on $A \cap U$. Choose a superharmonic function $s > 0$ on Ω such that $s(y) = +\infty$ for every y in the polar set $\{y \in A \cap U : v(y) \neq u(y)\} \cup \bigcup_{j>0} \{y \in A \cap U : \widehat{R}_u^{A_j \cap \Delta(U)} \neq R_u^{A_j \cap \Delta(U)}\}$. For any $\delta > 0$ we then have $u'_k + \delta s \geqslant v + \delta s \geqslant u$ on $A \cap U$. Because $v \geqslant \widehat{R}_u^{A_j} \geqslant \widehat{R}_u^{A_j \cap \Delta(U)}$ we obtain for $j \geqslant k$

$$u'_k \geqslant \widehat{R}_u^{A_j \cap \Delta(U)} + (v_j - \widehat{R}_u^{A_j \cap \Delta(U)}) = v_j,$$

and hence $u'_k \geqslant v_j \geqslant u$ on $W_j \cap U$ for $j \geqslant k$. Altogether, $u'_k + \delta s \geqslant u$ on $A \cap U$ and on $W_j \cap U$. It follows by Definition 3.4.3 that $u'_k + \delta s \geqslant \widehat{R}_u^A$, and hence for $\delta \to 0$ that $u'_k \geqslant \widehat{R}_u^A$ q.e. and actually everywhere on U. But $u'_k \downarrow v$ q.e. (namely at each point where u'_1 is finite). Consequently $v \geqslant \widehat{R}_u^A$ q.e. on U and so indeed everywhere on U.

(b) is easily deduced from (a) applied with A_j replaced by $A_1 \cup \ldots \cup A_j$, in view of (1) in Theorem 3.4.5 (extended to finite unions). There is also a simple direct proof, cf. [AG, Lemma 8.2.2 (i)] for the case of a Euclidean Green domain U. □

Theorem 3.4.8 *For any $A \subset \overline{U}$ we have $\widehat{R}_u^A = \widehat{R}_u^{A \cap \Delta(U)} + \widehat{R}_v^{A \cap U}$, where $v := u - \widehat{R}_u^{A \cap \Delta(U)}$.*

Proof Let $s \in \mathcal{S}_+(U)$, $s \geqslant u$ on $A \cap U$ and on a neighborhood of $A \cap \Delta(U)$. Then $s \geqslant \widehat{R}_u^{A \cap \Delta(U)}$, which is invariant by Theorem 3.4.6, and so $\widehat{R}_u^{A \cap \Delta(U)} \preccurlyeq s$. Furthermore, $s - \widehat{R}_u^{A \cap \Delta(U)} \geqslant u - \widehat{R}_u^{A \cap \Delta(U)} = v$ on $A \cap U$. It follows that $s - \widehat{R}_u^{A \cap \Delta(U)} \geqslant \widehat{R}_v^{A \cap U}$, and so $s \geqslant \widehat{R}_u^{A \cap \Delta(U)} + \widehat{R}_v^{A \cap U}$. This shows that $\widehat{R}_u^A \geqslant \widehat{R}_u^{A \cap \Delta(U)} + \widehat{R}_v^{A \cap U}$. For the opposite inequality let $w \in \mathcal{S}_+(U)$, $w \geqslant u$ on $A \cap U$ and on a neighborhood of $A \cap \Delta(U)$. Then $w \geqslant \widehat{R}_u^{A \cap \Delta(U)}$ on $A \cap U$, and since $\widehat{R}_u^{A \cap \Delta(U)}$ is invariant as noted above, we have $w - \widehat{R}_u^{A \cap \Delta(U)} \in \mathcal{S}_+(U)$. By hypothesis this function majorizes v on $A \cap U$, and we therefore get $w \geqslant \widehat{R}_u^{A \cap \Delta(U)} + \widehat{R}_v^{A \cap U}$. By varying w this leads to the remaining inequality $\widehat{R}_u^A \geqslant \widehat{R}_u^{A \cap \Delta(U)} + \widehat{R}_v^{A \cap U}$. □

We say that a (positive Radon) measure μ on $\overline{U}$ represents a function $u \in \mathcal{S}_+(U)$ if $u = K\mu$.

Corollary 3.4.9 *Let H be a compact subset of $\overline{U}$ and μ a Radon measure on $\overline{U}$ carried by H. Then $K\mu$ is invariant on $U \setminus H$.*

Proof By Theorem 3.4.8 there is a function $v \in \mathcal{S}_+(U)$ such that $\widehat{R}^H_{K\mu} = \widehat{R}^{H\cap\Delta(U)}_{K\mu} + \widehat{R}^{H\cap U}_v$. By Theorem 3.4.6 the former term on the right is invariant, and the latter term is invariant on $U \setminus H$ according to Lemma 2.9.7. □

Corollary 3.4.10 *Let $A \subset \overline{U}$ and $u \in \mathcal{S}_+(U)$. Then $\widehat{R}^A_u$ from Definition 3.4.3 is invariant on $U \setminus \overline{A}$.*

Proof For $A \subset U$ this follows from Lemma 2.9.7, and for $A \subset \Delta(U)$ it follows from Theorem 3.4.6. For general $A \subset \overline{U}$ it therefore follows right away by application of Theorem 3.4.8. □

Theorem 3.4.11 *Let $A \subset \overline{U}$ and $u \in \mathcal{S}_+(U)$. Then there exists a measure on $\overline{U}$ representing $\widehat{R}^A_u$ and carried by $\overline{A}$.*

Proof We may suppose that $\widehat{R}^A_u \neq 0$, in particular $u > 0$, the case $\widehat{R}^A_u = 0$ being trivial. For any probability measure ν on $\mathbf{B}$ we denote in this proof by $b(\nu)$ the barycenter of ν.

Suppose first that $A \subset U$. Let p be a fine potential > 0 on U. For any natural number $k \geq 1$ there exists a non-zero Radon measure σ_k on $\overline{U}$ representing the fine potential $\widehat{R}^A_{u\wedge kp} > 0$ on U, and σ_k is carried by U according to Theorem 3.2.22 and Corollary 3.2.24. In view of equations (3.2.1) and (3.2.2) we have

$$\widehat{R}^A_{u\wedge kp} = K\sigma_k = \int_U K(\cdot, y)d\sigma_k(y) = \int_U G_U(\cdot, y)d\tau_k(y)$$

with $d\tau_k(y) = \Phi(G_U(\cdot, y))^{-1}d\sigma_k(y)$, see paragraph preceding Theorem 3.2.17. By Corollary 3.4.10 $\widehat{R}^A_{u\wedge kp}$ is invariant on $U \setminus \overline{A}$, and hence τ_k is carried by $\overline{A}$ according to Theorem 2.9.11. It follows that σ_k likewise is carried by $\overline{A}$.

Consider for each k the probability measure ν_k on the chosen compact base $\mathbf{B}$ of the cone $\mathcal{S}_+(U)$, defined by $\nu_k(E) = \sigma_k(E \cap U)/\sigma_k(U)$ for any Borel subset E of $\overline{U}$. Clearly, ν_k is carried by $\overline{A}$ along with σ_k. The sequence (ν_k) has a subsequence (ν_{k_j}) which converges vaguely to a probability measure ν on $\overline{U}$, necessarily carried by $\overline{A}$. On the other hand, $\widehat{R}^A_{u\wedge kp} \to \widehat{R}^A_u$ pointwise and increasingly for $k \to +\infty$. It follows by Theorem 2.2.15(a) and Theorem 3.1.3 that $\widehat{R}^A_{u\wedge kp} \to \widehat{R}^A_u$ in the natural topology on $\mathcal{S}_+(U)$ as $k \to +\infty$, and hence $\Phi(\widehat{R}^A_{u\wedge kp}) \to \Phi(\widehat{R}^A_u) \in]0, +\infty[$ because Φ is naturally continuous on $\mathcal{S}_+(U)$. Identifying as usual ν_k and ν with probability measures on $\mathbf{B}$ we infer that

$$\frac{1}{\Phi(\widehat{R}^A_u)}\widehat{R}^A_u = \lim_{j\to\infty} \frac{1}{\Phi(\widehat{R}^A_{u\wedge k_j p})}\widehat{R}^A_{u\wedge k_j p} = \lim_{j\to\infty} b(\nu_{k_j}) = b(\nu) = K\nu,$$

the third equality holds by [Ph, p. 1 and Proposition 1.1]. Hence $\widehat{R}_u^A = K\mu$, where $\mu := \Phi(\widehat{R}_u^A)\nu$ (now again considered as a measure on $\overline{U}$) is carried by $\overline{A}$ along with ν.

Next, let $A \subset \Delta(U)$. According to Remark 3.4.4 there is a decreasing sequence of open sets W_j (depending on u) such that $A \subset \bigcap_j W_j \subset \bigcap_j \overline{W}_j \subset \overline{A}$ and $\widehat{R}_u^A = \widehat{\inf}_j \widehat{R}_u^{W_j \cap U} = \lim_j \widehat{R}_u^{W_j \cap U}$ (natural limit, again by Theorem 3.1.3). There is a sequence of reals $\alpha_j > 0$ and a real $\alpha > 0$ such that $\alpha_j \widehat{R}_u^{W_j \cap U} \in \mathbf{B}$ and $\alpha \widehat{R}_u^A \in \mathbf{B}$. The sequence (α_j) converges to α because the sequence $(\widehat{R}_u^{W_j \cap U})$ converges naturally to $\widehat{R}_u^A$. For any index j there exists, as shown in the preceding paragraph, a probability measure μ_j on $\mathbf{B}$ with the barycenter $\alpha_j \widehat{R}_u^{W_j \cap U}$ such that μ_j (when viewed as a measure on $\overline{U}$) is carried by $\overline{W}_j$. After passing to a subsequence we may suppose that μ_j converges to a probability measure μ on $\mathbf{B}$ which (again when viewed as a measure on $\overline{U}$) necessarily is carried by $\bigcap_j \overline{W}_j \subset \overline{A}$. The sequence $(b(\mu_j)) = (\alpha_j \widehat{R}_u^{W_j})$ of barycenters of the μ_j therefore converges to the barycenter $b(\mu)$ of μ, see [Ph, p. 1 and Proposition 1.1], whence $K\mu = b(\mu) = \alpha \widehat{R}_u^A$, and $\widehat{R}_u^A$ is represented by the measure $\frac{1}{\alpha}\mu$ carried by $\overline{A}$.

In the general case where just $A \subset \overline{U}$ we have by Theorem 3.4.8 $\widehat{R}_u^A = \widehat{R}_u^{A\cap\Delta(U)} + \widehat{R}_v^{A\cap U}$, where $v \in \mathcal{S}_+(U)$. As shown in the third paragraph of the present proof there exists a measure μ_1 on $\overline{U}$ representing $\widehat{R}_v^{A\cap U}$ and carried by $\overline{A}$. And as shown in the preceding paragraph there exists a measure μ_2 on $\overline{U}$ representing $\widehat{R}_u^{A\cap\Delta(U)}$ and likewise carried by $\overline{A}$. The measure $\mu = \mu_1 + \mu_2$ therefore represents $\widehat{R}_u^A$ and is carried by $\overline{A}$. □

Theorem 3.4.12 *Let $A \subset \overline{U}$. Then*

(i) *If $A \subset \Delta(U)$ then $\widehat{R}_p^A = 0$ for any $p \in \mathcal{P}(U)$.*

(ii) *If $A \subset \Delta(U)$ and $Y \in U \cup \Delta_1(U)$ then either $\widehat{R}_{K(\cdot,Y)}^A = 0$ or $\widehat{R}_{K(\cdot,Y)}^A = K(\cdot, Y)$. If moreover $Y \notin A$ then $\widehat{R}_{K(\cdot,Y)}^A = 0$.*

(iii) *If $Y \in \Delta_1(U) \setminus \overline{A}$ then $\widehat{R}_{K(\cdot,Y)}^A \neq K(\cdot, Y)$.*

Proof It follows from Theorem 3.4.6 that $\widehat{R}_p^A$ and $\widehat{R}_{K(\cdot,Y)}^A$ are invariant if $A \subset \Delta(U)$. This establishes (i) because $\widehat{R}_p^A$ is also a fine potential (along with p).

For the former assertion (ii) we have $\widehat{R}_{K(\cdot,Y)}^A \preccurlyeq K(\cdot, Y)$, again by Theorem 3.4.6, and since $K(\cdot, Y)$ is extreme there is a constant $c \geqslant 0$ such that $\widehat{R}_{K(\cdot,Y)}^A = cK(\cdot, Y)$. Hence it follows from (6) in Theorem 3.4.5 with $A = B$ that $c = 0$ or $c = 1$.

For the latter assertion (ii) suppose first that $Y \notin \overline{A}$ (the natural closure of A in $\overline{U}$). Suppose by contradiction that $\widehat{R}_{K(\cdot,Y)}^A = K(\cdot, Y)$. According to Theorem 3.4.11 there exists a measure λ on $\overline{U}$ carried by $\overline{A}$ such that

$$\widehat{R}_{K(\cdot,Y)}^A = \int_{\overline{U}} K(\cdot, Z)d\lambda(Z). \tag{3.4.2}$$

It follows that $K(\cdot, Y) = \int_{\overline{U}} K(\cdot, Z)d\lambda(Z)$, and so λ is a probability measure. Denote μ the probability measure on $\mathbf{B}$ defined by $\mu(E) = \lambda(E \cap \overline{U})$ for every Borel set

$E \subset \mathbf{B}$. Then μ has the barycenter $K(\cdot, Y)$. Since $K(\cdot, Y)$ is an extreme point of $\mathbf{B}$ we infer by [Al, Corollary I.2.4, p. 15] that $\mu = \varepsilon_Y$, and hence $Y \in \overline{A}$, which is contradictory. Thus $\widehat{R}^A_{K(\cdot,Y)} \neq K(\cdot, Y)$, and consequently $\widehat{R}^A_{K(\cdot,Y)} = 0$ according to the former assertion (ii) above.

It remains to consider the case where we just have $Y \notin A$. In this case A may be written as the union of an increasing sequence of subsets A_j of A with $Y \notin \overline{A}_j$ for any j. By Theorem 3.4.7(a) we then have $\widehat{R}^A_{K(\cdot,Y)} = \sup_j \widehat{R}^{A_j}_{K(\cdot,Y)} = 0$ and hence $\widehat{R}^A_{K(\cdot,Y)} = 0$, as claimed.

For (iii), suppose by contradiction that $\widehat{R}^A_{K(\cdot,Y)} = K(\cdot, Y)$. Again, there exists by Theorem 3.4.11 a probability measure λ on $\overline{U}$ carried by $\overline{A}$ such that (3.4.2) holds, and hence $\lambda = \varepsilon_Y$ as in the proof of the latter assertion of (ii) because $K(\cdot, Y) \in \mathrm{Ext}(\mathbf{B})$. It then follows that $Y \in \overline{A}$, which is contradictory. □

Actually, in Theorem 3.4.12(ii), if $Y \in A$ and hence $Y \notin U$ then $Y \in \Delta_1(U)$, and it follows that $\widehat{R}^A_{K(\cdot,Y)} = K(\cdot, Y)$, see Theorem 3.5.6 below.

The following result extends (4) in Theorem 3.4.5 to infinite sums.

Theorem 3.4.13 *Let $A \subset \overline{U}$. Let (μ_j) be a sequence of measures on $\overline{U}$ such that $\sum_j \int d\mu_j < +\infty$, and let $\mu = \sum_j \mu_j$. Then*

$$\widehat{R}^A_{K\mu} = \sum_j \widehat{R}^A_{K\mu_j}.$$

Proof Since $\mu(\overline{U}) < +\infty$, it follows that $K\mu \in \mathcal{S}_+(U)$, see paragraph following the proof of Theorem 3.1.8. If $A \subset U$ the stated equality holds by Theorem 2.2.15(a) and Corollary 2.2.19(b). Next, let $A \subset \Delta(U)$. The inequality "$\geq$" being trivial, let us prove the opposite inequality. Suppose first that $\widehat{R}^A_{K\mu} = K\mu$. For integers $k > 0$ we have by (4) in Theorem 3.4.5 (extended to sums of finitely many measures)

$$\widehat{R}^A_{K\mu} = \sum_{j \leqslant k} \widehat{R}^A_{K\mu_j} + \widehat{R}^A_{K(\sum_{j>k}\mu_j)}.$$

Since $\sum_j K\mu_j = K\mu < +\infty$ q.e. we have

$$\widehat{R}^A_{K(\sum_{j>k}\mu_j)} \leqslant K\sum_{j>k}\mu_j = \sum_{j>k} K\mu_j \downarrow 0$$

q.e. as $k \to \infty$. We thus have $\widehat{R}^A_{K\mu} \leqslant \sum_{j>0} \widehat{R}^A_{K\mu_j}$ with equality q.e., and indeed everywhere, both members of the inequality being of class $\mathcal{S}_+(U)$.

Without the temporary hypothesis $\widehat{R}^A_{K\mu} = K\mu$, we have $\widehat{R}^A_{K\mu} \preccurlyeq K\mu$ because $\widehat{R}^A_{K\mu}$ is invariant according to Theorem 3.4.6. Thus there exists a measure ν on $\overline{U}$ with $\nu \leqslant \mu$ and hence $K\nu \in \mathcal{S}_+(U)$ such that $K\mu = \widehat{R}^A_{K\mu} + K\nu$. After sweeping on A while invoking (6) in Theorem 3.4.5 we obtain $\widehat{R}^A_{K\nu} = 0$. Thus $K(\mu - \nu) = \widehat{R}^A_{K(\mu-\nu)} = \widehat{R}^A_{K\mu}$. Similarly, $K\mu_j = \widehat{R}^A_{K\mu_j} + K\nu_j$ with $\nu_j \leqslant \mu_j$ and $K(\mu_j - \nu_j) = \widehat{R}^A_{K\mu_j}$. As shown above it follows that $\widehat{R}^A_{K(\mu-\nu)} = \sum_j \widehat{R}^A_{K(\mu_j-\nu_j)}$ and hence

$$\widehat{R}^A_{K\mu} = \widehat{R}^A_{K(\mu-\nu)} = \sum_j \widehat{R}^A_{K(\mu_j-\nu_j)} = \sum_j \widehat{R}^A_{K\mu_j}.$$

The general case $A \subset \overline{U}$ follows immediately by application of Theorem 3.4.8. □

Remark 3.4.14 Even in the classical case $U = \Omega$, sweeping (and reduction) of a function $u \in \mathcal{S}_+(U)$ on an arbitrary set $A \subset \Delta(U)$ lacks the following two properties, valid when $A \subset U$. Fix a point $Y \in A \cap \Delta_1(U)$ and note that $\widehat{R}^A_{K(\cdot,Y)} = K(\cdot, Y)$ by Theorem 3.5.6 below. In the classical case, $K(\cdot, Y) \wedge c$ is a potential (hence a fine potential) for any constant $c > 0$, as noted in [Do4, Observation, p. 74] for the purpose of showing that the following Property 1. fails when $A = \Delta(U)$:

1. For any increasing sequence of functions $u_j \in \mathcal{S}_+(U)$ with pointwise supremum $u \in \mathcal{S}_+(U)$, we should have $\widehat{R}^A_u = \sup_j \widehat{R}^A_{u_j}$. This holds when $A \subset U$, by Theorem 2.2.15, but fails (classically) for $A = \Delta(U)$ and $u_j = K(\cdot, Y) \wedge j$ in view of the above. It does hold, however, for any sequence $(u_j) \subset \mathcal{S}_+(U)$ which is increasing in the specific order; this is a reformulation of Theorem 3.4.13 above.

2. For any $x \in U$, the affine function $u \mapsto \widehat{R}^A_u(x)$ on $\mathcal{S}_+(U)$ should be (naturally) l.s.c., that is, for any sequence of functions $u_j \in \mathcal{S}_+(U)$ converging (naturally) to $u \in \mathcal{S}_+(U)$,

$$\liminf \widehat{R}^A_{u_j}(x) \geq \widehat{R}^A_u(x).$$

But Property 2. fails (classically) for $A = \Delta(U)$ and $u_j = K(\cdot, Y) \wedge jp$, where $p > 0$ is a fine potential on U, hence $u = K(\cdot, Y)$, in view of the above.

3.5 Minimal Fine Topology

We introduce the minimal-fine topology (mf-topology) on $\overline{U}$ associated with the minimal thinness of a subset of U at a point of $\Delta_1(U)$, see Definition 3.3.2. This mf-topology is finer than the natural topology on $\overline{U}$, and induces on U the fine topology there. Next we study the pole of a non-zero minimal invariant function on U. The main result in this section is Theorem 3.5.7 providing an integral representation of the sweeping of a function $u \in \mathcal{S}_+(U)$ onto a set $A \subset \overline{U}$, whereby u is expressed canonically as $u = K\mu$ in terms of a unique measure μ on $U \cup \Delta_1(U)$, cf. Theorem 3.2.22. This result seems to be new even in the case where U is a Euclidean domain.

Like in classical potential theory, we define the minimal-fine (mf-)topology on $\overline{U}$ as follows:

Definition 3.5.1 A set $W \subset \overline{U}$ is said to be a minimal-fine neighborhood of a point $Y \in \overline{U}$ if

(a) $W \cap U$ is a fine neighborhood of Y in the usual sense, in case $Y \in U$,
(b) W contains the point Y and $U \setminus W$ is minimal-thin at Y, in case $Y \in \Delta_1(U)$,
(c) W contains the point Y, in case $Y \in \Delta(U) \setminus \Delta_1(U)$.

In the sequel, we will denote by mf-lim inf the lim inf in the sense of the mf-topology.

According to (a) above, the mf-topology on $\overline{U}$ induces on U the fine topology there, and U is mf-open in $\overline{U}$, that is, $\Delta(U)$ is mf-closed in $\overline{U}$ (since $\varnothing$ is minimal-thin at any point $Y \in \Delta_1(U)$). According to (c), the mf-topology on $\overline{U}$ induces the discrete topology on $\Delta(U) \setminus \Delta_1(U)$. In view of (c), (b), and Definition 3.3.2, $\Delta_1(U)$ is the set of mf-limit points of U in $\overline{U}$.

Theorem 3.5.2 *The* mf*-topology on $\overline{U}$ is finer than the natural topology (and is therefore Hausdorff).*

Proof Let W be a natural open neighborhood of a point $Y \in \overline{U}$. If $Y \in U$ then $W \cap U$ is a usual fine neighborhood of Y in U according to Corollary 3.2.15, and hence an mf-neighborhood of Y in $\overline{U}$ by Definition 3.5.1(a) above. If $Y \in \Delta(U) \setminus \Delta_1(U)$ there is nothing to prove in view of (c) in that definition. In the remaining case where $Y \in \Delta_1(U)$ we show that $U \setminus W$ is minimal-thin at Y, cf. Definition 3.5.1(b), which means that $\widehat{R}^{U\setminus W}_{K(\cdot,Y)} \neq K(\cdot, Y)$. Suppose that, on the contrary, $\widehat{R}^{U\setminus W}_{K(\cdot,Y)} = K(\cdot, Y)$. Writing $U \setminus W = A$ we have $Y \in W \subset \complement\overline{A}$ (complement relative to $\overline{U}$) because W is (naturally) open. It follows by Theorem 3.4.12(iii) that $\widehat{R}^{A}_{K(\cdot,Y)} \neq K(\cdot, Y)$, which is contradictory. □

Definition 3.5.3 Let h be a non-zero minimal invariant function. A point $Y \in \overline{U}$ is termed a pole of h if $\widehat{R}^{\{Y\}}_h = h$.

Remark 3.5.4 Any pole Y of h belongs to $\Delta(U)$, for if $Y \in U$ then $\widehat{R}^{\{Y\}}_h = 0$ because $\{Y\}$ is polar.

Theorem 3.5.5 *Every non-zero minimal invariant function h on U has precisely one pole. For any $Y \in \Delta_1(U)$ the pole of $K(\cdot, Y)$ is Y.*

Proof We may assume without loss of generality that $h \in \mathbf{B}$. Recall from Theorem 3.2.6 and Theorem 3.2.18 that h is of the form $K(\cdot, Y)$ for a (unique) $Y \in \Delta_1(U)$. Consider the family $\mathcal{C}$ of all (necessarily non-void) compact subsets C of $\overline{U}$ such that $\widehat{R}^{C}_{K(\cdot,Y)} = K(\cdot, Y)$, and note that $\mathcal{C}$ is non-void, for $\overline{U} \in \mathcal{C}$ because $\widehat{R}^{\overline{U}}_{K(\cdot,Y)} = K(\cdot, Y)$, an easy consequence of Definition 3.4.3. Equip $\mathcal{C}$ with the order defined by the inverse inclusion "$\supset$". For any totally ordered subfamily $\mathcal{C}^*$ of $\mathcal{C}$ the intersection C^* of $\mathcal{C}^*$ satisfies $\widehat{R}^{C^*}_{K(\cdot,Y)} = K(\cdot, Y)$ in view of the fine fundamental convergence theorem (Theorem 2.1.34), and hence $\mathcal{C}$ has a minimal element C_0 according to Zorn's lemma. The natural topology is Hausdorff, so if C_0 contains two distinct points Z_1 and Z_2 then there are compact subsets C_1 and C_2 of C_0 such that $C_0 = C_1 \cup C_2$, $Z_1 \in C_0 \setminus C_2$

and $Z_2 \in C_0 \setminus C_1$. Since $K(\cdot, Y)$ is extreme it then follows by Property of natural decomposition that either $\widehat{R}^{C_1}_{K(\cdot,Y)} = K(\cdot, Y)$ or $\widehat{R}^{C_2}_{K(\cdot,Y)} = K(\cdot, Y)$. In other words, either C_1 or C_2 belongs to $\mathcal{C}$, say $C_1 \in \mathcal{C}$. By minimality of C_0 we would then have $C_1 = C_0$ which contradicts $Z_2 \in C_0 \setminus C_1$. This shows that indeed $C_0 = \{Z\} \in \mathcal{C}$ for a certain $Z \in \overline{U}$, that is, $\widehat{R}^{\{Z\}}_{K(\cdot,Y)} = K(\cdot, Y)$. Thus Z is a pole of $K(\cdot, Y)$. Since $\{Z\}$ is a closed set it follows by Theorem 3.4.11 that $\widehat{R}^{\{Z\}}_{K(\cdot,Y)} = K\mu$ for some probability measure μ on $\mathbf{B}$ carried by $\{Z\}$, that is, for $\mu = \varepsilon_Z$. Thus $K(\cdot, Y) = K\mu = K(\cdot, Z)$, and so indeed $Y = Z$. □

Theorem 3.5.6 *Let h be a non-zero minimal invariant function on U with pole $Y \in \Delta_1(U)$ and let $A \subset \Delta(U)$. Then $\widehat{R}^A_h = h$ or 0 if $Y \in A$ or $Y \notin A$, respectively.*

Proof It follows from Theorems 3.2.18 and 3.5.5 that $h = \alpha K(\cdot, Y)$ for some constant $\alpha > 0$. By Theorem 3.4.6 the function $\widehat{R}^A_h$ is invariant and $\preccurlyeq h$, hence of the form ch for some constant $c \geq 0$. It follows by (6) in Theorem 3.4.5 with $A = B$ that $c = 0$ or $c = 1$. If $A \subset \Delta(U)$ contains the pole Y of h then, according to Theorem 2.1.31, $h \succcurlyeq \widehat{R}^A_h \succcurlyeq \widehat{R}^{\{Y\}}_h = h$, and so $\widehat{R}^A_h = h$. The rest follows from Theorem 3.4.12(ii). □

The following integral representation of the sweeping of a function of class $\mathcal{S}_+(U)$ on arbitrary sets $A \subset \overline{U}$ is based on Theorem 3.5.6, which in turn depended on Theorem 3.4.6.

Theorem 3.5.7 *For any set $A \subset \overline{U}$ and any Radon measure μ on $\overline{U}$ carried by $U \cup \Delta_1(U)$ we have*

$$\widehat{R}^A_{K\mu} = \int^* \widehat{R}^A_{K(\cdot,Y)} d\mu(Y). \tag{3.5.1}$$

If A is μ-measurable then the upper integral becomes a true integral.

Proof For any subset A of U this integral representation was established in Lemma 3.2.26 with the upper integral replaced by the integral. For $A \subset \Delta(U)$ it suffices to consider the case where μ is carried by $\Delta_1(U)$, for if ν denotes the restriction of μ to U then $K\nu$ and $K(\cdot, Y)$ (for $Y \in U$) are fine potentials according to Corollary 3.2.24, and so $\widehat{R}^A_{K\nu} = \widehat{R}^A_{K(\cdot,Y)} = 0$ by Theorem 3.4.12(i).

Proof that the inequality "$\geq$" holds in (3.5.1) *for* $A \subset \Delta(U)$. We may assume that $\widehat{R}^A_{K\mu} \not\equiv +\infty$ or, equivalently, $K_\mu \not\equiv +\infty$, that is, $K_\mu \in \mathcal{S}_+(U)$ (and $\widehat{R}^A_{K\mu} \in \mathcal{S}_+(U)$). By Remark 3.4.4 there is a decreasing sequence (W_j) of sets of class $\mathcal{W}(A)$ such that it suffices in Definition 3.4.3 to take for $W \in \mathcal{W}(A)$ the sets W_j. We show that the following equations and inequality hold quasi-everywhere on U:

$$\begin{aligned}\widehat{R}^A_{K\mu} &\underset{1}{=} \widehat{\inf_j R^{W_j \cap U}_{K\mu}} \underset{2}{=} \inf_j \widehat{R}^{W_j \cap U}_{K\mu} \underset{3}{=} \inf_j \int \widehat{R}^{W_j \cap U}_{K(\cdot,Y)} d\mu(Y) \\ &\underset{4}{=} \int \inf_j \widehat{R}^{W_j \cap U}_{K(\cdot,Y)} d\mu(Y) \underset{5}{=} \int \widehat{\inf_j R}^{W_j \cap U}_{K(\cdot,Y)} d\mu(Y) \underset{6}{\succcurlyeq} \int^* \widehat{R}^A_{K(\cdot,Y)} d\mu(Y).\end{aligned}$$

When these relations have been established quasi-everywhere on U, the desired resulting inequality holds everywhere on U. In fact, $\widehat{R}^A_{K\mu} \in \mathcal{S}_+(U)$ along with $K\mu$; and by Theorem 3.5.6 we have since μ is carried by $\Delta_1(U)$

$$\begin{aligned}\int^* \widehat{R}^A_{K(\cdot,Y)} d\mu(Y) &= \int^* K(\cdot, Y) 1_A(Y) d\mu(Y)\\ &= \int K(\cdot, Y) 1_{A^*}(Y) d\mu(Y) = K(1_{A^*}\mu) \in \mathcal{S}_+(U),\end{aligned}$$

where $A^* \subset \overline{U}$ denotes a G_δ set containing A such that $\mu^*(A^* \setminus A) = 0$, cf. Theorem 3.2.22. Equation 1 and inequality 6 hold everywhere on U by Definition 3.4.3. Equation 2 holds quasi-everywhere by the "fine" convergence theorem, see Theorem 2.1.34. Equation 3 holds by Lemma 3.2.26. Equation 4 is obvious (Lebesgue) at points x at which $\widehat{R}^A_{K\mu}(x) < +\infty$, that is, q.e. In the first place, the function $Y \mapsto \widehat{R}^{W_j \cap U}_{K(\cdot,Y)}(x)$ on $\overline{U}$ is l.s.c. according to Corollary 3.1.6 and Theorem 3.2.2(i). Secondly, the integrals are finite, being majorized by $\int K(\cdot, Y) d\mu(Y) = K\mu < +\infty$ at points x where $K\mu(x) < +\infty$. Concerning the remaining Eq. 5, note that for each k the function $\widehat{R}^{W_k \cap U}_{K(\cdot,Y)}$ is invariant on $U \setminus \widetilde{W}_k$ according to Lemma 2.9.7. For any j and any $k \geqslant j$ we have $W_k \subset W_j$, and $\widehat{R}^{W_k \cap U}_{K(\cdot,Y)}$ is therefore invariant on each $U \setminus \widetilde{W}_j$, and hence on their union according to Theorem 2.9.12(a), (b). It follows by Theorem 2.9.12(c) that $\widehat{\inf} \widehat{R}^{W_j \cap U}_{K(\cdot,Y)}$ is invariant on the finely open set U. For any point $x \in U$ such that $K\mu(x) < +\infty$ the set

$$E_x := \{Y \in \overline{U} : \widehat{R}^A_{K(\cdot,Y)}(x) = +\infty\}$$

is μ-null. Again by Theorem 2.9.12(c) we obtain

$$\widehat{\inf_j} \widehat{R}^{W_j \cap U}_{K(\cdot,Y)}(x) = \inf_j \widehat{R}^{W_j \cap U}_{K(\cdot,Y)}(x) \quad \mu\text{-a.e. for } Y \in \overline{U},$$

which implies Eq. 5 at points $x \in U$ with $K\mu(x) < \infty$. We have thus shown that $\widehat{R}^A_{K\mu} \geqslant \int^* \widehat{R}^A_{K(\cdot,Y)} d\mu(Y)$ also for $A \subset \Delta(U)$.

The asserted equality in case $A \subset \Delta(U)$. Recall from Lemma 3.2.14(c) the countable cover (V_k) of U by finely open sets with natural closures $\overline{V}_k$ in $\overline{U}$ contained in U. We may suppose that the sequence (V_k) is increasing. The complements $D_k := \complement\overline{V}_k$ (relative to $\overline{U}$) form a decreasing sequence of open subsets of $\overline{U}$ with the intersection $\Delta(U)$, and such that the mf-closures (that is, closures in mf-topology) $\widetilde{D}_k$ likewise have the intersection $\Delta(U)$. Consider first the case where $A = C \cap \Delta(U)$, C compact in $\overline{U}$. Choose a decreasing sequence of open subsets C_k of $\overline{U}$ containing C such that $\bigcap_k \overline{C}_k = C$. Suppose to begin with that μ is carried by some compact set $E \subset \Delta(U) \setminus A = \Delta(U) \setminus C$. We may assume that $E \cap C_k = \varnothing$. The decreasing open sets $W_k := C_k \cap D_k \supset A$ are of class $\mathcal{W}(A)$ and hence

$$\widehat{R}^A_{K\mu} \leqslant \widehat{R}^{W_k\cap U}_{K\mu} = \int_E \widehat{R}^{W_k\cap U}_{K(\cdot,Y)} d\mu(Y). \tag{3.5.2}$$

For each $Y \in E$ the functions $p_k := \widehat{R}^{W_k\cap U}_{K(\cdot,Y)}$ are fine potentials on U according to Theorem 3.4.12 (iii) and Lemma 3.3.1 because the mf-closure $\widetilde{W}_k$ of W_k is contained in $\widetilde{C}_k \cap \widetilde{D}_k \subset \overline{C}_k \cap \widetilde{D}_k$ according to Theorem 3.5.2, and hence does not meet E and in particular does not contain Y. It follows that $p := \widehat{\inf}_k p_k$ is a fine potential on U. By Lemma 2.9.7 the restriction of p_k to the finely open set $U \setminus \widetilde{W}_k$ is invariant. By Theorem 2.9.12(a) it follows that the restriction of p_k to the finely open set $U \setminus \widetilde{W}_l$ for $k \geq l$, and so is the restriction of p to $U \setminus \widetilde{W}_l$ by Theorem 2.9.12(c). Since $A \subset \Delta(U)$, we have

$$\bigcup\nolimits_{l\geqslant k}(U \setminus \widetilde{W}_l) = U \setminus \bigcap\nolimits_l \widetilde{W}_l \supset U \setminus \bigcap\nolimits_l (\overline{C}_l \cap \widetilde{D}_l) = U \setminus A = U$$

and hence $\bigcup_l (U \setminus \widetilde{W}_l) = U$. By Theorem 2.9.12(b) we infer that p is itself invariant on U, and being also a fine potential p must be 0. It therefore follows by (3.5.2) that $\widehat{R}^A_{K\mu} \leqslant \int_E \widehat{R}^{W_k\cap U}_{K(\cdot,Y)} d\mu(Y) \downarrow 0$ pointwise on U, and hence $\widehat{R}^A_{K\mu} = 0 \leqslant \int \widehat{R}^A_{K(\cdot,Y)} d\mu(Y)$. In combination with the opposite inequality in (3.5.1) obtained above (now with a true integral) this establishes equality in (3.5.1) in the present case where $A = C \cap \Delta(U)$ with C compact in $\overline{U}$ and μ carried by a compact set $E \subset \Delta(U) \setminus A$.

Next, replace the latter assumption on μ by the weaker temporary assumption that $\mu(A) = 0$. Choose an increasing sequence of compact sets $E_j \subset \Delta(U) \setminus A$ such that $\mu(E_j) \uparrow \mu(\Delta(U))$, and denote by μ_j the part of μ on E_j. By Theorem 3.4.13 it follows that

$$\widehat{R}^A_{K\mu} = \sup_j \widehat{R}^A_{K\mu_j} = \sup_j \int \widehat{R}^A_{K(\cdot,Y)} d\mu_j = \int \widehat{R}^A_{K(\cdot,Y)} d\mu(Y) = 0$$

according to Theorem 3.4.12(ii).

Without any such temporary assumption on μ we denote by μ_A and μ' the parts of μ on A and on $\overline{U} \setminus A$, respectively. Then $\mu'(A) = 0$ and hence

$$\widehat{R}^A_{K\mu} = \widehat{R}^A_{K\mu_A} + \widehat{R}^A_{K\mu'} \leqslant K\mu_A + \int \widehat{R}^A_{K(\cdot,Y)} d\mu'(Y) = \int \widehat{R}^A_{K(\cdot,Y)} d\mu(Y).$$

When combined with the inequality "$\geq$" in (3.5.1) obtained above (with an upper integral) this leads to equality in (3.5.1) (with a true integral) for arbitrary μ when $A = C \cap \Delta(U)$ with C compact.

More generally, if $A = C \cap \Delta(U)$ and if C is just the union of an increasing sequence of compact sets $C_j \subset \overline{U}$, then

$$\widehat{R}^{C_j\cap\Delta(U)}_{K\mu} = \int \widehat{R}^{C_j\cap\Delta(U)}_{K(\cdot,Y)} d\mu(Y) \leqslant \int \widehat{R}^A_{K(\cdot,Y)} d\mu(Y).$$

For $j \to \infty$ it follows by (a) in Theorem 3.4.7 that $\widehat{R}^A_{K\mu} \leqslant \int \widehat{R}^A_{K(\cdot,Y)} d\mu(Y)$. Together with the opposite inequality obtained above this leads to (3.5.1) (with a true integral) for any set $A = C \cap \Delta(U)$ with C a K_σ subset of $\overline{U}$. This applies in particular to $A = C \cap \Delta(U)$ with C open in $\overline{U}$.

Next, let A be any G_δ subset of $\Delta(U)$. Since $\Delta(U)$ is itself a G_δ in $\overline{U}$ this means that $A = C \cap \Delta(U)$ for some G_δ subset C of $\overline{U}$. Thus C is the intersection of a decreasing sequence of open sets $C_j \subset \overline{U}$. Denote by μ_j the part of μ on $\Delta(U) \setminus A_j$ where $A_j = C_j \cap \Delta(U)$. Then $\mu_j(A_j) = 0$, and again, since C_j is an open subset of $\overline{U}$,

$$\widehat{R}^A_{K\mu_j} \leqslant \widehat{R}^{A_j}_{K\mu_j} = \int \widehat{R}^{A_j}_{K(\cdot,Y)} d\mu_j(Y) = 0$$

according to Theorem 3.4.12(ii). But $\widehat{R}^A_{K\mu_j} \uparrow \widehat{R}^A_{K\mu} = 0$ in the specific order by Theorem 3.4.13, whence the assertion.

Finally, let A be an arbitrary subset of $\Delta(U)$. Then A can be extended by a μ-nullset to a G_δ set $A^* \subset \Delta(U)$ because $\Delta(U)$ is itself a G_δ. We obtain the missing inequality "$\leq$" as follows:

$$\begin{aligned}\widehat{R}^A_{K\mu} \leq \widehat{R}^{A^*}_{K\mu} &= \int \widehat{R}^{A^*}_{K(\cdot,Y)} d\mu(Y) = \int K(\cdot, Y) 1_{A^*}(Y) d\mu(Y) \\ &= \int^* K(\cdot, Y) 1_A(Y) d\mu(Y) = \int^* \widehat{R}^A_{K(\cdot,Y)} d\mu(Y)\end{aligned}$$

according to Theorem 3.5.6. We have thus shown that (3.5.1) holds for any set $A \subset \Delta(U)$. According to the last equality the upper integral in the above display becomes a true integral if the subset A of $\Delta(U)$ is μ-measurable.

The general case of the theorem. The equality (3.5.1) being proved above in case $A \subset \Delta(U)$, it follows by Theorem 3.4.8 that for general $A \subset \overline{U}$

$$\widehat{R}^A_{K\mu} = \widehat{R}^{A\cap\Delta(U)}_{K\mu} + \widehat{R}^{A\cap U}_{v} = \int^* \widehat{R}^{A\cap\Delta(U)}_{K(\cdot,Y)} d\mu(Y) + \widehat{R}^{A\cap U}_{v},$$

where

$$\begin{aligned} v := K\mu - \widehat{R}^{A\cap\Delta(U)}_{K\mu} &= \int K(\cdot, Y) d\mu(Y) - \int^* \widehat{R}^{A\cap\Delta(U)}_{K(\cdot,Y)} d\mu(Y) \\ &= \int_* K(\cdot, Y)(1 - 1_{A^*\cap\Delta_1(U)}(Y)) d\mu(Y) \ = \ K\lambda,\end{aligned}$$

with $\lambda := (1 - 1_{A^*\cap\Delta_1(U)})\mu$, A^* denoting again a G_δ set containing A such that $\mu^*(A^* \setminus A) = 0$. Since $\lambda \leqslant \mu$, λ is carried by $\Delta_1(U)$. It follows that

$$\widehat{R}^{A\cap U}_{v} = \int \widehat{R}^{A\cap U}_{K(\cdot,Y)} d\lambda(Y) = \int \widehat{R}^{A\cap U}_{K(\cdot,Y)} (1 - 1_{A^*\cap\Delta_1(U)}(Y)) d\mu(Y),$$

$$\widehat{R}^A_{K\mu} = \int \big(K(\cdot, Y)1_{A^*\cap\Delta_1(U)}(Y) + \widehat{R}^{A\cap U}_{K(\cdot,Y)}(1-1_{A^*\cap\Delta_1(U)}(Y))\big)d\mu(Y). \qquad (3.5.3)$$

Similarly, for any $Y \in \Delta_1(U)$,

$$\widehat{R}^A_{K(\cdot,Y)} = \widehat{R}^{A\cap\Delta(U)}_{K(\cdot,Y)} + \widehat{R}^{A\cap U}_{v_Y} = 1_{A\cap\Delta_1(U)}(Y)K(\cdot, Y) + \widehat{R}^{A\cap U}_{v_Y},$$

$$v_Y := K(\cdot, Y) - \widehat{R}^{A\cap\Delta(U)}_{K(\cdot,Y)} = (1 - 1_{A\cap\Delta_1(U)}(Y))K(\cdot, Y) \; = \; K\lambda_Y$$

with $\lambda_Y := (1 - 1_{A^*\cap\Delta_1(U)}(Y))\varepsilon_Y$ carried by $\Delta_1(U)$. It follows that

$$\widehat{R}^{A\cap U}_{v_Y} = \int \widehat{R}^{A\cap U}_{K(\cdot,Z)}d\lambda_Y(Z) = (1 - 1_{A\cap\Delta_1(U)}(Y))\widehat{R}^{A\cap U}_{K(\cdot,Y)},$$

$$\widehat{R}^A_{K(\cdot,Y)} = 1_{A\cap\Delta_1(U)}(Y)K(\cdot, Y) + (1 - 1_{A\cap\Delta_1(U)}(Y))\widehat{R}^{A\cap U}_{K(\cdot,Y)}.$$

The upper integral of this expression for $\widehat{R}^A_{K(\cdot,Y)}$ with respect to $d\mu(Y)$ is just the right hand member of (3.5.3) because $1_{A\cap\Delta_1(U)} = 1_{A^*\cap\Delta_1(U)}$ μ-a.e. This proves that indeed $\widehat{R}^A_{K\mu} = \int^* \widehat{R}^A_{K(\cdot,Y)}d\mu(Y)$. □

Corollary 3.5.8 *Let $u \in \mathcal{S}_+(U)$, let $A \subset \Delta(U)$, and let μ denote the (unique) representing measure for the invariant function $\widehat{R}^A_u$ carried by $\Delta_1(U)$. Then μ is carried by A (in the sense that $\mu^*(\complement A) = 0$) if and only if $\widehat{R}^A_u = u$.*

Proof By the above theorem together with Theorem 3.5.6 we have

$$\widehat{R}^A_u = \int^* \widehat{R}^A_{K(\cdot,Y)}d\mu(Y) = \int 1_{A^*}(Y)K(\cdot, Y)d\mu(Y) = K(1_{A^*}\mu),$$

and by uniqueness this equals $u = K\mu$ if and only if $1_{A^*}\mu = \mu$, which means that μ shall be carried by A. □

The above corollary, sharpening Theorem 3.4.11 (in the present setting), is an analogue of Theorem 3.2.17, where μ is carried only by the closure $\overline{A}$ of A, $\overline{A}$ supposed to be contained in U.

Remark 3.5.9 We refer to [EKF2, Definition 3.14] for a definition of sweeping on subsets of $\overline{U}$ relative to the minimal-fine topology. It is shown there that sweeping on A relative to the mf-topology coincides with sweeping on A relative to the natural topology as defined in Definition 3.4.3.

3.6 The Dirichlet Problem at the Martin Boundary of a Fine Domain

By way of preparation for the PWB method with respect to the mf-topology, we have the following minimal-fine boundary minimum property, followed by its consequence for the natural topology. The requirement $Y \in \Delta_1(U)$ (rather than $Y \in \Delta(U)$) is motivated by the fact that $\Delta_1(U)$ is the set of all mf-limit points of U in $\overline{U}$, as noted after Definition 3.5.1, and so the stated mf-$\liminf_{x \to Y}$ is defined for $Y \in \Delta_1(U)$ only.

Theorem 3.6.1 *Let v be finely superharmonic on U, let $h > 0$ be finely harmonic on U, and suppose that*

$$\text{mf-}\liminf_{x \to Y, x \in U} \frac{v(x)}{h(x)} \geq 0 \quad \textit{for every } Y \in \Delta_1(U).$$

If moreover $v \geq -s$ on U for some $s \in \mathcal{S}_+(U)$ then $v \geq 0$ on U.

Proof For given $\varepsilon > 0$ and $Y \in \Delta(U)$ there exists by the assumed boundary inequality an mf-fine open mf-neighborhood $W_Y \subset \overline{U}$ of Y such that $v > -\varepsilon h$ on $W_Y \cap U$. (Take for example $W_Y = \{Y\}$ for $Y \in \Delta(U) \setminus \Delta_1(U)$, cf. Definition 3.5.1(c).) In terms of the mf-open set $W := \bigcup\{W_Y : Y \in \Delta(U)\}$ containing $\Delta(U)$ we infer that $v > -\varepsilon h$ on $W \cap U = \bigcup\{W_Y \cap U : Y \in \Delta(U)\}$. The set $E := \overline{U} \setminus W$ is mf-closed and contained in U, and so E is finely closed, as noted before Theorem 3.5.2. Furthermore, E is minimal-thin at Y in view of Definition 3.5.1(b), and hence $\widehat{R}^E_{K(\cdot,Y)} \neq K(\cdot, Y)$ by Definition 3.3.2, or equivalently by Lemma 3.3.1, that $\widehat{R}^E_{K(\cdot,Y)}$ is a fine potential on U for each $Y \in \Delta_1(U)$. We show that $\widehat{R}^E_s$ likewise is a fine potential. Write $s = p + q$ with p a fine potential and q finely invariant. Then $\widehat{R}^E_s = \widehat{R}^E_p + \widehat{R}^E_q$ with $\widehat{R}^E_p \leqslant p$ and hence $\widehat{R}^E_p$ is a fine potential. To prove that $\widehat{R}^E_s$ is a fine potential it therefore remains to show that so is $\widehat{R}^E_q$.

Decompose $\widehat{R}^E_q = p_1 + k$ with p_1 a fine potential and k invariant. By sweeping again on E we obtain $\widehat{R}^E_q = \widehat{R}^E_{p_1} + \widehat{R}^E_k$, and hence in terms of the representing measure μ for k

$$\int_{\Delta_1(U)} K(\cdot, Y) d\mu(Y) = k = \widehat{R}^E_k = \int_{\Delta_1(U)} \widehat{R}^E_{K(\cdot,Y)} d\mu(Y), \tag{3.6.1}$$

by Theorem 3.5.7. Denoting by U_j the countably many fine components of the finely open set $U \setminus E$ we choose for each j a point $x_j \in U_j$ such that $k(x_j) < +\infty$. Then $\widehat{R}^E_{K(\cdot,Y)}(x_j)$ is finite for μ-almost every $Y \in \Delta_1(U)$, say for $Y \in \Delta_1(U) \setminus A_j$ with A_j μ-negligible. For every $Y \in \Delta_1(U) \setminus \bigcup_j A_j$ we have $\widehat{R}^E_{K(\cdot,Y)}(x_j) < +\infty$ for every j, and hence by (3.6.1)

$$\int_{\Delta_1(U)} \big(K(x_j, Y) - \widehat{R}^E_{K(\cdot,Y)}(x_j)\big) d\mu(Y) = 0.$$

For μ-almost every Y it therefore follows that $K(x_j, Y) = \widehat{R}^E_{K(\cdot,Y)}(x_j)$ and so $K(\cdot, Y) = \widehat{R}^E_{K(\cdot,Y)}$ on each U_j because $\widehat{R}^E_{K(\cdot,Y)}$ is invariant on U_j according to Lemma 2.9.7 and hence $\widehat{R}^E_{K(\cdot,Y)}|_{U_j} \preceq K(\cdot, Y)|_{U_j}$ on U_j by Lemma 2.9.4 so that the positive finely superharmonic function $K(\cdot, Y)|_{U_j} - \widehat{R}^E_{K(\cdot,Y)}|_{U_j}$, null at x_j, is indeed null on all of U_j by Corollary 2.1.27. For μ-almost every Y we likewise infer that the equality $K(\cdot, Y) = \widehat{R}^E_{K(\cdot,Y)}$ holds on $U \setminus E$, and actually q.e. on U since the equality $K(\cdot, Y) = \widehat{R}^E_{K(\cdot,Y)}$ holds q.e. on E. Finally, this equality holds everywhere on U for μ-almost every Y by fine continuity of $K(\cdot, Y)$ and $\widehat{R}^E_{K(\cdot,Y)}$. As noted above, $\widehat{R}^E_{K(\cdot,Y)} \neq K(\cdot, Y)$ for every $Y \in \Delta_1(U)$, and consequently $\mu = 0$ and so $k = K\mu = 0$, which implies that $\widehat{R}^E_q = p_1$ indeed is a fine potential, and so is therefore $\widehat{R}^E_s$.

The (possibly empty) fine interior V of E has relative fine boundary $U \cap \partial_f V \subset U \setminus V \subset U \cap \widetilde{W}$. We have $v + \varepsilon h \geq 0$ on $U \cap W$ and hence by fine continuity on $U \cap \widetilde{W} \supset U \cap \partial_f V$. Furthermore, $v + \widehat{R}^E_s = v + s$ on $U \cap \mathfrak{b}(E) \supset V$, where $\mathfrak{b}(E)$ denotes the base of E in Ω; and so $v + \varepsilon h \geq -\widehat{R}^E_s$ on V. Altogether, it follows by the relative fine boundary minimum property (Theorem 2.4.14) applied to the fine potential $p = \widehat{R}^E_s$ that $v + \varepsilon h \geq 0$ on V. As noted above, the same inequality holds on $U \cap \widetilde{W} \supset U \setminus V$ and thus on all of U. By varying ε we conclude that indeed $v \geq 0$ on all of U. □

In view of Theorem 3.5.2, we have the following weaker Martin boundary minimum property relative to the natural topology on $\overline{U}$ (to be used in last part of the present section).

Corollary 3.6.2 *Let v be finely superharmonic on U, let $h > 0$ be finely harmonic on U, and suppose that*

$$\liminf_{x \to Y,\, x \in U} \frac{v(x)}{h(x)} \geq 0 \quad \textit{for any } Y \in \Delta(U).$$

If moreover v/h is lower bounded then $v \geq 0$ on U.

The upper and lower PWB^h *classes of a function on* $\Delta(U)$. We shall study the *Dirichlet problem at* $\Delta(U)$ relative to a fixed finely harmonic function $h > 0$ on U. We denote by μ_h the measure on $\Delta(U)$ carried by $\Delta_1(U)$ and representing h, that is, $h = \int K(\cdot, Y) d\mu_h(Y) = K\mu_h$. A function u on U (or on some finely open subset of U) is said to be *finely h-hyperharmonic, finely h-superharmonic, h-harmonic, h-invariant, or a fine h-potential*, respectively, if it has the form $u = v/h$, where v is finely hyperharmonic, finely superharmonic, harmonic, invariant, or a fine potential, respectively.

Let f be a function on $\Delta(U)$ with values in $\overline{\mathbb{R}}$. A finely h-hyperharmonic function $u = v/h$ on U is said to belong to the *upper* PWB^h *class* for f, denoted by $\overline{\mathcal{U}}^h_f$, if u is lower bounded and if

$$\liminf_{x \to Y,\, x \in U} u(x) \geqslant f(Y) \quad \text{for every } Y \in \Delta(U),$$

where lim inf is taken relative to the natural topology on $\overline{U}$. We define

$$\dot{H}^h_f = \inf \overline{\mathcal{U}}^h_f, \quad \overline{H}^h_f = \widehat{\dot{H}^h_f} = \widehat{\inf} \overline{\mathcal{U}}^h_f \ (\leqslant \dot{H}^h_f), \tag{3.6.2}$$

where $\widehat{u}$ for any function $u : U \to \overline{\mathbb{R}}$ denotes the finely l.s.c. regularization (that is, the greatest finely l.s.c. minorant) of u. For any family $\mathcal{U}$ of functions $U \to \overline{\mathbb{R}}$, $\widehat{\inf} \mathcal{U}$ stands for $\widehat{\inf \mathcal{U}}$. Clearly, $\dot{H}^h_f$ is finely u.s.c. on all of U, and $\overline{H}^h_f$ is finely l.s.c. there.

Both functions $\overline{H}^h_f$ and $\dot{H}^h_f$, of which the former serves as the *upper solution* of the Dirichlet problem, are needed here, unlike the classical case where we have the Harnack convergence theorem and hence $\overline{H}^h_f = \dot{H}^h_f$, cf. Example 3.6.3 below. In our setup, $\dot{H}^h_f$ may be neither finely h-hyperharmonic nor identically $-\infty$, but only nearly finely h-hyperharmonic on the finely open set $\{\overline{H}^h_f > -\infty\}$ which can be a non-void proper subset of U, see [Fu8, p. 118 bottom]. In Example 3.6.3 below it is shown that the finely open set $\{\overline{H}^h_f > -\infty\}$ can be a non-void proper subset of U.

According to the "fine" convergence theorem (Theorem 2.1.34), see also Lemma 2.2.3, $\overline{H}^h_f$ is finely h-hyperharmonic on $\{\overline{H}^h_f > -\infty\}$, and we have $\overline{H}^h_f = \dot{H}^h_f$ quasi-everywhere (q.e.)there; furthermore, since $\overline{\mathcal{U}}^h_f$ is lower directed, there is a decreasing net of functions $(u_\alpha) \subset \overline{\mathcal{U}}^h_f$ such that $\overline{H}^h_f = \widehat{\inf}_\alpha u_\alpha = \widehat{\dot{H}^h_f}$, cf. (3.6.2). If f is bounded then $\overline{H}^h_f = \dot{H}^h_f$, see Theorem 3.6.7(c) below.

The lower PWB^h class $\underline{\mathcal{U}}^h_f$ is defined similarly, or directly by $\underline{\mathcal{U}}^h_f = -\overline{\mathcal{U}}^h_{-f}$, and we have the functions $\underline{H}^h_f = -\overline{H}^h_{-f}$ and $\underset{\cdot}{H}^h_f = -\dot{H}^h_{-f}$, of which the former is the *lower solution* of the Dirichlet problem. For $f = 0$ we have $\dot{H}^h_0 \leq 0$ because $0 \in \overline{\mathcal{U}}^h_0$, and here equality prevails according to Corollary 3.6.2. Consequently, $\dot{H}^h_0 = 0$ and hence $\underset{\cdot}{H}^h_0 = 0$, which implies that $\overline{H}^h_0 = \underline{H}^h_0 = 0$. It follows that if $f \geq 0$ then every function of class $\overline{\mathcal{U}}^h_f$ is $\geqslant \dot{H}^h_f \geqslant \dot{H}^h_0 = 0$. Moreover, $\dot{H}^h_{\alpha f+\beta} = \alpha \dot{H}^h_f + \beta$ and hence $\overline{H}^h_{\alpha f+\beta} = \alpha \overline{H}^h_f + \beta$ for real constants $\alpha \geq 0$ and β (when 0 times $\pm\infty$ is defined to be 0).

The following example shows that it can occur that $\overline{H}^h_f$ takes the value $-\infty$ without being identically $-\infty$.

Example 3.6.3 In $\Omega = \mathbb{R}^3$ with the Green kernel $G(x, y) = |x - y|^{-1}$ let ω and U be as in Example 3.2.8. Choose a point $z \in \partial_f \omega$ and a sequence (z_j) of points of $\complement\overline{\omega}$ such that $|z_j - z| \leq 2^{-j}$. Then $u := \sum_j 2^{-j} G(\cdot, z_j)$ is infinite at z, but finite and harmonic on ω. Furthermore, $u = \sup_k u_k$, where $u_k := \sum_{j \leqslant k} 2^{-j} G(\cdot, z_j)$ is harmonic on ω and bounded on $\overline{\omega}$ ($\subset \mathbb{R}^3$). It follows that $u_k|_U$ is of class $\underline{\mathcal{U}}^h_f$, where $f := u|_{\Delta(U)}$ and $h \equiv 1$ on U. In fact,

$$\lim_{x \to Y, x \in U} u_k(x) = u_k(Y) \leqslant u(Y) = f(Y)$$

for $Y \in \Delta(U)$ (natural limit on $U \cup \Delta(U)$). Thus $\underset{\cdot}{H}{}^h_f \geq u_k|_U$, and hence

$$\underline{H}^h_f(z) \geq \underset{\cdot}{H}{}^h_f(z) \geqslant \sup_k u_k(z) = u(z) = +\infty.$$

To show that $\underline{H}^h_f < +\infty$ on $U \setminus \partial_i\omega \,(= \omega)$, let $v \in \underline{\mathcal{U}}^h_f$. Being upper bounded on the bounded open set ω, v is subharmonic on ω by Theorem 2.1.22, and so is therefore $v - u$. For any $Y \in \Delta(U)$ ($\cong (\Delta(\omega) \setminus \partial_i\omega)$ we have

$$\limsup_{x \to Y,\, x \in \omega} v(x) \leqslant f(Y) < +\infty$$

and hence

$$\limsup_{x \to Y,\, x \in \omega} (v(x) - u(x)) \leq 0.$$

with Euclidean limit for $Y \in \partial\omega \setminus \partial_i\omega$, because $\partial\omega \subset \Delta(\omega)$ in view of [Ai, Theorem 3.1 and Theorem 1.5 (iv)]. Since $\partial_i\omega$ is polar and $v - u \leqslant v$ is upper bounded, it follows by a boundary minimum principle that $v - u \leq 0$, that is, $v \leqslant u$ on ω. By varying $v \in \underline{\mathcal{U}}^h_f$ we conclude that $\underset{\cdot}{H}{}^h_f \leqslant u$ on $\omega \cong U \setminus \partial_i\omega$ and hence by regularization that $\underline{H}^h_f \leqslant u < +\infty$ on $U \setminus \partial_i\omega$. Altogether, $\overline{H}^h_{-f} = -\underline{H}^h_f$ equals $-\infty$ at z, but is finite on $U \setminus \{z\}$. And $\underline{H}^h_{-f} = -\overline{H}^h_f \leq 0$. Note also that $\underline{H}^h_f = \overline{H}^h_f = u$, as obtained from $u_k \leqslant \underline{H}^h_f \leqslant \overline{H}^h_f \leqslant u$ by taking supremum over k.

Henceforth we fix the finely harmonic function $h > 0$ on U, relative to which we shall study the Dirichlet problem at $\Delta(U)$.

As noted above, $\overline{H}^h_f$ is finely h-hyperharmonic on the finely open set $\{\overline{H}^h_f > -\infty\}$ (which may be void, e.g. if $f \equiv -\infty$). As expected from the classical theory, this can be improved.

Theorem 3.6.4 *Let $f : \Delta(U) \to \overline{\mathbb{R}}$. The upper solution $\overline{H}^h_f$ is either identically $+\infty$ or else finely h-harmonic q.e. on the finely open set $\{x \in U : \overline{H}^h_f(x) > -\infty\}$. Similarly, $\underline{H}^h_f$ is either identically $-\infty$ or finely h-harmonic q.e. on the finely open set $\{\underline{H}^h_f < +\infty\}$.*

Proof For the former assertion we may suppose that $\overline{H}^h_f \not\equiv +\infty$. We may also assume that $\overline{H}^h_f \not\equiv -\infty$, or in other words that the finely open set $U' := \{\overline{H}^h_f > -\infty\}$ is non-void. For any $u \in \overline{\mathcal{U}}^h_f$ and any relatively finely closed subset A of U consider the function $u^A : U \to \overline{\mathbb{R}}$ defined by

$$u^A(x) = \frac{1}{h}\int_U hu\, d\varepsilon_x^{A \cup (\complement U)}, \quad x \in U, \tag{3.6.3}$$

(sweeping of measures being understood relative to Ω). If $u \geq 0$ then $u^A = \frac{1}{h}\widehat{R}^A_{hu}$ (sweeping of functions being everywhere taken relative to U), see Theorem 1.4.44

and Theorem 2.3.11. In any case, u^A is finely h-superharmonic on U' according to Theorem 2.1.39 applied to the base B of $A \cup (\complement U)$ and to $f = hu$ (extended to Ω by 0 off U), noting that $u^- \leqslant c$ for some positive constant c, and so $f^- = hu^- \leqslant ch$. Thus $\int f^- d\varepsilon_x^B \leqslant c \int h \, d\varepsilon_x^B \leqslant ch(x) < +\infty$ by Theorem 2.1.14. Choose a decreasing net $(u_\alpha) \subset \overline{\mathcal{U}}_f^h$ with finely l.s.c. regularization $\overline{H}_f^h$ of $\inf_\alpha u_\alpha$. Now take $A = U \setminus V_k$ for some k, see Lemma 3.2.14. Then $A \cup (\complement U) = \complement V_k$, and $\varepsilon_x^{A \cup (\complement U)} = \varepsilon_x^{\complement V_k}$ is a probability measure for any $x \in V_k$ (the fine harmonic measure at x on the fine boundary of V_k in Ω, or equivalently in U since V_k has (natural) compact closure and hence fine closure in U), see Corollary 2.1.17, and proof of Theorem 3.1.8 for compacity of the (natural) closure $\overline{V_k}$ of V_k. With a constant $c \geq 0$ such that $u_\alpha + c \geq 0$ we then have $(u_\alpha + c)^A \geq 0$ and hence for any $x \in U$

$$u_\alpha(x) + c \geq (u_\alpha + c)^A(x) = \frac{1}{h} \int_U (hu_\alpha + ch) \, d\varepsilon_x^{\complement V_k} = \frac{1}{h} \int_U hu_\alpha \, d\varepsilon_x^{\complement V_k} + c,$$

and hence $u_\alpha \geqslant u_\alpha^A$ on V_k. Since $u_\alpha^A = u_\alpha$ q.e. on $A = U \setminus V_k$, which is a deleted natural neighborhood relative to U of any point $Y \in \Delta(U)$, we infer that $\overline{H}_f^h(x) = \widehat{\inf}_\alpha u_\alpha^A(x)$ for every $x \in V_k \cap U'$. Because each $(u_\alpha + c)^A$ and hence each u_α^A are finely h-harmonic q.e. on $V_k \cap U'$, $\overline{H}_f^h(x)$ is finely h-harmonic there according to Theorem 2.1.31, and therefore finely h-harmonic q.e. on $\bigcup_k V_k \cap U' = U' = \{\overline{H}_f^h > -\infty\}$.—For the latter assertion of the proposition, replace f by $-f$ in the above proof. □

Similarly to the classical case, cf. [Do4, p. 108], we pose the following definition, denoting by 1_A the indicator function of a set $A \subset \Delta(U)$:

Definition 3.6.5 A subset A of $\Delta(U)$ is said to be *h-harmonic measure null* if $\overline{H}_{1_A}^h = 0$.

It will be shown in Corollary 3.6.20 below that A is h-harmonic measure null if and only if A is μ_h-measurable with $\mu_h(A) = 0$.

Theorem 3.6.6 (a) *Every countable union A of h-harmonic measure null sets $A_j \subset \Delta(U)$ is h-harmonic measure null.*

(b) *A set $A \subset \Delta(U)$ is h-harmonic measure null if and only if there exists a lower bounded finely h-superharmonic function u (positive if we like) on U such that $\lim_{x \to Y, x \in U} u(x) = +\infty$ for every $Y \in A$.*

(c) *If a function $f : \Delta(U) \to [0, +\infty]$ has $\overline{H}_f^h = 0$ then $\{f > 0\}$ is h-harmonic measure null.*

(d) *If a function $f : \Delta(U) \to [0, +\infty]$ has $\overline{H}_f^h < +\infty$ then $\{f = +\infty\}$ is h-harmonic measure null.*

(e) *For two functions $f, g : \Delta(U) \to [0, +\infty]$, if $f \leqslant g$ off some h-harmonic measure null set then $\dot{H}_f^h \leqslant \dot{H}_g^h$ q.e. on U and $\overline{H}_f^h \leqslant \overline{H}_g^h$ everywhere on U.*

Proof We adapt the proof in [Do4, pp. 108, 111] for the classical case.

(a) Fix a point x_0 of the co-polar and hence non-void subset $\bigcap_j \{\dot{H}^h_{1_{A_j}} = 0\}$ of U. For given $\varepsilon > 0$ and integers $j > 0$ there are functions $u_j \in \overline{\mathcal{U}}^h_{1_{A_j}}$ with $u_j(x_0) < 2^{-j}\varepsilon$. It follows that the function $u := \sum_j u_j$ is of class $\overline{\mathcal{U}}^h_{1_{A_j}}$ for each j (along with each u_j) because $u(x_0) < \varepsilon < +\infty$. Hence $u \in \overline{\mathcal{U}}^h_{1_A}$, and so $\overline{H}^h_{1_A}(x_0) \leqslant \dot{H}^h_{1_A}(x_0) \leqslant u(x_0) < \varepsilon$. The positive finely h-hyperharmonic function $\overline{H}^h_{1_A}$ therefore equals 0 at x_0 and so indeed everywhere on U.

(b) If $\overline{H}^h_{1_A} = 0$ then we apply (a) to the sequence (A_j) with every $A_j = A$. The function u from (a) is then finely superharmonic of class $\overline{\mathcal{U}}^h_{(+\infty)1_A}$ and may be taken to be positive. Conversely, if there exists a function u as described in (b), we may arrange that $u \geq 0$ after adding a constant. Then $\varepsilon u \in \overline{\mathcal{U}}^h_{1_A}$ for every $\varepsilon > 0$. It follows that $\dot{H}^h_{1_A} \leqslant \varepsilon u$ and by varying ε that $\dot{H}^h_{1_A} = 0$ off the polar set of infinities of u, and hence q.e. on U. Consequently, $\overline{H}^h_f = 0$.

(c) For integers $j \geq 1$ let f_j denote the indicator function on U for the set $\{f > 1/j\}$. Then $0 = \overline{H}^h_f \geqslant \overline{H}^h_{f_j}/j \geq 0$, so the sets $\{f > 1/j\}$ are h-harmonic measure null, and so is by (a) the union $\{f > 0\}$ of these sets.

(d) Since $\overline{H}^h_f < +\infty$ there exists a point $x_0 \in U$ such that $\dot{H}^h_f(x_0) < +\infty$. Hence there is $u \in \overline{\mathcal{U}}^h_f$ with $u(x_0) < +\infty$, showing that u is finely h-superharmonic. Then $\lim_{x\to Y,\, x\in U} u(x) = +\infty$ for every $Y \in A := \{f = +\infty\}$. After adding a constant we arrange that the finely h-hyperharmonic function u is positive. According to (b) it follows that indeed $\overline{H}^h_{1_A} = 0$.

(e) Let $v \in \overline{\mathcal{U}}^h_g$ and let u be a positive h-superharmonic function on U with natural limit $+\infty$ at every point of the subset $\{f > g\}$ of $\Delta(U)$ with h-harmonic measure null, cf. (b). Then

$$\liminf_{x\to Y}\,(v(x) + \varepsilon u(x)) \geqslant g(Y) \begin{cases} \geqslant f(Y), & Y \in \{f \leqslant g\}, \\ = +\infty \geqslant f(Y), & Y \in \{f > g\}. \end{cases}$$

Thus $v + \varepsilon u \in \overline{\mathcal{U}}^h_f$ and hence $\dot{H}^h_f \leqslant v + \varepsilon u$. By varying v, it follows that $\dot{H}^h_f \leqslant \dot{H}^h_g + \varepsilon u$. For $\varepsilon \to 0$ we obtain indeed $\dot{H}^h_f \leqslant \dot{H}^h_g$ q.e., viz. off the polar set $E := \{u = +\infty\}$.—On the other hand, from the above inequality $\dot{H}^h_f \leqslant v + \varepsilon u$ it follows for $\varepsilon \to 0$ that $\dot{H}^h_f \leqslant v$ off E, and so $\overline{H}^h_f \leqslant v$ everywhere on U because E is finely discrete. By varying v it follows by finely l.s.c. regularization that $\overline{H}^h_f \leqslant \dot{H}^h_g$ and indeed $\overline{H}^h_f \leqslant \overline{H}^h_g$ because $\overline{H}^h_f$ is finely l.s.c. □

Theorem 3.6.7 *Let f be a function on $\Delta(U)$ with values in $\overline{\mathbb{R}}$.*

(a) $\dot{H}^h_f \geq \underset{\cdot}{H}^h_f$ *and hence* $\overline{H}^h_f \geq \underset{\cdot}{H}^h_f$ *and* $\dot{H}^h_f \geqslant \underline{H}^h_f$.

(b) $\overline{H}^h_f \geqslant \underline{H}^h_f$ *on* $\{\overline{H}^h_f > -\infty\} \cup \{\underline{H}^h_f < +\infty\}$.

(c) *If f is lower bounded then $\overline{H}^h_f(x) = \dot{H}^h_f(x)$ ($\geqslant \underline{H}^h_f$) at any point $x \in U$ at which $\dot{H}^h_f(x) < +\infty$. If $f \geq 0$ on $\Delta(U)$ then $\overline{H}^h_f$ (≥ 0) is either identically $+\infty$ or h-invariant on U.*

Proof Clearly, $\overline{\mathcal{U}}^h_f$ is lower directed and $\underline{\mathcal{U}}^h_f$ is upper directed. The constant function $+\infty$ belongs to $\overline{\mathcal{U}}^h_f$. If $+\infty$ is the only function of class $\overline{\mathcal{U}}^h_f$ then obviously $\dot{H}^h_f = +\infty$ and hence $\overline{H}^h_f = +\infty$. In the remaining case, it suffices to consider finely h-superharmonic functions in the definition of $\dot{H}^h_f$ and hence of $\overline{H}^h_f$.

(a) Let $u \in \overline{\mathcal{U}}^h_f$ and $v \in \underline{\mathcal{U}}^h_f$. Then $u - v$ is well defined, finely h-hyperharmonic, and lower bounded on U, and

$$\begin{aligned}\liminf_{x\to Y,\,x\in U}(u(x)-v(x)) &\geq \liminf_{x\to Y,\,x\in U} u(x) - \limsup_{x\to Y,\,x\in U} v(x)\\ &\geq f(Y) - f(Y) = 0\end{aligned}$$

if $f(Y)$ is finite; otherwise $\liminf u(x) - \limsup v(x) = +\infty \geq 0$, for if for example $f(Y) = +\infty$ then $\liminf u = +\infty$ whereas $\limsup v < +\infty$ since v is upper bounded. By the natural boundary minimum property given in Corollary 3.6.2 applied to the finely superharmonic function $hu - hv$ (if $\not\equiv +\infty$) it then follows that $u - v \geq 0$, and hence $u \geqslant v$. By varying u and v in either order we obtain $\dot{H}^h_f \geq \underset{\cdot}{H}^h_f$. Since $\underset{\cdot}{H}^h_f = \sup \underline{\mathcal{U}}^h_f$ is finely l.s.c. it follows that $\overline{H}^h_f \geq \underset{\cdot}{H}^h_f$, and similarly $\dot{H}^h_f \geq \underline{H}^h_f$.

(b) Consider first any point x_0 of the finely open set $V = \{\overline{H}^h_f > -\infty\}$. Since $\overline{H}^h_f$ is finely h-hyperharmonic and hence finely continuous on V we obtain by (a)

$$\overline{H}^h_f(x_0) = \text{f-}\lim_{x\to x_0,\,x\in V} \overline{H}^h_f(x) \geqslant \text{f-}\limsup_{x\to x_0,\,x\in V} \underset{\cdot}{H}^h_f(x) = \underline{H}^h_f(x_0),$$

the latter equality by regularization. The case $x_0 \in \{\underline{H}^h_f < +\infty\}$ is treated similarly, or by replacing f with $-f$.

(c) For any constant $c \geq 0$ we have $\dot{H}^h_{f+c} = \dot{H}^h_f + c$ and $\overline{H}^h_{f+c} = \overline{H}^h_f + c$. The former assertion in (c) therefore reduces to the case $f \geq 0$, whereby $\overline{\mathcal{U}}^h_f$ consists of positive functions because $\overline{H}^h_0 = 0$. We may assume that $\overline{H}^h_f \not\equiv +\infty$, and hence $h\overline{\mathcal{U}}^h_f \subset \mathcal{S}_+(U)$. Consider the cover of U by the finely open sets V_k from Lemma 3.2.14. Then $h\overline{\mathcal{U}}^h_f$ is a Perron family in the sense of Definition 3.4.1. In fact, if $u = v/h \in \overline{\mathcal{U}}^h_f$ then, for any $Y \in \Delta(U)$,

$$\liminf_{x\in U,\,x\to Y} \frac{1}{h}\widehat{R}_v^{U\setminus V_k}(x) = \liminf_{x\in U,\,x\to Y} u(x) \geqslant f(Y) \tag{3.6.4}$$

because $\widehat{R}_v^{U\setminus V_k} = v$ on $U \setminus \mathfrak{r}(V_k)$, where the regularized $\mathfrak{r}(V_k)$ of V_k (the least regular finely open set containing V_k) is the union of V_k and the polar set of all finely isolated points of $U \setminus V_k$. But V_k and $\mathfrak{r}(V_k)$ have the same fine closure in U and hence by Theorem 3.5.2 the same natural closure in $\overline{U}$. By Lemma 3.2.14 this natural closure of $\mathfrak{r}(V_k)$ is a naturally compact set contained in U, so $\overline{U} \setminus \overline{V}_k$ is a natural neighborhood of Y (in $\overline{U}$) on which we have $\widehat{R}_v^{U\setminus V_k} = v$, and that implies (3.6.4). It therefore follows by Theorem 3.4.2 that $\overline{H}^h_f = \widehat{\inf}\,\overline{\mathcal{U}}^h_f$ indeed is h-invariant, and that $\overline{H}^h_f(x) = \dot{H}^h_f(x)$ at any point $x \in U$ at which $\dot{H}^h_f(x) < +\infty$. □

Theorem 3.6.8 *Let f, g be two functions on $\Delta(U)$ with values in $\overline{\mathbb{R}}$.*

(1) *If $(f+g)(Y)$ is defined arbitrarily at points Y of indetermination then $\dot{H}^h_{f+g} \leqslant \dot{H}^h_f + \dot{H}^h_g$ and $\underset{\cdot}{H}^h_{f+g} \geq \underset{\cdot}{H}^h_f + \underset{\cdot}{H}^h_g$ hold at any point of U at which the right-hand member is well defined.*

(2) *If $(f+g)(Y)$ is defined arbitrarily at points Y of indetermination then $\overline{H}^h_{f+g} \leqslant \overline{H}^h_f + \overline{H}^h_g$ and $\underline{H}^h_{f+g} \geqslant \underline{H}^h_f + \underline{H}^h_g$ hold at any point of U at which the right hand member is well defined.*

(3) *For any point $x \in U$ we have $\dot{H}^h_f(x) < +\infty$ if and only if $\dot{H}^h_{f\vee 0}(x) < +\infty$.*

(4) *Let (f_j) be an increasing sequence of lower bounded functions $\Delta(U) \to]-\infty, +\infty]$. Writing $f = \sup_j f_j$ we have $\overline{H}^h_f = \sup_j \overline{H}^h_{f_j}$. If $\overline{H}^h_{f_j} = \underline{H}^h_{f_j}$ for every j then $\overline{H}^h_f = \underline{H}^h_f$.*

Proof For Assertions (1), (2), and (4) we proceed much as in [Do4, 1.VIII.7, Proofs of (c), (b), and (e)]. Assertion (1) for lower solutions reduces to that for upper solutions by replacing f, g by $-f, -g$. For upper solutions, suppose first that $f, g < +\infty$ (and so $f+g$ is well defined). Given two functions $u \in \overline{\mathcal{U}}^h_f$ and $v \in \overline{\mathcal{U}}^h_g$, we have $u + v \in \overline{\mathcal{U}}^h_{f+g}$ and hence $\dot{H}^h_{f+g}(x) \leqslant u(x) + v(x)$. Now suppose that $\dot{H}^h_f + \dot{H}^h_g$ is well defined at some point x, and (to avoid triviality) that $\dot{H}^h_f(x) + \dot{H}^h_g(x) < \infty$. By varying u and v we obtain $\dot{H}^h_{f+g}(x) \leqslant \dot{H}^h_f(x) + \dot{H}^h_g(x)$. In the general case define functions $f_0 < +\infty$, resp. $g_0 < +\infty$, which equal f, resp. g, except on the set $\{f = +\infty\}$, resp. $\{g = +\infty\}$. We may assume that these two exceptional sets are h-harmonic measure null, for if, e.g. $\{f = +\infty\}$ is not h-harmonic measure null then $\overline{H}^h_f \equiv +\infty$ by Theorem 3.6.6(d), hence $\dot{H}^h_f \equiv +\infty$, and so Assertion (1) holds trivially at any point of U at which $\dot{H}^h_f + \dot{H}^h_g$ is well defined. It therefore follows in view of Theorem 3.6.6(a), (e) that $f + g = f_0 + g_0$ off the h-harmonic measure null set $\{f = +\infty\} \cup \{g = +\infty\}$ and hence that

$$\dot{H}^h_{f+g} = \dot{H}^h_{f_0+g_0} \leqslant \dot{H}^h_{f_0} + \dot{H}^h_{g_0} \leqslant \dot{H}^h_f + \dot{H}^h_g$$

at any point where $\dot{H}^h_f + \dot{H}^h_g$ is well defined, the latter inequality because $f_0 \leqslant f$ and $g_0 \leqslant g$.

Assertion (2) reduces in the same way as Assertion (1) to the case of upper solutions and $f, g < +\infty$. As in the proof of Assertion (1) (with $f, g < +\infty$) we have $\dot{H}^h_{f+g} \leqslant u + v$ for any $u \in \overline{\mathcal{U}}^h_f$ and $v \in \overline{\mathcal{U}}^h_g$. By varying v we obtain $\dot{H}^h_{f+g} \leqslant u + \dot{H}^h_g$. Here and in the rest of the proof an inequality with a sum on the right-hand side is understood to hold in U wherever that sum is well defined. By finely l.s.c. regularization we infer that

$$\overline{H}^h_{f+g}(x) = \text{f-}\liminf_{y\to x,\, y\in U} \dot{H}^h_{f+g}(y) \leqslant \text{f-}\limsup_{y\to x,\, y\in U} u(y) + \text{f-}\liminf_{y\to x,\, y\in U} \dot{H}^h_g(y) = u(x) + \overline{H}^h_g(x)$$

because u is finely continuous. By varying $u \in \overline{\mathcal{U}}^h_f$ we get $\overline{H}^h_{f+g} \le \dot{H}^h_f + \overline{H}^h_g$. Since $\overline{H}^h_{f+g}$ and $\overline{H}^h_g$ are finely l.s.c. we finally obtain by finely l.s.c. regularization $\overline{H}^h_{f+g} \le \overline{H}^h_f + \overline{H}^h_g$ wherever the sum on the right is well defined.

For Assertion (3) let $x \in U$ be given, and suppose that $\dot{H}^h_f(x) < +\infty$. There is then $u \in \overline{\mathcal{U}}^h_f$ with $u(x) < +\infty$, u being finely h-superharmonic $\geqslant -c$ for some constant $c \ge 0$. It follows that $u + c \in \overline{\mathcal{U}}^h_f$ and even $u + c \in \overline{\mathcal{U}}^h_{f\vee 0}$ because $u + c \ge 0$. Consequently, $\dot{H}^h_{f\vee 0}(x) \leqslant u(x) + c < +\infty$. The converse follows from $\dot{H}^h_f \leqslant \dot{H}^h_{f\vee 0}$.

Assertion (4) reduces easily to the case of positive functions f_j. Then $\overline{H}^h_f$ and each $\overline{H}^h_{f_j}$ are positive and hence finely h-hyperharmonic by Theorem 3.6.7(c). The inequality $\overline{H}^h_f \geqslant \sup_j \overline{H}^h_{f_j}$ is obvious, and we may therefore assume that the positive finely h-hyperharmonic function $\sup_j \overline{H}^h_{f_j}$ is not identically $+\infty$, and therefore is h-invariant, again according to Theorem 3.6.7(c). Denote by E_j the polar subset $\{\overline{H}^h_{f_j} < \dot{H}^h_{f_j}\} \cup \{\overline{H}^h_{f_j} = \infty\}$ of U and write $E := \bigcup_j E_j$ (polar). For a fixed $x \in U \setminus E$ and for given $\varepsilon > 0$ choose functions $u_j \in \overline{\mathcal{U}}^h_{f_j}$ so that

$$u_j(x) < \dot{H}^h_{f_j}(x) + 2^{-j}\varepsilon = \overline{H}^h_{f_j}(x) + 2^{-j}\varepsilon. \tag{3.6.5}$$

In particular, u_j is finely h-superharmonic. Define a finely h-hyperharmonic function u by

$$u = \sup_j \overline{H}^h_{f_j} + \sum_j (u_j - \overline{H}^h_{f_j}) \ge \overline{H}^h_{f_k} + (u_k - \overline{H}^h_{f_k}) = u_k \tag{3.6.6}$$

for any index k. Then

$$\liminf_{\xi \to Y,\, \xi \in U} u(\xi) \geqslant \liminf_{\xi \to Y,\, xi \in U} u_k(\xi) \geqslant f_k(Y)$$

for every $Y \in \Delta(U)$ and every index k. Thus $u \in \overline{\mathcal{U}}^h_f$ and $\overline{H}^h_f \le \dot{H}^h_f \leqslant u$. In particular, by the former equality (3.6.6) and by (3.6.5),

$$\overline{H}^h_f(x) \leqslant u(x) \leqslant \sup_j \overline{H}^h_{f_j}(x) + \varepsilon,$$

and hence the finely h-hyperharmonic function $\overline{H}^h_f$ is finely h-superharmonic. Because $\sup_j \overline{H}^h_{f_j}$ is h-invariant and majorized by $\overline{H}^h_f$, there is by Lemma 2.9.4 a finely superharmonic function $s \ge 0$ on U such that $\overline{H}^h_f = s + \sup_j \overline{H}^h_{f_j}$, and hence $s(x) \leqslant \varepsilon$. For $\varepsilon \to 0$ we get $s(x) = 0$ and hence $s = 0$, and so indeed $\overline{H}^h_f = \sup_j \overline{H}^h_{f_j}$.

Finally, if each $\overline{H}^h_{f_j} = \underline{H}^h_{f_j}$ then by Theorem 3.6.7(b)

$$\overline{H}^h_f \geqslant \underline{H}^h_f \geqslant \sup_j \underline{H}^h_{f_j} = \sup_j \overline{H}^h_{f_j} = \overline{H}^h_f,$$

where thus equality prevails, and hence $\overline{H}^h_f = \underline{H}^h_f$. □

In what follows we define resolutivity of a numerical function on $\Delta(U)$, necessarily without requiring fine harmonicity of H^h_f everywhere, but only quasi-everywhere, and we show that resolutivity is equivalent to an apparently weaker, but technically supple concept called quasi-resolutivity, which possibly has not been considered before in the literature (for the classical case where U is Euclidean open). Theorem 3.6.22 below implies the corresponding known result for the classical case, cf. [Do4, Theorem 1.VIII.8].

Definition 3.6.9 A function f on $\Delta(U)$ with values in $\overline{\mathbb{R}}$ is said to be *h-resolutive* if $\overline{H}^h_f = \underline{H}^h_f$ on U and if this function, also denoted by H^h_f, is neither identically $+\infty$ nor identically $-\infty$.

It follows that H^h_f is finely continuous on U and finely h-harmonic on the finely open set $\{\overline{H}^h_f > -\infty\} \cap \{\underline{H}^h_f < +\infty\} = \{-\infty < H^h_f < +\infty\}$.

For any function $f : \Delta(U) \to \overline{\mathbb{R}}$ we consider the following two subsets of U:

$$E^h_f = \{\overline{H}^h_f = -\infty\} \cup \{\underline{H}^h_f = +\infty\} \cup \{\overline{H}^h_f \neq \underline{H}^h_f\},$$

$$P^h_f = \{\dot{H}^h_f > \overline{H}^h_f > -\infty\} \cup \{\underset{\cdot}{H}^h_f < \underline{H}^h_f < +\infty\},$$

of which P^h_f is polar, by the fundamental convergence theorem. Note that $E^h_{-f} = E^h_f$ and $P^h_{-f} = P^h_f$.

Definition 3.6.10 A function f on $\Delta(U)$ with values in $\overline{\mathbb{R}}$ is said to be *h-quasi-resolutive* if E^h_f is polar, or equivalently if the relations $\overline{H}^h_f > -\infty$, $\underline{H}^h_f < +\infty$, and $\overline{H}^h_f = \underline{H}^h_f$ hold quasi-everywhere on U.

When f is h-quasi-resolutive on U the functions $\overline{H}^h_f$ and $\underline{H}^h_f$ are finely h-hyperharmonic and finely h-hypoharmonic, respectively, off the polar set $\{\overline{H}^h_f = -\infty\} \cup \{\underline{H}^h_f = +\infty\}$, and they are actually equal and hence finely h-harmonic off the larger polar set E^h_f. We then denote by H^h_f the common restriction of $\overline{H}^h_f$ and $\underline{H}^h_f$ to $U \setminus E^h_f$. Since $\overline{H}^h_f$ is finely h-hyperharmonic on $U \setminus E_f$, and $\underline{H}^h_f$ is finely h-hypoharmonic there, it follows by Theorem 3.6.7(a) (still assuming f to be h-quasi-resolutive) that

$$\dot{H}^h_f = \underset{\cdot}{H}^h_f = \overline{H}^h_f = \underline{H}^h_f = H^h_f \quad \text{on } U \setminus E^h_f.$$

Lemma 3.6.11 *Every h-resolutive function f is h-quasi-resolutive.*

Proof The disjoint subsets $E_+ := \{H^h_f = +\infty\}$ and $E_- := \{H^h_f = -\infty\}$ of U are finely closed. For any fine component V of $U \setminus E_-$ such that $V \cap E_+$ is nonpolar we have $V \cap E_+ = V$, that is, $V \subset E_+$, because $\overline{H}^h_f$ is finely h-hyperharmonic on V. Denote by W the union of these fine components V, and by W^* the (countable) union of the remaining fine components V^* of $U \setminus E_-$. Then $W \subset E_+$ whereas the set $P := W^* \cap E_+$ is polar along with each $V^* \cap E_+$. Since $E_+ \cap E_- = \varnothing$ we obtain

$$E_+ = (U \setminus E_-) \cap E_+ = (W \Cup W^*) \cap E_+ = (W \cap E_+) \Cup (W^* \cap E_+) = W \Cup P,$$

$\Cup$ denoting disjoint union. Now, $(U \setminus P) \cap E_+ = E_+ \setminus P$ is finely closed relative to the non-void fine domain $U \setminus P$ (cf. Corollary 2.1.46), but also finely open, being equal to W as seen from the above display. Thus either $W = U \setminus P$ or $W = \varnothing$. But $W = U \setminus P$ would imply $U = W \Cup P = E_+$, contradicting $H^h_f \not\equiv +\infty$, and so actually $E_+ = P$ (polar). Similarly (or by replacing f with $-f$) it is shown that E_- is polar, and so altogether f is h-quasi-resolutive. □

Remark 3.6.12 In view of Lemma 3.6.11 an h-quasi-resolutive function f is h-resolutive if and only if $E^h_f = \varnothing$ (any polar subset of U such as for example $\{\overline{H}^h_f = -\infty\}$ being a proper subset). It will be shown in Corollary 3.6.21 that h-resolutivity and h-quasi-resolutivity are actually identical concepts.

Theorem 3.6.13 *Let $f, g : \Delta(U) \to \overline{\mathbb{R}}$ be h-resolutive, resp. h-quasi-resolutive. Then*

(1) *For $\alpha \in \mathbb{R}$ we have $E^h_{\alpha f} \subset E^h_f$ and hence αf is h-resolutive, resp. h-quasi-resolutive. Furthermore, $H^h_{\alpha f} = \alpha H^h_f$ on $U \setminus E^h_f$, and hence everywhere, resp. q.e., on U.*

(2) *If $f + g$ is defined arbitrarily at points of indetermination then $f + g$ is h-quasi-resolutive, and $H^h_{f+g} = H^h_f + H^h_g$ everywhere, resp. q.e., on U.*

(3) *$f \vee g$ and $f \wedge g$ are h-resolutive, resp. h-quasi-resolutive. In case $g = 0$ we have $H^h_{f\vee 0} = \frac{1}{h}\widehat{R}_{h\overline{H}^h_f \vee 0}$ everywhere on U.*

Proof For Assertion (1), consider separately the cases $\alpha > 0$, $\alpha < 0$, and $\alpha = 0$, and apply Remark 3.6.12.

For Assertion (2), choose any definition of $f + g$ at points of indetermination of that sum. We then have everywhere, resp. q.e., on U:

$$\overline{H}^h_{f+g} \geq \underset{\cdot}{H}^h_{f+g} \geqslant H^h_f + H^h_g \geq \overline{H}^h_{f+g} \tag{3.6.7}$$

the first inequality (valid everywhere on U) by Theorem 3.6.7(a), and the remaining inequalities everywhere, resp. q.e., by Assertions (1) and (2), respectively, in Theorem 3.6.8. Similarly,

$$\underline{H}^h_{f+g} \leq \dot{H}^h_{f+g} \leqslant H^h_f + H^h_g \leq \underline{H}^h_{f+g} \tag{3.6.8}$$

everywhere, resp. q.e. Clearly, all these inequalities in (3.6.7) and (3.6.8) are equalities. Since f and g are h-resolutive, resp. h-quasi-resolutive we conclude that $f+g$ indeed is h-resolutive, resp. h-quasi-resolutive, and that $H^h_{f+g} = H^h_f + H^h_g$ everywhere, resp. q.e., on U.

For Assertion (3) we adapt the proof in [Do4, 1.VIII.7 (d)]. Since $f \wedge g = -[(-f) \vee (-g)]$ and $f \vee g = [(f-g) \vee 0] + g$ it follows by (1) and (2) that the former claim in (3) reduces to the h-resolutivity, resp. h-quasi-resolutivity, of $f^+ = f \vee 0$. For given $x \in U \setminus (E^h_f \cup P^h_f)$ and integers $j > 0$ choose $u_j \in \overline{\mathcal{U}}^h_f$ with $u_j(x) \leqslant \dot{H}^h_f(x) + 2^{-j} = \overline{H}^h_f(x) + 2^{-j}$. The series $\sum_{j=k}^{\infty}(u_j - H^h_f)$ of positive finely h-superharmonic functions on $U \setminus (E^h_f \cup P^h_f)$ (u_j being likewise restricted to $U \setminus (E^h_f \cup P^h_f)$) has a positive finely h-superharmonic sum, finite at x. Recall that H^h_f is defined and finely h-harmonic on $U \setminus E^h_f$ and in particular on $U \setminus (E^h_f \cup P^h_f)$. Consequently, $H^h_f \vee 0$ is finely h-subharmonic (and positive) on $U \setminus (E^h_f \cup P^h_f)$ and majorized there by $\overline{H}^h_{f\vee 0}$, which is finite valued on $U \setminus (E^h_f \cup P^h_f)$ by (3) in Theorem 3.6.8 because $\overline{H}^h_f < +\infty$ on $U \setminus (E^h_f \cup P^h_f)$ and because $\overline{H}^h_f = \dot{H}^h_f$ there. Now apply Theorem 2.4.19 with f replaced by $h\overline{H}^h_f \vee 0$ on U, which is finely subharmonic on $U \setminus (E^h_f \cup P^h_f)$. Using the notion of sweeping relative to U of a function $U \to [0, +\infty]$, see Definition 2.2.6, it follows that $\frac{1}{h}\widehat{R}_{\overline{H}^h_f \vee 0}$ is finely h-harmonic on $U \setminus (E^h_f \cup P^h_f)$, being majorized there by $\overline{H}^h_f \vee 0$ and hence by the larger finely h-superharmonic function $\overline{H}^h_{f\vee 0} < +\infty$ on $U \setminus (E^h_f \cup P^h_f)$. The positive function

$$\frac{1}{h}\widehat{R}_{h\overline{H}^h_f \vee 0} + \sum_{j=k}^{\infty}(u_j - H^h_f) \tag{3.6.9}$$

restricted to $U \setminus (E^h_f \cup P^h_f)$ is therefore finely h-superharmonic. Moreover, this positive finely h-superharmonic function on $U \setminus (E^h_f \cup P^h_f)$ majorizes $u_k \in \overline{\mathcal{U}}^h_f$ there (being $\geqslant \frac{1}{h}\widehat{R}_{h\overline{H}^h_f \vee 0} + (u_k - H^h_f) \geqslant u_k$ there), and this majorization remains in force after extension by fine continuity to U, cf. Theorem 2.1.43. Thus the extended positive function (3.6.9) belongs to $\overline{\mathcal{U}}^h_{f\vee 0}$. For $k \to \infty$ it follows that $\overline{H}^h_{f\vee 0} \leqslant \frac{1}{h}\widehat{R}_{h\overline{H}^h_f \vee 0}$ q.e. on U, and hence everywhere on U, by fine continuity. In fact, being $\geq 0 > -\infty$, $\overline{H}^h_{f\vee 0}$ is finely h-hyperharmonic on U.

On the other hand it follows from Theorem 3.6.4 that $\underline{H}^h_{f\vee 0}$ is finely h-harmonic q.e. on $\{\underline{H}^h_{f\vee 0} < +\infty\}$, which is a finely open subset of $\{\dot{H}^h_{f\vee 0} < +\infty\} = \{\dot{H}^h_f < +\infty\}$ in view of Theorem 3.6.7(a) together with (3) in Theorem 3.6.8. Since f is h-quasi-resolutive, cf. Lemma 3.6.11, it follows that $\underline{H}^h_{f\vee 0}$ is finely h-harmonic, in particular finely h-superharmonic, q.e. on U. And $\underline{H}^h_{f\vee 0}$ majorizes both $\underline{H}^h_f$ and 0, hence majorizes $\underline{H}^h_f \vee 0$, so $\underline{H}^h_{f\vee 0} \geqslant \frac{1}{h}\widehat{R}_{h\underline{H}^h_f \vee 0}$ q.e. and in fact everywhere on U because $\underline{H}^h_{f\vee 0}$ is finely u.s.c. We have thus established the second inequality and by Theorem 3.6.7(b) the last inequality in the following string

$$\overline{H}^h_{f\vee 0} \leqslant \frac{1}{h}\widehat{R}_{h\overline{H}^h_f \vee 0} = \frac{1}{h}\widehat{R}_{h\underline{H}^h_f \vee 0} \leqslant \underline{H}^h_{f\vee 0} \leq \overline{H}^h_{f\vee 0} \tag{3.6.10}$$

everywhere on U. The first inequality here was established at the end of the preceding paragraph. The equality holds, again by Theorem 3.6.7(b), because $\underline{H}^h_f = \overline{H}^h_f$ on $U \setminus (E^h_f \cup P^h_f)$ and hence q.e. on U. Thus (3.6.10) holds, and all inequalities are equalities. As shown above, $\underline{H}^h_{f\vee 0}$ is finely harmonic q.e. and hence finite q.e. on U. Consequently, $f \vee 0$ is indeed h-resolutive, resp. h-quasi-resolutive, cf. Remark 3.6.12, and $H^h_{f\vee 0} = \frac{1}{h}\widehat{R}_{h\overline{H}^h_f \vee 0}$ holds everywhere on U in either case. □

Lemma 3.6.14 *Let f be an h-quasi-resolutive function on $\Delta(U)$. If f^+ and f^- are h-resolutive then so is f.*

Proof According to Lemma 3.6.11, f^+ and f^- (and hence $-f^+$ and $-f^-$ by (1) in Theorem 3.6.13) are h-quasi-resolutive (besides being h-resolutive). It follows that the finely hyperharmonic function $\overline{H}^h_{f^+} \geq 0$ on U is not $\equiv +\infty$ and hence is $< +\infty$ q.e., and similarly $\overline{H}^h_{-f^-} > -\infty$ q.e. Since $-f^- \leqslant f \leqslant f^+$ it follows that

$$-\infty < \overline{H}^h_{-f^-} \leqslant \overline{H}^h_f \leqslant \overline{H}^h_{f^+} < +\infty \quad \text{q.e.},$$

and similarly the same with $\overline{H}$ replaced by $\underline{H}$ throughout (corresponding to replacing f by$-f$, again in view of (1) in Theorem 3.6.13). Applying (1) and (2) in Theorem 3.6.8, the latter to the sum $f = f^+ + (-f^-)$ which clearly is well defined on $\Delta(U)$, we obtain

$$\overline{H}^h_f \leqslant \overline{H}^h_{f^+} - \underline{H}^h_{f^-} = \underline{H}^h_{f^+} - \overline{H}^h_{f^-} \leqslant \underline{H}^h_f \quad \text{q.e.}$$

It follows that $\overline{H}^h_f = \underline{H}^h_f$ q.e. because $\overline{H}^h_f \geqslant \underline{H}^h_f$ q.e. by Theorem 3.6.7(b) since $\overline{H}^h_f > -\infty$ q.e. by quasi-resolutivity of f. □

Corollary 3.6.15 *Let (f_j) be an increasing sequence of lower bounded h-resolutive, resp. h-quasi-resolutive functions $\Delta(U) \to]-\infty, +\infty]$, and let $f = \sup_j f_j$. If $\overline{H}^h_f \not\equiv +\infty$ then f is h-resolutive, resp. h-quasi-resolutive.*

Proof By adding a constant to all f_j we reduce the claims to the case $f_j \geq 0$ in view of (2) in Theorem 3.6.13. For every j we then have $\underline{H}^h_f \geqslant \underline{H}^h_{f_j} = \overline{H}^h_{f_j} \geq 0$ everywhere, resp. q.e., the equality everywhere, resp. q.e. by definition of h-resolutivity and h-quasi-resolutivity, and the inequalities by Theorem 3.6.7(b). Hence $\underline{H}^h_f \geqslant \sup_j \overline{H}^h_{f_j} = \overline{H}^h_f$ everywhere, resp. q.e., according to (4) in Theorem 3.6.8. Again by Theorem 3.6.7(b) we have $\underline{H}^h_f \leqslant \overline{H}^h_f$ on $\{\overline{H}^h_f > -\infty\} = U$, and we conclude that $\underline{H}^h_f = \overline{H}^h_f$ everywhere, resp. q.e. Under the extra hypothesis $\overline{H}^h_f \not\equiv +\infty$, the finely hyperharmonic function $\overline{H}^h_f \geq 0$ is finite q.e. on U. Consequently, f is indeed h-resolutive, resp. h-quasi-resolutive. □

Recall that μ_h denotes the unique measure on $\overline{U}$ carried by $\Delta_1(U)$ and representing h (which is invariant, cf. Lemma 2.9.2 and Theorem 3.2.20), that is, $h = K\mu_h = \int K(\cdot, Y)d\mu_h(Y)$.

Theorem 3.6.16 *For any μ_h-measurable subset A of $\Delta(U)$ the indicator function 1_A is h-resolutive, and*

$$H^h_{1_A} = \frac{1}{h}\int_A K(\cdot, Y)d\mu_h(Y) = \frac{1}{h}\widehat{R}^A_h \tag{3.6.11}$$

on U. In particular, the constant function 1 on $\Delta(U)$ is h-resolutive and hence h-quasi-resolutive, and $H^h_1 = 1$.

Proof Because $h = K\mu_h$ and because μ_h is carried by $\Delta_1(U)$ we may apply Theorem 3.5.7 and Theorem 3.5.6, the latter with h there replaced by the minimal invariant function $K(\cdot, Y)$ with pole $Y \in \Delta_1(U)$, cf. Definition 3.5.3 and Theorem 3.5.5. Furthermore, we invoke the notion of sweeping $\widehat{R}^A_u$ of a function $u \in \mathcal{S}_+(U)$ on a subset A of $\Delta(U)$, defined by

$$\widehat{R}^A_u = \widehat{\inf}\{\widehat{R}^{W\cap U}_u : W \in \mathcal{W}(A)\},$$

cf. Definition 3.4.3 and subsequent text. Here $\mathcal{W}(A)$ denotes the family of all naturally open subsets of $\overline{U}$ such that $W \supset A$. For any $Y \in \Delta_1(U)$ and any set $A \subset \Delta(U)$, we have

$$\widehat{R}^A_{K(\cdot,Y)} = 1_A(Y)K(\cdot, Y),$$

again by Theorem 3.5.6. For the proof of Theorem 3.6.16, we have

$$\widehat{R}^A_h = \widehat{R}^A_{K\mu_h} = \int_{\Delta_1(U)} \widehat{R}^A_{K(\cdot,Y)}d\mu_h(Y) = \int_{\Delta(U)} K(\cdot, Y)1_A(Y)d\mu_h(Y).$$

Now consider any finely h-hyperharmonic function $u = v/h \geq 0$ on U such that $u \geq 1$ on some open set $W \in \mathcal{W}(A)$ as defined above. Then $u \in \overline{\mathcal{U}}^h_{1_A}$ and hence $u \geqslant \dot{H}^h_{1_A} \geqslant \overline{H}^h_{1_A}$. A function u as just described is $u = 1$ (hence $v = h$). By varying W it follows by Definition 3.4.3 as recalled above, that $\frac{1}{h}\widehat{R}^A_h \geqslant \overline{H}^h_{1_A}$. We have

$$\frac{1}{h}\int_{\Delta(U)} K(\cdot, Y)1_A(Y)d\mu_h(Y) = \frac{1}{h}\widehat{R}^A_h \geq \overline{H}^h_{1_A}. \tag{3.6.12}$$

Applying this inequality to the μ_h-measurable set $\Delta(U) \setminus A$ in place of A we obtain

$$\frac{1}{h}\int_{\Delta(U)} K(\cdot, Y)1_{\Delta(U)\setminus A}(Y)d\mu_h(Y) \geqslant \overline{H}^h_{1_{\Delta(U)\setminus A}}. \tag{3.6.13}$$

By adding the left hand, resp. right hand, members of (3.6.12) and (3.6.13) this leads by (2) in Theorem 3.6.8 to

$$1 = \frac{1}{h}\int_{\Delta_1(U)} K(\cdot, Y)d\mu_h(Y) \geq \overline{H}^h_{1_A} + \overline{H}^h_{1_{\Delta(U)\setminus A}} \geq \overline{H}^h_1 = 1. \tag{3.6.14}$$

Thus equalities prevail throughout in (3.6.12), (3.6.13), and (3.6.14). It follows altogether that

$$\begin{aligned}\underline{H}^h_{1_A} &= -\overline{H}^h_{-1_A} = 1 - \overline{H}^h_{1-1_A} = 1 - \overline{H}^h_{1_{\Delta(U)\setminus A}} \\ &= \overline{H}^h_{1_A} = \frac{1}{h}\widehat{R}^A_h = \frac{1}{h}\int_A K(\cdot, Y)d\mu_h(Y),\end{aligned}$$

so that indeed 1_A is h-resolutive and hence h-quasi-resolutive, and (3.6.11) holds. □

For any function $f : \Delta(U) \to \overline{\mathbb{R}}$ we define $f(Y)K(x, Y) = 0$ at points (x, Y) where $f(Y) = 0$ and $K(x, Y) = +\infty$. If f is μ_h-measurable then so is $Y \mapsto f(Y)K(x, Y)$ for each $x \in U$ because $K(x, Y) > 0$ is μ_h-measurable (even l.s.c.) as a function of $Y \in \Delta(U)$ according to Theorem 3.2.2(i).

Theorem 3.6.17 *Let f be a μ_h-measurable lower bounded function on $\Delta(U)$. Then*

$$\overline{H}^h_f = \frac{1}{h}\int f(Y)K(\cdot, Y)d\mu_h(Y) > -\infty,$$

and $\overline{H}^h_f$ is either identically $+\infty$ or the sum of an h-invariant function and a constant ≤ 0. Furthermore f is h-quasi-resolutive if and only if f is h-resolutive, and that holds if and only if $\frac{1}{h}\int_{\Delta_1(U)} f(Y)K(\cdot, Y)d\mu_h(Y) < +\infty$ q.e. on U, or equivalently: at some point of U. In particular, every bounded μ_h-measurable function $f : \Delta(U) \to \mathbb{R}$ is h-resolutive and h-quasiresolutive.

Proof Consider first the case of a positive μ_h-measurable function f. Then f is the pointwise supremum of an increasing sequence of positive μ_h-measurable step functions f_j (that is, finite valued functions f_j taking only finitely many values, each on some μ_h-measurable set; in other words: affine combinations of indicator functions of μ_h-measurable sets). For any index j it follows by Theorem 3.6.16 and by (1) and (2) in Theorem 3.6.13 (the latter extended to finite sums) that each f_j is h-resolutive and that

$$H^h_{f_j} = \frac{1}{h}\int f_j(Y)K(\cdot, Y)d\mu_h(Y)$$

on U, whence

$$\begin{aligned}0 &\leq \frac{1}{h}\int f(Y)K(\cdot, Y)d\mu_h(Y) \\ &= \frac{1}{h}\sup_j \int f_j(Y)K(\cdot, Y)d\mu_h(Y) = \sup_j H^h_{f_j} = \overline{H}^h_f = \underline{H}^h_f\end{aligned}$$

by (4) in Theorem 3.6.8. Thus $f \geq 0$ is h-quasi-resolutive if and only if $H^h_f < +\infty$ q.e. on U, that is, iff $\int f(Y)K(\cdot, Y)d\mu_h(Y) < +\infty$ q.e. on U, or equivalently iff $\int f(Y)K(\cdot, Y)d\mu_h(Y) \not\equiv +\infty$, noting that H^h_f is finely hyperharmonic on the fine domain U. For a general lower bounded μ_h-measurable function f on $\Delta(U)$ there is a constant $c \geq 0$ such that $g := f + c \geq 0$ and hence $\overline{H}^h_g = \overline{H}^h_f + c \geq 0$ by Theorem 3.6.6(e) and the 5th paragraph after Corollary 3.6.2. It follows that

$$\overline{H}^h_f = \frac{1}{h}\int g(Y)K(\cdot, Y)d\mu_h(Y) - c = \frac{1}{h}\int f(Y)K(\cdot, Y)d\mu_h(Y) > -\infty$$

and hence by Theorem 3.6.7(c) applied to g that $\overline{H}^h_f = \overline{H}^h_g - c$ has the asserted form.

Next, consider a bounded μ_h-measurable function f on $\Delta(U)$. As just shown, we have

$$\overline{H}^h_f = \frac{1}{h}\int f(Y)K(\cdot, Y)d\mu_h(Y)$$

and the same with f replaced by $-f$, whence $\overline{H}^h_f = \underline{H}^h_f$, finite valued because f is bounded. Thus f is h-resolutive. Let $c \geq 0$ be a constant such that $|f| \leqslant c$. Then $\overline{H}^h_f = c - \underline{H}^h_{c-f}$ which is finely h-harmonic because $c - f \geq 0$ and so H^h_{c-f} is h-invariant by Theorem 3.6.7(c) and hence finely h-harmonic, being finite valued.

Returning to a general lower bounded μ_h-measurable function f, suppose first that f is h-quasi-resolutive. Then, as shown in the first paragraph of the proof, $\frac{1}{h}\int f(Y)K(\cdot, Y)d\mu_h(Y) = \overline{H}^h_f$ is finite q.e. on U, and in particular $\overline{H}^h_f \not\equiv +\infty$. Thus Corollary 3.6.15 applies to the increasing sequence of bounded μ_h-measurable and hence h-resolutive functions $f \wedge j$ converging to f, and we conclude that f is h-resolutive (in particular h-quasi-resolutive) and hence that $\frac{1}{h}\int f(Y)K(\cdot, Y)d\mu_h(Y) = \overline{H}^h_f$ is finite q.e. on U. □

Corollary 3.6.18 *Let $f : \Delta(U) \to \overline{\mathbb{R}}$ be μ_h-measurable. Then f is h-resolutive if and only if $|f|$ is h-resolutive.*

Proof If f is h-resolutive then so is $|f| = f \vee (-f)$ according to (3) and (1) in Theorem 3.6.13. Since $|f|$ is lower bounded (and μ_h-measurable) then by Theorem 3.6.17 $|f|$ is even h-resolutive and $|f|K(x, \cdot)$ is μ_h-integrable for quasi-every $x \in U$. So are therefore $f^+K(x, \cdot)$ and $f^-K(x, \cdot)$, and it follows, again by Theorem 3.6.17, that f^+ and f^- are h-resolutive. So is therefore $f = f^+ - f^-$ by Lemma 3.6.14. □

Theorem 3.6.19 *Every h-quasi-resolutive function $f : \Delta(U) \to \overline{\mathbb{R}}$ is μ_h-measurable.*

Proof We begin by proving this for $f = 1_A$, the indicator function of a subset A of $\Delta(U)$, cf. [Do4, 1.VIII.6 (f)]. Clearly, $0 \leq \underset{\cdot}{H}^h_f(x) \leqslant \dot{H}^h_f \leq 1$ by Corollary 3.6.2 and Theorem 3.6.7(a), and hence $\overline{H}^h_f = \dot{H}^h_f$ and $\underline{H}^h_f = \underset{\cdot}{H}^h_f$ according to Theorem 3.6.7(c). Choose a point $x_0 \in U$. For any integer $j \geq 1$ there exists $u'_j \in \overline{\mathcal{U}}^h_f$ such that $u'_j(x_0) < \overline{H}^h_f(x_0) + 1/j$. Since $\overline{\mathcal{U}}^h_f$ is lower directed and $1 \in \overline{\mathcal{U}}^h_f$ we may successively replace u'_j by some $u_j \in \overline{\mathcal{U}}^h_f$ with $u_j \leq 1 \wedge u'_1 \wedge \ldots \wedge u'_j$ and thereby arrange that $u_j \leq 1$, the sequence $(u_j) \subset \overline{\mathcal{U}}^h_f$ is decreasing, and $u_j(x_0) < \overline{H}^h_f(x_0) + 1/j$. Thus $\overline{H}^h_f(x_0) = \inf_j u_j(x_0)$. Denote by g_j the function defined on $\overline{U}$ by

$$g_j(Y) = \liminf_{z \to Y,\, z \in U} u_j(z)$$

for any $Y \in \overline{U}$. Clearly, g_j is l.s.c. on $\overline{U}$ and $1_A \leqslant g_j \leq 1$ on $\Delta(U)$. Write $f_2 = \inf_j g_j$ (restricted to $\Delta(U)$). Then f_2 is Borel measurable and $1 \geqslant f_2 \geqslant f = 1_A$, whence $\overline{H}^h_{f_2} \geqslant \overline{H}^h_f$. For the opposite inequality note that $u_j \in \overline{\mathcal{U}}^h_{f_2}$ because $g_j \geqslant f_2$. Hence $\inf_j u_j \geqslant \overline{H}^h_{f_2} \geqslant \overline{H}^h_f$ with equality at x_0. Because $0 \leqslant f \leqslant f_2 \leqslant 1$, $\overline{H}^h_f$ is invariant according to Theorem 3.6.7(c), and hence finely harmonic by Lemma 2.9.2. It therefore follows that the function $\overline{H}^h_{f_2} - \overline{H}^h_f$ is positive and finely h-superharmonic on U. Being 0 at x_0 it is identically 0 according to Corollary 2.1.27, and so $\overline{H}^h_{f_2} = \overline{H}^h_f$. Similarly there is a positive Borel measurable function $f_1 \leqslant f$ such that $\underline{H}^h_{f_1} = \underline{H}^h_f$. Since f is h-quasi-resolutive we obtain from Theorem 3.6.7(a), (b) q.e. on U

$$H^h_f = \underline{H}^h_{f_1} \leqslant \overline{H}^h_{f_1} \leqslant H^h_f \leqslant \underline{H}^h_{f_2} \leqslant \overline{H}^h_{f_2} = H^h_f,$$

thus with equality q.e. all through. Hence f_1 and f_2 are h-quasi-resolutive, and so is therefore $f_2 - f_1$ by Assertions (1) and (2) of Theorem 3.6.13, which also shows that $H^h_{f_2 - f_1} = H_{f_2} - H_{f_1} = 0$ q.e. Because $f_2 - f_1$ is positive and Borel measurable on $\Delta(U)$ it follows by Theorem 3.6.17 that $\int (f_2(Y) - f_1(Y))K(\cdot, Y)d\mu_h(Y) = 0$, and hence $f_1 = f_2$ μ_h-a.e. because $K(\cdot, Y) > 0$ is l.s.c. by Theorem 3.2.2(i). It follows that $f = 1_A$ is μ_h-measurable, and so is therefore A.

Next we treat the case of a finite valued h-quasi-resolutive function f on $\Delta(U)$. Adapting the proof given in [Do4, p. 115] in the classical setting we consider the space $\mathcal{C}(\overline{\mathbb{R}}, \mathbb{R})$ of continuous (hence bounded) functions $\overline{\mathbb{R}} \to \mathbb{R}$, and denote by Φ the space of functions $\varphi \in \mathcal{C}(\overline{\mathbb{R}}, \mathbb{R})$ such that $\varphi \circ f$ is h-quasi-resolutive. In view of Theorem 3.6.13, Φ is a vector lattice, closed under uniform convergence because $|\varphi_j - \varphi| < \varepsilon$ implies $|H^h_{\varphi_j \circ f} - \overline{H}^h_{\varphi \circ f}| \leqslant \varepsilon$ and $|H^h_{\varphi_j \circ f} - \underline{H}^h_{\varphi \circ f}| \leqslant \varepsilon$ on $U \setminus E^h_f$, and so $|\overline{H}^h_{\varphi \circ f} - \underline{H}^h_{\varphi \circ f}| \leq 2\varepsilon$ on $U \setminus E^h_f$. We infer that $\overline{H}^h_{\varphi \circ f} = \underline{H}^h_{\varphi \circ f}$ (finite values) q.e. on U, and so $\varphi \circ f$ is indeed h-quasi-resolutive. Furthermore, Φ includes the functions $\varphi_p : t \mapsto (1 - |t - p|) \vee 0$ on $\mathbb{R}$ for integers $p \geq 1$, again by Theorem 3.6.13. These functions separate points of $\mathbb{R}$. In fact, for distinct $s, t \in \mathbb{R}$, say $s < t$, take $p = [s]$ (that is, $p \leqslant s < p + 1$). If also $p \leqslant t < p + 1$ then clearly $\varphi_p(t) < \varphi_p(s) \leq 1$, and in the remaining case $t \geqslant p + 1$ we have $\varphi_p(t) = 0 < \varphi_p(s)$. It therefore follows by the lattice version of the Stone-Weierstrass theorem that $\Phi = \mathcal{C}(\overline{\mathbb{R}}, \mathbb{R})$. Next, the class Ψ of (not necessarily continuous) functions $\psi : \mathbb{R} \to \mathbb{R}$ for which $\psi \circ f$ is h-quasi-resolutive is closed under bounded monotone convergence, by Corollary 3.6.15 (adapted to a bounded monotone sequence of functions $f_j \to f$). Along with the continuous functions $\mathbb{R} \to \mathbb{R}$, Ψ therefore includes every bounded Borel measurable function $\mathbb{R} \to \mathbb{R}$. In particular, the indicator function 1_J of an interval $J \subset \mathbb{R}$ belongs to Ψ, and hence $1_J \circ f$ is h-quasi-resolutive. We conclude by the first part of the proof that the h-quasi-resolutive indicator function $1_J \circ f = 1_{f^{-1}(J)}$ is μ_h-measurable for every interval $J \subset \mathbb{R}$, and so is therefore the function f.

Finally, for an arbitrary h-quasi-resolutive function $f : \Delta(U) \to \overline{\mathbb{R}}$, write $A_+ := \{f = +\infty\}$ and $A_- := \{f = -\infty\}$. By (3) in Theorem 3.6.13, $f \vee 0$ is h-quasi-resolutive and

$$\overline{H}^h_{(+\infty)1_{A+}} = \overline{H}^h_{f1_{A+}} \leqslant \overline{H}^h_{f\vee 0} < +\infty \tag{3.6.15}$$

on $U \setminus (E^h_f \cup P^h_f)$, cf. paragraph preceding Definition 3.6.10. It follows that $2\overline{H}^h_{(+\infty)1_{A+}} = \overline{H}_{(+\infty)1_{A+}} < +\infty$ q.e. on U, hence $\overline{H}^h_{(+\infty)1_{A+}} = 0$ q.e. (and hence everywhere) on U. Since $-f$ likewise is h-quasi-resolutive we have $\underline{H}^h_{(+\infty)1_{A-}} = 0$ q.e. Furthermore, $0 \leqslant \underline{H}^h_{(+\infty)1_{A+}} \leqslant \overline{H}^h_{(+\infty)1_{A+}} = 0$ q.e. by Theorem 3.6.7(b), and similarly $\underline{H}^h_{(+\infty)1_{A+}} = \overline{H}^h_{(+\infty)1_{A-}} = 0$ q.e. Writing $A = A_+ \cup A_- = \{|f| = +\infty\}$ we infer from (2) in Theorem 3.6.13 that $1_A = 1_{A_+} + 1_{A_-}$ is h-quasi-resolutive and hence so is $(+\infty)1_A$ in view of (3.6.15) and Corollary 3.6.15. Furthermore, $\overline{H}^h_{(+\infty)1_A} = \overline{H}^h_{(+\infty)1_{A+}} + \overline{H}^h_{(+\infty)1_{A-}} = 0$ q.e. As shown in the first paragraph of the proof it follows that A is μ_h-measurable. Define $g : \Delta(U) \to \mathbb{R}$ by $g = f$ except that $g = 0$ on $A = \{|f| = +\infty\}$. Then $g \leqslant f + (+\infty)1_A$ and $f \leqslant g + (+\infty)1_A$, and hence by Theorem 3.6.8

$$\overline{H}^h_g \leqslant H^h_f + \overline{H}^h_{(+\infty)1_A} = H^h_f \leqslant \underline{H}^h_g + \underline{H}^h_{(+\infty)1_A} = \underline{H}^h_g$$

quasi-everywhere, actually with equalities q.e., and hence g is h-quasi-resolutive along with f. As established in the second paragraph of the proof, g is μ_h-measurable and so is f. In fact, $\{f \neq g\} = A$ is μ_h-measurable with $\mu_h(A) = 0$ by the following corollary. □

Corollary 3.6.20 (cf. [Do4, p. 114].) *A set $A \subset \Delta(U)$ is h-harmonic measure null if and only if A is μ_h-measurable with $\mu_h(A) = 0$.*

Proof If A is μ_h-measurable with $\mu_h(A) = 0$ then

$$\overline{H}^h_{1_A} = \frac{1}{h}\int 1_A(Y)K(\cdot, Y)d\mu_h(Y) = 0$$

according to Theorem 3.6.17, and hence A is μ_h-harmonic null (Definition 3.6.5). Conversely, if $\overline{H}^h_{1_A} = 0$ then $0 \leqslant \underline{H}^h_{1_A} \leqslant \overline{H}^h_{1_A} = 0$ and so 1_A is h-resolutive. It follows by Theorem 3.6.19 that 1_A is μ_h-measurable, and again by Theorem 3.6.17 that $\frac{1}{h}\int 1_A(Y)K(\cdot, Y)d\mu_h(Y) = \overline{H}^h_{1_A} = 0$. Since $K(x, \cdot) > 0$ is l.s.c. by Theorem 3.2.2(i) we conclude that indeed $\mu_h(A) = 0$. □

Corollary 3.6.21 *A function f on $\Delta(U)$ with values in $\overline{\mathbb{R}}$ is h-resolutive if and only if f is h-quasi-resolutive.*

Proof The "only if" part is contained in Lemma 3.6.11. For the "if" part, suppose that f is h-quasi-resolutive and hence μ_h-measurable, by Theorem 3.6.19. If $f \geq 0$ then f is h-resolutive according to Theorem 3.6.17. For arbitrary $f : \Delta(U) \to \overline{\mathbb{R}}$ this applies to f^+ and f^-, which are h-quasi-resolutive according to (3) in Theorem 3.6.13. Consequently, $f = f^+ - f^-$ is likewise h-resolutive by (1) and (2) in Theorem 3.6.13. □

Theorem 3.6.22 *A function $f : \Delta(U) \to \overline{\mathbb{R}}$ is h-resolutive if and only if the function $Y \mapsto f(Y)K(x, Y)$ on $\Delta(U)$ is μ_h-integrable for quasi-every $x \in U$. In the affirmative case, we have quasi-everywhere on U*

$$H^h_f = \frac{1}{h}\int f(Y)K(\cdot, Y)d\mu_h(Y).$$

Proof Suppose first that f is h-resolutive. By Theorem 3.6.19 f is then μ_h-measurable. According to Assertions (1) and (3) in Theorem 3.6.13 the function $|f| = f \vee (-f)$ is also h-resolutive, and it follows by Theorem 3.6.17 that $\overline{H}^h_{|f|}(x) = \frac{1}{h}\int |f(Y)|K(x, Y)d\mu_h(Y) < +\infty$ for every quasi-every $x \in U$ (or equivalently: for some point x in U). Conversely, suppose that $fK(x, \cdot)$ is μ_h-integrable for quasi-every $x \in U$. For any $x \in U$, $K(x, \cdot) > 0$ is l.s.c. and hence μ_h-measurable on $\overline{U}$ by Theorem 3.2.2(i), and so f^+ and f^- must be μ_h-measurable. By Theorem 3.6.17, f^+ and f^- are therefore h-quasi-resolutive, that is, h-resolutive by Corollary 3.6.21, and so is therefore $f = f^+ - f^-$ by (1) and (2) in Theorem 3.6.13. □

Remark 3.6.23 In the case where U is Euclidean open it follows by the Harnack convergence theorem for harmonic functions (not extendable to finely harmonic functions) for any numerical function f on $\Delta(U)$ that $\dot{H}^h_f$ is h-hyperharmonic on U (in particular $> -\infty$) and hence equal to $\overline{H}^h_f$ (except if $\overline{H}^h_f \equiv -\infty$). It follows by Theorem 3.6.7(a) that $\overline{H}^h_f = \dot{H}^h_f \geq \underset{.}{H}^h_f = \underline{H}^h_f$. If f is h-resolutive then by Definition 3.6.9 equality prevails on all of U. Summing up, Theorem 3.6.22 is a (proper) extension of the corresponding classical result, cf., e.g. [Do4, Theorem 1.VIII.8]. In the present setting where U is only a fine domain, note that even if $f \geq 0$ is h-resolutive, H^h_f is h-invariant, but finely h-harmonic only if finite valued, and that need not be the case, as it can be seen from Example 3.6.3. In fact, μ_h is a bounded measure (restriction to $\Delta_1(U)$ of a Radon measure on the compact space $\overline{U}$), and each function u_k is bounded on $\overline{\omega} \supset \overline{U}$. It follows that $u_k|_{\Delta_1(U)}$ is μ_h-integrable and therefore h-resolutive by Theorem 3.6.22.

Remark 3.6.24 For a brief discussion of an alternative, but compatible concept of h-resolutivity based on the minimal-fine (mf) topology, cf. Definition 3.5.1, instead of the natural topology, and therefore using Theorem 3.6.1 instead of its corollary, see [EKF3, Section 5].

Notes: El Kadiri [EK2] introduced the topology defined in Sect. 3.1 on the cone $\mathcal{S}_+(U)$ of the positive finely superharmonic functions on a fine domain U and proved that $\mathcal{S}_+(U)$ has a compact base and established the integral representation in $\mathcal{S}_+(U)$. The definition of the Martin boundary and the Martin kernel of a fine domain was given by El Kadiri and Fuglede in [EKF1], where they obtained the integral representation of invariant functions and established a Fatou-Naïm-Doob theorem for finely positive superharmonic functions. They also introduced the minimal-fine topology and studied the sweeping on subsets of the Martin space of a fine domain U and solved the Dirichlet problem at the Martin boundary of U in [EKF2, EKF3].

Chapter 4
Further Developments

This chapter deals with some further developments in fine potential theory. In Sect. 4.1, we prove that any fine domain $U \subset \mathbb{R}^N$ ($N \geq 2$) is polygonally connected and we apply this to a result about the existence of certain asymptotic paths (see Theorem 4.1.7).

In Sect. 4.2 we are interested in the relationship between finely superharmonic functions on a finely open subset U of a Greenian domain $\Omega \subset \mathbb{R}^N$ ($N \geq 2$) and the Dirichlet space $\widehat{\mathcal{D}}(\Omega)$ (the completion of $\mathcal{D}(\Omega) = C_0^\infty(\Omega)$ for the Dirichlet norm). We prove that the space $\mathcal{H}(\Omega, U)$ formed by the functions in $\widehat{\mathcal{D}}(\Omega)$ that are finely harmonic quasi-everywhere in U is the same as the set of the limits in $\widehat{\mathcal{D}}(\Omega)$ of functions in $\widehat{\mathcal{D}}(\Omega)$ that are harmonic in open sets containing U (see Theorem 4.2.21). At the end, we identify the apparent singularities for $\mathcal{H}(\Omega, U)$ to bounded point evaluations for $\mathcal{H}(\Omega, U)$, or equivalently, to points x of U such that $\varepsilon_x^{\complement U}$ is of finite energy (Theorem 4.2.25).

In Sect. 4.3, we define the Dirichlet Laplacian $A(U)$ on a bounded finely open set $U \subset \mathbb{R}^N$ and prove that $A(U)$ is a positive selfadjoint operator on $L^2(\mathfrak{r}(U))$ ($\mathfrak{r}(U)$ is the regularizer of U) and has a compact resolvent. Denoting by $\lambda_n(U)$ the nth eigenvalue of $A(U)$ (repeated according to multiplicity), we prove that for any decreasing sequence of bounded finely open sets U_i, $\lambda_n(U_i) \to \lambda_n(U)$ as $i \to \infty$, $n \in \mathbb{N}$, where $U = \mathrm{int}_f U_\infty$, the fine interior of $U_\infty = \bigcap_i U_i$.

In Sect. 4.4, we define and study the notion of finely harmonic morphisms. This extends the concept of harmonic morphisms between open subsets of $\mathbb{R}^N$ and $\mathbb{R}^{N'}$, cf. [CC2, Fu13, BW], to finely open sets. We prove that every nonconstant finely harmonic morphism of a fine domain $U \subset \mathbb{R}^N$ into $\mathbb{R}^{N'}$ ($N' \geq 2$) has a finely open range and determines a finely open mapping. Moreover, the pre-image of any polar subset of $\mathbb{R}^{N'}$ is polar.

In Sect. 4.5, we study the notion of a finely holomorphic function on a finely open subset of $\mathbb{C}$. This notion extends in a natural way the classical notion of holomorphic function on a Euclidean open subset of $\mathbb{C}$. By using the theory of BLD functions

M. El Kadiri and B. Fuglede, *Classical Fine Potential Theory*, Springer Monographs in Mathematics, https://doi.org/10.1007/978-981-96-0430-2_4

and the Cauchy transform of square integrable functions (with compact support) we prove some equivalent properties and some results on differentiation of finely holomorphic functions. In particular, the Taylor series of a finely holomorphic function f represents f asymptotically in a suitable fine neighborhood of a given point, and f is completely determined by this series. We prove that the set of zeros of a nonzero finely holomorphic function is at most countable. At the end we study the subspaces $\mathcal{O}^p(U)$ of $L^p(U)$ $(2 < p \leq +\infty)$ formed by the functions in $L^p(U)$ that are finely holomorphic in the finely open set $U \subset \mathbb{C}$.

4.1 Polygonal Connectivity of Fine Domains

Theorem 4.1.1 *Any two points x, y of a fine domain $U \subset \mathbb{R}^N$, $N \geqslant 2$, can be joined by a finite polygonal path contained in U.*

The planar case $N = 2$ reduces immediately to the case $N = 3$ applied to the fine domain $U \times \mathbb{R} \subset \mathbb{R}^3$ and the points $(x, 0)$ and $(y, 0)$ and any other point $(z, t) \in U \times \mathbb{R}$ by $(x, 0)$. We therefore consider the case $N \geqslant 3$ only.

We denote by $\mathbf{S}$ the unit sphere $\{x \in \mathbb{R}^N : |x| = 1\}$ in $\mathbb{R}^N$, and by $\mathrm{i} : \mathbb{R}^N \cup \{\infty\} \to \mathbb{R}^N \cup \{\infty\}$ the inversion in $\mathbf{S}$, given by $\mathrm{i}(\infty) = 0$, $\mathrm{i}(0) = \infty$, and $\mathrm{i}(x) = \frac{x}{|x|^2}$ for $x \in \mathbb{R}^N \setminus \{0\}$. A set $A \subset \mathbb{R}^N$ is said to be thin at ∞ if $\mathrm{i}(A)$ is thin at 0. For $N \geqslant 3$ this is equivalent to $\widehat{R}_1^A(0) \neq 1$, see, for example, [AG, Theorem 7.6.2]. The following Lemma is well known.

Lemma 4.1.2 *For $N \geqslant 3$ a set $A \subset \mathbb{R}^N$ is thin at ∞ if and only if $\widehat{R}_1^A$ is the Newtonian potential of a measure $\mu \geqslant 0$ on $\mathbb{R}^N$.*

Proof Suppose that A is thin at ∞, and write $\widehat{R}_1^A = p + u$ where p is a potential and u a harmonic function $\geqslant 0$ on $\mathbb{R}^N$. Then u is constant according to a version of Liouville's theorem, see Sect. 1.1.2, say $u = c$ with $0 \leqslant c < 1$ (for if $c = 1$ then $p = 0$ and hence $\widehat{R}_1^A = 1$, contradicting). Thus $p = \widehat{R}_1^A - c \geqslant (1 - c)\widehat{R}_1^A$, showing that $(1 - c)\widehat{R}_1^A$ is a potential, and so is therefore $\widehat{R}_1^A$. The converse is obvious. □

Lemma 4.1.3 *For any set $A \subset \mathbb{R}^N \setminus \{0\}$ $(N \geqslant 3)$ we have*

$$m^*(\pi(A)) \leqslant 2^{N-2}\widehat{R}_1^A(0),$$

where π denotes central projection onto the unit sphere $\mathbf{S}$ in $\mathbb{R}^N$, that is, $\pi(x) = \frac{x}{|x|}$ for $x \in \mathbb{R}^N \setminus \{0\}$, while m is the normalized surface measure on $\mathbf{S}$.

Proof The following intermediate relations will be obtained in the course of the proof:

$$m^*(\pi(A)) \leqslant \operatorname{cap}^* \pi(A) = \widehat{R}_1^{\pi(A)}(0) \leqslant 2^{N-2}\widehat{R}_1^A(0). \tag{4.1.1}$$

In view of Theorem 1.3.6(a) it suffices to consider the case of a bounded set A. Then A is thin at ∞, and hence by Lemma 4.1.2

$$\widehat{R}_1^A = h * \mu = \int h(\cdot - y)\, d\mu(y), \ h(x) = |x|^{2-N}, \tag{4.1.2}$$

where $\mathrm{cap}^*(A)$ denotes the outer Newtonian capacity of a set $A \subset \mathbb{R}^N$ and μ is the capacitary measure for A, see [Do4, 1.XIII.10–13], [AG, Sections 5.4–5.5]. The first two relations in (4.1.1) are simple and more or less known. In fact, writing $m_{\pi(A)}$ for the restriction of m to $\pi(A) \subset \mathbf{S}$ we have

$$h * m_{\pi(A)} \leqslant h * m = h \wedge 1 \leqslant 1,$$

(see [AG, Example 4.2.9, p. 99]) and hence $m^*(\pi(A)) \leqslant \mathrm{cap}^*\, \pi(A)$, see, for example, [AG, Theorem 5.5.5(i)]. Next, if λ denotes the outer capacitary measure for $\pi(A)$, then λ is carried by $\mathbf{S}$ and

$$\mathrm{cap}^*\, \pi(A) = \int_{\mathbf{S}} d\lambda = \int_{\mathbf{S}} h\, d\lambda = (h * \lambda)(0) = \widehat{R}_1^{\pi(A)}(0),$$

see, for example, [Do4, 1.XIII.11-12].

Concerning the latter inequality in (4.1.1), the outer capacitary measure μ for the bounded set A is carried by $\mathbb{R}^N \backslash \{0\}$ because $\int h\, d\mu \leqslant 1$ by (4.1.2) and $h(0) = +\infty$. Let ν denote the image of $h\mu$ under $\pi : \mathbb{R}^N \backslash \{0\} \to \mathbf{S}$. The total mass of ν is

$$\int d\nu = \int h\, d\mu = \widehat{R}_1^A(0),$$

again by (4.1.2). Below we show that

$$h * \nu \geqslant 2^{2-N} \text{ q.e. on } \pi(A). \tag{4.1.3}$$

This will imply that $h * \nu \geqslant 2^{2-N} \widehat{R}_1^{\pi(A)}$ on all of $\mathbb{R}^N$ by Theorem 1.3.2, in particular at 0, whence the latter inequality (4.1.1) as follows:

$$\widehat{R}_1^A(0) = \int h\, d\nu = (h * \nu)(0) \geqslant 2^{2-N} \widehat{R}_1^{\pi(A)}(0).$$

Thus it remains only to establish (4.1.3). For $x \in A$ we have

$$\begin{aligned}
(h * \nu)(\pi(x)) &= \int_{\mathbf{S}} \left|\pi(x) - y\right|^{2-N} d\nu(y) \\
&= \int_{\mathbb{R}^N \backslash \{0\}} \left|\pi(x) - \pi(y)\right|^{2-N} \left|y\right|^{2-N} d\mu(y).
\end{aligned}$$

It therefore suffices to prove that

$$\big|\pi(x) - \pi(y)\big|\big|y\big| \leqslant 2\big|x - y\big| \text{ for } x, y \in \mathbb{R}^{N-2} \tag{4.1.4}$$

since this will lead to (4.1.3) as follows:

$$(h * \nu)(\pi(x)) \geqslant 2^{2-N}(h * \mu)(x) = 2^{2-N}\widehat{R}_1^A(x) = 2^{2-N}$$

q.e. for $x \in A$.

Finally, to verify the inequality (4.1.4), write $z = |y|\pi(x)$. Then $|z| = |y|$, and the left hand member of (4.1.4) equals

$$|z - y| \leqslant |x - z| + |x - y| \leqslant 2|x - y|$$

because

$$|x - z| = |x - |y|\pi(x)| = ||x| - |y|||\pi(x)| = ||x| - |y|| \leq |x - y|.$$

This completes the proof of (4.1.1) and hence of Lemma 4.1.3. While the constant 2 in (4.1.4) is the best possible, the question whether 2^{N-2} is the best possible constant in Lemma 4.1.3 remains open. □

The following Theorem will easily lead to Theorem 4.1.1.

Theorem 4.1.4 *Every point of a fine domain U in $\mathbb{R}^N$ ($N \geqslant 2$) has a fine neighborhood $V \subset U$ such that any two distinct points $x, y \in V$ can be joined by a polygonal path in U consisting of just two straight segments $[x, z]$ and $[z, y]$ of equal length and such that the angle between $[x, y]$ and $[x, z]$ is less than prescribed number $\alpha > 0$.*

The idea of the proof of this proposition is that Lemma 4.1.3, when applied to $A = B\backslash U$ for a suitable ball B of radius r and center 0, controls the surface measure of the set $\pi(A)$ of directions ξ (in the unit sphere $\mathbf{S}$ in $\mathbb{R}^N$) for which the segment $[x, x + r\xi]$ of length r and direction ξ meets $\complement U$, that is, $[0, r\xi]$ is not contained in U.

Proof of Theorem 4.1.4. As in the case of Theorem 4.1.1 we may assume that $N \geqslant 3$. For the given point of U we may take the origin 0 in $\mathbb{R}^N$. Consider on the unit sphere $\mathbf{S}$ in $\mathbb{R}^N$ a cap of spherical radius α (supposed $< \frac{\pi}{3}$). Let σ denote the normalized surface measure of such a cap. According to Lemma 1.3.9(b) there is a number $r > 0$ small enough that

$$\widehat{R}_1^{B(0,2r)\backslash U}(0) < 2^{-N}\sigma. \tag{4.1.5}$$

By fine continuity of $\widehat{R}_1^{B(0,2r)\backslash U}$ this inequality holds at any point x (in place of 0) from a suitable fine neighborhood V of 0 in $\mathbb{R}^N$, and we may take $V \subset B(0, \frac{r}{2})$. Now consider two distinct $x, y \in V$ and the hyperplane $H \subset \mathbb{R}^N$ consisting of all points $z \in \mathbb{R}^N$ of equal distance from x and y. For any $z \in H$ the segment $[y, z]$ is contained in U if and only if $[x, z]$ is contained in the image U^* of U under reflection in H. We shall thus prove that there are points $z \in H$ such that $[x, z] \subset U \cap U^*$ and that the angle at x between $[x, y]$ and $[x, z]$ is $< \alpha$. The latter requirement means that $\frac{z-x}{|z-x|}$ lies on the (open) spherical cap C_α on $\mathbf{S}$ of spherical radius α and center $\frac{y-x}{|y-x|}$, hence of normalized measure σ by the definition of σ in the beginning of the proof.

Because $x, y \in V \subset B(0, \frac{r}{2})$ we have $|x - y| < r$ and $B(x, r), B(y, r) \subset B(0, 2r)$, and hence by (4.1.5)

$$\widehat{R}_1^{B(x,r)\setminus U}(x) < 2^{-N}\sigma, \ \widehat{R}_1^{B(y,r)\setminus U}(y) < 2^{-N}\sigma. \tag{4.1.6}$$

Reflection in H maps y to x and $B(y, r)\setminus U$ onto $B(x, r)\setminus U^*$. It follows by (4.1.6) that

$$\widehat{R}_1^{B(x,r)\setminus U^*}(x) = \widehat{R}_1^{B(y,r)\setminus U}(y) < 2^{-N}\sigma;$$

and hence

$$\widehat{R}_1^{B(x,r)\setminus(U\cap U^*)}(x) < 2^{1-N}\sigma.$$

After translation by $-x$ this writes

$$\widehat{R}_1^A(0) < 2^{1-N}\sigma$$

with

$$A = (B(x, r)\setminus(U \cap U^*)) - x.$$

Note that $x \in U \cap U^*$ and hence $A \subset \mathbb{R}^N\setminus\{0\}$. Furthermore, $[x, x + r\xi] \subset U \cap U^*$ for any $\xi \notin \pi(A)$. By Lemma 4.1.3,

$$m^*(C_\alpha \cap \pi(A)) \leqslant m^*(\pi(A)) \leqslant 2^{N-2}2^{1-N}\sigma = \frac{1}{2}\sigma = \frac{1}{2}m(C_\alpha),$$

and hence $C_\alpha\setminus\pi(A) \neq \varnothing$. Choose $\xi \in C_\alpha\setminus\pi(A)$, and consider the right triangle with vertices x, $\frac{1}{2}(x + y)$, and z, where z denotes the intersection of H and the line through x with direction ξ. The angle at x in that triangle is $< \alpha < \frac{\pi}{3}$ and the side $[x, \frac{1}{2}(x + y)]$ has length $\frac{1}{2}|y - x| < \frac{1}{2}r$. The side $[x, z]$ therefore has length $< r$, and hence $[x, z] \subset [x, x + r\xi] \subset U \cap U^*$. This completes the proof. □

Proof of Theorem 4.1.1. For given $x \in U$ let U_x denote the set of all points of U which can be joined to x by a finite polygonal line contained in U. Theorem 4.1.4 implies that U_x and $U\setminus U_x$ are finely open and that $U_x \neq \varnothing$. Since U is finely connected we conclude that indeed $U_x = U$. □

Thus fine connectivity of the finely open set $U \subset \mathbb{R}^N$ implies polygonal connectivity. We show that the converse is true in dimension $N = 2$, but fails for $N \geqslant 3$.

Theorem 4.1.5 *A finely open set $U \subset \mathbb{R}^2$ is finely connected if (and only if) U is polygonally (or just arcwise) connected.*

Proof Consider any fine component V of polygonally connected finely open set U. In particular, V is (Euclidean) connected, and V is finely closed relative to U. We show that V is even closed relative to U. In terms of the closure $\overline{V}$ of V consider any point $x \in U \cap (\overline{V} \setminus V)$. Since $U \setminus V$ is finely open and $x \in U \setminus V$ there exists a sequence of circles C_n centered at x and of radius $r_n \downarrow 0$ such that each $C_n \subset U \setminus V$, see [AG, Theorem 7.3.9]. Being connected, V meets at most one of the annuli between circles C_n and C_{n+1} but this contradicts $x \in \overline{V}$. Thus $U \cap \overline{V} = V$.

Since the fine topology is locally connected the fine components of U are finely open, and their cardinality is countable in view of the quasi-Lindelöf principle (any polar and finely open set being void). Let $U_0, U_1, \dots$ denote the fine components of U (if there are more than one of them), and let $x_0 \in U_0$, $x_1 \in U_1$. Let $t \mapsto x(t)$ be a continuous mapping of $[0, 1]$ into U such that $x(0) = x_0$ and $x(1) = x_1$. Since U_j is closed relative to U, each set $E_j := \{t \in [0, 1] : x(t) \in U_j\}$ is closed relative to U. Thus $[0, 1]$ has been decomposed into countable family of disjoint closed subsets $E_0, E_1, \dots$, and $0 \in E_0$, $1 \in E_1$. But this is impossible by a simple argument similar to that which serves to prove that the Cantor set is uncountable. □

Example 4.1.6 Let U_0 be a domain in $\mathbb{R}^3$ forming a Lebesgue spin near a boundary point $0 \in \mathbb{R}^3$, and let $U_1 = \complement \overline{U}_0$. Then U_0 and $U_1 \cup \{0\}$ are disjoint and finely open, so their union U is finely open, but finely disconnected. Because U_0 contains a segment of a line L in $\mathbb{R}^3$, U is polygonally connected (via a larger segment of L meeting also U_1).

We proceed to apply Theorem 4.1.1 to a result about existence of certain asymptotic paths. In his thesis from 1914 Felix Iversen [Iv] showed that, given a nonconstant entire function f, there exists a continuous path in the complex plane leading infinity and along which $f(z) \to \infty$. More generally, he proved the following: Let f be holomorphic on an open set U in $\mathbb{C}$. If $|f(z)| \leqslant \alpha$ on the boundary ∂U for some constant α, then either $|f(z)| \leqslant \alpha$ on all of U, or else there exists a path in U leading to ∞ and along which $f(z) \to \infty$ (and hence $z \to \infty$).

Iversen obtained this result in the course of his study of the inverse of a meromorphic function. It has long been known that there is a simpler proof based on the Phragmén-Lindelöf principle, cf. e.g. Bieberbach [Bi, p. 272]. As noted by Hayman [HK] this proof carries over so as to produce an asymptotic path (e.g. an infinite sectionally polygonal line) for any *continuous* subharmonic function u defined for example on $\mathbb{R}^2$, thus replacing $\log |f|$ in Iversen's theorem.

Naturally, one wants to dispense with the requirements of continuity of the subharmonic function u. The sets $\{u > \lambda\}$ ($\lambda \in \mathbb{R}$) are then no longer open, and Hayman and his student M.N.M. Talpur considered instead the closed sets $\{u \geqslant \lambda\}$ and their component which are continua. Using some deep analytical results of Hayman

[Hay] they settled the 2-dimensional case with a sectionally polygonal path. In higher dimensions their method led to a chain of continua rather than a path.

However, the sets $\{u > \lambda\}$ are finely open because u is finely continuous. We proceed to formulate and prove the following "fine" version of Iversen's theorem. In Corollary 4.1.8 below we thereby a full generalization of Iversen's theorem in which $\log|f|$ (f holomorphic) is replaced by an arbitrary (not necessarily continuous) subharmonic function u on an unbounded domain in $\mathbb{R}^N$ ($N \geqslant 2$).

Theorem 4.1.7 *Let u be finely subharmonic on an unbounded fine domain U in $\mathbb{R}^N$ ($N \geqslant 2$), and write $\sup_{x\in U} u(x) = \gamma$ ($\leqslant +\infty$). Suppose there is a constant $\alpha < \gamma$ such that*

$$\text{f-}\limsup_{x\to y,\, x\in U} u(x) \leqslant \alpha \quad \textit{for every } y \in \partial_f U. \tag{4.1.7}$$

Then there exists an injective sectionally polygonal path in U along which $u(x) \to \gamma$.

Corollary 4.1.8 *Let u be subharmonic on an unbounded domain U in $\mathbb{R}^N$ ($N \geqslant 2$), and write $\sup_{x\in U} u(x) = \gamma(\leqslant +\infty)$. Suppose there is a constant $\alpha < \gamma$ such that*

$$\text{f-}\limsup_{x\to y,\, x\in U} u(x) \leqslant \alpha \quad \textit{for every } y \in \partial U. \tag{4.1.8}$$

Then there exists an injective sectionally polygonal path Γ in U converging to ∞ along which $u(x) \to \gamma$. If $N = 2$ and $U = \mathbb{R}^2$ then necessarily $\gamma = +\infty$ except if u is constant.

For the proof of Theorem 4.1.7 (and hence of most of Corollary 4.1.8) we shall need the following "fine" version of Phragmén-Lindelöf principle mentioned above.

Theorem 4.1.9 *Let v be finely subharmonic and bounded from above [upper bounded] on a finely open proper subset V of $\mathbb{R}^N$ ($N \geqslant 2$). Suppose that*

$$\text{f-}\limsup_{x\to y,\, x\in V} v(x) \leqslant 0 \quad \textit{for every} \quad y \in \partial_f V.$$

If $N \geqslant 3$ suppose in addition that $\mathbb{R}^N \setminus V$ is not thin at ∞. Then $v \leqslant 0$.

Proof Let $c > 0$ denote a constant such that $v \leqslant c$, and define

$$w = \begin{cases} v_+ & \text{on } V \\ 0 & \text{on } \mathbb{R}^N \setminus V. \end{cases}$$

According to Lemma 2.1.51 and Theorem 2.1.22(a), $w \geqslant 0$ is subharmonic on $\mathbb{R}^N$, and $w \leqslant c$ because $v \leqslant c$. In the planar case $N = 2$ it follows that w is constant, cf. e.g. [AG, Corollary 3.5.4], and hence $w \equiv 0$ because $V \neq \mathbb{R}^2$. In the case $N \geqslant 3$, since $c - w$ is superharmonic $\geqslant 0$ on $\mathbb{R}^N$ and $c - w \geqslant c$ on $\mathbb{R}^N \setminus V$ we have $c - w \geqslant \widehat{R}_c^{\complement V} = c$ on $\mathbb{R}^N$ (complement and sweeping relative to $\mathbb{R}^N$), that is, $\widehat{R}_1^{\complement V} = 1$ on $\mathbb{R}^N$, by Lemma 4.1.2. Thus $w \leqslant 0$, and so indeed $v \leqslant v_+ = w \leqslant 0$ on V. □

Proof of Theorem 4.1.7. As in the case of Theorem 4.1.1 we may assume that $N \geqslant 3$. The crucial step in the proof of Theorem 4.1.7 is to show that there is a constant β with $\alpha \leqslant \beta < \gamma$ such that

$$\sup_{x \in V} u(x) = \gamma \tag{4.1.9}$$

for every fine component V of any set $\{u > \lambda\}$ with $\beta < \lambda < \gamma$. We prove this by contradiction.

If there is no such constant β then there exists a number μ with $\alpha < \mu < \gamma$ such that

$$\sup_{x \in W} u(x) < \gamma \tag{4.1.10}$$

for some fine component W of $\{u > \mu\}$. Then W is a proper fine subdomain of U, for if $W = U$ then $\{u > \mu\} = U$ in contradiction with (4.1.7) because $\alpha < \mu$. It follows that $U \cap \partial_f W \neq \varnothing$ because U is a fine domain and $W \neq \varnothing, U$. Clearly $u = \mu$ on $U \cap \partial_f W$ because W is a fine component of $\{u > \mu\}$. Applying Theorem 4.1.9 with V and v replaced by W and the bounded finely subharmonic function $(u - \mu)|_W > 0$, cf. (4.1.10), we find that $\mathbb{R}^N \backslash W$ must be thin at ∞, for otherwise $u - \mu \leqslant 0$ on $W \subset \{u > \mu\}$. Consider any number λ such that

$$\sup_{x \in W} u(x) < \lambda < \gamma, \tag{4.1.11}$$

and any fine component V of $\{u > \lambda\}$. For the proof that $\sup_{x \in V} u(x) = \gamma$ we may assume that u is bounded on V. We have $V \subset \{u > \lambda\} \subset \mathbb{R}^N \backslash W$. Consequently $\mathbb{R}^N \backslash V$ is unthin at ∞. It therefore follows by Theorem 4.1.9, now with v replaced by the bounded finely subharmonic $u - \lambda$ on V, that $u \leqslant \lambda < \gamma$ on V, which contradicts $V \subset \{u > \lambda\}$. Thus actually $\sup_{x \in V} u(x) = \gamma$ for any fine component V of $\{u > \lambda\}$ with λ satisfying (4.1.11). This contradicts the non-existence of β in (4.1.9).

We have thus shown that there does exist β with $\alpha \leqslant \beta < \gamma$ such that (4.1.9) holds for every fine component V of any set $\{u > \lambda\}$ with $\beta < \lambda < \gamma$. For the construction of the desired sectionally polygonal path L we choose a strictly increasing sequence of numbers λ_n with $\beta < \lambda_n < \gamma$ such that $\lambda_n \to \gamma$. Next we define recursively a decreasing sequence of fine components V_n of $\{u > \lambda_n\}$ and a sequence of points $x_n \in V_n \cap \{u > \lambda_{n+1}\}$, noting that the supremum of u over V_n equals $\gamma > \lambda_{n+1}$ because $\lambda_n > \beta$, and so x_n exists as stated. To pass from (V_n, x_n) to (V_{n+1}, x_{n+1}), define V_{n+1} as the fine component of x_n in $\{u > \lambda_{n+1}\} \ni x_n$, and x_{n+1} as any point of the non-void set $V_{n+1} \cap \{u > \lambda_{n+1}\} \subset \{u > \lambda_n\}$ and $x_n \in V_n$, so that V_n is the fine component of x_n in $\{u > \lambda_n\}$.

Now join $x_n \in V_n$ with $x_{n+1} \in V_{n+1} \subset V_n$ by polygonal path Γ_n in the fine domain $V_n \subset \mathbb{R}^N$, cf. Theorem 4.1.1. Putting these paths together one after the other we obtain a sectionally polygonal path Γ along which $u(x) \to \gamma$ because $u > \lambda_n$ on $\Gamma_n \cup \Gamma_{n+1} \cup \ldots$, and $\lambda_n \uparrow \gamma$.

In proving that Γ may be taken to be simple (that is, injective) we may assume that $u < \gamma$ in V_1 (and hence in every V_n), the alternative being the trivial case $u \equiv \gamma$ in V_1 according to Corollary 2.1.27 applied to $\gamma - u$. Being u.s.c., u has on each Γ_n a maximal value $\gamma_n (< \gamma)$. With this in mind it is easy to modify in a sequence of steps the preliminary path Γ (formed by the paths Γ_n) so as to obtain an injective path (formed by suitable subarcs of the arcs Γ_n), necessarily converging to ∞. □

Proof of Corollary 4.1.8. The existence of an injective sectionally polygonal path Γ in U along which $u(x) \to \infty$ is a particular case of Theorem 4.1.7. We now prove that Γ converges to ∞. Otherwise Γ had a finite limit point $x \in \overline{U}$ then $x \in \{u > \lambda_n\}$ for any n, and hence $x \in U$ in view of (4.1.7) (now with $\partial_f U$ replaced by ∂U). Being subharmonic, u is u.s.c. on U and hence there is a sequence (x_n) in U with $u(x_n) > \lambda_n$ such that $u(x) \geqslant \lim_n u(x_n) \geqslant \lim_n \lambda_n = \gamma$. Since $u \leqslant \gamma$ on the domain U it follows by the maximum principle that $u \equiv \gamma$.

Finally, if $N = 2$ and $U = \mathbb{R}^2$ we have $\gamma = +\infty$ unless u is bounded from above and hence constant, see, for example, [AG, Corollary 3.5.4]. □

4.2 Finely Superharmonic Functions in Dirichlet Space

For any open subset Ω of $\mathbb{R}^N$, $N \geqslant 2$, let $\mathcal{D}(\Omega)$ denote the space of all infinitely differentiable functions $\Omega \to \mathbb{C}$ of compact support. Following [DeL] we denote by $\mathrm{BL}(\Omega)$ the set of all (complex) distributions $T \in \mathcal{D}'(\Omega)$ such that $\partial T/\partial x_i \in L^2(\Omega)$ (with the restriction to Ω of Lebesgue measure ℓ_N) for every $i \in \{1, \dots, N\}$ (BL stands for the mathematician Beppo Levi). For $S, T \in \mathrm{BL}(\Omega)$ define their (Dirichlet) inner product

$$\langle S, T\rangle = \int_\Omega \sum_{i=1}^N \partial_i S \overline{\partial_i T} d\ell_N,$$

where $\partial_i = \partial/\partial x_i$, and the corresponding norm

$$\|T\|^2 = \langle T, T\rangle = \int_\Omega \sum_{i=1}^N |\partial_i T|^2 d\ell_N.$$

we shall henceforth always assume that Ω is connected. Then $\|T\| = 0$ if and only if T is constant. Let $\mathrm{BL}^*(\Omega)$ denote the quotient space $\mathrm{BL}(\Omega)/\{\text{constants}\}$, which is a complex Hilbert space according to [DeL, Théorème 1.1 (p. 308)]. The norm $\|\cdot\|^*$ on $\mathrm{BL}^*(\Omega)$ is given by

$$\|T^*\|^* = \inf\{\|T\| : T \in T^*\},\ T^* \in \mathrm{BL}^*(\Omega).$$

We refer to [DeL, Théorème 2.1 (p. 310)] for the following theorem and its proof (in our setting):

Theorem 4.2.1 *A distribution $T \in \mathrm{BL}(\Omega)$ is a function of class $L^q_{loc}(\Omega)$ with $\frac{1}{q} = \frac{1}{2} - \frac{1}{N}$ if $N \geqslant 3$, and of class $L^r_{loc}(\Omega)$ for any finite r if $N = 2$.*

The functions of class $\mathrm{BL}(\Omega)$ are called Beppo Levi functions. In particular every Beppo Levi function is of class $L^2_{loc}(\Omega)$.

In analogy with the above we have on $\mathcal{D}(\Omega)$ the Dirichlet inner product

$$\langle u, v\rangle = \int_\Omega \nabla u.\overline{\nabla v} d\ell_N, \quad u, v \in \mathcal{D}(\Omega),$$

where ∇u denotes the gradient of u, and where $\|u\| = \langle u, u\rangle^{1/2}$ now is an actual norm. Denote by $\widehat{\mathcal{D}}(\Omega)$ the Hilbert space completion of the pre-Hilbert space $\mathcal{D}(\Omega)$. As pointed out in [DeL, p. 318], the space $\widehat{\mathcal{D}}(\Omega)$ is only useful if its elements are distributions on Ω. We shall see that this is always the case for $N \geqslant 3$, but for $N = 2$ it holds if and only if Ω is Greenian.

For $N \geqslant 3$, a finite number $q > 2$ is defined by $\frac{1}{q} = \frac{1}{2} - \frac{1}{N}$. We denote by $\mathcal{E}_q(\Omega)$ the set of functions (distributions) $u \in L^q(\Omega)$ such that $\partial_i u (= \partial u/\partial x_i)$ is of class $L^2(\Omega)$ (with respect to Lebesgue measure ℓ_N) for every $i \in \{1, \dots, N\}$. Then $\mathcal{E}_q(\Omega)$ is a Banach space with the norm

$$\|u\|_{\mathcal{E}_q(\Omega)} = \|u\|_{L^q(\Omega)} + \|u\|.$$

Theorem 4.2.2 *For $N \geqslant 3$, $\widehat{\mathcal{D}}(\Omega)$ equals the norm closure $\mathcal{D}_q(\Omega)$ of $\mathcal{D}(\Omega)$ in $\mathcal{E}_q(\Omega)$, and the elements of $\widehat{\mathcal{D}}(\Omega)$ are therefore distributions on Ω, more precisely functions of class $L^r_{loc}(\Omega)$. For $N = 2$ the elements of $\widehat{\mathcal{D}}(\Omega)$ are distributions if and only if Ω is Greenian. They are then functions of class $L^r_{loc}(\Omega)$ which after extension by 0 on $\mathbb{R}^2 \backslash \Omega$ become functions of class $L^r_{loc}(\mathbb{R}^2)$ for any finite r.*

For the proof we refer to [DeL, Théorème 4.4 (p. 322)] for $N \geqslant 3$, and [DeL, Théorème 2.1 (p. 350)] for $N = 2$.

Henceforth we therefore suppose (in the absence of other indication) that Ω *is a Greenian domain* in $\mathbb{R}^N$, $N \geqslant 2$.

Theorem 4.2.3 $\widehat{\mathcal{D}}(\Omega) \subset \mathrm{BL}(\Omega)$.

Proof Given $u \in \widehat{\mathcal{D}}(\Omega)$, there is a sequence (u_n) in $\mathcal{D}(\Omega)$ such that $\|u_n - u\| \to 0$. Then $\|u_n - u_p\| \to 0$ as $n, p \to \infty$. Since $\|u_n^* - u_p^*\| \leqslant \|u_n - u_p\|$ there exists $v^* \in \mathrm{BL}^*(\Omega)$ such that $\|u_n^* - v^*\| \to 0$. Choose $v \in v^*$. For given $\varepsilon > 0$ there exists $n \in \mathbb{N}$ such that $\|u_n - u\| < \varepsilon$ and $\|u_n^* - v^*\| < \varepsilon$. There exists $v_n \in v^*$ such that $\|u_n - v_n\| < \varepsilon$ and hence $\|u^* - v^*\| \leqslant \|u - v_n\| < 2\varepsilon$. Thus $u^* = v^*$ and so $u - v$ is constant, whence indeed $u \in \mathrm{BL}(\Omega)$ along with v. □

As shown in [DeL, Théorème 4.5 (p. 322)], every $u \in \mathrm{BL}(\Omega)$ has a unique decomposition $u = u_0 + h$ where $u_0 \in \widehat{\mathcal{D}}(\Omega)$ and h harmonic. Hence 0 is the only harmonic function in $\mathrm{BL}(\Omega)$ of class $\widehat{\mathcal{D}}(\Omega)$.

As in Sect. 1.1.14 we denote by $C^*(A)$ the outer Green capacity of a set $A \subset \Omega$. For A open (or just capacitable: $C^*(A) = C_*(A)$) we may write $C(A)$ in place of $C^*(A)$ and speak about the (Green) capacity of A.

Definition 4.2.4 A function $f : \Omega \to \overline{\mathbb{R}}$ is said to be quasi-continuous if

$$\inf\{C^*(A) : A \subset \Omega \text{ with } f|_{\complement A} \text{ continuous}\} = 0.$$

It clearly suffices to consider open sets A. This concept of quasi-continuity is identical with that of Definition 1.5.2 in which the outer Green capacity $C^*(A)$ of a subset A of Ω is replaced by $\widehat{R}_p^A$ ($p \in \mathcal{P}^c$) as a function of A with values in the cone of all positive hyperharmonic functions on Ω. The identity of the two concepts of quasi-continuity holds because Corollary 1.5.14 (with "u.s.c." replaced by "continuous" if we like) likewise holds with quasi-continuity according to Definition 4.2.4 above (in place of Definition 1.5.2); this is essentially the theorem of Choquet [Ch3]. We shall, however, not use this identity of the two concepts of quasi-continuity.

Theorem 4.2.5 *Every function $u \in \mathrm{BL}(\Omega)$, resp. $\widehat{\mathcal{D}}(\Omega)$, equals almost everywhere a quasi-continuous function (necessarily of class $\mathrm{BL}(\Omega)$, resp. $\widehat{\mathcal{D}}(\Omega)$).*

Proof According to the preceding theorem it suffices to consider the case $u \in \mathrm{BL}(\Omega)$, for which we refer to the proof of [DeL, Theorem 3.1 (p. 354)]. □

A quasi-continuous Beppo Levi function is termed "Fonction BL précisée" in [DeL], and is now called BLD function (Beppo Levi and Deny). The set of BLD functions on Ω is denoted by $\mathrm{BLD}(\Omega)$. Concerning $\widehat{\mathcal{D}}(\Omega)$ we shall henceforth let $\widehat{\mathcal{D}}(\Omega)$ stands for the set of all quasi-continuous functions which are of class $\widehat{\mathcal{D}}(\Omega)$ in the previous sense. The decomposition theorem mentioned after the proof of Theorem 4.2.3 leads to

Theorem 4.2.6 *Every $u \in \mathrm{BLD}(\Omega)$ has a unique decomposition $u = u_0 + h$ where $u_0 \in \widehat{\mathcal{D}}(\Omega)$ (in the new sense) and h is (complex) harmonic on Ω.*

Lemma 4.2.7 *For any (quasi-continuous) function $f \in \widehat{\mathcal{D}}(\Omega)$ and $\alpha \in]0, +\infty[$,*

$$C^*(\{|f| > \alpha\}) \leqslant \frac{\|f\|^2}{a_N \alpha^2},$$

where a_N is the Poisson constant (see Sect. 1.1.8).

Proof In the case $\Omega = \mathbb{R}^N$, $N \geqslant 3$, this is [DeL, Lemme 4.1 (p. 356)], the proof of which was given for $f \in \mathcal{D}(\mathbb{R}^N)$ in [DeL, Lemme 1.1 (p. 347)]. Altogether, the proof of the present Lemma 4.2.7 is quite similar. □

Theorem 4.2.8 *Let (f_n) be a norm-convergent sequence in $\widehat{\mathcal{D}}(\Omega)$ and let $f \in \widehat{\mathcal{D}}(\Omega)$ be its limit. There exists a subsequence (f_{n_j}) of (f_n) which converges pointwise q.e. in Ω to f, and such that the convergence is uniform on $\complement A$ for suitable open sets $A \subset \Omega$ of arbitrarily small capacity $C(A)$.*

Proof We show that it suffices to choose (n_j) so that

$$\sum_{j=1}^{\infty} 4^j \| f_{n_{j+1}} - f_{n_j} \| < +\infty.$$

Writing

$$E_j := \{|f_{n_{j+1}} - f_{n_j}| > 2^{-j}\}, \ A_k = \bigcup_{j=k}^{\infty} E_j,$$

we have from Lemma 4.2.7 with $\alpha = 2^{-j}$

$$C^*(A_k) \leqslant \sum_{j=k}^{\infty} C^*(E_j) \leqslant \frac{1}{a_N} \sum_{j=k}^{\infty} 4^j \| f_{n_{j+1}} - f_{n_j} \|^2$$

and hence $C^*(A_k) \to 0$ as $k \to +\infty$. But the sequence (f_{n_j}) converges uniformly to f on $\complement A_k$. Here we can replace A_k by a larger open set likewise of small capacity. □

Theorem 4.2.9 *In order that a function f of class* BLD(Ω) *be of class $\widehat{\mathcal{D}}(\Omega)$ it is necessary and sufficient that f have the fine limit* 0 *at quasi-every each point of $\partial_f \Omega$, and also at infinity if $N \geqslant 3$.*

For the proof and further information, see [DeL, p. 359].

We denote by $\mathcal{E}_+(\Omega)$ the convex cone of measures μ on Ω of finite energy

$$\begin{aligned} G_\Omega(\mu, \mu) &= \int_\Omega G_\Omega \mu \, d\mu = \int_{\Omega \times \Omega} G_\Omega d(\mu \otimes \mu) \\ &= \int_\Omega \int_\Omega G(x, y) \, d\mu(x) \, d\mu(y). \end{aligned}$$

Theorem 4.2.10 (a) *For any $\mu \in \mathcal{E}_+(\Omega)$ we have $G\mu \in \widehat{\mathcal{D}}_+(\Omega)$ and*

$$\int_\Omega f \, d\mu = \frac{1}{a_N} \langle f, G_\Omega \mu \rangle \quad \textit{for any } f \in \widehat{\mathcal{D}}(\Omega).$$

(b) *For any $f \in \widehat{\mathcal{D}}_+(\Omega)$ there exists $\lambda \in \mathcal{E}_+(\Omega)$ such that $G_\Omega \lambda \geqslant f$ q.e. on Ω.*

(c) *The potentials $G_\Omega \mu$, $\mu \in \mathcal{E}_+(\Omega)$, form a total subset of $\widehat{\mathcal{D}}(\Omega)$.*

Proof For (a), see [De3, p. 160 and Théorème (p. 170)]. For (b), see [De3, Lemme 3 (p. 161)] with $\omega = X$ (the present Ω). If (c) were false, there would exist

$f \in \widehat{\mathcal{D}}(\Omega)\setminus\{0\}$ such that $\langle f, G_\Omega\mu\rangle = 0$ for every $\mu \in \mathcal{E}_+(\Omega)$, by (a). We may assume that f is quasi-continuous and real valued, and furthermore that $\{f > 0\}$ is nonpolar. Then there exists $\varepsilon > 0$ such that the quasi-closed set $A = \{f \geqslant \varepsilon\}$ is nonpolar and therefore, has a nonpolar compact subset K. Let μ denote the equilibrium measure on K; then $\mu(K) = C(K) > 0$, but $\varepsilon\mu(K) \leqslant \int f\,d\mu = 0$, which is contradictory. □

Remark 4.2.11 For any $\mu \in \mathcal{E}_+(\Omega)$ we have $G_\Omega\mu < +\infty$ μ-a.e., that is, $G_\Omega\mu$ is semibounded (also called stable), cf. Definition 1.5.16. Equivalently, μ does not charge the polar sets.

As already mentioned it is understood that, in the absence of other indication we consider quasi-continuous representatives of elements of $\widehat{\mathcal{D}}(\Omega)$ or equivalently representatives which are continuous q.e. on Ω. And we mostly do not distinguish between such a representative and its equivalence class modulo polar sets. We denote by for example $\widehat{\mathcal{D}}_+(\Omega)$ the set of positive functions of class $\widehat{\mathcal{D}}(\Omega)$. The fine interior of a set $E \subset \Omega$ is denoted by E'.

For any set $E \subset U$ define

$$\widehat{\mathcal{D}}(\Omega, E) = \{f \in \widehat{\mathcal{D}}(\Omega) : f = 0 \text{ q.e. on } \complement E\},$$

$$\mathcal{S}(\Omega, E) = \{f \in \widehat{\mathcal{D}}(\Omega) : f \text{ is finely superharmonic q.e. on } E'\},$$

$$\mathcal{H}(\Omega, E) = \{f \in \widehat{\mathcal{D}}(\Omega) : f \text{ is (complex) finely harmonic q.e. on } E'\},$$

$$\mathcal{P}(\Omega, E) = \{f \in \widehat{\mathcal{D}}(\Omega) : f|_E \text{ is a fine potential on } E'\}.$$

In the definition of $\mathcal{S}(\Omega, E)$ we may equivalently replace "finely superharmonic" by "finely hyperharmonic". Clearly $\widehat{\mathcal{D}}(\Omega, E)$ and $\mathcal{H}(\Omega, E)$ are linear subspaces of the Hilbert space $\widehat{\mathcal{D}}(\Omega)$ whereas $\widehat{\mathcal{D}}_+(\Omega, E)$, $\mathcal{S}(\Omega, E)$, $\mathcal{S}_+(\Omega, E)$, and $\mathcal{P}(\Omega, E)$ are convex cones in $\widehat{\mathcal{D}}(\Omega)$. Note that $\mathcal{P}(\Omega, E) \subset \mathcal{S}_+(\Omega, E)$. Every element of $\mathcal{S}_+(\Omega, E)$ has a representative which is finely superharmonic on all of E', by Theorem 2.1.43. Every element g of $\mathcal{P}(\Omega, E)$ has a representative whose restriction to E' is a semibounded fine potential on E'. In fact, let f be a representative of g such that $f|_{E'}$ is a fine potential, there exists by Theorem 4.2.10 and Remark 4.2.11 a positive measure $\lambda \in \mathcal{E}_+(\Omega)$ with $f \le G_\Omega\lambda$ and $G_\Omega\lambda$ is semibounded, that is, λ does not charge the polar sets (Definition 1.5.16). It then follows by Theorem 2.4.13(a) that $f|_{E'}$ is semibounded.

Let us note that $\widehat{\mathcal{D}}(\Omega, E) = \widehat{\mathcal{D}}(\Omega, E')$. Indeed, if $f \in \widehat{\mathcal{D}}(\Omega)$ equals 0 q.e. on $\Omega\setminus E$, then it equals 0 q.e. on $\Omega\setminus E'$ $(= \widetilde{\Omega\setminus E})$ because f is finely continuous outside a polar set, see Theorems 4.2.5 and 1.5.15. We show later that not every function of class $\mathcal{H}(\Omega, E)$ can be represented by functions which are finely harmonic everywhere on E' (see Theorem 4.2.25).

Theorem 4.2.12 *Let f be a finely superharmonic function on a finely open set $U \subset \mathbb{R}^N$. Every point $x \in U$ at which $f(x) < +\infty$ has a relatively compact regular finely open fine neighborhood V in U such that $\overline{V} \subset \Omega$ and $f|_V$ can be extended to* *a finely continuous function* $\widehat{f} \in \mathcal{S}(\Omega, V)$, Ω *denoting any Greenian domain in* $\mathbb{R}^N$ *containing* $\overline{V}$. *If* f *is finely harmonic on* U *it may be arranged that* $\widehat{f} \in \mathcal{H}(\Omega, V)$.

Proof By the local extension property (Theorem 2.1.25) there exists a bounded regular finely open fine neighborhood V of x such that $\overline{V} \subset U$, and two locally bounded potentials $u = G_\Omega\mu$, $v = G_\Omega\nu$ such that $f = u - v$ on V. Since $\overline{V}$ is bounded we may take Ω bounded. Replacing u and v by their sweepings $\widehat{R}_u^V = G_\Omega\mu^V$ and $\widehat{R}_v^V = G_\Omega\nu^V$, where μ^V and ν^V have compact supports contained in $\overline{V}$, we achieve that μ and ν are supported by $\overline{V}$ and hence $\int G_\Omega\mu\, d\mu < +\infty$ and $\int G_\Omega\nu\, d\nu < +\infty$. According to Theorem 4.2.10(a) it follows that $G_\Omega\mu, G_\Omega\nu \in \widehat{\mathcal{D}}_+(\Omega)$, and hence $u - v = G_\Omega\mu - G_\Omega\nu \in \widehat{\mathcal{D}}(\Omega)$. Thus $f|_V = (u - v)|_V$ can be extended to $\widehat{f} := u - v \in \mathcal{S}(\Omega, V)$, and even $\widehat{f} \in \mathcal{H}(\Omega, V)$ if f is finely harmonic on U. □

Theorem 4.2.13 *We have* $\mathcal{S}(\Omega, E) \cap \widehat{\mathcal{D}}(\Omega, E) = \mathcal{P}(\Omega, E)$ *and* $\mathcal{H}(\Omega, E) \cap \widehat{\mathcal{D}}(\Omega, E) = \{0\}$.

Proof The latter assertion follows from the former because $\{0\} \subset \mathcal{H}(\Omega, E) \cap \widehat{\mathcal{D}}(\Omega, E) \subset \mathcal{S}(\Omega, E) \cap \widehat{\mathcal{D}}(\Omega, E)$ and $\mathcal{H}(\Omega, E) \cap \mathcal{P}(\Omega, E) = \{0\}$. For the non-trivial part $\mathcal{S}(\Omega, E) \cap \widehat{\mathcal{D}}(\Omega, E) \subset \mathcal{P}(\Omega, E)$ of the former assertion let $f \in \mathcal{S}(\Omega, E) \cap \widehat{\mathcal{D}}(\Omega, E)$, in particular f is real valued. Since $|f| \in \widehat{\mathcal{D}}_+(\Omega, E)$ in view of [DeL, p. 316] it follows by Theorem 4.2.10(b) that $|f| \leqslant G_\Omega\lambda$ q.e. on Ω for some positive measure λ on Ω of finite energy. By remark 4.2.11, the potential $p := G_\Omega\lambda$ is semibounded. By hypothesis, f is finely superharmonic on $E' \backslash e$ for some polar set $e \subset E'$, and f extends finely continuously by 0 q.e. on $\partial_f(E' \backslash e)$ (relative to Ω) by the definition of $\widehat{\mathcal{D}}(\Omega, E)$. Since $f \geqslant -|f| \geqslant -p$ it therefore follows by the fine boundary minimum principle (Theorem 2.1.10) that $f \geqslant 0$ everywhere on $E' \backslash e$. As noted just before Theorem 4.2.12 we may therefore assume that $f \geqslant 0$ on all of E'.

For the proof that $f|_{E'}$ is a fine potential it remains to verify that every finely subharmonic minorant $v \geqslant 0$ of $f|_{E'}$ equals 0. And that follows by the same fine boundary principle, now applied to the finely superharmonic function $-v$ on E'. In fact, $0 \leqslant f \leqslant p$ everywhere on E' and p is a semibounded potential on Ω, and finally

$$\text{f-}\limsup_{x \to y,\, x \in E'} v(x) \leqslant \text{f-}\limsup_{x \to y,\, x \in E'} f(x) = f(y) = 0$$

q.e. for $y \in \partial_f E'$ because $f \in \widehat{\mathcal{D}}(\Omega, E) = \widehat{\mathcal{D}}(\Omega, E')$. □

Remark 4.2.14 Clearly $\widehat{\mathcal{D}}(\Omega, E)$, $\mathcal{S}(\Omega, E)$, and $\mathcal{H}(\Omega, E)$ only depend on the equivalence class of E modulo polar sets. The same therefore applies to $\mathcal{P}(\Omega, E) = \mathcal{S}(\Omega, E) \cap \widehat{\mathcal{D}}(\Omega, E)$. For any $f \in \mathcal{P}(\Omega, E)$ the fine potential $f|_{E'}$ is semibounded, for if e is a polar subset of E' then $f \in \mathcal{P}(\Omega, E' \backslash e)$ and hence $f|_{E' \backslash e}$ is a fine potential.

Theorem 4.2.15 *Each of the sets* $\widehat{\mathcal{D}}_+(\Omega)$, $\widehat{\mathcal{D}}(\Omega, E)$, $\mathcal{S}(\Omega, E)$, $\mathcal{H}(\Omega, E)$, *and* $\mathcal{P}(\Omega, E)$ *is normclosed in* $\widehat{\mathcal{D}}(\Omega)$, *and so are therefore* $\widehat{\mathcal{D}}_+(\Omega, E)$ *and* $\mathcal{S}_+(\Omega, E)$.

Proof Let (f_n) be a sequence in $\widehat{\mathcal{D}}(\Omega)$ converging to $f \in \widehat{\mathcal{D}}(\Omega)$. After passing to subsequence we may by Theorem 4.2.8 assume that $f_n \to f$ pointwise q.e. on Ω and uniformly on $\complement A$ for some $A \subset \Omega$ of capacity as small as we please. We may arrange that A is finely closed relative to Ω. (Otherwise replace A by its fine closure $\widetilde{A}$ relative to Ω, noting that $C^*(\widetilde{A}) = C^*(A)$, an easy consequence of [AG, Theorem 5.5.5 (ii)] or [Do4, 1.XIII.13] because $\widetilde{A \setminus e} = \widetilde{A}$ for any polar, hence finely isolated set $e \subset \Omega$.)

If every f_n is of class $\widehat{\mathcal{D}}(\Omega, E)$ then $f_n = 0$ q.e. on $\complement E$ and hence $f = 0$ q.e. on $\complement E \setminus A$ and hence q.e. on $\complement E$, so that indeed $f \in \widehat{\mathcal{D}}(\Omega, E)$. If instead every $f_n \in \mathcal{S}(\Omega, E)$, that is, f_n is finely superharmonic on the finely open set $E' \setminus e_n$ for some polar set e_n, in particular $(E' \setminus A) \setminus e_n$, then the uniform limit f is finely superharmonic on $(E' \setminus A) \setminus e$ with $e := \bigcup_n e_n$ polar. Here we may take for A a sequence of finely closed sets $A_k \subset \Omega$ with $C^*(A_k) \to 0$. Then f becomes finely superharmonic q.e. on $E' \setminus A_k$, hence q.e. on $E' \setminus \bigcap_k A_k$, and indeed q.e. on E' because $C^*(\bigcap_k A_k) = 0$, that is, $\bigcap_k A_k$ is polar, in view of [AG, Corollary 5.5.7].

We have thus shown that $\widehat{\mathcal{D}}(\Omega, E)$ and $\mathcal{S}(\Omega, E)$ are closed in $\widehat{\mathcal{D}}(\Omega)$. The same therefore hold for $\mathcal{P}(\Omega, E)$ in view of Theorem 4.2.13, and for $\mathcal{H}(\Omega, E) = \mathcal{H}_{\mathbb{R}}(\Omega, E) + i\mathcal{H}_{\mathbb{R}}(\Omega, E)$ where $\mathcal{H}_{\mathbb{R}}(\Omega, E) = \mathcal{S}(\Omega, E) \cap (-\mathcal{S}(\Omega, E))$ denotes the set of real-valued functions of class $\mathcal{H}(\Omega, E)$. Finally $\widehat{\mathcal{D}}_+(\Omega, E) = \widehat{\mathcal{D}}(\Omega, E) \cap \widehat{\mathcal{D}}_+(\Omega)$ and $\mathcal{S}_+(\Omega, E) = \mathcal{S}(\Omega, E) \cap \widehat{\mathcal{D}}_+(\Omega)$ are closed in $\widehat{\mathcal{D}}(\Omega)$ because $\widehat{\mathcal{D}}_+(\Omega)$ is so. In fact $f_n \in \widehat{\mathcal{D}}_+(\Omega)$ and $f_n \to f \in \widehat{\mathcal{D}}(\Omega)$ then, as in Theorem 4.2.8, $0 \leqslant f_{n_j} \to f$ q.e., so $f \in \widehat{\mathcal{D}}_+(\Omega)$. □

If Γ is a convex cone in the Hilbert space $\widehat{\mathcal{D}}(\Omega)$ we denote by Γ° the dual of the convex cone $-\Gamma$ in $\widehat{\mathcal{D}}(\Omega)$:

$$\Gamma^\circ = \{f \in \widehat{\mathcal{D}}(\Omega) : \mathfrak{Re}\langle f, \varphi \rangle \geqslant 0 \quad \text{for every} \varphi \in \Gamma\}.$$

It is well known that $\Gamma^{\circ\circ} = \overline{\Gamma}$, the norm closure of Γ in $\widehat{\mathcal{D}}(\Omega)$. If Γ is a linear subspace of $\widehat{\mathcal{D}}(\Omega)$ then so is Γ° which equals the orthogonal complement of Γ in $\widehat{\mathcal{D}}(\Omega)$, as shown by replacing φ by $-\varphi$ and by $i\varphi$.

For any Euclidean open subset ω of our Greenian domain $\Omega \subset \mathbb{R}^N$, $N \geqslant 2$, we identify a function of class $\mathcal{D}(\omega)$ with its extension to Ω by the value 0 on $\complement\omega$.

Theorem 4.2.16 *Let ω be an open subset of Ω. Then $\widehat{\mathcal{D}}(\Omega, \omega) = \overline{\widehat{\mathcal{D}}(\omega)}$ and $\widehat{\mathcal{D}}_+(\Omega, \omega) = \overline{\widehat{\mathcal{D}}_+(\omega)}$ (norm closures in the Hilbert space $\widehat{\mathcal{D}}(\Omega)$). Furthermore*

$$\mathcal{S}(\Omega, \omega) = \{f \in \widehat{\mathcal{D}}(\Omega) : f|_\omega \ \textit{ is superharmonic}\},$$

$$\mathcal{H}(\Omega, \omega) = \{f \in \widehat{\mathcal{D}}(\Omega) : f|_\omega \ \textit{ is (complex) harmonic}\},$$

$$\mathcal{P}(\Omega, \omega) = \{f \in \widehat{\mathcal{D}}(\Omega) : f|_\omega \ \textit{ is (usual) potential}\}.$$

Finally, $\mathcal{H}(\Omega, \omega)$ and $\widehat{\mathcal{D}}(\Omega, \omega)$ are orthogonal complements of each other in $\widehat{\mathcal{D}}(\Omega)$, and $\mathcal{S}(\Omega, \omega) = \widehat{\mathcal{D}}_+(\Omega, \omega)^\circ$, $\mathcal{D}_+(\Omega, \omega) = \mathcal{S}(\Omega, \omega)^\circ$.

Proof We first show that if ω is connected then

$$\widehat{\mathcal{D}}(\Omega,\omega)=\widehat{\mathcal{D}}(\omega) \tag{4.2.1}$$

and hence $\widehat{\mathcal{D}}_+(\Omega,\omega)=\widehat{\mathcal{D}}_+(\omega)$. It is understood that a function $f\in\widehat{\mathcal{D}}(\omega)$ is identified with its extension f_* to Ω by the value 0 on $\complement\omega$. If $f\in\widehat{\mathcal{D}}(\Omega,\omega)\subset \mathrm{BLD}(\Omega)$ (Theorem 4.2.3) then $f|_\omega\in \mathrm{BLD}(\omega)$, and even $f|_\omega\in\widehat{\mathcal{D}}(\omega)$ because $f|_\omega$ is quasi-continuous along with f. Conversely, suppose that $f\in\widehat{\mathcal{D}}(\omega)$. Then f_* is finely continuous on Ω according to Corollary 1.5.14 (with "u.s.c." replaced by "continuous"), which also shows that f_* is quasi-continuous. It remains to show that $f_*\in\widehat{\mathcal{D}}(\omega)$ and hence $f_*\in\widehat{\mathcal{D}}(\Omega,\omega)$. Choose a sequence $(f_n)\subset\widehat{\mathcal{D}}(\Omega,\omega)$ so that $\|f_n-f_*\|\to 0$. Then (f_n) is Cauchy both in $\widehat{\mathcal{D}}(\omega)$ and $\widehat{\mathcal{D}}(\Omega)$, so (f_n) converges in the Hilbert space $\widehat{\mathcal{D}}(\Omega)$ to some $g\in\widehat{\mathcal{D}}(\Omega)$ which we may take quasi-continuous on Ω according to Theorem 4.2.5. After passing to a subsequence we may by Theorem 4.2.8 assume that $f_n\to g$ pointwise q.e. on Ω. Since $f_n=0$ on $\complement\omega$ we have $g=0$ q.e. on $\complement\omega$, and so $g\in\widehat{\mathcal{D}}(\Omega,\omega)$. Finally, $g=f_*$ q.e., both on $\complement\omega$ and on ω, the latter by renewed application of Theorem 4.2.8, now with Ω replaced by ω, whereby we may assume that $f_n\to f=f_*$ pointwise q.e. on ω. Consequently, $f_*\in\widehat{\mathcal{D}}(\Omega,\omega)$.

Dropping the above requirements that ω be connected, $\mathcal{D}(\omega)$ may be viewed as a subset of $\widehat{\mathcal{D}}(\Omega,\omega)$, which is closed in $\widehat{\mathcal{D}}(\Omega)$ by Theorem 4.2.15 applied to $E=\omega$. Hence $\overline{\mathcal{D}(\omega)}\subset\widehat{\mathcal{D}}(\Omega,\omega)$, and similarly $\overline{\mathcal{D}_+(\omega)}\subset\widehat{\mathcal{D}}_+(\Omega,\omega)$. Conversely, let $f\in\widehat{\mathcal{D}}(\Omega,\omega)$, and denote by ω_j, $j\in J\subset\mathbb{N}$ ($J=\{1,\cdots,k\}$ finite or $J=\mathbb{N}$), the components of ω. Put $f_j=1_{\omega_j}f=$ the extension of $f|_{\omega_j}$ to Ω by 0 on $\complement\omega_j$. Since the linear subspaces $\mathcal{D}(\omega_j)$ of $\widehat{\mathcal{D}}(\Omega)$ are pairwise orthogonal, so are their closures $\widehat{\mathcal{D}}(\omega_j)=\widehat{\mathcal{D}}(\Omega,\omega_j)$ by (4.2.1). Hence $(f_j)_{j\in J}$ is an orthogonal family in $\widehat{\mathcal{D}}(\Omega)$. If $J=\mathbb{N}$, every partial sum

$$s_n=f_1+\cdots+f_n$$

is the orthogonal projection of f on $\sum_{j=1}^n\widehat{\mathcal{D}}(\Omega,\omega_j)$. In fact, $f-s_n=0$ q.e. on ω_j, $j\leqslant n$, whence $\langle f-s_n,\varphi\rangle=0$ for every $\varphi\in\widehat{\mathcal{D}}(\Omega,\omega_j)=\overline{\mathcal{D}(\omega_j)}$, $j\leqslant n$, by (4.2.1) applied with ω replaced by ω_j. In particular, $\sum_{i\leq n}||f_i||^2=||s_n||^2\leq||f||^2(<\infty)$ and hence $||s_p-s_q||^2\to 0$ as $p,q\to+\infty$, that is, (s_n) is a Cauchy sequence in $\widehat{\mathcal{D}}(\Omega)$, and hence $\|s_n-s\|\to 0$ for some $s\in\widehat{\mathcal{D}}(\Omega)$, and after passing to a subsequence we may assume according to Theorem 4.2.8 that $s_n\to s$ q.e. on Ω, whence $s=f$ q.e. on Ω. Since $f_n\in\overline{\mathcal{D}(\omega_n)}\subset\overline{\mathcal{D}(\omega)}$ we have $s\in\overline{\mathcal{D}(\omega)}$, and so $\widehat{\mathcal{D}}(\Omega,\omega)\subset\overline{\mathcal{D}(\omega)}$, and altogether $\widehat{\mathcal{D}}(\Omega,\omega)=\overline{\mathcal{D}(\omega)}$. We also have $\widehat{\mathcal{D}}_+(\Omega,\omega)\subset\overline{\mathcal{D}_+(\omega)}$ and hence $\widehat{\mathcal{D}}_+(\Omega,\omega)=\overline{\mathcal{D}_+(\omega)}$. In fact, for any $\varepsilon>0$, there exists $n\in\mathbb{N}$ such that $\|s_n-f\|<\varepsilon$, and for $j\leqslant n$ there exists $\varphi_j\in\mathcal{D}(\Omega)$ such that $\|f_j-\varphi_j\|<\varepsilon/n$. It follows that $\varphi:=\varphi_1+\cdots+\varphi_n\in\mathcal{D}(\omega)$ and $\|f-\varphi\|<2\varepsilon$. By varying ε we conclude that indeed $f\in\overline{\mathcal{D}(\omega)}$. If $f\geqslant 0$ we can take $\varphi_j\geqslant 0$ here according to [De3, p. 140] applied to the normal contraction $Tz=(\mathfrak{Re}z)_+$, $z\in\mathbb{C}$, and so $f\in\overline{\mathcal{D}_+(\omega)}$. If instead of $J=\mathbb{N}$ we have $J=\{1,\ldots,n\}$ for some $n\in\mathbb{N}$ and if again $f\in\widehat{\mathcal{D}}(\Omega,\omega)$ then $s_n=f$. With φ_j as before we obtain $\|f-\varphi\|<\varepsilon$, etc. This leads again to

$f \in \overline{\mathcal{D}(\omega)}$, and if $f \in \mathcal{D}_+(\Omega, \omega)$ then $f \in \overline{\mathcal{D}_+(\omega)}$. This completes the proof that $\widehat{\mathcal{D}}(\Omega, \omega) = \overline{\mathcal{D}(\omega)}$ and $\widehat{\mathcal{D}}_+(\Omega, \omega) = \overline{\mathcal{D}_+(\omega)}$.

For the proof of the stated expression for $\mathcal{H}(\Omega, \omega)$ we begin by showing that the closed linear subspaces $\mathcal{H}(\Omega, \omega)$ and $\widehat{\mathcal{D}}(\Omega, \omega)$ of $\widehat{\mathcal{D}}(\Omega)$ (cf. Theorem 4.2.15) are the orthogonal complement of each other. A vector $h \in \widehat{\mathcal{D}}(\Omega)$ is orthogonal to $\widehat{\mathcal{D}}(\Omega, \omega) = \overline{\mathcal{D}(\omega)}$ if and only if

$$0 = \langle h, \varphi \rangle = \int_\Omega \nabla h . \nabla \overline{\varphi} d\ell_N = -\int_\Omega h \Delta \overline{\varphi} d\ell_N = -\int_\omega h \Delta \overline{\varphi} d\ell_N$$

for every $\varphi \in \mathcal{D}(\omega)$. Replacing φ by $\overline{\varphi} \in \mathcal{D}(\omega)$ this means that $\Delta(h|_\omega) = 0$ in the distributional sense, in other words that $h|_\omega$ almost everywhere equals a (complex) harmonic function. Thus every $f \in \widehat{\mathcal{D}}(\Omega)$ has a unique orthogonal decomposition $f = g + h$ with $g \in \widehat{\mathcal{D}}(\Omega, \omega)$ and h harmonic on ω and of class $\widehat{\mathcal{D}}(\Omega)$. In particular, for $f \in \mathcal{S}(\Omega, \omega)$ we obtain $g = f - h \in \widehat{\mathcal{D}}(\Omega, \omega) \cap \mathcal{S}(\Omega, \omega) = \mathcal{P}(\Omega, \omega)$, by Theorem 4.2.13 (because $h \in \mathcal{S}(\Omega, \omega)$). The non-trivial inclusion relation for $\mathcal{S}(\Omega, \omega)$ therefore reduces to the one for $\mathcal{P}(\Omega, \omega)$ (h being even harmonic on ω), which in turn holds because potentials and fine potentials on ω are the same according to Theorem 2.4.22.

The non-trivial inclusion relation for $\mathcal{H}(\Omega, \omega)$ reduces to showing that every real-valued function h of class $\mathcal{H}(\Omega, \omega)$ is harmonic on ω, and that is indeed the case because h and $-h$ are in $\mathcal{S}(\Omega, \omega)$ and hence superharmonic on ω, as shown in the preceding paragraph. We have therefore also shown $\widehat{\mathcal{D}}(\Omega, \omega)$ and $\mathcal{H}(\Omega, \omega)$ are the orthogonal complements of each other.

Finally, we show that $\mathcal{S}(\Omega, \omega) = \widehat{\mathcal{D}}_+(\Omega, \omega)^\circ$, and hence $\widehat{\mathcal{D}}_+(\Omega, \omega) = \mathcal{S}(\Omega, \omega)^\circ$ because $\mathcal{S}(\Omega, \omega)$ is closed in $\widehat{\mathcal{D}}(\Omega)$, by Theorem 4.2.15. For given $s \in \widehat{\mathcal{D}}(\Omega)$ we have $s \in \mathcal{D}_+(\Omega, \omega)^\circ$ if and only if

$$0 \leqslant \langle s, \varphi \rangle = \int_\Omega \nabla s . \nabla \varphi d\ell_N = -\int_\Omega s \Delta \varphi d\ell_N = -\int_\omega s \Delta \varphi d\ell_N$$

for $\varphi \in \mathcal{D}_+(\omega)$. This means that $\Delta(s|_\omega) \leqslant 0$ in the distributional sense, and since $s \in \widehat{\mathcal{D}}(\Omega) \subset L^1_{loc}(\Omega)$ we have $s|_\omega \in L^1_{loc}(\omega)$. We conclude that $s \in \mathcal{D}_+(\Omega, \omega)^\circ$ if and only if $s|_\omega$ almost everywhere equals a superharmonic function. With s and hence $s|_\omega$ quasi-continuous this means that $s|_\omega$ quasi-everywhere equals a superharmonic function, that is, $s \in \mathcal{S}(\Omega, \omega)$. □

Definition 4.2.17 A set $E \subset \Omega$ is called quasi-analytic (resp. quasi-coanalytic) if E differs by a polar set from an analytic (resp. coanalytic) subset of Ω.

Every finely closed set $A \subset \Omega$ is quasi-analytic, for A differs only by a polar set from its base, which is a G_δ-set. Every finely open subset of Ω is therefore quasi-analytic (even quasi-Borel).

Theorem 4.2.18 *For any quasi-analytic set $E \subset \Omega$ we have*

$$\widehat{\mathcal{D}}(\Omega, E) = \bigcap_{\omega \supset E} \widehat{\mathcal{D}}(\Omega, \omega), \quad \widehat{\mathcal{D}}_+(\Omega, E) = \bigcap_{\omega \supset E} \widehat{\mathcal{D}}_+(\Omega, \omega),$$

where ω ranges over all open subsets of Ω containing E.

Proof It suffices to prove that $f \in \widehat{\mathcal{D}}(\Omega, E)$ for any function f such that $f \in \widehat{\mathcal{D}}(\Omega, \omega)$ for every open set $\omega \supset E$. Since f is quasi-continuous there is a polar G_δ-set e such that $f|_{\complement e}$ is finely continuous, and hence Borel measurable according to Theorem 1.4.36. The set

$$A := \{x \in \complement E \backslash e : f(x) \neq 0\}$$

differs only by a polar set from the Borel set $B := \{x \in \complement \mathfrak{r}(E') \backslash E : f(x) \neq 0\} \subset A$. By Choquet's capacitability theorem it follows that A is C-capacitable, for $C^*(A) = C^*(B) = C_*(B) \leqslant C_*(A)$. Every compact set $K \subset A$ is polar because $\omega := \complement K$ is open and contains $\complement A$ and hence E, whence $f \in \widehat{\mathcal{D}}(\Omega, \omega)$, by hypothesis. We conclude that A itself is polar, and by the definition of A that $f = 0$ off the polar set $A \cup e$. This shows that indeed $f \in \widehat{\mathcal{D}}(\Omega, E)$. □

Corollary 4.2.19 *For any $\mu \in \mathcal{E}_+(\Omega)$ and any quasi-analytic set $E \subset \Omega$ the sweeping $\widehat{R}^{\complement E}_{G_\Omega \mu} = G_\Omega \mu^{\complement E}$ is the orthogonal projection of $G_\Omega \mu$ on $\mathcal{H}(\Omega, E)$.*

Proof In fact, $G_\Omega \mu^{\complement E} = \widehat{R}^{\complement E}_{G_\Omega \mu} \leqslant G_\Omega \mu$ (see Theorem 1.4.24 for the equality) and hence

$$\int G_\Omega \mu^{\complement E} \, d\mu^{\complement E} \leqslant \int G_\Omega \mu \, d\mu^{\complement E} = \int G_\Omega \mu^{\complement E} \, d\mu \leqslant \int G_\Omega \mu \, d\mu < +\infty,$$

the equality by the reciprocity theorem, cf. e.g. [Hel, Theorem 6.14] or Sect. 1.1.5. It follows by Theorem 4.2.10(a) that $G_\Omega \mu^{\complement E} \in \widehat{\mathcal{D}}(\Omega)$, and more precisely $G_\Omega \mu^{\complement E} \in \mathcal{H}(\Omega, E) = \mathcal{H}(\Omega, E')$ because $\widehat{R}^{\complement E}_{G_\Omega \mu}$ is finely harmonic q.e. on E'. According to (2.6.5) in Lemma 2.6.3 we have

$$G_\Omega \mu = G_{E'} \mu + \widehat{R}^{\complement E'}_{G \Omega \mu}.$$

Hence $G_{E'} \mu \in \widehat{\mathcal{D}}(\Omega)$ along with $G_\Omega \mu$ and $G_\Omega \mu^{\complement E}$, and more precisely $G_{E'} \mu \in \widehat{\mathcal{D}}(\Omega, E)$ since $G_{E'} \mu = 0$ q.e. on $\complement E' \supset \complement E$. It only remains to apply Theorem 4.2.18. □

For any family (E_α) of linear subspaces [resp. convex subcones] E_α of $\widehat{\mathcal{D}}(\Omega)$ [resp. $\widehat{\mathcal{D}}_{\mathbb{R}}(\Omega)$] we denote, by abuse of notation, by $\sum_\alpha E_\alpha$ the least linear subspace [resp. least convex subcone] of $\widehat{\mathcal{D}}(\Omega)$ containing $\bigcup_\alpha E_\alpha$. The elements of $\sum_\alpha E_\alpha$ are all finite sum of sums of vectors from various E_α. By $\overline{\sum_\alpha E_\alpha}$ we denote of course the (norm) closure of $\sum_\alpha E_\alpha$ in $\widehat{\mathcal{D}}(\Omega)$ [resp. $\widehat{\mathcal{D}}_{\mathbb{R}}(\Omega)$].

Theorem 4.2.20 *For any family $\mathcal{V}$ of finely open sets with the union U we have*

$$\mathcal{S}(\Omega, U) = \bigcap_{V \in \mathcal{V}} \mathcal{S}(\Omega, V), \ \mathcal{H}(\Omega, U) = \bigcap_{V \in \mathcal{V}} \mathcal{H}(\Omega, V),$$

$$\widehat{\mathcal{D}}(\Omega, U) = \overline{\sum_{V \in \mathcal{V}} \widehat{\mathcal{D}}(\Omega, V)}, \ \widehat{\mathcal{D}}_+(\Omega, U) = \overline{\sum_{V \in \mathcal{V}} \widehat{\mathcal{D}}_+(\Omega, V)}.$$

Proof This becomes obvious if the four equality signs are replaced by "$\supset$", cf. Theorem 4.2.15. By the quasi-Lindelöf principle there exists a sequence $(V_n) \subset \mathcal{V}$ and a polar set e_0 such that $U = e_0 \cup \bigcup_n V_n$. If $f \in \mathcal{S}(\Omega, V)$ for every $V \in \mathcal{V}$ then f is finely superharmonic on $V_n \backslash e_n$ for every $n \geqslant 1$ (e_n a polar subset of V_n) and hence on

$$U \backslash \bigcup_{n \geqslant 0} e_n \subset \bigcup_{n \geqslant 1} (V_n \backslash e_n),$$

thus quasi-everywhere on U, that is, $f \in \mathcal{S}(\Omega, U)$. This establishes the remaining inclusion "$\subset$" for $\mathcal{S}(\Omega, U)$. The proof for $\mathcal{H}(\Omega, U)$ is similar.

It will be shown in the next theorem that $\widehat{\mathcal{D}}(\Omega, U)^\perp = \mathcal{H}(\Omega, U)$ (and similarly with U replaced by any $V \in \mathcal{V}$). By duality we therefore obtain the asserted expression for $\widehat{\mathcal{D}}(\Omega, U)$, and from that we now deduce the asserted expression for $\widehat{\mathcal{D}}_+(\Omega, U)$. We first show that

$$\varphi \in \sum_{V \in \mathcal{V}} \widehat{\mathcal{D}}(\Omega, V) \Rightarrow \varphi^+ \in \sum_{V \in \mathcal{V}} \widehat{\mathcal{D}}_+(\Omega, V).$$

By definition, φ is a finite sum $\varphi = \sum_{j=1}^p \varphi_j$ of functions $\varphi_j \in \widehat{\mathcal{D}}(\Omega, V_j)$, $V_j \in \mathcal{V}$. According to [DeL, p. 316], $\widehat{\mathcal{D}}(\Omega)$ is a Riesz space (a vector lattice), and since $\varphi^+ \leqslant \sum_{j=1}^p \varphi_j^+$ there exist by the Riesz decomposition property vectors $\psi_1, \dots, \psi_p \in \widehat{\mathcal{D}}_+(\Omega)$ such that $\psi_j \leqslant \varphi_j^+$ (whence $\psi_j \in \widehat{\mathcal{D}}_+(\Omega, V_j)$) and that

$$\varphi^+ = \sum_{j-1}^p \psi_j \in \sum_{V \in \mathcal{V}} \widehat{\mathcal{D}}_+(\Omega, V).$$

This being done, let $f \in \mathcal{D}_+(\Omega, U)$, and choose a sequence $(f_n) \subset \sum_{V \in \mathcal{V}} \widehat{\mathcal{D}}(\Omega, V)$ so that $\| f_n - f \| \to 0$. It follows that $\| f_n^+ - f \| \to 0$, by [De3, p. 140] applied to the normal contraction T of $\mathbb{C}$ defined by $Tz = (\Re z)_+$ for $z \in \mathbb{C}$. Thus f belongs to the closure of $\sum_{V \in \mathcal{V}} \widehat{\mathcal{D}}_+(\Omega, V)$. This completes the proof of Theorem 4.2.20 except for the proof that $\widehat{\mathcal{D}}(\Omega, U)^\circ = \mathcal{H}(\Omega, U)$ given in the next theorem. □

We are now prepared for the main results of this section.

Theorem 4.2.21 *For any quasi-coanalytic subset E of a Greenian domain $\Omega \subset \mathbb{R}^N$, $N \geqslant 2$, we have*

$$\widehat{\mathcal{D}}(\Omega, E)^\perp = \mathcal{H}(\Omega, E) = \overline{\sum_{\omega \supset E} \mathcal{H}(\Omega, \omega)} = \overline{\bigcup_{\omega \supset E} \mathcal{H}(\Omega, \omega)},$$

$$\widehat{\mathcal{D}}_+(\Omega, E)^\circ = \mathcal{S}(\Omega, E) = \overline{\sum_{\omega \supset E} \mathcal{S}(\Omega, \omega)} = \overline{\bigcup_{\omega \supset E} \mathcal{S}(\Omega, \omega)},$$

where ω ranges over all Euclidean open subsets of Ω containing E. In other words, for $f \in \widehat{\mathcal{D}}(\Omega)$ the following three statements are equivalents:

(i) $\langle f, \varphi \rangle = 0$ *for every* $\varphi \in \widehat{\mathcal{D}}(\Omega, E)$ [*resp.* $\langle f, \varphi \rangle \geqslant 0$ *for every* $\varphi \in \widehat{\mathcal{D}}_+(\Omega, E)$],

(ii) $f \in \mathcal{H}(\Omega, E)$ [*resp.* $f \in \mathcal{S}(\Omega, E)$],

(iii) f *is the norm limit of a sequence of functions* $f_n \in \widehat{\mathcal{D}}(\Omega)$ *such that each* f_n *is harmonic* [*resp. superharmonic*] *in some open set* $\omega_n \supset U$.

Proof In view of Theorem 4.2.18 we deduce from the last assertion of Theorem 4.2.16 that

$$\widehat{\mathcal{D}}_+(\Omega, E)^\circ = \overline{\sum_{\omega \supset E} \mathcal{D}_+(\Omega, \omega)^\circ} = \overline{\bigcup_{\omega \supset E} \mathcal{S}(\Omega, \omega)} = \mathcal{S}(\Omega, E),$$

$$\widehat{\mathcal{D}}(\Omega, E)^\perp = \overline{\sum_{\omega \supset E} \mathcal{D}(\Omega, \omega)^\perp} = \overline{\bigcup_{\omega \supset E} \mathcal{H}(\Omega, \omega)} = \mathcal{H}(\Omega, E).$$

For the third equality in either line, note that the family $\{\omega \supset E\}$ is upper directed and that the map $\omega \mapsto \mathcal{H}(\Omega, \omega)$, resp. $\omega \mapsto \mathcal{S}(\Omega, \omega)$, is decreasing. For the last equality in either line, the corresponding inclusion relation "$\subset$" is obvious because $\mathcal{S}(\Omega, E)$ and $\mathcal{H}(\Omega, E)$ are closed in $\widehat{\mathcal{D}}(\Omega)$ according to Theorem 4.2.15. For the relation "$\supset$" concerning $\mathcal{H}(\Omega, E)$, let $f \in \mathcal{H}(\Omega, E)$ be given. We have the orthogonal decomposition $f = g + h$, $g \in \widehat{\mathcal{D}}(\Omega, E)$, $h \in \widehat{\mathcal{D}}(\Omega, E)^\perp$, hence $h \in \mathcal{H}(\Omega, E)$, and so $g = f - h \in \mathcal{H}(\Omega, E) \cap \widehat{\mathcal{D}}(\Omega, E) = \{0\}$, whence indeed $f = h \in \widehat{\mathcal{D}}(\Omega, E)^\perp$. This also completes the Proof of Theorem 4.2.20. Finally let $f \in \mathcal{S}(\Omega, E)$ and let us prove that $f \in \widehat{\mathcal{D}}(\Omega, E)^\circ$. By hypothesis f is finely superharmonic and finite valued on $U := E' \backslash e$ for some polar set $e \subset E'$. With the notation from the Proof of Theorem 4.2.12, every $x \in E'$ with $f(x) < +\infty$ has a relatively compact regular finely open fine neighborhood V with $\tilde{V} \subset U$ on which f has a representation $f = G_\Omega \mu - G_\Omega \nu$, where the measures μ and ν on Ω have finite G_Ω-energy and $G_\Omega \nu$ is finely harmonic on V, whence $\nu(V) = 0$ (see Theorem 2.6.4). For any $\varphi \in \widehat{\mathcal{D}}(\Omega, V)$ it follows by Theorem 4.2.10(a) that

$$\langle f, \varphi \rangle = \int_\Omega \nabla f . \nabla \varphi d\ell_N = \int_\Omega \nabla G_\Omega(\mu - \nu) . \nabla \varphi d\ell_N = a_N \int_\Omega \varphi d(\mu - \nu).$$

Here $\int_\Omega \varphi \, d\nu = \int_V \varphi \, d\nu = 0$ because $\nu(V) = 0$. Consequently, $\langle f, \varphi \rangle \geqslant 0$ for every $\varphi \in \widehat{\mathcal{D}}_+(\Omega, V)$, and so $f \in \widehat{\mathcal{D}}_+(\Omega, V)^\circ$. By the quasi-Lindelöf principle there is a polar set $E_0 \subset U$ and an increasing sequence of finely open sets V_n with the property of V above, such that $U = E_0 \cup \bigcup_n V_n$. It follows by Theorem 4.2.20 and Remark 4.2.14 that

$$f \in \bigcap_n (\widehat{\mathcal{D}}_+(\Omega, V_n))^\circ = \overline{\sum_n \widehat{\mathcal{D}}_+(\Omega, V_n)}^\circ = \overline{\bigcup \widehat{\mathcal{D}}_+(\Omega, V_n)}^\circ$$
$$= \widehat{\mathcal{D}}_+(\Omega, \bigcup_n V_n)^\circ = \widehat{\mathcal{D}}_+(\Omega, E)^\circ.$$

For the second equality note that the sequence (V_n) is increasing and that the map $V \mapsto \widehat{\mathcal{D}}(\Omega, V)$ (V finely open set in Ω) is decreasing. □

Corollary 4.2.22 *For every subset E of Ω and any function $f \in \widehat{\mathcal{D}}(\Omega, E)$ we have $\nabla f = 0$ a.e. on $\complement E'$.*

Proof E' is the union of the upper directed family $\mathcal{V}$ of all finely open sets V such that $\overline{V} \subset E'$. Since $f \in \widehat{\mathcal{D}}(\Omega, E')$ (see 3th paragraph after Remark 4.2.11) there exists by Theorem 4.2.20 an increasing sequence $(V_n) \subset \mathcal{V}$ and a sequence of functions $f_n \in \widehat{\mathcal{D}}(\Omega, V_n)$ such that $f_n \to f$ in $\widehat{\mathcal{D}}(\Omega)$ and hence $\nabla f_n \to \nabla f$ in $L^2(\Omega)$. The assertion therefore holds because $\nabla f_n = 0$ a.e. on $\complement \overline{V_n} \supset \complement E'$. □

Corollary 4.2.23 *For any quasi-coanalytic set $E \subset \Omega$ we have the following two orthogonal decompositions into a sum of two linear subspaces of $\widehat{\mathcal{D}}(\Omega)$, resp. into a sum of the convex subcone $\mathcal{P}(\Omega, E)$ and the real-linear subspace $\mathcal{H}_{\mathbb{R}}(\Omega, E)$:*

$$\widehat{\mathcal{D}}(\Omega) = \widehat{\mathcal{D}}(\Omega, E) \oplus \mathcal{H}(\Omega, E), \quad \mathcal{S}(\Omega, E) = \mathcal{P}(\Omega, E) \oplus \mathcal{H}_{\mathbb{R}}(\Omega, E).$$

The former decomposition amounts to the relation $\mathcal{H}(\Omega, E) = \widehat{\mathcal{D}}(\Omega, E)^\perp$ in Theorem 4.2.21, and the latter decomposition follows easily because $\mathcal{P}(\Omega, E) = \mathcal{S}(\Omega, E) \cap \widehat{\mathcal{D}}(\Omega, E)$, by Theorem 4.2.13, and because $\mathcal{H}_{\mathbb{R}}(\Omega, E) \subset \mathcal{S}(\Omega, E)$.

Corollary 4.2.24 (Spectral synthesis) *For any quasi-coanalytic set $E \subset \Omega$, the cone $\mathcal{P}_E$ of all potentials $G_\Omega\mu$ of measures $\mu \in \mathcal{E}_+(\Omega)$ with $\mu(E') = 0$ is total in $\mathcal{H}(\Omega, E)$*[1].

In fact, $\{G_\Omega\mu : \mu \in \mathcal{E}_+(\Omega)\}$ is total in $\widehat{\mathcal{D}}(\Omega)$ according to Theorem 4.2.10(c). By orthogonal projection on the closed linear subspace $\mathcal{H}(\Omega, E)$ it follows from Corollary 4.2.19(a) that $\{G_\Omega\mu^{\complement E} : \mu \in \mathcal{E}_+(\Omega)\}$ is total in $\mathcal{H}(\Omega, E) = \mathcal{H}(\Omega, E')$, and so is therefore the larger set $\{G_\Omega\nu : \nu \in \mathcal{E}_+(\Omega),\ \nu(E') = 0\} \subset \mathcal{H}(\Omega, E)$, noting that $\mu^{\complement E}(E') = 0$ because the measure $\nu := \mu^{\complement E}$ is carried by $b(\complement E) \subset \complement E'$, see Theorem 1.4.23(a).

When applied to an open set $E \subset \Omega$ and hence a closed set $F := \complement E$, Corollary 4.2.24 reduces to the spectral synthesis theorem in [De3, p. 168], the proof of which is a model for the above proof.

We now pass to the question of more precisely where in E' a suitable representative of an element $f \in \mathcal{H}(\Omega, E)$ is defined.

[1] The Proof of Corollary 4.2.24 given in [Fu20, Corollary (p. 146)] is incomplete.

Theorem 4.2.25 *For any quasi-coanalytic set $E \subset \Omega$ and any point $x \in E$ the following three propositions are equivalent.*

(a) *Every element $f \in \mathcal{H}(\Omega, E)$ has representatives which are finely harmonic on some finely open set containing x.*

(b) *The linear form $f \mapsto f(x)$, defined on the linear span*

$$\sum_{\omega \supset E} \{\mathcal{H}(\Omega, \omega) : \omega \text{ open}\}$$

(dense in $\mathcal{H}(\Omega, E)$ by Theorem 4.2.21) is continuous.

(c) *The swept measure $\varepsilon_x^{\complement E}$ (seen as harmonic measure on E' at x) has finite energy*

$$\begin{aligned}\int G_\Omega \varepsilon_x^{\complement E}\, d\varepsilon_x^{\complement E} &= G_\Omega \varepsilon_x^{\complement E}(x) = \int G_\Omega \varepsilon_x \, d\varepsilon_x^{\complement E} \\ &= \widehat{R}^{\complement E}_{G_\Omega \varepsilon_x}(x) < +\infty.\end{aligned}$$

(sweepings and complements relative to Ω).

If E and x have these three equivalent properties then we have for every $f \in \mathcal{H}(\Omega, E)$

$$f(x) = x(f) = \int f \, d\varepsilon_x^{\complement E} = \frac{1}{a_N} \langle G_\Omega \varepsilon_x^{\complement E}, f \rangle,$$

where $f(x)$ denotes the common value at the point x of representations mentioned in (a), *whereas $x(f)$ denotes the value at f of the extension to $\mathcal{H}(\Omega, E)$ by continuity of the linear form $h \mapsto h(x)$ on $\bigcup_{\omega \supset E} \mathcal{H}(\Omega, \omega)$ from* (b). *The operator norm of this extension therefore equals*

$$\|x(\cdot)\| = \frac{1}{\sqrt{a_N}} \|G_\Omega \varepsilon_x^{\complement E}\|.$$

Proof By way of preparation, consider an open set $\omega \subset \Omega$. For any element $f \in \mathcal{H}(\Omega, \omega)$, represented by a (quasi-continuous) function of class $\widehat{\mathcal{D}}(\Omega)$ which is harmonic on ω, cf. Theorem 4.2.16, denote by f^ω the orthogonal projection of the element $|f| \in \widehat{\mathcal{D}}(\Omega)$ ($\widehat{\mathcal{D}}(\Omega)$ being a Riesz space, see [DeL, p. 316] and Theorem 4.2.9) on $\mathcal{H}(\Omega, \omega)$. Then $-|f| \in \mathcal{S}(\Omega, \omega)$ and $-|f|^\omega$ is the orthogonal projection of $-|f|$ on $\mathcal{H}(\Omega, \omega)$. Hence $-f^\omega \in \mathcal{H}_\mathbb{R}(\Omega, \omega)$ by the orthogonal decomposition $\mathcal{S}(\Omega, \omega) = \mathcal{P}(\Omega, \omega) \oplus \mathcal{H}_\mathbb{R}(\Omega, \omega)$, cf. Corollary 4.2.23. Thus $f^\omega \in \mathcal{H}_\mathbb{R}(\Omega, \omega)$. Clearly $\|f^\omega\| \leqslant \||f|\| \leqslant \|f\|$ in view of [DeL, Theorem 3.2 (p. 316) and Remark 3.3 (p.p. 316–317]. Finally,

$$f^\omega \geqslant |f| \geqslant 0 \quad \text{on } \omega \tag{4.2.2}$$

because $-f^\omega$ is the harmonic part of the superharmonic function $-|f|$ on ω, as noted above.

For the proof that (a)⇒(b), suppose that (a) holds true whereas (b) fails. Then there exists a sequence of open sets $\omega_n \supset E$ and corresponding functions $f_n \in \mathcal{H}(\Omega, \omega_n)$ such that $\|f_n\| < 2^{-n}$ and $|f_n(x)| \geqslant 1$. As shown above, we have $\|f_n^{\omega_n}\| < 2^{-n}$ and $f_n^{\omega_n}(x) \geqslant 1$. The series $\sum_n f_n^{\omega_n}$ converges in $\widehat{\mathcal{D}}(\Omega)$ to an element $h \in \mathcal{H}(\Omega, E)$ because $\mathcal{H}(\Omega, \omega_n) \subset \mathcal{H}(\Omega, E)$ and because $\mathcal{H}(\Omega, E)$ is closed by Theorem 4.2.15. After passing to subsequence we may arrange by Theorem 4.2.8 that $\sum_n f_n^{\omega_n} = h$ pointwise q.e. in Ω. But $f_n^{\omega_n}$ is finely harmonic on $E' \subset \omega_n$, and hence the pointwise sum $\sum_n f_n^{\omega_n}$ is a finely hyperharmonic function h' on all of E'. And $h'(x) = +\infty$ because $f_n^{\omega_n}(x) \geqslant 1$. Since h' represents h on E' it follows from Theorem 4.2.9 that every representative of h has fine limit $\text{f-}\lim_{y \to x,\, y \neq x} h(y) = +\infty$, in contradiction with (a). This shows that (a)⇒(b).

For the converse, consider first a point $x \in E'$. For given $f \in \mathcal{H}(\Omega, E)$ there exists by Theorem 4.2.21(iii) a sequence of functions $s_n \in \mathcal{H}(\Omega, \omega_n)$ such that $\|s_n - f\| < 2^{-n-1}$. Replacing ω_n by $\omega_1 \cap \cdots \cap \omega_n$ we achieve that the sequence (ω_n) is decreasing. Writing $f_n = s_n - s_{n-1}$ (with $s_0 = 0$) we obtain $f_n \in \mathcal{H}(\Omega, \omega_n)$ and $s_n = f_1 + \cdots + f_n$. According that (b) holds there is a constant c such that

$$f_n^{\omega_n}(x) \leqslant c\|f_n^{\omega_n}\| \leqslant c\|f_n\| = c\|s_n - s_{n-1}\| \leqslant c2^{-n}.$$

The series $\sum_n f_n^{\omega_n}$ therefore converges in $\widehat{\mathcal{D}}(\Omega)$ to an element $h \in \mathcal{H}(\Omega, E)$ and (after passing to a subsequence) pointwise q.e. on E' to a finely hyperharmonic function $h' : E' \to [0, +\infty]$; and h' is even finely harmonic on the finely open set $\{y \in E' : h'(y) < +\infty\}$, (which contains x), see Theorem 2.1.31. In particular, h' is bounded in some finely open set V such that $x \in V \subset E'$. The series $\sum_n f_n$ itself converges in $\widehat{\mathcal{D}}(\Omega)$ to the given element $f \in \mathcal{H}(\Omega, E)$ and (after passing to a subsequence) pointwise q.e. on E' to a finely hyperharmonic function $f' : E' \to [0, +\infty]$. Since $f_n \in \mathcal{H}(\Omega, \omega_n)$ we have by (4.2.2) $|f_n| \leqslant f_n^{\omega_n}$. Hence

$$|f'| \leqslant \sum_n |f_n| \leqslant \sum_n f_n^{\omega_n} = h' \quad \text{q.e. on } E'.$$

Thus f has a representative f' which is bounded on the above V (along with h') and hence finely harmonic on V, by Theorem 2.1.45 since $f \in \mathcal{H}(\Omega, E)$.

In general case $x \in E$ consider the regularized $\mathfrak{r}(E)$ which is an F_σ-set equivalent to E', and so $E \cup \mathfrak{r}(E)$ is quasi-coanalytic. In view of Remark 4.2.14 it follows that

$$\mathcal{H}(\Omega, E) = \mathcal{H}(\Omega, E') = \mathcal{H}(\Omega, \mathfrak{r}(E)) = \mathcal{H}(\Omega, E \cup \mathfrak{r}(E)).$$

For the last equality the inequality "⊃" is trivial, and

$$\mathcal{H}(\Omega, E \cup \mathfrak{r}(E)) = \mathcal{H}(\Omega, (E \cup \mathfrak{r}(E))') \supset \mathcal{H}(\Omega, \mathfrak{r}(E))$$

because $(E \cup \mathfrak{r}(E))' \subset \mathfrak{r}(E)$. For the given point $x \in E \subset E \cup \mathfrak{r}(E)$ the property (b) holds a fortiori with E replaced by the larger set $E \cup \mathfrak{r}(E)$ which is likewise quasi-coanalytic. Hence property (a) holds for $E \cup \mathfrak{r}(E)$ and therefore a fortiori for

E itself. From the proof of (b) $\Rightarrow$ (c) below we deduce that $x \in \mathfrak{r}(E)$, and hence $x \in (E \cup \mathfrak{r}(E))'$ (because otherwise $\varepsilon_x^{\complement E} = \varepsilon_x$ would be of infinite energy), whence the statement (a).

For the proof that (b)$\Rightarrow$(c) suppose that (c) fails to hold. Let us first compute the energy $\int G_\Omega(z,t)d\varepsilon_x^{\complement E}(z)\,d\varepsilon_x^{\complement E}(t)$ of $\varepsilon_x^{\complement E}$. Putting $\mu = \varepsilon_x$ and $A = \complement E$, we have by Theorems 1.4.24, 1.4.30, the definition of μ^A and Corollary 1.4.10 and Fubini's theorem

$$\begin{aligned}\mathcal{E}(\mu^A) &= \int G_\Omega(z,t)d\mu^A(z)d\mu^A(t) = \int G\mu^A d\mu^A = \int \widehat{R}^A_{G_\Omega\mu} d\mu^A \\ &= \int G\mu d(\mu^A)^A = \int G_\Omega\mu d\mu^A = \int G_\Omega\mu^A d\mu.\end{aligned}$$

and hence the equalities in (c). For any compact set $K \subset \complement E$ write $\omega_K = \complement K$ and

$$f_K = G_\Omega\varepsilon_x^K = \widehat{R}^K_{G_\Omega\varepsilon_x} \in \mathcal{H}(\Omega, \omega_K),$$

noting that $G_\Omega\varepsilon_x$ is bounded on $K \subset \complement\{x\}$ and so $\varepsilon_x^K \in \mathcal{E}_+(\Omega)$, whence $G_\Omega\varepsilon_x^K \in \mathcal{H}(\Omega, \omega_K)$ by Theorem 4.2.10(a). Since $\complement E$ is supposed to be quasi-analytic by Choquet's capacitability theorem

$$\sup_K f_K(x) = \sup_K \widehat{R}^K_{G_\Omega\varepsilon_x}(x) = \widehat{R}^{\complement E}_{G_\Omega\varepsilon_x}(x) = +\infty$$

because (c) fails. By Theorem 4.2.10(a) we have

$$\frac{1}{a_N}\|f_K\|^2 = \int G_\Omega\varepsilon_x^K\, d\varepsilon_x^K = \widehat{R}^K_{G_\Omega\varepsilon_x}(x) = f_K(x),$$

and hence (b) fails to hold:

$$\sup_K \frac{f_K(x)^2}{\|f_K\|^2} = \frac{1}{a_N}\sup_K f_K(x) = +\infty.$$

For the proof of the opposite implication (c)$\Rightarrow$(b) and the rest of the theorem, denote by $\mathcal{H}_x(\Omega, E)$ the linear subspace of $\mathcal{H}(\Omega, E)$ formed by those elements of $\mathcal{H}(\Omega, E)$ which have a representative which is finely harmonic on a finely open set containing x. For any such representative f we have

$$f(x) = \int f\, d\varepsilon_x^{\complement E} = \frac{1}{a_N}\langle G_\Omega\varepsilon_x^{\complement E}, f\rangle. \tag{4.2.3}$$

The latter equality follows from (c) in view of Theorem 4.2.10(a). For the former equality, f is defined, finite, and finely continuous on $\complement e$ for some polar set $e \subset \Omega$, and f is finely harmonic on $E'\backslash e$ by fine continuity (cf. Theorem 2.1.45). Since (b) $\Longleftrightarrow$ (a) we may assume that $x \notin e$, for otherwise we may replace e by $e\backslash\{x\}$.

The function

$$\underline{f}(z) = \begin{cases} f(z) & \text{for } z \in \complement e \\ -\infty & \text{for } z \in e \end{cases}$$

is finely harmonic along with f on the finely open set $E'\backslash e$. Since $|f| \in \widehat{\mathcal{D}}(\Omega)$ (cf. [DeL, Théorème 3.2, p. 316] and Theorem 4.2.9) there exists by Theorem 4.2.10(b) a potential $p = G_\Omega\lambda$, $\lambda \in \mathcal{E}_+(\Omega)$, such that $f \geqslant -p$ q.e. on Ω; and p is semi-bounded by Remark 4.2.11. Because e is polar we even have $\underline{f} \geqslant -p$ q.e. on Ω. Thus $f + p \geqslant 0$ q.e. on $\complement e$ and everywhere on $E'\backslash e$ because $f + p$ is finely superharmonic on $E'\backslash e$. Altogether, $\underline{f}$ is finely l.s.c. on Ω, and $\underline{f}$ is finely superharmonic and $\geqslant -p$ on the finely open set $E'\backslash e$. It therefore follows by Theorem 2.1.14 that

$$f(x) = \underline{f}(x) \geqslant \int^* f\, d\varepsilon_x^{\complement(E'\backslash e)} = \int^* f\, d\varepsilon_x^{\complement E}$$

because $\varepsilon_x^{\complement(E\backslash e)} = \varepsilon_x^{\complement E'} = \varepsilon_x^{\complement E}$. Replacing f by $-f \in \mathcal{H}(\Omega, E)$ we conclude that indeed $f(x) = \int f\, d\varepsilon_x^{\complement E}$.

We have thus deduced from (c) that if $f \in \mathcal{H}_x(\Omega, E)$ (that is, if an element $f \in \mathcal{H}(\Omega, E)$ has a representative, now denoted by f_x, which is finely harmonic in some fine neighborhood of x), then (4.2.3 holds, and hence

$$|f_x(x)| \leqslant \frac{1}{a_N} \|G_\Omega\varepsilon_x^{\complement E}\| \|f_x\|.$$

This applies in particular to any finite sum of elements of $\mathcal{H}(\Omega, \omega)$ for variable open sets $\omega \supset E$. In other terms, $\mathcal{H}_x(\Omega, E)$ contains $\sum_{\omega\supset E} \mathcal{H}(\Omega, \omega)$ which is dense in $\mathcal{H}(\Omega, E)$ according to Theorem 4.2.21. The above inequality therefore extends with $f_x \in \mathcal{H}_x(\Omega, E)$ replaced by any $f \in \mathcal{H}(\Omega, E)$, and we have thus shown that (c)⇒(b), and altogether that (a), (b), and (c) are indeed equivalent. And we have likewise established the rest of Theorem 4.2.25. □

Corollary 4.2.26 *For any set $E \subset \Omega$ and any point $x \in \Omega$, property (c) in Theorem 4.2.25 holds if and only if there exists $f \in \mathcal{H}(\Omega, E)$ such that $f = G_\Omega\varepsilon_x$ q.e. on $\complement E$. In the affirmative case we have $f = \widehat{R}_{G_\Omega\varepsilon_x}^{\complement E} = G_\Omega\varepsilon_x^{\complement E}$.*

Proof In fact, if (c) holds then $f := G_\Omega\varepsilon_x^{\complement E} - \widehat{R}_{G_\Omega\varepsilon_x}^{\complement E}$ is of class $\widehat{\mathcal{D}}(\Omega)$ by Theorem 4.2.10(a), and even of class $\mathcal{H}(\Omega, E)$ with $f = G_\Omega\varepsilon_x$ q.e. on $\complement E$, then we may assume that $x \in E$ (otherwise replace E by $E \cup \{x\}$), which leaves $\mathcal{H}(\Omega, E)$ unchanged in view of Remark 4.2.14. As in the proof of (b)⇒(c) in Theorem 4.2.25 write $f_K = G_\Omega\varepsilon_x^K$ for compact $K \subset \Omega$, and write $\omega_K = \complement K$. Then $f - f_K \in \widehat{\mathcal{D}}(\Omega, \omega_K) = \mathcal{H}(\Omega, \omega_K)^\perp$. Thus f_K is the orthogonal projection of f on $\mathcal{H}(\Omega, \omega_K)$, and so $\|f_K\| \leqslant \|f\|$. Since $\complement E'$ differs by a polar set from the G_δ-set $b(\complement E')$ we have

$$\widehat{R}^{\complement E}_{G_\Omega\varepsilon_x}(x) \leqslant \widehat{R}^{\complement E'}_{G_\Omega\varepsilon_x}(x) \leqslant \sup_{K\subset\complement E'} \widehat{R}^{K}_{G_\Omega\varepsilon_x}(x)$$
$$= \frac{1}{a_N}\sup_{K\subset\complement E'} \|f_K\|^2 \leqslant \frac{1}{a_N}\|f\|^2 < +\infty$$

because $f_K(x) = \frac{1}{a_N}\|f_K\|^2$ by Theorem 4.2.10(a) applied to $\mu = \varepsilon_x$ and $f = f_K = G_\Omega\varepsilon_x^K$. Thus (c) holds. □

Remark 4.2.27 (Kolsrud [Ko]). In the case $\Omega = \mathbb{R}^N$ if $N \geqslant 3$, $\Omega =$ the open unit disk if $N = 2$, and again arbitrary $E \subset \Omega$ and $x \in \Omega$, (c) further holds if and only if the Wiener type criterion

$$\sum_n \frac{C^*(A_n(x)\setminus E)}{C(A_n(x))^2} < +\infty$$

is fulfilled. Here $A_n(x) = \{y \in \Omega : 2^{-n-1} < |y - x| < 2^{-n}\}$. This contradiction clearly implies that $\complement E$ is thin at x, that is, E (or equivalently E') is a fine neighborhood of x. For the proof see [Ko].

4.3 The Dirichlet Laplacian on Finely Open Sets

Consider first a (Euclidean) open set $\Omega \subset \mathbb{R}^N$, $N \geqslant 2$, and the complex Hilbert space $L^2(\Omega)$ (with the restriction to Ω of the Lebesgue measure ℓ_N on $\mathbb{R}^N$), and furthermore the Sobolev space

$$W^{1,2}(\Omega) = \{u \in L^2(\Omega) : \nabla u \in (L^2(\Omega))^N\}$$

where the gradient $\nabla u = (\partial_1 u, \dots, \partial_N u)$ is understood in the distributional sense. For information about Sobolev spaces we refer to [Ad]. Equipped with the inner product

$$\langle u, v\rangle = \int_\Omega (u\overline{v} + \nabla u\overline{\nabla v})d\ell_N$$

in terms of Lebesgue measure ℓ_N on $\mathbb{R}^N$, $W^{1,2}(\Omega)$ becomes a (complex) Hilbert space with norm $\|u\|_1$ given by

$$\|u\|_1^2 = \int_\Omega (|u|^2 + |\nabla u|^2)d\ell_N.$$

Thus $W^{1,2}(\Omega)$ resembles the Beppo Levi space $\mathrm{BL}(\Omega)$ studied in the preceding section. Clearly

$$W^{1,2}(\Omega) \subset \mathrm{BL}(\Omega).$$

We shall mainly work with the closure $W_0^{1,2}(\Omega)$ of $\mathcal{D}(\Omega)$ in $W^{1,2}(\Omega)$, where $\mathcal{D}(\Omega)$ denotes the space of all infinitely differentiable of compact support complex valued functions on Ω. Since $\mathcal{D}(\Omega)$ is a linear subspace of the Hilbert space $W^{1,2}(\Omega)$, the closure $W_0^{1,2}(\Omega)$ of $\mathcal{D}(\Omega)$ in $W^{1,2}(\Omega)$ is also (a representation of) the completion of $\mathcal{D}(\Omega)$. We may abbreviate $W_0^{1,2}(\mathbb{R}^N)$, $\mathcal{D}(\mathbb{R}^N)$, etc, by omitting $\mathbb{R}^N$ from the notations.

On a bounded open set $\Omega \subset \mathbb{R}^N$ the Dirichlet Laplacian $A(\Omega)$ is the Friederichs extension of the usual Laplacian Δ (acting on $\mathcal{D}(\Omega)$) to a selfadjoint operator on the Hilbert space $L^2(\Omega)$ associated with the Dirichlet form

$$Q(u) = \int_\Omega |\nabla u|^2 d\ell_N$$

defined on the Sobolev space $W_0^{1,2}(\Omega)$. For the meaning of this definition see [Da] (or the analogous definition of the Dirichlet Laplacian on a bounded finely open set $U \subset \mathbb{R}^N$ given below). It is well known that $A(\Omega)$ has a compact inverse, see again [Da] (or as a generalized below to the fine topology setting). Denote by $\lambda_n(\Omega)$ $(n \geq 1)$ the eigenvalues of the positive selfadjoint operator $A(\Omega)$ (repeated according to multiplicity). Consider now a decreasing sequence of non-void bounded open sets $\Omega_i \subset \mathbb{R}^N$ with intersection $\Omega_\infty = \bigcap_i \Omega_i$, and denote by $\Omega = int\Omega_\infty$ the interior of Ω_∞. As shown in [Fu29, Theorem 2] or Theorem 4.3.7 below, the increasing sequence $(\lambda_n(\Omega_i))_{i\in\mathbb{N}}$ converges to $\lambda_n(\Omega)$ for every n if and only if Ω_∞ is stable in the sense of Keldyŝ, that is, $\complement\Omega_\infty$ and $(\complement\Omega_\infty)^\circ$ shall be thin at the same points of $\mathbb{R}^N$. This provides an answer to a question raised in [Da, p. 129]. The stated result was deduced in [Fu30] from an abstract version thereof, involving variationally defined selfadjoint operators on Hilbert space.

In the present section we obtain a similar more general result, now for any decreasing sequence of bounded *finely* open sets $U_i \subset \mathbb{R}^N$ (Theorem 4.3.7 below), as desired in [Fu29] from [Fu30]. The resulting convergence property

$$\lambda_n(U_i) \to \lambda_n(U) \text{ as } i \to +\infty, n \in \mathbb{N}, \tag{4.3.1}$$

where now $U = int_f U_\infty$, the interior of $U_\infty := \bigcap_i U_i$ in the fine topology, holds without any additional assumption like stability. In [Fu30], this convergence property (4.3.1) is deduced from the above-mentioned abstract result. Here we shall obtain (4.3.1) by reformulating its proof in [Fu30] now in the setting of fine potential theory on $\mathbb{R}^N$. From this "fine" convergence property (4.3.1) one easily recovers the corresponding "Euclidean" convergence property that was our starting point, simply because stability of a set $E \subset \mathbb{R}^N$ is equivalent to $E'\backslash E^\circ$ being a polar set. We are thus faced with yet another situation where a stability condition, entering crucially in a result involving the Euclidean topology, disappears when passing to a (more general) fine topology version.

In Theorem 4.3.10, etc. we generalize Theorem 4.3.7 by replacing the decreasing sequence (U_i) by an order convergence sequence.

Recall from Theorem 1.5.15 that a function $f : \Omega \to \overline{\mathbb{R}}$ (Ω open in $\mathbb{R}^N$) is quasi-continuous if and only if it is finely continuous q.e. The following regularity properties of functions of class $W_0^{1,2}(\Omega)$ (Ω bounded and open in $\mathbb{R}^N$), due to Deny [De1, DeL] are instrumental in what follows.

(a) Every function $u \in W_0^{1,2}(\Omega)$ can be redefined, or corrected, on a Lebesgue nullset so as to become quasi-continuous. Two functions of class $W_0^{1,2}(\Omega)$ which are equal Lebesgue almost everywhere (a.e.) become equal even q.e. after such a correction. Since $W_0^{1,2}(\Omega) \subset \mathrm{BL}(\Omega)$, this follows from [DeL, p. 353], where "corrected" is called "précisée". (See also Theorem 4.2.5 above.) In what follows we shall usually understand implicitly that functions of class $W_0^{1,2}(\Omega)$ have been corrected in this way.

(b) Every sequence of functions $u_i \in W_0^{1,2}(\Omega)$ converging in Sobolev norm $\|\cdot\|_1$ to some $u \in W^{1,2}(\Omega)$ has a subsequence which converges pointwise q.e. to u, see [DeL, Théorème 4.1 (p. 357)] or Theorem 4.2.8 above.

(c) A function $u \in W_0^{1,2}(\mathbb{R}^N) = W_0^{1,2}$ belongs to $W_0^{1,2}(\Omega)$ if and only if $u(x) = 0$ q.e. for $x \in \complement\Omega$. See [DeL, Théorème 5.1 (p. 359)], [FL2, Corollaire 16].

For an arbitrary set $E \subset \mathbb{R}^N$ we define

$$W_0^{1,2}(E) = \{u \in W_0^{1,2} : u = 0 \text{ q.e. on } \complement E\}, \tag{4.3.2}$$

cf. [FL2, Definition 1]. See also the analogous definition of $\widehat{\mathcal{D}}(\Omega, E)$ in the first display after Remark 4.2.11 in Sect. 4.2. In view of (c) above, this definition agrees with the standard definition of $W_0^{1,2}(E)$ in the particular case where E is a Euclidean open set. Clearly, $W_0^{1,2}(E_1) = W_0^{1,2}(E_2)$ if E_1 and E_2 differ by polar set.

$W_0^{1,2}(E)$ is *closed* as a (linear) subspace of $(W_0^{1,2}, \|\cdot\|_1)$. This follows from (b) above because the union of a sequence of polar sets is polar. Note that

$$W_0^{1,2}(E) = W_0^{1,2}(int_f E). \tag{4.3.3}$$

In fact, if $u \in W_0^{1,2}(E)$ then u (corrected) is finely continuous off some polar set, and $u = 0$ holds in $\complement E$ off some bigger polar set e. Hence $u = 0$ in the fine closure of $\complement(e \cup E)$ relative to $\complement e$, and this fine closure equals $\complement(e \cup E')$ because the polar set e is finely closed.

Lemma 4.3.1 (Gårding [Gar]) *For any bounded open set $\Omega \subset \mathbb{R}^N$ we have*

$$\int_\Omega u^2 d\ell_N \leqslant (\text{diam } \Omega)^2 \int_\Omega |\nabla u|^2 d\ell_N$$

for any $u \in \mathcal{D}(\Omega)$.

Proof We only use that Ω is contained in a band, for example $\Omega \subset \{x \in \mathbb{R}^N : a < x_1 < b\}$, where $a, b \in \mathbb{R}$ and $a < b$, in which case the diameter diam Ω in the inequality can be replaced by the width $b - a$ of the band. After extending u to $\mathbb{R}^N$ by $u = 0$ off Ω we have, writing $x' = (x_2, \ldots, x_N) \in \mathbb{R}^{N-1}$, hence $x = (x_1, x')$,

$$u(x) = \int_a^{x_1} \partial_1 u(t, x')dt,$$

$$u(x)^2 \leqslant (x_1 - a) \int_a^{x_1} \left|\partial_1 u(t, x')\right|^2 dt, \; x_1 \in [a, b],$$

$$\int_{\mathbb{R}^{N-1}} u(x_1, x')^2 d\ell_{N-1}(x') \leqslant (x_1 - a) \int_a^b (\int_{\mathbb{R}^{N-1}} \left|\nabla u(t, x')\right|^2 d\ell_{N-1}(x'))dt$$

whence

$$\int_\Omega u(x)^2 d\ell_N(x) \leq (b-a)^2 \int_\Omega \left|\nabla u(x)\right|^2 d\ell_N(x)$$

after integration with respect to $x_1 \in [a, b]$. □

From Sect. 4.2 we have the Hilbert space completion $\widehat{\mathcal{D}}(\Omega)$ of the pre-Hilbert space $\mathcal{D}(\Omega)$ with the Beppo Levi norm

$$\|u\| := (\int_\Omega |\nabla u|^2 d\ell_N)^{\frac{1}{2}}, \; u \in \widehat{\mathcal{D}}(\Omega).$$

According to Lemma 4.3.1 we have

$$\widehat{\mathcal{D}}(\Omega) = W_0^{1,2}(\Omega)$$

(as sets), and the Beppo Levi norm $\|\cdot\|$ on $\widehat{\mathcal{D}}(\Omega)$ is equivalent to the Sobolev norm $\|\cdot\|_1$. We may therefore apply results about $\widehat{\mathcal{D}}(\Omega)$ from Sect. 4.2 to functions of class $W_0^{1,2}(\Omega)$. Note also that if $\overline{E} \subset \Omega$, then $W_0^{1,2}(E) = \mathcal{D}(\Omega, E)$ by identifying each $u \in W_0^{1,2}(E)$ with its restriction to Ω, see (c) above.

Let $\wedge$ and $\vee$ denote infimum and supremum in the complete lattice Sub $W_0^{1,2}$ of all closed subspaces of the Hilbert space $(W_0^{1,2}, \|\cdot\|_1)$ under inclusion. Thus $\bigwedge_i \mathcal{U}_i$ is the intersection and $\bigvee_i \mathcal{U}_i$ is the closed (linear) span of the union of a family $(\mathcal{U}_i) \subset$ Sub $W_0^{,2}$.

Lemma 4.3.2 (i) *For any increasing sequence of non-void finely open sets* $U_i \subset \mathbb{R}^N$ *with* $\bigcup U_i$ *bounded we have*

$$\bigvee_i W_0^{1,2}(U_i) = W_0^{1,2}(\bigcup_i U_i).$$

(ii) *For any sequence of sets* $E_i \subset \mathbb{R}^N$ *we have*

$$\bigwedge_i W_0^{1,2}(E_i) = W_0^{1,2}\Big(\bigcap_i E_i\Big) = W_0^{1,2}(int_f \bigcap_i E_i).$$

Proof Let $U = \bigcup_i U_i$ and let Ω be an open ball containing $\overline{U}$. We obviously have $W_0^{1,2}(U) \supset \overline{\bigcup_i W_0^{1,2}(U_i)}$ (the closure in $W_0^{1,2}$). Conversely, let $f \in W_0^{1,2}(U) = \widehat{\mathcal{D}}(\Omega, U)$ (see the paragraph following the Proof of Lemma 4.3.1) be orthogonal to each $W_0^{1,2}(U_i)$, then for any measure μ of finite energy with compact support contained in U we have by Theorem 4.2.10(a) $\int f d\mu = \int f d\mu^{\complement U_i}$ for every i (because $G_\Omega\mu - G_\Omega\mu^{\complement U_i} = G_\Omega\mu - \widehat{R}^{\complement U_i}_{G_\Omega\mu} \in W_0^{1,2}(\mathfrak{r}(U_i)) = W_0^{1,2}(U_i)$, see the first paragraph of the proof of Theorem 4.3.5). But the sequence of the measures $\mu^{\complement U_i}$ admits a weak* cluster limit in energy ν carried by $\complement U$. Hence $\int f d\mu = \int f d\nu = 0$ because $f \in W_0^{1,2}(U)$. It then follows that $f = 0$ q.e. and therefore $W_0^{1,2}(U) \subset \overline{\bigcup_i W_0^{1,2}(U_i)}$. In part (ii) the former equality holds because the union of a sequence of polar sets is polar and the later follows from (4.3.3). □

Note that $\bigcup_i U_i$ is the supremum and $\left(\bigcap_i U_i\right)'$ is the infimum in the lattice of finely open subsets of $\mathbb{R}^N$.

Lemma 4.3.3 *For any set $E \subset \mathbb{R}^N$ and $u \in W_0^{1,2}(E)$ we have $\nabla u = 0$ a.e. on $\complement E'$.*

Proof Like any finely open set, $E'(= int_f E)$ is the union of the family $\mathcal{V}$ of all finely open subsets V with compact fine closure $\overline{V} \subset E'$, see Sect. 1.2. The fine topology being quasi-Lindelöf, see Sect. 1.2, there is an increasing sequence of finely open sets $V_i \in \mathcal{V}$ whose union U differs only by polar set from E'. In view of (4.3.3) it follows that $W_0^{1,2}(E) = W_0^{1,2}(E')$ is the closure of $\bigcup_i W_0^{1,2}(V_i)$ in $W_0^{1,2}$, by Lemma 4.3.2(i). Thus there is a sequence of functions $u_i \in W_0^{1,2}(V_i)$ such that $u_i \to u$ in $W_0^{1,2}$, hence $\nabla u_i \to \nabla u$ in L^2; and since $\nabla u_i = 0$ a.e. on the open set $\complement \overline{V_i} \supset \complement E'$, the assertion is proved. □

In Sect. 2.5, we have introduced the Green kernel G_U for a *bounded* finely pen set $U \subset \mathbb{R}^N$. Recall from the paragraph following the proof of Lemma 1.4.27 that the fine limit points of $\complement U$ are called *regular* fine boundary points for U, and the remaining, finely isolated points of $\complement U$ are called the *irregular* fine boundary points for U. The irregular points form a finely discrete, that is, a polar subset $\mathfrak{i}(U)$ of the fine boundary $\partial_f U$ of U. A finely open set U is said to be *regular* if $\mathfrak{i}(U) = \emptyset$. For any finely open set $U \subset \mathbb{R}^N$ the set

$$\mathfrak{r}(U) = U \cup \mathfrak{i}(U)$$

is finely open Euclidean F_σ-set, and moreover regular because the operator $\mathfrak{r}$ is idempotent $\mathfrak{r}(\mathfrak{r}(U)) = \mathfrak{r}(U)$ just like the base operator $\mathfrak{b}$ defined by $\mathfrak{b}(\complement U) = \complement\mathfrak{r}(U)$, see (1.4.20). We call $\mathfrak{r}(U)$ the *regularizer* of U; it is the smallest regular finely open set containing U. Note that the regularization of a Euclidean open set is not necessarily a Euclidean open set but is necessarily finely open. Recall that the Green kernel G_U for a finely open set U equals the Green kernel $G_{\mathfrak{r}(U)}$ for $\mathfrak{r}(U)$. By Definition 2.5.10 and subsequent text we have

$$G_U(x, y) = G_\Omega(x, y) - \widehat{R}^{\complement U}_{G_\Omega(\cdot, y)}(x), \ x, y \in \Omega \ (x \neq y),$$

and $G_U(x, x) = +\infty$, resp. 0, if $x \in \mathfrak{r}(U)$, resp. $x \notin \mathfrak{r}(U)$, where G_Ω is the Green kernel of an open ball Ω in $\mathbb{R}^N$ containing U (or any Green domain containing U). As shown in Remark 2.5.12 the restriction of $G_U(\cdot, y)$ $(y \in U)$ to U (also denoted G_U) does not depend on the choice of Ω. Recall that the kernel G_U is symmetric: $G_U(x, y) = G_U(y, x)$, and Borel measurable (Theorem 2.5.11 and Lemma 2.5.13).

Theorem 4.3.4 *Let $U \subset \mathbb{R}^N$ be a bounded finely open set. On the real Hilbert space $L^2(\mathfrak{r}(U))$, a bounded linear operator injective selfadjoint integral operator, denoted $\mathcal{G}_U$ is defined for $f \in L^2(\mathfrak{r}(U))$ by*

$$\mathcal{G}_U f(x) = \int_{\mathfrak{r}(U)} G_U(x, y) f(y) \, d\ell_N(y), \ x \in U. \tag{4.3.4}$$

Proof We shall often tacitly extend f from $\mathfrak{r}(U)$ to Ω by 0 off $\mathfrak{r}(U)$. Suppose first that $f \geqslant 0$. Denoting by ℓ_N the Lebesgue measure on $\mathbb{R}^N$, fm stands for the m-absolutely continuous measure on Ω with density f with respect to the restriction m of ℓ_N to Ω. Denoting by G the Green kernel on $\Omega \times \Omega$ we have

$$\mathcal{G}_U f = G_U(fm) \leqslant G(fm) \leqslant h * \bar{f},$$

where $\bar{f}$ is the extension of f by 0 on $\mathbb{R}^N \backslash \mathfrak{r}(U)$, $h(x) = |x|^{2-N}$ if $N \geqslant 3$ and $h(x) = |\log|x|| + C$ for some suitable constant $C > 0$ if $N = 2$. Since $h \in L^1_{loc}$ $(= L^1_{loc}(\mathbb{R}^N))$ and $f \in L^2_c$ (c stands for compact support), we have $h * f \in L^2_{loc}$ and hence $\mathcal{G}_U f \in L^2(\mathfrak{r}(U))$, $\mathfrak{r}(U)$ being bounded. For general $f \in L^2(U)$, apply the above to f^+ and f^-. □

The operator $\mathcal{G}_U$ will be called the *Green operator* for U.

We now pass to the Dirichlet Laplacian $A(U)$ on the bounded finely open set $U \subset \mathbb{R}^N$. Let $\mathcal{H}$ denote the closure of $W_0^{1,2}(U)$ in $L^2(\mathfrak{r}(U))$. As expected from the Euclidean case we show in Theorem 4.3.5(a) below that $\mathcal{H} = L^2(\mathfrak{r}(U))$. Consider the following symmetric bilinear form

$$Q(u, v) = \langle \nabla u, \nabla v \rangle = \int \nabla u . \nabla v \, d\ell_N, \ u, v \in W_0^{1,2}, \tag{4.3.5}$$

cf. (4.3.2) and Lemma 4.3.3 applied to the finely open set $E = U$. For any $u \in W_0^{1,2}(U) = W_0^{1,2}(\mathfrak{r}(U))$ with $Q(u, u) = 0$ we have $\nabla u = 0$ a.e. on U, and since $\nabla u = 0$ a.e. on $\complement U$ by Lemma 4.3.3 we conclude that u is constant a.e. on $\mathbb{R}^N$ because $\mathbb{R}^N$ is connected, and indeed $u = 0$ q.e. on $\mathbb{R}^N$ because $u = 0$ q.e. on $\complement U$, in particular $u = 0$ q.e. on U, by the regularity property (a) above. This shows that the symmetric positive semidefinite bilinear form Q on the Hilbert space $W_0^{1,2}(U)$ is *strictly positive definite*. Then the form Q is furthermore *closed* in the sense that

$$Q(u, u) + \|u\|^2_{L^2(U)} = \|u\|_1^2$$

for $u \in W_0^{1,2}(U)$, with $W_0^{1,2}(U)$ complete in the Sobolev norm $\|\cdot\|_1$. We may therefore define the Dirichlet Laplacian on U as the positive selfadjoint operator $A(U)$ on the closure $\mathcal{H}$ of $W_0^{1,2}(U)$ associated with closed form Q, cf. Friederichs [Fr], Lax-Milgram [LM], Davies [Da]. It follows that the operator $A(U)^{\frac{1}{2}}$ has the domain $W_0^{1,2}(U) = W_0^{1,2}(\mathfrak{r}(U))$ and

$$\langle A(U)^{\frac{1}{2}}u, A(U)^{\frac{1}{2}}v\rangle = \langle \nabla u, \nabla v\rangle_{L^2(\mathfrak{r}(U))}, \quad u, v \in W_0^{1,2}(U).$$

The domain of $A(U)$, written dom $A(U)$, consists of all those $u \in W_0^{1,2}(U)$ for which there exists $f \in \mathcal{H}(U)$ such that

$$\int_{\mathfrak{r}(U)} \nabla u.\nabla v d\ell_N = \int_{\mathfrak{r}(U)} f v d\ell_N, \quad v \in W_0^{1,2}(U), \tag{4.3.6}$$

and we then have $A(U)u = f$. If $A(U)u = 0$, then $Q(u,u) = \langle \nabla u, \nabla u\rangle_{L^2(U)} = 0$ and hence $u = 0$, as shown above. Then $A(U)$ is *injective*.

We proceed to establish that $\mathcal{H} = L^2(\mathfrak{r}(U))$, as already announced. Since U shall be bounded we henceforth choose Ω as a *bounded* domain in $\mathbb{R}^N$ $(N \geqslant 2)$ containing $\mathfrak{r}(U)$ (for example an open ball).

Theorem 4.3.5 *Let $U \subset \mathbb{R}^N$ be a bounded finely open set.*
(a) *$W_0^{1,2}(U)$ is dense in $L^2(\mathfrak{r}(U))$, that is, $\mathcal{H} = L^2(\mathfrak{r}(U))$.*
(b) *$A(U)^{-1} = \mathcal{G}_U$.*

Proof Choose an open ball $\Omega \supset U$. The Green potential $G_\Omega\mu = \int G_\Omega(\cdot, y)d\mu(y)$ of a (positive) measure μ on Ω belongs to $W_0^{1,2}(\Omega)$ if and only if μ has finite energy

$$\int G_\Omega \mu d\mu = \int G_\Omega(x, y) d\mu(x) d\mu(y) < +\infty,$$

[De1], and we shall now assume that μ is a *positive* measure on Ω of finite energy. Consider the sweeping $\mu^{\complement U}$ of μ on $\complement U$ relative to Ω and its potential $G_\Omega\mu^{\complement U} = \widehat{R}^{\complement U}_{G_\Omega\mu}$ according to Theorem 1.4.24. We have by Theorem 1.4.8 and symmetry of G_Ω

$$\begin{aligned}\int G_\Omega\mu^{\complement U} d\mu^{\complement U} &= \int \widehat{R}^{\complement U}_{G_\Omega\mu} d\mu^{\complement U} \le \int G_\Omega\mu d\mu^{\complement U} \\ &= \int G_\Omega\mu^{\complement U} d\mu \le \int G_\Omega \mu d\mu < +\infty,\end{aligned}$$

so that $\widehat{R}^{\complement U}_{G_\Omega\mu} = G_\Omega\mu^{\complement U} \in W_0^{1,2}(\Omega)$. Because $\widehat{R}^{\complement U}_{G_\Omega\mu} = G_\Omega\mu$ q.e. on $\complement U$ we obtain by (4.3.2)

$$G_U\mu = G_\Omega\mu - \widehat{R}^{\complement U}_{G_\Omega\mu} \in W_0^{1,2}(\mathfrak{r}(U)), \tag{4.3.7}$$

cf. Definition 2.5.10 with (2.5.7) integrated with respect to μ and application of (1.4.2). (Alternatively, apply (2.6.5) in Lemma 2.6.3.)

We now apply (4.3.7) to a measure μ having a density $f \in L^2(\mathfrak{r}(U))$, $f \geqslant 0$, with respect to the restriction m of ℓ_N to $\mathfrak{r}(U)$. According to Theorem 4.3.4, $\mathcal{G}_U f \in L^2(\mathfrak{r}(U))$, and hence the energy $\int(\mathcal{G}_U f) f dm$ is finite, so $\mathcal{G}_U f \in W_0^{1,2}(\mathfrak{r}(U))$ (see the beginning of the present proof). For brevity write

$$p = G_\Omega f,\ q = \widehat{R}^{\complement U}_{G_\Omega f},\ u = \mathcal{G}_U f = p - q \tag{4.3.8}$$

by (4.3.7). Then $\Delta p = -f$ in the distributional sense in Ω, and hence (with $L^2(\Omega)$ inner products in the rest of the proof):

$$\langle \nabla p, \nabla v\rangle = -\langle \Delta p, v\rangle = \langle f, v\rangle,$$

first for $v \in \mathcal{D}(\Omega)$ and for any $v \in W_0^{1,2}(\Omega)$. Moreover, $\langle \nabla q, \nabla v\rangle = 0$ for $v \in W_0^{1,2}(U)$ according to (i), (ii) in Theorem 4.2.21 because q is finely harmonic q.e. on U by Corollary 2.1.30 and hence $q \in \mathcal{H}(\Omega, U)$. From (4.3.6) and (4.3.8) we thus obtain

$$\langle \nabla \mathcal{G}_U f, \nabla v\rangle = \langle f, v\rangle \text{ for } v \in W_0^{1,2}(U) \tag{4.3.9}$$

and any $f \in L^2(\mathfrak{r}(U))$, the previous requirement $f \geqslant 0$ being superfluous here.

If f is orthogonal to $W_0^{1,2}(U)$ in $L^2(\mathfrak{r}(U))$ then $\langle \nabla u, \nabla v\rangle = 0$ for any $v \in W_0^{1,2}(U)$ by (4.3.9), hence $A(U)u = 0$, by (4.3.6) and the subsequent two lines. It follows that $u = \mathcal{G}_U f = 0$, and finally $f = 0$ by the injectivity of $\mathcal{G}_U$ on $L^2(\mathfrak{r}(U))$, cf. above. This establishes (a).

Leaving out the trivial case $U = \varnothing$, $L^2(\mathfrak{r}(U))$ is infinite dimensional. Denote by

$$\lambda_1(U) \leqslant \lambda_2(U) \leqslant \ldots \leqslant \lambda_n(U) \leq \ldots.$$

the eigenvalues of $A(U)$ repeated according to multiplicity. We have $\lambda_1(U) > 0$ because $A(U)$ is injective. Furthermore $\lambda_n(U) \to +\infty$ since $A(U)$ is compact. Clearly dim $W_0^{1,2}(U) = +\infty$ since $U \neq \varnothing$. For two open sets U, U' with $U \subset U'$ we have $\lambda_n(U) \geqslant \lambda_n(U')$ for every n in view of the Rayleigh-Ritz formula, cf. e.g. [Da, pp. 88, 91]. If $U = \varnothing$, we put $\lambda_n(U) = +\infty$ for any $n \in \mathbb{N}$.

For any interval $J \subset \mathbb{R}$ (possibly a singleton) we denote by $E(J, U)$ the spectral measure of J for $A(U)$. If J is bounded then range of $E(J, U)$ is contained in dom $A(U)^{\frac{1}{2}} = W_0^{1,2}(U)$. Because

$$\left\|A(U)^{\frac{1}{2}} u\right\|^2 = Q(u) := Q(u, u),\ u \in W_0^{1,2}(U),$$

we have

$$\lambda_n(U) = \min\{Q(u) : u \in W_0^{1,2}(U),\ \|u\| = 1,\ E([\lambda_1(U), \lambda_n(U)[, U)u = 0\}, \tag{4.3.10}$$

the minimizing vectors being precisely the $L^2(\mathbb{R}^N)$-normalized eigenfunctions of $A(U)$ corresponding to $\lambda_n(U)$. In fact, every competing function u is orthogonal in

$L^2(\Omega)$ to all eigenvectors for $A(U)$ corresponding to eigenvalues $< \lambda_n(U)$, and hence u is the $L^2(\mathbb{R}^N)$-limit of a sequence of finite linear combinations of eigenvectors for $A(U)$ corresponding to eigenvalues $> \lambda_n(U)$. It follows that

$$\begin{aligned} Q(u) &= \langle A(U)^{\frac{1}{2}}u, A(U)^{\frac{1}{2}}u\rangle_1 \\ &\geq \lambda_n(U)\langle u, u\rangle_1 = \lambda_n(U). \end{aligned}$$

Hence every eigenvector for $A(U)$ corresponding to the eigenvalue $\lambda_n(U)$ is indeed minimizing in (4.3.10), which proves (4.3.10).

For arbitrary $f \in L^2(\mathfrak{r}(U))$ we similarly derive from (4.3.9) and (4.3.6):

$$A(U)\mathcal{G}_U f = f, \; f \in L^2(\mathfrak{r}(U)).$$

Thus

$$A(U) : L^2(\mathfrak{r}(U)) \to L^2(\mathfrak{r}(U))$$

is bijective with the inverse $\mathcal{G}_U$. □

The injection

$$(W_0^{1,2}(U), \|\cdot\|_1) \hookrightarrow L^2(U) \text{ is compact,} \tag{4.3.11}$$

by a result of Rellich [Re] applied to an open ball Ω containing U. (We may of course extend $\mathcal{G}_U$ and $A(U)$ linearly to $L^2(\Omega)$ or to $L^2 = L^2(\mathbb{R}^N)$ by 0 on $L^2(\Omega\backslash\mathfrak{r}(U))$ or $L^2(\complement\mathfrak{r}(U))$).

The positive selfadjoint operator $A(U)$ on $L^2(\mathfrak{r}(U))$ has a compact resolvent in view of (4.3.11). Thus $\mathcal{G}_U$ is a *positive compact* operator on $L^2(\mathfrak{r}(U))$ (and hence on $L^2(\mathbb{R}^N)$). The compactness of $\mathcal{G}_U$ was obtained differently by Hansen [Han, p. 128].

Lemma 4.3.6 *For any bounded finely open sets U, V, $U \subset V$, we have $\lambda_n(U) \geqslant \lambda_n(V)$. Moreover, the following are equivalent*

(i) $W_0^{1,2}(U) = W_0^{1,2}(V)$,
(ii) $V\backslash U$ *is polar,*
(iii) $\lambda_n(U) = \lambda_n(V)$,
(iv) $A(U) = A(V)$.

Proof For any $n \in \mathbb{N}$ it follows by the Rayleigh-Ritz characterization of $\lambda_n(U)$ that $\lambda_n(U) \geqslant \lambda_n(V)$, e.g. [Da, pp. 88, 89].

(i)⇔(ii). Let 1_V denote the indicator function for V. Then $1_V \in L^2(\mathbb{R}^N)$, and the function $\mathcal{G}_V 1_V$ belongs to $W_0^{1,2}(V) = W_0^{1,2}(U)$ by (4.3.7) and satisfies $\mathcal{G}_V 1_V > 0$ on V (because every fine component ω of V has non-zero Lebesgue measure, and $G_V > 0$ on $\omega \times \omega$). By (4.3.2) this implies that indeed $V\backslash U$ is polar. The converse (ii)⇒(i) is obvious, and so (i)⇔(ii).

(iii)⇔(iv). By induction with respect to n one shows that the eigenspaces of $A(U)$ and $A(V)$ corresponding to the common eigenvalue $\lambda_n(U) = \lambda_n(V)$ are the same in view of (4.3.10). By the spectral theorem for selfadjoint operators this implies that $A(U) = A(V)$. The converse is obvious, so (iii)⇔(iv).

(iv)⇒(i) because

$$W_0^{1,2}(U) = \text{dom } A(U)^{\frac{1}{2}} = \text{dom } A(V)^{\frac{1}{2}} = W_0^{1,2}(V).$$

Conversely, (i)⇒(iv) is obvious. □

Theorem 4.3.7 *Let $(U_i)_{i\in\mathbb{N}}$ be a decreasing sequence of bounded finely open subsets of $\mathbb{R}^N$, and denote by U the fine interior of $U_\infty = \bigcap_i U_i$. For any $n \in \mathbb{N}$ we then have $\lambda_n(U_i) \uparrow \lambda_n(U)$ as $i \uparrow \infty$.*

Proof We have $\lambda_n(U) \uparrow +\infty$ as $n \to \infty$ because $A(U)^{-1}$ is compact. Fix $n \in \mathbb{N}$ so that $\lambda_n(U) > \lambda_{n-1}(U)$ (if $n > 1$). Denote by n' the smallest integer $\nu > n$ such that $\lambda_\nu(U) > \lambda_{\nu-1}(U)$,

$$\lambda_{n-1}(U) < \lambda_n(U) = \ldots = \lambda_{n'-1}(U) < \lambda_{n'}(U). \tag{4.3.12}$$

We show that

$$\lambda_\nu(U_i) \uparrow \lambda_\nu(U) \text{ as } i \to +\infty, n \leqslant \nu < n', \tag{4.3.13}$$

and that

$$E(J_n(U_i), U_i) \to E(J_n(U), U) \text{ as } i \to \infty, \tag{4.3.14}$$

where we write

$$J_n(U_i) = [\lambda_n(U_i), \lambda_{n'}(U_i)[\tag{4.3.15}$$

cf. (4.3.12) with U replaced by U_i and where the convergence of operators is in the operators norm topology for $L^2(\mathbb{R}^N)$.

Clearly, the theorem has been proved once (4.3.13) has been established for every n as stated and for any ν as stated. Here it suffices to consider $\nu = n$ since for $n < \nu < n'$ we obtain by (4.3.10), (4.3.13) for $i \to +\infty$

$$\lambda_\nu(U) \geqslant \lambda_\nu(U_i) \geqslant \lambda_n(U_i) \uparrow \lambda_n(U) = \lambda_\nu(U).$$

We prove (4.3.13) with $\nu = n$ and (4.3.14) simultaneously by induction. In case $n > 1$ we suppose that (4.3.13) and (4.3.14) hold with n replaced by any m from finite set

$$M = \{m \in \mathbb{N} : 1 \leqslant m < n, \text{ and } \lambda_m(U) > \lambda_{m-1}(U) \text{ if } m > 1\} \tag{4.3.16}$$

If $\lambda_n(U_i) \to +\infty$ as $i \to \infty$ then $\lambda_n(U) = +\infty$, and hence (4.3.13) holds for $\nu = n$ while (4.3.14) drops out. We may assume that *the increasing sequence $(\lambda_n(U_i))_{i\in\mathbb{N}}$ is bounded*, say $\sup_i \lambda_n(U_i) < +\infty$.

Consider the $(n-1)$-dimensional spectral space $\mathcal{F}_i \subset W_0^{1,2}(U_i)$ of the interval $[\lambda_1(U_i), \lambda_n(U_i)[$ and the orthogonal projection operator F_i of $W_0^{1,2}(U_i)$ on that interval:

$$F_i = E([\lambda_1(U_i), \lambda_n(U_i)[, U_i) = \sum_{m \in M} E(J_m(U_i), U_i) \tag{4.3.17}$$

with $J_m(U_i) := [\lambda_m(U_i), \lambda_{m'}(U_i)[$, m' denoting the smallest integer $\mu > m$ such that $\lambda_\mu(U_i) > \lambda_{\mu-1}(U_i)$, similarly to the definition of n' and $J_n(U_i)$, cf. (4.3.12) and (4.3.15). Here we have used that the intervals $J_m(U_i)$, $m \in M$, are pairwise disjoint and together cover $[\lambda_1(U_i), \lambda_n(U_i)[$ because 1 is the smallest number in M, and when m is the largest number in M we have $m' = n$, that is, $\lambda_m(U) = \ldots = \lambda_{n-1}(U) < \lambda_n(U)$. In the case $n = 1$ it is understood that $M = \varnothing$ and $F_i = 0$ (the projection of $L_2(\mathbb{R}^N)$ on $\{0\}$).

By the hypothesis of induction we have for each $m \in M$

$$E(J_m(U_i), U_i) \to E(\{\lambda_m(U)\}, U) \text{ as } i \to \infty,$$

and hence by summation over M while invoking (4.3.17)

$$F_i \to F := E([\lambda_1(U), \lambda_n(U)[) = \sum_{m \in M} E(\{\lambda_m(U)\}, U) \tag{4.3.18}$$

in the operator norm topology, that is, $\|F_i - F\| \to 0$.

Fix for a while $\nu \in \mathbb{N}$, $\nu \geqslant n$, such that the sequence $(\lambda_\nu(U_i))_{i\in\mathbb{N}}$ is bounded:

$$\sup_{i\in\mathbb{N}} \lambda_\nu(U_i) < +\infty, \tag{4.3.19}$$

for example $\nu = n$, and consider for each $i \in \mathbb{N}$ an $L^2(\mathbb{R}^N)$-normalized eigenvector u_i for $A(U_i)$ corresponding to the eigenvalue $\lambda_\nu(U_i)$ for $A(U_i)$, or equivalently: corresponding to eigenvalue $\lambda_\nu(U_i)^{1/2}$ for $A(U_i)^{1/2}$ on $W_0^{1,2}$. Because $u_i \in W_0^{1,2}(U_i) \subset W_0^{1,2}(U_1)$, and $Q(u_i) = \lambda_\nu(U_i)$ remains bounded, by (4.3.19), we may assume in view of the compactness of the injection $(W_0^{1,2}(U_1), \|\cdot\|) \hookrightarrow L^2(\mathbb{R}^N)$ that there is a subsequence, still denoted $(u_i)_{i\in\mathbb{N}}$, which converges in $L^2(\mathbb{R}^N)$ to some $u \in L^2(\mathbb{R}^N)$.

For $i < j$ we have $u_i, u_j \in W_0^{1,2}(U_i)$, hence by (4.3.17) $F_i u_j \in \mathcal{F}_i \subset W_0^{1,2}(U_i)$. Thus

$$u_{ij} = u_i + u_j - F_i u_j \in W_0^{1,2}(U_i). \tag{4.3.20}$$

Note that $F_i u_i = F_j u_j = 0$ because $\nu \geqslant n$ and hence $\lambda_\nu(U_i) \geqslant \lambda_n(U_i)$, etc. It follows by (4.3.18) that

$$F_i(u_j) = (F_i - F_j)(u_j) \to 0 \text{ in } L^2(\mathbb{R}^N) \text{ as } i, j \to \infty \tag{4.3.21}$$

because $F_i - F_j \to F - F = 0$ in operator norm topology and $\|u_j\| = 1$. Moreover $F_i u_{ij} = F_i u_i = 0$. Invoking (4.3.10) applied to $u = u_{ij}/\|u_{ij}\|$ (if $u_{ij} \neq 0$) and $U = U_i$ we obtain from (4.3.17)

$$E([\lambda_1(U_i), \lambda_n(U_i)[, U_i)u_{ij} = F_i u_{ij} = 0$$

(trivially if $u_{ij} = 0$), and hence

$$Q(u_{ij}) \geqslant \lambda_n(U_i)\|u_{ij}\|^2 = 4\lambda_n(U_i) + o(1) \tag{4.3.22}$$

as $i, j \to \infty$ because $\lambda_n(U_i) \uparrow \sup_i \lambda_n(U_i) < +\infty$ as $n \to \infty$ according to (4.3.19), and because $u_{ij} \to 2u$ in $L^2(\mathbb{R}^N)$, hence $\|u_{ij}\|^2 \to 4\|u\|^2 = 4$, by (4.3.20), (4.3.21) in view of $u_i \to u$ in $L^2(\mathbb{R}^N)$ (see end of the preceding paragraph). By the parallelogram identity for the closed positive semidefinite bilinear form Q on $W_0^{1,2}(U)$ given by (4.3.5) we obtain since $u_i + u_j = F_i u_j$

$$\begin{aligned} Q(u_i - u_j) &\leq 2Q(u_i) + 2Q(u_j) - Q(u_{ij} + F_i u_j) \\ &\leq 2\lambda_\nu(U_i) + 2\lambda_\nu(U_j) - 4\lambda_n(U_i) + o(1). \end{aligned} \tag{4.3.23}$$

This is because $F_i u_j \in \mathcal{F}_i$, an $(n-1)$-dimensional subspace of $W_0^{1,2}(\mathbb{R}^N)$, so the $L^2(\mathbb{R}^N)$-norm $\|\cdot\|$ and the $W_0^{1,2}$-norm $\|\cdot\|_1$ are equivalent on $\mathcal{F}_i$. Furthermore, Q is continuous on $(W_0^{1,2}(\mathbb{R}^N), \|\cdot\|_1)$. Hence $-Q(u_{ij} + F_i u_j) = -Q(u_{ij}) + o(1) \leqslant -4\lambda_n(U_j) + o(1)$ as $i, j \to \infty$ by (4.3.22).

In the particular case $\nu = n$ the estimate (4.3.23) reads

$$Q(u_i - u_j) \leqslant 2\lambda_n(U_j) - 2\lambda_n(U_i) + o(1) = o(1)$$

as $i, j \to \infty$ by the convergence of the bounded increasing sequence $(\lambda_n(U_i))_{i\in\mathbb{N}}$, cf. (4.3.19). Thus (u_i) is a Cauchy sequence in $(W_0^{1,2}, \|\cdot\|_1)$, and therefore converges there to its limit u in $L^2(\mathbb{R}^N)$. And since each $u_i \in W_0^{1,2}(U_i)$ we have $u = 0$ q.e. on $\complement U_i$ for each i, hence altogether $u = 0$ q.e. on $\bigcup_i \complement U_i = \complement U_\infty$. Thus $u \in W_0^{1,2}(U_\infty) = W_0^{1,2}(int_f U_\infty) = W_0^{1,2}(U)$ by (4.3.3) and the definition of U in the theorem. Moreover, $Fu = \lim F_l u_l = 0$ in view of (4.3.18). This leads to (4.3.13) in the decisive case $\nu = n$ as follows in view of (4.3.10):

$$\lambda_n(U) \leqslant Q(u) = \lim_i Q(u_i) = \lim_i \lambda_n(U_i) \leqslant \lambda_n(U),$$

first for the subsequence of (u_i) chosen in the paragraph containing (4.3.19) above, and next obviously for the entire original sequence (u_i) converging to u. This implies that

$$\lambda_n(U) < +\infty \tag{4.3.24}$$

by the boundedness of $(\lambda_n(U_i))_{i\in\mathbb{N}}$, see (4.3.19). Moreover, (4.3.19) holds with ν replaced by any integer $\nu \in [n, n'[$ because $\lambda_\nu(U_i) \leqslant \lambda_\nu(U) = \lambda_n(U)$ by (4.3.12). It follows from (4.3.12) and (4.3.13) for $\nu = n$ that for i large enough

$$\lambda_n(U_i) > \lambda_{n-1}(U) \geqslant \lambda_{n-1}(U_i). \tag{4.3.25}$$

We may of course assume that (4.3.25) holds for all $i \in \mathbb{N}$.

In order next to establish (4.3.14) under the inductive hypothesis from the paragraph containing (4.3.16) we write for brevity

$$E = E(\{\lambda_n(U)\}, U) \tag{4.3.26}$$

$$E_i = E([\lambda_n(U_i), \lambda_{n'-1}(U_i)], U_i) \tag{4.3.27}$$

cf. (4.3.15), and $\mathcal{E}, \mathcal{E}_i$ for the ranges of E, E_i respectively. Clearly $\dim\, \mathcal{E}_i \geqslant n' - n = \dim\, \mathcal{E}$, cf. (4.3.12), and $\mathcal{E}_i \supset$ range $E(J_n(U_i), U_i)$. We have $E_i = E(J_n(U_i), U_i)$ if and only if $\lambda_{n'}(U_i) > \lambda_{n'-1}(U_i)$, or equivalently: $\dim \mathcal{E}_i = n' - n$. Put $p = n'$ if these three conditions are fulfilled for all sufficiently large $i \in \mathbb{N}$, and put $p = n' + 1$ otherwise. After passing to a subsequence of (U_i), again denoted by (U_i), we may suppose in the case $p = n' + 1$ that $\lambda_{n'}(U_i) = \lambda_{n'-1}(U_i)$ holds for all $i \in \mathbb{N}$. For every $\nu = n, \ldots, p - 1$ we therefore have by (4.3.24) in either case $p = n'$ or $p = n' + 1$

$$\lambda_\nu(U_i) \uparrow \lambda_n(U)(< +\infty). \tag{4.3.28}$$

In fact, for $n \leqslant \nu < n'$ we have $\lambda_\nu(U_i) \uparrow \lambda_\nu(U) = \lambda_n(U)$ by (4.3.13) and (4.3.12). And if $\nu = n'$, hence $p = n' + 1$, we have from the lines following (4.3.27): $\lambda_{n'}(U_i) = \lambda_{n'-1}(U_i) \uparrow \lambda_{n'-1}(U) = \lambda_n(U)$, again in view of (4.3.12).

In either case $p = n'$ or $p = n' + 1$ we further have $\dim\, \mathcal{E}_i \geqslant p - n$ and $E_i = E([\lambda_n(U_i), \lambda_{p-1}(U_i)], U_i)$. We may therefore choose an orthonormal family of $p - n$ vectors

$$e_n(U_i), \ldots, e_{p-1}(U_i) \in \mathcal{E}_i(\subset W_0^{1,2}(U_i))$$

such that

$$A(U_i)e_\nu(U_i) = \lambda_\nu(U_i)e_\nu(U_i), \; n \leqslant \nu < p.$$

In view of (4.3.28) we may now apply the results obtained earlier to $u_i = e_\nu(U_i)$, $n \leqslant \nu < p$, cf. the paragraph containing (4.3.19). After passing to yet another subsequence of $(U_i)_{i \in \mathbb{N}}$ (cf. the preceding paragraph) we may thus assume that there exist $e_1, \ldots, e_{p-1} \in L^2(\mathbb{R}^N)$ such that

$$e_\nu(U_i) \to e_\nu \text{ in } L^2(\mathbb{R}^N) \text{ as } i \to \infty \tag{4.3.29}$$

for $n \leqslant \nu < p$. And (4.3.23) together with $\lambda_n(U_i) \geqslant \lambda_\nu(U_i)$ lead to

$$Q(e_\nu(U_j) - e_\nu(U_i)) \leqslant 2\lambda_\nu(U_i) + 2\lambda_\nu(U_j) - 4\lambda_\nu(U_i) + o(1) = o(1)$$

as $j > i \to \infty$, according to (4.3.28). Each Cauchy sequence $(e_\nu(U_i))_{i \in \mathbb{N}}$ in $(W_0^{1,2}(\mathbb{R}^N), \|\cdot\|_n)$ converges there to its limit e_ν in $L^2(\mathbb{R}^N)$, cf. (4.3.29), and since $e_\nu(U_i) \in W_0^{1,2}(U_i)$ we obtain

$$e_\nu \in W_0^{1,2}(U), \; n \leqslant \nu < p. \tag{4.3.30}$$

According to (4.3.29) we have $\langle e_\mu, e_\nu\rangle = \delta_{\mu\nu}$ for $n \leqslant \mu, \nu < p$, and by (4.3.18), (4.3.17)

$$Fe_\nu = \lim_i F_i e_\nu(U_i) = 0. \tag{4.3.31}$$

From (4.3.28) we proceed to derive that in terms of the $W_0^{1,2}(U_i)$-norm $\|\cdot\|_1$

$$Q(e_\nu) = \lim_i Q(e_\nu(U_i)) = \lim_i \lambda_\nu(U_i) = \lambda_n(U) < +\infty \tag{4.3.32}$$

for $\nu \in [n, p[$. Here $e_\nu(U_i) \in \mathcal{E}_i \subset$ range $E([\lambda_n(U_i), \lambda_\nu(U_i)], U_i)$, and so dim $(\mathcal{E}_i) \leqslant \nu < +\infty$. It follows that the $L^2(\mathbb{R}^N)$-norm and $W_0^{1,2}(\mathbb{R}^N)$-norm on $\mathcal{E}_i$ are equivalent, which proves (4.3.32).

From (4.3.30), (4.3.31), (4.3.32) together with (4.3.10) and the definition of F in (4.3.18) we obtain

$$A(U)e_\nu = \lambda_n(U),\ n \leqslant \nu < p,$$

and so the orthonormal family of $p - n$ vectors e_ν is contained in the range $\mathcal{E}$ of E, cf. (4.3.26). Since $p - n \leqslant n' - n = \dim\ \mathcal{E}$ we conclude that actually $p = n'$ and so

$$\dim\ \mathcal{E}_i = n' - n = \dim\ \mathcal{E} \tag{4.3.33}$$

holds for all sufficiently large $i \in \mathbb{N}$ from the original sequence, see the paragraph containing (4.3.28). For these i we therefore have $\lambda_{n'}(U_i) > \lambda_{n'-1}(U_i)$ and

$$E_i = E(J_n(U_i), U_i). \tag{4.3.34}$$

For given $f \in L^2(\mathbb{R}^N)$ with $\|f\| = 1$ we proceed to show (again for the original sequence (U_i)) that

$$E_i f \to Ef \text{ in } L^2(\mathbb{R}^N) \text{ as } i \to \infty, \tag{4.3.35}$$

cf. (4.3.26), (4.3.27). We have $E_i f \in$ dom $A(U_i)^{1/2} = W_0^{1,2}(U_i) \subset W_0^{1,2}(U_1)$ and

$$Q(E_i f) \leqslant \lambda_{n'-1}(U_i)\|E_i f\|^2 \leqslant \lambda_{n'-1}(U_i),$$

a bounded sequence because $\lambda_q(U_i) \leqslant \lambda_q(U) < +\infty$ for every $q \in \mathbb{N}$ by Lemma 4.3.6. In view of the compactness of the injection $(W_0^{1,2}(U_1), \|\cdot\|_1) \hookrightarrow L^2(\mathbb{R}^N)$ the set $\{E_i f\}$ is relatively compact in $L^2(\mathbb{R}^N)$. This implies that (4.3.35) will be established once it has been shown that Ef is the limit in $L^2(\mathbb{R}^N)$ of any convergent subsequence of the sequence $(E_i f)_{i\in\mathbb{N}}$. To achieve this we choose $e_n(U_i), \dots, e_{n'-1}(U_i)$ as in the paragraph containing (4.3.29) above (now with $p = n'$). After passing to a suitable further subsequence we may assume that there exist $e_n, \dots, e_{n'-1} \in L^2(\mathbb{R}^N)$ (perhaps not the same as before) such that (4.3.29) holds for $n \leqslant \nu < n'$ for this new subsequence (again denoted by $(e_\nu(U_i))_{i\in\mathbb{N}}$). As shown in the preceding paragraph the e_ν again form an orthonormal base of $\mathcal{E}$, and we obtain by (4.3.29)

$$E_i f = \sum_{\nu=n}^{n'-1} \langle f, e_\nu(U_i)\rangle e_\nu(U_i) \to \sum_{\nu=n}^{n'-1} \langle f, e_\nu\rangle e_\nu = Ef$$

as $i \to \infty$ (after the last extraction of a subsequence). As pointed out above, this establishes (4.3.35), even for the original sequence.

Finally, $E_i \to E$ in the operator norm topology as $i \to \infty$ according to Lemma 4.3.8 below because dim $\mathcal{E}_i =$ dim $\mathcal{E} < +\infty$ by (4.3.33). In view of (4.3.34) and (4.3.26) this completes the simultaneous inductive proof of (4.3.13) and (4.3.14), and hence of the theorem. □

The following lemma is presumably known:

Lemma 4.3.8 *On a Hilbert space $\mathcal{H}$ let P and P_n, $n \in \mathbb{N}$, be selfadjoint projections of equal and finite rank r, and such that*

$$P_n f \to Pf \text{ for every } f \in \mathcal{H} \text{ as } n \to \infty.$$

Then $\|P_n - P\| \to 0$.

Proof We may assume that $r > 0$. Let $e \in$ range P, $\|e\| = 1$, and write $P_e f = \langle f, e\rangle e$. Then $Q := P - P_e$ is a self-adjoint projection rank $r - 1$. Since $P_n e \to Pe = e$ we may assume that $P_n e \neq 0$. Writing $e_n = \|P_n e\|^{-1} P_n e$ we have $\|e_n\| = 1$, $e_n \to e$, and

$$\begin{aligned} P_{e_n} f &:= \langle f, e_n\rangle e_n \to \langle f, e\rangle e = P_e f, \\ Q_n f &:= P_n f - P_{e_n} f \to Pf - P_e f = Qf. \end{aligned} \tag{4.3.36}$$

Because Q_n and Q both have rank $r - 1$ we may assume by induction

$$\|Q_n - Q\| \to 0 \tag{4.3.37}$$

For any $f \in \mathcal{H}$ with $\|f\| = 1$ we have

$$(P_{e_n} - P_e)f = \langle f, e_n - e\rangle e_n + \langle f, e\rangle(e_n - e),$$

hence

$$\|P_{e_n} - P_e\| \leqslant 2\|e_n - e\| \to 0. \tag{4.3.38}$$

Finally, by (4.3.36), $P_n - P = (P_{e_n} - P_e) + (Q_n - Q)$, and hence by (4.3.37), (4.3.38)

$$\|P_n - P\| \leqslant \|P_{e_n} - P_e\| + \|Q_n - Q\| \to 0.$$

□

We proceed to extend and expand Theorem 4.3.7 by replacing the decreasing sequence (U_i) by any *order convergent* sequence of finely open sets, now supposed uniformly bounded. Recall that we denote by E' the fine interior of a set $E \subset \mathbb{R}^N$.

Definition 4.3.9 A sequence of uniformly bounded finely open sets $U_i \subset \mathbb{R}^N$ is said to be order convergent with the order limit U if

$$U = \Big(\bigcap_i \bigcup_{j \geqslant i} U_j\Big)' = \bigcup_i \Big(\bigcap_{j \geqslant i} U_j\Big)'.$$

In particular, any decreasing sequence of bounded finely open sets $U_i \subset \mathbb{R}^N$ is order convergent with order limit $\left(\bigcap_i U_i\right)'$. Any increasing sequence of uniformly bounded sets $U_i \subset \mathbb{R}^N$ is order convergent with the order limit $\bigcup_i U_i$.

Theorem 4.3.10 *Let $(U_i)_{i\in\mathbb{N}}$ be an order convergent sequence of uniformly bounded open subsets of $\mathbb{R}^N$ with the order limit U. Then*

(1) $\lambda_n(U_i) \to \lambda_n(U)$ as $i \to \infty$, $n \in \mathbb{N}$,

(2) $E(J, U_i) \to E(J, U)$ as $i \to \infty$ for every bounded interval $J \subset \mathbb{R}$ whose endpoints are not eigenvalues of $A(U)$,

(3) $\mathcal{G}_{U_i} \to \mathcal{G}_U$ as $i \to \infty$,

the convergence in (2) *and* (3) *being in the norm topology for bounded operators on the Hilbert space $L^2(\mathbb{R}^N)$.*

Proof (1) For a decreasing sequence (U_i) this holds by Theorem 4.3.7. Next, consider an increasing sequence (U_i) of uniformly bounded finely open sets U_i and its order limit $U = \bigcup_i U_i$. For given $n \in \mathbb{N}$ the sequence $(\lambda_n(U_i))$ is decreasing. Let $\alpha \in \mathbb{R}$, $\alpha > \inf_i \lambda_n(U_i)$. By Lemma 4.3.2(i), $\bigcup_i W_0^{1,2}(U_i)$ is dense in $(W_0^{1,2}(U), \|\cdot\|_1)$, and the Rayleigh-Ritz formula shows that there exists an n-dimensional (linear) subspace $\mathcal{L}$ of $\bigcup_i W_0^{1,2}(U_i)$ such that

$$\lambda(\mathcal{L}) = \sup\{Q(u) : u \in \mathcal{L}, \|u\| = 1\} < \alpha,$$

see, for example, [Da, p. 91]. Since $(W_0^{1,2}(U_i))_{i\in\mathbb{N}}$ is increasing we have $\mathcal{L} \subset W_0^{1,2}(U_i)$ for some i, hence, according to (4.3.10),

$$\begin{aligned}\lambda_n(U_i) &= \min\{Q(u) : u \in W_0^{1,2}(U_i), \|u\| = 1, E([\lambda_1(U_i), \lambda_n(U_i)]) = 0\}\\ &\leq \min\{Q(u) : u \in \mathcal{L}, \|u\| = 1, E([\lambda_1(U_i), \lambda_n(U_i)]) = 0\}\\ &\leq \lambda(\mathcal{L}) < \alpha.\end{aligned}$$

which proves (1) for an increasing sequence $(U_i)_{i\in\mathbb{N}}$.

For an arbitrary order convergent sequence $(U_i)_{i\in\mathbb{N}}$ with the limit U we consider the decreasing, resp. increasing sequence, of uniformly bounded finely open sets

$$U_i^* = \bigcup_{j \geqslant i} U_j, \; U_{*i} = \left(\bigcap_{j \geq i} U_j \right)' \tag{4.3.39}$$

with

$$U = \left(\bigcap U_i^* \right)' = \bigcup U_{*i}$$

and hence by Theorem 4.3.7 and Lemma 4.3.6

$$\lambda_n(U) = \sup_i \lambda_n(U_i^*) \leq \sup_i \inf_{j \geqslant i} \lambda_n(U_j) = \liminf_i \lambda_n(U_i)$$

and

$$\lambda_n(U) = \inf_i \lambda_n(U_{*i}) \geq \inf_i \sup_{j \geqslant i} \lambda_n(U_j) = \limsup_i \lambda_n(U_i).$$

This establishes (1).

(2) We decompose the given bounded interval $J \subset \mathbb{R}$ into finitely many pairwise disjoint intervals J_m of the same kind and such that each J_m contains at most one eigenvalue of $A(U)$. Then $E(J, U) = \sum_m E(J_m, U)$ and similarly with U replaced by any U_i. This reduces (2) to the case where J itself contains at most one eigenvalue of $A(U)$, necessarily lying in the interior of J. Denote by a and $b(>a)$ the endpoints of J. Let $\lambda_p(U)$ be the greatest eigenvalue of $A(U)$ less than a, and let $\lambda_q(U)$ be the least eigenvalue of $A(U)$ exceeding b. In view of (1), for i sufficiently large, $\lambda_p(U_i)$ is the greatest eigenvalue of $A(U_i)$ less than a, and $\lambda_q(U_i)$ is the least eigenvalue of $A(U_i)$ exceeding b. In case J contains no eigenvalue of $A(U)$ then likewise no eigenvalue of $A(U_i)$, so $E(J, U_i) = 0 \to 0 = E(J, U)$ as $i \to \infty$. In the remaining case where J contains exactly one (perhaps multiple) eigenvalue $\lambda = \lambda_n(U)$ of $A(U)$ (that we may suppose that $> \lambda_{n-1}$) we have $p + 1 \leqslant q - 1$ and

$$\lambda_p(U) < \lambda_{p+1}(U) = \ldots = \lambda_{q-1}(U) < \lambda_q(U).$$

Hence

$$J_n(U_i) = [\lambda_n(U_i), \lambda_{n'-1}(U_i)] \subset J, \; \lambda_{n'}(U_i) \notin J$$

for large i by (1). This amounts to (4.3.12), now with $n = p + 1$ and $n' = q$. For $i \to \infty$ we therefore obtain by (4.3.27), (4.3.34), and (4.3.14)

$$\begin{aligned} E(J, U_i) &= E([\lambda_n(U_i), \lambda_{n'-1}(U_i)], U_i) \\ &= E_i = E(J_n(U_i), U_i) \to E(\{\lambda_n(U)\}, U) = E(J, U) \end{aligned}$$

which establishes (2).

(3) As shown in the paragraph containing (4.3.6), $A(U)$ is injective, that is, $\lambda_n(U) \neq 0$. In view of (1) for $n = 1$ there is therefore a real constant $\tau > 0$ such that $\lambda_1(U_i) > \tau$ for all i. Denote by $E(\cdot)$, $E_i(\cdot)$ the spectral measure for $A(U)$, $A(U_i)$

respectively. From (2) follows

$$\int \varphi(t)E_i(dt) \to \int \varphi(t)E(dt) \tag{4.3.40}$$

in the operator norm topology for every continuous function $\varphi : [0, +\infty] \longrightarrow \mathbb{R}$ tending to 0 at infinity. (This is shown in standard way by approximating φ uniformly by finite linear combinations of indicator functions of bounded intervals whose endpoints are not eigenvalues of $A(U)$.) To obtain (3), apply (4.3.40) to $\varphi(t) = 1/\max\{t, \tau\}$. This completes the proof of Theorem 4.3.10. □

Remark 4.3.11 A perturbation theorem of Rellich [Re] shows directly that conversely, (3) implies (2). The present proof of Theorem 4.3.10, adapted to the more general setting of [Fu30, Theorem 2.1], is simpler than that proof given there.

Theorem 4.3.12 *Let $U \subset \mathbb{R}^N$ be a non-void bounded finely open.*

(a) *Every eigenfunction $u \in L^2(U)$ for $A(U)$ can be redefined on a polar set so as it become bounded and finely continuous everywhere on U and to have fine limit at every regular fine boundary point for U.*

(b) *If U is finely connected, the lowest eigenvalue $\lambda_1(U)$ of the positive selfadjoint operator $A(U)$ is simple, and the corresponding (finely continuous) eigenfunctions has constant sign.*

Proof (a) Suppose that $A(U)u = \lambda u$, that is, $\mathcal{G}_U u = \lambda^{-1}u$ by Theorem 4.3.5(b). It follows that $\mathcal{G}_U^n u = \lambda^{-n}u$ for $n \in \mathbb{N}$. As before, let $\Omega = \mathbb{R}^N$ if $N \geqslant 3$ and Ω be for example an open ball containing U if $N = 2$. Let $\mathcal{G} = \mathcal{G}_\Omega$ be the corresponding Green operator. By Theorem 4.3.4, $\mathcal{G}_U$ is a bounded selfadjoint operator on $L^2(\mathfrak{r}(U))$. The Riesz kernel of (real) order $\alpha \in]0, N[$ on $\mathbb{R}^N$ is $h_\alpha(x-y)$, $x, y \in \mathbb{R}^N$, with

$$h_\alpha(x) = |x|^{\alpha-N}, \quad x \in \mathbb{R}^N.$$

The Riesz composition formula, cf. [Ri], reads

$$h_\alpha * h_\beta \approx h_{\alpha+\beta} \text{ for } \alpha + \beta < N,$$

where $\approx$ indicates equality up to some multiplicative constant factor. By repeated, say $n - 1$ ($\geqslant 0$) applications, we have for $2n < N$

$$h_2^{*n} := h_2 * h_2 * \ldots * h_2 \ (n \text{ factors}) \approx h_{2n},$$

and hence $\mathcal{G}_U^n u^\pm \leqslant h_2^{*n} * u^\pm \approx h_{2n} * u^\pm$ for $u \in L^2(\mathbb{R}^N)$ and $2n < N$. We also need that $4n > N$ to secure that $h_{2n} \in L^2_{loc}(\mathbb{R}^N)$, and hence $h_{2n} * u$ be bounded (and continuous) on $\mathbb{R}^N$. (Note that $u \in L^2_c$ since u is bounded, and that the convolution of class $L^2(\mathbb{R}^N)$ is bounded and uniformly continuous.) If $N \neq 2$ and $N \neq 4$ then there exists at least one natural number n such that $2n < N < 4n$, and hence

$$h_n * h_n = h_{2n} = h_2^{*n}.$$

Then $\mathcal{G}_U u^\pm \leqslant h_{2n} * u^\pm$ is bounded, and so is therefore $\mathcal{G}_U u$. In the remaining cases $N = 2$ or $N = 4$ the best we can do is to take $n = N/2 = 1$ or 2. Since $N < 4n$ we still have $h_{2n} \in L^2_{loc}(\mathbb{R}^N)$, but we no longer have $h_n * h_n \approx h_{2n}$. Instead, $h_n * h_n$ is approximately the logarithmic kernel $\log |x|$, see [Fu2, Lemma 3.1] for a precise estimate, which amply suffices for concluding that again $h_n * h_n \in L^2_{loc}(\mathbb{R}^N)$ (since $N > 0$).

The extension $\lambda G_U(u^+ m) - \lambda G_U(u^- m)$ of $u = \lambda \mathcal{G}_U u$ from $\mathfrak{r}(U)$ to $\mathbb{R}^N$ (m denoting the restriction of the Lebesgue measure to U) is bounded and finely continuous in view of Theorem 2.1.26 and equals 0 on $\complement \mathfrak{r}(U)$, hence has indeed the fine limit 0 at $\partial_f \mathfrak{r}(U) = \partial_{f,r} U$, the regular part of $\partial_f U$.

(b) In the particular case $\lambda = \lambda_1 (= \lambda_1(U))$ consider any normalized eigenvector u for λ_1, that is, $u = \lambda_1 \mathcal{G}_U u$. Then $|u|$ is normalized as well and $\mathcal{G}_U |u| \in L^2(\mathfrak{r}(U))$. Clearly, $\mathcal{G}_U |u| \geqslant |\mathcal{G}_U u|$. In terms of the inner product $\langle \cdot, \cdot \rangle$ on $L^2(\mathfrak{r}(U))$ we have

$$\langle \mathcal{G}_U |u|, |u| \rangle \geqslant \langle \mathcal{G}_U u, u \rangle = \lambda_1$$

with equality because $\lambda_1^{-1} = \max \langle \mathcal{G}_U f, f \rangle$ over all normalized vectors $f \in L^2(\mathfrak{r}(U))$. The subsets $V^+ = \{u > 0\}$ and $V^- = \{u < 0\}$ of $\mathfrak{r}(U)$ are finely open and $V^+ \cup V^- = \mathfrak{r}(U)$, $V^+ \cap V^- = \varnothing$ so when U or equivalently $\mathfrak{r}(U)$ is finely connected then either $V^+ = \mathfrak{r}(U)$ or $V^- = \mathfrak{r}(U)$. We may assume that $V^+ = \mathfrak{r}(U)$, that is, $u > 0$. Finally, λ_1 must be simple, for otherwise there would be two orthogonal eigenvectors u and v for $\mathcal{G}_U$ corresponding to the eigenvalue λ_1^{-1}, and we may assume that $u > 0$ and $v > 0$, hence $\langle u, v \rangle = \int_{\mathfrak{r}(U)} uv dm > 0$, a contradiction. □

4.4 Finely Harmonic Morphisms

We shall extend the concept of a harmonic morphisms between open subsets of $\mathbb{R}^N$ and $\mathbb{R}^{N'}$, cf. [CC2, Fu13, BW], by allowing the domain to be finely open.

Let U denote a finely open subset of $\mathbb{R}^N$, $N \geqslant 1$, and let $N' \geqslant 1$ be given. Recall that the fine topology on $\mathbb{R}^1$ equals the usual Euclidean topology. In this section we do not require that U be a subset of a Greenian set Ω, cf. Remark 2.1.9. The following definition turns out to be equivalent to the perhaps more natural (and apparently more general) definition, in which the Euclidean topology in (i) above is replaced by the fine topology, and u' in (ii) is finely harmonic on a finely open subset V' of $\mathbb{R}^{N'}$.

Definition 4.4.1 A map $\varphi : U \to \mathbb{R}^{N'}$ is called a finely harmonic morphism if

(i) φ is continuous from U with the fine topology into $\mathbb{R}^{N'}$ with the usual Euclidean topology, and

(ii) for any harmonic function u' on an open subset V' of $\mathbb{R}^{N'}$, $u' \circ \varphi$ is finely harmonic on the finely open subset $\varphi^{-1}(V')$ of U (cf. Remark 2.1.9).

Condition (ii) is regarded as automatically fulfilled if $\varphi^{-1}(V') = \varnothing$, that is, if $\varphi(U) \cap V' = \varnothing$. Here are some immediate consequences of Definition 4.4.1:

(a) The sheaf property: for any finely open subset V of U, the restriction $\varphi|_V$ is a finely harmonic morphism, and for any family (U_α) of finely open subsets of $\mathbb{R}^N$ and any map $\varphi : \bigcup_\alpha U_\alpha \to \mathbb{R}^{N'}$ such that $\varphi|_{U_\alpha} : U_\alpha \to \mathbb{R}^{N'}$ is a finely harmonic morphism, φ is itself a finely harmonic morphism. It follows that, for any open subset V' of $\mathbb{R}^{N'}$ and any finely harmonic morphism $\varphi : U \to \mathbb{R}^{N'}$, the restriction of φ to $\varphi^{-1}(V')$ (a finely open subset of U) is a finely harmonic morphism of $\varphi^{-1}(V')$ into V' (and likewise into $\mathbb{R}^{N'}$).

(b) Writing $\varphi = (\varphi_1, \dots, \varphi_{N'})$, each coordinate function $\varphi_n : U \to \mathbb{R}$ is finely harmonic by (ii) because the coordinate x'_n of $x' \in \mathbb{R}^{N'}$ is a harmonic function of x' on $\mathbb{R}^{N'}$.

(c) If $N' = 1$ (whereby harmonic functions on open subsets of $\mathbb{R}^{N'}$ are the same as affine-linear functions) then a finely harmonic morphism $U \to \mathbb{R}$ is the same as a finely harmonic function on U.

(d) If $U' = \mathbb{C}$ (identified with $\mathbb{R}^2$) then $\varphi : U \to \mathbb{C}$ is a finely harmonic morphism if and only if, for every $n = 0, 1, 2, \dots$, φ^n is a complex finely harmonic function.

(e) If U is open in $\mathbb{R}^N$ then every *harmonic morphism* $\varphi : U \to \mathbb{R}^{N'}$ in the sense of [CC2] (there named harmonic map) is also a finely harmonic morphism. It can be shown that the two concepts are actually identical if $N' \geqslant 2$, cf. Laine [Lai], but clearly not for $N \geqslant 3$, $N' = 1$, cf. (b) above, because a finely harmonic function on $U \subset \mathbb{R}^N$ need not be harmonic when $N \geqslant 3$, see Remark 2.1.23.

Theorem 4.4.2 *If $\varphi : U \to \mathbb{R}^{N'}$ is a nonconstant finely harmonic morphism (U fine domain in $\mathbb{R}^N$) then $N' \leqslant N$.*

Proof According to the last assertion of Theorem 4.2.12, U can be covered by bounded regular finely open sets $V \subset U$ such that each coordinate function φ_n ($n = 1, \dots, N'$) can be extended to a bounded function $\widehat{\varphi}_n \in \mathcal{H}(\Omega, V) \subset \widehat{\mathcal{D}}(\Omega)$ (Ω an open ball in $\mathbb{R}^N$. To achieve boundedness, replace each V by $\{x \in V : |\varphi(x)| < l\}$ and $\widehat{\varphi}_n$ by $\max\{-l, \min\{\widehat{\varphi}_n, l\}\}$; $l \in \mathbb{N}$). For any bounded finely harmonic function $u \in \mathcal{H}(\Omega, V)$ we have $\int \nabla u . \nabla \lambda d\ell_N = 0$ for any $\lambda \in \widehat{\mathcal{D}}(\Omega, V)$, according to Theorem 4.2.21. Since the functions $x'_j x'_k$ and $x_j'^2 - x_k'^2$ with distinct $j, k \in \{1, \dots, N'\}$ are harmonic functions of $x' = (x'_1, \dots, x'_N) \in \mathbb{R}^{N'}$, the functions $\varphi_j \varphi_k$ and $\varphi_j^2 - \varphi_k^2$ are finely harmonic on U. Since $\widehat{\varphi}_j$ and $\widehat{\varphi}_k$ are bounded and of class $\widehat{\mathcal{D}}(\Omega)$, so are $\widehat{\varphi}_j \widehat{\varphi}_k$ and $\widehat{\varphi}_j^2 - \widehat{\varphi}_k^2$, and they are thus of class $\mathcal{H}(\Omega, V)$. We therefore have for any bounded $\lambda \in \widehat{\mathcal{D}}(\Omega, V)$

$$
\begin{aligned}
0 &= \int_\Omega \nabla(\widehat{\varphi}_j\widehat{\varphi}_k).\nabla\lambda d\ell_N \\
&= \int_\Omega \widehat{\varphi}_j\nabla\widehat{\varphi}_k.\nabla\lambda d\ell_N + \int_\Omega \widehat{\varphi}_k\nabla\widehat{\varphi}_j.\nabla\lambda d\ell_N \\
&= \int_\Omega \nabla\widehat{\varphi}_k(\nabla(\widehat{\varphi}_j\lambda) - \lambda.\nabla\widehat{\varphi}_j)d\ell_N + \int_\Omega \nabla\widehat{\varphi}_j(\nabla(\widehat{\varphi}_k\lambda) - \lambda.\nabla\widehat{\varphi}_k)d\ell_N \\
&= -2\int_\Omega (\nabla\widehat{\varphi}_j.\nabla\widehat{\varphi}_k)\lambda d\ell_N \qquad (4.4.1)
\end{aligned}
$$

because $\widehat{\varphi}_j \in \widehat{\mathcal{D}}(\Omega, V)$ and $\widehat{\varphi}_k \in \mathcal{H}(\Omega, V)$, etc. Similarly, $\widehat{\varphi}_j^2 - \widehat{\varphi}_k^2$ is finely harmonic on V and indeed of class $\mathcal{H}(\Omega, V)$ because ${x'_j}^2 - {x'_k}^2$ is a harmonic function of $x' = (x'_1, \dots, x'_{N'}) \in \mathbb{R}^{N'}$. We this time obtain for any bounded $\lambda \in \widehat{\mathcal{D}}(\Omega, V)$

$$
\begin{aligned}
0 &= \int_\Omega \nabla(\widehat{\varphi}_j^2 - \widehat{\varphi}_k^2)\nabla\lambda\, d\ell_N \\
&= 2\int_\Omega \nabla(\varphi_j\lambda)\nabla\varphi_j\, d\ell_N \\
&\quad -2\int_\Omega \nabla(\varphi_k\lambda)\nabla\varphi_k\, d\ell_N - 2\int_\Omega (\|\nabla\varphi_k\|^2 - \|\nabla\varphi_k\|^2)\lambda\, d\ell_N \\
&= -2\int_\Omega (\|\nabla\widehat{\varphi}_j\|^2 - \|\nabla\widehat{\varphi}_k\|^2)\lambda\, d\ell_N. \qquad (4.4.2)
\end{aligned}
$$

Choosing $\lambda > 0$ we shall deduce from (4.4.1) and (4.4.2) by Lebesgue's differentiation theorem that

$$
\nabla\widehat{\varphi}_j.\nabla\widehat{\varphi}_k = 0 \text{ and } \|\nabla\widehat{\varphi}_j\|^2 - \|\nabla\widehat{\varphi}_k\|^2 = 0 \qquad (4.4.3)
$$

ℓ_N-a.e. on V for distinct $j, k \in \{1, \dots, N'\}$. Thus $\nabla\widehat{\varphi}_1, \dots, \nabla\widehat{\varphi}_{N'}$ are at ℓ_N-a.e. point of V mutually orthogonal vectors in $\mathbb{R}^N$ of equal norm. If $N' > N$ then these equal norms must all be identically 0 ℓ_N-a.e. on V, and indeed everywhere on V by fine continuity. Since U is supposed to be finely connected, this contradicts the hypothesis that φ is nonconstant, and so actually $N' \leqslant N$.

To establish (4.4.3) at ℓ_N-a.e. point $x \in V$, note that we may in (4.4.1) and (4.4.2) replace $\lambda \in \widehat{\mathcal{D}}(\Omega, V)$ by $\kappa\lambda \in \widehat{\mathcal{D}}(\Omega, V)$ for any bounded function $\kappa \in \widehat{\mathcal{D}}(\Omega)$. If f denotes one of the functions $\nabla\widehat{\varphi}_j.\nabla\widehat{\varphi}_k$ or $\|\nabla\widehat{\varphi}_j\|^2 - \|\nabla\widehat{\varphi}_k\|^2$ with distinct $j, k \in \{1, \dots, N'\}$ and if λ is bounded and of class $\widehat{\mathcal{D}}(\Omega)$, we thus have from (4.4.1) and (4.4.2) $\int_\Omega f\lambda\kappa\, d\ell_N = 0$ for every bounded $\kappa \in \widehat{\mathcal{D}}(\Omega)$, in particular for any $\kappa \in \mathcal{D}(\Omega)$. Thus $f\lambda = 0$ as a distribution on Ω of class $L^1(\ell_N)$, and hence $f\lambda = 0$ ℓ_N-a.e. For the proof of (4.4.3) it therefore suffices to show that there exists a bounded function $\lambda \in \widehat{\mathcal{D}}(\Omega, V)$ such that $\lambda > 0$ ℓ_N-a.e. on V. For this we state

$$
\lambda := G_\Omega\mu - \widehat{R}^{\complement V}_{G_\Omega\mu} = G_\Omega\mu - G_\Omega\mu^{\complement V}
$$

for some measure $\mu \in \mathcal{E}_+(\Omega)$, that is, μ shall have finite energy $\int G_\Omega\mu\, d\mu$ whence $G_\Omega\mu \in \widehat{\mathcal{D}}(\Omega)$ according to Theorem 4.2.10(a). It suffices to choose μ as a bounded measure on Ω with a bounded potential $G_\Omega\mu$. Then $\mu^{\complement V}$ has the same two properties. It follows that λ is bounded and that

$$\lambda = G_\Omega\mu - G_\Omega\mu^{\complement V} \in \widehat{\mathcal{D}}_+(\Omega, V)$$

because $G_\Omega\mu^{\complement V} = \widehat{R}^{\complement V}_{G_\Omega\mu}$ equals $G_\Omega\mu$ q.e. on $\complement V$. It remains to choose μ so that $\widehat{R}^{\complement V}_{G_\Omega\mu} < G_\Omega\mu$ ℓ_N-a.e. on V, and this holds (even everywhere on V) according to Theorem 1.4.17 if the bounded potential $G_\Omega\mu$ is *strict* (Definition 1.4.14), V being regular. And since the ball Ω is bounded we may, according to Theorem 1.4.15, simply take $\mu = m$ (Lebesgue measure on Ω), for, if for example $N \geq 3$, $G_\Omega m$ is majorized on Ω by $G_{\mathbb{R}^N} m$ (m being identified with the measure $1_\Omega \cdot \ell_N$ on $\mathbb{R}^N$) which is a finite and continuous Newtonian potential, hence bounded on the compact ball $\overline{\Omega}$. (This also holds for $N = 2$ if the diameter of Ω is $< +\infty$.) □

In (ii) in Definition 4.4.1, "harmonic" can equivalently be replaced (twice) by "hyperharmonic":

Theorem 4.4.3 *A map $\varphi : U \to \mathbb{R}^{N'}$ is a finely harmonic morphism if and only if*

(i) *φ is fine-to-usual continuous, and*

(ii) *for any hyperharmonic function u' on an open subset U' of $\mathbb{R}^{N'}$ (such that $\varphi^{-1}(U') \neq \varnothing$), $u' \circ \varphi$ is finely hyperharmonic on $\varphi^{-1}(U')$.*

Proof The "if" part is obvious. For the "only if" part it is easily seen that we may assume u' to be superharmonic. For any regular domain $V' \Subset U'$ (that is, $V' \subset \overline{V'} \subset U'$) write

$$u'_{V'} = \begin{cases} H^{V'}_{u'} & \text{on } V' \\ u' & \text{on } U' \setminus V'. \end{cases} \tag{4.4.4}$$

Then $u'_{V'}$ is superharmonic (see Sects. 1.1.7, 1.1.16 and 1.1.18), and $u'_{V'} \leqslant u'$ on U', and $u'_{V'}$ is harmonic on V'. By the Brelot property (Theorem 1.5.29), every $x \in U$ has a compact fine neighborhood $K \subset U$ such that $\varphi|_K : K \to \mathbb{R}^{N'}$ is usual-to-usual continuous and hence $\varphi(K)$ is compact. For any finite set $\mathcal{V}'$ of regular domains $V' \Subset U'$ write

$$u'_{\mathcal{V}'} = \min_{V' \in \mathcal{V}'} u'_{V'}.$$

Then $u'_{\mathcal{V}'}$ is superharmonic $\leqslant u'$ on U', and continuous on $\bigcup \mathcal{V}'$. By induction we show that $u'_{\mathcal{V}'} \circ \varphi$ is finely hyperharmonic on $\varphi^{-1}(V')$ for any $V' \in \mathcal{V}'$, and hence on $\bigcup \mathcal{V}'$. Write $\mathcal{V}' = \{V'_1, \dots, V'_m\}$ for some $m \in \mathbb{N}$ and some regular domains $V'_1, \dots, V'_m \Subset U'$. For any regular domain $V'_{m+1} \Subset U'$ we shall prove that the function

$$(u'_{\mathcal{V}'} \wedge u'_{V'_{m+1}}) \circ \varphi = (u'_{\mathcal{V}'} \circ \varphi) \wedge (u'_{V'_{m+1}} \circ \varphi)$$

is finely hyperharmonic for example on $\varphi^{-1}(V'_{m+1})$ under the hypothesis that $u'_{\mathcal{V}'} \circ \varphi$ is finely hyperharmonic on $\varphi^{-1}(V'_j)$, $j \in \{1, \dots, m\}$. The function $u'_{V'_{m+1}} \circ \varphi$

is even finely harmonic on $\varphi^{-1}(V'_{m+1})$ and $u'_{\mathcal{V}'_{m+1}}$ is finely l.s.c. on U', and so $u'_{\mathcal{V}'_{m+1}}$ is the supremum of a sequence of continuous functions $> -\infty$ on U', whence $u'_{\mathcal{V}'_{m+1}} \circ \varphi$ is the supremum of a sequence of finely continuous functions $> -\infty$ on $\varphi^{-1}(U')$. Finally, $u'_{\mathcal{V}'_{m+1}} \circ \varphi = u' \circ \varphi \geqslant u'_{\mathcal{V}} \circ \varphi$ on $U \setminus \varphi^{-1}(V'_{m+1})$ and consequently $(u'_{\mathcal{V}'} \circ \varphi) \wedge (u'_{\mathcal{V}'_{m+1}} \circ \varphi)$ is indeed finely hyperharmonic on U according to Lemma 2.1.51.

The finite covers $\mathcal{V}'$ of $\varphi(K)$ by regular domains $V \Subset U'$ form a directed set under the pre-order relation

$$\mathcal{V}'_1 \leqslant \mathcal{V}'_2 \iff \forall V'_2 \in \mathcal{V}'_2,\ \exists V'_1 \in \mathcal{V}'_1 : V'_1 \supset V'_2.$$

In fact, the family $\mathcal{V}'_3$ defined by

$$\mathcal{V}'_3 := \{V'_1 \cap V'_2 : V'_1 \in \mathcal{V}'_1,\ V'_2 \in \mathcal{V}'_2\}$$

is likewise a finite cover of $\varphi(K)$; and we clearly have $\mathcal{V}'_1 \leqslant \mathcal{V}'_3$, etc. The net $(u'_{\mathcal{V}'})$, indexed by this directed set, is obviously pointwise increasing. It remains to show that $u' = \sup_{\mathcal{V}'} u'_{\mathcal{V}'}$ pointwise on U', for then $u' \circ \varphi = \sup_{\mathcal{V}'} u'_{\mathcal{V}'} \circ \varphi$ is finely hyperharmonic on U along with each $u'_{\mathcal{V}'} \circ \varphi$, as shown above. Each $u'_{\mathcal{V}'}$ is hyperharmonic and hence l.s.c. For fixed $x' \in \varphi(K)$ and $\varepsilon > 0$ there exists $\delta > 0$ such that $u' \geqslant u'(x') - \varepsilon$ on the open ball $B(x', \delta) \subset U'$. Let $\mathcal{V}'$ denote a finite cover of $\varphi(K)$ by regular domains of diameter $< \delta$. For any $V' \in \mathcal{V}'$ such that $x' \in V'$ we have $V' \subset B(x', \delta)$ and hence $u' \geqslant u'(x') - \varepsilon$ on V'. It follows that $H_{u'}^{V'} \geqslant u'(x') - \varepsilon$ on V' and hence $u'_{\mathcal{V}'} \geqslant u'(x') - \varepsilon$ on all of U', in particular $u'_{\mathcal{V}'}(x') \geqslant u'(x') - \varepsilon$. For any remaining $V' \in \mathcal{V}'$ we have $x' \in U' \setminus V'$ and hence $u'_{\mathcal{V}'}(x') = u'(x')$. Consequently, $u'_{\mathcal{V}'}(x') \geqslant u'(x') - \varepsilon$, which was to be proved. □

Remark 4.4.4 The proof of Theorem 4.4.3 for usual harmonic morphisms given in [CC2, Theorem 3.1] adapts to yield an alternative proof of Theorem 4.4.3, using Lyon's Theorem (Remark 2.1.2).

We proceed to characterize finely harmonic morphisms by use of the fine topology on both Ω and $\mathbb{R}^{N'}$.

Theorem 4.4.5 *A map $\varphi : U \to \mathbb{R}^{N'}$ is a finely harmonic morphism if and only if*

(i) *φ is fine-to-fine continuous on U, and*

(ii) *for any finely harmonic (resp. finely hyperharmonic) function u' on a finely open subset V' of $\mathbb{R}^{N'}$, $u' \circ \varphi$ is finely harmonic (resp. finely hyperharmonic) on $\varphi^{-1}(V')$.*

Proof When applied to $\pm u'$, the finely hyperharmonic version of Theorem 4.4.5 clearly will imply the finely harmonic version. Next, if the finely harmonic version of Theorem 4.4.3 holds then clearly φ is a finely harmonic morphism (Definition 4.4.1). Finally, if φ is a finely harmonic morphism then $u' \circ \varphi$ is finely hyperharmonic and hence finely continuous on U for every hyperharmonic function u' on some open subset of $\mathbb{R}^{N'}$ and this implies that φ is fine-to-fine continuous by definition of fine topology on $\mathbb{R}^{N'}$.

It remains to establish the finely hyperharmonic version of (ii) of Theorem 4.4.5 under the assumption that $\varphi : U \to \mathbb{R}^{N'}$ is a finely harmonic morphism. We may assume that $u' < +\infty$ on V'; otherwise represent u' as the pointwise supremum of the increasing sequence of finite valued $u' \wedge n, n \in \mathbb{N}$, on V'. For given $x \in \varphi^{-1}(V')$ (in order to cover the case $N' = 2$) choose an open ball Ω' in $\mathbb{R}^{N'}$ containing $\varphi(x)$, and replace V' by $V' \cap \Omega'$. By the finely local extension property (Theorem 2.1.25) there exists a finely open set W' with $\varphi(x) \in W' \subset \widetilde{W}' \subset U'$ such that

$$u' = p' - q' \text{ on } W',$$

where p' and q' are locally bounded potential on Ω', and q' is finely harmonic in W'. According to Corollary 2.8.2 there exists a compact fine neighborhood K' of $\varphi(x)$ with $K' \subset W'$ and a sequence of harmonic functions u'_n (each defined on some open set $\omega'_n \subset \Omega'$ with $\omega'_n \supset K'$) such that $u'_n \to q'$ uniformly on K'. It follows that $u'_n \circ \varphi \to q' \circ \varphi$ uniformly on V where V denotes the fine interior of $\varphi^{-1}(K')$, which is a fine neighborhood of x in view of property (i) established in the preceding paragraph. By hypothesis, each $u'_n \circ \varphi$ is finely harmonic on V, and hence so is $q' \circ \varphi$ on V by virtue of Corollary 2.1.21.

On the other hand, $p' \circ \varphi$ is finely hyperharmonic on V by (ii) of Theorem 4.4.3, and we conclude that $u' \circ \varphi = p' \circ \varphi - q' \circ \varphi$ is finely hyperharmonic on V, and hence on all of $\varphi^{-1}(U')$. □

Corollary 4.4.6 *If $\varphi : U \to \mathbb{R}^{N'}$ and $\psi : U' \to \mathbb{R}^{N''}$ (with $U' \subset \varphi(U)$ a finely open subset of $\mathbb{R}^{N'}$ and $N'' \geqslant 1$) are finely harmonic morphisms then so is $\psi \circ \varphi : U \to \mathbb{R}^{N''}$.*

The following theorem is likewise based on Theorem 4.4.5.

Theorem 4.4.7 *Suppose that the finely open set $U \subset \mathbb{R}^N$ is finely connected. Let $\varphi : U \to \mathbb{R}^{N'}$ be a finely harmonic morphism; nonconstant if $N' \geqslant 2$. Then $\varphi^{-1}(E')$ is a polar subset of U for any polar subset E' of $\mathbb{R}^{N'}$.*

Proof (cf. [CC2, Theorem 3.1]). We may assume that $N' \geqslant 2$, for otherwise $E' = \varnothing$ and hence $\varphi^{-1}(E') = \varnothing$. Let $E = U \cap \mathfrak{b}(\varphi^{-1}(E'))$ be the base of $\varphi^{-1}(E')$ relative to U, cf. Sect. 1.4. Suppose by contradiction that $\varphi^{-1}(E')$ is nonpolar, that is, $E \neq \varnothing$. Let $x \in E$ and let s' be a positive superharmonic function on an open bounded neighborhood V' of $\varphi(x)$ such that $s' = +\infty$ on $V' \cap E'$. Let $y' \in V' \setminus \{\varphi(x)\}$ and let V'_1, V'_2 be disjoint open subsets of V' such that $\varphi(x) \in V'_1$, $y' \in V'_2$. Then $\varphi(x) \in V' \setminus V'_2$, $y' \in V' \setminus V'_1$ and

$$s' \leqslant {}^{V'}\widehat{R}_{s'}^{V' \setminus V'_1} + {}^{V'}\widehat{R}_{s'}^{V' \setminus V'_2} \text{ on } V'. \tag{4.4.5}$$

Let V be the fine component of x in the finely open set $\varphi^{-1}(V')$, a subset of U containing x. Then $V \cap \varphi^{-1}(E')$ is non polar because $x \in E$. Being infinite on $V \cap$

$\varphi^{-1}(E')$ and finely hyperharmonic on V, $s' \circ \varphi$ is infinite on V. Being finite at x, ${}^{V'}\widehat{R}_{s'}^{V'\setminus V_1'} \circ \varphi$ is superharmonic on V, and hence by (4.4.5) ${}^{V'}\widehat{R}_{s'}^{V'\setminus V_2'} \circ \varphi$ cannot be superharmonic on V. Thus ${}^{V'}\widehat{R}_{s'}^{V'\setminus V_2'} \circ \varphi = +\infty$ on V. Since $y' \in V_2'$ we have ${}^{V'}\widehat{R}_{s'}^{V'\setminus V_2'}(y') < +\infty$ and hence $y' \notin \varphi(V)$. By varying $y' \in V'\setminus\{\varphi(x)\}$ we infer that $\varphi(V) = \{\varphi(x)\}$. By (i) in Theorem 4.4.5, $\varphi^{-1}(E')$ is finely closed relative to U because the polar subset E' of $\Omega \subset \mathbb{R}^N$ is finely closed. It follows that $E \subset \varphi^{-1}(E')$ so $\varphi(x) \in \varphi(E) \subset E'$. For any point $y \in V$ we have $\varphi(y) = \varphi(x) \in E'$, $y \in V \subset \varphi^{-1}(E')$, and hence $V \subset U \cap \mathfrak{b}(V) \subset U \cap \mathfrak{b}(\varphi^{-1}(E')) = E$. Thus E is finely open and φ is locally constant on E. By hypothesis, U is finely connected, and since $E \neq \varnothing$ is finely open and finely closed relative to U (because $\mathfrak{b}(\varphi^{-1}(E'))$ is finely closed in $\mathbb{R}^N$) it follows that $E = U$ and φ constant, against hypothesis. Our assumption that $E \neq \varnothing$ therefore is false, and we conclude that actually $E = \varnothing$, that is, $\varphi^{-1}(E')$ is indeed polar. □

Corollary 4.4.8 *Let U be finely connected and let $\varphi : U \to \mathbb{R}^{N'}$ be a finely harmonic morphism, non constant if $N' \geqslant 2$. If s' is finely superharmonic on a finely open $V' \subset \mathbb{R}^{N'}$ the function $s' \circ \varphi$ is finely superharmonic on $\varphi^{-1}(V')$.*

Theorem 4.4.9 (cf. [CC2, Theorem 3.5]). *Let U be finely connected and let $\varphi : U \to \mathbb{R}^{N'}$ be a nonconstant finely harmonic morphism. If $N' \geqslant 2$ then $\varphi(U)$ is finely open in $\mathbb{R}^{N'}$ and the map φ is fine-to-fine open.*

Proof Let $x \in U$, by the Brelot property (Theorem 1.5.29) there is a fine domain $E \subset U$ such that $x \in E$ and $\varphi|_E$ is continuous on E (with the Euclidean topology). We shall prove that $\varphi(E)$ is a fine neighborhood of $x' := \varphi(x)$ in $\mathbb{R}^{N'}$. According to Lemma 1.3.9(b) there is an open ball V in $\mathbb{R}^N$ centered at x such that $R_1^{\overline{V}\setminus E}(x) < 1$. Indeed, let F be a compact fine neighborhood of x such that $F \subset E$, Lemma 1.3.9 tells us that there is an open ball B in $\mathbb{R}^N$ centered at x such that $R_1^{B\setminus F}(x) = \widehat{R}_1^{B\setminus F}(x) < 1$, see Theorem 1.3.3(b) for the former equality. If V is an open ball centered at x such that $\overline{V} \subset B$, then $R_1^{\overline{V}\setminus E}(x) \leq R_1^{B\setminus F}(x) < 1$. Hence there is a superharmonic function $s \geqslant 0$ on an open neighborhood of $\overline{B}$ such that $s > 1$ on $\overline{V}\setminus E$ and $s(x) < 1$. It follows that

$$G := \{y \in \overline{V} : s(y) > 1\} \supset \overline{V}\setminus E,$$

that is, $\overline{V}\setminus G \subset E$. Since $N' \geqslant 2$, $\{x'\}$ is polar in $\mathbb{R}^{N'}$ and hence $\varphi^{-1}(\{x'\})$ is a polar subset of $\mathbb{R}^N$ by Theorem 4.4.7, in particular of zero harmonic measure with respect to V. Let K be a compact subset of $(\partial V)\setminus\varphi^{-1}(\{x'\})$ such that

$$H_f^V(x) + s(x) < 1,$$

where $f := 1_{(\partial V)\setminus K}$. Since $x' \notin \varphi(K)$ and since $\varphi(K)$ is compact there exists an open ball V' such that $x' \in V' \subset \overline{V'} \subset \mathbb{R}^{N'}\setminus\varphi(K)$. The sets

$$W := V \cap \varphi^{-1}(V'),\ G' := V'\setminus\varphi(\overline{W}\setminus G)$$

are finely open in $\mathbb{R}^N$, resp. open in $\mathbb{R}^{N'}$, the former because φ is fine-to-usual continuous and the latter because $\overline{W}$ is compact along with $\overline{V}$ whereas G is open so that $\overline{W}\backslash G$ is a compact subset of E, and hence $\varphi(\overline{W}\backslash G)$ is compact because $\varphi|_E$ is continuous in Euclidean topology. Since

$$V'\backslash G' \subset \varphi(\overline{W}\backslash G) \subset \varphi(\overline{V}\backslash G) \subset \varphi(E)$$

it suffices to prove that G' is thin at x', for then $V'\backslash G'$ is a fine neighborhood of $x' \in \mathbb{R}^N$.

Let K' be a compact subset of G' and let $s' = {}^{V'}\widehat{R}_1^{K'}$. Then s' is harmonic on $V'\backslash K'$, see Theorem 1.3.4(a), and hence $s' \circ \varphi$ is finely harmonic on the finely open set

$$\varphi^{-1}(V'\backslash K') = \varphi^{-1}(V')\backslash\varphi^{-1}(K') \supset W\backslash\varphi^{-1}(K').$$

Since H_f^V is continuous on $\overline{V}$ and s is l.s.c. on $\overline{V} \supset \overline{W} \supset \overline{W\backslash\varphi^{-1}(K')}$, and since $s' \circ \varphi \leqslant 1$ on $\varphi^{-1}(V') \supset W \supset W\backslash\varphi^{-1}(K')$, the lower limit of $H_f^V + s - s' \circ \varphi$ is $\geqslant 0$ at each point of $\partial_f(W\backslash\varphi^{-1}(K'))$. Moreover $H_f^V + s - s' \circ \varphi \geqslant -1$ on the bounded finely open set $W\backslash\varphi^{-1}(K') \subset W \subset V$, with compact closure in $\overline{V} \subset \mathbb{R}^N$. It therefore follows by the fine boundary minimum principle (Theorem 2.1.10) in view of Remark 2.1.18 that $H_f^V + s - s' \circ \varphi \geqslant 0$ on $W\backslash\varphi^{-1}(K')$, that is,

$$s'(x') = {}^{V'}\widehat{R}_1^{K'}(x') \leqslant H_f^V(x) + s(x)$$

and taking for K' an increasing sequence of compact sets $K'_n \subset G'$ with union G' we conclude that

$${}^{V'}\widehat{R}_1^{G'}(x') = \sup_n {}^{V'}\widehat{R}_1^{K'_n}(x') \leqslant H_f^V(x) + s(x) < 1$$

which by Lemma 1.3.9(a) shows that indeed G' is thin at x'. □

Theorem 4.4.10 *Let $\varphi : U \to \mathbb{R}^{N'}$ be an injective finely harmonic morphism, $N' \geqslant 2$. Then $N = N'$, $\varphi(U)$ is finely open, and $\varphi^{-1} : \varphi(U) \to \mathbb{R}^N$ is a finely harmonic morphism.*

Proof Being injective, φ is clearly nonconstant, and hence $\varphi(U)$ is finely open and φ^{-1} is finely continuous according to the preceding theorem. Let V be a finely open subset of U and let $u : V \to \mathbb{R}$ be finely harmonic. We shall prove that $u \circ \varphi^{-1}$ is finely harmonic on $\varphi(V)$, see Theorem 4.4.5. Let $V' \subset U'$ be a regular fine domain with $\widetilde{V'} \subset \varphi(V)$, and let

$$V_n := \{x \in V : |x| < n \text{ and } |u(x)| < n\} \text{ for } n \in \mathbb{N}.$$

Then each V_n is a bounded finely open subset of V and $V_n \uparrow V$. We may therefore assume from the beginning that V and u are bounded. According to Theorems 2.3.1

or 2.3.7, $v' = H^{V'}_{u\circ\varphi^{-1}}$ is finely harmonic on V' and its extension by $u \circ \varphi^{-1}$ to $\partial_f V' \subset \varphi(V)$ is finely continuous. Hence $v' \circ \varphi$ is finely harmonic on $\varphi^{-1}(V')$ and finely continuous on $\widetilde{\varphi^{-1}(V')} = \varphi^{-1}(\widetilde{V'}) \subset V$. Since $\varphi^{-1}(V')$ is bounded along with V, and since $v' \circ \varphi$ is bounded on $\varphi^{-1}(V')$, there is a bounded potential p on an open ball $\Omega \subset \mathbb{R}^N$ containing $\varphi^{-1}(V')$ such that $|v' \circ \varphi| \leqslant p$ on $\varphi^{-1}(V')$. We conclude by Theorem 2.1.10 and Remark 2.1.18 that $v' \circ \varphi = u$, that is, $v' = u \circ \varphi^{-1}$ which thus indeed is finely harmonic on $\varphi(V)$. The equality $N = N'$ now follows from Theorem 4.4.2 applied to φ and φ^{-1}. □

4.5 Finely Holomorphic Functions

In view of the close relationship between holomorphic functions of a complex variable and harmonic functions of two real variables it seems natural to look for a concept of finely holomorphic functions defined on a finely open set of $\mathbb{C}$ (or of Riemann surface). Several natural definitions of such a concept present themselves, and they turn out to be equivalent all of them (Theorems 4.5.9, 4.5.14 and 4.5.17 below). We chose to begin with the following definition, inspired by the characterization Corollary 2.8.2 of fine harmonicity.

Definition 4.5.1 A function $f : U \to \mathbb{C}$ (U finely open of $\mathbb{C}$) is termed *finely holomorphic* if every point of U has a compact fine neighborhood $K \subset U$ such that $f|_K \in R(K)$.

Recall that for any compact $K \subset \mathbb{C}$, $R(K)$ denotes the uniform closure of the algebra of restrictions to K of holomorphic functions, or equivalently, of rational functions with poles off K, defined on open subsets of $\mathbb{C}$ containing K. For properties of $R(K)$ we refer to Zalcman [Zal]. Every function of class $R(K)$ is (complex) finely harmonic on the fine interior K' of K. Hence every finely holomorphic function on U is finely harmonic on U.

Several equivalent definitions of fine holomorphy will be given in the following Theorems 4.5.9 and 4.5.14 below.

The finely holomorphic functions on U clearly form a subalgebra of the algebra of all finely continuous functions $U \to \mathbb{C}$. We denote this subalgebra by $\mathcal{O}(U)$. In its dependance on the finely open set $U \subset \mathbb{C}$, $\mathcal{O}(U)$ has the usual sheaf property.

Theorem 4.5.2 *Let $f : U \to \mathbb{C}$ be finely holomorphic or the conjugate of a finely holomorphic function. Then f is finely harmonic morphism. In particular f is fine-to-fine continuous. If U is finely connected and f is nonconstant then f is a fine-to-fine open map $U \to \mathbb{C}$, and the pre-image of any polar subset of $\mathbb{C}$ is a polar set of U.*

The term "fine-to-fine" indicates that the fine topology is used both on U and on $\mathbb{C}$.

Proof of Theorem 4.5.2. For every natural number n, f^n is finely holomorphic, resp. antiholomorphic; in particular f^n is a finely harmonic morphism according to (d) after Definition 4.4.1. The remaining assertions therefore follow from Theorems 4.4.5, 4.4.7 and 4.4.9. □

Remark 4.5.3 It is a longstanding open problem whether, conversely, any finely harmonic morphism $f : U \to \mathbb{C}$ is either finely holomorphic or the conjugate of a finely holomorphic function. The answer is in the affirmative if U is a Euclidean domain or if f is of class C^1, see [Fu26]. Pyrh has proved that this result remains true if f is finely locally injective, see [Py]. However, it is not known if the condition of finely local injectivity (q.e.) is automatically fulfilled by any nonconstant finely harmonic morphism.

Definition 4.5.1 has the drawback that it does not allow for a simple proof that $\mathcal{O}(U)$ is closed under uniform convergence. This property, however, follows easily once finely holomorphic functions have been characterized as (complex) finely harmonic functions $f : U \to \mathbb{C}$ such that $z \mapsto zf(z)$ likewise is finely harmonic (Theorem 4.5.9). But that property does not lead easily to the main results to be obtained, such as the following.

Every finely holomorphic function $f : U \to \mathbb{C}$ is finely differentiable, and its fine derivative $f' : U \to \mathbb{C}$ is finely holomorphic (Theorems 4.5.14, 4.5.16). A finely holomorphic function on a fine domain is uniquely determined by its fine derivatives at any point (Theorem 4.5.23). A finely holomorphic function on a fine domain has at most countably many zeros (Theorem 4.5.26).

We shall begin by characterizing finely holomorphic functions in Dirichlet space, drawing on the results in Sect. 2.1. In what follows, Ω denotes a Greenian domain of $\mathbb{C}$ (a domain in $\mathbb{C}$ of nonpolar complement), and $\mathcal{D}(\Omega)$ denotes the space of all C^∞-functions $\Omega \to \mathbb{C}$ of compact support. As in Sect. 4.2 we denote by $\widehat{\mathcal{D}}(\Omega)$ the completion of $\mathcal{D}(\Omega)$ endowed with the Dirichlet norm $\|\cdot\|$ associated with the scalar product

$$\langle f, g\rangle = \int_\Omega \nabla f \overline{\nabla g}\, dm = 4\int_\Omega \partial f \overline{\partial g}\, dm = 4\int_\Omega \overline{\partial} f \partial \overline{g}\, dm, \tag{4.5.1}$$

where the gradient $\nabla = (\frac{\partial}{\partial x}, \frac{\partial}{\partial y})$ (in the distributional sense) is taken with respect to $x = \mathfrak{Re}\, z$, $y = \mathfrak{Im}\, z$ for $z \in \mathbb{C}$, m is the restriction to Ω of the Lebesgue measure ℓ_2 on $\mathbb{C} \cong \mathbb{R}^2$, and we write

$$\partial = \frac{1}{2}\Big(\frac{\partial}{\partial x} - i\frac{\partial}{\partial y}\Big),\quad \overline{\partial} = \frac{1}{2}\Big(\frac{\partial}{\partial x} + i\frac{\partial}{\partial y}\Big).$$

For the middle equality in (4.5.1) note that

$$\begin{aligned} 4\int_\Omega \partial f \overline{\partial g}\, dm &= \int (f_x - i f_y)\overline{(g_x - i g_y)}\, dm \\ &= \int (f_x \overline{g_x} + f_y \overline{g_y})\, dm = \int \nabla f \overline{\nabla g}\, dm \end{aligned}$$

because

$$\int (f_x\overline{g_y} - f_y\overline{g_x})\, dm = \int (-f_{yx}\overline{g} + f_{yx}\overline{g})\, dm = 0,$$

first for $f \in \mathcal{D}(\Omega)$, and next for $f \in \widehat{\mathcal{D}}(\Omega)$ by continuity of the linear form $f \mapsto \int (f_x\overline{g_x} + f_y\overline{g_y})\, dm$ on $\widehat{\mathcal{D}}(\Omega)$ for fixed $g \in \widehat{\mathcal{D}}(\Omega)$.

As in Sect. 4.2 the elements of the Hilbert space $\widehat{\mathcal{D}}(\Omega)$ are equivalence class represented by functions which are finite and finely continuous q.e. on Ω. For any finely open set $U \subset \Omega$ we write (in part as in Sect. 2.1)

$$\begin{aligned}
\widehat{\mathcal{D}}(\Omega, U) &= \{f \in \widehat{\mathcal{D}}(\Omega) : f = 0 \text{ q.e. on } \complement U\},\\
\mathcal{H}(\Omega, U) &= \{f \in \widehat{\mathcal{D}}(\Omega) : f \text{ is finely harmonic q.e. on } U\},\\
\mathcal{O}(\Omega, U) &= \{f \in \widehat{\mathcal{D}}(\Omega) : \overline{\partial} f = 0 \text{ a.e. on } U\}.
\end{aligned}$$

These three linear subspaces of $\widehat{\mathcal{D}}(\Omega)$ are closed. For $\mathcal{O}(\Omega, U)$ this is because the linear map $f \mapsto \nabla f$ of $\widehat{\mathcal{D}}(\Omega)$ into $L^2(\Omega)$ (with respect to m) is bounded. For $\widehat{\mathcal{D}}(\Omega, U)$ and $\mathcal{H}(\Omega, U)$, see Theorem 4.2.15. Recall that by Theorem 4.2.21 or Corollary 4.2.23 we have the orthogonal decomposition

$$\widehat{\mathcal{D}}(\Omega) = \widehat{\mathcal{D}}(\Omega, U) \oplus \mathcal{H}(\Omega, U). \tag{4.5.2}$$

The space

$$\mathcal{O}^\infty(\Omega, U) := \mathcal{O}(\Omega, U) \cap L^\infty(\Omega)$$

of all bounded functions of class $\mathcal{O}(\Omega, U)$ is an algebra (subalgebra of the algebra of all bounded functions of class $\widehat{\mathcal{D}}(\Omega)$). In fact, for $f, g \in \mathcal{O}^\infty(\Omega, U)$, $\partial(fg) = f\partial g + g\partial f$ and $\overline{\partial}(fg) = f\overline{\partial} g + g\overline{\partial} f$ are of class $L^2(\Omega)$, and $\overline{\partial}(fg) = 0$ a.e. along with $\overline{\partial} f$ and $\overline{\partial} g$.

Lemma 4.5.4 *For any $f \in \widehat{\mathcal{D}}(\Omega)$ and any finely open set V with compact closure $\overline{V}$ in Ω the following three properties are equivalent:*

(i) $\overline{\partial} f = 0$ *a.e. on V (that is, $f \in \mathcal{O}(\Omega, V)$).*

(ii) *The functions f and $z \mapsto zf(z)$ are finely harmonic q.e. on V.*

(iii) $\int f\overline{\partial g} = 0$ *for every $g \in \widehat{\mathcal{D}}(\Omega, V)$.*

Proof According to Theorem 4.2.2, $\widehat{\mathcal{D}}(\Omega) \subset L^2_{loc}(\Omega)$. It follows that $f|_V \in L^2(V)$. There exists $\varphi \in \mathcal{D}(\Omega)$ such that $\varphi(z) = z$ on some set ω, $V \subset \omega \subset \Omega$.

(i)⇒(ii). For any $g \in \widehat{\mathcal{D}}(\Omega, V)$ we have $\partial\overline{g} = 0$ a.e. on $\complement V$ by Corollary 4.2.22, and hence by (4.5.1)

$$\langle f, g\rangle = 4\int \overline{\partial} f \partial\overline{g} = 0.$$

It follows by (4.5.2) applied to V in place of U that $f \in \widehat{\mathcal{D}}(\Omega, V)^\perp = \mathcal{H}(\Omega, V)$. Furthermore $\varphi f \in \widehat{\mathcal{D}}(\Omega)$ and

$$\overline{\partial}(\varphi f) = (\overline{\partial}\varphi) f + \varphi\overline{\partial} f = 0 \text{ a.e. on } V$$

because $\overline{\partial}\varphi = \overline{\partial}z = 0$ on $\omega \supset V$. As above it follows that $\varphi f \in \mathcal{H}(\Omega, V)$. Altogether (ii) holds.

(ii)⇒(iii). We suppose that $f, \varphi f \in \mathcal{H}(\Omega, V)$. For any $g \in \widehat{\mathcal{D}}(\Omega, V)$ we obtain by (4.5.1), (4.5.2)

$$4\int_\Omega \partial f \overline{\partial g} = \langle f, g\rangle = 0,\ 4\int_\Omega \partial(\varphi f)\overline{\partial g} = \langle \varphi f, g\rangle = 0.$$

Because $\partial\varphi = 1$ on $\omega \supset V$ and $\partial g = 0$ a.e. on $\complement V$ it follows that

$$\begin{aligned}\int_\Omega f\overline{\partial g} &= \int_\Omega f\partial\varphi\overline{\partial g} = \int_\Omega (\partial(\varphi f) - \varphi\partial f)\overline{\partial g}\\ &= -\int_\Omega \varphi\partial f\overline{\partial g} = -\int_\Omega \partial f\overline{\overline{\varphi}\partial g} = -\int_\Omega \partial f\overline{\partial(\overline{\varphi} g)}\\ &= \langle f, \overline{\varphi} g\rangle\end{aligned}$$

since $g\partial\overline{\varphi} = 0$ a.e. on ω and $g = 0$ a.e. on $\Omega\backslash V$, again by Corollary 4.2.22, and finally (4.5.2) applies because $f \in \mathcal{H}(\Omega, V)$ and $\overline{\varphi} g \in \widehat{\mathcal{D}}(\Omega, V)$.

(iii)⇒ (i). The identity

$$\int_\Omega (\overline{\partial} f)\overline{g} = -\int_\Omega f\overline{\partial g}$$

for $g \in \mathcal{D}(\Omega)$ extends by continuity to $g \in \widehat{\mathcal{D}}(\Omega)$ and in particular to the present $g \in \widehat{\mathcal{D}}(\Omega, V)$. Replacing g by ψg for $\psi \in \mathcal{D}(\Omega)$, whence $\psi g \in \widehat{\mathcal{D}}(\Omega, V)$, we infer from (iii) that $\int_\Omega (\overline{\partial} f)\overline{\psi g} = 0$ and hence by varying ψ

$$(\overline{\partial} f)\overline{g} = 0 \text{ a.e. on } \Omega.$$

The measure $\mu = m$ has finite energy $\int G_\Omega\mu\, d\mu$ and the potential $G_\Omega\mu$ is therefore strict according to Theorem 1.4.15. It follows that the function $g := G_\Omega\mu - \widehat{R}^{\complement V}_{G_\Omega\mu}$ (sweeping relative to Ω) is of class $\widehat{\mathcal{D}}(\Omega, V)$, see Theorem 4.2.10, and since $g > 0$ on V we conclude that indeed $\overline{\partial} f = 0$ a.e. on V. □

The proofs of Theorems 4.5.9 and 4.5.14 below depend much more on the properties of the Cauchy transform

$$f(z) = \int \frac{1}{z-\zeta}\varphi(\zeta)\, d\ell_2(\zeta),\ z \in \mathbb{C}, \tag{4.5.3}$$

of function $\varphi \in L^2_c(\mathbb{C})$ (square integrable with respect to Lebesgue measure ℓ_2 on $\mathbb{C}$, and of compact support). We may briefly write $f = \frac{1}{z} * \varphi$.

Lemma 4.5.5 *The Cauchy transform $f = \frac{1}{z} * \varphi$ of a function $\varphi \in L^2_c(\mathbb{C})$ is defined and finite q.e. for $z \in \mathbb{C}$. Furthermore, f is a Beppo Levi function which is finely continuous q.e., that is, a* BLD *function.*

Proof For BLD functions see the beginning of Sect. 4.2. Since $\zeta \mapsto \frac{1}{z-\zeta}$ is locally integrable, f is well defined a.e. (with respect to ℓ_2), and $f \in L^1_{loc}(\mathbb{C})$. The distributional derivatives of f are

$$\overline{\partial} f = \pi\varphi \tag{4.5.4}$$

$$\partial f = -\lim_{\delta\to 0}\int_{|\zeta-z|>\delta}\frac{\varphi(\zeta)}{(z-\zeta)^2}\,d\ell_2(\zeta). \tag{4.5.5}$$

The equation (4.5.4) is derived from

$$f = \frac{1}{z} * \varphi = 2(\partial \log|z|) * \varphi = 2\partial[(\log|z|) * \varphi]$$

by distributional differentiation

$$\begin{aligned}\overline{\partial} f &= 2\overline{\partial}\partial[(\log|z| * \varphi] = \frac{1}{2}\Delta[(\log|z|) * \varphi]\\ &= \frac{1}{2}(\Delta \log|z|) * \varphi = \pi\varepsilon * \varphi,\end{aligned}$$

where $\Delta = 4\overline{\partial}\partial$ denotes the Laplacian on $\mathbb{C}$ and ε is the Dirac measure at 0 (unit mass at 0), and so $\Delta\log|z| = 2\pi\varepsilon$ by Poisson's formula in the plane.

In (4.5.5), the right hand member is the Cauchy principal value of the singular integral $\int_{\mathbb{C}}\frac{-\varphi(\zeta)}{(z-\zeta)^2}\,d\ell_2(\zeta)$ obtained by formal differentiation of (4.5.3). In other terms (4.5.5) expresses that f is a Hilbert transform of φ, and (4.5.5) therefore holds as particular case of Calderon-Zygmund [CZ, Theorems 1 and 2], see also Schwartz [Sc, p. 48]. Furthermore, (4.5.5) implies that $\partial f \in L^2(\mathbb{C})$ according to [CZ], and (4.5.4) implies of course that $\overline{\partial} f \in L^2(\mathbb{C})$.

Then f is a Beppo Levi function. We proceed to show that f is defined and finite even q.e. and that f is quasi-continuous (or equivalently: finely continuous q.e., cf. Theorem 1.5.15), and f is thus altogether a BLD function.

For the proof that $f = \frac{1}{z} * \varphi$ is well defined q.e. and has a quasi-continuous representative we shall need a substitute for M. Riesz's composition formula

$$\int_{\mathbb{R}^N}|x-\xi|^{\alpha-N}|y-\xi|^{\beta-N}\,d\ell_2(\xi) = C_{\alpha,\beta,N}|x-y|^{\alpha+\beta-N},$$

valid for $\alpha, \beta > 0$ and $\alpha < N$, $c_{\alpha,\beta,N} > 0$ being an explicit constant depending on α, β and N (see [Ri] or [Lan, p. 45]). In our case we have $\alpha = \beta = 1$ and $N = 2$; and so $\alpha + \beta = N$, not permitted above. Instead, let $0 < \rho \leqslant \frac{1}{2}$ be given and denote by D_ρ and $D_{3\rho}$ two concentric open disks in $\mathbb{C}$ of radii ρ and 3ρ, respectively in Frostman [Fr, p. 61] and proved in [Fu2, Lemma 3.1] we have

$$\frac{1}{2\pi}\int_{D_{3\rho}} \frac{1}{|z-\zeta|}\frac{1}{|w-\zeta|}\,d\ell_2(\zeta) = \log\frac{1}{|z-w|} + F_\rho(z,w), \tag{4.5.6}$$

where F_ρ is bounded and continuous on $\overline{D}_\rho \times \overline{D}_\rho$, say

$$|F_\rho(z,w)| \leqslant M(\rho) \text{ for } z, w \in \overline{D}_\rho. \tag{4.5.7}$$

By integrating (4.5.6) with respect to $d\mu(z)d\mu(w)$, where μ denotes a probability measure on D_ρ we obtain after interchanging the order of integration

$$\frac{1}{2\pi}\int_{D_{3\rho}} \left(\frac{1}{|z|} * \mu\right)^2 d\ell_2(z) \leqslant \int_{D_\rho}\int_{D_\rho} \log\frac{1}{|z-w|}\,d\mu(z)d\mu(w) + M_\rho. \tag{4.5.8}$$

For any compact set $K \subset D_\rho$ the capacity cap K with respect to the kernel $-\log|z-w| > 0$ considered on $D_\rho \times D_\rho$ is defined by

$$\log\frac{1}{\text{cap } K} = \inf_\mu \int_K\int_K \log\frac{1}{|z-w|}\,d\mu(z)\,d\mu(w), \tag{4.5.9}$$

where μ ranges over the probability measures on K. The inner capacity $\text{cap}_* E$ and the outer capacity $\text{cap}^* E$ of a set $E \subset D_\rho$ are defined by

$$\text{cap}_* E = \sup\{\text{cap } K : K \text{ compact in } D_\rho,\ K \subset E\},$$

$$\text{cap}^* E = \inf\{\text{cap}_* G : G \text{ open in } D_\rho,\ G \supset E\}.$$

For a Borel set or an analytic set $E \subset D_\rho$ it can be shown by Choquet's capacitability theorem that $\text{cap}^* E = \text{cap}_* E$, but we shall not need this, except that it holds trivially for open subsets $E \subset D_\rho$.

In the rest of the proof, Ω denotes the open unit disk in $\mathbb{C}$. We need to compare the outer logarithmic capacity $c^*(E)$ of a set $E \subset D_\rho$ ($\rho \leqslant \frac{1}{2}$), or just $E \subset \frac{1}{2}\Omega$, with the outer Green capacity $C^*(E)$ with respect to Ω. The Green kernel on $\Omega \times \Omega$ is given by

$$G_\Omega(z,w) = \log\frac{|z-w'|}{|z-w||w'|}, \quad z, w \in \Omega, \tag{4.5.10}$$

where $w' = \frac{1}{\overline{w}}$ and (in the limit) $G_\Omega(z,0) = -\log|z|$. In the sequel we assume that $|z| \leqslant \frac{1}{2}$, $|w| \leqslant \frac{1}{2}$ and hence $|w'| \geqslant 2$, $\frac{|z|}{|w'|} \leqslant \frac{1}{4}$, and $\frac{3}{4}|w'| \leqslant |z-w'| \leqslant \frac{5}{4}|w'|$. It follows that

$$\log\frac{3/4}{|z-w|} \leqslant G_\Omega(z,w) \leqslant \log\frac{5/4}{|z-w|}, \tag{4.5.11}$$

and for any probability measure μ on $\frac{1}{2}\Omega$

$$\log\frac{3}{4}\leqslant\int\int G_\Omega(z,w)\,d\mu(z)\,d\mu(w)-\int\int\log\frac{1}{|z-w|}\,d\mu(z)\,d\mu(w)\leqslant\log\frac{5}{4}$$

(understood to imply that the double integrals are either both infinite). Taking infimum over all probability measures on $\frac{1}{2}\Omega$ in either double integral we infer that

$$\log\frac{3}{4}+\log\frac{1}{\text{cap }K}\leqslant\frac{1}{C(K)}\leqslant\log\frac{5}{4}+\log\frac{1}{\text{cap }K} \tag{4.5.12}$$

and hence similarly with K replaced by any set $E\subset\frac{1}{2}\Omega$, and cap K and $C(K)$ either by $\text{cap}_*\,E$ and $C_*(E)$ or by $\text{cap}^*\,E$ and $C^*(E)$.

For the proof that the integral in (4.5.3) exists and finite q.e. in $\mathbb{C}$ we consider for given $t>0$ the open set

$$G_t=\{z\in D_\rho:\int_{D_{3\rho}}\frac{1}{|z-\zeta|}|\varphi(\zeta)|\,d\ell_2(\zeta)>t\}.$$

For any compact $K\subset G_t$ and any probability measure μ on K we have after interchanging integrations

$$\begin{aligned}t&\leqslant\textstyle\int_{D_{3\rho}}|\varphi(\zeta)|(\int\frac{1}{|z-\zeta|}\,d\mu(z))\,d\ell_2(\zeta),\\ t^2&\leqslant\textstyle\int_{D_{3\rho}}|\varphi|^2\,d\ell_2\int_K(\frac{1}{|z|}*\mu)^2\,d\ell_2(z),\\ \tfrac{t^2}{2\pi}&\leqslant\textstyle\int_{D_{3\rho}}|\varphi|^2\,d\ell_2(\int_K\int_K\log\frac{1}{|z-w|}\,d\mu(z)\,d\mu(w)+M(\rho)).\end{aligned}$$

according to (4.5.8). Taking infimum over all probability measures μ on K we obtain by (4.5.9)

$$\frac{t^2}{2\pi}\leqslant\left(\log\frac{1}{\text{cap }K}+M(\rho)\right)\int_{D_{3\rho}}|\varphi|^2\,d\ell_2,$$

and hence by varying K the same with cap K replaced by $\text{cap}^*\,G_t$. Since $G_t\subset D_\rho$ with $\rho\leqslant\frac{1}{2}$ it follows by the extended latter inequality (4.5.12)

$$\frac{t^2}{2\pi}\leqslant\left(\frac{1}{C^*(G_t)}-\log\frac{3}{4}+M(\rho)\right)\int_{D_{3\rho}}|\varphi|^2\,d\ell_2. \tag{4.5.13}$$

For $t\to+\infty$ this implies that $C^*(G_t)\to 0$. If E denotes $\{z\in D_\rho:\int_{D_{3\rho}}\frac{|\varphi(\zeta)|}{|z-\zeta|}\,d\ell_2(\zeta)=+\infty\}$ we have $E\subset G_t$ for every $t>0$ and hence $C^*(E)=0$, which means that E is polar. Since countably many disks D_ρ with suitable centers cover $\mathbb{C}$ we conclude that indeed $\int_{\mathbb{C}}\frac{|\varphi(\zeta)|}{|z-\zeta|}\,d\ell_2(\zeta)<+\infty$ q.e. on $\mathbb{C}$.

For the proof that f is finely continuous q.e., or equivalently that f is C-quasicontinuous, choose a sequence of finite continuous functions φ_n on $\mathbb{C}$ with compact support so that

$$\int|\varphi_n-\varphi|^2\,d\ell_2\leqslant\frac{1}{8^n}. \tag{4.5.14}$$

The Cauchy transforms $f_n = \frac{1}{z} * \varphi_n$ are finite and continuous on $\mathbb{C}$ and

$$|f - f_n| \leqslant \int \frac{1}{|z-\zeta|} |\varphi_n(\zeta) - \varphi(\zeta)|\, d\ell_2(\zeta).$$

The outer capacity of the open set

$$E_n := \{z \in D_\rho : \int \frac{1}{|z-\zeta|} |\varphi_n(\zeta) - \varphi(\zeta)|\, d\ell_2(\zeta) > \frac{1}{2^n}\}$$

can be estimated in the same way as cap* G_t in (4.5.13):

$$\frac{1}{2\pi 4^n} \leqslant \left(\frac{1}{C^*(E_n)} - \log\frac{3}{4} + M(\rho)\right) \int |\varphi_n - \varphi|^2\, d\ell_2. \tag{4.5.15}$$

Inserting (4.5.14) we obtain

$$C^*(E_n) \leqslant 4\pi.2^{-n}$$

for n large enough that $\log\frac{4}{3} + M(\rho) \leqslant \frac{2^n}{4\pi}$, say for $n \geqslant n_0$. For each $n \geqslant n_0$ consider the open set

$$\omega_n = \bigcup_{j>n} E_j.$$

Then $C(\omega_n) \leqslant 4\pi.2^{-n}$ and for $z \in \complement\omega_n = \bigcap_{j>n} \complement E_j$ we have

$$|f_j(z) - f(z)| \leqslant \int \frac{1}{|z-\zeta|} |\varphi_n(\zeta) - \varphi(\zeta)|\, d\ell_2(\zeta) \leqslant \frac{1}{2^j}$$

for $j > n$. This shows that $f_j(z) \to f(z)$ uniformly on $\complement\omega_n$, and hence f is continuous relative to $\complement\omega_n$. Consequently, f is quasi-continuous on $\mathbb{C}$, that is, f is indeed finely continuous q.e. □

As a kind of converse to Lemma 4.5.5 we have the following generalization of the Cauchy formula:

Lemma 4.5.6 *Let $f \in \widehat{\mathcal{D}}(\Omega)$ have compact support in the Greenian domain $\Omega \subset \mathbb{C}$. Then*

$$f(z) = \frac{1}{\pi} \int \frac{1}{z-\zeta} \overline{\partial} f(\zeta)\, dm(\zeta) \text{ q.e. for } z \in \Omega,$$

the integral being absolutely convergent q.e. for $z \in \Omega$.

Proof For the absolute convergence of the integral q.e. for $z \in \Omega$, see [DeL, Lemma 6.1 (p. 361)]. Denote by $F(z)$ the right hand member of the stated equality. In the case $f \in \mathcal{D}(\Omega)$ the assertion of the lemma is contained in the classical Cauchy formula, obtained by application of Stokes' theorem to the above integral after deleting from

Ω a small disk centered at z, see, for example, [Ho, Theorem 1.2.1]. In the general case, since $\mathcal{D}(\Omega)$ is dense in $\widehat{\mathcal{D}}(\Omega)$ there exists a sequence of functions $f_n \in \mathcal{D}(\Omega)$, converging to f in $\widehat{\mathcal{D}}(\Omega)$, that is, $\nabla f_n \to \nabla f$ in $L^2(\Omega)$, and in particular $\overline{\partial} f_n \to \overline{\partial} f$ in $L^2(\Omega)$.

Suppose to begin with that such a sequence (f_n) can be chosen so that, in addition, their supports are contained in a single compact set $K \subset \Omega$. For a given point z_0 consider an open disk $D_\delta(z_0)$ such that $\overline{D_\delta(z_0)} \subset \Omega$, and define

$$\varphi(z) = \frac{-1}{z} \text{ for } z \in K - \overline{D_\delta(z_0)}$$

and $\varphi(z) = 0$ for the remaining points of $\mathbb{C}$. Then $K - \overline{D_\delta(z_0)}$ is compact and hence $\varphi \in L^1(\mathbb{C})$, $\frac{1}{z}$ being locally integrable function of $z \in \mathbb{C}$. It follows that the convolution $\varphi * \overline{\partial} f \in L^2(\mathbb{C})$, and hence $\varphi * \overline{\partial} f_n \to \varphi * \overline{\partial} f$ in $L^2(\mathbb{C})$ because $\overline{\partial} f_n \to \overline{\partial} f$ in $L^2(\mathbb{C})$ when we extend any function of class $L^2(\Omega)$ to all of $\mathbb{C}$ by 0 in $\mathbb{C}\backslash\Omega$. In particular we find that, as functions of $z \in D_\delta(z_0)$,

$$\frac{1}{\pi}\int_\Omega \frac{1}{z-\zeta}\overline{\partial} f_n(\zeta) \to \frac{1}{\pi}\int_\Omega \frac{1}{z-\zeta}\overline{\partial} f(\zeta) = F(z)$$

in $L^2(D_\delta(z_0))$ as $n \to +\infty$. On the other hand

$$\frac{1}{\pi}\int_\Omega \frac{1}{z-\zeta}\overline{\partial} f_n(\zeta) = f_n(z) \to f(z)$$

in $L^2(D_\delta(z_0))$ because $f_n \in \mathcal{D}(\Omega)$. Thus $F(z) = f(z)$ a.e. for $z \in D_\delta(z_0) \subset \Omega$, where δ depends on z_0. Countably many such disks cover Ω, and so indeed $F(z) = f(z)$ a.e. for $z \in \Omega$, and hence for q.e. $z \in \Omega$ by q.e. fine continuity of f and F according to Theorem 4.2.5 and Lemma 4.5.5.

To settle the general case, let just $(f_n) \subset \mathcal{D}(\Omega)$ converge to f in $\widehat{\mathcal{D}}(\Omega)$. Choose a function $\psi \in \mathcal{D}(\Omega)$ so that $\psi = 1$ on Supp f (compact in Ω, by hypothesis). Then $\psi f \in \widehat{\mathcal{D}}(\Omega)$ because $\nabla(\psi f) = \psi \nabla f + f \nabla \psi$, which also shows that $\psi f_n \to \psi f = f$ in $\widehat{\mathcal{D}}(\Omega)$. Since $\text{Supp}(\psi f) \subset K :=$ Supp ψ we may replace f_n by ψf_n in the above part of the proof.[2] □

In Theorem 4.5.14 below we characterize fine holomorphy by properties involving fine differentiability. For the proofs we need the lemma below, which in turn depends on the following lemma—a kind of uniform Wiener criterion.

Lemma 4.5.7 *For any constant $\alpha > 0$ and any point $z_0 \in U$ there exists a fine neighborhood U_α of z_0 in U and a natural number N_α such that*

$$c^*(A_n(z)\backslash U) \leqslant 2^{-n\alpha}$$

[2] The hypothesis in Lemma 4.5.6 that $f \in \widehat{\mathcal{D}}(\Omega)$ have compact support in Ω can be omitted by a more demanding proof, see [DeL, Théorème 6.2, p. 362].

for any $z \in U_\alpha$ *and any* $n \geqslant N_\alpha$.

Here c^* denotes outer logarithmic capacity see, for example, [Do4, 1.XIII.18], [AG, Section 5.8], and $A_n(z)$ is the annulus

$$A_n(z) = \{\zeta \in \mathbb{C} : 2^{-n-1} \leq |z - \zeta| < 2^{-n}\}.$$

Proof We may assume that $0 \in U$ and $z_0 = 0$. Let Ω denote the open unit disk in $\mathbb{C}$. Since $0 \in U$, $\Omega \backslash U$ is thin at 0, and so is therefore a suitable open set $E \subset \Omega$ with $E \supset \Omega \backslash U$. In fact, there is a compact set H such that $0 \in H' \subset H \subset U$, and hence $E := \Omega \backslash H$ is open and contains $\Omega \backslash U$, but $\widetilde{E} = \Omega \backslash H'$ does not contain 0. According to [AG, Section 5.4], for any compact set $K \subset E \cap A_n(z)$, the sweeping $\widehat{R}_1^K$ relative to Ω equals the equilibrium potential $G_\Omega \lambda$ on K. The total mass of λ equals the Green capacity $C(K)$. It follows by (4.5.11) and (4.5.12) that

$$\begin{aligned} \widehat{R}_1^K(z) &= \int_\Omega G_\Omega(z, w)\, d\lambda(w) \geqslant \int_\Omega \log \frac{3/4}{|z - w|}\, d\lambda(w) \qquad (4.5.16) \\ &\geqslant \log(\frac{3}{4}2^n) C(K) \geqslant \log(\frac{3}{4}2^n) / \log \frac{5/4}{c(K)}. \end{aligned}$$

Since E is thin at 0 there exists by Lemma 1.3.9(b) a constant $r = r_\alpha > 0$ small enough that

$$\widehat{R}_1^{E \cap 2r\Omega}(0) < \frac{1}{4\alpha}.$$

We take $r < \frac{1}{2}$. Since $\widehat{R}_1^{E \cap 2r\Omega}$ is finely continuous on U there is a fine neighborhood U_α of 0 on Ω such that

$$\widehat{R}_1^{E \cap 2r\Omega}(z) < \frac{1}{2\alpha} \text{ for every } z \in U_\alpha.$$

We take $U_\alpha \subset r\Omega \subset \frac{1}{2}\Omega$. For $z \in U_\alpha$ and $2^{-n} < r$ we have

$$\widehat{R}_1^K(z) < \frac{1}{2\alpha}.$$

From this together with (4.5.16) we conclude that

$$c(K) \leqslant \frac{5}{4}(\frac{4}{3}2^{-n})^{2\alpha} \leqslant 2^{-n\alpha},$$

and hence $c^*(E \cap A_n(z)) \leqslant 2^{-n\alpha}$ for $z \in U_\alpha$ and n large enough that $2^{-n} < r = r_\alpha$ and $2^{n\alpha} > \frac{5}{4}(\frac{4}{3})^{2\alpha}$. Since $E \cap A_n(z)$ is a Borel set it follows that $c^*(A_n(z) \backslash U) \leqslant c(E \cap A_n(z)) \leqslant 2^{-n\alpha}$ for $z \in U_\alpha$ and n large enough as stated (depending only on α). □

Lemma 4.5.8 *Let K be a compact subset of $\mathbb{C}$ and let $U \subset \mathbb{C}$ be finely open. For any constant $\alpha > 0$ the function h_α on U defined by*

$$h_\alpha(z) = \int_{K\setminus U} |z - \zeta|^{-\alpha}\, d\ell_2(\zeta),\ z \in U,$$

is finite and finely continuous everywhere on U. Every point $z_0 \in U$ therefore has a compact fine neighborhood $V \subset U$ on which each h_α is bounded and continuous:

$$c_\alpha := \sup_{z\in V} h_\alpha(z) < +\infty. \tag{4.5.17}$$

Proof For $\alpha < 2$ the function $z \mapsto |z|^{-\alpha}$ is locally ℓ_2-integrable, and h_α is therefore finite and continuous even in the Euclidean topology, and we may take for V any compact fine neighborhood of z_0 contained in U. In the remaining case $\alpha \geqslant 2$ we may assume that U is regular and hence an F_σ set. With the notation from Lemma 4.5.7 we have $|z - \zeta| \geqslant 2^{-n-1}$ for $z \in U_\alpha$, $\zeta \in A_n(z)$, and hence

$$\begin{aligned}\int_{A_n(z)\setminus U} |z-\zeta|^{-\alpha}\, d\ell_2(\zeta) &\leqslant 2^{(n+1)\alpha}\ell_2(A_n(z)\setminus U)\\ &\leqslant 2^{(n+1)\alpha}\pi\big[\operatorname{cap}(A_n(z)\setminus U)\big]^2 \leqslant 2^{(n+1)\alpha}2^{-2n\alpha}\pi\\ &= 2^{-(n-1)\alpha}\pi\end{aligned}$$

for $z \in U_\alpha$, $n \geqslant N_\alpha$. Here we have used the classical inequality

$$\ell_2(E) \leqslant \pi(\operatorname{cap}(E))^2$$

for every bounded Borel set $E \subset \mathbb{C}$, expressing that among all bounded Borel subsets of $\mathbb{C}$ of given logarithmic capacity, the disk has the largest area, see [PS]. By addition over $n \geqslant N_\alpha$ we obtain

$$\int_{|z-\zeta|<2^{-N_\alpha}} |z-\zeta|^{-\alpha}\, d\ell_2(\zeta) \leqslant \sum_{n\geqslant N_\alpha} 2^{-(n-1)\alpha}\pi \leqslant 2\pi\cdot 2^{-(N_\alpha-1)\alpha}$$

for $z \in U_\alpha$. On the other hand we obviously have

$$\int_{|z-\zeta|\geqslant 2^{-N_\alpha},\,\zeta\in K} |z-\zeta|^{-\alpha}\, d\ell_2(\zeta) \leqslant 2^{\alpha N_\alpha}\ell_2(K),$$

whence by adding the two integrals

$$h_\alpha(z) \leqslant 2\pi\cdot 2^{-(N_\alpha-1)\alpha} + 2^{\alpha N_\alpha}\ell_2(K),\ z \in U_\alpha,$$

showing that h_α is bounded on U_α. This holds for any $\alpha > 0$. In particular, $h_{2\alpha}$ is bounded on $U_{2\alpha}$, and so is therefore h_α.

Choose a decreasing sequence of open sets ω_j such that $U \subset \omega_j \subset \mathbb{C}$, and that $\ell_2(K \cap \omega_j) \to \ell_2(K \cap U)$ and hence $\ell_2(K \cap \omega_j \setminus U) \to 0$ as $j \to \infty$. Consider the functions $h_{\alpha,j}$ defined on ω_j by

$$h_{\alpha,j}(z) = \int_{K\setminus\omega_j} |z-\zeta|^{-\alpha}\, d\ell_2(\zeta),\ z \in \omega_j.$$

Then $h_{\alpha,j} \leqslant h_\alpha < +\infty$ on $U_{2\alpha}$. And

$$\begin{aligned}0 \leqslant h_\alpha(z) - h_{\alpha,j}(z) &= \int_{K\cap\omega_j\setminus U} |z-\zeta|^{-\alpha}\, d\ell_2(\zeta)\\ &\leqslant \Big(\int_{K\setminus U} |z-\zeta|^{-2\alpha}\, d\ell_2(\zeta)\Big)^{\frac{1}{2}} \Big(\int_{K\cap\omega_j\setminus U} d\ell_2(\zeta)\Big)^{\frac{1}{2}}\\ &\leqslant \sup_{z\in U_{2\alpha}} (h_{2\alpha}(z))^{\frac{1}{2}} \ell_2(K\cap\omega_j\setminus U)^{\frac{1}{2}} < +\infty\end{aligned}$$

because $h_{2\alpha}$ is bounded on $U_{2\alpha}$ as noted above. Since $\ell_2(K \cap \omega_j \setminus U) \to 0$ as $j \to \infty$ we have thus shown that $h_{\alpha,j} \to h_\alpha$ uniformly on $U_{2\alpha}$. Furthermore, $h_{\alpha,j}$ is continuous on ω_j, for every point $z_0 \in U$ has a compact fine neighborhood K_0 contained in U, and $h_{\alpha,j}$ is clearly finite and continuous on K_0 because $|z-\zeta|$ ($z \in K_0$, $\zeta \in K\setminus\omega_j$) is minorized by the positive distance between the disjoint compacts sets K_0 and $K\setminus\omega_j$. We conclude that the uniform limit $h_\alpha : U_{2\alpha} \to \mathbb{R}$ of the sequence of finite continuous restrictions of $h_{\alpha,j}$ to $U_{2\alpha}$ is finite and continuous on a fine neighborhood $U_{2\alpha}$ of each point $z_0 \in U$, cf. Lemma 4.5.8. Thus h_α is finite and finely continuous on U.

According to the Brelot property (Theorem 1.5.29) every point of U has a compact fine neighborhood V in U such that each $h_\alpha|_V$ with α rational is (finite and) continuous in the Euclidean topology, and hence bounded. To show that the same is true for any $\alpha > 0$, suppose first that the compact set $V - K \subset \mathbb{C}$ is contained in the unit disk. Then $\alpha \mapsto h_\alpha(z)$ is increasing for each $z \in V$. For any $\alpha > 0$ and any increasing sequence (α_n) of rationals converging to α we have $h_\alpha(z) = \sup_n h_{\alpha_n}(z)$ for every $z \in V$, and hence $h_\alpha|_V$ is l.s.c. Similarly $h_\alpha|_V$ is also u.s.c. and altogether $h_\alpha|_V$ is continuous. In the remaining case write $k = \sup\{|z| : z \in V - K\} > 0$. The above argument then shows that each $k^\alpha h_\alpha|_V$ is continuous because $k/|z-\zeta| \geq 1$ for $z, \zeta \in V - K$. □

Theorem 4.5.9 *Each of the following three properties is necessary and sufficient for a function $f : U \to \mathbb{C}$ to be finely holomorphic:*

(i) *f and $z \mapsto zf(z)$ are finely harmonic on U.*

(ii) *f is finely harmonic on U, and there exists a cover of U by bounded finely open sets $V \subset U$ and a function $\widehat{f} \in \widehat{\mathcal{D}}(\Omega)$, where $\Omega \supset V$ is a bounded open subset of $\mathbb{C}$ (depending on V) such that $\widehat{f} = f$ on V and $\overline{\partial}\widehat{f} = 0$ a.e. on V (that is, $\widehat{f} \in \mathcal{O}(\Omega, V)$).*

(iii) *Every point of U has a finely open fine neighborhood $V \subset U$ on which f coincides with the Cauchy transform of some function $\varphi \in L^2(\mathbb{C})$ (or equivalently well: $\varphi \in L^p(\mathbb{C})$ for some $p > 1$) of compact support such that $\varphi = 0$ a.e. on V:*

$$f(z) = \int_{\mathbb{C}} \frac{1}{z-\zeta}\varphi(\zeta)\, d\ell_2(\zeta),\ z \in V.$$

Note that this integral exists and is finite for each $z \in \mathbb{C}$, by Hölder inequality.

Proof If f is finely holomorphic then (i) holds because fine harmonicity is preserved under uniform convergence, by Corollary 2.1.21.

(i)⇒(ii). For brevity write $h(z) = zf(z)$, $z \in U$. According to Theorem 4.2.12, U can be covered by bounded regular finely open sets $V \subset U$ such that $f|_V$ and $h|_V$ can be extended to finely continuous functions $\widehat{f}, \widehat{h} \in \mathcal{H}(\Omega, V)$, where Ω denotes a bounded domain in $\mathbb{C}$ depending on V and containing $\overline{V}$. It now follows from Lemma 4.5.4 that (i) and (ii) above are equivalent.

(ii)⇒(iii). For a given point $z_0 \in U$ let V be a bounded finely open subset of U containing z_0 and such that $\overline{V} \subset U$, let Ω be a bounded open subset of $\mathbb{C}$ such that $\Omega \supset \overline{V}$, and let $\widehat{f} \in \mathcal{O}(\Omega, V)$ be such that $\widehat{f} = f$ on V. It then follows from Lemma 4.5.6 that

$$f(z) = \widehat{f}(z) = \frac{1}{\pi}\int_{\Omega} \frac{1}{z-\zeta}\overline{\partial}\widehat{f}(\zeta)\, d\ell_2(\zeta) \text{ q.e. on } \Omega.$$

Here $\varphi := \frac{1}{\pi}\overline{\partial}\widehat{f} \in L^2_c(\Omega) \subset L^2_c(\mathbb{C})$, and in particular $\varphi = 0$ a.e. on $V \subset \Omega \cap U$.

(iii) implies fine holomorphy. We may assume that f itself is the restriction to U of the Cauchy transform of some function $\varphi \in L^p_c(\mathbb{C})$ with $p > 1$ and $\varphi = 0$ a.e. on U:

$$f(z) = \int_{\mathbb{C}} \frac{1}{z-\zeta}\varphi(\zeta)\, d\ell_2(\zeta), \quad z \in U. \tag{4.5.18}$$

Let now V denote any bounded fine neighborhood of a given point of U such that $\overline{V} \subset U$. Define $\psi(z) = \frac{1}{z}$ for $z \in \overline{V} - \operatorname{Supp}\varphi$ (compact) and $\psi(z) = 0$ elsewhere on $\mathbb{C}$. Then $f = \varphi * \psi$ on V, and since $\varphi \in L^p(\mathbb{C})$ for some $p > 1$ and $1_{\complement U}\psi(z - \cdot) \in L^q(\mathbb{C})$ for every $z \in U$ and $q > 1$ according to Lemma 4.5.8, in particular for $\frac{1}{p} + \frac{1}{q} = 1$, $\varphi * \psi$ is well defined and finite valued on $\mathbb{C}$, by Hölder's inequality. Hence f is well defined on V.

Choose a decreasing sequence of open sets ω_n with compact closures in $\mathbb{C}$ and with the compact intersection $\overline{V}$. Writing $\varphi_n = 1_{\mathbb{C}\setminus\omega_n}\varphi$ consider the function f_n defined on $\mathbb{C}$ by

$$f_n(z) = \int_{\mathbb{C}\setminus\omega_n} \frac{1}{z-\zeta}\varphi(\zeta)\, d\ell_2(\zeta) = \int_{\mathbb{C}} \frac{1}{z-\zeta}\varphi_n(\zeta)\, d\ell_2(\zeta). \tag{4.5.19}$$

Defining now $\psi(z) = \frac{1}{z}$ for $z \in \overline{\omega}_1 - \operatorname{Supp}\varphi$ (compact) and $\psi(z) = 0$ elsewhere on $\mathbb{C}$, we have $f_n = \psi * \varphi_n$ on ω_n. By Hölder's inequality

$$\sup_{z \in V} \left| f(z) - f_n(z) \right| \leqslant \|\psi\|_{L^q} \|\varphi_n - \varphi\|_{L^p} \to 0$$

as $n \to +\infty$, and so $f_n \to f$ uniformly on V. Each function f_n is holomorphic on ω_n, and we conclude that indeed $f \in R(\overline{V})$. □

Corollary 4.5.10 *On a (Euclidean) open set U, the holomorphic and the finely holomorphic functions are the same.*

In view of the same with "holomorphic" replaced by "harmonic" in Theorem 2.7.2, this follows by (i).

Corollary 4.5.11 *The algebra $\mathcal{O}(U)$ of all finely holomorphic functions on U is closed under finely locally uniform convergence.*

In view of (i) in Theorem 4.5.9 this holds because the space of all finely harmonic functions $U \to \mathbb{C}$ is closed under finely locally uniform convergence, see Corollary 2.1.21.

Corollary 4.5.12 *Let f be a finely holomorphic on $U \setminus E$, where E denotes a polar subset of the finely open set $U \subset \mathbb{C}$. If f is bounded on some deleted fine neighborhood of each point of E then f extends by fine continuity to a (unique) finely holomorphic function on all U.*

In view of (i) in Theorem 4.5.9 this follows immediately from Theorem 2.1.45 applied to the real and imaginary parts of f and $z \mapsto zf(z)$.

Remark 4.5.13 For any compact set $K \subset \mathbb{C}$ every function $f \in R(K)$ is finely holomorphic on the fine interior K' of K, in view of Definition 4.5.1 and Corollary 4.5.11. On the other hand, there exists a compact set K and a function $f \in C(K) \setminus R(K)$ such that f is finely holomorphic on K' because $K' = \varnothing$. This is seen by example in McKissick [McKi] of a compact $K \subset \mathbb{C}$ with $R(K) \neq C(K)$ for which $R(K)$ is *normal* on K, i.e., for every pair of disjoints closed sets $K_0, K_1 \subset K$ there exists a function $f \in R(K)$ such that $f(K_0) = \{0\}$ and $f(K_1) = \{1\}$, for that implies that $K' = \varnothing$. In fact, if $K' \neq \varnothing$ we choose in some fine component V of K' distinct points z_0, z_1, and take $K_1 = \{z_1\}$, K_0= a compact fine neighborhood of z_0 contained in the finely open set $K' \setminus \{z_1\} = K' \setminus K_1$. The above function $f \in R(K)$ is then identically 0 on K_0 and hence on all of V according to Theorem 4.5.23 below (which is independent of the present Remark 4.5.13), but this contradicts $f(z_1) = 1$. Consequently we have indeed $K' = \varnothing$.

Theorem 4.5.14 *Either of the following properties* (iv), (v) *is necessary and sufficient for a function $f : U \to \mathbb{C}$ (U finely open in $\mathbb{C}$) to be finely holomorphic:*

(iv) *Every point of U has a finely open fine neighborhood $V \subset U$ such that $f|_V$ extends to a C^1-function $\widehat{f}$ on some open set $\omega \subset \mathbb{C}$ containing V with $\partial \widehat{f} = f'$ and $\overline{\partial} \widehat{f} = 0$ on V.*

(v) *f is a fine C^1-function on U in the complex variable sense.*

Saying that f is a fine C^1-function in the complex variable sense on U shall mean that, when U is given the fine topology and $\mathbb{C}$ the Euclidean topology, then $\frac{f(w)-f(z)}{w-z} \in \mathbb{C}$ has for each $z \in U$ a limit in $\mathbb{C}$ as $w \to z$ finely in U, and that this (fine) limit, denoted $f'(z)$ and termed the fine derivative of f at z, is continuous as a function of $z \in U$ with values in $\mathbb{C}$.

Proof of Theorem 4.5.14. Suppose that f is finely holomorphic and thus has property (iii) in Theorem 4.5.9. We may assume that f is itself the restriction to U of the Cauchy transform of some function $\varphi \in L^p_c(\mathbb{C})$ with $p > 1$ and $\varphi = 0$ a.e. on U, as in (4.5.18). Consider a compact fine neighborhood $V \subset U$ of a given point of U as in Lemma 4.5.8. As in the last paragraph of the proof of Theorem 4.5.9, consider a decreasing sequence of open sets ω_n with compact closures in $\mathbb{C}$ and with the intersection V. Again write $\varphi_n = 1_{\mathbb{C}\backslash\omega_n}\varphi$ and define f_n as in (4.5.19), whereby f_n is holomorphic on ω_n with the derivative

$$f_n'(z) = \int_{\mathbb{C}\backslash\omega_n} \frac{1}{(z-\zeta)^2}\varphi(\zeta)\, d\ell_2(\zeta) = \int_{\mathbb{C}} \frac{1}{(z-\zeta)^2}\varphi_n(\zeta)\, d\ell_2(\zeta).$$

Now define preliminarily $f'(z)$ by formal differentiation of (4.5.18) under integral sign while taking into account that $\varphi = 0$ a.e. on U by hypothesis:

$$f'(z) = \int_{S\backslash U} \frac{-1}{(z-\zeta)^2}\varphi(\zeta)\, d\ell_2(\zeta) \tag{4.5.20}$$

where S denotes the closed support of φ. The integral exists and is finite for $z \in V$ by Cauchy-Schwarz because $\varphi \in L^2_c(\mathbb{C})$ and $\int_{S\backslash U} |z-\zeta|^{-4}\, d\ell_2(\zeta) \leqslant c_4$ by Lemma 4.5.8.

Consider any two points $z, w \in V$. With integration over $S\backslash U$ we have (omitting $d\ell_2(\zeta)$)

$$f(w) - f(z) = \int \Big(\frac{1}{w-\zeta} - \frac{1}{z-\zeta}\Big)\varphi(\zeta) = (w-z)\int \frac{-1}{(w-\zeta)(z-\zeta)}\varphi(\zeta),$$

$$\big|f(w) - f(z)\big|^2 \leqslant |w-z|^2 \int \frac{1}{\big|w-\zeta\big|^2\big|z-\zeta\big|^2} \int |\varphi(\zeta)|^2,$$

$$\big|f(w) - f(z)\big| \leqslant c_4\|\varphi\|_{L^2}\big|w-z\big| \tag{4.5.21}$$

by two applications of Cauchy-Schwarz and by Lemma 4.5.8 with $\alpha = 4$. Similarly by (4.5.20)

$$f(w) - f(z) - (w-z)f'(z) = (w-z)^2 \int \frac{1}{(z-\zeta)^2(w-\zeta)}\varphi(\zeta)\, d\ell_2(\zeta),$$

$$\left|f(w) - f(z) - (w-z)f'(z)\right| \leqslant \sqrt[4]{c_4c_8}\|\varphi\|_{L^2}\left|w-z\right|^2,$$

$$f'(w) - f'(z) \tag{4.5.22}$$

$$= (w-z)\int \Big(\frac{1}{(z-\zeta)(w-\zeta)^2}\varphi(\zeta) + \frac{1}{(z-\zeta)^2(w-\zeta)}\varphi(\zeta)\Big)\, d\ell_2(\zeta),$$

$$\left|f'(w) - f'(z)\right| \leqslant 2\sqrt[4]{c_4c_8}\|\varphi\|_{L^2}\left|w-z\right|. \tag{4.5.23}$$

In view of these three estimations (4.5.21), (4.5.22), (4.5.23), which are uniform in $z, w \in V$, it follows that from Whitney's extension theorem [Wh] (written in complex form) that $f|_V$ can be extended to a C^1-function $\widehat{f} : \mathbb{C} \to \mathbb{C}$ of compact support such that $\partial\widehat{f} = f'$ and $\overline{\partial}\widehat{f} = 0$ on V. This establishes (iv).

(iv)$\Rightarrow$(v). For any $z \in V$ and any $\varepsilon > 0$ there exists a number $\delta > 0$ such that

$$\left|\widehat{f}(w) - \widehat{f}(z) - (w-z)\partial\widehat{f}(z) - (\overline{w}-\overline{z})\overline{\partial}\widehat{f}(z)\right| \leqslant \varepsilon\left|w-z\right| \text{ for } \left|w-z\right| < \delta.$$

Since $\widehat{f} = f$ on V and $\overline{\partial}\widehat{f}(z) = 0$ it follows by division by $w - z$ for $w \in V\backslash\{z\}$, in particular for $w \in V'\backslash\{z\}$, that

$$\left|\frac{f(w) - f(z)}{w-z} - \partial\widehat{f}(z)\right| \leqslant \varepsilon$$

holds for every $w \in V'$ with $|w - z| < \delta$, and the set of these w is a fine neighborhood of z. Thus f is finely differentiable on V' with the fine derivative $f' = \partial\widehat{f}$, which is continuous relative to V' and in particular finely continuous on V'. We conclude that f is a fine C^1-function in the complex variable sense on V', a fine neighborhood of z.

For the remaining assertion that (v) implies f being finely holomorphic we need the following

Lemma 4.5.15 *Suppose that $f : U \to \mathbb{C}$ is fine C^1-function in the complex variable sense and that f and its fine derivative f' are (Euclidean) continuous on U. Let $z_0, z_1 \in U$, and suppose that the point $z_t = (1-t)z_0 + tz_1$ belongs to U for every $t \in [0, 1]$. Then*

$$f(z_1) - f(z_0) = (z_1 - z_0)\int_0^1 f'(z_t)dt.$$

Proof For given $\varepsilon > 0$ let E_ε denote the set of numbers $t \in [0, 1]$ such that

$$\left|f(z_t) - f(z_0) - (z_1 - z_0)\int_0^t f'(z_\tau)\, d\tau\right| \leqslant \varepsilon|z_1 - z_0|t. \tag{4.5.24}$$

Then E_ε is closed because $f(z_t)$ and $f'(z_t)$ are continuous functions of $t \in [0, 1]$, and E_ε is non-void since $0 \in E_\varepsilon$. Denote $t = \max E_\varepsilon$, and suppose $t < 1$. By our hypothesis (v) there exists a fine neighborhood U_t of z_t such that

$$\left|f(w) - f(z_t) - (w - z_t)f'(z_t)\right| \leqslant \frac{1}{2}\varepsilon\left|w - z_t\right|$$

for every $w \in U_t$. According to Brelot [Br2, n° 12] there exists a decreasing sequence of numbers $t_n \in [0, 1]$ tending to t as $n \to \infty$ and such that the circles Γ_n centered at z_t and passing through z_{t_n} are contained in U_t. Hence $z_{t_n} - z_t = (z_1 - z_0)(t_n - t)$ and by the above display

$$\left|f(z_{t_n}) - f(z_t) - (z_{t_n} - z_t)f'(z_t)\right| \leqslant \frac{1}{2}\varepsilon\left|z_1 - z_0\right|(t_n - t). \tag{4.5.25}$$

By continuity of f' at z_t

$$\left|f'(z_\tau) - f'(z_t)\right| \leqslant \frac{1}{2}\varepsilon \text{ for } \tau \in [t, t_n]$$

and a suitable n. It follows that

$$\begin{aligned}\left|\int_t^{t_n} f'(z_\tau)\,d\tau - (t_n - t)f'(z_t)\right| &\leqslant \int_t^{t_n}\left|f'(z_t) - f'(z_\tau)\right| d\tau \\ &\leqslant \frac{1}{2}\varepsilon(t_n - t),\end{aligned}$$

and we may therefore replace $(z_{t_n} - z_t)f'(z_t) = (z_1 - z_0)(t_n - t)f'(z_t)$ on the left of (4.5.25) by $(z_1 - z_0)\int_t^{t_n} f'(z_\tau)\,d\tau$ at the expense of replacing $\frac{1}{2}\varepsilon$ by ε on the right of (4.5.25). With this change in (4.5.25) we obtain after adding (4.5.24) the inequality (4.5.24) with t replaced by t_n. This means that $t_n \in E_\varepsilon$, contradicting the maximality of t, and so actually (4.5.24) holds for $t = 1$. Letting $\varepsilon \to 0$ we conclude that (4.5.24) holds with εt replaced by 0, and this proves the lemma. □

Returning to the proof of Theorem 4.5.14, suppose conversely that f is a fine C^1-function in the complex variable sense. Henceforth, f' denotes the fine derivative of f. Since f and f' are finely continuous (as mappings into $\mathbb{C}$ with the Euclidean topology) it follows by the Brelot property that every point of U has a compact fine neighborhood V in U such that f and f' are (Euclidean) continuous relative to V. By Theorem 4.1.4 we may further assume that, for any two points z, w of the fine interior V' of V, there exists a point $z' \in V'$ such that straight segments $[z, z']$ and $[z', w]$ lie in V' and $|z - z'| = |z' - w| \leqslant 2|z - w|$. According to Lemma 4.5.15 applied to U replaced with V' and z_0, z_1 by either z', z or z', w we have after subtraction

$$f(w) - f(z) = \int_L f'(\zeta)\,d\zeta,$$

where the path L is the (straight) segment from z to z' followed by the segment from z' to w. Since f and f' are uniformly continuous on V (compact) there corresponds to any $\varepsilon > 0$ a number $\delta = \delta(\varepsilon) > 0$ such that

$$|f(z) - f(\zeta)|,\ |f'(z) - f'(\zeta)| \leqslant \varepsilon \text{ for any } z, \zeta \in V \text{ with } |z - \zeta| \leqslant \delta.$$

Denoting by $|L|$ the length of the above path L from z to w in V' (so $|L| \leqslant 2|w - z|$) we obtain

$$\begin{aligned}\big|f(w) - f(z) - (w - z)f'(z)\big| &= \Big|\int_L (f'(\zeta) - f'(z))\,d\zeta\Big| \\ &\leqslant |L|\varepsilon \leqslant 2\varepsilon|w - z|\end{aligned}$$

provided that $[z - z'|(= |w - z'|) \leqslant \delta$, and hence in particular if $|w - z| \leqslant \delta$ (for then $|z - z'| = \frac{1}{2}|L| \leqslant |w - z| \leqslant \delta$). This obviously extends by continuity to the case where z and w belong to the Euclidean closure $\overline{V'}$ of V' (and $|w - z| \leqslant \delta$). Proceeding as in the proof that (iii)⇒(iv) we find that f extends to a C^1-function $\mathbb{C} \to \mathbb{C}$ of compact support such that $\partial f = f'$ and $\overline{\partial} f = 0$ on $\overline{V'}$. Defining $\varphi = \overline{\partial} f \in L^2(\mathbb{C})$ in terms of the above extension f we have

$$f(z) = \frac{1}{\pi}\int_{\mathbb{C}} \frac{1}{z - \zeta}\varphi(\zeta)\,d\ell_2(\zeta)$$

by the classical Cauchy formula, see the proof of Lemma 4.5.6. □

Theorem 4.5.16 *If $f : U \to \mathbb{C}$ is finely holomorphic then so is its fine derivative f'. Consequently every finely holomorphic function is infinitely finely differentiable and its fine derivatives of all orders are finely holomorphic.*

Proof According to Theorem 4.5.9(iii) we may assume that $f = z^{-1} * \varphi$ on U for some $\varphi \in L^2_c(\mathbb{C})$ with $\varphi = 0$ a.e. on U. As shown in the proof that f is C^1-function, the fine derivative f' equals $-z^{-2} * \varphi$ on U and is thus obtained by usual differentiation under integral sign. As in the beginning of the proof of Theorem 4.5.14, consider a compact fine neighborhood V of a given point of U as in Lemma 4.5.8 with $\alpha = 2$, and a decreasing sequence of open sets ω_n of compact closures in $\mathbb{C}$ and with the intersection V. Again write $\varphi_n = 1_{\mathbb{C}\setminus\omega_n}\varphi$ and define $f_n = z^{-1} * \varphi_n$ as in (4.5.19). Then $f_n' = -z^{-2} * \varphi_n$ is holomorphic on ω_n and $f_n' \to f'$ uniformly on V. In fact

$$f_n'(z) - f'(z) - \int_{S\setminus U} \frac{-1}{(z - \zeta)^2}(\varphi_n(\zeta) \quad \varphi(\zeta))\,d\ell_2(\zeta)$$

with $S = \operatorname{Supp}\varphi$. By Cauchy-Schwarz,

$$|f_n'(z) - f'(z)|^2 \leqslant \int_{S\setminus U} \frac{d\ell_2(\zeta)}{|z - \zeta|^4}\int_{\mathbb{C}} |\varphi_n - \varphi|^2\,d\ell_2,$$

and the right hand members converge to 0 as $n \to \infty$, uniformly on V according to Lemma 4.5.8. Thus the restriction of f' to V is of class $R(V)$, and so f' is indeed finely holomorphic on U, cf. Definition 1.2.1. □

Theorem 4.5.17 *A function $f : U \to \mathbb{C}$ (U finely open set in $\mathbb{C}$) is finely holomorphic if and only if f is fine-to-fine C^1-function in the complex variable sense.*

Saying that f is fine-to-fine C^1-function only differs from f being fine C^1-function by using the fine topology also on target $\mathbb{C}$.

Proof of Theorem 4.5.17. The "if part" is obvious because the fine topology is finer than the Euclidean one. For the "only if" part, suppose that f is finely holomorphic. For given $z \in U$ consider the function $g : U \to \mathbb{C}$ defined by $g(z) = f'(z)$ and

$$g(w) = \frac{f(w) - f(z)}{w - z} \text{ for } w \in U \setminus \{z\}.$$

Since $\mathcal{O}(U)$ is an algebra, $w \mapsto f(w) - f(z)$ and $w \mapsto \frac{1}{w-z}$ are finely holomorphic on $U \setminus \{z\}$ and so is g and g is finely continuous even at z. By Theorem 4.5.9, g and $w \mapsto wg(w) = f(w) - f(z) + zg(w)$ are finely harmonic on $U \setminus \{z\}$, and even at z in view of Theorem 2.1.45; and so g is finely holomorphic on all of U. According to Theorem 4.5.2, g therefore is fine-to-fine continuous at z, which means that f indeed is fine-to-fine differentiable at z. Likewise, f' is fine-to-fine continuous, being finely holomorphic according to Theorem 4.5.16. □

Theorem 4.5.18 (a) *If $f : U \to \mathbb{C}$ and $g : V \to \mathbb{C}$ are finely holomorphic, then $g \circ f$ is finely holomorphic on $U \cap f^{-1}(V)$ and*

$$(g \circ f)' = (g' \circ f) f' \text{ on } U \cap f^{-1}(V).$$

(b) *If $f : U \to \mathbb{C}$ is injective and finely holomorphic then $g := f^{-1}$ is finely holomorphic on $f(U)$ (finely open), $f' \neq 0$ on U and*

$$g' = \frac{1}{f' \circ g} \text{ on } U.$$

(c) *Let $f : U \to \mathbb{C}$ be finely holomorphic and let $z_0 \in U$. In order that the restriction of f to a suitable fine neighborhood V of z_0 in U be injective it is necessary and sufficient that $f'(z_0) \neq 0$.*

Proof (a) Note that $U \cap f^{-1}(V)$ is finely open because f is fine-to-fine continuous by Theorem 4.5.2. Given a point $z_0 \in U \cap f^{-1}(V)$ the point $w_0 = f(z_0) \in V$ has by Theorem 4.5.14 a finely open fine neighborhood $V_2 \subset V$ such that $g|_{V_2}$ extends to C^1-function $\widehat{g} \in C^1_c(\mathbb{C})$ with $\partial\widehat{g} = g'$ and $\overline{\partial}\widehat{g} = 0$ on V_2. Next, z_0 has a finely open fine neighborhood $V_1 \subset U \cap f^{-1}(V_2)$ such that $f|_{V_1}$ extends to C^1-function $\widehat{f} \in C^1_c(\mathbb{C})$ with $\partial\widehat{f} = f'$ and $\overline{\partial}\widehat{f} = 0$ on V_1. It follows that $\widehat{g} \circ \widehat{f} \in C^1_c(\mathbb{C})$ and $\widehat{g} \circ \widehat{f}$ extends $g \circ f|_{V_1}$. Furthermore, $\overline{\partial}(\widehat{g} \circ \widehat{f}) = [(\partial\widehat{g}) \circ \widehat{f})]\overline{\partial}\widehat{f} + [(\overline{\partial}\widehat{g}) \circ \widehat{f}]\overline{\partial \widehat{f}} = 0$ on

V_1. Consequently, by Theorem 4.5.14, $g \circ f$ is finely holomorphic on V_1 and hence on all of the finely open set $U \cap f^{-1}(V)$, and $(g \circ f)' = \partial(\widehat{g} \circ \widehat{f}) = (g' \circ f)f'$ on V_1, whence $(g \circ f)' = (g' \circ f)f'$ on U.

(b) By Theorem 4.5.2, f is a finely harmonic morphism and hence so is $g = f^{-1} : f(V) \to \mathbb{C}$ according to Theorem 4.4.10. The functions $z \mapsto z$ and, by Theorem 4.5.9(i), $z \mapsto zf(z)$ are finely harmonic on U. Their compositions $w \mapsto g(w)$ and $w \mapsto g(w)w$ with the finely harmonic morphism g are therefore finely harmonic on $f(U)$ (Theorem 4.4.5), and so g is finely holomorphic on $f(U)$. From $g \circ f = \mathrm{id}$ (the identity map $U \to U$) we infer by (a) above that $(g' \circ f)f' = 1$ and hence $f' \neq 0$ and $g' \circ f = 1/f'$, or equivalently $g' = (1/f') \circ f^{-1} = (1/f') \circ g = 1/(f' \circ g)$.

(c) If the restriction of f to some (finely open) fine neighborhood V of z_0 is injective then $f'(z_0) \neq 0$ according to (b). Conversely, if $f'(z_0) \neq 0$ then, by Theorem 4.5.14, there is a fine neighborhood U_0 of z_0 in U such that $f|_{U_0}$ extends to some $\widehat{f} \in C^1_c(\mathbb{C})$ with $\partial\widehat{f}(z_0) = f'(z_0) \neq 0$ and hence $\widehat{f}$ is injective in some Euclidean neighborhood V of z_0. It follows that $f = \widehat{f}|_{U_0}$ is injective on the fine neighborhood $V \cap U_0$ of z_0 in U. □

The regularity properties (iv) and (v) of a finely holomorphic function given in Theorem 4.5.14 extend to derivation of higher order as follows.

Theorem 4.5.19 *If $f : U \to \mathbb{C}$ (U finely open) is finely holomorphic then every point of U has a compact fine neighborhood $V \subset U$ with the following three properties:*

(a) *There exists a sequence of holomorphic functions $f_j : \omega_j \to \mathbb{C}$ (ω_j open in $\mathbb{C}$, $\omega_j \supset V$) such that, for any integer $n \geqslant 0$, the ordinary n'th derivatives $f_j^{(n)}$ converge uniformly to $f^{(n)}$, the fine n'th derivative of f, on V as $j \to +\infty$.*

(b) *For any integers m, n the quotient*

$$\frac{f^{(n)}(w) - \sum_{k=0}^{m-1} \frac{1}{k!}(w-z)^k f^{(n+k)}(z)}{(w-z)^m}$$

(understood as $f^{(n)}(w)$ for $m = 0$) is bounded as function of $(z, w) \in V \times V$.

(c) *There exists an ordinary C^∞-function $\widehat{f} : \mathbb{C} \to \mathbb{C}$ such that $\widehat{f} = f$ on V and hence*

$$\partial^n \widehat{f} = f^{(n)} \text{ and } \partial^n \overline{\partial}^{n'} \widehat{f} = 0 \text{ on } V$$

for any integers $n \geqslant 0$ and $n' \geqslant 1$.

Proof In view of Theorem 4.5.9(iii) we may assume for given p, $1 < p \leqslant +\infty$, that f equals the Cauchy transform of some function $\varphi \in L^p(\mathbb{C})$ of compact support S such that $\varphi = 0$ a.e. on U:

$$f(z) = \int_{\mathbb{C}} \frac{1}{z - \zeta} \varphi(\zeta)\, d\ell_2(\zeta), \ z \in U.$$

The proof for that case will be given in three steps. In steps 1 and 2, $f^{(n)} : U \to \mathbb{C}$ will be understood as obtained by differentiation under the integral sign:

$$f^{(n)}(z) = (-1)^n n! \int \frac{1}{(z-\zeta)^{n+1}} \varphi(\zeta)\, d\ell_2(\zeta), \ z \in U, \tag{4.5.26}$$

with integration over $S\backslash U$ here and in the rest of the proof. In step 3, an application of Whitney's theorem will show that $f^{(n)}$ in that sense equals the n'th fine derivative of f in U. Let $z_0 \in U$ and $V \subset U$ a compact fine neighborhood of z_0.

Step 1. Choose a decreasing sequence of open sets ω_j having the intersection V. As in the end of the proof of Theorem 4.5.9 put $\varphi_j = 1_{\complement\omega_j}\varphi$ and $f_j = z^{-1} * \varphi_j$. Then f_j is holomorphic on ω_j with the derivatives

$$f_j^{(n)}(z) = (-1)^n n! \int \frac{1}{(z-\zeta)^{n+1}} \varphi_j(\zeta)\, d\ell_2(\zeta), \ z \in \omega_j. \tag{4.5.27}$$

From (4.5.26), (4.5.27), and (4.5.17) it follows by Hölder's inequality, writing $q = p/(p-1)$, that

$$\sup_{z\in V} \left| f_j^{(n)}(z) - f^{(n)}(z) \right| \leqslant n!(c_{(n+1)q})^{1/q} \|\varphi_j - \varphi\|_{L^p} \to 0$$

as $j \to \infty$. This establishes (a), though with $f^{(n)}$ defined by (4.5.26).

At the same time we infer by Corollary 4.5.11 that each $f^{(n)}$ from (4.5.26 is finely holomorphic in the fine interior V' of V and hence on all of U, because each $f_j^{(n)}$ from (4.5.27) is finely holomorphic on ω_j, in particular on V'. For $n = 0$ we recover (iii) in Theorem 4.5.9.

Step 2. For integers $m, n \geqslant 0$, and consider the m'th remainder in the Taylor expansion of the function $z \mapsto z^{-n-1}$. For $z, w \in \mathbb{C}\backslash\{0\}$ this remainder term equals

$$\begin{aligned} R(z,w) &= \ w^{-n-1} - \sum_{k=0}^{m-1} \frac{1}{k!}(w-z)^k \partial^k z^{-n-1} \\ &= \ z^{-m-n} w^{-n-1} P(z,w), \end{aligned}$$

where P is a polynomial in z, w. If we regard P as a polynomial in z and $w - z$ we obtain an expansion in a finite sequence

$$P(z,w) = \sum_{i\geqslant 0} P_i(z)(w-z)^i$$

in terms of certain polynomials P_i of a single variable. According to Taylor's formula the limit

$$\lim_{w\to z} \frac{R(z,w)}{(w-z)^m} = \lim_{w\to z} z^{-m-n} w^{-n-1} \sum_{i\geqslant 0} P_i(z)(w-z)^{i-m}$$

exists finitely for any $z \neq 0$. It follows that

$$P_i = 0 \text{ for } i < m,$$

and hence (for $w \neq z, 0$), writing $i = m + j$,

$$\begin{aligned}\frac{R(z,w)}{(w-z)^m} &= z^{-m-n} w^{-n-1} \sum_{j \geqslant 0} P_{m+j}(z)(w-z)^j \\ &= z^{-m-n} w^{-n-1} \sum_{i,j \geqslant 0} a_{i,j} z^i w^j \\ &= \sum_{r,s} b_{r,s} z^{-r} w^{-s}\end{aligned}$$

which is a finite sum with constant coefficients $b_{r,s}$.

This applies with z, w replaced by $z - \zeta$, $w - \zeta$ provided that z, w and ζ are distinct. Take $z, w \in V$ and $\zeta \in \complement U$. After multiplication by $(-1)^n n! \varphi(\zeta)\, d\ell_2(\zeta)$ and integration over $S \backslash U$ this yields by (4.5.26)

$$\begin{aligned}&\frac{1}{(w-z)^m}\Big(f^{(n)}(w) - \sum_{k=0}^{m-1} \frac{1}{k!}(w-z)^k f^{(n+k)}(z)\Big) \\ &= \int (-1)^n n! \frac{R(z-\zeta, w-\zeta)}{[(w-\zeta)-(z-\zeta)]^m} \varphi(\zeta)\, d\ell_2(\zeta) \\ &= (-1)^n n! \sum_{r,s} b_{r,s} \int_{S\backslash U} (z-\zeta)^{-r}(w-\zeta)^{-s} \varphi(\zeta)\, d\ell_2(\zeta).\end{aligned}$$

This finite sum is bounded as function of $z, w \in V$ (for fixed m, n). In fact, by Hölder inequality,

$$\begin{aligned}&\Big|\int_{S\backslash U} (z-\zeta)^{-r}(w-\zeta)^{-s}\varphi(\zeta)\, d\ell_2(\zeta)\Big| \\ &\leqslant \|\varphi\|_{L^p}\Big(\int_{S\backslash U} |z-\zeta|^{-rq} |w-\zeta|^{-sq}\, d\ell_2(\zeta)\Big)^{\frac{1}{q}} \\ &\leqslant \|\varphi\|_{L^p} h_{2rq}(z)^{\frac{1}{2q}} h_{2sq}(w)^{\frac{1}{2q}},\end{aligned}$$

which is bounded function of $z, w \in V$ for each of the finitely many relevant values of (r, s) (Lemma 4.5.8). We conclude that

$$\Big|f^{(n)}(w) - \sum_{k=0}^{m-1} \frac{1}{k!}(w-z)^k f^{(n+k)}(z)\Big| \leqslant A_{m,n} \|\varphi\|_{L^p} |w-z|^m \tag{4.5.28}$$

for $z, w \in V$ and integers $m, n \geqslant 0$, the constants $A_{m,n}$ being independent of $z, w \in V$ and of the function φ. And this amounts to assertion (b) of Theorem 4.5.19 for the present case and with the $f^{(n)}$ defined by (4.5.26).

For $m = 2$ and a given point $z \in V'$ it follows from (4.5.28) for $w \in V'$ $(w \neq z)$ that

$$\left|\frac{f^{(n)}(w) - f^{(n)}(z)}{w - z} - f^{(n+1)}(z)\right| \leqslant A_{2,n}\|\varphi\|_{L^p}\left|w - z\right|.$$

Thus each $f^{(n)}$ as defined in (4.5.26) is finely differentiable at any point $z \in V'$ and hence on all of U, with the fine derivation $f^{(n+1)}$ likewise defined in agreement with (4.5.26). It follows by induction that the function $f = z^{-1} * \varphi$ is infinitely finely differentiable on U, and its n'th fine derivative is the function $f^{(n)}$ defined by (4.5.26).

Step 3. For integers $n, n' \geqslant 0$ and for any $z \in V$, write

$$f^{(n,n')}(z) = \begin{cases} f^{(n)}(z) & \text{for } n' = 0 \\ 0 & \text{for } n' > 0. \end{cases} \tag{4.5.29}$$

Then the estimate (4.5.28) may be written as

$$\left|f^{(n,n')}(w) - \sum_{k+k' \leqslant m-1} \frac{1}{k!k'!}(w - z)^k(\overline{w} - \overline{z})^{k'} f^{(n+k,n'+k')}(z)\right| \leqslant A_{m,n}\|\varphi\|_{L^p}\left|w - z\right|^m$$

for $z, w \in V$ (compact) and $m, n, n' \geqslant 0$. It therefore follows by Whitney's theorem that $\widehat{f} = f$ on V for some C^∞-function $\widehat{f} : \mathbb{C} \to \mathbb{C}$ and hence

$$\partial^n \overline{\partial}^{n'} \widehat{f} = f^{(n,n')} \text{ on } V$$

for $n, n' \geqslant 0$. In view of (4.5.29) this amounts to assertion (c) of Theorem 4.5.19 for the present typical case. □

Remark 4.5.20 The formal Taylor series of finely holomorphic function f is generally divergent, but it represents f asymptotically in the sense of Theorem 4.5.19.

Corollary 4.5.21 *A finely holomorphic function f on a fine domain $U \subset \mathbb{C}$ is constant if its derivative f' is identically 0 on U.*

Proof According to Theorem 4.5.14(iv), U can be covered by fine domains V for each of which $f|_V$ extends to usual C^1-function $\widehat{f}$ on some open set $\omega \subset \mathbb{C}$ containing V such that $\partial \widehat{f} = f' = 0$ and $\overline{\partial} \widehat{f} = 0$ on V. According to Theorem 4.1.1, any two points z_1, z_2 of such a fine domain V can be joined by a finite polygonal path contained in $V \subset \omega$. Since $\widehat{f} \in C^1(\omega)$ has $\partial \widehat{f} = \overline{\partial} \widehat{f} = 0$ it follows that $\widehat{f}(z_1) = \widehat{f}(z_2)$, that is, $f z_1) = f(z_2)$. Thus f is finely locally constant on U. For every constant $c \in \mathbb{C}$ the set $f^{-1}(c)$ is therefore finely open in U, and finely closed relative to U because f is finely continuous. Consequently, $f^{-1}(c) = U$ or $\varnothing$, and there is precisely one constant c such that $f^{-1}(c) = U$. □

The fine topology is locally connected, but not locally simply connected. That explains why a finely holomorphic function generally has no primitive. If such a primitive exists (and is finely holomorphic), then it is uniquely determined up to an additive constant according to the above corollary. We bring an example of a finely holomorphic function having no primitive. Examples of this kind have been used by Borel [Bor2] and later by Doob [Do2] and Fuglede [Fu19].

Example 4.5.22 Let (a_n) be a sequence of numbers $a_n > 0$ with $\sum_n a_n < +\infty$. Write $z_n = 2^{-n}$ ($n \in \mathbb{N}$) and

$$f(z) = \sum_{n\in\mathbb{N}} \frac{a_n}{z - z_n}. \tag{4.5.30}$$

This series converges absolutely and locally uniformly on the open set $\Omega := \mathbb{C}\backslash(\{z_n\}_{n\in\mathbb{N}} \cup \{0\})$, and f is therefore well defined and holomorphic on Ω, and meromorphic on $\mathbb{C}\backslash\{0\}$. Let (r_n) be a sequence of numbers such that $0 < r_n < |z_n|$ and

$$\sum_n \frac{n}{\log(1/r_n)} < +\infty.$$

By the Wiener criterion, see, for example, [AG, Theorem 7.7.3], the union A of closed disks $\{|z - z_n| \leqslant r_n\}$ is thin at $0 \notin A$ (and closed relative to $\mathbb{C}\backslash\{0\}$) and therefore $U = \mathbb{C}\backslash A$ is a fine neighborhood of 0. Suppose now that the a_n have been chosen so that $\sum_n a_n/z_n < +\infty$. The series (4.5.30) then converges on the finely open set U containing 0, and f furthermore becomes finely holomorphic on the fine domain $\mathbb{C}\backslash\{z_n\}_{n\in\mathbb{N}}$ $(= \complement\{z_n\}_{n\in\mathbb{N}})$ (and "finely meromorphic" on $\mathbb{C} \cup \{\infty\}$, cf. [Fu27, Section 5]). We show that f has no primitive on any fine neighborhood of 0. Suppose on the contrary that f is derivative of a finely holomorphic function F on a finely open set U_1 such that $0 \in U_1 \subset \complement\{z_n\}_{n\in\mathbb{N}}$. According to Theorem 4.5.9(ii) applied to F on the finely open set U_1 there is a bounded finely open set V such that $0 \in V \subset \overline{V} \subset U_1$ and that $F|_V$ extends to a function $\widehat{f} \in C^1(\mathbb{C})$, whence $\partial\widehat{f} = F' = f$ and $\overline{\partial}\widehat{f} = 0$ on V. It follows that

$$\int f(z)\,dz = \int (\partial\widehat{f} + \overline{\partial}\widehat{f})\,dz = 0$$

along any circle $\{|z| = r\}$ contained in V (and such a circle exists). But that is impossible for

$$\int_{\{|z|=r\}} f(z)\,dz = \sum_n \int_{\{|z|=r\}} \frac{a_n}{z - z_n}\,dz = 2\pi i \sum_{|z_n|<r} a_n \neq 0$$

in view of the uniform convergence of the series (4.5.30) on $V \subset U = \mathbb{C}\backslash A$.

Theorem 4.5.23 (Uniqueness) *A finely holomorphic function f on a fine domain $U \subset \mathbb{C}$ is completely determined on a fine domain by the sequence of its fine deriva-*

tives $f^{(n)}(z_0)$ *of all orders* $n \geqslant 0$ *at a prescribed point* $z_0 \in U$. *In particular,* f *is uniquely determined by its restrictions to any non-void finely open subset of* U.

Proof We may assume that $z_0 = 0$ and all $f^{(n)}(0) = 0$. According to Theorem 4.5.19(b) (with $n = 0$) there is a fine neighborhood V of 0 in U such that $f(z)/|z|^n$ is bounded on V for any integer $n \geqslant 0$:

$$|f(z)| \leqslant a_n |z|^n, \ z \in V. \tag{4.5.31}$$

We may assume moreover that V is a fine domain contained in the unit disk $\Omega \subset \mathbb{C}$ and that $|f| \leqslant 1$ on V.

Denote by $G_\Omega(\cdot, 0)$, resp. $G_V(\cdot, 0)$, the Green function with pole at 0 for Ω, resp. for V, cf. Sect. 2.5. Defining $\log 0 = -\infty$ the function $-\log|f(z)| \geqslant 0$ is finely hyperharmonic by Theorem 4.5.2, and we obtain from (4.5.31)

$$\begin{aligned} -\log|f(z)| &\geqslant -n\log|z| - \log a_n \\ &\geqslant nG_V(z, 0) - \log a_n \end{aligned}$$

on V since $-\log|z| = G_\Omega(z, 0) \geqslant G_V(z, 0)$. Because $G_V(\cdot, 0)$ is finely continuous and $G_V(0, 0) = +\infty$ there is a finely closed fine neighborhood W_n of 0 in V such that

$$\frac{1}{2} n G_V(z, 0) \geqslant \log a_n, \ z \in W_n.$$

The function

$$u(z) := -\log|f(z)| - \frac{1}{2} n G_V(z, 0) \tag{4.5.32}$$

on V is therefore positive on W_n, and $\geqslant -\frac{1}{2} n G_V(z, 0)$ on the finely open set $V \setminus W_n$. Since $G_V(\cdot, 0)$ is a fine potential on V and $u \geqslant 0$ on $V \cap \partial_f (V \setminus W_n)$ ($\subset \partial_{\mathcal{L}} W_n$) it follows by the relative fine boundary minimum principle (applied to u on $V \setminus W_n$) that $u \geqslant 0$ on $V \setminus W_n$ and hence on all of V. Because $G_V(z, 0) > 0$ for $z \in V$ we infer from (4.5.32) for $n \to +\infty$ that $f(z) = 0$ for $z \in V$. But then $f(z) = 0$ for all $z \in U$, for otherwise f would be nonconstant, and hence the finely open set $V = f^{-1}(0)$ would be polar according to the last assertion of Theorem 4.5.2. This proves the former assertion of the present theorem. The latter assertion follows because $f^{(n)}(z_0)$ only depends on the restriction of f to any fine neighborhood of z_0 in U. □

Remark 4.5.24 Contrary to Theorem 4.5.23, a finely *harmonic* function f on a fine domain $U \subset \mathbb{C}$ (or $U \subset \mathbb{R}^N$, $N \geqslant 2$) is not uniquely determined by its restrictions to a non-void finely open subset of U, see Lyons [Ly4].

Theorem 4.5.25 *Each zero* z_0 *of a nonconstant finely holomorphic function* f *of a fine domain* U *has a finite multiplicity* $n(z_0) = n \in \mathbb{N}$, *characterized by any one of the following three equivalent properties*

(i) $f^{(k)}(z_0) = 0$ *for* $0 \leqslant k < n$ *and* $f^{(n)}(z_0) \neq 0$.

(ii) *There exists a finely holomorphic function* g *on* U *with* $g(z_0) \neq 0$ *such that*

$$f(z) = (z - z_0)^n g(z), \; z \in U.$$

(iii) *There exists a finely open fine neighborhood* V_0 *of* z_0 *in* U *and an injective finely holomorphic function* h *on* V_0 *such that* $h'(z_0) \neq 0$ *and*

$$f(z) = h(z)^n, \; z \in V_0.$$

In the affirmative case there is a basis of finely open fine neighborhoods V *of* $z_0 \in V_0$ *from* (iii) *such that* f *is n-to-1 on* $V \setminus \{z_0\}$. [3]

Proof The existence of $n \in \mathbb{N}$ satisfying (i) is the content of the former assertion of the uniqueness theorem 4.5.23.

(i)⇒(ii). Define $g_0(z) = (z - z_0)^{-n} f(z)$ for $z \in U \setminus \{z_0\}$. Thus g_0 is finely holomorphic on $U \setminus \{z_0\}$ along with f and $z \mapsto (z - z_0)^{-n}$, $\mathcal{O}(U)$ being an algebra. According to (b) in Theorem 4.5.19 with $n = 0$ and m replaced by n, there is a finely open fine neighborhood U_1 of z_0 in U such that g_0 is bounded on $U_1 \setminus \{z_0\}$ and hence extends by fine continuity to a finely holomorphic function g_1 on U_1. Define $g = g_0$ on $U \setminus \{z_0\}$ and $g = g_1$ on U_1, noting that $g_0 = g_1$ on $(U \setminus \{z_0\}) \cap U_1 = U_1 \setminus \{z_0\}$. Then g is finely holomorphic on U and $f(z) = (z - z_0)^n g(z)$ on all of U by fine continuity.

(ii)⇒(iii). Since $g(z_0) \neq 0$ there is an open disk ω_0 in $\mathbb{C} \setminus \{0\}$ centered at $g(z_0)$ and a single-valued holomorphic branch of $w \mapsto w^{1/n}$ defined on ω_0. The function $h(z) := (z - z_0) g(z)^{1/n}$ is well defined on the finely open fine neighborhood $V_0 := g^{-1}(\omega_0)$ of z_0 in U, and $h'(z_0) = g(z_0)^{1/n} \neq 0$. Clearly $h^n = f$ on V_0.

(iii)⇒(i) follows by repeated fine differentiation of $f = h^n$ because $h(z_0) = 0$ and $h'(z_0) \neq 0$.

Now suppose that (i), (ii), or (iii) holds. According to Theorem 4.5.19(c) with $n = 1$ every fine neighborhood of z_0 contains a finely open fine neighborhood V_1 of z_0 in U for which there exists a C^∞-function $\widehat{h} : \mathbb{C} \to \mathbb{C}$ such that $\widehat{h} = h$ on V_1 and $\partial \widehat{h}(z_0) = h'(z_0)$ $(\neq 0)$. By the implicit function theorem it follows that $\widehat{h}$ is injective on some open disk ω_1 centered at z_0, and hence h is injective on the finely open fine neighborhood $V_2 := V_1 \cap \omega_1$ of z_0 in U, and in particular $h(z) \neq h(z_0)$ for $z \in V_2 \setminus \{z_0\}$. With $\varepsilon := \exp(2\pi i/n)$ define

$$V := V_2 \cap h^{-1}\Big(\bigcap_{k=1}^{n} \varepsilon^k h(V_2)\Big),$$

which is a finely open fine neighborhood of z_0 in V_2, noting that $h(V_2)$ is finely open according to Theorem 4.5.2 applied to h and Theorem 4.4.9. To see that f is n-to-1 on $V \setminus \{z_0\}$ let $z \in V \setminus \{z_0\}$. Then $h(z) = \varepsilon^k h(z_k)$ $(\neq 0)$ for every $k \in \{1, \ldots, n\}$ and a unique $z_k \in V_2 \setminus \{z_0\}$. Actually, $z_k \in V \setminus \{z_0\}$ with $z_k \in V_2$. Thus $f(z) = h(z)^n = \varepsilon^{nk} h(z_k)^n = f(z_k)$ since $\varepsilon^{nk} = 1$, showing that $V \cap f^{-1}(f(z))$ contains $z_1, \ldots, z_n$.

[3] The formulation of this last paragraph represents a correction to (iv) in [Fu27, Proposition 4.3].

For any point $\zeta \in V\backslash\{z_0\}$ with $f(\zeta) = f(z)$ we have $h(\zeta)^n = h(z)^n \neq 0$ and hence $h(\zeta) = \varepsilon^{-k} h(z) = h(z_k)$ for some $k \in \{1, \ldots, n\}$. By injectivity of h on $V\backslash\{z_0\}$ we conclude that $\zeta = z_k$, so $f|_{V\backslash\{z_0\}}$ is indeed n-to-1. □

It follows already from the last assertion of Theorem 4.5.2 that the zeros of a nonconstant finely holomorphic function on a fine domain U form a polar set. We can now show that $f^{-1}(0)$ is even countable.

Theorem 4.5.26 *The set $f^{-1}(0)$ of zeros of a finely holomorphic function $f \not\equiv 0$ on a fine domain $U \subset \mathbb{C}$ is countable.*

Proof The Greenian domain Ω can be covered by a sequence of open disks Ω_j of diameter < 1. Consider the (countably many) fine components of each of the finely open sets

$$\Omega_j \cap \{z \in U : |f(z)| < k\}, \ j, k \in \mathbb{N}.$$

By the uniqueness Theorem 4.5.23 we have $f \not\equiv 0$ on each of these fine components. We may therefore assume that U is contained in an open disk Ω of diameter < 1, and so the Euclidean closure $\overline{U}$ is contained in a larger open disk of diameter < 1, and furthermore that $|f| \leqslant 1$ on U (after multiplying f by a small constant > 0). Now consider any finite set $\{z_1, \ldots, z_n\}$ of distinct zeros of f in U, and let m_j denote the multiplicity of z_j, $j \in \{1, \ldots, n\}$, cf. Theorem 4.5.25. The function

$$F(z) := f(z)(z - z_1)^{-m_1} \ldots (z - z_n)^{-m_n} \tag{4.5.33}$$

is well defined on $U\backslash\{z_1, \ldots, z_n\}$ and extends by fine continuity to a finely holomorphic function, likewise denoted F, on all of U, as seen by repeated application of (ii) in Theorem 4.5.25.

For any $j \in \{1, \ldots, n\}$ the function

$$h_j(z) = -\log|z - z_j|, \ z \in \Omega,$$

is > 0 because Ω has diameter < 1. By Riesz decomposition,

$$h_j(z) = G_\Omega(z, z_j) + g_j(z), \ z \in \Omega, \tag{4.5.34}$$

where g_j is harmonic > 0 on Ω. By sweeping on $\complement U$ (sweeping and complement relative to Ω),

$$\widehat{R}_{h_j}^{\complement U} = \widehat{R}_{G_\Omega(\cdot, z_j)}^{\complement U} + g_j, \ z \in \Omega, \tag{4.5.35}$$

because $\widehat{R}_{g_j}^{\complement U} = g_j$ on Ω. In fact, the function $u := \widehat{R}_{g_j}^{\complement U} - g_j$ is well defined and finely superharmonic on Ω, and $u = 0$ on the base $\mathfrak{b}(\complement U) = \complement\mathfrak{r}(U)$, $\mathfrak{r}(U)$ denoting the regularized finely open set of U. It follows that u has the fine limit 0 at $\partial_f \mathfrak{r}(U) \subset \widetilde{U} \subset \overline{U} \subset \Omega$. By Theorem 2.1.10 applied to the bounded finely open set $\mathfrak{r}(U)$ we infer that $u \geqslant 0$ on $\mathfrak{r}(U)$, that is, $\widehat{R}_{g_j}^{\complement U} = g_j$ on $\mathfrak{r}(U)$, and hence indeed on $\mathfrak{r}(U) \cup \mathfrak{b}(\complement U) = \Omega$.

Subtracting (4.5.35) from (4.5.34) leads by Definition 2.5.10 to

$$h_j - \widehat{R}_{h_j}^{\complement U} = G_\Omega(\cdot, z_j) - \widehat{R}_{G_\Omega(\cdot,z_j)}^{\complement U} = G_U(\cdot, z_j). \qquad (4.5.36)$$

By (4.5.33), (4.5.36), and the inequality $|f| \leqslant 1$, it follows that

$$\begin{aligned} 0 \leqslant -\log|f| &= \sum_{j=1}^{n} m_j h_j - \log|F| \\ &= \sum_{j=1}^{n} m_j G_U(\cdot, z_j) + \sum_{j=1}^{n} m_j \widehat{R}_{h_j}^{\complement U} - \log|F|. \end{aligned}$$

But $p := \sum_{j=1}^{n} m_j G_U(\cdot, z_j)$ is a fine potential on U along with each $G_U(\cdot, z_j)$. The finely subharmonic minorant $\log|F| - \sum_{j=1}^{n} m_j \widehat{R}_{h_j}^{\complement U}$ of p therefore is $\leqslant 0$, that is,

$$\log|F| \leqslant \sum_{j=1}^{n} m_j \widehat{R}_{h_j}^{\complement U} \text{ on } U.$$

By the preceding equality we deduce the Jensen type inequality

$$\sum_{j=1}^{n} m_j G_U(z, z_j) \leqslant -\log|f(z)|, \ z \in U.$$

Because $G_U(z, z_j) > 0$ this inequality, when applied to a point $z \in U$ such that $f(z) \neq 0$, shows that $f^{-1}(0)$ indeed is countable, any uncountable family of strictly positive reals having the sum $+\infty$. □

For any real number $p \in [1, +\infty[$ we have the Banach space $L^p(U)$ (as before with respect to Lebesgue measure ℓ_2 on the finely open set $U \subset \mathbb{C}$ and the L^p-norm $\|\cdot\|_{L^p}$). Note that the set U is ℓ_2-measurable because U only differs by a polar and hence ℓ_2-null set from its regularized set $\mathfrak{r}(U)$ which is a F_σ-set. We define

$$\mathcal{O}^p(U) = \mathcal{O}(U) \cap L^p(U).$$

Recall furthermore that two finely continuous functions on U are identical if they are equal ℓ_2-a.e. For $p = +\infty$, $\mathcal{O}^p(U)$ is therefore the space of bounded finely holomorphic functions with the sup norm.

We proceed to show that the regularity properties in Theorem 4.5.19 hold uniformly in question as long as $f \in \mathcal{O}^p(U)$ has bounded L^p-norm for some $p \in]2, +\infty]$. For this we shall need the following two lemmas.

Lemma 4.5.27 *If $f \in \mathcal{O}^2(U)$ then $\int_U f\overline{\partial}\varphi = 0$ for every $\varphi \in \mathrm{BLD}(\mathbb{C})$ with compact support such that $\varphi = 0$ q.e. on $\complement U$.*

Proof Choose a Greenian domain $\Omega \subset \mathbb{C}$ containing the support of φ. Replacing the given finely open set $U \subset \mathbb{C}$ by $\Omega \cap U$ we arrange that $U \subset \Omega$ because $\varphi = 0$ off $\Omega \cap U$. Since f is finely holomorphic on U we may by Theorem 4.5.9(ii) cover U by finely open sets V with compact closure $\overline{V} \subset U$ such that $f|_V$ extends to a function $f_V \in \mathcal{O}(\Omega, V)$, that is, $f_V \in \widehat{\mathcal{D}}(\Omega)$ and $\overline{\partial} f_V = 0$ a.e. on V (see text in paragraphs containing (4.5.1) and (4.5.2)). For any $\varphi \in \widehat{\mathcal{D}}(\Omega, V)$, (that is, $\varphi \in \widehat{\mathcal{D}}(\Omega)$ and $\varphi = 0$ q.e. on $\complement V$), we have $\overline{\partial}\varphi = 0$ a.e. on $\complement V$ according to Corollary 4.2.22, and hence

$$\int_U f\overline{\partial}\varphi = \int_V f\overline{\partial}\varphi = \int_V f_V\overline{\partial}\varphi = 0$$

by Lemma 4.5.4, (i)⇒(iii). Since $f \in L^2(U)$ the equation $\int f\overline{\partial}\varphi = 0$ extends by linearity and continuity to any $\varphi \in \widehat{\mathcal{D}}(\Omega, U)$ in view of Theorem 4.2.20. That completes the proof because the restriction to Ω of the given function φ is of class $\widehat{\mathcal{D}}(\Omega)$ since $\varphi = 0$ q.e. on $\complement\Omega (\subset \complement U)$ in view of Theorem 4.2.9. □

Remark 4.5.28 The hypothesis that φ have compact support can be omitted, see [Fu18, Remarque 1 (p. 81)].

Lemma 4.5.29 *Let* $2 < p \leqslant +\infty$. *For any* $f \in \mathcal{O}^p(U)$ *and any* $\varphi \in \mathrm{BLD}(\mathbb{C})$ *with compact support such that* $\varphi = 0$ *q.e. on* $\complement U$ *we have*

$$\int_U \frac{f(z) - f(\zeta)}{z - \zeta}\overline{\partial}\varphi(\zeta)\, d\ell_2(\zeta) = 0 \text{ for } z \in U, \tag{4.5.37}$$

and hence

$$f(z)\varphi(z) = \frac{1}{\pi}\int_U \frac{1}{z - \zeta} f(\zeta)\overline{\partial}\varphi(\zeta)\, d\ell_2(\zeta) \text{ q.e. for } z \in U. \tag{4.5.38}$$

Proof For fixed $z \in U$ consider the function $g : U \to \mathbb{C}$ defined by

$$g(\zeta) = \frac{f(z) - f(\zeta)}{z - \zeta} \text{ for } U\setminus\{z\}$$

and by $g(z) = f'(z)$ (the fine derivative of f at z). This function is finely holomorphic according to Corollary 4.5.12, being bounded on a deleted neighborhood V of z in U, and furthermore $g \in L^2(U)$ because

$$\int_{U\setminus V} |g(\zeta)|^2\, d\ell_2(\zeta)$$
$$\leqslant \Big(\int_{U\setminus V} |z - \zeta|^{-\frac{2p}{p-2}}\, d\ell_2(\zeta)\Big)^{1-\frac{2}{p}} \Big(\int_U |f(z) - f(\zeta)|^p\, d\ell_2(\zeta)\Big)^{\frac{2}{p}} < +\infty$$

by Hölder inequality and Lemma 4.5.8 since we may take U bounded. The former relation (4.5.37) now holds by Lemma 4.5.27 applied with f replaced by g.

For (4.5.38) recall from Lemma 4.5.6 the Cauchy identity

$$\varphi(z) = \frac{1}{\pi}\int_{\mathbb{C}} \frac{1}{z-\zeta}\overline{\partial}\varphi(\zeta)\, d\ell_2(\zeta) \text{ q.e. on } \Omega \tag{4.5.39}$$

with an absolutely convergent integral, cf. Lemma 4.5.6. As at the end of the proof of the preceding lemma, the present φ does belong to $\widehat{\mathcal{D}}(\Omega)$ (with Ω a Greenian domain in $\mathbb{C}$). After multiplying (4.5.39) with $f(z)$ (which is finite for quasi-every $z \in U$) and subtracting (4.5.37) we obtain (4.5.38). □

Theorem 4.5.30 *Let $2 < p \leqslant +\infty$. Then $\mathcal{O}^p(U)$ is closed in $L^p(U)$ and hence a Banach space. For any given point $z_0 \in U$ the compact fine neighborhood V of z_0 mentioned in Theorem 4.5.19 can be so chosen that the assertions* (a), (b), (c) *of that theorem hold uniformly in f on any norm bounded subset of $\mathcal{O}^p(U)$. In particular, this uniform version of* (b) *states that*

$$\left| f^{(n)}(w) - \sum_{k=0}^{m-1} \frac{1}{k!}(w-z)^k f^{(n+k)}(z) \right| \leqslant a_{m,n} \|f\|_{L^p(U)} |w-z|^m \tag{4.5.40}$$

for any $f \in \mathcal{O}^p(U)$, any $(z, w) \in V \times V$, and any $m, n = 0, 1, \ldots,$ with constant $a_{m,n}$ independent of f, z, and w.

Proof We may suppose that U is bounded. There exists a function $\varphi \in \mathrm{BLD}(\mathbb{C})$ such that $\varphi = 1$ on a fine neighborhood U_0 of z_0 contained in U and that $\varphi = 0$ q.e. on $\complement U$. For this, choose an open disk Ω in $\mathbb{C}$ so that $U \subset \Omega$, and a strict potential $G_\Omega\mu$ of a measure μ on Ω of finite G_Ω-energy (Definition 1.4.14 and Theorem 1.4.15), for example $\mu =$ Lebesgue measure on Ω. Then take $\varphi = \inf\{1, a\psi\}$ with

$$\psi = G_\Omega\mu - \widehat{R}^{\complement U}_{G_\Omega\mu} \in \widehat{\mathcal{D}}_+(\Omega, U)$$

(sweeping relative to Ω), and with the constant a satisfying $a\psi(z_0) > 1$. With this choice of φ it follows for any $f \in \mathcal{O}^p(U)$ and quasi-every $z \in U_0$ by Lemma 4.5.29 that

$$f(z) = f(z)\varphi(z) = \frac{1}{\pi}\int_U \frac{1}{z-\zeta} f(\zeta)\overline{\partial}\varphi(\zeta)\, d\ell_2(\zeta),$$

the integral being absolutely convergent. This integral is the Cauchy transform of the function $f\overline{\partial}\varphi$ of class $L^{2p/(p+2)}(U)$ because $f \in L^p(U)$ and $\varphi \in \mathrm{BL}(\mathbb{C})$, hence $\overline{\partial}\varphi \in L^2(\mathbb{C})$. The function $f\overline{\partial}\varphi$ equals 0 a.e. on $\complement U$ according to Corollary 4.2.22. Since $p > 2$ we have $\frac{2p}{p+2} > 1$, and it follows by Theorem 4.5.9(iii) that f is q.e. equal on U_0 to a finely holomorphic function which therefore equals f everywhere on U_0. As in the proof of Theorem 4.5.14 we choose a compact fine neighborhood V of z_0 in U as in Lemma 4.5.8, that is, so that (4.5.17) holds, but now also so that $V \subset U_0$. According to Lemma 4.5.29,

$$f(z) = \frac{1}{\pi} \int_U \frac{1}{z - \zeta} f(\zeta)\overline{\partial}\varphi(\zeta)\, d\ell_2(\zeta) \text{ q.e. on } V$$

for every $f \in \mathcal{O}^p(U)$. Since $f \in L^p(U)$ with $p > 2$ and $\overline{\partial}\varphi \in L^2(\mathbb{C})$ we have by Hölder's inequality $f\overline{\partial}\varphi \in L^q$ with $q = \frac{2p}{p+2} > 1$. In the beginning of the proof of Theorem 4.5.19 we may therefore replace φ by the present $\frac{1}{\pi} f\overline{\partial}\varphi$ and U by U_0. Hence $\|\varphi\|_{L^p}$ shall be replaced by $\frac{1}{\pi}\|f\|_{L^p}\|\overline{\partial}\varphi\|_{L^2}$ in the rest of the proof of Theorem 4.5.19. Since $\varphi \in \mathrm{BLD}(\mathbb{C})$ (see beginning of proof of Theorem 4.5.30), we have $\overline{\partial}\varphi \in L^2(\mathbb{C})$, and so $\|\overline{\partial}\varphi\|_{L^2(\mathbb{C})}$ is indeed a finite constant. As in (b) in Theorem 4.5.19 the stated quotient is now majorized in absolute value by $\frac{1}{\pi} A_{m,n}\|f\|_{L^p}\|\overline{\partial}\varphi\|_{L^2}$, cf. (4.5.28), which means that (4.5.40) in Theorem 4.5.30 holds with $a_{m,n} = \frac{1}{\pi} A_{m,n}\|\overline{\partial}\varphi\|_{L^2}$. Taking $m, n = 0$ in (4.5.40) we obtain

$$\sup_{w \in V} |f(w)| \leqslant a_{0,0}\|f\|_{L^p}$$

for any $f \in \mathcal{O}^p(U)$. Every sequence $(f_j) \subset \mathcal{O}^p(U)$ which converges in $L^p(U)$ with limit $f \in L^p(U)$ therefore is a Cauchy sequence on $\mathcal{O}^p(U)$ with the seminorm $f \mapsto \sup_{w \in V} |f(w)|$. On the fine interior V' of V the sequence (f_j) hence converges uniformly to a finely holomorphic function $f_{V'}$ on V' according to Corollary 4.5.11. By varying the given point z_0 we thus obtain a cover of U by finely open sets V' and corresponding finely holomorphic functions $f_{V'}$ on V' such that $f_{V'} = f$ a.e. on V'. For two such open sets V_1', V_2' we therefore have $f_{V_1'} = f_{V_2'}$ a.e. and hence everywhere on $V_1' \cap V_2'$ by fine continuity. By the sheaf property of fine holomorphy there is a finely holomorphic function on U extending all the $f_{V'}$. By the quasi-Lindelöf principle this function coincides with f q.e. on U and hence everywhere on U, again by fine continuity. □

Corollary 4.5.31 *For any $z \in U$ and $n = 0, 1, 2, \ldots$ the linear form $D_z^{(n)}$ on $\mathcal{O}^p(U)$ $(2 < p \leqslant +\infty)$ defined by $D_z^{(n)} f = f^{(n)}(z)$ is continuous. Its norm $\|D_z^{(n)}\|$ is a finely continuous function on U.*

Proof For the former assertion take $m = 0$, $w = z_0$ (and z arbitrary) in (4.5.40). For the latter assertion take $m = 1$, $z, w \in V$:

$$\begin{aligned} &\big|\|D_w^{(n)}\| - \|D_z^{(n)}\|\big| \\ &\quad \leqslant \|D_w^{(n)} - D_z^{(n)}\| = \sup\{|f^{(n)}(w) - f^{(n)}(z)| : \|f\|_{L^p} \leqslant 1\} \leqslant a_{1,n}|w - z|. \end{aligned}$$

□

Remark 4.5.32 The continuous linear form $D_z^{(n)}$ on $\mathcal{O}^p(U)$ may be termed the n'th order (fine) *point derivation on* $\mathcal{O}^p(U)$. In the case U is the fine interior K' of a compact set $K \subset \mathbb{C}$ and where $p = +\infty$, the former assertion of Corollary 4.5.31 extends an analogous result by Debiard and Gaveau [DG1] on subalgebra $A_f(K)$ of $R(K)$ consisting of all continuous functions on K with fine interior K'. The fact that when $D_z^{(1)}$ acts on $A_f(K)$, its norm is finely locally bounded is due to Lyons [Ly2].

Remark 4.5.33 For a usual holomorphic function f on a disk $|z| < r$ in $\mathbb{C}$ we have $f^{(p)}(0) = O(p!r^{-p})$ because the radius of convergence of the Taylor series is > 0. For finely holomorphic functions f on a fine domain $U \subset \mathbb{C}$ there is no such limit of growth of the fine derivative $f^{(p)}$ at a given point of U, as we see from the following example continuing Example 4.5.22.

Example 4.5.34 For any prescribed increasing sequence of numbers $c_p > 1$ there is a finely holomorphic function f on a fine domain $U \subset \mathbb{C}$ containing 0 such that $|f^{(p)}(0)|/c_p \to +\infty$ as $p \to \infty$. As in Example 4.5.22 take

$$f(z) = \sum_{n\in\mathbb{N}} \frac{a_n}{z - z_n}$$

now with $z_n = 1/c_{(n+1)^2}, r_n = (z_n)^{n^2}, a_n = \frac{r_n}{n^2}$. The union A of the disks $\{|z - z_n| \leqslant r_n\}$ is thin at 0 by the Wiener criterion because $\sum_{n\geqslant 2} \frac{\log z_n}{\log r_n} = \sum_{n\geqslant 2} \frac{1}{n^2} < +\infty$. Hence $\complement A$ is a finely open set containing 0 because $r_n < |z_n|$ and the fine component U of 0 in $\complement A$ is a fine domain. The given series converges uniformly on U because $a_n/|z - z_n| \leqslant a_n/r_n = 1/n^2$ for $z \in U$. For each p we have

$$f^{(p)}(0) = -(p!) \sum_{n\geqslant 2} a_n/(z_n)^{p+1}$$

according to Corollary 4.5.31 applied to the p'th order continuous point derivative $D_0^{(p)}$ on the Banach space $\mathcal{O}^\infty(U)$ of bounded holomorphic functions on U with the L^∞-norm (sup-norm, by fine continuity). Hence

$$|f^{(p)}(0)| \geqslant p!a_n(z_n)^{-p-1} = p!n^{-2}(z_n)^{n^2-p-1}$$

for any $n \geqslant 2$. For $p \geqslant 4$ we may take $n = [\sqrt{p}]$, that is, $(n+1)^2 > p \geqslant n^2$, and hence indeed

$$|f^{(p)}(0)| \geqslant (p-1)!(z_n)^{n^2-p-1} \geqslant (p-1)!z_n^{-1} = (p-1)!c_{(n+1)^2} \geqslant (p-1)!c_p.$$

We close this section by a brief discussion on fine domain of existence in $\mathbb{C}$, roughly speaking, fine domain in $\mathbb{C}$ that admits a finely holomorphic function that cannot be extended as a finely holomorphic function at any fine boundary point.

A well-known result by Weierstrass asserts that every domain $D \subset \mathbb{C}$ is a domain of existence, roughly speaking, it admits a holomorphic function f that cannot be extended analytically at any boundary point of D. In his thesis [Bor1] Borel showed, however, that it may be that f can be (uniquely) extended to a strictly larger set X in an "analytic" way, albeit that X is no longer Euclidean open. This eventually led Borel to the introduction of his Cauchy domains and monogenic functions, cf. [Bor2]. Finely holomorphic functions on fine domains in $\mathbb{C}$ are the natural extension and setting for Borel's ideas, see [Fu19], also for some historic remarks on Borel's work.

For fine domains the results are different from both the classical Weierstrass case of one variable and the classical several variables case, where Hartogs showed that there exist domains $D \subsetneq D^*$ in $\mathbb{C}^n$ with the property that every holomorphic function on D extends to the larger domain D^*. Indeed, it was shown by Fuglede, Groot, and Wiegerinck that every fine domain that is (1) a Euclidean F_σ as well as a Euclidean G_δ, or (2) fine regular, is a fine domain of existence, see [FuGW, Theorem 3.9]. Euclidean domains are of the form (1) and therefore are fine domains of existence. Noting that on Euclidean domains holomorphic and finely holomorphic functions are the same, the above result includes Weierstrass' theorem. However, they show that fine domains V with the property that their complement contains a non-empty polar set E that is of the first Baire category in its Euclidean closure K (in particular, E has no Euclidean isolated points) and $K \backslash E \subset V$, are not fine domains of existence.

Notes: Except Lemma 4.3.3, which is due to Feyel and Lapradelle [FL1], all the material and results of this chapter are essentially due to Fuglede [Fu13, Fu17, Fu18, Fu20, Fu29, Fu30].

Chapter 5
Fine Complex Potential Theory

This chapter is devoted to the theory of $\mathcal{F}$-plurisubharmonic and $\mathcal{F}$-holomorphic functions, that is, the plurisubharmonic and holomorphic functions relative to the coarsest topology $\mathcal{F}$ on $\mathbb{C}^n$ making continuous all plurisubharmonic functions.

In Sect. 5.1, we recall some background on plurisubharmonic functions to be used along the chapter. In Sect. 5.2 we prove that the topology $\mathcal{F}$ is locally connected and has the (complex) quasi-Lindelöf property. As applications of this topology we prove in Sect. 5.3 a Brelot property for plurisubharmonic functions on an open subset of $\mathbb{C}^n$ (Theorem 5.3.1), and establish some weak convergence results in $\mathcal{F}$-topology (Corollary 5.3.4) for the Monge-Ampère operator, and an equality on Monge-Ampère measures on $\mathcal{F}$-open sets (Corollary 5.3.8). Applied to the "slicing", this allows to give an expression of the slice of an $(n-1, n-1)$-current T with respect to a hyperplane $z_1 = a$ (current on $z_1 = a$), see Proposition 5.3.13.

In Sects. 5.4 and 5.5 we define and study two concepts of $\mathcal{F}$-plurisubharmonic (resp. $\mathcal{F}$-holomorphic) functions, and we prove a fundamental convergence theorem for $\mathcal{F}$-plurisubharmonic functions (Theorem 5.4.13) and removable singularity theorems for weakly $\mathcal{F}$-plurisubharmonic functions (Theorem 5.4.13), and for $\mathcal{F}$-holomorphic functions (Corollary 5.5.5).

In Sect. 5.6 we show that the composition with weakly $\mathcal{F}$-holomorphic maps preserves $\mathcal{F}$-plurisubharmonicity and $\mathcal{F}$-holomorphy (Theorem 5.6.4).

In Sect. 5.7 we use, as in Sect. 4.2, duality with respect to pairing of Hilbert spaces and prove several density results (Theorem 5.7.10 and Corollary 5.7.11) for $\mathcal{F}$-plurisubharmonic functions. As a consequence, we prove a local approximation result for $\mathcal{F}$-plurisubharmonic functions (Theorem 5.7.17).

In Sect. 5.8 we define the Monge-Ampère measure $(dd^c u)^n$ associated with a finite $\mathcal{F}$-plurisubharmonic function u on an $\mathcal{F}$-open set, and we prove that the Monge-Ampère operator $u \mapsto (dd^c u)^n$ is continuous on decreasing sequences of finite $\mathcal{F}$-plurisubharmonic functions.

M. El Kadiri and B. Fuglede, *Classical Fine Potential Theory*, Springer Monographs in Mathematics, https://doi.org/10.1007/978-981-96-0430-2_5

In Sect. 5.9 we define the notions of maximal and $\mathcal{F}$-locally maximal $\mathcal{F}$-plurisubharmonic functions on an $\mathcal{F}$-open set and study their properties, extending the classical ones on a Euclidean open subset of $\mathbb{C}^n$. In Sect. 5.10 we prove that the Monge-Ampère measure associated with a finite maximal plurifinely plurisubharmonic function is zero, and we discuss the converse of this result.

5.1 Background in Plurisubharmonic Functions Theory

Plurisubharmonic functions on open subsets of $\mathbb{C}^n$ were introduced in 1942 by Lelong as "les objets souples de l'analyse complexe" and independently by Oka at the same time. The present section is devoted to background material on the theory of plurisubharmonic functions and the Monge-Ampère operator on open sets of $\mathbb{C}^n$. The reader is referred to the excellent book of Klimek [Kl] on the subject which we often follow in the beginning.

5.1.1 *Complex Differential Forms*

Let V and W be vector spaces over $\mathbb{R}$ and k an integer ≥ 1. A mapping $f : \underbrace{V \times \ldots \times V}_{k-times} \to W$ is said to be *k-linear* if it is $\mathbb{R}$-linear in each variable when the other variables are fixed. It is called *alternating* if moreover $f(v_1, \ldots, v_k) = 0$ whenever $v_j = v_{j+1}$ for some j, $1 \leq j < k$. The set of such mappings is denoted by $\Lambda^k(V, W)$ and naturally endowed with the structure of a real vector space. The space $\Lambda^1(V, W)$ is often denoted by $\mathcal{L}(V, W)$. It is convenient to define $\Lambda^0(V, W) = W$. If $f_1, \ldots, f_k \in \mathcal{L}(V, \mathbb{C})$, we define the *wedge product* or the *exterior product* $f_1 \wedge \ldots \wedge f_k$ by the formula

$$f_1 \wedge \ldots \wedge f_k(v_1, \ldots, v_k) = \det(f_i(v_j))_{i,j \leq k} \tag{5.1.1}$$

for all $v_1, \ldots, v_k \in V$. Of course $f_1 \wedge \ldots \wedge f_k \in \Lambda^k(V, \mathbb{C})$. Moreover, if $\{f_1, \ldots, f_m\} \subset \mathcal{L}(V, \mathbb{C})$ is a basis (over $\mathbb{C}$) of $\mathcal{L}(V, \mathbb{C})$, then

$$\{f_{j_1} \wedge \ldots \wedge f_{j_k} : 1 \leq j_1 < \ldots < j_k \leq m\} \tag{5.1.2}$$

is a basis (over $\mathbb{C}$) of $\Lambda^k(V, \mathbb{C})$. If $p, q \in \mathbb{N}$ are such that $p + q \leq m$, the *exterior multiplication* is the unique bilinear mapping

$$\wedge : \Lambda^p(V, \mathbb{C}) \times \Lambda^q(V, \mathbb{C}) \to \Lambda^{p+q}(V, \mathbb{C})$$

that conforms with (5.1.1) (with $k = p + q$) on the basic vectors (5.1.2) in $\Lambda^p(V, \mathbb{C})$ and $\Lambda^q(V, \mathbb{C})$ (with $k = p$ and $k = q$, respectively).

Suppose now $V = \mathbb{R}^m$ and Ω is an open subset of $\mathbb{R}^m$. A *differential form of degree* $k \geq 0$, or a *k-form* on Ω, is a mapping

$$\omega : \Omega \to \Lambda^k(\mathbb{R}^m, \mathbb{C}).$$

For $j = 1, \ldots, m$, we denote by dx_j the 1-form defined on $\mathbb{R}^m$ by $dx_j(y) = y_j$ for $y = (y_1, \ldots, y_m) \in \mathbb{R}^m$.

A multi-index is an element $\alpha = (\alpha_1, \ldots, \alpha_k)$ of the set $\mathbb{N}^k$, and we put $|\alpha| = k$. We say that this multi-index is *increasing* if $1 \leq \alpha_1 < \ldots < \alpha_k$.

If ω is a differential form of degree k on Ω, it can be written as

$$\omega = \sum_{|\alpha|=k}{}' \omega_\alpha dx_\alpha,$$

where

$$dx_\alpha = dx_{\alpha_1} \wedge \ldots \wedge dx_{\alpha_k},$$

ω_α are complex functions on Ω, and the "prime" indicates that the sum is taken over increasing multi-indices $\alpha = (\alpha_1, \ldots, \alpha_k)$ where $\alpha_k \leq m$. When the coefficients ω_α belong to a particular class $\mathcal{G}(\Omega, \mathbb{C})$ of complex functions (e.g. $\mathcal{C}(\Omega, \mathbb{C})$, $\mathcal{C}^\infty(\Omega, \mathbb{C})$, etc.), we write $\omega \in \mathcal{G}(\Omega, \Lambda^k(\mathbb{R}^m, \mathbb{C}))$.

The *exterior multiplication* of differential forms ω and η on Ω of degree k and l respectively is the form $\omega \wedge \eta$ of degree $k + l$ defined on Ω by

$$(\omega \wedge \eta)(x) = \omega(x) \wedge \eta(x).$$

If $f : \Omega \to \mathbb{C}$ is a differentiable function on Ω, the differential of f is the differential form of degree 1 on Ω given by

$$df = \sum_{j=1}^{m} \frac{\partial f}{\partial x_j} dx_j.$$

This can be extended to differential forms on Ω with differentiable coefficients as follows:

$$d\omega = \sum_{|\alpha|=k}{}' d\omega_\alpha \wedge dx_\alpha$$

whenever $\omega = \sum'_{|\alpha|=k} \omega_\alpha \wedge dx_\alpha$, $k \geq 1$. Clearly, $d\omega$ is of degree $k + 1$.

The operator d is called operator of *exterior differentiation* (to be distinguished from the differential d). In particular $d^2 = dd = 0$ (e.g. Cartan [Ca4]).

In $\mathbb{C}^n$ we will use the following standard notations:

Writing $z = (z_1, \ldots\ldots, z_n) \in \mathbb{C}^n, z_j = x_j + iy_j \in \mathbb{C}, x_j, y_j \in \mathbb{R}$, we put $dz_j = dx_j + idy_j, d\bar{z}_j = dx_j - idy_j$, and

$$\frac{\partial}{\partial z_j} = \frac{1}{2}\left(\frac{\partial}{\partial x_j} - i\frac{\partial}{\partial y_j}\right),$$

$$\frac{\partial}{\partial \bar{z}_j} = \frac{1}{2}\left(\frac{\partial}{\partial x_j} + i\frac{\partial}{\partial y_j}\right).$$

For an open subset Ω of $\mathbb{C}^n$ and a differentiable function $f : \Omega \to \mathbb{C}$, we have

$$df = \sum_{j=1}^{n}\left(\frac{\partial f}{\partial x_j}dx_j + \frac{\partial f}{\partial y_j}dy_j\right),$$

$$\partial f = \sum_{j=1}^{n}\frac{\partial f}{\partial z_j}dz_j, \quad \bar{\partial} f = \sum_{j=1}^{n}\frac{\partial f}{\partial \bar{z}_j}d\bar{z}_j.$$

If $\alpha = (\alpha_1, \ldots, \alpha_k) \in \mathbb{N}^k, k \leq n$, we put

$$dz_\alpha = dz_{\alpha_1} \wedge \ldots \wedge dz_{\alpha_k}, \quad d\bar{z}_\alpha = d\bar{z}_{\alpha_1} \wedge \ldots \wedge d\bar{z}_{\alpha_k}.$$

If $p, q \in \mathbb{Z}_+$ are such that $p + q = r \in \mathbb{N}$, then we denote by $\Lambda^{p,q}(\mathbb{C}^n, \mathbb{C})$ the (complex) subspace of $\Lambda^r(\mathbb{C}^n)$ generated by the forms $dz_\alpha \wedge d\bar{z}_\beta$, where α and β are increasing multi-indices with $|\alpha| = p$ and $|\beta| = q$ (the subspace of $\Lambda^r(\mathbb{C}^n)$ generated by the forms dz_α ($|\alpha| = r$), resp. $d\bar{z}_\beta$ ($|\beta| = r$), is denoted by $\Lambda^{r,0}(\mathbb{C}^n, \mathbb{C})$, resp. $\Lambda^{0,r}(\mathbb{C}^n, \mathbb{C})$) and define $\wedge^{0,0}(\mathbb{C}^n, \mathbb{C}) = \mathbb{C}$. Then one has the direct sum decomposition

$$\Lambda^r(\mathbb{C}^n, \mathbb{C}) = \bigoplus_{p+q=r} \Lambda^{p,q}(\mathbb{C}^n, \mathbb{C}).$$

Let ω be a differential form on an open set $\Omega \subset \mathbb{C}^n$. If ω can be written as

$$\omega = \sum_{|\alpha|=p}\sum_{|\beta|=q} \omega_{\alpha,\beta}\, dz_\alpha \wedge d\bar{z}_\beta,$$

then it is called a *differential form of type* (p, q) or a (p, q)*-form*. The $(p, 0)$-forms, resp $(0, q)$-forms are the forms $\sum_{|\alpha|=p} \omega_\alpha dz_\alpha$, resp. $\sum_{|\beta|=q} \omega_\beta d\bar{z}_\beta$.

We extend the differential operators d, ∂ and $\bar{\partial}$ on differentiable functions on Ω to differential forms (with differentiable coefficients) as follows:

$$d\omega = \sum_{\alpha,\beta} d\omega_{\alpha,\beta} \wedge dz_\alpha \wedge d\bar{z}_\beta,$$

$$\partial\omega = \sum_{\alpha,\beta} \partial\omega_{\alpha,\beta} \wedge dz_\alpha \wedge d\bar{z}_\beta,$$

$$\bar{\partial}\omega = \sum_{\alpha,\beta} \bar{\partial}\omega_{\alpha,\beta} \wedge dz_\alpha \wedge d\bar{z}_\beta.$$

The forms $\partial\omega$ and $\bar{\partial}\omega$ are of type $(p+1, q)$ and $(p, q+1)$ respectively. Furthermore, $d = \partial + \bar{\partial}$ and hence

$$0 = d^2\omega = \partial^2\omega + (\partial\bar{\partial} + \bar{\partial}\partial)\omega + \bar{\partial}^2\omega.$$

Since all three terms are of different types, we conclude that

$$\partial^2 = \partial\partial = 0,$$

$$\bar{\partial}^2 = \bar{\partial}\bar{\partial} = 0,$$

$$\partial\bar{\partial} = -\bar{\partial}\partial.$$

Another important differential operator that will be used in this section is the operator d^c defined by

$$d^c = i(\bar{\partial} - \partial).$$

Note that

$$dd^c = 2i\partial\bar{\partial} \tag{5.1.3}$$

and that, if $u \in C^2(\Omega)$, then

$$dd^c u = 2i \sum_{j,k=1}^{n} \frac{\partial^2 u}{\partial z_j \partial \bar{z}_k} dz_j \wedge d\bar{z}_k.$$

The *complex Monge-Ampère operator* in $\mathbb{C}^n$ is defined as the n-th exterior power of dd^c, i.e.

$$(dd^c)^n = \underbrace{dd^c \wedge \ldots \wedge dd^c}_{n-times}.$$

We close this subsection by a description of the notion of positivity in $\Lambda^r(\mathbb{C}^n, \mathbb{C})$.

A form $\omega \in \Lambda^r(\mathbb{C}^n, \mathbb{C})$ belongs to $\Lambda^r(\mathbb{C}^n, \mathbb{R})$ if and only if $\omega = \bar{\omega}$, the natural conjugate of ω. We then say that ω is real. The *standard Kähler form* in $\mathbb{C}^n$ is defined as

$$\beta = \beta_n = \frac{i}{2} \sum_{j=1}^{n} dz_j \wedge d\bar{z}_j. \tag{5.1.4}$$

This form is invariant with respect to unitary isomorphisms of $\mathbb{C}^n$.

The *standard volume form* in $\mathbb{C}^n$ will be denoted by $dV = dV_n$. We have

$$dV_n = (n!)^{-1}(\beta_n)^n = (n!)^{-1} \underbrace{\beta_n \wedge \ldots \wedge \beta_n}_{n-times}. \tag{5.1.5}$$

A $2n$-form ω (i.e. an element of $\Lambda^{2n}(\mathbb{C}^n, \mathbb{C})$) is called *positive* if $\omega = \tau dV$ for some positive number τ.

To define positive forms of other degrees, we need to introduce the notion of strongly positive forms.

A form $\omega \in \Lambda^{p,p}(\mathbb{C}^n, \mathbb{R})$ is said to be *elementary strongly positive* if there are linearly independent $\mathbb{C}$-linear mappings $\eta_j : \mathbb{C}^n \to \mathbb{C}$, $j = 1, \ldots, p$, such that

$$\omega = \frac{i}{2}\eta_1 \wedge \bar{\eta}_1 \wedge \ldots \wedge \frac{i}{2}\eta_p \wedge \bar{\eta}_p. \tag{5.1.6}$$

A form $\omega \in \Lambda^{p,p}(\mathbb{C}^n, \mathbb{R})$ is called *strongly positive* if it belongs to the convex cone in $\Lambda^{p,p}(\mathbb{C}^n, \mathbb{R})$ generated by the elementary strongly positive forms; that is, $\omega = \sum_j \lambda_j \omega_j$ for some positive reals $\lambda_1, \ldots, \lambda_m$ and some elementary strongly positive forms $\omega_1, \ldots, \omega_m$ $(m \geq 1)$.

A form $\omega \in \Lambda^{p,p}(\mathbb{C}^n, \mathbb{C})$ is said to be *positive* if for any elementary strongly positive $\eta \in \Lambda^{n-p,n-p}(\mathbb{C}^n, \mathbb{R})$ (and hence for any strongly positive form $\eta \in \Lambda^{n-p,n-p}(\mathbb{C}^n, \mathbb{R})$), the form $\omega \wedge \eta$ is positive. The family of all positive (p, p)-forms on $\mathbb{C}^n$ is denoted by $\Lambda_+^{p,p}(\mathbb{C}^n)$. Note that $\Lambda_+^{p,p}(\mathbb{C}^n)$ is a convex cone in $\Lambda^{p,p}(\mathbb{C}^n, \mathbb{R})$.

5.1.2 Currents

Let Ω be an open set in $\mathbb{C}^n$. The family of all differential forms of bidegree (p, q) whose coefficients belong to $\mathcal{C}_0(\Omega, \mathbb{C})$, respectively $\mathcal{C}_0^\infty(\Omega, \mathbb{C})$, will be denoted by $\mathcal{D}_0^{p,q}(\Omega)$, respectively $\mathcal{D}^{p,q}(\Omega)$. That is,

$$\mathcal{D}_0^{p,q}(\Omega) = \mathcal{C}_0(\Omega, \Lambda^{p,q}(\mathbb{C}^n, \mathbb{C})), \quad \mathcal{D}^{p,q}(\Omega) = \mathcal{C}_0^\infty(\Omega, \Lambda^{p,q}(\mathbb{C}^n, \mathbb{C})).$$

The elements of $\mathcal{D}^{p,q}(\Omega)$ are called the *test forms.*

Note that if $p = q = 0$ then $\mathcal{D}_0^{p,q}(\Omega) = \mathcal{C}_0(\Omega, \mathbb{C})$ and $\mathcal{D}^{p,q}(\Omega) = \mathcal{C}_0^\infty(\Omega) = \mathcal{D}(\Omega)$, the space of test functions as studied in the theory of distributions.

We equip the space $\mathcal{D}^{p,q}(\Omega)$ with the topology for which a sequence $(\varphi^j)_{j\geq 0}$ of elements of $\mathcal{D}^{p,q}(\Omega)$, say

$$\varphi^j = \sum_{|I|=p,|J|=q}{}' \varphi_{I,J}^j \, dz_I \wedge d\bar{z}_J,$$

converges to $\varphi = \sum'_{|I|=p,|J|=q} \varphi \, dz_I \wedge d\bar{z}_J$ if and only if the following conditions are satisfied:

(1) there is a compact set $K \subset \Omega$ such $\mathrm{Supp}(\varphi^j_{I,J}) \subset K$ for all j, I, J;
(2) $D^\alpha \varphi^j_{I,J} \to D^\alpha \varphi_{I,J}$ uniformly as $j \to \infty$ for all I, J and $\alpha \in (\mathbb{Z}_+)^{2n}$.

The elements of the dual space $(\mathcal{D}^{n-p,n-q}(\Omega))'$ are called *currents of bidegree* (p, q) or (p, q)*-current*. The elements of the dual space $(\mathcal{D}_0^{n-p,n-q}(\Omega))'$ are called *currents of order* 0 *and bidegree* (p, q). Clearly, the currents of bidegree (n, n) are just the distributions on Ω, whereas the currents of bidegree $(0, 0)$ and order 0 are the Radon measures on Ω.

Let ψ be a (p, q)-form whose coefficients are locally integrable in Ω. One can associate with ψ the current T_ψ defined by

$$T_\psi(\varphi) = \int_\Omega \psi \wedge \varphi \tag{5.1.7}$$

for any $(n-p, n-q)$-test form φ on Ω.

Let T be a (p, q)-current in Ω. Let $I = (i_1, \dots, i_p)$ and $J = (j_1, \dots, j_q)$ with $p, q \le n$ be two increasing multi-indices and denote $K = (k_1, \dots, k_{n-p})$ and $L = (l_1, \dots, l_{n-q})$, respectively, the increasing complements of I and J, respectively, in $\{1, \dots, n\}$. Define

$$T_{I,J}(\varphi) = \varepsilon_{I,J} T(\varphi dz_K \wedge d\bar{z}_L),$$

where $\varphi \in C_0^\infty(\Omega, \mathbb{C})$ and $\varepsilon_{I,J}$ is chosen so that

$$\varepsilon_{I,J} dz_I \wedge d\bar{z}_J \wedge dz_K \wedge d\bar{z}_L = dV_n$$

(the volume form). Then $T_{I,J}$ is a distribution, and the current T can be regarded as a differential form with distributional coefficients $T_{I,J}$. Therefore we can write

$$T = \sum_{I,J}{}' T_{I,J} dz_I \wedge d\bar{z}_J,$$

where the summation is taken over increasing multi-indices only, of length p and q, respectively. Therefore,

$$T(\varphi) = \sum_{I,J}{}' (\varepsilon_{I,J})^{-1} T_{I,J}(\varphi_{K,L}),$$

where

$$\varphi = \sum_{K,L}{}' \varphi_{K,L} \, dz_K \wedge d\bar{z}_L$$

is a test form. If T is a current of order 0, the distributions $T_{I,J}$ are Radon measures. In this case it is convenient to write T as if it were the current T_ψ induced by a differential form ψ. Namely we write

$$\int_\Omega T \wedge \varphi$$

instead of $T(\varphi)$. If $T = T_\psi$ with

$$\psi := \sum_{I,J}{}' \psi_{I,J}\, dz_I \wedge d\bar{z}_J,$$

then

$$T_{I,J}(\varphi) = \int_\Omega \psi_{I,J}\varphi\, d\ell_{2n}$$

(ℓ_{2n} is the Lebesgue measure on $\mathbb{C}^n \cong \mathbb{R}^{2n}$). Observe that if T is a (p, q)-current on Ω, if ψ is a (k, l)-form on Ω with coefficients in $\mathcal{C}^\infty(\Omega, \mathbb{C})$, and if $\max\{p + k, q + l\} \le n$, then the formula

$$(T \wedge \psi)(\varphi) = T(\psi \wedge \varphi) \quad (\varphi \in \mathcal{D}^{n-p-k,\, n-q-l}(\Omega))$$

defines a $(p + k, q + l)$-current.

We furnish the space of all currents (respectively, currents of order 0) of bidegree (p, q) with the weak*-topology. A sequence of (p, q)-currents is convergent in this topology if it converges pointwise on $\mathcal{D}^{n-p,\,n-q}(\Omega)$ (respectively, $\mathcal{D}_0^{n-p,n-q}(\Omega)$).

A current T is said to be *positive of degree* p if it is a (p, p)-current such that for any $\omega \in \mathcal{D}_0^{n-p,\,n-p}(\Omega)$ we have $T(\omega) \ge 0$ whenever ω is a strongly positive form. In other words, T is positive if $T \wedge \omega$ is a positive distribution (hence a measure, by [Kl, Proposition 3.3.2]) for any strongly positive form $\omega \in \mathcal{D}_0^{n-p,\,n-p}(\Omega)$. Any positive current is real, that is, $T(\omega) = \overline{T(\overline{\omega})}$, and, following [Kl, Proposition 3.3.4], it has measure coefficients (i.e. is of order 0).

5.1.3 Plurisubharmonic Functions

Let Ω be an open subset of $\mathbb{C}^n$. An u.s.c. function $u : \Omega \to [-\infty, +\infty[$ is said to be *plurisubharmonic* if u is not identically $-\infty$ on any component of Ω and if, for each $a \in \Omega$ and $b \in \mathbb{C}^n$, the function $\lambda \mapsto u(a + \lambda b)$ is subharmonic or identically $-\infty$ on every component of the open set $\{\lambda : a + \lambda b \in \Omega\}$ of $\mathbb{C}$. The set of plurisubharmonic functions on Ω is denoted by PSH(Ω).

The following theorem (see [Kl, Theorem 2.9.14]) gathers the important properties of plurisubharmonic functions that will often be used in this chapter.

Theorem 5.1.1 *Let Ω be an open subset of $\mathbb{C}^n$.*

(i) *The family* PSH(Ω) *is a convex cone, i.e. if α, β are positive numbers and $u, v \in$* PSH(Ω)*, then $\alpha u + \beta v \in$* PSH(Ω).

(ii) *If Ω is connected and if $(u_j)_{j\in\mathbb{N}}$ is a decreasing sequence of functions in* PSH(Ω)*, then $u := \lim_{j\to\infty} u_j \in$* PSH($\Omega$) *or $u \equiv -\infty$.*

(iii) *If $u : \Omega \to \mathbb{R}$ and if $(u_j)_{j\in\mathbb{N}}$ is a sequence of functions in* PSH(Ω) *converging to u uniformly on compact subsets of Ω, then $u \in$* PSH(Ω).

(iv) *Let* $\{u_\alpha\}_{\alpha\in A}$ *be a family of functions in* PSH(Ω) *such that* $u := \sup_{\alpha\in A} u_\alpha$ *is locally upper bounded. Then the u.s.c. regularization* u^* *of* u *is in* PSH(Ω).

The convex subcone of PSH(Ω) formed by the negative plurisubharmonic functions on Ω is denoted by $\mathrm{PSH}_-(\Omega)$.

According to [Kl, Corollary 2.9.6], every plurisubharmonic function on Ω is a subharmonic function on Ω considered as a $\mathbb{R}^{2n}$-open set. For further properties of plurisubharmonic functions that will be used in this chapter we refer the reader to the book of Klimek [Kl].

The following result is analogous to the convergence theorem for sub/superharmonic functions. Corresponding to the reduction R_1^E in potential theory, the *relative extremal function* or *pluriharmonic measure* $u = u_{E,\Omega}$ (or simply u_E) of a subset E of an open set Ω in $\mathbb{C}^n$, cf. [Kl, p. 158], is defined by

$$u = u_{E,\Omega} := \sup\{\psi \in \mathrm{PSH}_-(\Omega) : \psi \leq -1 \text{ on } E\}. \tag{5.1.8}$$

According to Theorem 5.1.1(iv) the u.s.c. regularization u^* of u:

$$u^*(z) = \limsup_{\Omega\ni\zeta\to z} u(\zeta)$$

is plurisubharmonic in Ω.

For any function $u \in \mathrm{PSH}(\Omega)$, $dd^c u$ is a positive $(1, 1)$-current with measures coefficients in view of [Kl, Proposition 3.3.5] and, if $u \in \mathcal{C}^2(\Omega)$, we have

$$dd^c u = 2i \sum_{1\leq j,k\leq n} \frac{\partial^2 u}{\partial z_j \partial \bar{z}_k} dz_j \wedge d\bar{z}_k$$

(see [Kl, p. 15]).

5.1.4 The Complex Monge-Ampère Operator

Let Ω be an open subset of $\mathbb{C}^n$, $n \geq 1$, and u a plurisubharmonic function on Ω. If u is of class $\mathcal{C}^2$, the Monge-Ampère measure associated with u is the measure

$$\det\left(\left(\frac{\partial^2 u}{\partial z_j \partial \bar{z}_k}\right)_{1\leq j,k\leq n}\right) V$$

on Ω, where

$$dV = \left(\frac{i}{2}\right)^n dz_1 \wedge \bar{z}_1 \wedge \ldots \wedge dz_n \wedge d\bar{z}_n$$

is the usual volume form on $\mathbb{C}^n$ (restricted to Ω). This measure obviously coincides with the current $(dd^c u)^n$.

If Ω is a bounded open subset of $\mathbb{C}^n$ with smooth boundary then, following [Kl, Proposition 3.4.1], for any real valued C^2-functions $v_1, \ldots, v_k$ on a neighborhood of $\overline{\Omega}$ and any smooth $(n-k, n-k)$-form φ on a neighborhood of $\overline{\Omega}$ such that $\varphi = 0$ on $\partial\Omega$, one has

$$\int_\Omega dd^c v_1 \wedge \ldots \wedge dd^c v_k \wedge \varphi =$$

$$\int_\Omega v_j dd^c v_1 \wedge \ldots \wedge dd^c v_{j-1} \wedge dd^c v_{j+1} \wedge \ldots \wedge dd^c v_k \wedge dd^c \varphi \tag{5.1.9}$$

As noted above, for any function $u \in \mathrm{PSH}(\Omega)$ the $(1, 1)$-current $dd^c u$ is positive. Bedford and Taylor [BT2] observed that this property combined with the formula (5.1.9) for integration by parts can serve as a basis for an inductive definition of the (generalized) complex Monge-Ampère operator acting on locally bounded plurisubharmonic functions as follows:

Let $u_1, \ldots, u_n \in \mathrm{PSH}(\Omega) \cap L^\infty_{loc}(\Omega)$. If $1 \le k \le n$, then $dd^c u_1 \wedge \ldots \wedge dd^c u_k$ can be defined inductively as a positive (k, k)-current of order 0 by the formula

$$\int_\Omega dd^c u_1 \wedge \ldots \wedge dd^c u_k \wedge \chi = \int_\Omega u_k dd^c u_1 \wedge \ldots \wedge dd^c u_{k-1} \wedge dd^c \chi, \tag{5.1.10}$$

where χ is a test form in Ω of bidegree $(n-k, n-k)$ (see [Kl, p. 113]). If $u_1 = \ldots = u_n = u$, the measure (current of bidegree (n, n)) $dd^c u_1 \wedge \ldots \wedge dd^c u_n$ is denoted by $(dd^c u)^n$.

By [Kl, Proposition 3.3.4], $dd^c u_1 \wedge \ldots \wedge dd^c u_k$ is a current of order 0, i.e. it has measure coefficients. For complete details see [Kl, p. 113].

The operator $(dd^c \cdot)^n$, acting on locally bounded plurisubharmonic functions, is called the *(complex) Monge-Ampère operator*. Sometimes, the operator

$$(u_1, \ldots, u_k) \mapsto dd^c u_1 \wedge \ldots \wedge dd^c u_k$$

is also referred to as the Monge-Ampère operator.

The Monge-Ampère operator was extended to a larger class of plurisubharmonic functions, called the *domain of definition of the Monge-Ampère operator*, by Cegrell [Ce2] and Błocki [Bl3].

One of the most important properties of the Monge-Ampère operator is its continuity on decreasing sequences, cf. [Kl, Theorem 3.4.3]:

If $(u_1^j)_{j\in\mathbb{N}}, \ldots, (u_k^j)_{j\in\mathbb{N}}$ are decreasing sequences of plurisubharmonic functions converging pointwise on Ω to locally bounded plurisubharmonic functions $u_1, \ldots, u_n$, respectively, then

$$\lim_{j\to+\infty} dd^c u_1^j \wedge \ldots \wedge dd^c u_k^j = dd^c u_1 \wedge \ldots \wedge dd^c u_k$$

in the sense of weak*-convergence of currents of order 0.

5.1.5 Monge-Ampère Capacity

Let Ω be an open set in $\mathbb{C}^n$. Following [BT2] we define the *capacity* of a compact subset of Ω as

$$C(K, \Omega) = C(K) = \sup\left\{\int_K (dd^c u)^n : u \in \mathrm{PSH}(\Omega),\ 0 < u < 1\right\},$$

and the (inner) capacity of any set $E \subset \Omega$ by

$$C(E, \Omega) = \sup\{C(K) : K \text{ is a compact subset of } E\}.$$

It follows by use of (5.1.9) that $C(K, \Omega) < +\infty$. By the above Definition we have for any Borel set $E \subset \Omega$

$$C(E, \Omega) = \sup\left\{\int_E (dd^c u)^n : u \in \mathrm{PSH}(\Omega), 0 < u < 1\right\}.$$

We have the following result (see [Kl, Proposition 3.5.1]).

Proposition 5.1.2 *Let $\Omega \subset \mathbb{C}^n$ and $C(E) = C(E, \Omega)$.*
(1) *If $E_1 \subset E_2 \subset \Omega$ then $C(E_1) \leq C(E_2)$.*
(2) *If $E \subset \Omega_1 \subset \Omega_2$, then $C(E, \Omega_1) \geq C(E, \Omega_2)$.*
(3) $C(\bigcup_{j=l}^{\infty} E_j) \leq \sum_{j=1}^{\infty} C(E_j)$.
(4) *If Ω is hyperconvex, and if (K_j) is a decreasing sequence of compact subsets of Ω with $K = \bigcap K_j$ then*

$$\lim_{j\to\infty} C(K_j, \Omega) = C(K, \Omega).$$

(5) *If $E_1 \subset E_2 \subset \ldots$ are Borel sets in Ω, then $C(\bigcup_j E_j) = \lim_{j\to\infty} C(E_j)$.*

The *outer capacity* $C^*(E, \Omega)$ (relative to Ω) of a set $E \subset \Omega$ is defined by

$$C^*(E, \Omega) = \inf\{C(\omega, \Omega) : \omega \supset E,\ \text{open in } \Omega\}.$$

The properties (1), (2) and (3) in the above proposition remain true if C is replaced by C^*, and it follows from (4) that if Ω is hyperconvex then $C^*(K, \Omega) = C(K, \Omega)$, see [Kl, Proposition 4.7.1].

Definition 5.1.3 A function f on Ω is said to be *quasi-continuous* if for any $\varepsilon > 0$ there is an open subset $\mathcal{O}$ of Ω such that $C(\mathcal{O}) = C(\mathcal{O}, \Omega) < \varepsilon$ and f is continuous on $\Omega \setminus \mathcal{O}$ (in the Euclidean topology).

Definition 5.1.4 A set $A \subset \Omega$ is said to be *quasi-open* (*quasi-closed*) if

$$\inf\{C^*(A\Delta E, \Omega) : E \text{ open } \ (E \text{ closed })\} = 0.$$

This is equivalent to the fact that for every $\varepsilon > 0$, there is an open (closed) set $E \subset \Omega$ such that $C^*(A\Delta E) \leq \varepsilon$.

The proof of the following lemma is analogous to that of Lemma 1.5.5.

Lemma 5.1.5 *A set $A \subset \Omega$ is quasi-open, resp. quasi-closed, if (and only if)*

$$\inf\{C^*(E \setminus A, \Omega) : E \text{ open in } \Omega, A \subset E\} = 0,$$

resp.

$$\inf\{C^*(A \setminus E, \Omega) : E \text{ closed in } \Omega,\ E \subset A\} = 0.$$

The proposition is an easy consequence of Lemma 5.1.5, the properties (b) and (c) from Proposition 5.1.6 and [Kl, Theorem 4.7.5].

Proposition 5.1.6 (i) *A set that differs from a quasi-open set by a pluripolar set (see Definition 5.1.9 below) is quasi-open.*
(ii) *A countable union of quasi-open subsets of Ω is quasi-open.*

Theorem 5.1.7 ([Fu5, Lemma 3.3]) *A function $f : \Omega \to \overline{\mathbb{R}}$ is quasi-continuous if and only if for any open set $V \subset \overline{\mathbb{R}}$, the set $f^{-1}(V)$ is quasi-open.*

Proof Suppose that f is quasi-continuous. For given $\varepsilon > 0$, there is an open set $\mathcal{O} \subset \Omega$ such that $C(\mathcal{O}, \Omega) < \varepsilon$ and f is continuous on $\Omega \setminus \mathcal{O}$. This means that $f^{-1}(V) \cap (\Omega \setminus \mathcal{O})$ is open relative to $\Omega \setminus \mathcal{O}$, that is, there is an open set $\omega \subset \Omega$ such that $f^{-1}(V) \cap (\Omega \setminus \mathcal{O}) = \omega \cap (\Omega \setminus \mathcal{O})$. It then follows that $f^{-1}(V) \subset \omega \cup \mathcal{O}$ and $C^*((\omega \cup \mathcal{O}) \setminus f^{-1}(V), \Omega) \leq \varepsilon$. Since ε is arbitrarily chosen, we infer by Lemma 5.1.5 that $f^{-1}(V)$ is quasi-open.

Conversely, let (V_n) be a sequence of open sets forming a basis of the topology of $\mathbb{R}$ and suppose that $f^{-1}(V_n)$ is quasi-open for every n. For a given $\varepsilon > 0$ there is for every $n \in \mathbb{N}$ an open set $\omega_n \subset \Omega$ with $C(\omega_n, \Omega) < \varepsilon/2^n$ such that $(\Omega \setminus f^{-1}(V_n)) \setminus \omega_n$ is closed in Ω. In terms of the open set $\omega := \bigcup_n \omega_n$ it follows *a fortiori* that $(\Omega \setminus f^{-1}(V_n)) \setminus \omega$ is closed in Ω for every $n \in \mathbb{N}$, and hence $f|_{\complement\omega}$ is continuous. Since such an open set ω with $C(\omega, \Omega) < \varepsilon$ has been found for every $\varepsilon > 0$ we conclude that f indeed is quasi-continuous. □

By using capacity properties the following analogue of Cartan's theorem for subharmonic functions was obtained by Bedford and Taylor [BT2].

Theorem 5.1.8 *Plurisubharmonic functions are quasi-continuous.*

For the proof see [Kl, Theorem 3.5.5] for bounded plurisubharmonic functions. For the extension to general plurisubharmonic functions see [Kl, p. 173, paragraph before Proposition 4.7.12].

Definition 5.1.9 A subset E of Ω is said to be *pluripolar* if for any point $p \in E$ there is an open neighborhood $\mathcal{O}$ of p and a plurisubharmonic function u on $\mathcal{O}$ such that $\mathcal{O} \cap E \subset \{u = -\infty\}$. The set E is *(pluri)negligible* if there is a locally bounded family $\{u_\alpha : \alpha \in A\}$ of plurisubharmonic functions on Ω such that $E \subset \{u^* > u\}$, where $u(z) = \sup\{u_\alpha(z) : \alpha \in A\}$ is the upper envelope of the family and $u^*(z) = \limsup_{\zeta \to z} u(\zeta)$ is the u.s.c. regularization of u.

It is a basic theorem, due to Josefson [J], that the notion of pluripolar sets can be given by global functions. That is, if E is pluripolar then there exists a plurisubharmonic function $u \not\equiv -\infty$ on $\mathbb{C}^n$, such that $E \subset \{u = -\infty\}$, cf. [Kl, Theorem 4.7.4].

The pluripolar sets can be characterized in terms of outer capacities.

Theorem 5.1.10 *Let Ω be a hyperconvex domain in $\mathbb{C}^n$. A subset E of Ω is pluripolar if and only if $C^*(E, \Omega) = 0$.*

The following theorem is due to Bedford and Taylor [BT2]:

Theorem 5.1.11 *A set $E \subset \Omega$ is negligible if and only if it is pluripolar.*

For the proofs of Theorems 5.1.10 and 5.1.11 see [Kl, Theorem 4.7.5] and [Kl, Theorem 4.7.6].

5.2 The Pluri-Fine Topology on $\mathbb{C}^n$

Definition 5.2.1 The *pluri-fine topology* or topology $\mathcal{F}$ (or also $\mathcal{F}$-topology) on an open subset Ω of $\mathbb{C}^n$ is the weakest topology in which all plurisubharmonic functions on Ω are continuous.

The topology $\mathcal{F}$ was briefly introduced by Fuglede in [Fu25] and was further studied by Bedford and Taylor in [BT3], and by El Marzguioui and Wiegerinck in [EMW1]. The concept is analogous to the H. Cartan fine topology on $\mathbb{R}^N$, in which all subharmonic functions are continuous. In particular, on $\mathbb{C} \cong \mathbb{R}^2$, fine and pluri-fine topology coincide.

Notions pertaining to the pluri-fine topology are indicated with the prefix pluri-fine (plurifinely) or $\mathcal{F}$-, e.g. plurifinely open set or $\mathcal{F}$-open set means open set in the pluri-fine topology. The $\mathcal{F}$-closure (pluri-fine closure), resp. Euclidean closure, of a set $A \subset \Omega$ will be denoted by $\overline{A}^{\mathcal{F}}$, resp. $\overline{A}$, and its $\mathcal{F}$-boundary (pluri-fine boundary), resp. Euclidean boundary, by $\partial_{\mathcal{F}} A$, resp. ∂A. The $\mathcal{F}$-interior (pluri-fine interior) of a set $A \subset \Omega$ is denoted by A''. The lim and lim sup relative to the pluri-fine topology shall be denoted by $\mathcal{F}$- lim and $\mathcal{F}$- lim sup respectively. Topological properties not explicitly referring to the pluri-fine topology $\mathcal{F}$ or the Cartan fine topology are tacitly understood to refer to the Euclidean topology. Complements of subsets of Ω are taken relative to Ω.

In this section, we shall outline some basic facts about the pluri-fine topology that will be used in the rest of this chapter.

Definition 5.2.2 A subset E of Ω is *pluri-thin* at $z \in \Omega$ if and only if either $z \notin \overline{E}$ or there is an open set $\omega \ni z$ and a plurisubharmonic function u on ω such that

$$\limsup_{\zeta \to z,\, \zeta \in (E \setminus \{z\}) \cap \omega} u(\zeta) < u(z). \tag{5.2.1}$$

A subset E is *pluri-thin* if it is pluri-thin at every point $z \in \Omega$.

As in the classical case, a basic fact is that the "jump" in (5.2.1) can be infinite when E is pluri-thin at z.

Proposition 5.2.3 *The set E is pluri-thin at $z \in \overline{E}$ if and only if there exists a plurisubharmonic function v on $\mathbb{C}^n$, locally bounded outside a neighborhood of z, such that $v(z) > -\infty$ and*

$$\limsup_{\zeta \to z,\, \zeta \in E} v(\zeta) = -\infty.$$

Proof By hypothesis there exists an open set $\omega \ni z$ and $\psi \in \mathrm{PSH}(\omega)$ such that

$$\limsup_{\zeta \to z,\, \zeta \in (E \setminus \{z\}) \cap \omega} \psi(\zeta) < \psi(z).$$

Without loss of generality, we may suppose that $z = 0$, $\omega = \{|\zeta| < 1\}$, $\psi < -1$ on $\omega \cap E$, and $\psi(0) = -\eta > -1$. By the upper semicontinuity of ψ, we may choose $\delta_j > 0$ (for j sufficiently large) such that $\delta_j < 1$ and $\psi(\zeta) < -\eta + 2^{-j-1}$ for $|\zeta| < \delta_j$. It follows that the function defined on $\mathbb{C}^n$ by

$$v_j(\zeta) = \begin{cases} \max\{\frac{2^{-j-1}}{|\log \delta_j|} \log |\zeta|, \psi + \eta - 2^{-j}\} & \text{if } |\zeta| < \delta_j \\ \frac{2^{-j-1}}{|\log \delta_j|} \log |\zeta| & \text{if } |\zeta| \geq \delta_j \end{cases}$$

is plurisubharmonic on $\mathbb{C}^n$. Further, $v_j(0) = -2^{-j}$ and $v_j(\zeta) \leq -1 + \eta - 2^{-j}$ for $\zeta \in E$, ζ sufficiently close to 0. Thus $v = \sum v_j$ is the desired function. □

It is with the notion of pluri-thinness that the difference between the topology $\mathcal{F}$ and the Cartan fine topology arises. Namely, in the classical Potential theory the (everywhere) thin sets coincide with the polar sets. But that is not the case here. Pluri-thin sets are pluripolar, but not conversely. Indeed, the set $E = \{z_1 = 0\}$ in $\mathbb{C}^2$ is pluripolar but not pluri-thin, cf. [Kl, p. 175].

The study of pluri-thin sets appears to be very delicate. Sadullaev [Sa] has given a discussion of pluri-thinness and provided many very interesting examples and open questions concerning this notion.

Since pluripolar sets are not necessarily pluri-thin, it is not possible to give a necessary and sufficient criterion for pluri-thinness in terms of capacity. The *unregularized* functions u_K, however, may be used to yield something analogous to the Wiener criterion (cf. [Do4, p. 249]), where the specific values $r_j = \alpha^j$ are used, for some $\alpha \in]0, 1[$. The following pluri-thinness criterion is due to Bedford and Taylor [BT3].

Theorem 5.2.4 *Let $A \subset \Omega \subset \mathbb{C}^n$ be given, Ω open, and let z_0 be a point of $\overline{A} \cap \Omega$. Then A is pluri-thin at z_0, if and only if there exist $r_1 > r_2 > r_3 > \dots$ (all > 0) such that*

$$\sum_{j=1}^{\infty} u_{A_j}(z_0) > -\infty \tag{5.2.2}$$

where $A_j = A \cap \{r_j > |z - z_0| \geq r_{j+1}\}$, and $u_{A_j} = u_{A_j,\Omega}$ is the relative (unregularized) extremal function of A_j.

Proof If (5.2.2) holds, then there exist $v_j \in \mathrm{PSH}(\Omega)$, $v_j < 0$, with $v_j \leq -1$ on A_j and

$$v_j(z_0) > u_{A_j}(z_0) - 2^{-j-1}.$$

It follows, then, that $V_J := \sum_{j=J}^{\infty} v_j \in \mathrm{PSH}(\Omega)$, and $-2^{-J} + \sum_{j=J}^{\infty} u_{A_j}(z_0) < V_J(z_0)$. If J is chosen large enough, then $V_J(z_0) > -1$, but

$$\limsup_{A \ni \zeta \to z_0} V_J(\zeta) \leq -1$$

and so A is pluri-thin at z_0.

Conversely, if A is pluri-thin at z_0 we let $v \in \mathrm{PSH}(\Omega)$ be the function given by Proposition 5.2.3. Without loss of generality we may assume that $v < 0$. Now let us choose $r_1 > r_2 > r_3 > \dots$ (all > 0), such that

$$\sup_{\{|\zeta - z_0| < r_j\} \cap A} v(\zeta) < -2^j.$$

It then follows that

$$u_{A_j}(z_0) \geq 2^{-j} v(z_0)$$

and so (5.2.2) holds. □

Theorem 5.2.5 *Finite intersections of sets of the form*

$$\mathcal{B}_{u,V,\alpha} = \{z \in V : u(z) > \alpha\}$$

where $\alpha \in \mathbb{R}$, $V \subset \Omega$ is open and $u \in \mathrm{PSH}(V)$, constitute a basis for the pluri-fine topology on Ω.

Proof It is clear that if $\alpha \in \mathbb{R}$ and $u \in \mathrm{PSH}(\Omega)$, then the sets $\{u < \alpha\}$ and $\{u > \alpha\}$ are $\mathcal{F}$-open. In particular, for any $a \in \Omega$ the function $z \mapsto \|z - a\|$ is a plurisubharmonic function on Ω, and so $B(a, r)$, $r < d(a, \mathbb{C}^n \setminus \Omega)$, is an $\mathcal{F}$-open. Consequently, all open sets in Ω are $\mathcal{F}$-open. Consider a real α, an open set $V \subset \Omega$ and $u \in \mathrm{PSH}(V)$. Take $a \in B_{u,V,\alpha}$ and choose $r > 0$ such that $\overline{B}(a, r) \subset V$. Note that

$$a \in \{z \in B(a, r) : u(z) > \alpha\} \subset B_{u,V,\alpha}.$$

Take $\rho > r$ such that $\overline{B}(a, \rho) \subset V$ and set $M = \sup u(\overline{B}(a, \rho))$. Choose $C, D \in \mathbb{R}$ such that $C \log r + D = \alpha$ and $C \log \rho + D = M$, and define

$$v(z) = \begin{cases} \max\{C \log \|z - a\| + D, u(z)\} & \text{if } z \in B(a, \rho) \\ C \log \|z - a\| + D & \text{if } z \in \Omega \setminus B(a, \rho). \end{cases}$$

Then $v \in \mathrm{PSH}(\Omega)$ by [Kl, Corollary 2.9.15], and in $B(a, r)$, $u > \alpha$ if and only if $v > \alpha$. Thus

$$\{z \in B(a, r) : u(z) > \alpha\} = B(a, r) \cap \{z \in \Omega : v(z) > \alpha\},$$

and since both sets at the right-hand side are $\mathcal{F}$-open, it follows that a is in the $\mathcal{F}$-interior of $B_{u,V,\alpha}$. Consequently $B_{u,V,\alpha}$ is $\mathcal{F}$-open.

Any topology on Ω with respect to which all plurisubharmonic functions are continuous must contain the sets $B_{u,V,\alpha}$, where $V \subset \Omega$ is open, $u \in \mathrm{PSH}(V)$ and $\alpha \in \mathbb{R}$. Therefore the topology generated by finite intersections of those sets is the topology $\mathcal{F}$. □

It follows easily from Theorem 5.2.5 that a nonvoid $\mathcal{F}$-open set has non-zero Lebesgue measure, although it may have empty Euclidean interior.

Corollary 5.2.6 *If $\Omega_1 \subset \Omega_2$ are open subsets of $\mathbb{C}^n$ then the $\mathcal{F}$-topology on Ω_1 is the same as the topology induced on Ω_1 by the $\mathcal{F}$-topology on Ω_2.*

Following Corollary 5.2.6 a set $U \subset \mathbb{C}^n$ is $\mathcal{F}$-open in $\mathbb{C}^n$ if and only if it is $\mathcal{F}$-open in any open set $\Omega \subset \mathbb{C}^n$ containing U. One consequence of this is that the definition of the $\mathcal{F}$-open sets in Ω is independent of Ω. Hence, if $\mathcal{F}(\Omega)$ is the plurifine topology on Ω, then a subset S of Ω belongs to $\mathcal{F}(\Omega)$ if and only if $S = T \cap \Omega$ for some $T \in \mathcal{F}(\mathbb{C}^n)$. Furthermore, S is an $\mathcal{F}$-open subset of Ω if and only if it is locally $\mathcal{F}$-open.

Theorem 5.2.7 *Let $z \in \Omega$ and $V \subset \Omega$ be such that $z \in V$. Then V is a $\mathcal{F}$-neighborhood of z if and only if $\complement V$ is pluri-thin at z.*

Proof The proof is a straightforward adaptation of the proof of Theorem 1.2.4. □

Theorem 5.2.8 *The sets of the form $V_{B,u} = \{z \in B : u(z) > 0\}$, where $B \subset \Omega$ is an open ball and u is a plurisubharmonic function on B, form a base for the topology $\mathcal{F}$.*

Proof The sets $V_{B,u} := \{z \in B : u(z) > 0\}$, where $B \subset \Omega$ is an open ball and u is a plurisubharmonic function on B, are $\mathcal{F}$-open by Theorem 5.2.5 and the definition of the topology $\mathcal{F}$. Let $U \subset \Omega$ be a non-empty $\mathcal{F}$-open set and let $a \in U$. We will prove that U contains an $\mathcal{F}$-neighborhood of a of the form $V_{B,u}$ as stated in the theorem. This is trivial if a belongs to the Euclidean interior of U. Consider the case when a

is an accumulation point of $\complement U$. By Theorem 5.2.7 there is a ball B centered at a, a plurisubharmonic function v on B and a real c such that $\limsup_{\zeta\to z,\zeta\in \complement U\cap B} v(\zeta) < c < v(z)$. This implies that $a \in V_{B,v-c} = \{z \in B : v(z) - c > 0\} \subset U$. □

Remark 5.2.9 We may replace 0 in Theorem 5.2.8 by any real number α.

Corollary 5.2.10 *Every $\mathcal{F}$-neighborhood of a point z of $\mathbb{C}^n$ contains an $\mathcal{F}$-neighborhood of z which is compact in the Euclidean topology.*

Proof Let V be an $\mathcal{F}$-neighborhood of z. By Theorem 5.2.8 there is a ball $B = B(z, \varepsilon)$ and a plurisubharmonic function u on B such that $u(z) > 0$ and that $\{u > 0\} \subset V$. The set $\overline{B}(z, \frac{\varepsilon}{2}) \cap \{u \geq \frac{u(z)}{2}\}$ is an $\mathcal{F}$-neighborhood of z contained in V and is compact, for u is u.s.c. on B. □

Corollary 5.2.11 *The topology $\mathcal{F}$ is completely regular, that is, for any $\mathcal{F}$-closed set $A \subset \Omega$ and any point $a \in \complement A$ there exists an $\mathcal{F}$-continuous real valued function f on Ω such that $f|_A \equiv 0$ and $f(a) \neq 0$.*

Proof By Theorem 5.2.7 there is an open ball $B = B(a, r)$ $(r > 0)$ and a function $u \in \mathrm{PSH}(B)$ such that $a \in \{z \in B : u > 0\} \subset \complement A$. By replacing u by u^+, we may suppose $u \geq 0$. Let χ be a continuous function on Ω of compact support contained in B such that $\chi(a) = 1$, and let f be the function $\chi \cdot u$ extended by 0 outside of B. Then f is $\mathcal{F}$-continuous on Ω and we have $f = 0$ on A and $f(a) = u(a) \neq 0$. □

Proposition 5.2.12 *The topology $\mathcal{F}$ is biholomorphically invariant, that is, a subset U of $\mathbb{C}^n$ is $\mathcal{F}$-open if and only if $h(U)$ is $\mathcal{F}$-open for any biholomorphic function h on an open neighborhood of U into its image.*

Proof The proposition is an obvious consequence of the biholomorphic invariance of the class of plurisubharmonic functions by composition with biholomorphic maps, see [Kl, Corollary 2.9.5]. □

A slight variation in Theorem 5.2.8 shows that we can also use a neighborhood subbasis of the topology $\mathcal{F}$ of the form

$$\{\zeta : |\zeta - z| \leq \varepsilon, u(\zeta) \geq 0\}$$

($\varepsilon > 0$ and u a plurisubharmonic function on an open neighborhood of $\overline{B}(z, \varepsilon)$), see the proof of Corollary 5.2.10. Since these sets are compact, we may argue as on p. 167 of [Do4] to obtain the Baire property for the fine topology: *if $\mathcal{O}_1, \mathcal{O}_2, \ldots,$ are $\mathcal{F}$-open and $\mathcal{F}$-dense, then $\bigcap_j \mathcal{O}_j$ is $\mathcal{F}$-dense.*

A direct consequence of the equivalence of negligible sets and pluripolar sets (Theorem 5.1.11) is the following property of the pluri-fine topology (see [Do4, p. 181] for an analogous proof).

Theorem 5.2.13 (Quasi-Lindelöf property) *An arbitrary union of $\mathcal{F}$-open subsets differs from a countable subunion by at most a pluripolar set.*

The following result contains Theorem 5.1.8.

Theorem 5.2.14 *Every $\mathcal{F}$-continuous function $f : \Omega \to \overline{\mathbb{R}}$ is quasi-continuous.*

Proof Let $V \subset \overline{\mathbb{R}}$ be open. The set $f^{-1}(V)$ is $\mathcal{F}$-open, and it follows by Theorem 5.2.8 and the quasi-Lindelöf property of the topology $\mathcal{F}$ (Theorem 5.2.13) that there is a pluripolar set $E \subset \Omega$ and, for every integer $j \geq 1$, an open ball $B_j \subset \Omega$ and a function $\varphi_j \in \mathrm{PSH}(B_j)$ such that $f^{-1}(V) = (\bigcup_j \{\varphi_j > 0\}) \cup E$. Since φ_j is quasi-continuous by Theorem 5.1.8, the set $\{\varphi_j > 0\}$ is quasi-open in B_j, and hence in Ω in view of Proposition 5.1.2(2). It follows by Proposition 5.1.6 that $f^{-1}(V)$ is quasi-open. By Theorem 5.1.7 we conclude that f is quasi-continuous on Ω. □

The following proposition marks an important difference between the pluri-fine topology and the (Cartan) fine topology.

Proposition 5.2.15 (a) *The pluri-fine topology on $\mathbb{C}^n$ induces on every $\mathbb{C}$-linear subspace $L \cong \mathbb{C}^k$ of $\mathbb{C}^n$ the pluri-fine topology on L. Explicitly, for any $\mathcal{F}$-open set $U \subset \mathbb{C}^n$ the intersection $L \cap U$ is $\mathcal{F}$-open in L, and so is the orthogonal projection of Ω on L.*

(b) *A set $\omega \subset \mathbb{C}^k$ is $\mathcal{F}$-open in $\mathbb{C}^k$ if and only if $\omega \times \mathbb{C}^{n-k}$ is $\mathcal{F}$-open in $\mathbb{C}^n$.*

Proof For $z = (z_1, \ldots, z_n) \in \mathbb{C}^n$ write $z' = (z_1, \ldots, z_k)$, $z'' = (z_{k+1}, \ldots, z_n)$. For (a) it suffices to consider the particular subspace $L_0 = \{(z', 0'') : z' \in \mathbb{C}^k\}$ which we identify with $\mathbb{C}^k$. For any $\mathcal{F}$-open set $U \subset \mathbb{C}^n$ denote by ω the part of U in L_0. Consider a point $a' \in \omega$. According to Theorem 5.2.8 there exist (Euclidean) open neighborhoods U' of a' in $\mathbb{C}^k$ and U'' of $0''$ in $\mathbb{C}^{n-k}$ and a plurisubharmonic function ψ on $U' \times U'' \subset \mathbb{C}^n \cong \mathbb{C}^k \times \mathbb{C}^{n-k}$ such that

$$(a', 0'') \in \{(z', z'') \in U' \times U'' : \psi(z', z'') > 0\} \subset U. \tag{5.2.3}$$

Define $\varphi : U' \to [-\infty, +\infty[$ by $\varphi(z') = \psi(z', 0'')$. Then φ is plurisubharmonic and

$$a' \in \{z' \in U' : \varphi(z') > 0\} \subset \omega. \tag{5.2.4}$$

Thus ω is indeed an $\mathcal{F}$-neighborhood of a' in $\mathbb{C}^k$. For each $t \in \mathbb{C}^{n-k}$ the translate $U_t = U - (0', t)$ of Ω is $\mathcal{F}$-open in $\mathbb{C}^n$. It follows that $U_t \cap L_0$ is $\mathcal{F}$-open in L_0, and so is therefore the union of the $U_t \cap L_0$, that is, the projection of U on L_0.

For (b) we have just shown, in particular, that if $U := \omega \times \mathbb{C}^{n-k}$ is $\mathcal{F}$-open in $\mathbb{C}^n$ then ω is $\mathcal{F}$-open in $\mathbb{C}^k$. To establish the converse, suppose that ω is $\mathcal{F}$-open in $\mathbb{C}^k$ and let us prove that every point $a = (a', a'')$ of $\omega \times \mathbb{C}^{n-k}$ is an $\mathcal{F}$-inner point of that set. Since ω is an $\mathcal{F}$-neighborhood of a' in $\mathbb{C}^k$ there exists again by Theorem 5.2.8 a plurisubharmonic function φ on a (Euclidean) neighborhood U' of a' in $\mathbb{C}^k$ such that (5.2.4) holds. The function ψ defined on $\mathbb{C}^n$ by $\psi(z', z'') = \varphi(z')$ is plurisubharmonic –an easy and well-known consequence of the definition of plurisubharmonicity. Furthermore, (5.2.3) holds (with $U = \omega \times \mathbb{C}^{n-k}$ and with $(a', 0'')$ replaced by a) for any neighborhood U'' of a'' in $\mathbb{C}^{n-k}$. Thus $\omega \times \mathbb{C}^{n-k}$ is indeed an $\mathcal{F}$-neighborhood of a in $\mathbb{C}^n$. □

Theorem 5.2.16 *The (Cartan) fine topology on $\mathbb{C}^n \cong \mathbb{R}^{2n}$, $n \geq 2$, is strictly finer than the pluri-fine topology.*

Proof It is clear that the (Cartan) fine topology on $\mathbb{C}^n \cong \mathbb{R}^{2n}$ is finer than the pluri-fine topology $\mathcal{F}$, the plurisubharmonic functions on $\mathbb{C}^n$ being subharmonic on $\mathbb{C}^n$ by [Kl, Theorem 2.9.12]. The complement of a Lebesgue spine at 0 in $\mathbb{C}^n$ is a (Cartan) fine neighborhood of 0, but it is not a pluri-fine $\mathcal{F}$-neighborhood of 0 because its intersection with some complex line containing 0 is not (Cartan) finely open in $\mathbb{C}$ since it does not contain any circle around 0. Thus the (Cartan) fine topology on $\mathbb{C}^n \cong \mathbb{R}^{2n}$ is strictly finer than the pluri-fine topology $\mathcal{F}$. □

Since a pluripolar set in $\mathbb{C}^n \cong \mathbb{R}^{2n}$ is a polar set, it follows from Theorem 5.2.16 that it has empty $\mathcal{F}$-interior.

Due to the fact that the fine topology on $\mathbb{C}^n \cong \mathbb{R}^{2n}$ is finer than the pluri-fine topology $\mathcal{F}$ on $\mathbb{C}^n$, the following two theorems are immediate consequences of Corollarys 2.1.24 and 2.1.46:

Theorem 5.2.17 *An open set $U \subset \mathbb{C}^n$ is $\mathcal{F}$-connected if and only if U is connected in the Euclidean topology.*

Theorem 5.2.18 *Let $U \subset \mathbb{C}^n$ be an $\mathcal{F}$-connected $\mathcal{F}$-open set. If $E \subset U$ is an $\mathcal{F}$-closed pluripolar set, then $U \setminus E$ is $\mathcal{F}$-connected.*

Contrary to classical fine topology the set $U \setminus E$ in Theorem 5.2.18 is not $\mathcal{F}$-open in general. Indeed, following the paragraph after the proof of Proposition 5.2.3 there exists a pluripolar set $E \subset \mathbb{C}^n$, $(n \geq 2)$, which is not pluri-thin at a point a, and hence necessarily $a \in \overline{E}$ and $E \setminus \{a\}$ is not pluri-thin at a, so that $\mathbb{C}^n \setminus (E \setminus \{a\})$ is not an $\mathcal{F}$-neighborhood of a and therefore $U \setminus (E \setminus \{a\})$ is not $\mathcal{F}$-open.

Lemma 5.2.19 *Let $B = B(0, 1)$ (the unit ball in $\mathbb{C}^n$) and $U := \{\varphi > 0\}$, where $\varphi \in \mathrm{PSH}(B)$ is such that $0 \leq \varphi \leq 1$. If $U = V \cup W$, where V and W are nonempty $\mathcal{F}$-open sets such that $V \cap W = \varnothing$, then the following function is plurisubharmonic:*

$$\varphi_V(z) = \begin{cases} \varphi(z) & \text{if } z \in B \setminus W \\ 0 & \text{if } z \in W. \end{cases}$$

Proof U is $\mathcal{F}$-open because φ is plurisubharmonic. Let $\zeta \in B, \zeta \neq 0$, denote by L_ζ the complex line through 0 and ζ and by φ_ζ the restriction of φ to $L_\zeta \cap B$. It follows from Proposition 5.2.15 that $W_\zeta := W \cap L_\zeta$ is a finely open subset of $L_\zeta \cap B$ which may be empty. The restriction of φ_V to W_ζ is clearly equal to the function

$$\psi(z) = \begin{cases} \varphi_\zeta(z) & \text{if } z \in V \cap L_\zeta \\ 0 & \text{if } z \in B \cap L_\zeta \setminus V \end{cases}$$

on $L_\zeta \cap B$. Since $\limsup_{z \in V \cap L_\zeta, z \to \partial_f(V \cap L_\zeta)} \psi(z) = 0$ and $\varphi \geq 0$ on B, it follows that ψ is finely subharmonic on $B \cap L_\zeta$ according to Lemma 2.1.51, and thus subharmonic on $B \cap L_\zeta$ by Theorem 2.7.2. According to [Lel2] it follows that φ_V is plurisubharmonic. □

Theorem 5.2.20 *The topology $\mathcal{F}$ on an open set Ω in $\mathbb{C}^n$ is locally connected.*

Proof Let $z_0 \in \Omega$ and let $D \subset \Omega$ be an $\mathcal{F}$-open $\mathcal{F}$-neighborhood of z_0. By Theorem 5.2.8 there exists an open ball $B(z_0, \varepsilon) \subset \overline{B}(0, \varepsilon) \subset \Omega$ and a function $\varphi \in \mathrm{PSH}(B(z_0, \varepsilon))$, such that the set $\{z \in B(z_0, \varepsilon) : \varphi(z) > 0\}$ is an $\mathcal{F}$-open $\mathcal{F}$-neighborhood of z_0 contained in D. In view of Corollary 5.2.6 and the fact that the pluri-fine topology is biholomorphically invariant, see Proposition 5.2.12, there is no loss of generality if we assume that $z_0 = 0$, $\varepsilon = 1$ and that $0 \leq \varphi \leq 1$ on $B(0, 1)$. To prove the theorem we will find an $\mathcal{F}$-open $\mathcal{F}$-neighborhood of 0 which is $\mathcal{F}$-connected and contained in $V_\varphi := \{z \in B(0, 1) : \varphi(z) > 0\}$. We can assume that V_φ is not $\mathcal{F}$-connected since otherwise there is nothing to prove. Denote by $\mathcal{G}$ the (nonempty) set of all $\mathcal{F}$-open sets V which contain 0 and for which there exists a nonempty $\mathcal{F}$-open set W_V such that $V_\varphi = V \cup W_V$ and $V \cap W_V = \varnothing$. It follows from Lemma 5.2.19 that the function φ_V is plurisubharmonic for all $V \in \mathcal{G}$. Moreover, the family $(\varphi_V)_{V \in \mathcal{G}}$ is lower directed and lower bounded by 0, hence its infimum ψ is plurisubharmonic on $B(0, 1)$ according to [Hel, Theorem 4.15]. Now we claim that the set $U = \{z \in V_\varphi : \psi(z) > 0\}$ is an $\mathcal{F}$-connected $\mathcal{F}$-open $\mathcal{F}$-neighborhood of 0. To prove the claim observe first that for every V, $\psi(0) = \varphi_V(0) = \varphi(0) > 0$, which means that U is an $\mathcal{F}$-open $\mathcal{F}$-neighborhood of 0. Next, from the definition of φ_V in Lemma 5.2.19 we see that $\varphi_V(z) = \varphi(z) > 0$ on V and $\varphi_V(z) = 0$ on $B(0, 1)\backslash V$. Consequently, $U = \bigcap_{V \in \mathcal{G}}\{z \in B(0, 1) : \varphi_V(z) > 0\} = \bigcap_{V \in \mathcal{G}} V$ and $V_\varphi = (\bigcap_{V \in \mathcal{G}} V) \cup (\bigcup_{V \in \mathcal{G}} W_V)$ and $(\bigcap_{V \in \mathcal{G}} V) \cap (\bigcup_{V \in \mathcal{G}} W_V) = \varnothing$. Therefore U is an element of $\mathcal{G}$. It is minimal in the sense that it cannot be split into two disjoint non-empty $\mathcal{F}$-open sets, which proves the claim and the theorem. □

Corollary 5.2.21 *Every $\mathcal{F}$-component of an $\mathcal{F}$-open set $U \subset \mathbb{C}^n$ is $\mathcal{F}$-open. Moreover, the set of these components is at most countable.*

The proof of this corollary is the same as in the classical case and uses the quasi-Lindelöf property (Theorem 5.2.13).

Let Ω be an open set in the complex plane $\mathbb{C}$ and let $E \subset \Omega$. Subharmonic functions on Ω are denoted by $\mathrm{SH}(\Omega)$, while $\mathrm{SH}_-(\Omega) = \{u \in \mathrm{SH}(\Omega) : u \leq 0\}$. Recall that the *harmonic measure* (or the *relative extremal function*) of E (relative to Ω) at $z \in \Omega$ is defined as follows (see, for example, [Kl, p. 158]):

$$u_{E,\Omega}(z) = \omega(z, E, \Omega) := \sup\{u(z) : u \in \mathrm{SH}_-(\Omega), u|_E \leq -1\}.$$

that is, $\omega(z, E, \Omega) = -R_1^E(z)$. This function need not be subharmonic in Ω, but its u.s.c. regularization

$$\omega(z, E, \Omega)^* = -\widehat{R}_1^E(z) = \limsup_{\Omega \ni v \to z} \omega(v, E, \Omega) \geqslant \omega(z, E, \Omega)$$

is subharmonic, see Sect. 1.1.13. If E is a closed subset of Ω, then $\omega(\cdot, E, \Omega)$ coincides with the Perron solution of the Dirichlet problem in $\Omega\backslash E$ with boundary values -1 on $\partial E \cap \Omega$ and 0 on $\partial\Omega\backslash\partial E$. Recall the following result, cf. [Br7, Ch4].

Theorem 5.2.22 *Let Ω be a bounded open subset of $\mathbb{C}$. If $E \subset \Omega$ is a Borel set, then there exists an increasing sequence of compact sets $K_j \subset E$ such that $\omega(z, K_j, \Omega)^* \downarrow \omega(z, E, \Omega)^*$.*

Let $\mathbb{D} = \{z \in \mathbb{C} : |z| < 1\}$ be the open unit disk in $\mathbb{C}$ and $E \subset \mathbb{D}$. We associate to E its *circular projection*

$$E^\circ = \{|z| : z \in E\}.$$

To prove the next lemma, we need the following classical theorem of A. Beurling and R. Nevanlinna related to the Carleman-Milloux problem, cf. [Beu] and [Nev]. See also [Bet].

Theorem 5.2.23 *Let $F \subset \mathbb{D}$ be compact. Then*

$$\omega(z, F, \mathbb{D}) \leq \omega(-|z|, F^\circ, \mathbb{D})$$

for all $z \in \mathbb{D} \backslash F$.

It is well known following [AG, Theorem 7.3.9] that any finely open subset of $\mathbb{C} \cong \mathbb{R}^2$ contains circles with arbitrarily small radius around any point of it. The following proposition is an extension in some sense of this result to $\mathcal{F}$-open sets of $\mathbb{C}^n$. To prove it, we need the following Lemma:

Lemma 5.2.24 *For every $d < c < 0$, there exists $\kappa > 0$ such that for every $\varphi \in \mathrm{SH}_-(\mathbb{D})$ with $\varphi(0) > c$ and for every point a in the $\mathcal{F}$-open set*

$$V = \{z \in D(0, 1/8) : \varphi(z) > c\},$$

the set $\Omega = \{z \in \mathbb{D} : \varphi(z) \geq d\}$ contains a circle $C(a, \delta_{\varphi,a})$ with radius $\delta_{\varphi,a} > \kappa$, $D(a, r)$ denoting the disk of center a and radius $r > 0$ in $\mathbb{C}$.

Proof After multiplying φ by a positive constant we can assume that $d = -1$. Moreover, we may assume that the set $E = \{z \in \mathbb{D} : \varphi(z) < d\}$ is nonempty since otherwise the lemma trivially holds. Let $a \in V$ be fixed. We will first prove the following estimate

$$\varphi(a) \leq \omega(a, E_a^\circ, D(a, 3/4))^*, \tag{5.2.5}$$

where $E_a^\circ = \{a + |z - a| : z \in E \cap D(a, 3/4)\}$. Let f be the function $f(z) = z + a$. Note that the circular projection commutes with f^{-1}, i.e. $f^{-1}(E_a^\circ) = (f^{-1}(E \cap D(a, \frac{3}{4})))^\circ$. Hence, to prove (5.2.5) it is enough, in view of the conformal invariance of the harmonic measure, to prove that the estimate (5.2.5) holds for the particular point $a = 0$, i.e.

$$\varphi(0) \leq \omega(0, E^\circ, D(0, 3/4))^*. \tag{5.2.6}$$

By Theorem 5.2.22, there is an increasing sequence of compact subsets K_j of E° such that

$$\omega(0, K_j, D(0, 3/4))^* = \omega(0, K_j, D(0, 3/4)) \downarrow \omega(0, E^\circ, D(0, 3/4))^*. \quad (5.2.7)$$

The equality in (5.2.7) holds because K_j is compact and $0 \notin K_j$, see Theorem 1.3.3(a). Let $\varepsilon > 0$. It follows from (5.2.7) that there exists a natural number j_0 such that

$$\omega(0, K_{j_0}, D(0, 3/4)) \leqslant \omega(0, E^\circ, D(0, 3/4))^* + \varepsilon. \quad (5.2.8)$$

We can find a compact set $L \subset E$ such that $L^\circ = K_{j_0}$. By Theorem 5.2.23 together with inequality (5.2.8) we get

$$\omega(0, L, D(0, 3/4)) \leq \omega(0, K_{j_0}, D(0, 3/4)) \leq \omega(0, E^\circ, D(0, 3/4))^* + \varepsilon. \quad (5.2.9)$$

Since $L \subset E$, and $\varphi(z) < -1$ for all $z \in E$, inequality (5.2.9) implies the following estimate

$$\varphi(0) \leqslant \omega(0, E^\circ, D(0, 3/4))^* + \varepsilon. \quad (5.2.10)$$

As ε is arbitrary, the estimate (5.2.6), and therefore also (5.2.5), follows. Let now $\alpha \in]0, 1/4[$ be a constant such that

$$I = \{z \in D(a, 3/4) : \mathfrak{Im} z = \mathfrak{Im} a \text{ and } \mathfrak{Re} a + \alpha \leqslant \mathfrak{Re} z \leqslant 1/2\} \subset E_a^\circ.$$

Then by (5.2.5)

$$\varphi(a) \leq \omega(a, E_a^\circ, D(a, 3/4))^* \leq \omega(a, I, D(a, 3/4)). \quad (5.2.11)$$

Again by the conformal invariance, (5.2.11) yields

$$\varphi(a) \leq \omega(0, f^{-1}(I), D(0, 3/4)). \quad (5.2.12)$$

Since $f^{-1}(I) = [\alpha, 1/2 - \mathfrak{Re} a]$, it follows that

$$\varphi(a) \leq \omega(0, [\alpha, 3/8], D(0, 3/4)). \quad (5.2.13)$$

Let $\alpha_j \downarrow 0$ be a sequence decreasing to 0. Since $j \mapsto \omega(0, [\alpha_j, 3/8], D(0, 3/4))$ decreases to -1 (see, for example, [Hel, Theorem 8.38]), there exists a constant $0 < \kappa < 3/8$ depending only on c but not on the function φ such that

$$\omega(0, [\kappa, 3/8], D(0, 3/4)) < c. \quad (5.2.14)$$

The last inequality together with (5.2.13) hence show that for all $a \in V$, the interval $J = \{z \in D(a, 3/4) : \mathfrak{Im} z = \mathfrak{Im} a \text{ and } \kappa + \mathfrak{Re} a \leq \mathfrak{Re} z \leq 3/8\}$ cannot be a subset of E_a°. We conclude that there exists a $\delta_{\varphi,,a} \in [\kappa, 1/2]$ such that

$$\{z : |z-a| = \delta_{\varphi,a}\} \subset \Omega = \{z \in D : \varphi(z) \geq d\}. \tag{5.2.15}$$

□

Lemma 5.2.25 *Let $\varphi \in \mathrm{SH}(\mathbb{D})$ such that $0 \leq \varphi \leq 1$ and $U = \{z \in \mathbb{D} : \varphi > 0\}$. Suppose that there exists a piecewise-$\mathcal{C}^1$ Jordan curve $\gamma \subset \mathbb{D}$ such that $\gamma \subset U$. Let Γ be the bounded component of the complement of γ. Then $W = U \cap \Gamma$ is polygonally connected and, hence finely connected.*

Proof For each $z \in \mathbb{C}$ we denote by $Q(z, d)$ the (open) square of center z and diameter d, with sides parallel to coordinates axis and by $S(z, d)$ the boundary of $Q(z, d)$. The proof of [Hel, Theorem 10.14] can be adapted to show that every finely open subset $V \subset \mathbb{C}$ contains boundaries of squares centered at a fixed point $z \in V$ and with arbitrarily small diameters. Let j be an integer ≥ 1. For every $z \in \gamma$ there exists a real $0 < d_z < \frac{1}{j}$ such that $S(z, d_z) \subset U$ and $\varphi > \varphi(z)/2$ on $S(z, d_z)$. The squares $Q(z, d_z)$, $z \in \gamma$, cover γ and, since γ is compact, we can find a minimal finite set $J_j = \{1, \ldots, m_j\} \subset \mathbb{N}$ such that $\gamma \subset \bigcup_{k \in J_j} Q(z_k, d_{z_k})$. Now $\Omega_j = \bigcup_{k \in J_j} Q(z_k, d_{z_k})$ is an open neighborhood of γ and its boundary is contained in

$$\{\varphi > \frac{1}{2} \min\{\varphi(z_k) : 1 \leq k \leq m_j\}\}.$$

Since γ is locally connected, the boundary of Ω_j will consist of two polygonal curves if j is sufficiently large. One of these components, say, γ_j is contained in Γ. Denote by Γ_j the bounded component of the complement of γ_j. Let $0 < \varepsilon < \frac{1}{2} \min\{\varphi(z_k) : 1 \leq k \leq m_j\}$ and $K_\varepsilon^j = \{\varphi \geq \varepsilon\} \cap \overline{\Gamma}_j$. Then K_ε^j is a compact subset of $\mathbb{D}$. Since $\partial\Gamma_j (= \gamma_j)$ is contained in K_ε^j, an easy application of the maximum principle shows that K_ε^j is connected. Let $z_1, z_2 \in K_\varepsilon^j$, then, by repeating the above argument, we can find for every $\delta > 0$ a polygonal curve C contained in $K_{\varepsilon/2}^j$ such that $\max\{(d(z_1, C), d(z_2, C)\} < \delta$. A well-known result, cf. [Ran, Corollary 5.4.4], states that for $z \in \mathbb{D}$ and almost $\theta \in [0, 2\pi]$

$$\lim_{r \to 0} \varphi(z + re^{i\theta}) = \varphi(z).$$

Hence there exists a polygonal line in $K_{\varepsilon/2}^j$ that connects z_1 and z_2. Letting $\varepsilon \to 0$ we conclude that $U \cap \overline{\Gamma}_j = \bigcup_{\varepsilon > 0} K_j^\varepsilon$ is polygonally connected. Since every interval is finely connected so is $U \cap \overline{\Gamma}_j$. Finally, since $W = \bigcup_{j \geq 1} U \cap \overline{\Gamma}_j$, we conclude that W is finely connected. □

Proposition 5.2.26 *Let $U \subset \mathbb{C}^n$ be an $\mathcal{F}$-open set and let $a \in U$. Then there exists a constant $\kappa = \kappa(U, a)$ and an $\mathcal{F}$-neighborhood $V \subset U$ of a with the property that for any complex line L through $v \in V$ the $\mathcal{F}$-component of the $\mathcal{F}$-open set $U \cap L$ that contains v, contains a circle about v with radius at least κ.*

Proof Let $a \in U$. By Theorem 5.2.8 there exists a constant $r > 0$, and a plurisubharmonic function $\varphi \in \mathrm{PSH}(B(a,r))$ such that $\{z \in B(a,r) : \varphi(z) > -1\}$ is an $\mathcal{F}$-neighborhood of a contained in U. Since the pluri-fine topology is biholomorphically invariant, there is no loss of generality if we assume that $r = 2$, $\varphi \leq 0$, $a = 0$ and $\varphi(0) = -1/4$. Let

$$V = \{z \in B(0, 1/8) : \varphi(z) > -1/2\}.$$

Let $v \in V$ and let L be a complex line through v. The restriction φ_L of φ to $B(v,1) \cap L$ is subharmonic and satisfies the conditions of Lemma 5.2.24 with $c = -1/2$. Consequently, there exists a constant κ (not depending on φ_L), such that the set $\{z \in B(v,1) : \varphi(z) \geq -1\} \cap L$, and therefore $U \cap L$, contains a circle with radius $\delta_{\varphi_L, v} \in]\kappa, \frac{1}{2}[$ about v. It follows from Lemma 5.2.25 that the set $U \cap L \cap B(v, \delta_L)$ is $\mathcal{F}$-connected. This completes the proof of the proposition. □

Notes: Proposition 5.2.3, Theorems 5.2.4, 5.2.8 and its corollaries, and Theorem 5.2.13 are due to Bedford and Taylor [BT3]. Theorem 5.2.14 is new. Theorems 5.2.17, 5.2.18, Lemma 5.2.19, Theorem 5.2.20 and Corollary 5.2.21, Lemmas 5.2.24 and 5.2.25 and Proposition 5.2.26 are due to El Marzguioui and Wiegerinck [EMW1]. The other results of this section are due to El Kadiri, Fuglede, and Wiegerinck [EKFW].

5.3 Applications of the Pluri-Fine Topology to the Pluripotential Theory

In this section, we give some applications of the pluri-fine topology to pluripotential theory on $\mathbb{C}^n$. The first one concerns the quasi-Brelot property for plurisubharmonic functions. The following theorem extends up to pluripolar sets the Brelot property for subharmonic functions on an open subset of $\mathbb{R}^N$ (Theorem 1.5.29).

Theorem 5.3.1 (Quasi-Brelot property) *Let u be a plurisubharmonic function on an open set $\Omega \subset \mathbb{C}^n$. Then there exists a pluripolar set $E \subset \Omega$ such that for every $z \in \Omega \backslash E$ we can find an $\mathcal{F}$-neighborhood $\mathcal{O}_z \subset \Omega$ of z such that u is continuous in the usual sense in $\mathcal{O}_z$.*

Proof Without loss of generality we may assume that Ω is a Euclidean ball B and that u is continuous near the boundary of B. By the quasi-continuity property, see Theorem 5.1.8, we can select a sequence of relatively compact open subsets ω_j of B, such that the Monge-Ampère capacity $C(\omega_j, B) < \frac{1}{j}$, and u is continuous on $B \backslash \omega_j$. Denote by $\widetilde{\omega}_j$ the $\mathcal{F}$-closure of ω_j. The pluriharmonic measure $u^*_{\omega_j, B}$ is equal to the pluriharmonic measure $u^*_{\widetilde{\omega}_j, B}$ because for a plurisubharmonic function ψ on B the set $\{\psi \leq -1\}$ is $\mathcal{F}$-closed, thus $\psi|_{\omega_j} \leq -1 \Leftrightarrow \psi|_{\widetilde{\omega}_j} \leq -1$. Now, using [Kl, Proposition 4.7.2]

$$C(\omega_j, B) = C^*(\omega_j, B) = \int_\Omega (dd^c u^*_{\omega_j,B})^n = \int_\Omega (dd^c u^*_{\widetilde{\omega}_j,B})^n = C^*(\widetilde{\omega}_j, B). \tag{5.3.1}$$

Let $E = \bigcap_j \widetilde{\omega}_j$. By Proposition 5.1.2(1) and (5.3.1), $C^*(E, B) \le C^*(\omega_j, B) \le \frac{1}{j}$, for every j. Hence E is a pluripolar subset of B, according to [Kl, Theorem 4.7.5]. Let $z \notin E$. Then there exists j_0 such that $z \notin \widetilde{\omega}_{j_0}$. Clearly, the set $B \backslash \widetilde{\omega}_{j_0}$ is an $\mathcal{F}$-neighborhood of z. Since u is continuous on $B \backslash \omega_{j_0}$, it is also continuous on the smaller set $B \backslash \widetilde{\omega}_{j_0} (\subset B \backslash \omega_{j_0})$. □

The second application of the pluri-fine topology is the following result and its consequences:

Theorem 5.3.2 *Let (T_j) denote a sequence of positive currents of bidegree (k, k) such that $T_j \to T$ on an open set $\Omega \subset \mathbb{C}^n$, in the weak topology on the space of currents. Then the following are equivalent (where convergence is in the sense of currents of order 0).*

(1) *T has zero mass on any pluripolar set and $uT_j \to uT$ for every locally bounded plurisubharmonic function u on Ω,*

(2) *T has zero mass on any pluripolar set and $\psi T_j \to \psi T$ for every bounded, quasi-continuous function ψ on Ω,*

(3) *T has zero mass on any pluripolar set and $\psi_j T_j \to \psi T$ for each uniformly bounded sequence of quasi-continuous functions (ψ_j) which converge monotonically, either increasing or decreasing, to ψ quasi-everywhere,*

(4) *the sequence (T_j) puts uniformly small mass on sets of small capacity; i.e. if $\omega \Subset \Omega$, then for any sequence $(\mathcal{O}_j)$ of open subsets of ω, with $\lim_{j\to\infty} C(\mathcal{O}_j) = 0$, we have*

$$\lim_{j\to\infty} \sup\{|T|(\mathcal{O}_j), |T_k|(\mathcal{O}_j) : k = 1, 2, \dots\} = 0.$$

Proof That (3) ⇒ (2) ⇒ (1) is trivial. We will next show that (1) ⇒ (4). Assume not. Then we can find an open set $\omega \Subset \Omega$, open sets $\mathcal{O}_j \subset \omega$, $j = 1, 2, \dots$, and a number $\delta > 0$ such that $|T_j|(\mathcal{O}_j) = \int_{\mathcal{O}_j} \beta_{n-k} \wedge T_j \ge \delta > 0$ and $C(\mathcal{O}_j) \le 2^{-j}$. Here $\beta_j = (4^{-j}/j!)(dd^c|z|^2)^j$. The assertions (1) and (4) are local, so it is no loss of generality to assume that $\omega = \{|z| < r\} \Subset \{|z| < 1\} = \Omega$. Consider the extremal functions $u_j = u^*_{\mathcal{O}_j}$, for $\mathcal{O}_j$ relative to the unit ball Ω. Since $C(\mathcal{O}_j) \to 0$, the bounded functions $u_j \to 0$ almost everywhere in Ω. Replacing $\mathcal{O}_j$ by $\bigcup_{l\ge j} \mathcal{O}_l$, if necessary, we can assume that $\mathcal{O}_1 \supset \mathcal{O}_2 \supset \dots$, so that $u_1 \le u_2 \le \cdots$. Further, $u_j \to 0$ almost everywhere, hence also pointwise outside a set E of capacity zero, i.e. a pluripolar set (see Theorem 5.1.10). It then also follows that $u_j \to 0$ locally in L^1. Hence, we can further assume that $\sum |u_j|$ converges locally in L^1.

For each $l = 1, 2, \dots$, let $\widetilde{v}_l = \sum_{j\ge l} u_j$ and $v_l = \max\{\widetilde{v}_l, -1\}$. Then $\widetilde{v}_l$ and v_l are plurisubharmonic on Ω, $v_1 \le v_2 \le \cdots$, and $v_l \to 0$ locally in L^1. It therefore follows that $v_l \to 0$, except possibly on a set E of capacity zero according to Theorem 5.1.1(iv) and Theorem 5.1.11. Thus, if $v = \sup v_l$, then $v\beta_{n-k} \wedge T = 0$. But, from the bounded convergence theorem, $v_l\beta_{n-k} \wedge T \to 0$. Thus, since $v_l\beta_{n-k} \wedge T$ is a negative Borel measure, we can choose l so large that

$$-\frac{\delta}{2} < \int_\omega v_l \beta_{n-k} \wedge T \leq 0$$

For this fixed value of l, we apply the hypothesis (1) to the plurisubharmonic function v_l. Then

$$\begin{aligned}-\frac{\delta}{2} &< \lim_{j\to+\infty} \int_\omega v_l \beta_{n-k} \wedge T_j \leq \lim_{j\to+\infty} \int_{\mathcal{O}_j} u_j \beta_{n-k} \wedge T_j \\ &\leq - \lim_{j\to+\infty} \int_{\mathcal{O}_j} \beta_{n-k} \wedge T_j \leq -\delta\end{aligned}$$

which is a contradiction. Thus, (1) $\Rightarrow$ (4) is proved.

It remains to show that (4) $\Rightarrow$ (3). The capacity is subadditive, $C(\bigcup E_j) \leq \sum_j C(E_j)$. Therefore, if the ψ_j, ψ, are quasi-continuous and $\psi_j \to \psi$, monotonically, quasi-everywhere, then for each $\varepsilon > 0$, there exists an open set $\mathcal{O}$ with $C(\mathcal{O}) < \varepsilon$, ψ_j, ψ continuous on $\Omega\backslash\mathcal{O}$, and $\psi_j \to \psi$ monotonically on $\Omega\backslash\mathcal{O}$. It then follows from Dini's theorem that $\psi_j \to \psi$ uniformly on compact subsets of $\Omega\backslash\mathcal{O}$. Thus we can write $\psi_j = \widetilde{\psi}_j + \eta_j$, $\psi = \widetilde{\psi} + \eta$, where $\widetilde{\psi}_j$, $\widetilde{\psi}$ are continuous on Ω, $|\widetilde{\psi}_j| \leq \sup|\widetilde{\psi}_j| \leq M$, $\eta_j, \eta = 0$ on $\Omega\backslash\mathcal{O}$, and $\widetilde{\psi}_j \to \widetilde{\psi}$ uniformly on compact subsets of $\Omega\backslash\mathcal{O}$. Thus, if φ is any continuous $(n-k, n-k)$-form with compact support in Ω, then

$$\begin{aligned}&\limsup_{j\to\infty} \left| \int \varphi \wedge (\psi_j T - \psi T) \right| \\ &\quad \leq \limsup_{j\to\infty} \left| \int \varphi \wedge ((\widetilde{\psi}_j - \widetilde{\psi}) T_j) \right| \\ &\quad + \left| \int_\Omega \varphi \wedge (\widetilde{\psi}(T_j - T)) \right| + \left| \int_\Omega \varphi \wedge (\eta_j T_j - \eta T) \right|.\end{aligned}$$

Because $\widetilde{\psi}_j \to \widetilde{\psi}$ uniformly on $(\Omega\backslash\mathcal{O}) \cap \operatorname{Supp}\varphi$, the first term in the last expression does not exceed

$$\limsup_{j\to+\infty} \left| \int_{\mathcal{O}} \varphi \wedge ((\widetilde{\psi}_j - \widetilde{\psi}) T_j) \right| \leq C \limsup_{j\to+\infty} |T_j|(\mathcal{O}),$$

where C is a constant depending on a maximum of ψ_j, ψ and φ. This term thus tends to zero as $C(\mathcal{O}) \to 0$, by hypothesis (4). The second term $\to 0$, because $T_j \to T$ weakly as currents on Ω and φ, $\widetilde{\psi}$ are continuous on Ω. The integrand in the third term is uniformly bounded and equal to zero off of $\mathcal{O}$. Hence, the third term does not exceed const$(|T_j|(\mathcal{O}) + |T|(\mathcal{O}))$, which also tends to zero as $C(\mathcal{O}) \to 0$. Thus we have $\psi_j T_j \to T$, as asserted. This completes the proof. □

Let (u_j^0), ..., (u_j^n), $j = 1, 2, \ldots$, denote $n+1$ sequences of plurisubharmonic functions on Ω, and consider the following types of currents:

$$\begin{aligned} S_j &= dd^c u_j^1 \wedge \ldots \wedge dd^c u_j^n, \\ T_j &= u_j^0 dd^c u_j^1 \wedge \ldots \wedge dd^c u_j^n = u_j^0 S_j, \\ U_j &= du_j^0 \wedge d^c u_j^1 \wedge \ldots \wedge dd^c u_j^n. \end{aligned} \tag{5.3.2}$$

Suppose that the sequences (u_j^k) are uniformly bounded and converge monotonically almost everywhere, either increasing or decreasing, to plurisubharmonic functions $u^k, k = 0, 1, \ldots, n$. Let S, T, U denote the corresponding currents with u_j^k replaced by u^k. We then have the following corollary of Theorem 5.3.2.

Corollary 5.3.3 *Let (X_j), X denote any of the 3 sequences of currents S_j, S, etc. of (5.3.2). If $u_j^0 \geq 0$, then we have for any $\mathcal{F}$-open subset $\mathcal{O}$ of Ω,*

$$\int_{\mathcal{O}} X \leq \liminf_{j \to +\infty} \int_{\mathcal{O}} X_j.$$

Proof The currents X_j are all positive, and satisfy (4) of Theorem 5.3.2. Let $\mathcal{O} \subset \Omega$ be $\mathcal{F}$-open, $K \subset \mathcal{O}$ (Euclidean) compact and let $\varepsilon > 0$. The characteristic function $1_{\mathcal{O}}$ of $\mathcal{O}$ being $\mathcal{F}$-l.s.c., it follows from the regularity of the topology $\mathcal{F}$ (see Corollary 5.2.11), the quasi-Lindelöf property and Theorem 5.2.14 that there is an increasing sequence of quasi-continuous functions $f_j \geq 0$ on Ω such that $1_{\mathcal{O}} = \sup_j f_j$ q.e., and hence by Theorem 5.3.2(2) we have for every continuous function φ with compact support such that $1_K \leq \varphi \leq 1$

$$\int_K X \leq \sup_k \int \varphi f_k X \leq \int \varphi f_{k_0} X + \varepsilon = \lim_{j \to \infty} \int \varphi f_{k_0} X_j + \varepsilon \leq \liminf \int_{\mathcal{O}} X_j + \varepsilon.$$

for some integer k_0. By varying ε and K we obtain the stated inequality. □

Corollary 5.3.4 *Let (u_j^1), ..., (u_j^n) be monotone sequences of bounded plurisubharmonic functions on a Euclidean domain Ω, converging respectively to bounded plurisubharmonic functions u_1, ..., u_n. Then for every bounded $\mathcal{F}$-continuous function φ on Ω with compact support, we have*

$$\lim_{j \to +\infty} \int \varphi dd^c u_j^1 \wedge \ldots \wedge dd^c u_j^n = \int \varphi dd^c u_1 \wedge \ldots \wedge dd^c u_n.$$

Proof Since φ is quasi-continuous by Theorem 5.2.14, this follows immediately from Theorem 5.3.2(2). □

Corollary 5.3.5 *Let u be a negative locally bounded plurisubharmonic function on an open set $\Omega \subset \mathbb{C}^n$, and let $\mathcal{O} \subset \Omega$ be an $\mathcal{F}$-open set. If*

$$\widetilde{u}_{\Omega \setminus \mathcal{O}} = (\sup\{v \in \mathrm{PSH}_-(\Omega) : v \leq u \text{ on } \Omega \setminus \mathcal{O}\})^*$$

then $(dd^c \widetilde{u}_{\Omega \setminus \mathcal{O}})^n = 0$ on $\mathcal{O}$.

Proof By the quasi-Lindelöf property, we may assume that $\mathcal{O}$ is a Borel set (since no mass is put on a pluripolar set). Thus we may take open sets $\mathcal{O}_1 \supset \mathcal{O}_2 \dots$ such that $\widetilde{u}_{\Omega\setminus\mathcal{O}_j}$ increases to $\widetilde{u}_{\Omega\setminus\mathcal{O}}$ q.e. Since $(dd^c\widetilde{u}_{\Omega\setminus\mathcal{O}_j})^n$ puts no mass on $\mathcal{O}$, the result follows from Corollary 5.3.3. □

Lemma 5.3.6 *Let $(u_k^1), \dots, (u_k^n)$ and $(v_k^1), \dots, (v_k^n)$ be monotonically decreasing (or increasing) sequences of locally bounded plurisubharmonic functions on an open set $\Omega \subset \mathbb{C}^n$ that converge to plurisubharmonic functions $u_1, \dots, u_n$ and $v_1, \dots, v_n$ that are locally bounded on Ω. Let $\mathcal{O} \subset \Omega$ be an $\mathcal{F}$-open set. Suppose that for all k*

$$(dd^c u_1^k \wedge \dots \wedge dd^c u_n^k)|_{\mathcal{O}} = (dd^c v_1^k \wedge \dots \wedge dd^c v_n^k)|_{\mathcal{O}}.$$

Then

$$(dd^c u_1 \wedge \dots \wedge dd^c u_n)|_{\mathcal{O}} = (dd^c v_1 \wedge \dots \wedge dd^c v_n)|_{\mathcal{O}}.$$

Proof By Theorem 5.2.8 and the Proof of Theorem 5.2.5 and the quasi-Lindelöf property, we may write $\mathcal{O}$ (modulo a pluripolar set) as a countable union of sets of the form $B \cap \{\psi > 0\}$ where B is an open ball, $B \Subset \Omega$, and $\psi \in \mathrm{PSH}(\Omega)$. Since pluripolar sets have measure zero, it suffices to prove the result for $\mathcal{O} = B \cap \{\psi > 0\}$. Obviously there is an $\mathcal{F}$-continuous function $\widetilde{\psi}$ with compact support in Ω such that $\{z \in \Omega : \widetilde{\psi}(z) > 0\} = \mathcal{O}$. By Corollary 5.3.4,

$$\lim_{j\to+\infty} \int \widetilde{\psi} dd^c u_1^j \wedge \cdots \wedge dd^c u_n^j = \int \widetilde{\psi} dd_1^u \wedge \cdots \wedge dd^c u_n$$

and

$$\lim_{j\to+\infty} \int \widetilde{\psi} dd^c v_1^j \wedge \cdots \wedge dd^c v_n^j = \int \widetilde{\psi} dd^c v_1 \wedge \cdots \wedge dd^c v_n$$

This holds also if $\widetilde{\psi}$ is replaced by $f\widetilde{\psi}$ for any continuous f. We conclude that

$$\int f\widetilde{\psi} dd^c u_1 \wedge \cdots \wedge dd^c u_n = \int f\widetilde{\psi} dd^c v_1 \wedge \cdots \wedge dd^c v_n$$

holds for all $f \in \mathcal{C}(\Omega)$. Thus $dd^c u_1 \wedge \cdots \wedge u_n - dd^c v_1 \wedge \cdots dd^c v_n$ vanishes on $\{\widetilde{\psi} > 0\}$, which proves the lemma. □

Proposition 5.3.7 *Let $u_1, \dots, u_n, v_1, \dots, v_n$ be locally bounded plurisubharmonic functions on an open set $\Omega \subset \mathbb{C}^n$, let $1 \le m \le n$, and let $\mathcal{O} \subset \{u_1 > v_1\} \cap \cdots \cap \{u_m > v_m\}$ be $\mathcal{F}$-open. Then*

$$\begin{aligned}(dd^c \max(u_1, v_1) \wedge \cdots \wedge dd^c \max(u_m, v_m) \wedge dd^c u_{m+1} \wedge \cdots \wedge dd^c u_n)|_{\mathcal{O}} \\ = (dd^c u_1 \wedge \cdots \wedge dd^c u_n)|_{\mathcal{O}}.\end{aligned}$$

Proof In case $\mathcal{O}$ is a Euclidean open set the statement is obvious. Let $(u_1^k), \dots, (u_m^k)$ be decreasing sequences of smooth plurisubharmonic functions that converge, respectively, to $u_1, \dots, u_m$. We put $\mathcal{O}_k = \bigcap_{j=1}^m \{u_j^k > v_j\}$. Then for every k the set $\mathcal{O}_k$ is a Euclidean open set on which $\max\{u_j^k, v_j\} = u_j^k$, and $\mathcal{O} \subset \mathcal{O}_k$. Therefore

$$(dd^c \max(u_1^k, v_1) \wedge \cdots \wedge dd^c \max(u_m^k, v_m) \wedge dd^c u_{m+1} \cdots \wedge dd^c u_n)|_{\mathcal{O}} \\ = (dd^c u_1^k \wedge \cdots \wedge dd^c u_m^k \wedge dd^c u_{m+1} \cdots \wedge dd^c u_n)|_{\mathcal{O}}.$$

By letting $k \to \infty$ and using Lemma 5.3.6 this leads to

$$(dd^c \max(u_1, v_1) \wedge \cdots \wedge dd^c \max(u_m, v_m) \wedge dd^c u_{m+1} \cdots \wedge dd^c u_n)|_{\mathcal{O}} \\ = (dd^c u_1 \wedge \cdots \wedge dd^c u_n)|_{\mathcal{O}}.$$

□

Corollary 5.3.8 *Let $u_1, \dots, u_n, v_1, \dots, v_n$ be locally bounded plurisubharmonic functions on Ω and let $\mathcal{O} \subset \Omega$ be $\mathcal{F}$-open. If $u_1 = v_1, \dots, u_n = v_n$ on $\mathcal{O}$, then*

$$(dd^c u_1 \wedge \cdots \wedge dd^c u_n)|_{\mathcal{O}} = (dd^c v_1 \wedge \cdots \wedge dd^c v_n)|_{\mathcal{O}}.$$

Proof Let $k > 0$. As $u_j = \max\{u_j, v_j - \frac{1}{k}\}$, $j = 1, \dots, n$, on $\mathcal{O}$, we have by Proposition 5.3.7

$$(dd^c u_1 \wedge \cdots \wedge dd^c u_n)|_{\mathcal{O}} = (dd^c \max(u_1, v_1 - \frac{1}{k}) \wedge \cdots \wedge dd^c \max(u_n, v_n - \frac{1}{k}))|_{\mathcal{O}}.$$

Letting $k \to \infty$, we obtain from Lemma 5.3.6

$$(dd^c u_1 \wedge \cdots \wedge dd^c u_n)|_{\mathcal{O}} = (dd^c \max(u_1, v_1) \wedge \cdots \wedge dd^c \max(u_n, v_n))|_{\mathcal{O}}.$$

Similarly we find

$$(dd^c v_1 \wedge \cdots \wedge dd^c v_n)|_{\mathcal{O}} = (dd^c \max(u_1, v_1) \wedge \cdots \wedge dd^c \max(u_n, v_n))|_{\mathcal{O}},$$

which completes the proof. □

Corollary 5.3.9 *Let $\Omega \subset \mathbb{C}^n$ be open, and let $u, v \in \mathrm{PSH}(\Omega)$ be locally bounded. If $\mathcal{O} \subset \Omega$ is $\mathcal{F}$-open, and if $u = v$ on $\mathcal{O}$, then $(dd^c u)^n|_{\mathcal{O}} = (dd^c v)^n|_{\mathcal{O}}$, i.e. the measures agree on measurable subsets of $\mathcal{O}$.*

To prove Theorem 5.4.11 below we need some results on the notion of "*slicing*" following [BT4]. Recall that the *slice* of an $(n-1, n-1)$-current T with respect to a hyperplane $z_1 = a$ is the current on $z_1 = a$ given by the formula

$$\langle T, z_1, a\rangle(\Psi) = \lim_{\varepsilon \to 0} \frac{1}{\pi \varepsilon^2} \int_{|z_1 - a| \leq \varepsilon} \Psi(z_2, \dots, z_n) \frac{i}{2} dz_1 \wedge d\bar{z}_1 \wedge T \qquad (5.3.3)$$

where Ψ is a test form on $z_1 = a$. By [Fed, Section 4.3, p. 435] the slice of a normal (or flat) current exists for almost all $a \in \mathbb{C}$ and is characterized by the formula

$$\int \varphi(z_1)\Psi(z_2,\ldots,z_n)\wedge T = \int_{a\in\mathbb{C}} \varphi(a)\wedge\{\langle T,z_1,a\rangle(\Psi)\} \tag{5.3.4}$$

for all test forms $\Psi(z_2,\ldots,z_n)$ and $(1,1)$-test forms $\varphi(z_1)$. From (5.3.4) and the identity

$$\begin{aligned}\int \varphi(z_1)\Psi(z')w(z)(dd^c u)^{n-1} \qquad\qquad\qquad\qquad\qquad\qquad (5.3.5)\\ = \int_{z_1}\varphi(z_1)\wedge\left\{\int_{z'}\Psi(z')w(z_1,z')(dd^c u(z_1,z'))^{n-1}\right\}\end{aligned}$$

where $z'=(z_2,\ldots,z_n)$, it follows that when $T=w(dd^c u)^{n-1}$ with u,w bounded plurisubharmonic functions, then

$$\langle T,z_1,a\rangle = \frac{1}{2\pi}w(a,z')(dd^c u(a,z'))^{n-1}$$

for almost $a\in\mathbb{C}$. The formula (5.3.5) is obvious when u,w are smooth and holds by a limiting argument in the general case. Similar remark applies to the currents $u_1dd^cu_2\wedge\ldots\wedge dd^cu_n$ and $wddu_1\wedge d^cu_2\wedge dd^cu_3\wedge\ldots\wedge dd^cu_n$ and the u_j are locally bounded plurisubharmonic functions.

In fact, these slicing formulas hold for all hyperplanes $z_1=a$, not just almost all hyperplanes. This will be proved in Proposition 5.3.13.

Let $u_1,\ldots,u_k\in \mathrm{PSH}(\Omega)\cap L^\infty_{loc}(\Omega)$ and $v\in\mathrm{PSH}(\Omega)$. The current $vdd^cu_1\wedge\cdots\wedge dd^cu_k$ is of locally finite mass and the current $dd^cv\wedge dd^cu_1\wedge\cdots\wedge dd^cu_k = dd^c(vdd^cu_1\wedge\cdots\wedge dd^cu_k)$ is a closed positive current of bidegree $(k+1,k+1)$. For a proof of this fact we refer to [D].

Lemma 5.3.10 *Let $T=dd^cu_1\wedge\cdots\wedge dd^cu_{n-1}$, where $u_j\in\mathrm{PSH}(\Omega)\cap L^\infty_{loc}(\Omega)$. Let $v\in\mathrm{PSH}(\Omega)$ and $v_a=\max\{v,a\}$, $a\in\mathbb{R}$. Then for all $u\in\mathrm{PSH}(\Omega)\cap L^\infty_{loc}(\Omega)$*

$$\lim_{a\to-\infty} udd^cv_a\wedge T = udd^cv\wedge T.$$

Proof We have for any test function ψ on Ω,

$$\begin{aligned}\int\psi udd^cv_a\wedge T &= \int v_a dd^c(\psi uT)\\ &= \int v_a\psi dd^cu\wedge T + 2\int v_a d\psi\wedge d^cu\wedge T + \int v_a udd^c\psi\wedge T.\end{aligned}$$

The first and third integrals converge to the corresponding integrals with v_a replaced by v, as $a\to-\infty$. The second integral differs from the corresponding integral by

$$\left|\int (v_a - v)d\psi \wedge d^c u \wedge T\right|$$
$$\leq \left|\int (v_a - v)d\psi \wedge d^c\psi \wedge T\right|^{\frac{1}{2}} \left|\int_{\mathrm{Supp}(\psi)} (v_a - v)du \wedge d^c u \wedge T\right|^{\frac{1}{2}}$$
$$= I_1(a)I_2(a).$$

Now, $I_1(a) \to 0$ by the dominated convergence theorem since vT has locally finite mass. It is no loss of generality to assume $u \geq 0$ on a neighborhood of the support of ψ so that u^2 is plurisubharmonic and $2du \wedge d^c u \leq dd^c u^2$. But $vdd^c(u^2) \wedge T$ has locally finite mass so $I_2(a) \to 0$ as $a \to -\infty$. Thus

$$\lim_{a\to-\infty} \int \psi u dd^c v_a \wedge T = \int v dd^c(\psi u) \wedge T$$

which is the conclusion of the Lemma. □

Corollary 5.3.11 *Let* $T = dd^c u_1 \wedge \cdots \wedge dd^c u_{n-1}$, *where* $u_j \in \mathrm{PSH}(\Omega) \cap L^\infty_{loc}(\Omega)$. *Let* $v \in \mathrm{PSH}(\Omega)$ *and* $v_a = \max\{v, a\}$, $a \in \mathbb{R}$. *On the set* $\{v > a\}$ *we have*

$$dd^c v_a \wedge T = dd^c v \wedge T.$$

Proof If $a > b > -\infty$, then it follows from Corollary 5.3.8 that the currents $dd^c v_a \wedge T$ and $dd^c v_b \wedge T$ agree on the finely open set $\{v > a\}$. Thus,

$$(v - a)^+(dd^c v_a \wedge T - dd^c v_b \wedge T) \equiv 0.$$

By Lemma 5.3.10, this current has limit

$$(v - a)^+(dd^c v_a \wedge T - dd^c v \wedge T)$$

as $b \to -\infty$. Hence $dd^c v_a \wedge T = dd^c v \wedge T$ on the set $\{v > a\}$. □

Let $T = dd^c u_1 \wedge \cdots \wedge u_{n-1}$, where $u_1, \ldots, u_{n-1} \in \mathrm{PSH}(\Omega) \cap L^\infty_{loc}(\Omega)$. Let $v \in \mathrm{PSH}(\Omega)$ and $a \in \mathbb{R}$. We define

$$S(T; a, v) = S(T; a) = 1_{\{v \leq a\}} dd^c v_a \wedge T,$$

where $v_a = \max\{v, a\}$.

The *residual mass* of T on the set $\{v = -\infty\}$ is the (n, n)-current

$$S(T; v) = S(T) = 1_{\{v=-\infty\}} dd^c v \wedge T.$$

We clearly have

$$S(T; v) = \lim_{a\to-\infty} 1_{\{v\leq a\}} dd^c v \wedge T.$$

Proposition 5.3.12 *For any* $v \in \mathrm{PSH}(\Omega)$ *we have*

$$\lim_{a \to -\infty} S(T; a, v) = S(T; v).$$

Proof We have

$$dd^c v \wedge T = 1_{\{v \le a\}} dd^c v \wedge T + 1_{\{v > a\}} dd^c v \wedge T$$

and

$$dd^c v_a \wedge T = 1_{\{v \le a\}} dd^c v_a \wedge T + 1_{\{v > a\}} dd^c v_a \wedge T.$$

By Corollary 5.3.11 the last terms in each of these equalities are equal. From Lemma 5.3.10 with $u \equiv 1$, the limit as $a \to -\infty$ of $dd^c v_a \wedge T$ is $dd^c v \wedge T$. Thus

$$S(T, v) = \lim_{a \to -\infty} 1_{\{v \le a\}} dd^c v \wedge T = \lim_{a \to -\infty} 1_{\{v \le a\}} dd^c v_a \wedge T.$$

□

Proposition 5.3.13 *Let* u, w *be locally bounded plurisubharmonic functions on an open set* $\Omega \subset \mathbb{C}^n$. *If* T *denotes the current* $w(dd^c u)^{n-1}$, *then the slice of* T *with respect to the hyperplane* $z_1 = 0$ *exists. Further,*

$$\langle T, z_1, 0 \rangle = \frac{1}{2\pi} w(0, z')(dd^c u(0, z'))^{n-1}$$

where $z' = (z_2, \ldots, z_n)$. *The analogous formula holds for*

$$T = w du_1 \wedge d^c u_2 \wedge dd^c u_3 \wedge \ldots \wedge dd^c u_n$$

and

$$T = w dd^c u_2 \wedge \ldots \wedge dd^c u_n$$

where w *and the* u_i *are locally bounded plurisubharmonic functions.*

Proof We will prove the proposition only for the case $T = w(dd^c u)^n$. The other cases may be proved by exactly the same argument. Or, one can use a polarization identity to reduce to the case considered here, along with the identity $dd^c u^2 = 2du \wedge d^c u + 2u dd^c u$. For $\varepsilon > 0$, let $v_\varepsilon = \max\{\log|z|, \log\varepsilon\}$. Then we have for any test function ψ on Ω

$$\int \psi w dd^c v_\varepsilon \wedge (dd^c u)^{n-1} = \frac{1}{2\varepsilon^2} \int_{|z_1| \le \varepsilon} (\psi w) dd^c |z_1|^2 \wedge (dd^c u)^{n-1}$$
$$+ \frac{1}{2\varepsilon^2} \int_{|z_1| \le \varepsilon} (\varepsilon^2 - |z_1|^2) dd^c(\psi w) \wedge (dd^c u)^{n-1}.$$

(The formula is easily verified when w, u are smooth, and the general case is obtained from a limiting argument.) The first term on the right-hand side is 2π times the integral on the right-hand side of (5.3.3) when ψ is independent of z_1. The second term on the right-hand side tends to 0 as $\varepsilon \to 0$, since it involves a bounded function times the currents $dd^c\psi \wedge (dd^cu)^{n-1}$, $d\psi \wedge d^cw \wedge (dd^cu)^{n-1}$ or $\psi dd^cw \wedge (dd^cu)^{n-1}$. All of these put zero mass on the pluripolar set $\{z_1 = 0\}$ and, hence, have the total mass on the "collar" $|z_1| \leq \varepsilon$, which goes to zero as $\varepsilon \to 0$. On the left-hand side, we have

$$\begin{aligned} wdd^cv_\varepsilon \wedge (dd^cu)^{n-1} &= 1_{\{|z_1|\leq\varepsilon\}} wdd^cv_\varepsilon \wedge (dd^cu)^{n-1} \\ &= S(w(dd^cu)^{n-1}; \log\varepsilon, \log|z_1|). \end{aligned}$$

By Lemma 5.3.10 and Proposition 5.3.12, it follows that the limit of the left-hand side exists and equals

$$1_{\{z_1=0\}} wdd^c \log|z_1| \wedge (dd^cu)^{n-1}.$$

Thus, the slice of $w(dd^cu)^{n-1}$ on $\{z_1 = 0\}$ exists and is equal to $T_0 = wdd^c \log|z_1| \wedge (dd^cu)^{n-1}$. (Note that T_0, a current on $\mathbb{C}^n$, is really supported on $\{z_1 = 0\}$ and hence can be identified with a current on $\{z_1 = 0\}$.)

We still have to show that

$$wdd^c \log|z| \wedge (dd^cu)^{n-1} = w(0, z') \wedge (dd^cu(0, z'))^{n-1}. \tag{5.3.6}$$

When u and w are smooth, this is obvious. The general case then follows by taking smooth u_j, w_j decreasing to u, w. Both sides of (5.3.6) converge according to [D, Théorème 2.6] and this completes the proof. □

Notes: Theorem 5.3.1 was proved by El Marzguioui and Wiegerinck in [EMW3]. Theorem 5.3.2 and its corollaries, Lemma 5.3.6, which is a generalization of Corollary 5.3.9 (due to Bedford and Taylor [BT3, Lemma 4.1]), Proposition 5.3.7 and Corollary 5.3.8 were obtained by El Kadiri and Wiegerinck in [EKW]. The part of this section concerning the slicing is taken from [BT4].

5.4 Plurifinely Plurisubharmonic Functions

For a compact set $K \subset \mathbb{C}^n$ we denote by $\mathrm{PSH}_0(K)$ the convex cone of all restrictions to K of finite continuous plurisubharmonic functions defined on open subsets of $\mathbb{C}^n$ containing K, and by $\mathrm{PSH}(K)$ the closure of $\mathrm{PSH}_0(K)$ in $C(K, \mathbb{R})$ (the continuous functions $K \to \mathbb{R}$ with the uniform norm); then $\mathrm{PSH}(K)$ is likewise a convex cone.

Definition 5.4.1 Let U denote an $\mathcal{F}$-open (i.e. plurifinely open) subset of $\mathbb{C}^n$.

(i) A function $u : U \to \mathbb{R}$ is said to be C-strongly $\mathcal{F}$-plurisubharmonic (or $\mathcal{F}$-*cpsh* in abbreviated form) if every point of U has a compact $\mathcal{F}$-neighborhood K in U such that $u|_K \in \mathrm{PSH}(K)$.

(ii) A function $u : U \to [-\infty, +\infty[$ which is not identically $-\infty$ on any $\mathcal{F}$-component of U, is said to be *strongly $\mathcal{F}$-plurisubharmonic* if u is the pointwise limit of a decreasing net of $\mathcal{F}$-*cpsh* functions on U.

(iii) A function $u : U \to [-\infty, +\infty[$ which is not identically $-\infty$ on any $\mathcal{F}$-component of U is said to be *weakly $\mathcal{F}$-plurisubharmonic*, or simply $\mathcal{F}$-*plurisubharmonic*, if u is $\mathcal{F}$-u.s.c. and, for every complex line L in $\mathbb{C}^n$, the restriction of u to the finely open subset $L \cap U$ of L is *finely hypoharmonic*.

The concept of $\mathcal{F}$-cpsh functions, defined in (i), is an auxiliary one. Every strongly $\mathcal{F}$-plurisubharmonic function is $\mathcal{F}$-u.s.c. (even $\mathcal{F}$-continuous).

The set of C-strongly plurisubharmonic, resp. $\mathcal{F}$-plurisubharmonic functions on the $\mathcal{F}$-open $U \subset \mathbb{C}^n$ is denoted by $\mathcal{F}$-CPSH(U), resp. $\mathcal{F}$-PSH(U). If a function u defined on a subset A of $\mathbb{C}^n$ containing U, we write $u \in \mathcal{F}$-PSH(U) if $u|_U \in \mathcal{F}$-PSH(U).

If u is strongly, resp. weakly, $\mathcal{F}$-plurisubharmonic on U ($\mathcal{F}$-open in $\mathbb{C}^n$) then the restriction of u to $L \cap U$ (L a $\mathbb{C}$-linear subspace $\cong \mathbb{C}^k$ of $\mathbb{C}^n$) has the same property in every $\mathcal{F}$-component of $L \cap U$ on which u is not identically $-\infty$. This follows easily from Proposition 5.2.15(a). For $n = 1$, u is strongly, resp. weakly, $\mathcal{F}$-plurisubharmonic on U (finely open in $\mathbb{C}$) if and only if u is finely subharmonic on U. This is obvious in the weak case. In the strong case, suppose first that u is finite and finely subharmonic on U. By the Brelot property (Theorem 1.5.29), every point of U has a compact fine neighborhood K in U such that $u|_K \in \mathcal{C}(K, \mathbb{R})$ (u being finely continuous by Theorem 2.1.26). Because u is finite and finely hypoharmonic in the fine interior K' of K we have $u \in S(K)$ according to [BH2, Theorem 4.7], or Theorem 2.8.1, and so u is $\mathcal{F}$-cpsh on U. For a general finely subharmonic function u on U write $u = \inf_{n\in\mathbb{N}} \max\{u, -n\}$ and note that $\max\{u, -n\}$ is finite and finely hypoharmonic, cf. Corollary 2.1.21. Conversely, if u is strongly $\mathcal{F}$-plurisubharmonic we may assume by the same corollary that u is even $\mathcal{F}$-cpsh. For any compact set $K \subset \mathbb{C}$, every function of class PSH(K) is finite and finely subharmonic on K' according to Corollary 2.1.21. With K as in (i) this shows that u indeed is finite and finely subharmonic on U.

By an $\mathcal{F}$-domain we understand an $\mathcal{F}$-connected $\mathcal{F}$-open set.

The following theorem, which is an immediate consequence of Definition 5.4.1 and properties of finely hypoharmonic functions, gathers the first important properties of $\mathcal{F}$-plurisubharmonic functions that will often be used in this chapter.

Theorem 5.4.2 *Let U be an $\mathcal{F}$-open subset of $\mathbb{C}^n$.*

(i) *Sheaf property: The $\mathcal{F}$-plurisubharmonic functions on $\mathcal{F}$-open subsets of U form a sheaf.*

(ii) *The family $\mathcal{F}$-PSH(U) is a convex cone, that is, if α, β are positive numbers and $u, v \in \mathcal{F}$-PSH(U), then $\alpha u + \beta v \in \mathcal{F}$-PSH$(U)$.*

(iii) $u_1 \vee \ldots \vee u_k \in \mathcal{F}$-PSH$(U)$ *for any* $u_1, \ldots, u_k \in \mathcal{F}$-PSH$(U)$.

(iv) *If U is $\mathcal{F}$-connected and if $(u_j)_{j\in\mathbb{N}}$ is a decreasing sequence of functions in $\mathcal{F}$-PSH(U), then $u := \lim_{j\to\infty} u_j \in \mathcal{F}$-PSH$(U)$ or $u \equiv -\infty$.*

(v) *If $u : U \to \mathbb{R}$ and if $(u_j)_{j\in\mathbb{N}}$ is a sequence of functions in $\mathcal{F}$-PSH(U) converging to u uniformly on compact subsets of U, then $u \in \mathcal{F}$-PSH(U).*

Property (i) means that (1) if u is $\mathcal{F}$-plurisubharmonic on an $\mathcal{F}$-open set $U \subset \mathbb{C}^n$ then the restriction of u to any $\mathcal{F}$-open subset of U is $\mathcal{F}$-plurisubharmonic, and (2) a function u on an $\mathcal{F}$-open set $U \subset \mathbb{C}^n$ is $\mathcal{F}$-plurisubharmonic on U provided that u has $\mathcal{F}$-plurisubharmonic restrictions to each member of a family $\mathfrak{V}$ of $\mathcal{F}$-open sets having the union U. This property follows easily from (d) in Theorem 5.4.3 below and the properties of finely hypoharmonic functions on finely open subsets of $\mathbb{C}$.

The set $\mathcal{F}$-CPSH(U) has the same properties (i), (ii), (iii), (iv) and (v).

The cone $\mathcal{F}$-PSH(U) is furthermore stable under pointwise infimum for lower directed (possibly infinite) families whose pointwise infimum is not identically $-\infty$ on each $\mathcal{F}$-component of U, and closed under $\mathcal{F}$-locally uniform convergence in view of Corollary 2.1.21. For upper directed families of $\mathcal{F}$-plurisubharmonic functions, see Theorem 5.4.14 below.

Theorem 5.4.3 *Let u be an $\mathcal{F}$-plurisubharmonic function on an $\mathcal{F}$-open set $U \subset \mathbb{C}^n$.*

(a) *Every point $z_0 \in U$ such that $u(z_0) > -\infty$ has an $\mathcal{F}$-open $\mathcal{F}$-neighborhood $\mathcal{O} \subset U$ on which u can be represented as the difference $u = v_1 - v_2$ between two bounded plurisubharmonic functions v_1 and v_2 defined on some open ball $B(z_0, r)$ containing $\mathcal{O}$.*

(b) *r, $\mathcal{O}$, and v_2 can be chosen independently of u provided that u maps U into a prescribed bounded interval $]a, b[$.*

(c) *u is $\mathcal{F}$-continuous.*

(d) *If U is $\mathcal{F}$-connected then $\{z \in U : u(z) = -\infty\}$ is a relatively $\mathcal{F}$-closed, pluripolar subset of U.*

Proof (a) To begin with, suppose that u is *bounded*. We may then assume that $-1 < u < 0$, for u maps U into a bounded interval $]a, b[$, and hence $\frac{u-b}{b-a}$ maps U into $]-1, 0[$ and is likewise weakly $\mathcal{F}$-plurisubharmonic. Let $V \subset \mathbb{C}^n$ be a compact $\mathcal{F}$-neighborhood of z_0 such that z_0 is not a Euclidean interior point of V. Since the complement $\complement V$ of V is pluri-thin at z_0, there exists $0 < r < 1$ and a plurisubharmonic function v on $B(z_0, r)$ such that

$$\limsup_{z\to z_0, z\in B(z_0,r)\setminus V} v(z) < v(z_0).$$

Without loss of generality we may suppose that v is negative on $B(z_0, r)$ and

$$v(z) = -1 \text{ on } B(z_0, r)\setminus V \quad \text{and} \quad v(z_0) = -1/2.$$

Hence

$$u(z) + \lambda v(z) \le -\lambda \quad \text{for } z \in U \cap B(z_0, r)\setminus V \text{ and } \lambda > 0. \tag{5.4.1}$$

Now define a function u_λ on $B(z_0, r)$ by

$$u_\lambda(z) = \begin{cases} \max\{-\lambda, u(z) + \lambda v(z)\} & \text{if } z \in U \cap B(z_0, r) \\ -\lambda & \text{if } z \in B(z_0, r)\backslash V. \end{cases} \tag{5.4.2}$$

This definition makes sense because $\big(U \cap B(z_0, r)\big) \bigcup \big(B(z_0, r)\backslash V\big) = B(z_0, r)$, and the two definitions agree on $U \cap B(z_0, r)\backslash V$ in view of (5.4.1).

Clearly, u_λ is weakly $\mathcal{F}$-plurisubharmonic on $U \cap B(z_0, r)$ and on $B(z_0, r)\backslash V$, hence on all of $B(z_0, r)$ in view of the sheaf property, cf. Theorem 5.4.2(i). Since u_λ is bounded on $B(z_0, r)$, it follows from Theorem 2.1.22 that u_λ is subharmonic on each complex line where it is defined. It is well known that a bounded function, which is subharmonic on each complex line where it is defined, is plurisubharmonic, cf. [Lel5, Theorem 4, p. 30]. Thus, u_λ is plurisubharmonic on $B(z_0, r)$.

Since $v(z_0) = -1/2$, the set $\mathcal{O} = \{z \in B(z_0, r) : v(z) > -3/4\}$ is an $\mathcal{F}$-neighborhood of z_0, and because $v = -1$ on $B(z_0, r)\backslash V$ it is clear that $\mathcal{O} \subset V \subset U$.

Observe now that $-4 \le u(z) + 4v(z)$ for every $z \in \mathcal{O}$. Hence $u = v_1 - v_2$ on $\mathcal{O}$, with $v_1 = u_4$ and $v_2 = 4v$, both plurisubharmonic on $B(z_0, r)$. Thus v is weakly $\mathcal{F}$-plurisubharmonic on $\mathcal{O}$, which is an $\mathcal{F}$-neighborhood of z_0. It follows that u is $\mathcal{F}$-continuous on $\mathcal{O}$ along with v_1 and v_2, provided that u is bounded.

Without assuming that u be bounded, u remains $\mathcal{F}$-continuous in $\mathcal{O}$ according to (c), proven below. It follows that u is bounded on some $\mathcal{F}$-neighborhood W of z_0 in U, and we therefore have a decomposition of u as required, on some $\mathcal{F}$-neighborhood (replacing the above $\mathcal{O}$) of z_0 on $W \subset U$.

(b) Again we may assume that $-1 < u < 0$. The set V and the plurisubharmonic function v in the proof of (a) then do not depend on u, and that applies to $v_2 = 4v$ as well.

(c) In the remaining case where u may be unbounded (cf. the proof of (a) above), note that u is $\mathcal{F}$-u.s.c. and $< +\infty$. Choose $c, d \in \mathbb{R}$ with $d < c$. Then the set $U_c = \{z \in U : u(z) < c\}$ is $\mathcal{F}$-open. The function $\max\{u, d\}$ is bounded and weakly $\mathcal{F}$-plurisubharmonic on U_c, hence $\mathcal{F}$-continuous there. The set

$$\{z \in U : d < u(z) < c\} = \{z \in U_c : d < \max\{u(z), d\} < c\}$$

is therefore $\mathcal{F}$-open, and hence u is $\mathcal{F}$-continuous.

For (d) we refer to the proof given in [EMW3, Theorem 4.1]. □

The next proposition generalizes Lemma 2.1.51 to the pluri-fine situation.

Proposition 5.4.4 *Let V and U be $\mathcal{F}$-open sets in $\mathbb{C}^n$ such that $V \subset U$. Suppose that $u \in \mathcal{F}$-$\mathrm{PSH}(U)$, $v \in \mathcal{F}$-$\mathrm{PSH}(V)$, and $\mathcal{F}$-$\limsup_{z \in V, z \to \zeta} v(z) \le u(\zeta)$ for all $\zeta \in \partial_{\mathcal{F}} V \cap U$. Then the function*

$$w = \begin{cases} \max\{u, v\} & \text{on } V \\ u & \text{on } U\backslash V \end{cases}$$

is $\mathcal{F}$-plurisubharmonic on U.

Proof It follows easily from the hypotheses that the function w is $\mathcal{F}$-u.s.c. on U and not identically $-\infty$ on any $\mathcal{F}$-component of U because $w \geq u$ on U. Let L be a complex line of $\mathbb{C}^n$ such that $L \cap U \neq \varnothing$. If $L \cap V = \varnothing$ we have $w|_{L\cap U} = u|_{L\cap U}$ and the function $w|_{L\cap U}$ is finely hypoharmonic. Next suppose that $L \cap V \neq \varnothing$. The functions u and v are finely hypoharmonic on $L \cap U$ and $L \cap V$ respectively. Moreover, we have $\partial_f(L \cap V) \subset L \cap \partial_{\mathcal{F}} V$ and

$$\text{f-}\limsup_{z\in L\cap V, z\to\zeta} v(z) \leq \mathcal{F}\text{-}\limsup_{z\in V, z\to\zeta} v(z) \leq u(\zeta)$$

for any $\zeta \in \partial_f(L \cap V)$ because the $\mathcal{F}$-topology on U induces on $L \cap U$ the fine topology, according to Proposition 5.2.15(a). Hence the function $w|_{L\cap U}$ is finely hypoharmonic by Lemma 2.1.51. It then follows that w is $\mathcal{F}$-plurisubharmonic on U. □

Proposition 5.4.5 *Let u be a strongly $\mathcal{F}$-plurisubharmonic function on an $\mathcal{F}$-open set $U \subset \mathbb{C}^n$, then u is weakly $\mathcal{F}$-plurisubharmonic.*

Proof In view of Theorem 5.4.2(iv) we may assume that u is even $\mathcal{F}$-cpsh. Let (u_j) be a sequence of finite continuous plurisubharmonic functions on open sets Ω_j containing K from (i) in Definition 5.4.1 such that $u_j|_K \to u|_K$ uniformly. For any complex line L in $\mathbb{C}^n$, $u_j|_{|L\cap K''}$ is finely hypoharmonic. This uses Theorem 2.1.22 and the fact that the intersection of any $\mathcal{F}$-open subset of $\mathbb{C}^n$ with any complex line L is finely open (in L), by Proposition 5.2.15. It follows by Corollary 2.1.21 that $u|_{L\cap K''}$ is finely hypoharmonic, and in particular finely continuous. Consequently, u is indeed weakly $\mathcal{F}$-plurisubharmonic. □

Proposition 5.4.6 *Let U be a Euclidean open subset of $\mathbb{C}^n$. For a function $u : U \to [-\infty, +\infty[$ the following are equivalent:*

(i) *u is plurisubharmonic (in the ordinary sense).*
(ii) *u is strongly $\mathcal{F}$-plurisubharmonic on any component of U.*
(iii) *u is weakly $\mathcal{F}$-plurisubharmonic on any component of U.*

Proof Every finite continuous plurisubharmonic function on U is of course $\mathcal{F}$-cpsh. It follows that any plurisubharmonic function on U is strongly $\mathcal{F}$-plurisubharmonic (being locally the pointwise limit of a decreasing sequence of finite continuous plurisubharmonic functions). Conversely, if u is weakly $\mathcal{F}$-plurisubharmonic on U then u is plurisubharmonic on every component of U. To see this, we can assume without loss of generality that $U$is connected and hence $\mathcal{F}$-connected by Theorem 5.2.17. First observe that u is $\mathcal{F}$-locally bounded from above, so for every point $a \in U$ we can find an $\mathcal{F}$-neighborhood $V \subset U$ where $u < M_a$, say. According to Proposition 5.2.26 we may further arrange that there exists $\delta > 0$ such that, for every complex line L passing through a, the intersection $V \cap L$ contains a circle C_L about a with radius at least δ. Let D_L be the disk in L bounded by C_L. Then $D_L \subset U \cap L$ and C_L is the fine boundary of D_L. By the maximum principle for finely hypoharmonic functions on a planar domain (see Lemma 2.7.1), it follows that $u < M_a$ on the disks D_L, bounded by these circles, hence in particular on the ball $B(a, \delta) = \bigcup_L D_L$. For functions that are locally bounded from above in the Euclidean topology, the statement follows easily from Theorem 2.1.22(a) in view of [Lel5, Theorem 4, p. 30]. □

It follows from Proposition 5.4.6 that the family of $\mathcal{F}$-plurisubharmonic functions on a Euclidean open set $U \subset \mathbb{C}^n$ is the cone PSH(U).

Recall that by fine topology on $\mathbb{C}^n$ we mean the Cartan fine topology on $\mathbb{C}^n \cong \mathbb{R}^{2n}$ and that this topology is finer than the pluri-fine topology, see Theorem 5.2.16. A well-known characterization of plurisubharmonic functions by Lelong (see [Kl, Theorem 2.9.12]) may be adapted as follows, and will lead to further properties of weakly $\mathcal{F}$-plurisubharmonic functions (Theorems 5.4.13 and 5.4.14).

Lemma 5.4.7 *Let u_1, u_2 be bounded subharmonic functions on an open set $B \subset \mathbb{R}^n$, and consider the function $u = u_1 - u_2$ on B. Let U be a finely open Borel subset of B. Then $u|_U$ is finely subharmonic if and only if the signed Riesz measure Δu on B has a positive restriction to U.*

Proof Let μ_1 and μ_2 be the positive measures on B given by $\mu_1 = \Delta u_1$ and $\mu_2 = \Delta u_2$, see Sect. 1.1.6. Suppose first that $(\Delta u)|_U \geq 0$, and let us prove that u then is finely subharmonic on U. We may suppose that B is bounded, the general case is obtained by covering B by a countable set of bounded open sets. Writing $\Delta u := \mu = \mu_+ - \mu_-$ on B we have, by hypothesis, $(\mu_-)(U) = 0$ and $\mu = \mu_1 - \mu_2$ so that $\mu_- \leq \mu_2$ and thus $G\mu_- \leq G\mu_2 < +\infty$ because u_2 is bounded on B, G denoting the Green kernel on B. It then follows that the function $G\mu$ is finely hyperharmonic by Theorem 2.1.6, and indeed (being also bounded) finely superharmonic on U. By the Riesz representation theorem, $u = -G\mu + h$ on B, with h harmonic on B, in particular finely harmonic on U, and u is therefore likewise finely subharmonic on U. Conversely, suppose that $u|_U$ is finely subharmonic. Recall the corollary in [Do4, 1.XI.18] that if two subharmonic functions g_1 and g_2, defined on some open set, coincide on a set A, then their Riesz masses satisfy $\Delta g_1 = \Delta g_2$ on the fine interior of A. Hence the Riesz measure $\Delta u = \Delta u_1 - \Delta u_2$ on U is independent of the choice of u_1 and u_2. In fact, if $u = w_1 - w_2$ (with w_1 and w_2 subharmonic on U) then $u_1 + w_2 = u_2 + w_1$ on U, hence $\Delta u_1 + \Delta w_2 = \Delta u_2 + \Delta w_1$ on (the fine interior of) U, that is, $\Delta w_1 - \Delta w_2 = \Delta u_1 - \Delta u_2$ on U.

Since u is finely subharmonic on U it follows by the proof of Theorem 2.1.25 that every point $x \in U$ has a fine neighborhood $V_x \subset \overline{V}_x \subset U$ in which we can write $u = v_1 - v_2$, where v_1 and v_2 are superharmonic functions on some Euclidean open neighborhood B_0 of $\overline{V}_x$ in B. Replacing if necessary V_x by $U \cap \mathfrak{r}(V_x)$ we may suppose in view of Corollary 1.4.18 that V_x is a Borel set. Moreover, v_2 is the swept function on $B_0 \setminus V_x$ of a certain superharmonic function ≥ 0 on B_0. The Riesz mass of v_2 is concentrated on the fine boundary of the complement of V_x, cf. e.g. [Do4, Theorem 1.XI.14 (b)], hence on the fine boundary of V_x. It follows that the Riesz mass Δu of u is positive on V_x for every x. By the quasi-Lindelöf property, we can find countably many $x_j \in U$ such that $U = \bigcup_j V_{x_j} \cup E$, where E is polar. Clearly Δu is positive on $\bigcup_j V_{x_j}$. Because E is polar, the Riesz mass of a bounded subharmonic function does not charge E, so we have $\Delta u_i(E) = 0$ $(i = 1, 2)$. We conclude that Δu is positive on U. □

Lemma 5.4.8 *Let u be a bounded finely subharmonic function on an $\mathcal{F}$-open set $U \subset \mathbb{C}^n$ and suppose that for every $\mathbb{C}$-affine bijection h of $\mathbb{C}^n$ the function $u \circ h$ is finely subharmonic on $h^{-1}(U)$. Then every $z_0 \in U$ admits a (compact) $\mathcal{F}$-neighborhood K_{z_0} such that u can be written as*

$$u = u_1 - u_2, \quad on\ K_{z_0},$$

where u_1, u_2 are plurisubharmonic functions defined on a ball $B_{z_0} \supset K_{z_0}$.

Proof We proceed as in the proof of Theorem 5.4.3(a). We can assume that $-1 < u < 0$. There exists a compact $\mathcal{F}$-neighborhood V of z_0 and a negative plurisubharmonic function v on a ball $B(z_0, r) \supset V$ such that $v(z_0) = -1/2$ and $v \leq -1$ on $\complement V$. Then for every $\lambda > 0$ the function

$$u_\lambda(z) = \begin{cases} \max\{-\lambda, u(z) + \lambda v(z)\} & \text{for } z \in B(z_0, r) \cap U \\ -\lambda & \text{for } z \in B(z_0, r) \backslash U_{z_0} \end{cases}$$

is a bounded finely subharmonic function on $B(z_0, r)$, hence subharmonic on B_{z_0} by Theorem 2.1.22. Similarly for every $\mathbb{C}$-affine bijection h of $\mathbb{C}^n$ the function $u_\lambda \circ h$ is finely subharmonic, hence subharmonic on $h^{-1}(B(z_0, r))$. From this we conclude that u_λ is in fact plurisubharmonic. Taking $\lambda = 4$, we see that

$$u_4(z) = u(z) + 4v(z)$$

on the $\mathcal{F}$-neighborhood $K_{z_0} = \{v \geq -2/3\} \subset V \cap B(z_0, r)$ of z_0, and K_{z_0} is compact along with V. This proves the lemma. □

Corollary 5.4.9 *Keep the notation as above, Then u is $\mathcal{F}$-continuous on K_{z_0}, hence on U. Moreover, $dd^c u$ is the restriction to K_{z_0} of a well-defined closed Hermitian $(1, 1)$ current with measure coefficients on $B(z_0, r)$.*

For $\varepsilon = (\varepsilon_1, ..., \varepsilon_n)$, $(n \geq 2)$, we write $\varepsilon' = (\varepsilon_2, ..., \varepsilon_n)$, $\varepsilon'^2 = \Pi_{j=2}^n \varepsilon_j^2$ and $|z'| < \varepsilon'$ for $|z_j| < \varepsilon_j$, $j = 2, \cdots, n$.

Lemma 5.4.10 *Let $D \subset \mathbb{C}^n$ be a bounded domain, $\psi = \psi(z_1)$ a test function on $\{z \in D : z' = 0'\}$, and let w and u be bounded plurisubharmonic functions on D. Then*

$$\begin{aligned} &\int_{\{z_2=0,\dots,z_n=0\}} \psi(z_1) w(z_1, 0') dd^c u(z_1, 0') \\ &= \lim_{\varepsilon' \to 0'} \frac{1}{2^{n-1}\varepsilon'^2} \int_{\{|z'| < \varepsilon'\}} \psi(z_1) w(z) dd^c |z_2|^2 \wedge \ldots \wedge dd^c |z_n|^2 \wedge dd^c u. \end{aligned} \tag{5.4.3}$$

Proof Apply slicing (cf. Proposition 5.3.13) with respect to $z_2 = 0$ to the current $T = wdd^c|z_3|^2 \wedge \ldots \wedge dd^c|z_n|^2 \wedge dd^c u$ to obtain

$$\langle T, z_2, 0\rangle = \frac{1}{2\pi} w(z_1, 0, z_3, \ldots, z_n) dd^c|z_3|^2 \wedge \cdots \wedge dd^c|z_n|^2 \wedge dd^c u.$$

Next in $\{z_2 = 0\}$, apply slicing with respect to $z_3 = 0$ to the current $wdd^c|z_4|^2 \wedge \cdots \wedge dd^c|z_n|^2 \wedge dd^c u$. Continuing in this fashion we obtain (5.4.3). □

Theorem 5.4.11 *A function $u : U \to [-\infty, +\infty[$ (U is $\mathcal{F}$-open in $\mathbb{C}^n$) is a weakly $\mathcal{F}$-plurisubharmonic function if and only if u is $\mathcal{F}$-locally bounded from above and $u \circ h$ is $\mathbb{R}^{2n}$-finely subharmonic in the $\mathcal{F}$-open set $h^{-1}(U)$ for every $\mathbb{C}$-affine bijection h of $\mathbb{C}^n$.*

Proof Since u is not identically $-\infty$ on any $\mathcal{F}$-component of U it follows that $u \circ h$ is not identically $-\infty$ on any $\mathcal{F}$-component of $h^{-1}(U)$. Let us remark that $u \circ h = \inf_p u_p \circ h$ where $u_p = \max\{u, -p\}$ $(p \in \mathbb{N})$, and hence, according to Theorem 5.4.2(iv), it suffices to prove the theorem for u finite. We also may assume, without loss of generality, that U is $\mathcal{F}$-connected and, by the sheaf property and the fact that U is the union of $\mathcal{F}$-open Borel sets (see Theorem 5.2.8), that U is a Borel set.

Let us first prove the "only if part" of the theorem. According to Theorem 5.4.3(a), every point $a \in U$ has a bounded $\mathcal{F}$-open $\mathcal{F}$-neighborhood $\mathcal{O} \subset U$ on which u is representable as $u = u_1 - u_2$, where u_1 and u_2 are bounded plurisubharmonic functions, defined on an open ball B in $\mathbb{C}^n$ containing $\mathcal{O}$. Let V be an $\mathcal{F}$-open set and K a compact set such that $a \in V \subset K \subset \mathcal{O}$. Recall that such V and K exist by Corollary 5.2.10. By shrinking $\mathcal{O}$ and replacing V by $\mathfrak{r}(V)$ $(\subset K)$ we may suppose that $\mathcal{O}$ and V are a Borel sets. For every $j \in \{1, \ldots, n\}$ the distributions

$$M_{i,j} = M_{i,j}(z) = \frac{\partial^2 u_i(z_1, \ldots, z_n)}{\partial z_j \partial \bar{z}_j}, \qquad i = 1, 2,$$

are well defined positive measures on B. Below we show that $(M_{1,j} - M_{2,j})|_K \geq 0$ and hence $\Delta(u_1 - u_2) = 4\sum_{j=1}^n (M_{1,j} - M_{2,j}) \geq 0$ on V, where Δ denotes the Laplacian on $\mathbb{C}^n \cong \mathbb{R}^{2n}$. According to Lemma 5.4.7 this implies that $u_1 - u_2$ indeed is $\mathbb{R}^{2n}$-finely subharmonic on V, and hence actually on all of U, by varying a.

For the proof that $(M_{1,j} - M_{2,j})|_K \geq 0$ it is convenient to write points of $\mathbb{C}^n$ as (z, w), now with $z \in \mathbb{C}$ and $w \in \mathbb{C}^{n-1}$. Each of the above measures $M_{i,j}$ on B then takes the form

$$M_i = M_i(z, w) = \frac{\partial^2 u_i(z, w)}{\partial z \partial \bar{z}}, \qquad i = 1, 2,$$

and we shall prove that $(M_1 - M_2)|_K \geq 0$.

For each $w \in \mathbb{C}^{n-1}$ define

$$B(w) = \{z \in \mathbb{C} : (z, w) \in B\}$$

(and similarly with B replaced by other subsets of $\mathbb{C}^n$). The functions $u_i(\cdot, w)$, $i = 1, 2$, induce subharmonic functions $u_i(\cdot, w)$ on the open subset $B(w)$ of $\mathbb{C}$, and the distributions

$$\mu_{i,w} = \mu_{i,w}(z) = \frac{\partial^2 u_i(z, w)}{\partial z \partial \bar{z}}, \qquad i = 1, 2,$$

are therefore positive measures on the open set $B(w)$ (if non-empty). Being weakly $\mathcal{F}$-plurisubharmonic on $\mathcal{O}$, $u = u_1 - u_2$ induces the finely subharmonic function $u(\cdot, w) = u_1(\cdot, w) - u_2(\cdot, w)$ on the finely open set $\mathcal{O}(w)$. According to the planar version of Lemma 5.4.7 applied to the induced bounded subharmonic functions $u_i(\cdot, w)$ on $B(w)$, $i = 1, 2$, the Riesz measure $\mu_{1,w} - \mu_{2,w}$ of $u(\cdot, w)$ is positive on the finely open set $\mathcal{O}(w) \subset B(w)$.

Let V_z, V_w denote Lebesgue measure on $\mathbb{C}$, $\mathbb{C}^{n-1}$, respectively. For any test function $\varphi \in C_0^\infty(B)$ we have by Fubini's theorem

$$\begin{aligned}
\int_B \varphi \, dM_i &= \int_B \frac{\partial^2 \varphi(z, w)}{\partial z \partial \bar{z}} u_i(z, w) \, dV_z \, dV_w \\
&= \int_{\mathbb{C}^{n-1}} \Big(\int_{B(w)} \frac{\partial^2 \varphi(z, w)}{\partial z \partial \bar{z}} u_i(z, w) \, dV_z \Big) dV_w \\
&= \int_{\mathbb{C}^{n-1}} \Big(\int_{B(w)} \varphi(z, w) \, d\mu_{i,w}(z) \Big) dV_w .
\end{aligned}$$

There exists a decreasing sequence of functions $\varphi_k \in C_0^\infty(B)$ with $0 \le \varphi_k \le 1$ so that $\varphi_k = 1$ on K and $\varphi_k \downarrow 1_K$ (the indicator function for K) as $k \uparrow \infty$. Since M_i and $\mu_{i,w}$ are locally finite positive measures and B and φ_k are bounded, we obtain by the monotone convergence theorem

$$\begin{aligned}
M_i(K) &= \int_B 1_K \, dM_l = \lim_{k \to \infty} \int_B \varphi_k \, dM_i \\
&= \lim_{k \to \infty} \int_{\mathbb{C}^{n-1}} \Big(\int_{B(w)} \varphi_k(\cdot, w) \, d\mu_{i,w} \Big) dV_w \\
&= \int_{\mathbb{C}^{n-1}} \Big(\int_{B(w)} 1_{K(w)} \, d\mu_{i,w} \Big) \, dV_w \\
&- \int_{\mathbb{C}^{n-1}} \mu_{l,w}(K(w)) \, dV_w .
\end{aligned}$$

It follows that

$$M_1(K) - M_2(K) = \int_{\mathbb{C}^{n-1}} \big(\mu_{1,w}(K(w)) - \mu_{2,w}(K(w)) \big) \, dV_w \ge 0$$

because $\mu_{1,w} - \mu_{2,w} \ge 0$ on $\mathcal{O}(w) \supset K(w)$, and the proof of the "only if" part of the theorem is complete.

We shall now prove the "if" part of the theorem. We keep the notations of Lemmas 5.4.8 and 5.4.10 and proceed as follows. First we will show that $dd^c u$ is positive on the compact $K = K_{z_0}$. Next we apply Lemma 5.4.10 to show that the restriction of u to any complex line through z_0 is finely subharmonic in a fine neighborhood of z_0.

Let v be a plurisubharmonic function on a ball B, let h be a $\mathbb{C}$-affine bijection of $\mathbb{C}^n$ and $\varphi \in C_0^\infty(B)$ a test function. Then the action of the Riesz measure $\Delta(v \circ h)$ on $\varphi \circ h$ can be expressed as follows

$$\begin{aligned}4^{n-1}(n-1)! \int_{h^{-1}(B)} & \varphi \circ h(z) \Delta(v \circ h) \\ &= \int_{h^{-1}(B)} \varphi \circ h(z) dd^c(v \circ h) \wedge \left(dd^c \|z\|^2\right)^{n-1} \\ &= \int_B \varphi(\zeta) dd^c v(\zeta) \wedge \left(dd^c \|h^{-1}(\zeta)\|^2\right)^{n-1}.\end{aligned}$$

Returning to u, we have by Lemma 5.4.7 that the Riesz measure $\Delta(u \circ h)$ is positive on $h^{-1}(K_{z_0})$, hence (with $h^{-1} = g$) we obtain that

$$dd^c u(\zeta) \wedge \left(dd^c \|g(\zeta)\|^2\right)^{n-1} \tag{5.4.4}$$

is a positive measure on K_{z_0} for every $\mathbb{C}$-affine bijection g of $\mathbb{C}^n$, and by continuity also for $\mathbb{C}$-affine map $g : \mathbb{C}^n \to \mathbb{C}^n$.

To finish the proof we want to show that for any complex line L passing through z_0, the restriction of u to $L \cap U$ is finely subharmonic in a fine neighborhood of z_0 relative to L. We write $z = (z_1, z')$ and can assume that $z_0 = 0$ and that L is given by $z' = 0$. Because K is an $\mathcal{F}$-neighborhood of 0, there exists a bounded positive plurisubharmonic function w defined on a ball B_0 about 0 such that $w = 0$ on $B_0 \backslash K$, while $w(0) > 0$. Then $\{w > 0\}$ is an $\mathcal{F}$-open subset of K that contains 0. On K we have $u = u_1 - u_2$, where u_1, u_2 are plurisubharmonic on a ball containing K, say on B_0. We apply Lemma 5.4.10 to u_1 and u_2 separately and subtract to obtain from (5.4.4) (with $g(z) = g(z_1, z') = (0, z')$)

$$\begin{aligned}&\int_{\{z_2=0,\dots,z_n=0\}} \psi(z_1) w(z_1, 0')\, dd^c u(z_1, 0') \\ &= \lim_{\varepsilon' \downarrow 0} \frac{1}{2^{n-1}\varepsilon'^2} \int_{\{|z'|<\varepsilon'\}} \psi(z_1) w(z) dd^c |z_2|^2 \wedge \ldots \wedge dd^c |z_n|^2 \wedge dd^c u.\end{aligned}$$

If ε' is sufficiently small then the integrals occurring in the limit on the right-hand side are positive for every positive test function ψ on $B_0 \cap \{z' = 0\}$.

We conclude that the Riesz measure of $u|_{L\cap U}$ is positive on a fine neighborhood of z_0. By Lemma 5.4.7, $u|_{L\cap U}$ is finely subharmonic on this neighborhood. Varying z_0 over L and using the sheaf property, we find that $u|_{L\cap U}$ is finely subharmonic. □

Corollary 5.4.12 *Let u be a bounded weakly $\mathcal{F}$-plurisubharmonic function on an $\mathcal{F}$-domain $U \subset \mathbb{C}^n$ such that u admits the representation $u = u_1 - u_2$ of Lemma 5.4.8 on U. Then for $\mathbb{C}$-affine functions $l_1, \ldots, l_{n-1}$ on $\mathbb{C}^n$ and $K = K_{z_0}$ from Lemma 5.4.8*

$$\int_U 1_K(z) dd^c|l_1|^2 \wedge \ldots \wedge dd^c|l_{n-1}|^2 \wedge dd^c u \geq 0.$$

Proof This follows from (5.4.4) with $g(z) = (0, l_1(z), \ldots, l_{n-1}(z))$. □

From Theorem 5.4.11 we derive the following two results, one about removable singularities for weakly $\mathcal{F}$-plurisubharmonic functions, and the other about the supremum of a family of such functions.

Theorem 5.4.13 *Let $u : U \to [-\infty, +\infty[$ be $\mathcal{F}$-locally bounded from above on an $\mathcal{F}$-open set $U \subset \mathbb{C}^n$, and let E be an $\mathcal{F}$-closed pluripolar subset of U. If u is (weakly) $\mathcal{F}$-plurisubharmonic on $U \backslash E$ then u has a unique extension to a weakly $\mathcal{F}$-plurisubharmonic function on all of U, and this extension u^* is given by*

$$u^*(z) = \mathcal{F}\text{-}\limsup_{\zeta \in U \backslash E, \zeta \to z} u(\zeta), \quad z \in U.$$

Proof The stated function u^* (the $\mathcal{F}$-u.s.c. regularization of u) equals u on the $\mathcal{F}$-open set $U \backslash E$ because u is $\mathcal{F}$-u.s.c. on $U \backslash E$. Furthermore, u^* is $\mathcal{F}$-u.s.c. and $< +\infty$ on all of U (finiteness because u is $\mathcal{F}$-locally bounded from above). By the "only if part" of Theorem 5.4.11, for any $\mathbb{C}$-linear bijection h of $\mathbb{C}^n$, $u^* \circ h$ therefore is $\mathbb{R}^{2n}$-finely subharmonic in $h^{-1}(U \backslash E) = h^{-1}(U) \backslash h^{-1}(E)$ and $\mathcal{F}$-u.s.c. $< +\infty$ on $h^{-1}(U)$. In particular, $u^* \circ h < +\infty$ is $\mathbb{R}^{2n}$-u.s.c. on $h^{-1}(U)$. Because E is pluripolar so is $h^{-1}(E)$, which thus is $\mathbb{R}^{2n}$-polar. According to Theorem 2.1.43, $u^* \circ h$ is therefore $\mathbb{R}^{2n}$-finely subharmonic on all of $h^{-1}(U)$, and so u^* is indeed weakly $\mathcal{F}$-plurisubharmonic on U, by the "if part" of Theorem 5.4.11. Because the pluripolar set E has empty $\mathcal{F}$-interior, u^* is the only weakly $\mathcal{F}$-plurisubharmonic and hence $\mathcal{F}$-continuous extension of u to U. □

Theorem 5.4.14 *Let U denote an $\mathcal{F}$-open subset of $\mathbb{C}^n$. For any $\mathcal{F}$-locally upper bounded family of weakly $\mathcal{F}$-plurisubharmonic functions u_α on U, the least $\mathcal{F}$-u.s.c. majorant u^* of the pointwise supremum $u = \sup_\alpha u_\alpha$ is likewise weakly $\mathcal{F}$-plurisubharmonic on U, and $\{z \in U : u(z) < u^*(z)\}$ is pluripolar.*

Proof We may assume that the set A of indices α is upper directed, that the net $(u_\alpha)_{\alpha \in A}$ is increasing and, furthermore that U is $\mathcal{F}$-connected. For any function $v : U \to [-\infty, +\infty[$ which is $\mathcal{F}$-locally bounded from above, write

$$v^*(z) = \mathcal{F}\text{-}\limsup_{\zeta \to z} v(\zeta), \quad \check{v}(z) = \mathcal{F}\text{-}\limsup_{\zeta \to z} v(\zeta),$$

where $\mathcal{F}$- lim sup is the lim sup taken in the sense of the (Cartan) fine (or the $\mathbb{R}^{2n}$-fine) topology on $\mathbb{C}^n \cong \mathbb{R}^{2n}$. Then $\check{u}(z) \leq u^*(z) < +\infty$, the former inequality because the

$\mathbb{R}^{2n}$-fine topology is finer than the $\mathcal{F}$-topology. As in Theorem 5.4.11, let $h : \mathbb{C}^n \to \mathbb{C}^n$ be a $\mathbb{C}$-affine bijection, and note that

$$u \circ h = \sup_\alpha (u_\alpha \circ h), \quad (u \circ h)\check{} = \check{u} \circ h, \qquad \text{in } h^{-1}(U),$$

the latter equation because h is an $\mathbb{R}^{2n}$-fine homeomorphism. By Theorem 5.4.11, $u_\alpha \circ h$ is $\mathbb{R}^{2n}$-finely subharmonic. Now $u_\alpha \circ h \le u^* \circ h$. Furthermore, u^* and hence $u^* \circ h$ and $\check{u} \circ h$ are $\mathbb{R}^{2n}$-finely locally bounded from above. It follows by Lemma 2.2.3 that $\check{u} \circ h = (u \circ h)\check{}$ is $\mathbb{R}^{2n}$-finely subharmonic.

We proceed to show that $\check{u} = u^*$ in U, and hence that $\check{u}$ is $\mathcal{F}$-u.s.c. there. Invoking also Theorem 5.4.11 we shall thus altogether find that $\check{u} = u^*$ becomes $\mathcal{F}$-plurisubharmonic on U, and in particular $\mathcal{F}$-continuous there, by Theorem 5.4.3(c). Consider first a point $z_0 \in U$ such that $u(z_0) > -\infty$. Fix $\beta \in A$ such that $u_\beta(z_0) > -\infty$, and choose an $\mathcal{F}$-open $\mathcal{F}$-neighborhood V of z_0 such that $V \subset U$ and

$$u_\beta(z_0) - 1 < u_\beta \le u^* < u(z_0) + 1 \qquad \text{on } V,$$

noting that the $\mathcal{F}$-plurisubharmonic function u_β is $\mathcal{F}$-continuous and that u^* is $\mathcal{F}$-u.s.c. and $< +\infty$. Since $u_\beta \le u_\alpha \le u$ for every $\alpha \succcurlyeq \beta$ in A, any such u_α maps V into some fixed bounded interval. According to Theorem 5.4.3(a), (b) there exist $r > 0$, an $\mathcal{F}$-open set $\mathcal{O}$ such that $z_0 \in \mathcal{O} \subset B(z_0, r)$, and locally bounded ordinary plurisubharmonic functions φ_α and ψ on $B(z_0, r)$ such that $u_\alpha = \varphi_\alpha - \psi$ on $\mathcal{O}$ for every $\alpha \succcurlyeq \beta$ in A. The net (φ_α) is increasing, along with the given net (u_α). The plurisubharmonic functions φ_α and ψ are $\mathcal{F}$-continuous, in particular $\mathbb{R}^{2n}$-finely continuous. Writing $\sup_\alpha \varphi_\alpha = \varphi$ and denoting by $\bar{\varphi}$ the Euclidean $\mathbb{R}^{2n}$-subharmonic regularization of φ in $B(z_0, r)$, we therefore have $\check{\varphi} = \bar{\varphi}$ there, by Brelot-Cartan's fundamental convergence theorem, see Sect. 1.1.13. Because $\check{\varphi} \le \varphi^* \le \bar{\varphi}$ it follows that $\check{\varphi} = \varphi^*$ in $B(z_0, r)$, and consequently

$$\check{u} = (\varphi - \psi)\check{} = \check{\varphi} - \psi = \varphi^* - \psi = (\varphi - \psi)^* = u^* \quad \text{in } \mathcal{O}$$

since ψ is $\mathcal{F}$-continuous and hence $\mathbb{R}^{2n}$-finely continuous on the $\mathcal{F}$-open, hence $\mathbb{R}^{2n}$-finely open set $\mathcal{O} \subset B(z_0, r)$.

Next, the set $\{z \in \mathcal{O} : u(z) < u^*(z)\} = \{z \in \mathcal{O} : \varphi(z) < \varphi^*(z)\}$ is pluripolar, by Theorem 5.1.11. Writing

$$E = \{z \in U : u(z) < u^*(z)\}, \quad e = \{z \in U : u(z) = -\infty\},$$

we have thus found that every point $z_0 \in U \backslash e$ has an $\mathcal{F}$-neighborhood $\mathcal{O}$ for which $\mathcal{O} \cap E$ is pluripolar. Because $e = \bigcap_\alpha \{z \in V : u_\alpha(z) = -\infty\}$ is $\mathcal{F}$-closed relative to V, and pluripolar by Theorem 5.4.3 (d) (since some u_α being $\not\equiv -\infty$), we infer by the quasi-Lindelöf principle (Theorem 5.2.13) that indeed E is pluripolar. Finally, we have found that u^* is $\mathcal{F}$-plurisubharmonic on each $\mathcal{F}$-open set $\mathcal{O}$ as above (as z_0 varies), and hence on their union $U \backslash e$, by the sheaf property. Because u^* is $\mathcal{F}$-u.s.c.

and $< +\infty$ on U, and that e is pluripolar, we conclude from Theorem 5.4.13 above that indeed u^* is weakly $\mathcal{F}$-plurisubharmonic in all of U. □

Taking for U a *Euclidean* open set Ω we obtain in particular the following

Corollary 5.4.15 *For any family $\{u_\alpha\}$ of ordinary plurisubharmonic functions on a Euclidean open set $\Omega \subset \mathbb{C}^n$ such that $u := \sup_\alpha u_\alpha$ is locally bounded from above, the least plurisubharmonic majorant of u exists and can be expressed as the u.s.c. regularization of u in the Euclidean topology on $\mathbb{C}^n$, as well as in the $\mathcal{F}$-topology and in the $\mathbb{R}^{2n}$-fine topology; that is, $\bar{u} = u^* = \check{u}$.*

The version of this involving the *Euclidean* topology is due to Lelong [Lel2], or see [Lel4, p. 26] or [Kl, Theorem 2.9.10]. Being locally bounded from above, u is in particular $\mathcal{F}$-locally bounded from above, and hence so is u^*, being $\mathcal{F}$-plurisubharmonic by Theorem 5.4.14. Because Ω is Euclidean open, it follows by Proposition 5.4.6 that u^* even is an ordinary plurisubharmonic function. From $u \le u^* \le \bar{u}$ it therefore follows that $u^* = \bar{u}$. Similarly, $\check{u} = \bar{u}$ in view of Theorem 2.1.22(a).

The identity $u^* = \bar{u}$ is perhaps new even in the Euclidean case.

Proposition 5.4.16 *Let u be a bounded weakly $\mathcal{F}$-plurisubharmonic function on an $\mathcal{F}$-domain $U \subset \mathbb{C}^n$ such that u admits the representation $u = u_1 - u_2$ of Lemma 5.4.8 on U. Then for holomorphic functions $g_1, \ldots, g_{n-1}$ on $K = K_{z_0}$ from Lemma 5.4.8*

$$\int_U 1_K(z) dd^c|g_1|^2 \wedge \ldots \wedge dd^c|g_{n-1}|^2 \wedge dd^c u \ge 0. \tag{5.4.5}$$

Proof Corollary 5.4.12 yields that (5.4.5) is valid for compact sets $\widetilde{K} \subset K_{z_0}$ and $\mathbb{C}$-affine functions g_i. For arbitrary holomorphic functions g_j we have

$$\int_U 1_K(z) dd^c|g_1|^2 \wedge \ldots \wedge dd^c|g_{n-1}|^2 \wedge dd^c u \tag{5.4.6}$$

$$= \lim_{N\to+\infty} \sum_{j=1}^{N} \int_U 1_{E_j^N}(z) dd^c|l_1^{j,N}|^2 \wedge \ldots \wedge dd^c|l_{n-1}^{j,N}|^2 \wedge dd^c u$$

for suitable quasi Borel sets E_j^N and complex affine approximants $l_k^{j,N}$ of g_k on E_j^N $(k = 1, \ldots, n-1)$. Hence the right-hand side of (5.4.6) is indeed positive. □

Theorem 5.4.17 *Let u be a weakly $\mathcal{F}$-plurisubharmonic function on an $\mathcal{F}$-domain $U \subset \mathbb{C}^n$:*

(a) *For any $\mathcal{F}$-closed set $E \subset \{z \in U : u(z) = -\infty\}$, $U \setminus E$ is an $\mathcal{F}$-domain.*

(b) *If $u \le 0$ then either $u < 0$ or $u = 0$.*

Proof (a) According to (d) in Theorem 5.4.3 E is pluripolar and hence $U\backslash E$ is an $\mathcal{F}$-domain by Theorem 5.2.18, E being $\mathcal{F}$-closed.

(b) Let $v = \inf_n nu$. Then, by Theorem 5.4.2(iv), v is $\mathcal{F}$-plurisubharmonic function or $v \equiv -\infty$. In the former case, the set $\{v = -\infty\} = \{u < 0\}$ is $\mathcal{F}$-open and $\mathcal{F}$-closed subset of U, and hence $\{u < 0\} = \varnothing$ or $\{u < 0\} = U$, so that $u < 0$ on U or $u \equiv 0$. In the latter case we obviously have $u < 0$ on U. □

Next we give two characterizations of functions of class $\widehat{\mathcal{D}}(\mathbb{C}^n)$ (see Sect. 4.2, with $\mathbb{C}^n \cong \mathbb{R}^{2n}$) which are $\mathcal{F}$-plurisubharmonic on an $\mathcal{F}$-open set. The Lebesgue measure on $\mathbb{C}^n$ will be denoted dV instead of $d\ell_{2n}$ as in Sect. 4.2.

Let D be a Greenian domain of $\mathbb{C}^n$. (For $n \geq 2$, we can take $D = \mathbb{C}^n$, and for $n = 1$, any bounded domain D will do). According to Theorem 4.2.21 an $\mathbb{R}^{2n}$-finely continuous (hence quasi-continuous) function $u \in \widehat{\mathcal{D}}(D)$ is finely subharmonic quasi-everywhere (hence actually everywhere by Theorem 2.1.43) on a $\mathbb{R}^{2n}$-finely open set $U \subset D$, if and only if $u < +\infty$ and the inequality sign holds in:

$$\frac{1}{4}\int_D \nabla u.\nabla\varphi dV = \sum_{j=1}^{n}\int_D (\partial_j u)(\overline{\partial}_j\varphi) dV \leq 0 \tag{5.4.7}$$

for every $\varphi \in \widehat{\mathcal{D}}_+(D, U)$. (It suffices of course to integrate over U.)

Theorem 5.4.18 *Let U be $\mathcal{F}$-open in $\mathbb{C}^n$ ($n \geq 2$). Given an $\mathcal{F}$-continuous function $u \in \widehat{\mathcal{D}}(\mathbb{C}^n)$ with values in $[-\infty, +\infty[$, the following are equivalent:*

(a) *u is weakly $\mathcal{F}$-plurisubharmonic on U,*

(b) *for every $\varphi \in \widehat{\mathcal{D}}_+(\mathbb{C}^n, U)$ and every $\lambda = (\lambda_1, \dots, \lambda_n) \in \mathbb{C}^n$,*

$$\sum_{j,k=1}^{n} \lambda_j\overline{\lambda}_k \int_U (\partial_j u)(\overline{\partial}_k\varphi) dV \leq 0,$$

(c) *for every regular holomorphic map $h : \omega \to \mathbb{C}^n$ (ω is open in $\mathbb{C}^n$), $u \circ h$ is weakly $\mathcal{F}$-plurisubharmonic on $h^{-1}(U)$ ($\subset \omega$).*

Proof (a)⇒ (b). Using the characterization of weakly $\mathcal{F}$-plurisubharmonic functions given in Theorem 5.4.11, one may adapt the proof of the "only if part" of [Kl, Theorem 2.9.12] as follows. Suppose $u \in \widehat{\mathcal{D}}(\mathbb{C}^n)$ is weakly $\mathcal{F}$-plurisubharmonic on U, and so $u \circ T$ is $\mathbb{R}^{2n}$-finely subharmonic on $T^{-1}(U)$ for any $\mathbb{C}$-affine bijection T of $\mathbb{C}^n$. To prove (b) with constant $\lambda = (\lambda_1, \dots, \lambda_n) \in \mathbb{C}^n$, take

$$T_\varepsilon(z) = z_1\lambda + \varepsilon\sum_{l=2}^{n} z_l e_l, \ \varepsilon > 0,$$

where $(e_1, \dots, e_n)$ denotes the canonical basis of $\mathbb{C}^n$. From (5.4.7) we obtain (with integrations over $\mathbb{C}^n$), replacing U and φ, as we may, by $T_\varepsilon^{-1}(U)$ and $\varphi \circ T_\varepsilon \in \widehat{\mathcal{D}}_+(\mathbb{C}^n, T_\varepsilon^{-1}(U))$,

$$
\begin{aligned}
0 &\geq \sum_{l=1}^{n} \int \partial_l (u \circ T_\varepsilon) \overline{\partial}_l (\varphi \circ T_\varepsilon) dV \\
&= \sum_{j,k=1}^{n} \int [(\partial_j u) \circ T_\varepsilon][(\overline{\partial}_k \varphi) \circ T_\varepsilon](\lambda_j \overline{\lambda}_k + O(\varepsilon)) dV \\
&= |\det T_\varepsilon|^2 \Big(\sum_{j,k=1}^{n} \int (\partial_j u)(\overline{\partial}_k \varphi) \lambda_j \overline{\lambda}_k dV + O(\varepsilon)\Big).
\end{aligned}
$$

Here $\det T_\varepsilon$ denotes the Jacobian of T_ε. This leads to (b) after division by $|\det T_\varepsilon|^2$ when we make $\varepsilon \to 0$.

(b) $\Rightarrow$ (a). Consider any $\mathbb{C}$-affine bijection $T = (T_1, \dots, T_n)$ of $\mathbb{C}^n$, say

$$
T_l(z) = \sum_{j=1}^{n} c_{lj} z_j + d_l, \ l \in \{1, \dots, n\}, \ z = (z_1, \dots, z_n) \in \mathbb{C}^n,
$$

with $c_{lj}, d_l \in \mathbb{C}$ and $\det T \neq 0$. We obtain

$$
\begin{aligned}
\int \partial_l (u \circ T) \overline{\partial}_l (\varphi \circ T) dV &= \sum_{j,k=1}^{n} [(\partial_j u) \circ T][(\overline{\partial}_k \varphi) \circ T] c_{lj} \overline{c}_{lk} dV \\
&= |\det T|^2 \sum_{j,k=1}^{n} \int c_{lj} \overline{c}_{lk} (\partial_j u)(\overline{\partial}_k \varphi) dV \leq 0
\end{aligned}
$$

by (b) with $\lambda_j = c_{lj}$. After division by $|\det T|^2$ and summation over l this shows according to (5.4.7) and Theorem 5.4.11 that the $\mathcal{F}$-continuous function $u < +\infty$ indeed is $\mathcal{F}$-plurisubharmonic on U.

(c) $\Rightarrow$ (a). This is contained in the "if part" of Theorem 5.4.11 (even with h in (c) just a $\mathbb{C}$-affine bijection and with $u \circ h$ just $\mathbb{R}^{2n}$-finely subharmonic).

(a) $\Rightarrow$ (c). We may assume that $u > -\infty$ on U (otherwise replace u by $u^p := \max\{u, -p\}$, $p \in \mathbb{N}$, and let $p \to +\infty$). According to Theorem 5.4.3(a), every point $z_0 \in h^{-1}(U)$ then has an $\mathcal{F}$-open $\mathcal{F}$-neighborhood $\mathcal{O} \subset h^{-1}(U)$ on which $u = u_1 - u_2$, u_1 and u_2 being bounded plurisubharmonic on some open set $D \subset \mathbb{C}^n$ containing $\mathcal{O}$. In particular, U and $\mathcal{O}$ are $\mathbb{R}^{2n}$-finely open, and u_1 and u_2 are $\mathbb{R}^{2n}$-subharmonic on D. We may further assume that the Jacobian matrix $(\partial_j h_k)$ of the regular holomorphic map $h : \omega \to \mathbb{C}^n$ is bounded with determinant bounded away from 0. We have $-u \in \mathcal{S}(D, \mathcal{O})$ and hence by Theorem 4.2.21(i)

$$
\int_{\mathcal{O}} \nabla u . \nabla \varphi dV \leq 0 \text{ for } \varphi \in \widehat{\mathcal{D}}_+(D, \mathcal{O}).
$$

For any $\psi \in \widehat{\mathcal{D}}(\mathbb{C}^n)$ we have (by the properties of h required in (c)) $\psi \circ h \in \widehat{\mathcal{D}}(\omega)$. According to Theorem 5.4.11 it suffices to show that the $\mathcal{F}$-continuous function $u \circ h$ is $\mathbb{R}^{2n}$-finely subharmonic on $h^{-1}(\mathcal{O})$. For this it suffices by (5.4.7) to prove that, for every $j \in \{1, \dots, n\}$,

$$\int_{h^{-1}(\mathcal{O})} (\partial_j(u \circ h))(\bar{\partial}_j\psi)dV \leq 0 \text{ for every } \psi \in \widehat{\mathcal{D}}_+(\mathbb{C}^n, h^{-1}(\mathcal{O})),$$

and here ψ may be replaced equivalently by $\varphi \circ h$ with $\varphi \in \widehat{\mathcal{D}}_+(\mathbb{C}^n, \mathcal{O})$ (or just as well with $\varphi \in \widehat{\mathcal{D}}_+(D, \mathcal{O})$). We take $j = 1$ and write

$$dV = (i/2)^n dz_1 \wedge d\bar{z}_1 \wedge \dots \wedge dz_n \wedge d\bar{z}_n = (1/4)^n dd^c|z_1|^2 \wedge \dots \wedge dd^c|z_n|^2.$$

By abuse of notation we will write $h^{-1} = (h_1^{-1}, h_2^{-1}, \dots, h_n^{-1})$ for the components of the inverse h^{-1} of the map h. Then by the chain rule we obtain

$$\begin{aligned}
&\int_{h^{-1}(\mathcal{O})} (\partial_1(u \circ h))(\overline{\partial}_1(\varphi \circ h))dd^c|z_1|^2 \wedge \dots \wedge dd^c|z_n|^2 \\
&= \int d(u \circ h) \wedge d^c(\varphi \circ h) \wedge dd^c|z_2|^2 \wedge \dots \wedge dd^c|z_n|^2 \\
&= \int d(u \circ h) \wedge d^c(\varphi \circ h) \wedge dd^c|h_2^{-1} \circ h|^2 \wedge \dots \wedge dd^c|h_n^{-1} \circ h|^2 \\
&= \int du \wedge d^c\varphi \wedge dd^c|h_2^{-1}|^2 \wedge \dots \wedge dd^c|h_n^{-1}|^2 \qquad (5.4.8) \\
&= \int \varphi d(d^c u \wedge dd^c|h_2^{-1}|^2 \wedge \dots \wedge dd^c|h_n^{-1}|^2) \qquad (5.4.9) \\
&= -\int \varphi dd^c u \wedge dd^c|h_2^{-1}|^2 \wedge \dots \wedge dd^c|h_n^{-1}|^2.
\end{aligned}$$

The last three lines are in h-coordinates. Equality (5.4.8 is justified by approximating u and φ in $\widehat{\mathcal{D}}$ with functions in $\mathcal{D}$ and applying Stokes' theorem to the approximants. The final expression is negative because of Proposition 5.4.16, and we are done. □

Notes: The definition of (weakly) $\mathcal{F}$-plurisubharmonic function was first introduced and partially studied by El Kadiri [EK4]. Theorem 5.4.17 is due to El Marzguioui and Wiegerinck [EMW2]. Proposition 5.4.4 is due to El Kadiri and Smit [EKS]. Theorem 5.4.3 is essentially due to El Marzguioui and Wiegerinck [EMW2] and, in the more precise and improved form given in this section, it is due, along with the remaining results of this section, to El Kadiri, Fuglede, and Wiegerinck [EKFW].

5.5 Plurifinely Holomorphic Functions

The concept of finely holomorphic functions in one variable has been studied in Sect. 4.5 of the present book. In the present section we are going to study two concepts of plurifinely holomorphic functions in several variables: the *strongly $\mathcal{F}$-holomorphic* functions and the *weakly $\mathcal{F}$-holomorphic* functions.

For a compact set $K \subset \mathbb{C}^n$ we denote by $\mathcal{O}_0(K)$ the algebra of all restrictions to K of holomorphic functions defined in open subsets of $\mathbb{C}^n$ containing K, and by $\mathcal{O}(K)$ the closure of $\mathcal{O}_0(K)$ in $C(K, \mathbb{C})$ (the continuous functions $K \to \mathbb{C}$ with the uniform norm); then $\mathcal{O}(K)$ is likewise an algebra.

Definition 5.5.1 Let U denote an $\mathcal{F}$-open subset of $\mathbb{C}^n$.

(i) A function $f : U \to \mathbb{C}$ is said to be *strongly $\mathcal{F}$-holomorphic* if every point of U has a compact $\mathcal{F}$-neighborhood K in U such that $f|_K \in \mathcal{O}(K)$.

(ii) A function $f : U \to \mathbb{C}$ is said to be *weakly $\mathcal{F}$-holomorphic* (or simply $\mathcal{F}$-holomorphic) if f is $\mathcal{F}$-continuous and if, for every complex line L in $\mathbb{C}^n$, the restriction $f|_{L\cap U}$ is *finely holomorphic*.

Each of the concepts strongly and weakly $\mathcal{F}$-holomorphic is an $\mathcal{F}$-local one. The class of all strongly, resp. weakly, $\mathcal{F}$-holomorphic functions on U is an algebra, and the latter class is closed under $\mathcal{F}$-locally uniform convergence, in view of Corollary 4.5.11. Clearly, every strongly $\mathcal{F}$-holomorphic function is $\mathcal{F}$-continuous (on K'' from Definition 5.5.1(i), and so on all of U).

If f is strongly, resp. weakly, $\mathcal{F}$-holomorphic on U ($\mathcal{F}$-open in $\mathbb{C}^n$) then the restriction of f to $L \cap U$ (L a $\mathbb{C}$-linear subspace $L \cong \mathbb{C}^k$ of $\mathbb{C}^n$) has the same property in $L \cap U$. This follows easily from Proposition 5.2.15.

For $n = 1$, it follows from the definition of finely holomorphic functions (see Definition 4.5.1) that f is strongly (resp. weakly) $\mathcal{F}$-holomorphic on U (finely open in $\mathbb{C}$) if and only if f is finely holomorphic on U.

Remark 5.5.2 Similarly to Proposition 5.4.6, a function $f : U \to \mathbb{C}$, U Euclidean open subset of $\mathbb{C}^n$, is holomorphic if and only if f is strongly, or equivalently weakly, $\mathcal{F}$ holomorphic; the "if part" follows from Hartogs' theorem.

Proposition 5.5.3 *Every strongly $\mathcal{F}$-holomorphic function $f : U \to \mathbb{C}$ is weakly $\mathcal{F}$-holomorphic, and in particular $\mathcal{F}$-continuous.*

Proof For any K as in (i) from Definition 5.5.1 there exists a sequence of holomorphic functions f_ν defined on open sets containing K such that $f_\nu|_K \to f|_K$ uniformly. For every complex line L in $\mathbb{C}^n$ this shows that the finely holomorphic functions $f_\nu|_{L\cap K'}$ converge uniformly to $f|_{L\cap K'}$, which therefore is finely holomorphic by Corollary 4.5.11. Consequently $f|_{L\cap U}$ is finely holomorphic, and so f is indeed weakly $\mathcal{F}$-holomorphic, being also $\mathcal{F}$-continuous. □

The concept of *weakly* $\mathcal{F}$-holomorphic function can be characterized in terms of weakly $\mathcal{F}$-pluriharmonic functions (that is, functions $f : U \to \mathbb{C}$ such that $\pm\mathfrak{Re} f$ and $\pm\mathfrak{Im} f$ are weakly $\mathcal{F}$-plurisubharmonic on the $\mathcal{F}$-open set $U \subset \mathbb{C}^n$):

Proposition 5.5.4 *A function $f : U \to \mathbb{C}$ is weakly $\mathcal{F}$-holomorphic if and only if f and each of the functions $z \mapsto z_j f(z)$ ($j \in \{1, ..., n\}$) are weakly $\mathcal{F}$-pluriharmonic on U.*

Proof This reduces right away to the case $n = 1$ which is the assertion (i) in Theorem 4.5.9. □

Corollary 5.5.5 (Removable singularities for $\mathcal{F}$-holomorphic functions) *Let $U \subset \mathbb{C}^n$ be $\mathcal{F}$-open, let E be an $\mathcal{F}$-closed and pluripolar set in $\mathbb{C}^n$ and let $h : U \setminus E \to \mathbb{C}$ be a weakly $\mathcal{F}$-holomorphic function. If h is $\mathcal{F}$-locally bounded on some deleted $\mathcal{F}$-neighborhood of each point of $U \cap E$, then h extends uniquely to a weakly $\mathcal{F}$-holomorphic function $h^* : U \to \mathbb{C}$, given by*

$$h^*(z) = \mathcal{F}\text{-}\lim_{\zeta \to z, \zeta \in U \setminus E} h(\zeta), \; z \in U.$$

Proof The corollary is an immediate consequence of Proposition 5.5.4 and Theorem 5.4.13. □

For any $\mathcal{F}$-open set $U \subset \mathbb{C}^m$, an n-tuple $(h_1, \ldots, h_n)$ of strongly (resp. weakly) $\mathcal{F}$-holomorphic functions $h_j : U \to \mathbb{C}$ will be termed a strongly (resp. weakly) $\mathcal{F}$-*holomorphic map* $U \to \mathbb{C}^n$.

Proposition 5.5.6 *Let $U \subset \mathbb{C}^m$ be $\mathcal{F}$-open and let $h = (h_1, \ldots, h_n) : U \to \mathbb{C}^n$ be a strongly (resp. weakly) $\mathcal{F}$-holomorphic map.*

(a) *The map $h : U \to \mathbb{C}^n$ is continuous from U with the $\mathcal{F}$-topology on $\mathbb{C}^m$ to $\mathbb{C}^n$ with the Euclidean topology.*

(b) *For any plurisubharmonic function f on an open set Ω in $\mathbb{C}^n$, the function $f \circ h$ is strongly (resp. weakly) $\mathcal{F}$-plurisubharmonic or identically $-\infty$ on each $\mathcal{F}$-component of the $\mathcal{F}$-open set $h^{-1}(\Omega) = \{z \in U : h(z) \in \Omega\} \subset \mathbb{C}^m$.*

(c) *For any holomorphic function f on an open set Ω in $\mathbb{C}^n$, the function $f \circ h$ is strongly (resp. weakly) $\mathcal{F}$-holomorphic on the $\mathcal{F}$-open set $h^{-1}(\Omega) \subset \mathbb{C}^m$.*

Proof Assertion (a) holds because each h_j (whether strongly or weakly $\mathcal{F}$-holomorphic) is $\mathcal{F}$-continuous and that the Euclidean topology on $\mathbb{C}^n$ is the product of the Euclidean topology on each of n copies of $\mathbb{C}$.

For (b) with each h_j *strongly* $\mathcal{F}$-holomorphic we begin by showing that, if the plurisubharmonic function f on Ω is finite and continuous, then $f \circ h$ is even $\mathcal{F}$-*cpsh* on $h^{-1}(\Omega)$, which is $\mathcal{F}$-open according to (a). Every point $a \in h^{-1}(\Omega)$ has a compact $\mathcal{F}$-neighborhood K_j in $h^{-1}(\Omega)$ ($\subset U \subset \mathbb{C}^m$) such that $h_j|_{K_j} \in \mathcal{O}(K_j)$. Thus there exists a sequence $(h_j^\nu)_{\nu \in \mathbb{N}}$ of holomorphic functions h_j^ν on open sets U_j^ν in $\mathbb{C}^m$ containing K_j such that $h_j^\nu|_{K_j} \to h_j|_{K_j}$ uniformly as $\nu \to \infty$. Write $K = K_1 \cap \cdots \cap K_n$ and $h^\nu = (h_1^\nu, \ldots, h_n^\nu)$ on $U^\nu = U_1^\nu \cap \cdots \cap U_n^\nu$. Then $h_j|_{K_j} \in \mathcal{C}(K_j, \mathbb{C})$ and hence $h|_K \in \mathcal{C}(K, \mathbb{C}^n)$. It follows that $h(K)$ is a compact subset of $\Omega \subset \mathbb{C}^n$. Denoting by $\| \cdot \|$ the Euclidean norm on $\mathbb{C}^n$ and by B the closed unit ball in $\mathbb{C}^n$, there exists accordingly $\delta > 0$ such that $h(K) + \delta B \subset \Omega$. We may assume that $\|h^\nu(z) - h(z)\| < \delta$ for any ν and any $z \in K$. Under the present extra hypothesis, f is finite and uniformly continuous on the compact set $h(K) + \delta B$ containing any $h^\nu(K)$, and it follows that $f \circ h^\nu|_K \to f \circ h|_K$ uniformly as $\nu \to \infty$. Because $f \circ h^\nu$ is finite, continuous, and plurisubharmonic, on the open set $U^\nu \supset K$, we have $f \circ h|_K \in \mathrm{PSH}(K)$. By varying $a \in h^{-1}(\Omega)$ and hence the $\mathcal{F}$-neighborhood K of a in $h^{-1}(\Omega)$ we infer that $f \circ h$ is $\mathcal{F}$-cpsh on $h^{-1}(\Omega)$, cf. Definition 5.4.1(i).

If we drop the extra hypothesis that f be finite and continuous, f is the pointwise limit of a decreasing net of finite continuous plurisubharmonic functions f_ν on Ω, and $f \circ h$ is then the pointwise limit of the decreasing net of functions $f_\nu \circ h$ on $h^{-1}(\Omega)$ which we have just shown are *$\mathcal{F}$-cpsh*, and so, in view of Theorem 5.4.2(iv), $f \circ h$ is indeed strongly $\mathcal{F}$-plurisubharmonic or identically $-\infty$ in each $\mathcal{F}$-component of $h^{-1}(\Omega)$, cf. Definition 5.4.1(i), (ii) (with U replaced by $h^{-1}(\Omega)$).

Next suppose instead that each h_j is *weakly* $\mathcal{F}$-holomorphic on U, and consider a complex line L in $\mathbb{C}^m$; then $L \cap U$ is finely open in L. According to Definition 5.5.1(ii), $h_j|_{L\cap U}$ is then finely holomorphic, which is the same as strongly $\mathcal{F}$-holomorphic (see above for $n = 1$). As shown above (now with $m = 1$ and with U replaced by $L \cap U$) it follows that $f \circ h|_{L\cap h^{-1}(\Omega)}$ is strongly $\mathcal{F}$-plurisubharmonic, which is the same as finely subharmonic (because the dimension is 1). According to Definition 5.4.1(iii) this means that $f \circ h$ indeed is weakly $\mathcal{F}$-plurisubharmonic or identically $-\infty$ on each $\mathcal{F}$-component of $h^{-1}(\Omega)$, noting that $f \circ h$ is $\mathcal{F}$-u.s.c. in view of (a) because f is u.s.c.

For (c), suppose first that each h_j is *strongly* $\mathcal{F}$-holomorphic on U. Proceeding as in the first part of the proof of (b) we arrange that $f \circ h^\nu|_K \to f \circ h|_K$ uniformly as $\nu \to \infty$; but now $f \circ h^\nu$ is holomorphic on U^ν. We therefore conclude that $f \circ h|_K \in \mathcal{O}(K)$, and so $f \circ h$ is indeed strongly $\mathcal{F}$-holomorphic according to Definition 5.5.1(i).

If instead each h_j is *weakly* $\mathcal{F}$-holomorphic on U then, for every complex line L in $\mathbb{C}^m$, each $h_j|_{L\cap U}$ is again *strongly* $\mathcal{F}$-holomorphic. As just established, this implies that $f \circ h|_{L\cap h^{-1}(\Omega)}$ is strongly $\mathcal{F}$-holomorphic, or equivalently finely holomorphic. We conclude that indeed $f \circ h$ is *weakly* $\mathcal{F}$-holomorphic, according to Definition 5.5.1(ii), noting that $f \circ h$ is $\mathcal{F}$-continuous in view of (a). □

In the version of Proposition 5.5.6 with "strongly" in each of the three occurrences one may allow f in (b) to be just *strongly $\mathcal{F}$-plurisubharmonic* (in place of plurisubharmonic), and similarly f in (c) to be *strongly $\mathcal{F}$-holomorphic* (in place of holomorphic):

Theorem 5.5.7 *Let $U \subset \mathbb{C}^m$ be $\mathcal{F}$-open and let $h = (h_1, \ldots, h_n) : U \to \mathbb{C}^n$ be a weakly $\mathcal{F}$-holomorphic map.*

(a) *The map $h : U \to \mathbb{C}^n$ is continuous from U with the $\mathcal{F}$-topology on $\mathbb{C}^m$ to $\mathbb{C}^n$ with the $\mathcal{F}$-topology there.*

(b) *For any strongly $\mathcal{F}$-plurisubharmonic function f defined on an $\mathcal{F}$-open set Ω in $\mathbb{C}^n$, the function $f \circ h$ is weakly $\mathcal{F}$-plurisubharmonic or identically $-\infty$ on each $\mathcal{F}$-component of the $\mathcal{F}$-open set $h^{-1}(\Omega) = \{z \in U : h(z) \in \Omega\} \subset \mathbb{C}^m$.*

(c) *For any strongly $\mathcal{F}$-holomorphic function f defined on an $\mathcal{F}$-open set Ω in $\mathbb{C}^n$, the function $f \circ h$ is weakly $\mathcal{F}$-holomorphic on the $\mathcal{F}$-open set $h^{-1}(\Omega) \subset \mathbb{C}^m$.*

Proof For the present weakly $\mathcal{F}$-holomorphic functions assertion (a) is stronger than Proposition 5.5.6(a). We shall prove that $h^{-1}(\Omega)$ is $\mathcal{F}$-open in $\mathbb{C}^m$ for any $\mathcal{F}$-open set Ω in $\mathbb{C}^n$. Fix a point $a \in h^{-1}(\Omega)$ and write $h(a) = b$ $(\in \Omega)$. According to Theorem 5.2.8 there exists a plurisubharmonic function φ on an open ball $B(b, r)$ in $\mathbb{C}^n$ and a number $c < \varphi(b)$ such that the basic $\mathcal{F}$-neighborhood

$$W = \{w \in B(b, r) : \varphi(w) > c\}$$

of b in $\mathbb{C}^n$ is a subset of the $\mathcal{F}$-open set Ω in $\mathbb{C}^n$. Then $h^{-1}(W) \subset h^{-1}(\Omega)$ $(\subset U)$, and

$$h^{-1}(W) = \{z \in U : h(z) \in B(b, r) \text{ and } (\varphi \circ h)(z) > c\}$$

is $\mathcal{F}$-open in $\mathbb{C}^m$ because $h : U \to \mathbb{C}^n$ is $\mathcal{F}$-continuous by Proposition 5.5.6(a) and that $\varphi \circ h : h^{-1}(B(b, r)) \to [-\infty, +\infty[$ is weakly $\mathcal{F}$-plurisubharmonic by Proposition 5.5.6(b) (applied with Ω, f replaced by $B(b, r), \varphi$), and in particular $\mathcal{F}$-*continuous*, as noted at the end of Theorem 5.4.3(c). By varying $a \in h^{-1}(\Omega)$ we infer that indeed $h^{-1}(\Omega)$ is $\mathcal{F}$-open.

For (b) we may assume that f is even $\mathcal{F}$-*cpsh* on Ω. Let $K \subset \Omega$ be as in Definition 5.4.1(i), and let (f_j) be a sequence of finite continuous plurisubharmonic functions on open sets $\Omega_j \supset K$ such that $f_j|_K \to f|_K$ uniformly as $j \to \infty$. According to Proposition 5.5.6(a), (b) each $f_j \circ h$ is weakly $\mathcal{F}$-plurisubharmonic on the $\mathcal{F}$-open set $h^{-1}(\Omega_j)$. By (a), $h^{-1}(K'')$ is $\mathcal{F}$-open, and it follows that each $f_j \circ h|_{h^{-1}(K'')}$ likewise is weakly $\mathcal{F}$-plurisubharmonic, in particular $\mathcal{F}$-u.s.c.; and hence so is its uniform limit $f \circ h|_{h^{-1}(K'')}$ in view of Corollary 2.1.21. By varying $K \subset \Omega$ and hence K'' we conclude that indeed $f \circ h$ is weakly $\mathcal{F}$-plurisubharmonic on $h^{-1}(\Omega)$.

Finally, the proof of (c) is quite parallel to that of (b) in view of Proposition 5.5.6(a), (c), using Corollary 4.5.11 in place of Corollary 2.1.21. □

Corollary 5.5.8 *Let $h_j : U \to \mathbb{C}$ be finely holomorphic functions defined on a finely open set $U \subset \mathbb{C}$ $(j \in \{1, \dots, n\})$, and write $h = (h_1, \dots, h_n)$. For any strongly $\mathcal{F}$-plurisubharmonic (resp. strongly $\mathcal{F}$-holomorphic) function f defined on an $\mathcal{F}$-open set $\Omega \subset \mathbb{C}^n$, the function $f \circ h$ is finely hypoharmonic (resp. finely holomorphic) on the finely open set $h^{-1}(\Omega) = \{z \in U : h(z) \in \Omega\} \subset \mathbb{C}$.*

Remark 5.5.9 If it can be proved that the cone of $\mathcal{F}$-cpsh, resp. the algebra of strongly $\mathcal{F}$-holomorphic, functions on an $\mathcal{F}$-open subset Ω of $\mathbb{C}^n$ is closed under uniform convergence, then the proofs of Theorem 5.5.7(b), (c) easily show that 'weakly' can be replaced by "strongly" there. Indeed, with f $\mathcal{F}$-cpsh (resp. strongly $\mathcal{F}$-holomorphic) and h strongly $\mathcal{F}$-holomorphic, let $K \subset \Omega$ denote a compact $\mathcal{F}$-neighborhood of a point $a \in \Omega$, and let (f_j) denote a sequence of finite continuous plurisubharmonic (resp. holomorphic) functions, defined on open subsets Ω_j of $\mathbb{C}^n$ containing K, and such that $f_j \to f$ uniformly on K. Then $f_j \circ h$ is $\mathcal{F}$-cpsh (resp. strongly $\mathcal{F}$-holomorphic) on the $\mathcal{F}$-open set $h^{-1}(K') \subset h^{-1}(\Omega_j) \subset U \subset \mathbb{C}^m$, by Proposition 5.5.6 (see also the first two lines of the proof of (b) of Theorem 5.5.7). Under the stated extra hypothesis it will follow that the uniform limit $f \circ h$ of $(f_j \circ h)$ on $h^{-1}(K'')$ likewise is $\mathcal{F}$-cpsh (resp. strongly $\mathcal{F}$-holomorphic) on $h^{-1}(K'')$, and therefore on $h^{-1}(\Omega)$, by varying K and hence K''.—If f is merely *strongly* $\mathcal{F}$-plurisubharmonic or identically $-\infty$ on each $\mathcal{F}$-component of Ω, it follows as usual that indeed $f \circ h$ is likewise strongly $\mathcal{F}$-plurisubharmonic.

In the case $n = 1$ the hypothesis stated in the above remark is always fulfilled in view of Corollary 2.1.21 (resp. 4.5.11). We therefore have the following corollary of Theorem 5.5.7 for that case:

Corollary 5.5.10 *Let $h : U \to \mathbb{C}$ be a strongly $\mathcal{F}$-holomorphic function defined on an $\mathcal{F}$-open set $U \subset \mathbb{C}^m$. For any finely hypoharmonic (resp. finely holomorphic) function f defined on a finely open set $\Omega \subset \mathbb{C}$ the function $f \circ h$ is strongly $\mathcal{F}$-plurisubharmonic or identically $-\infty$ (resp. $\mathcal{F}$-holomorphic) on each $\mathcal{F}$ component of the $\mathcal{F}$-open set $h^{-1}(\Omega) \subset \mathbb{C}^m$.*

The same with "strongly" replaced by "weakly" at both places is simply the case $n = 1$ of Theorem 5.5.7 as it stands.

We close this section with an application to pluripolar hulls. Recall that the *pluripolar hull* P^*_Ω of a pluripolar set P relative to an open set Ω containing P is defined as the following set (closed relatively to Ω):

$$P^*_\Omega = \bigcap_u \{z \in \Omega : u(z) = -\infty\},$$

where the intersection is taken over all plurisubharmonic functions u defined in Ω and such that $u|_P \equiv -\infty$.

For any set $E \subset \mathbb{C}^m$, and any function $h : E \to \mathbb{C}$ we denote by $\Gamma_h(E) = \{(z, h(z)) : z \in E\}$ the graph of h ($\subset \mathbb{C}^{m+1}$), and by $\Gamma_h(E)^*_{\mathbb{C}^{m+1}}$ the pluripolar hull of $\Gamma_h(E)$ if $\Gamma_h(E)$ is pluripolar. It is clear that if $F \subset E$ then $\Gamma_{h|_F}(F)$, simply denoted by $\Gamma_h(F)$, is a subset of $\Gamma_h(E)$.

Proposition 5.5.11 *Let h be a weakly $\mathcal{F}$-holomorphic function on an $\mathcal{F}$-domain $U \subset \mathbb{C}^m$.*

(a) *If $h \not\equiv 0$, the set $h^{-1}(0)$ of zeros of h is pluripolar in $\mathbb{C}^m$.*

(b) *The graph $\Gamma_h(U)$ of h is pluripolar in $\mathbb{C}^{m+1}$.*

(c) *If E is a non-pluripolar subset of U then $\Gamma_h(E)$ is pluripolar, and $\Gamma_h(U) \subset \Gamma_h(E)^*_{\mathbb{C}^{m+1}}$.*

Proof (a) It follows from Proposition 5.5.6(b) that the function $\log|h(\cdot)|$ is weakly $\mathcal{F}$-plurisubharmonic or identically $-\infty$ on U. Since $\log|h(z)| = -\infty$ for $z \in h^{-1}(0)$, but $\log|h(\cdot)| \not\equiv -\infty$, we conclude from Theorem 5.4.3(d) that $h^{-1}(0)$ is pluripolar.

(b) Applying (a) to the function $\mathcal{F}$-holomorphic function $g : U \times \mathbb{C}$ defined by $g(z, w) = w - h(z)$ we see that $\Gamma_h(U) = g^{-1}(0)$ is pluripolar.

(c) The function $(z, w) \mapsto w - h(z)$ is weakly $\mathcal{F}$-holomorphic and $\not\equiv 0$ on the $\mathcal{F}$-open set $U \times \mathbb{C} \subset \mathbb{C}^{m+1}$. Again by Proposition 5.5.6(b) it follows that the function $(z, w) \mapsto \log|w - h(z)|$ is weakly $\mathcal{F}$-plurisubharmonic on $U \times \mathbb{C}$. Since this function equals $-\infty$ on $\Gamma_h(E)$ we conclude that $\Gamma_h(E)$ is pluripolar. By Josefson's theorem [J] there exists a plurisubharmonic function f on all of $\mathbb{C}^{m+1}$ such that $f(z, h(z)) = -\infty$ for every $z \in E$. Since E is non-pluripolar, it follows by Theorem 5.5.7(b) and Theorem 5.4.3(d) that $f(z, h(z)) = -\infty$ even for every $z \in U$, and hence $\Gamma_h(U)$ is pluripolar in $\mathbb{C}^{m+1}$. By definition of the pluripolar hull of $\Gamma_h(E)$ we conclude that indeed $\Gamma_h(U) \subset \Gamma_h(E)^*_{\mathbb{C}^{m+1}}$. □

With h supposed *strongly* $\mathcal{F}$-holomorphic on U, this proposition was obtained in [EMW3, Corollary 4.4 and Theorem 4.5] extending [EMW2, Theorem 6.4], and [EEW, Theorem 3.5].

Levenberg, Martin, and Poletsky in [LMP] conjectured that the pluripolar hull of a graph of an analytic function f is equal to the graph of the maximal analytic extension of f in the sense of Weierstrass. This conjecture was refuted in [EW1]. Other counter examples to this conjecture, cf. [Zwo, Si, PW]. Eventually Edlund and Jöricke, [EJ], made the connection with fine holomorphy, observing that in all the available counter examples the set E under consideration admits no analytic extension, but it does admit so called fine analytic extension. This connection motivates even more the study of the finely holomorphic functions.

Notes: The concept of strongly $\mathcal{F}$-holomorphic was first introduced by El Marzguioui and Wiegerinck in [EMW2, Definition 6.1] (where they called it $\mathcal{F}$-holomorphic). The weakly $\mathcal{F}$-holomorphic concept was introduced by El Kadiri, Fuglede, and Wiegerinck [EKFW]. All results of this section were obtained by these authors in [EKFW], Proposition 5.5.11 is due to El Marzguioui and Wiegerinck for strongly $\mathcal{F}$-holomorphic functions (see [EMW2, Theorem 6.4]).

5.6 Biholomorphic Invariance

We will show that the concept of weakly $\mathcal{F}$-plurisubharmonic functions is *biholomorphically invariant*. Recall from the text preceding Proposition 5.5.6 that an n-tuple $(h_1, \dots, h_n)$ of strongly/weakly $\mathcal{F}$-holomorphic functions $h_j : U \to \mathbb{C}$ (U is $\mathcal{F}$-open in some $\mathbb{C}^m$) is termed a strongly/weakly $\mathcal{F}$-holomorphic map (or curve if $m = 1$). The following concept of a strongly $\mathcal{F}$-biholomorphic map is an auxiliary one.

Definition 5.6.1 (Plurifinely biholomorphic map) *A strongly $\mathcal{F}$-biholomorphic map* h from an $\mathcal{F}$-open set $U \subset \mathbb{C}^n$ onto its image in $\mathbb{C}^n$ is an $\mathcal{F}$-homeomorphism with the property that there exists for every $z \in U$ a compact $\mathcal{F}$-neighborhood K_z of z in U and a C^∞-diffeomorphism F_z from an open neighborhood of K_z to its image in $\mathbb{C}^n$ such that $F_z|_{K_z} = h|_{K_z}$ and that $F_z|_{K_z}$ is a C^2-limit of biholomorphic maps defined on open sets containing K_z..

Proposition 5.6.2 *The composition $f \circ h$ of a weakly $\mathcal{F}$-plurisubharmonic function f on an $\mathcal{F}$-open set $\Omega \subset \mathbb{C}^n$ with a strongly $\mathcal{F}$-biholomorphic map $h : U \to \Omega$ (U is $\mathcal{F}$-open in $\mathbb{C}^n$) is weakly $\mathcal{F}$-plurisubharmonic on $h^{-1}(\Omega)(\subset \mathbb{C}^n)$.*

Proof For $n = 1$ this is contained in Theorem 4.4.5(ii) (in which h is any finely holomorphic function on U) and Theorem 4.5.2. Therefore suppose that $n \geq 2$. We may assume that Ω is $\mathcal{F}$-connected and that $f > -\infty$ on Ω (otherwise replace f by $\max\{f, -p\}$ and let $p \to +\infty$). According to Theorem 5.4.3(a), every point $z_0 \in \Omega$ then has an $\mathcal{F}$-open $\mathcal{F}$-neighborhood $\mathcal{O} \subset \Omega$ on which $f = f_1 - f_2$, f_1 and f_2 being bounded negative plurisubharmonic on some open ball $B = B(z_0, r)$ containing $\mathcal{O}$. In particular, f_1 and f_2 are $\mathbb{R}^{2n}$-subharmonic on B. We may further assume that $-f_1$ and $-f_2$ are $\mathbb{R}^{2n}$-potentials on B, for otherwise we may replace $-f_i$ for $i = 1, 2$ by its swept function (relative to B) $\widehat{R}^A_{-f_i}$ on $A = B(z_0, r/2)$ (and $\mathcal{O}$ by $\mathcal{O} \cap A$). In terms of the Green kernel G on B we therefore may write $-f_i = G\mu_i$ on B for some bounded positive measure μ_i of compact support in B. Since $-f_i$ is bounded, its G-energy $\int G\mu_i d\mu_i$ is finite, and hence $G\mu_i$ is of class $\widehat{\mathcal{D}}(B)$ according to Theorem 4.2.10(a). Hence we may further suppose that Ω is bounded in $\mathbb{C}^n$ and that f is bounded and of class $\widehat{\mathcal{D}}(D)$ for some bounded domain $D \subset \mathbb{C}^n$ containing Ω. Fix $z \in U$ and let K_z be a compact $\mathcal{F}$-neighborhood of z in U on which h has the properties described in Definition 5.6.1. It will be sufficient to see that the expression (5.4.8) is negative if $\varphi \in \widehat{\mathcal{D}}_+(D, \mathcal{O})$ for some $\mathcal{F}$-open set $\mathcal{O} \subset D$ with $z \in \mathcal{O} \subset K_z$. Notice that $df \wedge d^c\varphi$ is a form with L^1 coefficients that is supported on K_z. Thus let (h_m) be a sequence of biholomorphic maps on open sets containing K_z that converge in C^2 to h on K_z. Then

$$\lim_{m\to\infty} dd^c|h^{-1}_{m,2}|^2 \wedge \ldots \wedge dd^c|h^{-1}_{m,n}|^2 = dd^c|h^{-1}_2|^2 \wedge \ldots \wedge dd^c|h^{-1}_n|^2,$$

uniformly on K_z. Now the expression (5.4.9) is negative when we replace h by h_m. By Lebesgue's dominated convergence theorem we conclude that (5.4.9) is also negative for h a strongly $\mathcal{F}$ biholomorphic map. □

It is reasonable to expect that the concept of weakly $\mathcal{F}$-plurisubharmonic function is invariant even under composition with suitable weakly $\mathcal{F}$-biholomorphic mappings. Currently, there is no such thing as a fine inverse function theorem for weakly $\mathcal{F}$-holomorphic maps of several variables. In fact it is not even known if weakly $\mathcal{F}$-holomorphic functions have weakly $\mathcal{F}$-holomorphic partial derivatives. However, one can handle the special case of a map of the form

$$G(z) = g(z_1) + (0, z_2, \ldots, z_n), \tag{5.6.1}$$

where $g = (g_1, \ldots, g_n)$ is a finely holomorphic curve in $\mathbb{C}^n$, and that turns out to be sufficient.

Theorem 5.6.3 *The composition $f \circ g$ of a weakly $\mathcal{F}$-plurisubharmonic (resp. weakly $\mathcal{F}$-holomorphic) function f on an $\mathcal{F}$-open set $\Omega \subset \mathbb{C}^n$ with a finely holo-*

morphic curve g in $\mathbb{C}^n$ defined on a finely open set $U \subset \mathbb{C}$, is finely hypoharmonic (resp. finely holomorphic) on the finely open pre-image $g^{-1}(\Omega)$ ($\subset \mathbb{C}$).

Proof The theorem is known for $n = 1$, cf. Theorem 4.5.2 and Theorem 4.5.18(a), so we suppose that $n \geq 2$. According to Theorem 5.5.7(a), g is continuous from U with the fine topology to $\mathbb{C}^n$ with the pluri-fine topology. The pre-image $g^{-1}(\Omega)$ therefore is finely open in $\mathbb{C}$, and $f \circ g$ is finely continuous. We may of course assume that f is bounded, that U is finely connected, and that g is non-constant, for example that g_1 is non-constant, hence a fine-to-fine open map by Theorem 4.5.2. Given a point $z_0 \in g^{-1}(\Omega)(\subset U \subset \mathbb{C})$ with $g_1'(z_0) \neq 0$, cf. Corollary 4.5.21, there exists a finely open set $\mathcal{O} \subset \mathbb{C}$ such that $z_0 \in \mathcal{O} \subset g^{-1}(\Omega)$ and that g_1 is injective on $\mathcal{O}$ with $g_1'(z) \neq 0$ for every $z \in \mathcal{O}$, hence $(g_1|_{\mathcal{O}})^{-1}$ is finely holomorphic on the finely open set $g_1(\mathcal{O})$ by Theorem 4.5.18(b). We may further assume after shrinking $\mathcal{O}$ that there exists a C^∞-map $\varphi : \mathbb{C} \to \mathbb{C}^n$ such that $\varphi = g$ on $\mathcal{O}$ and hence $\partial\varphi = g', \bar{\partial}\varphi = 0$ on $\mathcal{O}$, see Theorem 4.5.19(c). Since $\partial\varphi_1(z_0) = g_1'(z_0) \neq 0$ we may arrange (by further shrinking $\mathcal{O}$) that f_1 is injective on some open set $\omega \subset \mathbb{C}$ containing the closure of $\mathcal{O}$ in $\mathbb{C}$, and hence that $\varphi_1|_\omega$ is a C^∞-diffeomorphism of ω onto $\varphi_1(\omega)$. Likewise, we may achieve that there exists a sequence of curves φ^νsuch that each coordinate φ_j^ν of φ^ν, $j \in \{1, \dots, n\}$, is a rational function defined on some open set $\mathcal{O}^\nu \subset \omega$ containing $\mathcal{O}$, and that $\varphi^\nu \to \varphi$, $(\varphi^\nu)' \to \varphi'(= \partial\varphi)$, uniformly on $\mathcal{O}$ as $\nu \to \infty$, cf. Theorem 4.5.19(a). We also require that φ_1^ν is injective with $\partial\varphi_1^\nu \neq 0$. With this final choice of $\mathcal{O}$ define $G : \mathcal{O} \times \mathbb{C}^{n-1} \to \mathbb{C}^n$ by (5.6.1), writing now $t \in \mathbb{C}^n$ in place of $z \in \mathbb{C}^n$. For a given point $z = (z_1, z') \in \mathcal{O} \times \mathbb{C}^{n-1}$ choose a compact fine neighborhood L_{z_1} of z_1 in $\mathbb{C}$ so that $L_{z_1} \subset \mathcal{O}(\subset \omega)$, and a number $c > \max\{|z_2|, \dots, |z_n|\}$. In analogy with (5.6.1) define

$$\Phi_z(t) = \varphi(t_1) + (0, t_2, \dots, t_n) \text{ for } t \in \omega \times \mathbb{C}^{n-1}.$$

It is easily verified that Φ_z is a C^∞-diffeomorphism of $\omega \times \mathbb{C}^{n-1}$ onto its image in $\mathbb{C}^n$. We have $\Phi_z = G$ on $\mathcal{O} \times \mathbb{C}^{n-1}$, and in particular on the compact $\mathcal{F}$-neighborhood $K_z := L_{z_1} \times [-c, c]^{n-1}$ of z in $\mathcal{O} \times \mathbb{C}^{n-1}$, and consequently $G : \mathcal{O} \times \mathbb{C}^{n-1} \to \mathbb{C}^n$ is a strongly $\mathcal{F}$-biholomorphic map.

Suppose first that f is weakly $\mathcal{F}$-plurisubharmonic on Ω. According to Proposition 5.6.2, $f \circ G$ is a weakly $\mathcal{F}$-plurisubharmonic map defined on the $\mathcal{F}$-open set $G^{-1}(\mathcal{O}) \times \mathbb{C}^{n-1}$. Now

$$G(t_1, 0, \dots, 0) = g(t_1) = (g_1(t_1), g_2(t_1), \dots, g_n(t_1)) \text{ for } t_1 \in \mathcal{O},$$

and hence $(f \circ G)(t_1, 0, \dots, 0) = (f \circ g)(t_1)$ for $t_1 \in \mathcal{O}$. It follows that $f \circ g$ is finely hypoharmonic on $\mathcal{O}$, hence so (by varying z_0) on $\{z \in g^{-1}(\Omega) : g_1'(z) \neq 0\}$, which differs only by a countable and hence polar set from $g^{-1}(\Omega)$, cf. Theorem 4.5.26. Because $f \circ g$ is finely continuous and $< +\infty$ on $g^{-1}(\Omega)$ ($\subset U$) we conclude by the removable singularities theorem (Theorem 2.1.43) that indeed $f \circ g$ is finely hypoharmonic on $g^{-1}(\Omega)$.

Finally, let instead f be weakly $\mathcal{F}$-holomorphic on Ω, in particular weakly (complex) $\mathcal{F}$-harmonic (that is, $f|_{L\cap\Omega}$ is (complex) finely harmonic for any complex line L in $\mathbb{C}^n$), by Theorem 4.5.9. As shown in the proof of Proposition 5.6.2 we may suppose that Ω is contained in some bounded domain $D \subset \mathbb{C}^n$ and that f is bounded and of class $\widehat{\mathcal{D}}(D)$. Each of the functions $z \mapsto z_j f(z)$, $j \in \{1, \dots, n\}$, therefore is bounded and of class $\widehat{\mathcal{D}}(D)$.

By the former part of the theorem, the bounded functions $\pm\mathfrak{Re}(f \circ g)$ and $\pm\mathfrak{Im}(f \circ g)$ are finely subharmonic on $g^{-1}(\Omega)$, and hence $f \circ g$ is (complex) finely harmonic there. Similarly, $(z_j f) \circ g = g_j.(f \circ g)$ is finely harmonic on $g^{-1}(\Omega)$ $(\subset U \subset \mathbb{C})$, $j \in \{1, \dots, n\}$. We therefore may further assume that U is contained in a bounded domain $D_1 \subset \mathbb{C}$ and that $f \circ g$ and each g_j are bounded and of class $\widehat{\mathcal{D}}(D_1)$. Some component of g, say g_1, is non-constant, and it therefore again follows from Theorem 4.5.18(c) that every point $z_0 \in g^{-1}(\Omega)$ $(\subset U)$ with $g_1'(z_0) \neq 0$ has a finely open fine neighborhood $\mathcal{O} \subset g^{-1}(\Omega)$ on which g_1 is injective with $g_1' \neq 0$; and the inverse $h_1 := (g_1|_{\mathcal{O}})^{-1}$ is finely holomorphic on the finely open set $g_1(\mathcal{O})$. Hence, so is the re-parameterized curve $g \circ h_1 : g_1(\mathcal{O}) \to \mathbb{C}^n$. The function h on $g_1(\mathcal{O}) \times \mathbb{C}^{n-1}$ defined by

$$h(z) = h(z_1, \dots, z_n) = h_1(z_1)$$

is weakly $\mathcal{F}$-holomorphic. Therefore, so is hf, and consequently $(hf) \circ g$ is finely harmonic on $\mathcal{O}$, like $f \circ g$ above. Note that $h \circ g = h_1 \circ g_1$. For $z \in \mathcal{O}$,

$$\begin{aligned}&[(hf) \circ g](z)\\ &\quad = [(h \circ g)(z)][(f \circ g)(z)] = [(h_1 \circ g_1)(z)][(f \circ g)(z)] = z \cdot (f \circ g)(z),\end{aligned}$$

and since this function $(hf) \circ g = z \cdot (f \circ g)$ of $z \in \mathcal{O}$ is finely harmonic it follows according to Theorem 4.5.9, that $f \circ g$ is finely holomorphic on Ω, hence (by varying z_0) quasi-everywhere on $\{t \in g^{-1}(\Omega) : g_1'(t) \neq 0\}$, and indeed everywhere on $g^{-1}(\Omega)$ by the removable singularities theorem for finely holomorphic functions, cf. Corollary 4.5.12. □

The following extension of Theorem 5.6.3 from finely holomorphic curves g to $\mathcal{F}$-holomorphic maps h is a strengthening of Theorem 5.5.7(b), (c), in which f was required to be strongly $\mathcal{F}$-plurisubharmonic (resp. strongly $\mathcal{F}$-holomorphic). Likewise, Theorem 5.6.4 (for a weakly $\mathcal{F}$-plurisubharmonic function f) extends Proposition 5.6.2 (in which $m = n$, and h is strongly biholomorphic).

Theorem 5.6.4 *The composition $u \circ h$ of a weakly $\mathcal{F}$-plurisubharmonic (resp. weakly $\mathcal{F}$-holomorphic) function u on an $\mathcal{F}$-open set $\Omega \subset \mathbb{C}^n$ with a weakly $\mathcal{F}$-holomorphic map $h : U \to \mathbb{C}^n$ (U is $\mathcal{F}$-open in $\mathbb{C}^m$) is weakly $\mathcal{F}$-plurisubharmonic or identically $-\infty$ (resp. weakly $\mathcal{F}$-holomorphic) on each $\mathcal{F}$-component of $h^{-1}(\Omega)$ $(\subset \mathbb{C}^m)$.*

Proof According to Theorem 5.5.7(a), h is continuous from $U \subset \mathbb{C}^m$ to $\mathbb{C}^n$ with their respective pluri-fine topologies. It follows that $h^{-1}(\Omega)$ is $\mathcal{F}$-open in $\mathbb{C}^m$, and that $u \circ h$ is $\mathcal{F}$-continuous. Next, we restrict $u \circ h$ to a complex line L in $\mathbb{C}^m$, and observe that $h|_{L\cap U}$ is a finely holomorphic curve. By Theorem 5.6.3, $u \circ h$ restricted to $L \cap U$ therefore is finely hypoharmonic (resp. finely holomorphic), and we are done. □

Note: The results of this section are due to El Kadiri, Fuglede and Wiegerinck [EKFW].

5.7 Local Approximation of $\mathcal{F}$-Plurisubharmonic Functions

In this section, we will prove several density results (Theorem 5.7.10 and Corollary 5.7.11) for $\mathcal{F}$-plurisubharmonic functions. Inspired by Sect. 4.2, we will employ duality with respect to a pairing $\mathcal{B}$ of Hilbert spaces. Contrary to the situation in Sect. 4.2, the pairing will not be *separated*– we will use throughout the terminology of [KeNa]—but this causes no problems. We denote by $\mathcal{M}_{n,m}$ the space of complex $n \times m$ matrices and by $\mathcal{H}_n$ the space of Hermitian matrices in $\mathcal{M}_{n,n}$. Next $\mathcal{H}_n^+$ denotes the cone in $\mathcal{H}_n$ consisting of positive Hermitian matrices. In this section $\Omega \subset \mathbb{C}^n \cong \mathbb{R}^{2n}$ will be a bounded, Euclidean open domain and U will be an $\mathcal{F}$-open subset of Ω.

Functions in $\widehat{\mathcal{D}}(\Omega)$ are equivalent if they are equal a.e., hence elements of $\widehat{\mathcal{D}}(\Omega, U)$ may be represented by functions which are everywhere 0 on $\Omega\backslash U$. We denote the cones of positive elements in $\widehat{\mathcal{D}}(\Omega)$ and $\widehat{\mathcal{D}}(\Omega, U)$ by $\widehat{\mathcal{D}}_+(\Omega)$ and $\widehat{\mathcal{D}}_+(\Omega, U)$, respectively, and we write $\widehat{\mathcal{D}}_-(\Omega) = -\widehat{\mathcal{D}}_+(\Omega)$ and $\widehat{\mathcal{D}}_-(\Omega, U) = -\widehat{\mathcal{D}}_+(\Omega, U)$.

Next we introduce cones of (plurifinely) plurisubharmonic functions:

$$\begin{aligned}\mathrm{PSH}(\Omega, U) &= \{f \in \widehat{\mathcal{D}}(\Omega) : f|_U \in \mathcal{F}\text{-}\,\mathrm{PSH}(U)\},\\ \mathrm{PSH}_-(\Omega, U) &= \mathrm{PSH}(\Omega, U) \cap \widehat{\mathcal{D}}_-(\Omega).\end{aligned}$$

For ω a Euclidean open in Ω, we introduce like spaces $\mathrm{PSH}(\Omega, \omega)$, $\mathrm{PSH}_-(\Omega, \omega)$ of functions that are, respectively, in $\mathrm{PSH}(\omega) \cap \widehat{\mathcal{D}}(\Omega)$ and in $\mathrm{PSH}(\omega) \cap \widehat{\mathcal{D}}_-(\Omega)$.

The following theorem is a different formulation of Theorem 5.4.18.

Theorem 5.7.1 *Let $u \in \widehat{\mathcal{D}}(\Omega)$ with values in $[-\infty, +\infty[$ and $\mathcal{F}$-continuous on U. Then the following conditions are equivalent:*

(a) *u is in* $\mathrm{PSH}(\Omega, U)$.

(b) *For every $M = (a_{ij}) \in \mathcal{H}_n^+$ and all $\varphi \in \widehat{\mathcal{D}}_+(\Omega, U)$*

$$-\sum_{j,k=1}^{n} a_{ij} \int_U \partial_j u \bar{\partial}_k \varphi dV \geq 0.$$

Proof (b)⇒(a). Let $\Lambda = (\lambda_1, \ldots, \lambda_n) \in \mathbb{C}^n$. We view Λ as an element of $\mathcal{M}_{n,1}$. Let Λ^* be the adjoint of Λ. Then $\Lambda\Lambda^* = (\lambda_i\overline{\lambda}_j) \in \mathcal{H}_n^+$. Now (b) implies that for all $\varphi \in \widehat{\mathcal{D}}_+(\Omega, U)$ one has

$$-\sum_{i,j=1}^{n} \lambda_i\overline{\lambda}_j \int_U \partial_i u \overline{\partial}_j \varphi dV \geq 0,$$

hence condition (a) is satisfied according to Theorem 5.4.18.

(a)⇒(b). Let $u \in \widehat{\mathcal{D}}(\Omega)$ belong to PSH(Ω, U), $\varphi \in \widehat{\mathcal{D}}_+(\Omega, U)$ and $M = (a_{ij}) \in \mathcal{H}_n^+$. Let $\alpha_1, \ldots, \alpha_n \geq 0$ be the eigenvalues of M, counted with multiplicity, and let $\Lambda^1, \ldots, \Lambda^n$ be an orthonormal basis of eigenvectors of M in $\mathbb{C}^n$ such that

$$M = \sum_{k=1}^{n} \alpha_k \Lambda^k \Lambda^{k*}.$$

If (a) holds, then

$$-\sum_{i,j} a_{ij} \int_U \partial_i u \overline{\partial}_j \varphi dV = -\sum_k \alpha_k \sum_{ij} \lambda_i^k \overline{\lambda_j^k} \int_U \partial_i u \overline{\partial}_j \varphi dV \geq 0$$

according to Theorem 5.4.18, hence assertion (b). □

Let X be the Hilbert space $\mathcal{H}_n \otimes \widehat{\mathcal{D}}(\Omega)$ endowed with the inner product inherited from $\widehat{\mathcal{D}}(\Omega)$ and let $\mathcal{B}$ be the $\mathbb{R}$-bilinear form on $X \times \widehat{\mathcal{D}}(\Omega)$ defined by

$$\mathcal{B}(M \otimes u, v) = -\sum_{j,k=1}^{n} a_{jk} \int \partial_j u \overline{\partial}_k v dV \tag{5.7.1}$$

for all $M \otimes u \in X$ and all $v \in \widehat{\mathcal{D}}(\Omega)$, where $M = (a_{jk})$. We have

Lemma 5.7.2 *The form $\mathcal{B}$ is continuous on $X \times \widehat{\mathcal{D}}(\Omega)$. The map $x \mapsto \mathcal{B}(x, \cdot)$ maps X continuously onto the dual space $\widehat{\mathcal{D}}(\Omega)^*$.*

Proof The first statement follows from

$$\left|\mathcal{B}(M \otimes u, v)\right| \leq \sum_{j,k=1}^{n} \left|a_{jk}\right| \int \left|\nabla u\right|\left|\nabla v\right| dV \leq n\left\|M\right\|_2\left\|u\right\|\left\|v\right\|.$$

For the second statement, just note that for every $u \in \widehat{\mathcal{D}}(\Omega)$ we have $\langle u, v\rangle = \mathcal{B}(I \otimes u, v)$. □

Remark 5.7.3 It follows immediately that if $\mathcal{B}(u, v) = 0$ for every $u \in X$ then $v = 0$, i.e. X *distinguishes points* in $\widehat{\mathcal{D}}(\Omega)$. But $\widehat{\mathcal{D}}(\Omega)$ does not distinguish points in X, hence $\mathcal{B}$ is not separated: Indeed, take any $g \in C_0^\infty(\Omega)$ not identically 0, $u_1 = \frac{\partial^2 g}{\partial z_1 \partial \overline{z}_1}$,

$u_2 = \frac{\partial^2 g}{\partial z_2 \partial \bar{z}_2}$, and let $m^1_{22} = 1$, $m^1_{ij} = 0$ if $(i, j) \neq (2, 2)$, and $m^2_{11} = 1$, $m^2_{ij} = 0$ if $(i, j) \neq (1, 1)$. Then for every $v \in \widehat{\mathcal{D}}(\Omega)$

$$\mathcal{B}(m^1 \otimes u_1 - m^2 \otimes u_2, v) = 0.$$

Let $\mathcal{B}$ be a pairing of topological vector spaces E and F. We denote by $w(E, F)$ the weak topology induced by F via $\mathcal{B}$ on E and similarly by $w(F, E)$ the weak topology induced by E via $\mathcal{B}$ on F. The dual of a cone $A \subset E$ is denoted by

$$A^* = \{f \in F : \forall x \in A,\ \mathcal{B}(x, f) \leq 0\}.$$

Similarly, the dual of a cone Z in F is the subset of $x \in E$ such that for all $f \in Z$, $\mathcal{B}(x, f) \leq 0$, and is denoted by Z_*. We will write $A^*{}_*$ for $(A^*)_*$.

For a set $A \subset E$ we define $co(A)$ as the smallest cone of E containing A (that is, the cone generated by A) and put

$$A^* = co(A)^*.$$

Then we have

$$A^* = \{f \in F : \forall x \in A,\ \mathcal{B}(x, f) \leq 0\}.$$

If $A \subset F$, the cone A_* is defined in a similar way.

We now adapt Theorem 16.3 from [KeNa] to our needs, and we provide its very similar proof for convenience of the reader.

Theorem 5.7.4 *Let $\mathcal{B}$ be a pairing of topological vector spaces E and F.*

(1) *For each cone $A \subset E$, A^* is an $w(F, E)$-closed, convex cone.*

(2) *If $A \subset B$ are cones in E then $B^* \subset A^*$.*

(3) *If $A \subset E$ is non-empty, then $A^*{}_*$ is the smallest convex, $w(E, F)$-closed cone in E that contains A.*

(4) For a nonempty set $A \subset E$ we have $A^*{}_*{}^* = A^*$.

(5) If $\{A_t\}$ is a family of cones in E, then $(\bigcup_t A_t)^* = \bigcap_t A_t^*$.

(6) If $\{A_t\}$ is a family of convex $w(E, F)$-closed cones in E, then $(\bigcap_t A_t)^*$ is the smallest convex, $w(F, E)$-closed cone in F that contains A_t^* for all t.

Proof Statements (1) and (2) are obvious. For (3), let C be the smallest $w(E, F)$-closed cone that contains A, then $C \subset A^*{}_*$. If $x \notin C$ then in view of the Hahn-Banach theorem, there exists a $w(E, F)$ continuous functional g on E such that $g|_C = 0$ and $g(x) = 1$. There exists $f \in F$ with $g(y) = B(y, f)$ for $y \in E$, cf. [KeNa, Theorem 16.2]. Then $f \in A^*$, hence $x \notin A^*{}_*$. Next, (4) follows immediately from (1) and the dual of (3). Now (5) is clear, as $B(x, f) \leq 0$ for $x \in \bigcup_t A_t$ if and only if $f \in \bigcap_t A_t^*$. Finally, by (3) $A_t = A^*_{t\,*}$, so that by (5)

$$\left(\bigcap_t A_t\right)^* = \left(\bigcap_t A^*_{t\,*}\right)^* = \left(\left(\bigcup_t A_t^*\right)_*\right)^*,$$

which equals the $w(E, F)$-closure of the convex hull of $\bigcup_t A_t^*$, and (6) is proven. □

Let $\mathcal{B}$ the pairing of the topological vector spaces X and $\mathcal{D}(\Omega)$ defined in 5.7.1.

Theorem 5.7.5 *For every quasi-analytic set $E \subset \Omega$, the cone $\widehat{\mathcal{D}}_+(\Omega, E)$ and the subspace $\widehat{\mathcal{D}}(\Omega, E)$ of $\widehat{\mathcal{D}}(\Omega)$ are $w(\widehat{\mathcal{D}}(\Omega), X)$-closed.*

Proof By Theorem 4.2.18 it suffices to prove that the result is true for $E = \omega$, an open subset of Ω. Let ω be an open subset of Ω and $u \in \overline{\widehat{\mathcal{D}}_+(\Omega, \omega)}$, the $w(\widehat{\mathcal{D}}(\Omega), X)$-closure of $\widehat{\mathcal{D}}_+(\Omega, \omega)$. Let (u_j) be a sequence of functions in $\widehat{\mathcal{D}}(\Omega, \omega)$ and $w(\widehat{\mathcal{D}}(\Omega), X)$-converging to $u \in \widehat{\mathcal{D}}(\Omega)$. For any bounded positive Borel measurable function φ on Ω such that $\varphi = 0$ on ω, we have $G_\Omega\varphi := G_\Omega\varphi m \in \widehat{\mathcal{D}}_+(\Omega)$ and

$$\int_\Omega u\varphi dm = \lim_j \frac{1}{a_{2n}} \mathcal{B}(I_n \otimes u_j, G_\Omega(\varphi m)) = \lim_j \int_\Omega u_j\varphi dm = 0,$$

by Theorem 4.2.10(a), where I_n is the unit complex matrix and m is the Lebesgue measure on Ω. This shows that $u = 0$ m-a.e. on $\complement\omega$. Since u is (or rather admits a representative) quasi-continuous on Ω (see Theorem 4.2.5), it follows that $u = 0$ q.e. on $\complement\omega$ so $u \in \widehat{\mathcal{D}}(\Omega, \omega)$. Suppose now that the u_j are in $\widehat{\mathcal{D}}_+(\Omega, \omega)$, then, $u \in \widehat{\mathcal{D}}(\Omega, \omega)$ as seen above and, for every bounded positive Borel measurable function φ on Ω, we have as above

$$\int_\Omega u\varphi dm = \lim_j \frac{1}{a_{2n}} \mathcal{B}(I_n \otimes u_j, G_\Omega(\varphi m)) = \lim_j \int_\Omega u_j\varphi dm \geq 0,$$

and hence $u \geq 0$ m-a.e., that is, $u \in \widehat{\mathcal{D}}_+(\Omega, \omega)$. □

Corollary 5.7.6 *For any quasi-analytic set $E \subset \Omega$ one has $\widehat{\mathcal{D}}(\Omega, E)_*{}^* = \widehat{\mathcal{D}}(\Omega, E)$ and $\widehat{\mathcal{D}}_+(\Omega, E)_*{}^* = \widehat{\mathcal{D}}_+(\Omega, E)$.*

Proof The corollary is an easy consequence of the above theorem and Theorem 5.7.4(3). □

We write $\Gamma(U)$ for the convex cone $\mathcal{H}_n^+ \otimes \widehat{\mathcal{D}}_+(\Omega, U) \subset X$ generated by the elements of X of the form $M \otimes u$ where $M \in \mathcal{H}_n^+$ and $u \in \widehat{\mathcal{D}}_+(\Omega, U)$. Next we introduce $C_\mathcal{F}(U)$ the set of (classes of) functions in $\widehat{\mathcal{D}}(\Omega)$ that are $\mathcal{F}$-continuous on U. Let $\Gamma(U)^*$ be the dual of the convex cone $\Gamma(U) \subset X$ relative to $\mathcal{B}$. Then according to Theorem 5.7.1 one has

$$\Gamma(U)^* \cap C_\mathcal{F}(U) = \mathrm{PSH}(\Omega, U). \tag{5.7.2}$$

Remark 5.7.7 In view of Theorem 4.2.21, respectively Theorem 4.2.16 and Theorem 5.7.1 the following hold:

(a) We have $\Gamma(U)^* \subset \mathcal{S}(\Omega, U)$, the cone of functions in $\widehat{\mathcal{D}}(\Omega)$ that are $\mathbb{R}^{2n}$-q.e. subharmonic in U.

(b) For $v \in \widehat{\mathcal{D}}(\Omega)$ The statements "v belongs to $\Gamma(\omega)^*$" and "for all $\Lambda = (\lambda_1, \dots, \lambda_n) \in \mathbb{C}^n$, the distribution $\sum_{j,k=1}^n \lambda_j \overline{\lambda_k} \frac{\partial^2 v}{\partial z_j \partial \bar{z}_k}$ is a positive measure on ω" are equivalent.

Recall from Theorem 4.2.18 that for finely open sets $U \subset \Omega$, hence a fortiori for $\mathcal{F}$-open set $U \subset \Omega$:

$$\widehat{\mathcal{D}}_+(\Omega, U) = \bigcap_\omega \widehat{\mathcal{D}}_+(\Omega, \omega). \tag{5.7.3}$$

Here the intersection runs over all (Euclidean) open sets ω with $U \subset \omega \subset \Omega$.

Proposition 5.7.8 *For all open sets $\omega \subset \Omega$, we have $\Gamma(\omega)^* = \mathrm{PSH}(\Omega, \omega)$.*

Proof The inclusion $\mathrm{PSH}(\Omega, \omega) \subset \Gamma(\omega)^*$ is contained in (5.7.2). For the other direction, let $f \in \Gamma(\omega)^*$. According to Remark 5.7.7(a) we have $\Gamma(\omega)^* \subset \mathcal{S}(\Omega, \omega)$, hence f is, or rather admits a representative which is u.s.c. in ω, therefore, in view of Remark 5.7.7(b), $f \in \mathrm{PSH}(\omega)$. We conclude that $f \in \mathrm{PSH}(\Omega, \omega)$. □

Proposition 5.7.9 *For all $\mathcal{F}$-open $U \subset \Omega$, we have $\Gamma(U) = \bigcap_\omega \Gamma(\omega)$, where the intersection is taken over all Euclidean open sets ω with $U \subset \omega \subset \Omega$.*

Proof The inclusion $\Gamma(U) \subset \bigcap_\omega \Gamma(\omega)$ is clear. For the other inclusion we proceed as follows. Let $x = \sum_{j\in J} M_j \otimes u_j \in \bigcap_\omega \Gamma(\omega) \subset X$, with J finite, $M_j \in \mathcal{H}_n^+\setminus\{0\}$, and $u_j \in \widehat{\mathcal{D}}_+(\Omega)$ for all $j \in J$. Let g be the $\mathbb{R}$-linear form on $\mathcal{H}_n$ defined by $g(M) = \mathrm{tr}(M) = \sum_{i=1}^n a_{ii}$, for all $M = (a_{ij}) \in \mathcal{H}_n$. Then $g(M) > 0$ for every $M \in \mathcal{H}_n^+\setminus\{0\}$. Now let $\omega \subset \Omega$ be open and such that $U \subset \omega$. Because $x \in \Gamma(\omega)$, we can write $x = \sum_{k\in K} L_k \otimes \widetilde{u}_k \in \Gamma(\omega)$ with K finite, $L_k \in \mathcal{H}_n^+\setminus\{0\}$, and $\widetilde{u}_k \in \widehat{\mathcal{D}}_+(\Omega, \omega)$ for all $k \in K$. For every linear continuous form h on $\widehat{\mathcal{D}}(\Omega)$, we have

$$h\Big(\sum_{j\in J} \mathrm{tr}(M_j)u_j\Big) = (g \otimes h)(x) = h\Big(\sum_{k\in K} \mathrm{tr}(L_k)\widetilde{u}_k\Big).$$

Therefore

$$\sum_{j\in J} \mathrm{tr}(M_j)u_j = \sum_{k\in K} \mathrm{tr}(L_k)\widetilde{u}_k.$$

As the functions $\widetilde{u}_k$ are q.e. 0 in $\Omega\setminus\omega$ and because $\mathrm{tr}(M_j), \mathrm{tr}(L_k) > 0$ and the functions u_j are positive, we infer that the latter are also 0 q.e. in $\Omega\setminus\omega$. Hence for all $j \in J$, we have $u_j \in \bigcap_\omega \widehat{\mathcal{D}}_+(\Omega, \omega)$, which equals $\widehat{\mathcal{D}}_+(\Omega, U)$ by (5.7.3). We conclude that $x \in \Gamma(U)$. Hence $\Gamma(U) = \bigcap_\omega \Gamma(\omega)$. □

Theorem 5.7.10

$$\mathrm{PSH}(\Omega, U) = \overline{\bigcup_\omega \mathrm{PSH}(\Omega, \omega)} \cap C_{\mathcal{F}}(U), \tag{5.7.4}$$

$$\mathrm{PSH}_-(\Omega, U) = \overline{\bigcup_\omega \mathrm{PSH}_-(\Omega, \omega)} \cap C_{\mathcal{F}}(U), \tag{5.7.5}$$

where the union is over all Euclidean open ω with $U \subset \omega \subset \Omega$, and the closure is in the sense of the strong (i.e. norm) topology in the Hilbert space $\widehat{\mathcal{D}}(\Omega)$.

Proof Proposition 5.7.9 gives $\Gamma(U) = \bigcap_\omega \Gamma(\omega)$. Invoking (5.7.2), Theorem 5.7.4 yields that

$$\begin{aligned}\mathrm{PSH}(\Omega, U) &= \Gamma(U)^* \cap \mathcal{C}_{\mathcal{F}}(U) = (\bigcap_\omega \Gamma(\omega))^* \cap \mathcal{C}_{\mathcal{F}}(U)\\ &= \overline{\bigcup_\omega \Gamma(\omega)^*} \cap \mathcal{C}_{\mathcal{F}}(U) = \overline{\bigcup_\omega \mathrm{PSH}(\Omega, \omega)} \cap \mathcal{C}_{\mathcal{F}}(U).\end{aligned}$$

Here at first the closure of the convex set $\bigcup_\omega \mathrm{PSH}(\Omega, \omega)$ is in $w(\widehat{\mathcal{D}}(\Omega), X)$, the topology induced by X on $\widehat{\mathcal{D}}(\Omega)$. This topology coincides with the usual weak topology on $\widehat{\mathcal{D}}(\Omega)$ because of Lemma 5.7.2 and next, by the Hahn-Banach theorem this closure equals the closure in the strong topology in $\widehat{\mathcal{D}}(\Omega)$. Hence (5.7.4) is proved.

The proof of (5.7.5) is a straightforward adaptation of the proof of (5.7.4).

$$\begin{aligned}\mathrm{PSH}_-(\Omega, U) &= \mathrm{PSH}(\Omega, U) \cap (-\widehat{\mathcal{D}}_+(\Omega))\\ &= (\Gamma(U)^* \cap (-\widehat{\mathcal{D}}_+(\Omega))) \cap \mathcal{C}_{\mathcal{F}}(U) = (\Gamma(U) \cup (-\widehat{\mathcal{D}}_+(\Omega))_*)^* \cap \mathcal{C}_{\mathcal{F}}(U)\\ &= (\bigcap_\omega \Gamma(\omega) \cup (-\widehat{\mathcal{D}}_+(\Omega))_*)^* \cap \mathcal{C}_{\mathcal{F}}(U)\\ &= (\bigcap_\omega (\Gamma(\omega) \cup (-\widehat{\mathcal{D}}_+(\Omega))_*))^* \cap \mathcal{C}_{\mathcal{F}}(U)\\ &= \overline{\bigcup_\omega (\mathrm{PSH}(\Omega, \omega) \cap (-\widehat{\mathcal{D}}_+(\Omega)))} \cap \mathcal{C}_{\mathcal{F}}(U) = \overline{\bigcup_\omega \mathrm{PSH}_-(\Omega, \omega)} \cap \mathcal{C}_{\mathcal{F}}(U),\end{aligned}$$

by Corollary 5.7.6 and Theorem 5.7.4(6), where the union is taken over all Euclidean open ω with $U \subset \omega \subset \Omega$. □

We next present more useful "bounded" versions of Theorem 5.7.10. In the next two corollaries we assume that U is compactly contained in Ω.

Corollary 5.7.11 *Suppose that $\overline{U} \subset \Omega$ and let $u \in \mathrm{PSH}(\Omega, U)$ be bounded from above by M on U. Then there exists a sequence (ω_j) of open sets $U \subset \omega_j \subset \Omega$ and functions $u_j \in \mathrm{PSH}(\Omega, \omega_j)$, $u_j \leq M$ on ω_j, such that (u_j) converges in the strong topology of $\widehat{\mathcal{D}}(\Omega)$ to a function $v \in PSH(\Omega, U)$ such that $v = u$ on U.*

Proof Let χ be a positive function in $\mathcal{C}_0^\infty(\Omega)$ such that $\chi = 1$ on a Euclidean open neighborhood of $\overline{U}$. Then, according to [DeL, p. 316], Theorem 4.2.3 and Theorem 4.2.9, and because $u \leq M$ on U, we have $\min\{u - M\chi, 0\} \in \mathrm{PSH}_-(\Omega, U)$ and $v := \min\{u - M\chi, 0\} + M\chi \in \mathrm{PSH}(\Omega, U)$ and satisfies $v = u$ on $\overline{U}$, and (5.7.5) in Theorem 5.7.10 gives a sequence of open sets $U \subset \widetilde{\omega}_j \subset \Omega$ and functions $f_j \in \mathrm{PSH}_-(\Omega, \widetilde{\omega}_j)$ converging strongly to $\min\{u - M\chi, 0\}$. Set $\omega_j = \omega \cap \widetilde{\omega}_j$. Then the functions $u_j = f_j + M\chi$ are in $\mathrm{PSH}(\Omega, \omega_j)$, $u_j \leq M$ on ω_j, and (u_j) converges strongly to v. □

Corollary 5.7.12 *Suppose that $\overline{U} \subset \Omega$ and let $u \in \mathcal{F}$-PSH(Ω, U) be bounded from above by M on U. Then there exists a sequence (ω_j) of open sets $U \subset \omega_j \subset \Omega$ and functions $u_j \in \mathrm{PSH}(\omega_j)$, $u_j \leq M$ such that (u_j) converges to u q.e. on U in the sense of the Green capacity with respect to Ω.*

Proof The corollary follows easily from the above corollary by applying Theorem 4.2.8. □

Lemma 5.7.13 *Let $(u_i)_{i\in I}$ be a family of $\mathcal{F}$-locally uniformly bounded $\mathcal{F}$-plurisubharmonic functions on U. Then for every $z \in U$ there exist an open ball $B = B(z, r)$, an $\mathcal{F}$-open $\mathcal{O}$, $z \in \mathcal{O} \subset B \cap U$, and, for all $i \in I$, a function $v_i \in \mathrm{PSH}(B, \mathcal{O}))$ such that $u_i = v_i$ on $\mathcal{O}$.*

Proof Let $z \in U$. Following the proof of Theorem 5.4.3, there exist an open ball $B = B(z, r)$, an $\mathcal{F}$-open $\mathcal{O}$ such that $z \in \mathcal{O} \subset B \cap U$, a family $(w_i)_{i\in I}$ of uniformly bounded plurisubharmonic functions on B, and a $\psi \in \mathrm{PSH}(B)$ such that for all $i \in I$, $u_i = w_i - \psi$ on $\mathcal{O}$. By adding the same constant to each of the functions $-w_i$ and $-\psi$ one may suppose that these are positive on B. After replacing B by $B' = B(z, \frac{r}{2})$ and $-w_i$ as well as $-\psi$ by their swept functions on B', and shrinking $\mathcal{O}$ if necessary, one can assume that $-w_i$ and $-\psi$ are potentials on B. Then one has $v_i := w_i - \psi \in \widehat{\mathcal{D}}(B)$, $u_i = v_i$ on $\mathcal{O}$ and hence $v_i \in \mathrm{PSH}(B, \mathcal{O})$. □

Theorem 5.7.14 *Let $u \in \mathcal{F}$-PSH(U) be finite. Then for all $z \in U$ there exist an $\mathcal{F}$-open $\mathcal{F}$-neighborhood $\mathcal{O} \subset U$ of z, a constant M, a sequence of open sets (ω_j) such that $\mathcal{O} \subset \omega_j \subset \Omega$, and a sequence of functions (u_j), $u_j \in \mathrm{PSH}(\omega_j)$ such that $u \leq M$ on $\mathcal{O}$, $u_j \leq M$ on ω_j, and (u_j) converges $\mathbb{R}^{2n}$-q.e. to u on $\mathcal{O}$.*

Proof Let $z \in U$. Because u is $\mathcal{F}$-continuous, there exists $M > 0$ such that $u \leq M$ on an $\mathcal{F}$-open $\mathcal{F}$-neighborhood $V \subset U$ of z. The preceding lemma applied to $u|_V$ provides us then with an open ball $B = B(z, r)$, an $\mathcal{F}$-open $\mathcal{F}$-neighborhood $\mathcal{O} \subset \overline{\mathcal{O}} \subset B \cap V$ of z and a function $v \in \mathrm{PSH}(B, \mathcal{O})$ such that $v \leq M$ and $u = v$ on $\mathcal{O}$. The result now follows immediately from Corollary 5.7.12 applied with Ω, U and u replaced by B, $\mathcal{O}$ and v. □

Corollary 5.7.15 *Let u be a finite function in $\mathcal{F}$-PSH(U). Then each point z of U admits a pluri-fine neighborhood $\mathcal{O} \subset U$ and a sequence (u_j) of plurisubharmonic functions defined and uniformly bounded on (Euclidean) neighborhoods of $\mathcal{O}$ such that $u = \inf_k[(\sup_{j\geq k} u_j)^*]$ in $\mathcal{O}$ and (u_j) converges to u $\mathbb{R}^{2n}$-q.e. on $\mathcal{O}$.*

Proof Let $z \in U$. By Theorem 5.7.14 there exist an $\mathcal{F}$-open set $\mathcal{O} \subset U$ containing z, a constant $M > 0$, a sequence of open sets (ω_j) with $\mathcal{O} \subset \omega_j \subset \Omega$, and functions $u_j \in \mathrm{PSH}(\omega_j)$ with $u_j \leq M$ on ω_j, such that (u_j) converges $\mathbb{R}^{2n}$-q.e. to u in $\mathcal{O}$. Therefore $u = \inf_k(\sup_{j\geq k} u_j)$ q.e. in $\mathcal{O}$. Because $\inf_k(\sup_{j\geq k} u_j) = \inf_k[(\sup_{j\geq k} u_j)^*]$ q.e. in $\mathcal{O}$ we conclude that $u = \inf_k[(\sup_{j\geq k} u_j)^*]$ $\mathbb{R}^{2n}$-q.e. in $\mathcal{O}$. As the functions u_j are bounded from above by M on $\mathcal{O}$, the function $\inf_k[(\sup_{j\geq k} u_j)^*]|_{\mathcal{O}}$ belongs to $\mathcal{F}$-PSH$(\mathcal{O})$ in view of Theorems 5.4.14 and 5.4.2(iv). It follows by $\mathcal{F}$-continuity of $\mathcal{F}$-plurisubharmonic functions that $u = \inf_k[(\sup_{j\leq k} u_j)^*]$ everywhere on $\mathcal{O}$, because an $\mathbb{R}^{2n}$-polar set in $\mathbb{C}^n$ has an empty fine interior and hence an empty $\mathcal{F}$-interior, the fine topology being finer than the $\mathcal{F}$-topology. □

Lemma 5.7.16 *Let (u_j) be a sequence of $\mathcal{F}$-plurisubharmonic functions that is $\mathcal{F}$-locally uniformly bounded from above on an $\mathcal{F}$-open $V \subset \Omega$. Then there exists a pluripolar set $E \subset V$ such that every point z of $V \setminus E$ admits a Euclidean compact $\mathcal{F}$-neighborhood K_z, such that for every j the restriction of u_j to K_z is continuous.*

Proof It is clear that the set $A = \bigcup_j \{u_j = -\infty\}$ is pluripolar (see Theorem 5.4.3(d)). Let $a \in V \setminus A$. Following the proof of Theorem 5.4.3, there exist an open ball $B = B(a, r)$ containing an $\mathcal{F}$-open $\mathcal{F}$-neighborhood $\mathcal{O}_a \subset V \cap B$ of a, a sequence (v_j) of uniformly bounded plurisubharmonic functions on B, and a bounded $\varphi \in \mathrm{PSH}_-(B)$, such that for all j, $u_j = v_j - \varphi$ on $\mathcal{O}_a$. Let C' be an upper bound for the v_j on B and C a real such that $C > C'$. Put $w_j = (v_j - C)/(C - v_j(a)) \leq 0$. By the quasi-Brelot property of plurisubharmonic functions (see Theorem 5.3.1) applied to the plurisubharmonic function $\varphi + \sum_j 2^{-j} w_j$, there exists a pluripolar set $E_a \subset V$ such that every point t of $B \setminus E_a$ admits an $\mathcal{F}$-neighborhood K_t that is compact in the Euclidean topology and such that the restriction of $\varphi + \sum_j 2^{-j} w_j$ to K_t is continuous. Because the functions w_j and φ are u.s.c., bounded and negative on K_t, we infer that the restriction of φ to K_t is continuous, and for every j the restriction of w_j, and hence of u_j, to K_t is continuous too. The quasi-Lindelöf property of the pluri-fine topology provides us with a pluripolar set F and a sequence of points (a_k) in V such that $V = \bigcup_k \mathcal{O}_{a_k} \cup F \cup A$. Now let E be the $\mathcal{F}$-closure in V of the pluripolar set $\bigcup_k E_{a_k} \cup F \cup A$. Then E is pluripolar and every point z of $V \setminus E$ admits an $\mathcal{F}$-neighborhood which is Euclidean compact and meets the conditions of the lemma. □

Theorem 5.7.17 *Let u be an $\mathcal{F}$-plurisubharmonic function in U. Then there exists a pluripolar $\mathcal{F}$-closed set E such that u is C-strongly $\mathcal{F}$-plurisubharmonic in $U \setminus E$.*

Proof Let $z \in U$. In view of Theorem 5.7.14 and its Corollary 5.7.15, we can find an $\mathcal{F}$-open $\mathcal{O}_z \subset U$ containing z and a sequence (u_j) of uniformly bounded from above plurisubharmonic functions defined on open neighborhoods of $\mathcal{O}_z$ such that $u = \inf_k [(\sup_{j \geq k} u_j)^*]$ on $\mathcal{O}_z$. By Theorem 5.4.14 there exists an $\mathcal{F}$-closed pluripolar subset P of $\mathbb{C}^n$ such that $\inf_k (\sup_{j \geq k} u_j) = \inf_k [(\sup_{j \geq k} u_j)^*]$ on $\mathcal{O}_z \setminus P$. By Lemma 5.7.16, one can also find an $\mathcal{F}$-closed pluripolar set Q such that every point $t \in (\mathcal{O}_z \setminus P) \setminus Q$ admits a compact $\mathcal{F}$-neighborhood K_t on which the restrictions of the functions u, $(\sup_{j \geq k} u_j)^*$, and $\max_{k \leq j \leq i} u_j$, $1 \leq k \leq i$, are Euclidean continuous. Let $t \in \mathcal{O}_z \setminus (P \cup Q)$ and let L_t be a compact $\mathcal{F}$ neighborhood of t such that $L_t \subset K_t \cap (\mathcal{O}_z \setminus (P \cup Q))$. The sequences $(\sup_{j \geq k} u_j)_k$ and $(\max_{k \leq j \leq i} u_j)_i$ are monotonic, hence by Dini's Theorem, we have that $\lim_{k \to +\infty} \sup_{j \geq k} u_j = u$, and for all k, $\lim_{i \to +\infty} \max_{k \leq j \leq i} u_j = \sup_{j \leq k} u_j$ uniformly on L_t. From this we infer that the restrictions to K_t of the functions $u_{k,i} = \max_{k \leq j \leq i} u_j$, $k \leq i$, which are plurisubharmonic in suitable neighborhoods of L_t, approximate u uniformly on L_t. □

Note: The results of this section are due to El Kadiri and Wiegerinck [EKW].

5.8 The Monge-Ampère Operator for $\mathcal{F}$-Plurisubharmonic Functions

Let Ω be an open set $\subset \mathbb{C}^n$, $U \subset \Omega$ an $\mathcal{F}$-domain and let u be the restriction to U of a function $v \in \mathrm{PSH}(\Omega) \cap L^\infty_{loc}(\Omega)$. Then u is $\mathcal{F}$-plurisubharmonic in U and it is natural to try and define on U the Monge-Ampère of u as the measure $(dd^c v)^n$ restricted to U. This definition makes sense because the measure $(dd^c u)^n$ thus defined does not depend on the choice of v according to Corollary 5.3.9. Then, because of the quasi-Lindelöf property of the $\mathcal{F}$-topology and the fact that the Monge-Ampère mass of a bounded plurisubharmonic function does not charge pluripolar sets, this result will enable us to define more generally $(dd^c u)^n$ for functions u that are $\mathcal{F}$-locally the restriction of functions v as above. In all that follows, Ω will be a domain in $\mathbb{C}^n$ $(n \geq 2)$ and U will be an $\mathcal{F}$-domain in Ω. We denote by $\mathrm{QB}(\mathbb{C}^n)$ the σ-algebra on $\mathbb{C}^n$ generated by the Borel sets and the pluripolar subsets of $\mathbb{C}^n$. It follows from Theorem 5.2.5 (or Theorem 5.2.8) and the quasi-Lindelöf property of the pluri-fine topology that any $\mathcal{F}$-open subset of $\mathbb{C}^n$ belongs to $\mathrm{QB}(\mathbb{C}^n)$. This allows us to define $\mathrm{QB}(U)$ as the trace of $\mathrm{QB}(\mathbb{C}^n)$ on U. Let u be a finite $\mathcal{F}$-plurisubharmonic function on a U. Then by Theorem 5.4.3(a), u is $\mathcal{F}$-locally the difference $w = v_1 - v_2$ of bounded plurisubharmonic functions v_1, v_2 defined on an open set in $\mathbb{C}^n$. As a consequence $\mathcal{F}$-plurisubharmonic functions are $\mathcal{F}$-locally bounded on U if and only if they are finite on U. Following Cegrell and Wiklund [CeW] one defines the (signed) Monge-Ampère mass $(dd^c w)^n$ associated with w as

$$(dd^c w)^n = \sum_{p=0}^{n} \binom{n}{p} (-1)^p (dd^c v_1)^p \wedge (dd^c v_2)^{n-p}. \tag{5.8.1}$$

Therefore we would like to define $(dd^c u)^n$ $\mathcal{F}$-locally by $(dd^c u)^n = (dd^c w)^n$. To do so we need some auxiliary results.

Proposition 5.8.1 *Let $f_1, g_1, f_2, g_2 \in \mathrm{PSH}(\Omega) \cap L^\infty_{loc}(\Omega)$ and let $\mathcal{O} = \{f_1 - g_1 > f_2 - g_2\}$. Then*

$$(dd^c \max\{f_1 - g_1, f_2 - g_2\})^n|_{\mathcal{O}} = (dd^c(f_1 - g_1))^n|_{\mathcal{O}}.$$

Proof We will use equation (5.8.1) and apply Proposition 5.3.7 with various m and for every j, $u_j = f_1 + g_2$ and $v_j = f_2 + g_1$. Then

$$
\begin{aligned}
&(dd^c \max\{f_1 - g_1, f_2 - g_2\})^n|_{\mathcal{O}} \\
&\quad =(dd^c(\max\{f_1 + g_2, f_2 + g_1\} - (g_1 + g_2)))^n|_{\mathcal{O}} \\
&\quad =\sum_{p=0}^{n}(-1)^p\binom{n}{p}(dd^c \max\{f_1 + g_2, f_2 + g_1\})^{n-p} \wedge (dd^c g_1 + g_2))^p|_{\mathcal{O}} \\
&\quad =\sum_{p=0}^{n}(-1)^p\binom{n}{p}(dd^c(f_1 + g_2))^{n-p} \wedge (dd^c(g_1 + g_2))^p|_{\mathcal{O}} \\
&\quad =(dd^c(f_1 - g_1))^n|_{\mathcal{O}}.
\end{aligned}
$$

□

Now we can prove a generalization of Corollary 5.3.9.

Theorem 5.8.2 *Suppose that u_1, u_2, v_1, v_2 are locally bounded plurisubharmonic functions on Ω. If $u_1 - u_2 = v_1 - v_2$ on an $\mathcal{F}$-open $\mathcal{O} \subset \Omega$, then $(dd^c(u_1 - u_2))^n|_{\mathcal{O}} = (dd^c(v_1 - v_2))^n|_{\mathcal{O}}$.*

Proof Let $k \in \mathbb{N}$. As $u_1 - u_2 > v_1 - v_2 - \frac{1}{k}$ in $\mathcal{O}$, we have in view of Proposition 5.8.1

$$
\begin{aligned}
&(dd^c(u_1 - u_2))^n|_{\mathcal{O}} = (dd^c \max\{u_1 - u_2, v_1 - v_2 - \frac{1}{k}\})^n|_{\mathcal{O}} \\
&= (dd^c(\max\{u_1 + v_2, u_2 + v_1 - \frac{1}{k}\}) - (u_2 + v_2)))^n|_{\mathcal{O}} \\
&= \sum_{p=0}^{n}\binom{n}{p}(-1)^p(dd^c \max\{u_1 + v_2, u_2 + v_1 - \frac{1}{k}\})^{n-p} \wedge (dd^c(u_2 + v_2))^p|_{\mathcal{O}}.
\end{aligned}
$$

We let $k \to \infty$ and obtain in view of Lemma 5.3.6

$$
\begin{aligned}
&(dd^c(u_1 \quad u_2))^n|_{\mathcal{O}} \\
&= \sum_{p=0}^{n}\binom{n}{p}(-1)^p(dd^c \max\{u_1 + v_2, u_2 + v_1\})^{n-p} \wedge (dd^c(u_2 + v_2))^p|_{\mathcal{O}} \\
&= (dd^c \max\{u_1 - u_2, v_1 - v_2\})^n|_{\mathcal{O}}.
\end{aligned}
$$

Similarly wc havc

$$
(dd^c(v_1 - v_2))^n|_{\mathcal{O}} = (dd^c \max\{u_1 - u_2, v_1 - v_2\})^n|_{\mathcal{O}},
$$

and the proof is complete. □

Theorem 5.8.2 will enable us to define the Monge-Ampère operator for finite $\mathcal{F}$-plurisubharmonic functions. We will show that the Monge-Ampère mass of a finite $\mathcal{F}$ plurisubharmonic function defined on an $\mathcal{F}$-open set can be defined and is positive.

The proof of the following lemma is inspired by that of Lemma 5.3.6.

Lemma 5.8.3 *Let $\mathcal{O} \subset \Omega$ be an $\mathcal{F}$-open set, and let (u_1^j) and (u_2^j) be two monotone sequences of plurisubharmonic functions that are bounded in Ω, each converging to a bounded plurisubharmonic function u_1, respectively u_2 on Ω. If $(dd^c(u_1^j - u_2^j))^n|_{\mathcal{O}} \geq 0$ for all j, then $(dd^c(u_1 - u_2))^n|_{\mathcal{O}} \geq 0$.*

Proof Observe that $(dd^c(u_1^j - u_2^j))^n$ and $(dd^c(u_1 - u_2))^n$ are defined on $\mathcal{O}$ by (5.8.1). By Theorem 5.2.8 and the quasi-Lindelöf property of the pluri-fine topology we can write $\mathcal{O}$ as a pluripolar set E joined with a countable union of $\mathcal{F}$-open sets of the form $B \cap \{\psi > 0\}$, where B is an open ball in $\mathbb{C}^n$, $B \subset \overline{B} \subset \Omega$, and ψ plurisubharmonic in a neighborhood of B. As pluripolar sets are negligible for the Monge-Ampère mass of bounded plurisubharmonic functions, it suffices to show the result for $\mathcal{O} = B \cap \{\psi > 0\}$. We proceed as follows. Multiplying ψ with a smooth cut-off function that is positive on B, we find an $\mathcal{F}$-continuous function $\widetilde{\psi} \geq 0$ with compact support such that $\mathcal{O} = \{z \in \Omega : \widetilde{\psi}(z) > 0\}$. Then, according to Theorem 5.3.2,

$$0 \leq \lim_{j \to \infty} \int \widetilde{\psi}(dd^c(u_1^j - u_2^j))^n = \int \widetilde{\psi}(dd^c(u_1 - u_2))^n.$$

When we replace $\widetilde{\psi}$ by $f\widetilde{\psi}$, where f is any $\mathcal{F}$-continuous functions ≥ 0 on Ω, this inequality remains valid. That is, $\int f\widetilde{\psi}(dd^c(u_1 - u_2))^n \geq 0$. Hence $\widetilde{\psi}(dd^c(u_1 - u_2))^n \geq 0$, and it follows that $(dd^c(u_1 - u_2))^n|_{\mathcal{O}} \geq 0$. □

Theorem 5.8.4 *Let $U \subset \Omega$ be a $\mathcal{F}$-open and u be in $\mathcal{F}$-$\mathrm{CPSH}(U)$. Then for each $z \in U$ there exists an open ball $B = B(z, r_z)$, an $\mathcal{F}$-open set $\mathcal{O}_z \subset B \cap U$ containing z and bounded functions $u_z, v_z \in \mathrm{PSH}(B)$ such that*

(a) $u = u_z - v_z$ on $\mathcal{O}_z$,

(b) $(dd^c(u_z - v_z))^n|_{\mathcal{O}_z} \geq 0$.

Proof Let $z \in U$. There exists a compact $\mathcal{F}$-neighborhood $K = K_z \subset U$ of z and functions $u_1, u_2, \ldots$ that are finite, continuous and plurisubharmonic on open neighborhoods $\mathcal{O}_1, \mathcal{O}_2, \ldots$ of K, such that the sequence $(u_j|_K)$ converges to u uniformly on K. Take any such K and u_j. The functions u_j are continuous, hence after shrinking the open $\mathcal{O}_j$ and modifying the functions u_j by small constants, we can assume that for every j, $\mathcal{O}_{j+1} \subset \mathcal{O}_j$ and $u_{j+1} \leqslant u_j$ on $\mathcal{O}_{j+1}$. Next it follows from the proof of Theorem 5.4.3(a), (b), that we can find an open set B containing z, a decreasing sequence (u_z^j) of bounded plurisubharmonic functions on B and finally two functions u_z and v_z in $\mathrm{PSH}(B)$ such that $u_z \leq u_z^j$ for every j on B and, on some $\mathcal{F}$-open $\mathcal{F}$-neighborhood $\mathcal{O}_z \subset B \cap U$ of z, one has $u_j = u_z^j - v_z$ and $u = u_z - v_z$. It is clear that (u_z^j) converges (pointwise) to u_z. For every j, we have $(dd^c(u_z^j - v_z))^n|_{\mathcal{O}_z} = (dd^c u_j)^n|_{\mathcal{O}_z} \geq 0$. As the sequence (u_z^j) decreases we conclude in view of Lemma 5.8.3 that $(dd^c(u_z - v_z))^n|_{\mathcal{O}_z} \geq 0$. □

Corollary 5.8.5 *Let $U \subset \Omega$ be $\mathcal{F}$-open and f be in $\mathcal{F}$-$\mathrm{CPSH}(U)$. Then for any $\mathcal{F}$-open set $\mathcal{O} \subset U$ on which $f = u - v$, where u and v are bounded plurisubharmonic functions on an open set $\subset \mathbb{C}^n$ containing $\mathcal{O}$, we have $(dd^c(u - v))^n|_{\mathcal{O}} \geq 0$.*

Proof According to Theorem 5.8.4 there exists for each $z \in \mathcal{O}$, an $\mathcal{F}$-open set $\mathcal{O}_z \subset \mathcal{O}$ and two bounded plurisubharmonic functions u_z and v_z on an open neighborhood of $\mathcal{O}_z$ such that $f = u_z - v_z$ on $\mathcal{O}_z$ and $(dd^c(u_z - v_z))^n|_{\mathcal{O}_z} \geq 0$. By the quasi-Lindelöf property of the $\mathcal{F}$ topology there is a sequence (z_j) of points of $\mathcal{O}$ and a pluripolar set $P \subset \mathcal{O}$ such that $\mathcal{O} = \bigcup_j \mathcal{O}_{z_j} \cup P$. Since $f = u - v$ on $\mathcal{O}$ and therefore on $\mathcal{O}_{z_j}$ for every j, it follows from Theorem 5.8.2 that $(dd^c(u - v))^n|_{\mathcal{O}_{z_j}} = (dd^c(u_{z_j} - v_{z_j}))^n|_{\mathcal{O}_{z_j}}$ and thus $(dd^c(u - v))^n|_{\mathcal{O}} \geq 0$, for $(dd^c(u - v))^n$ puts no mass on pluripolar sets. □

Now we can define the Monge-Ampère measure $(dd^c u)^n$ associated with a finite $\mathcal{F}$-plurisubharmonic function u as follows: Let first u be a C-strongly $\mathcal{F}$-plurisubharmonic on an $\mathcal{F}$-open set $U \subset \Omega$. By Theorem 5.8.4 and the quasi-Lindelöf property of the pluri-fine topology, there exists a sequence $(\mathcal{O}_j)$ of $\mathcal{F}$-open subsets of U and a pluripolar subset P of U such that $U = \bigcup_j \mathcal{O}_j \cup P$ and for each j, bounded plurisubharmonic functions u_j and v_j on an open neighborhood of $\mathcal{O}_j$ such that $u = u_j - v_j$ on $\mathcal{O}_j$. Because the measures $(dd^c(u_j - v_j))^n$ do not have mass on pluripolar sets and because $(dd^c(u_j - v_j))^n|_{\mathcal{O}_j}$ is positive for every j by Corollary 5.8.5, we can define a positive measure μ on U by putting for every $A \in \mathrm{QB}(U)$,

$$\mu(A) = \int_{\mathcal{O}_1} 1_A (dd^c(u_1 - v_1))^n + \sum_{k \geq 2} \int_{\mathcal{O}_k \setminus (\bigcup_{1 \leq j \leq k-1} \mathcal{O}_j)} 1_A (dd^c(u_k - v_k))^n. \tag{5.8.2}$$

By Theorem 5.8.2 the measure μ defined by (5.8.2) is independent of the choice of the sequence $(\mathcal{O}_j)$ and the functions u_j and v_j as described above. This measure is denoted by $(dd^c u)^n$.

Suppose now that $u \in \mathcal{F}\text{-}\mathrm{PSH}(U)$ is finite. Then, according to Theorem 5.7.17, there exists a pluripolar set $\mathcal{F}$-closed $F \subset \Omega$ such that u is C-strongly $\mathcal{F}$-plurisubharmonic on $U \setminus F$. Since the measure $(dd^c u|_{U \setminus F})^n$ on $U \setminus F$ does not charge the pluripolar sets, it can be extended to a positive measure on $(U, \mathrm{QB}(U))$, denoted by $(dd^c u)^n$, by putting $(dd^c u)^n(F) = 0$. The measure $(dd^c u)^n$ thus defined is independent of the choice of F and does not have mass on pluripolar sets, that is, a Borel measure on U which has no mass on pluripolar sets.

Definition 5.8.6 The measure $(dd^c u)^n$ thus defined on $(U, \mathrm{QB}(U))$ is the Monge-Ampère measure associated with u. The operator which associates to every finite function $u \in \mathcal{F}$-PSH(U) the measure $(dd^c u)^n$ will be called the Monge-Ampère operator on U.

The following theorem extends Corollary 5.3.9 to finite $\mathcal{F}$-plurisubharmonic functions on $\mathcal{F}$-open sets.

Theorem 5.8.7 *Let u and v be two finite $\mathcal{F}$-plurisubharmonic functions on an $\mathcal{F}$-open set $U \subset \Omega$. If $u = v$ on a $\mathcal{F}$-open $\mathcal{O} \subset U$, then*

$$(dd^c u)^n|_{\mathcal{O}} = (dd^c v)^n|_{\mathcal{O}}.$$

Proof Because of Theorem 5.4.3(a) and the quasi-Lindelöf property of the pluri-fine topology we can find bounded plurisubharmonic functions u_j^1, u_j^2, v_j^1, and v_j^2 defined on open neighborhoods of compact sets $K_j \subset U$, $(j = 1, 2, ...)$ and a pluripolar set P such that $\mathcal{O} = \bigcup_j K_j \cup P$, and for every j, $u = u_j^1 - u_j^2$ and $v = v_j^1 - v_j^2$ on an $\mathcal{F}$-open $\mathcal{F}$-neighborhood V_j of K_j. Now we have $(dd^c(u_j^1 - u_j^2))^n|_{K_j} = (dd^c(v_j^1 - v_j^2))^n|_{K_j}$ according to Theorem 5.8.2, and the proof is complete because $(dd^c u)^n(P) = (dd^c v)^n(P) = 0$. □

Now we can define as in [BT3] the non-polar part $\mathrm{NP}((dd^c u)^n)$ of $(dd^c u)^n$ for an $\mathcal{F}$-plurisubharmonic function u on an $\mathcal{F}$-open set $U \subset \mathbb{C}^n$. For every set $E \in \mathrm{QB}(U)$ we set

$$\mathrm{NP}(dd^c u)^n(E) = \lim_{j \to +\infty} \int_{E \cap \{u > -j\}} (dd^c u_j)^n,$$

where $u_j = \max\{u, -j\}$. This definition makes sense in view of Theorem 5.8.7. Indeed, it follows from this theorem that for each $j \in \mathbb{N}$ and $E \in \mathrm{QB}(U)$, we have $(dd^c u_j)^n = (dd^c u_{j+1})^n$ on $E \cap \{u > -j\}$ by Theorem 5.8.7, so the sequence $(\int_{E \cap \{u > -j\}} (dd^c u_j)^n)$ in $[0, +\infty]$ is increasing. It follows from this definition and Theorem 5.8.7 that if u, v are $\mathcal{F}$-plurisubharmonic functions on U that are equal on an $\mathcal{F}$-open set $V \subset U$, then $\mathrm{NP}(dd^c u)^n(E) = \mathrm{NP}(dd^c v)^n(E)$ for every set $E \in \mathrm{QB}(V)$ and that if u is finite on U, then $(dd^c u)^n = \mathrm{NP}(dd^c u)^n$ on $\mathrm{QB}(U)$. Moreover, $\mathrm{NP}(dd^c u)^n$ does not charge the pluripolar subsets of U.

If u is a plurisubharmonic function on Ω, the Borel measure $\mathrm{NP}(dd^c u)^n$ is not in general locally finite as shown by the following example, see [Ki, Exemple 3.1]:

$$u(z_1, z_2) := (|z_1|^2 - 1)(-\log|z_2|)^{\frac{1}{2}}$$

for $(z_1, z_2) \in \mathbb{C}^2$ near 0. The function u is plurisubharmonic near 0, it is smooth outside the $z_2 = 0$ axis and $\mathrm{NP}(dd^c u)^2|_{\{z_2 \neq 0\}} = (dd^c u|_{\{z_2 \neq 0\}})^2$, but the smooth measure $(dd^c(u|_{\{z_2 \neq 0\}}))^2$, is not locally finite near any point of $z_2 = 0$. This means that the positive Borel measure $\mathrm{NP}(dd^c u)^2$ is not locally finite.

Remark 5.8.8 Let u be a finite plurisubharmonic function on a Euclidean domain Ω in $\mathbb{C}^n$. Then u is in particular in $\mathcal{F}$- $\mathrm{PSH}(\Omega)$ and finite. The Monge-Ampère mass $(dd^c u)^n$ coincide with the non-polar part $\mathrm{NP}(dd^c u)^n$ of $(dd^c u)^n$ defined in [BT3, p. 236]. This measure is in general not a Radon measure, i.e. not Euclidean locally finite. It is, however, $\mathcal{F}$-locally finite. In general u may well not belong to $\mathcal{D}_\Omega$, the domain of the Monge-Ampère operator on Ω in the sense of Błocki (cf. [Bl1]) as shown by the following example due to Åhag, Cegrell and Hiep [ACH].

Example 5.8.9 Let Δ be the unit disk in $\mathbb{C}$ and let $\Omega \subset \Delta^n$ be a hyperconvex domain. Let u be defined on Δ^n by

$$u(z) = \sum_{j=1}^{\infty} \max\left\{\frac{1}{2^j} \log \frac{|z_1 - \frac{1}{2^j}|}{|1 - \frac{z_1}{2^j}|}, \frac{2^j}{j} \log|z_2|, \log|z_3|, \cdots, \log|z_n|, -2^j\right\}.$$

We start by proving that $u(z) > -\infty$ for all $z \in \Delta^n$. If $z_1 = 0$ then we have that

$$u(z) = \sum_{j=1} \frac{1}{2^j} \log \frac{1}{2^j} > -\infty.$$

If $z_1 \neq 0$ we choose j_0 be such that $|z_1| > \frac{1}{2^{j_0-1}}$. Hence

$$u(z) \geq \sum_{j=1}^{j_0} (-2^j) + \sum_{j=j_0+1}^{\infty} \frac{1}{2^j} \log \frac{|z_1 - \frac{1}{2^j}|}{|1 - \frac{z_1}{2^j}|} \geq \sum_{j=1}^{j_0} (-2^j) + \frac{\log |z_1|}{4} \sum_{j=j_0+1}^{\infty} \frac{1}{2^j} > -\infty.$$

Next, we shall show that $u \notin \mathcal{D}_W$ for all open neighborhoods W of 0 ($\mathcal{D}_W$ is the domain of definition of the Monge-Ampère operator on W). Set

$$u_k = \sum_{j=1}^{k} \max\left\{\frac{1}{2^j} \log \frac{|z_1 - \frac{1}{2^j}|}{|1 - \frac{z_1}{2^j}|}, \frac{2^j}{j} \log |z_2|, \log |z_3|, \cdots, \log |z_n|, -2^j\right\}.$$

We have $u_k \in \mathcal{D}_{\Delta^n}$ and $u_k \downarrow u$ as $k \to \infty$, and

$$(dd^c u_k)^n \geq \sum_{j=1}^{k}$$

$$(dd^c (\sum_{j=1}^{k} \max\left\{\frac{1}{2^j} \log \frac{|z_1 - \frac{1}{2^j}|}{|1 - \frac{z_1}{2^j}|}, \frac{2^j}{j} \log |z_2|, \log |z_3|, \cdots, \log |z_n|, -2^j\right\}))^n$$

$$= \sum_{j=1}^{k} \frac{1}{j} \sigma_{\{|z_1 - \frac{1}{2^j}| = e^{-4^j} |1 - \frac{z_1}{2^j}|\}} \times \sigma_{\{|z_2| = e^{-j}\}} \times \sigma_{\{|z_3| = e^{-2j}\}} \times \cdots \times \sigma_{\{|z_n| = e^{-2j}\}},$$

where $\sigma_{\{|z_j| - r\}}$ is the normalized surface measure on $\{|z_j| - r\}$. Hence

$$\int_W (dd^c u_k)^n \to \infty$$

as $k \to \infty$ for all neighborhood W of 0.

Definition 5.8.10 Let U be an $\mathcal{F}$-open set in $\mathbb{C}^n$ and let μ and μ_j, $j = 1, 2, \cdots$, be measures on $(U, \mathrm{QB}(U))$ that give measure zero to pluripolar sets. We say that (μ_j) *converges $\mathcal{F}$-locally vaguely* to μ on $(U, \mathrm{QB}(U))$ if for any $z \in U$ there exists an $\mathcal{F}$-open V such that $z \in V \subset U$ and $\lim_{j \to +\infty} \int \varphi d\mu_j = \int \varphi d\mu$ for every bounded $\mathcal{F}$-continuous function φ on U with compact support $\mathrm{Supp}(\varphi) \subset V$ (that is, φ is null outside a compact subset of V).

Remark 5.8.11 Let μ, ν and μ_j, $j = 1, 2, \cdots$, be measures on $(U, \mathrm{QB}(U))$ that give measure zero to pluripolar sets. Suppose that (μ_j) converges $\mathcal{F}$-locally vaguely to both μ and ν. We will show that $\mu = \nu$. Indeed, let $z \in U$ and pick V_z such that

$$\int \varphi d\mu = \lim_{j\to+\infty} \int \varphi d\mu_j = \int \varphi d\nu$$

for every bounded $\mathcal{F}$-continuous function φ on U with compact support $\text{Supp}(\varphi) \subset V_z$. According to Corollary 5.2.11 there is an $\mathcal{F}$-continuous φ_z on $\mathbb{C}^n$ with compact support within V_z such that $0 \leq \varphi_z \leq 1$ and $\varphi_z \equiv 1$ on an $\mathcal{F}$-open $\mathcal{F}$-neighborhood W_z $(\subset V_z)$ of z. Using the quasi-Lindelöf property of the pluri-fine topology, we can find a sequence (z_k) of points in U such that $U = (\bigcup_{k=1}^{\infty} W_{z_k}) \cup K$ where K is pluripolar. Define $\psi_1 = \varphi_{z_1}$, $\psi_2 = \min\{\varphi_{z_2}, 1 - \psi_1\}$, ..., $\psi_k = \min\{\varphi_{z_k}, 1 - (\psi_1 + \ldots + \psi_{k-1})\}, \ldots$. These functions ψ_k form a partition of unity for $U \setminus K$, and $\mu(K) = \nu(K) = 0$. If η is a positive $\mathcal{F}$-continuous function on U, we find that

$$\begin{aligned}\int_U \eta d\mu &= \int_{U\setminus K} \eta d\mu = \sum_{k=1}^{\infty} \int_{U\setminus K} \psi_k \eta d\mu = \sum_{k=1}^{\infty} \int_{U\setminus K} \psi_k \eta d\nu \\ &= \int_{U\setminus K} \eta d\nu = \int_U \eta d\nu.\end{aligned}$$

This implies that $\mu = \nu$ in view of the monotone class theorem.

Theorem 5.8.12 *Let (u_j) be a monotone sequence of finite, $\mathcal{F}$-plurisubharmonic functions on an $\mathcal{F}$-open set $U \subset \mathbb{C}^n$ that converges to a finite $\mathcal{F}$-plurisubharmonic function on U. Then the sequence of measures $(dd^c u_j)^n$ converges $\mathcal{F}$-locally vaguely to $(dd^c u)^n$.*

Proof As (u_j) is monotone, the function u and all of the functions u_j are $\mathcal{F}$-locally uniformly bounded on U. By Theorem 5.4.3(a), for every $z \in U$ there exists an open ball $B_z = B(z, r_z)$ in $\mathbb{C}^n$ and an $\mathcal{F}$-open $\mathcal{F}$-neighborhood $V_z \subset B_z \cap U$ of z, and functions v, w and v_j, $j \in \mathbb{N}$, that are plurisubharmonic and uniformly bounded on B_z such that $u_j = v_j - w$, $u = v - w$ on V_z and the sequence (v_j) is monotone. By (5.8.1) we have

$$(dd^c u_j)^n = \sum_{p=0}^{n} (-1)^p \binom{n}{p} (dd^c v_j)^p \wedge (dd^c w)^{n-p}$$

on V_z. Let φ be a bounded $\mathcal{F}$-continuous function on U that equals 0 outside of a compact $K \subset V_z$. Then according to Corollary 5.3.4

$$\begin{aligned}\lim_{j\to+\infty} \int \varphi (dd^c u_j)^n &= \lim_{j\to+\infty} \sum_{p=0}^{n} (-1)^p \binom{n}{p} \int \varphi (dd^c (v_j)^p \wedge (dd^c w)^{n-p} \\ &= \sum_{p=0}^{n} (-1)^p \binom{n}{p} \int \varphi (dd^c v)^p \wedge (dd^c w)^{n-p} = \int \varphi (dd^c u)^n.\end{aligned}$$

Hence the sequence of measures $(dd^c u_j)^n$ converges $\mathcal{F}$-locally vaguely to $(dd^c u)^n$. □

Corollary 5.8.13 *Let (u_j) be a monotone sequence of finite $\mathcal{F}$-plurisubharmonic functions on an $\mathcal{F}$-open set $U \subset \mathbb{C}^n$ that converges to a finite $\mathcal{F}$-plurisubharmonic function u on U. Then $(dd^c u)^n = 0$ if $(dd^c u_j)^n = 0$ for all j.*

Proof Applying Theorem 5.8.12, we see that $(dd^c u_j)^n$ converges $\mathcal{F}$-locally vaguely to $(dd^c u)^n$. However, by Definition 5.8.10 $(dd^c u_j)^n$ converges $\mathcal{F}$-locally vaguely to 0 as well. By Remark 5.8.11, this means that $(dd^c u)^n = 0$. □

Notes: Theorem 5.8.12 and its corollary are due to El Kadiri and Smit [EKS]. The other results are due to El Kadiri and Wiegerinck [EKW].

5.9 Maximal $\mathcal{F}$-Plurisubharmonic Functions

In analogy with *maximal* plurisubharmonic functions, which play a role in pluripotential theory comparable to that of harmonic functions in classical potential theory, we will introduce *$\mathcal{F}$-maximal* $\mathcal{F}$-plurisubharmonic functions. These relate similarly to finely harmonic functions and constitute the pluri-fine analog of maximal plurisubharmonic functions on Euclidean open sets.

Definition 5.9.1 Let $U \subset \mathbb{C}^n$ be an $\mathcal{F}$-open set and u an $\mathcal{F}$-plurisubharmonic function on U. We say that u is *$\mathcal{F}$-maximal* if for every bounded $\mathcal{F}$-open set $G \subset \overline{G} \subset U$, and for every bounded from above $\mathcal{F}$-u.s.c. function v on $\overline{G}^{\mathcal{F}}$ such that $v \in \mathcal{F}\text{-}\operatorname{PSH}(G)$ the following holds:

$$v \leq u \text{ on } \partial_{\mathcal{F}} G \Rightarrow v \leq u \text{ on } G.$$

We denote by $\mathcal{F}\text{-}\operatorname{MPSH}(U)$ the set of $\mathcal{F}$-maximal $\mathcal{F}$-plurisubharmonic functions on U and we say that a function $u \in \mathcal{F}\text{-}\operatorname{PSH}(U)$ is $\mathcal{F}$-maximal in an $\mathcal{F}$-open set $V \subset U$ if $u|_V \in \mathcal{F}\text{-}\operatorname{MPSH}(V)$. It follows easily from the definition that if $u \in \mathcal{F}\text{-}\operatorname{MPSH}(U)$ and $c \in \mathbb{R}$, then $u + c \in \mathcal{F}\text{-}\operatorname{MPSH}(U)$.

Proposition 5.9.2 *Let $U \subset \mathbb{C}^n$ be an $\mathcal{F}$-domain, and let (u_j) be a sequence of functions in $\mathcal{F}\text{-}\operatorname{MPSH}(U)$, decreasing to u. Then either $u \equiv -\infty$, or $u \in \mathcal{F}\text{-}\operatorname{MPSH}(U)$.*

Proof It follows from Theorem 5.4.2(iv) that either $u \equiv -\infty$ or else $u \in \mathcal{F}\text{-}\operatorname{PSH}(U)$. In the latter case, let G be a bounded $\mathcal{F}$-open set such that $\overline{G} \subset U$, and let v be an $\mathcal{F}$-u.s.c. and bounded from above function on $\overline{G}^{\mathcal{F}}$ such that $v \in \mathcal{F}\text{-}\operatorname{PSH}(G)$. Suppose that $v \leq u$ on $\partial_{\mathcal{F}} G$. Then for all j we have $v \leq u_j$ on $\partial_{\mathcal{F}} G$, and hence $v \leq u_j$ on G by $\mathcal{F}$-maximality of u_j. This implies that $v \leq u$ on G, so u is $\mathcal{F}$-maximal. □

Proposition 5.9.3 *Let $\Omega \subset \mathbb{C}^n$ be a Euclidean domain. A plurisubharmonic function u on Ω is $\mathcal{F}$-maximal if and only if u is maximal.*

Proof Suppose that $u \in \mathrm{PSH}(\Omega)$ is $\mathcal{F}$-maximal. Let G be a bounded open set in $\mathbb{C}^n$ such that $\overline{G} \subset \Omega$, and let v an u.s.c. function on $\overline{G}$ such that $v \in \mathrm{PSH}(G)$ and $v \le u$ on ∂G. Observe that $u \in \mathcal{F}\text{-}\mathrm{PSH}(\Omega)$ and $v \in \mathcal{F}\text{-}\mathrm{PSH}(G)$ by Proposition 5.4.6, and that v is bounded from above on $\overline{G}$. Then, because $\partial_{\mathcal{F}} G \subset \partial G$, we have $v \le u$ on $\partial_{\mathcal{F}} G$, and hence $v \le u$ on G by $\mathcal{F}$-maximality. In other words, u is maximal. For the converse, suppose that u is maximal. Let G be a bounded $\mathcal{F}$-open set with the property that $G \subseteq \overline{G} \subset \Omega$, and let v be a bounded from above $\mathcal{F}$-u.s.c. function on $\overline{G}$, such that $v \in \mathcal{F}\text{-}\mathrm{PSH}(G)$ and $v \le u$ on $\partial_{\mathcal{F}} G$. By Proposition 5.4.4, the function

$$w = \begin{cases} \max\{u, v\} & \text{on } G \\ u & \text{on } \Omega \backslash G \end{cases}$$

lies in $\mathcal{F}\text{-}\mathrm{PSH}(\Omega)$, and by Proposition 5.4.6 therefore also in $\mathrm{PSH}(\Omega)$. Obviously, $w \le u$ on $\Omega \backslash G$. Since $\overline{G}$ is compact, there is an open set W such that $\overline{G} \subset W \subset \overline{W} \subset \Omega$. Now we have $w \le u$ on $\partial_{\mathcal{F}} W$ and hence by maximality of u on Ω, we can conclude that $w \le u$ on W. This implies that $v \le u$ on G, so u is $\mathcal{F}$-maximal. □

Proposition 5.9.4 *Let $U \subset \mathbb{C}^n$ be an $\mathcal{F}$-open set, and $u \in \mathcal{F}\text{-}\mathrm{PSH}(U)$ a bounded function that is $\mathcal{F}$-maximal in $U \backslash F$ for some $\mathcal{F}$-closed pluripolar set $F \subset U$. Then u is $\mathcal{F}$-maximal in U.*

Proof By subtracting a constant from u if necessary, we may assume without loss of generality in view of the text following Definition 5.9.1, that $m < u < 0$ for some constant m. Let G be a bounded $\mathcal{F}$-open set such that $G \subseteq \overline{G} \subseteq U$, and let v be a bounded from above $\mathcal{F}$-u.s.c. function on $\overline{G}^{\mathcal{F}}$ such that $v \in \mathcal{F}\text{-PSH}(G)$, and satisfies $v \le u$ on $\partial_{\mathcal{F}} G$. As $u < 0$, we see that $v < 0$ on $\partial_{\mathcal{F}} G$. Since v is finely subharmonic on G (see Theorem 5.4.11), it follows that $v \le 0$ on G by the minimum principle, cf. Theorem 2.1.10, applied to $-v$. Since F is pluripolar, we can find by Josefson's theorem (cf. [Kl, Theorem 4.7.4]) a function $g \in \mathrm{PSH}(\mathbb{C}^n)$ such that $F \subseteq \{g = -\infty\}$. Since G is bounded, we can even pick a g that satisfies $g < 0$ on G. For $k \in \mathbb{N}$, set $G_k = \{z \in G : g(z) > km\} \subset G$. Since g is u.s.c. and $\mathcal{F}$-continuous, G_k is $\mathcal{F}$-open and $\overline{G_k} \subset \{z \in \overline{G} : g(z) \ge km\} \subset U \backslash F$. Note also that as $m < 0$, we have $G_k \subset G_{k+1}$. As $G_k = G \cap \{z \in \mathbb{C}^n : g(z) > km\}$, we find that $\partial_{\mathcal{F}} G_k \subseteq \partial_{\mathcal{F}} G \cup \partial_{\mathcal{F}} \{z \in \mathbb{C}^n : g(z) > km\}$, and $\partial_{\mathcal{F}} G_k \subseteq \overline{G}^{\mathcal{F}}$. The function $v + \frac{g}{k}$ is $\mathcal{F}$-plurisubharmonic on G_k, bounded from above by 0 on G_k, and $\mathcal{F}$-u.s.c. on $\overline{G_k}^{\mathcal{F}}$. On $\partial_{\mathcal{F}}$G we have $v + \frac{g}{k} \le v + 0 \le u$. As $\partial_{\mathcal{F}} \{z \in \mathbb{C}^n : g(z) > km\} \subset \{z \in \mathbb{C}^n : g(z) = km\}$ by $\mathcal{F}$-continuity of g, we find that on $\partial_{\mathcal{F}} \{z \in \mathbb{C}^n : g(z) > km\} \cap \overline{G}^{\mathcal{F}}$ we have $v + \frac{g}{k} = v + \frac{km}{k} \le m \le u$. Hence $v + \frac{g}{k} \le u$ on $\partial_{\mathcal{F}} G_k$. Since $\overline{G_k} \subset U \backslash F$, it follows by $\mathcal{F}$-maximality of u on $U \backslash F$ that $v + \frac{g}{k} \le u$ on G_k. As the sets G_k are increasing, the above inequality also implies that $v + \frac{g}{k} \le u$ on G_{k_0} for any $k \ge k_0$. Letting k tend to infinity, we find that $v \le u$ on G_{k_0}. Letting k_0 tend to infinity shows that $v \le u$ on $\bigcup_{k=1}^{\infty} G_k = G \backslash \{g = -\infty\}$. Since the pluripolar set $\{g = -\infty\}$ has empty $\mathcal{F}$-interior and u and v are $\mathcal{F}$-continuous, this implies that $v \le u$ on G. This proves the $\mathcal{F}$-maximality of u on U. □

We can also define an $\mathcal{F}$-local concept of $\mathcal{F}$-maximality:

Definition 5.9.5 An $\mathcal{F}$-plurisubharmonic function u on an $\mathcal{F}$-open set $U \subset \mathbb{C}^n$ is said to be *$\mathcal{F}$-locally $\mathcal{F}$-maximal* if each point of U has an $\mathcal{F}$-open $\mathcal{F}$-neighborhood $V \subset U$ such that the restriction of u to V is $\mathcal{F}$-maximal.

This definition can be used to formulate a corollary to Proposition 5.9.4:

Corollary 5.9.6 *Let $U \subset \mathbb{C}^n$ be an $\mathcal{F}$-open set, and $u \in \mathcal{F}\text{-}\mathrm{PSH}(U)$ a finite function that is $\mathcal{F}$-maximal in $U \backslash F$ for some $\mathcal{F}$-closed pluripolar set F. Then there exists an increasing sequence (V_j) of $\mathcal{F}$-open sets in U such that $\bigcup_{j \in \mathbb{N}} V_j = U$ and u is $\mathcal{F}$-maximal in each V_j. In particular, u is $\mathcal{F}$-locally $\mathcal{F}$-maximal.*

Proof For each $j \in \mathbb{N}$, set $V_j = \{z \in U : -j < u(z) < j\}$. By $\mathcal{F}$-continuity of u, the sets V_j are $\mathcal{F}$-open and we have $\bigcup_{j \in \mathbb{N}} V_j = U$ because u is finite on U. Since u is $\mathcal{F}$-maximal in $U \backslash F$, it will be bounded and $\mathcal{F}$-maximal in $V_j \backslash F$. By Proposition 5.9.4, u is $\mathcal{F}$-maximal in all of V_j. As the V_j cover U, u will be $\mathcal{F}$-locally $\mathcal{F}$-maximal in U. □

In the (complex) one-dimensional case, $\mathcal{F}$-maximality and $\mathcal{F}$-local $\mathcal{F}$-maximality are identical for finite finely subharmonic functions:

Proposition 5.9.7 *Let $U \subset \mathbb{C}$ be a finely open set, and let u be a finite finely subharmonic function on U. Then u is $\mathcal{F}$-maximal if and only if u is $\mathcal{F}$-locally $\mathcal{F}$-maximal, and also if and only if u is finely harmonic.*

Proof Note that the fine and pluri-fine topologies are equal in this setting. Suppose first that u is finely harmonic on U. Let G be a bounded finely open set such that $\overline{G} \subset U$. Let a function v on $\widetilde{G}$ be lower bounded from above and finely u.s.c., and such that $v|_G$ is a finely subharmonic on G and $v \leq u$ on $\partial_f G$. Then $u - v$ is finely superharmonic on G, and $\text{f-}\liminf_{\zeta \to z, \zeta \in G}(u - v)(\zeta) \geq 0$ for every $z \in \partial_f G$. Hence $u - v \geq 0$ on G by Theorem 2.7.3, whence $v \leq u$ on G. This proves the $\mathcal{F}$-maximality of u.

Now assume that u is $\mathcal{F}$-maximal in U. Since u is finely continuous, we can find a basis for the fine topology on U, consisting of bounded, regular sets V such that $V \subset \overline{V} \subset U$, and u is bounded on V. Let V be a set in such a basis and let $v := \overline{H}_u^V$. Since u is bounded on V it follows by Theorem 2.3.7 that v is finely harmonic on V. Since u is finely continuous and V is regular, one has $\text{f-}\lim_{y \to z} v(y) = u(z)$ for all $z \in \partial_f V$ according to Corollary 2.3.8. The $\mathcal{F}$-maximality of u now guarantees that $v \leq u$ on V. Since u is finite and finely continuous, the function u must be finely superharmonic by Theorem 2.3.7 and Definition 2.1.1, and hence finely harmonic because u is also finely subharmonic.

Finally, since being finely harmonic is an f-local property, we can conclude that being $\mathcal{F}$-maximal must also be a finely local property. As finely local and $\mathcal{F}$-local properties are the same in this setting, u is $\mathcal{F}$-maximal in U if and only if u is $\mathcal{F}$-locally $\mathcal{F}$-maximal in U. □

Proposition 5.9.8 *Let $U \subset \mathbb{C}$ be a finely open set, and $u \leq 0$ a finely subharmonic function on U such that $-u$ is invariant. Then u is $\mathcal{F}$-locally $\mathcal{F}$-maximal in U.*

Proof Working in $\mathbb{C}$, the fine and pluri-fine topologies are identical. Since we are looking for an $\mathcal{F}$-local result, we can cover U by bounded regular finely open subsets U_j and prove $\mathcal{F}$-local $\mathcal{F}$-maximality on each U_j. By Theorem 2.9.12(a) the function $-u|_{U_j}$ will likewise be invariant. Therefore, we can assume without loss of generality that U is bounded and regular. Since $-u$ is invariant, we can find an increasing sequence $(V_j)_{j\in\mathbb{N}}$ of finely open sets such that $\bigcup_{j=1}^{\infty} V_j = U$, $\widetilde{V}_j \subset U$ for all $j \in \mathbb{N}$, and $\widehat{R}_{-u}^{U\setminus V_j} = -u$, cf. Theorem 2.9.9. Using Theorem 2.2.15, (a) we can write $\widehat{R}_{-u}^{U\setminus V_j} = \lim_{k\to\infty} \widehat{R}_{\min\{-u,k\}}^{U\setminus V_j}$. As $\min\{-u, k\}$ is finite and finely superharmonic on U, the function $\widehat{R}_{\min\{-u,k\}}^{U\setminus V_j}$ is finely harmonic on V_j by Theorem 2.4.19(b). As a consequence, the function $-\widehat{R}_{\min\{-u,k\}}^{U\setminus V_j}$ is $\mathcal{F}$-maximal in V_j by Proposition 5.9.7. The function $u = -\widehat{R}_{-u}^{U\setminus V_j}$ is therefore the pointwise limit of a decreasing sequence of $\mathcal{F}$-maximal functions in V_j, hence $\mathcal{F}$-maximal in V_j by Proposition 5.9.2. Since the V_j cover U, it then follows that u is $\mathcal{F}$-locally $\mathcal{F}$-maximal in U. □

Theorem 5.9.9 *Let U be a regular finely open subset of $\mathbb{C}$ and u a positive finely superharmonic function on U. If $-u$ is $\mathcal{F}$-maximal then u is invariant.*

Proof We may suppose that U is a fine domain of $\mathbb{C}$. Suppose that $-u$ is $\mathcal{F}$-maximal. By the Riesz decomposition of u (see (2.9.1), there exists a fine potential p and an invariant function h on U such that $u = p + h$. By Theorem 2.9.9 and Corollary 3.2.28, there exists an increasing sequence (V_j) of finely open subsets of U such that $U = \bigcup V_j$, $\overline{V_j} \subset V_{j+1}$ for each j, $\inf_j \widehat{R}_p^{U\setminus V_j} = 0$ q.e. and $\widehat{R}_h^{U\setminus V_j} = h$ for each j. Let j be an integer ≥ 1 and s a positive finely superharmonic function on U such that $s \geq u$ on $U\setminus V_j$. Since $-u$ is $\mathcal{F}$-maximal we have $u \leq s$ on V_j. We then deduce that $u \leq R_u^{U\setminus V_j}$ and hence $u = R_u^{U\setminus V_j} = \widehat{R}_u^{U\setminus V_j}$. It follows that $p = \widehat{R}_p^{U\setminus V_j}$ for every j and, by passing to the infimum over j, we obtain $p = 0$ so that $u = h$ is invariant. □

Let $U \subset \mathbb{C}^n$ be $\mathcal{F}$-open, let u be a negative $\mathcal{F}$-plurisubharmonic function on an $\mathcal{F}$-open set V containing U, and let $A \subset U$. We denote by ${}^U u_A$, or simply u_A if there is no risk of ambiguity, the function defined on U by

$$u_A(z) = \sup\{v(z) : v \in \mathcal{F}\text{-}\operatorname{PSH}_-(U) \text{ and } v \leq u \text{ in } A\},$$

where $\mathcal{F}\text{-}\operatorname{PSH}_-(U)$ denotes the cone of negative $\mathcal{F}$-plurisubharmonic functions on U. By Theorem 5.4.14 the $\mathcal{F}$-u.s.c. regularization u_A^* of u_A is an $\mathcal{F}$-plurisubharmonic function on U. It will be called the *maximalized function* of u on A. If $U_1 \subset U_2 \subset U$ are two $\mathcal{F}$-open subsets of U and $A \subset U_1$, we denote by ${}^{U_1}u_A^*$ and ${}^{U_2}u_A^*$ the maximalized functions on A of the restrictions of u to U_1 and U_2, respectively. Then we have ${}^{U_2}u_A^* \leq {}^{U_1}u_A$ on U_1. This notion is somewhat similar to that of reduced and swept functions in the fine setting (see Sect. 2.2). Note that if U is Euclidean open then u_A^* is plurisubharmonic on U by Proposition 5.4.6.

Proposition 5.9.10 *Let U be a bounded $\mathcal{F}$-open subset of $\mathbb{C}^n$, $u \in \mathcal{F}\text{-}\mathrm{PSH}_-(U)$ and V an $\mathcal{F}$-open subset of U. Then $u^*_{U\setminus V}$ is $\mathcal{F}$-maximal in V.*

Proof Let G be an $\mathcal{F}$-open set such that $\overline{G} \subseteq V$ and let v be an upper bounded $\mathcal{F}$-u.s.c. function on $\overline{G}^{\mathcal{F}}$ such that $v \in \mathcal{F}$-PSH(G) and $v \le u^*_{U\setminus V}$ on $\partial_{\mathcal{F}} G$. Let us put

$$w = \begin{cases} \max\{u^*_{U\setminus V}, v\} & \text{on } G \\ u^*_{U\setminus V} & \text{on } U\setminus G. \end{cases}$$

Then $w \in \mathcal{F}\text{-}\mathrm{PSH}(U)$ according to Proposition 5.4.4, and we have $w = u^*_{U\setminus V}$ on $U\setminus V$ because $U\setminus V \subset U\setminus G$. Since u is $\mathcal{F}$-plurisubharmonic on U, one has $u_{U\setminus V} = u$ on $U\setminus V$. On the other hand, it follows from Theorem 5.4.14 that $u^*_{U\setminus V} = u_{U\setminus V}$ outside of some pluripolar set $A \subset U$. Since U is bounded, we find, using Josefson theorem (cf. [Kl, Theorem 4.7.4]), an $\mathcal{F}$-plurisubharmonic function $\psi < 0$ on U such that $A \subset \{\psi = -\infty\}$. Then for any $a > 0$, we have $w + a\psi \le u$ on $U\setminus V$, hence $w + a\psi \le u^*_{U\setminus V}$ on U. By letting a tend to 0, we obtain $w \le u^*_{U\setminus V}$ and hence $w = u^*_{U\setminus V}$ on $U\setminus\{\psi = -\infty\}$, so that the functions w and $u^*_{U\setminus V}$ can only differ on a pluripolar set by Theorem 5.4.3(d). However, as both are $\mathcal{F}$-continuous, and since pluripolar sets have empty $\mathcal{F}$-interiors, these functions must be identical on U. It follows then that $v \le u^*_{U\setminus V}$ and hence the restriction of $u^*_{U\setminus V}$ to V is $\mathcal{F}$-maximal. □

In the above proposition we have supposed that U is bounded, which guarantees for any pluripolar subset A of U the existence of a function $\psi \in \mathcal{F}\text{-}\mathrm{PSH}_-(U)$ such that $A \subset \{\psi = -\infty\}$. For a general U we have:

Proposition 5.9.11 *Let U be an $\mathcal{F}$-open subset of $\mathbb{C}^n$, $u \in \mathcal{F}\text{-}\mathrm{PSH}_-(U)$ and $C \subset U$ be an $\mathcal{F}$-closed subset relative to U. Then $u^*_{U\setminus C}$ $(= u_{U\setminus C})$ is $\mathcal{F}$-maximal in the $\mathcal{F}$-interior V of C.*

Proof Let G be an $\mathcal{F}$-open set such that $\overline{G} \subseteq V$ and let v be bounded from above $\mathcal{F}$-u.s.c. on $\overline{G}^{\mathcal{F}}$ such that $v \in \mathcal{F}$-PSH(G), and satisfies $v \le u^*_{U\setminus \overline{V}^{\mathcal{F}}}$ on $\partial_{\mathcal{F}} G$. Let us put

$$w = \begin{cases} \max\{u^*_{U\setminus C}, v\} & \text{on } G \\ u^*_{U\setminus C} & \text{on } U\setminus G. \end{cases}$$

By Proposition 5.4.4, $w \in \mathcal{F}\text{-}\mathrm{PSH}(U)$. As $G \subset V \subset C$, we have $w = u^*_{U\setminus C}$ on $U\setminus C$. But note that $u_{U\setminus C} = u$ on $U\setminus C$, and as $U\setminus C$ is $\mathcal{F}$-open and u is $\mathcal{F}$-u.s.c., we find that $u^*_{U\setminus C} = u$ on $U\setminus C$. Therefore, we have $w = u$ on $U\setminus C$, which implies that $w \le u^*_{U\setminus C}$ on U. It follows that $v \le u^*_{U\setminus C}$ and therefore the restriction of $u^*_{U\setminus C}$ to V is $\mathcal{F}$-maximal. □

In particular, if $V \subseteq U$ is an $\mathcal{F}$-open set, and $u \in \mathcal{F}\text{-}\mathrm{PSH}_-(U)$, then $u^*_{U\setminus \overline{V}^{\mathcal{F}}}$ and $u^*_{U\setminus \overline{V}}$ are $\mathcal{F}$-maximal in V.

The next proposition translates this back to information about open sets:

Proposition 5.9.12 *Let U be an $\mathcal{F}$-open subset of $\mathbb{C}^n$, $u \in \mathcal{F}$-$\mathrm{PSH}_-(U)$ and $C \subset U$ a subset that is relatively $\mathcal{F}$-closed with respect to U, with $\mathcal{F}$-interior V. Then $u^*_{U\setminus V} = u^*_{U\setminus C}$ which is therefore $\mathcal{F}$-maximal in V.*

Proof Since $V \subset C$ we immediately find that $u_{U\setminus V} \le u_{U\setminus C}$. Now suppose that $v \in \mathcal{F}$-$\mathrm{PSH}_-(U)$ satisfies $v \le u$ on $U\setminus C$. Since v and u are both $\mathcal{F}$-continuous, the set $\{v > u\}$ will be $\mathcal{F}$-open and contained in C. As V is the $\mathcal{F}$-interior of C, this means that $\{v > u\} \subset V$, and so $v \le u$ on $U\setminus V$. This proves that $u_{U\setminus V} = u_{U\setminus C}$ and hence $u^*_{U\setminus V} = u^*_{U\setminus C}$, which is $\mathcal{F}$-maximal in V by Proposition 5.9.11. □

For general $\mathcal{F}$-open $V \subset U$, however, it is not true that $u^*_{U\setminus V} = u^*_{U\setminus \overline{V}^{\mathcal{F}}}$:

Example 5.9.13 Let $U = B(0,3) \subset \mathbb{C}$, $V = B(0,2)\setminus C(0,1)$, where $C(0,1)$ denotes the circle of center 0 and radius 1, and define $u \in \mathcal{F}$-$\mathrm{PSH}_-(U)$ by $u(z) = |z|^2 - 10$. Then U and V are both Euclidean open, and $\overline{V}^{\mathcal{F}} = \overline{B}(0,2)$. Since U is Euclidean open in $\mathbb{C}$, subharmonic functions and finely subharmonic functions on U are the same, by Theorem 2.7.2. So we can use the maximum principle to see that $u^*_{U\setminus B(0,2)} = u_{U\setminus B(0,2)} = \max\{|z|^2 - 10, -6\}$. On the other hand, $u_{U\setminus V} = -9$ on $C(0,1)$, and by Theorem 5.4.14 the set $\{u^*_{U\setminus V} > u_{U\setminus V}\}$ is pluripolar. Since the functions $u_{U\setminus V}$ and $u^*_{U\setminus B(0,2)}$ are not equal at any point of $C(0,1)$, and $C(0,1)$ is not a pluripolar set, the functions $u^*_{U\setminus V}$ and $u^*_{U\setminus B(0,2)}$ cannot be equal.

In the case where U is bounded, the argument in the proof of Proposition 5.9.7 can be applied to get one more equality:

Proposition 5.9.14 *Let $U \subset \mathbb{C}^n$ be a bounded $\mathcal{F}$-open set, and let $A \subset B \subset U$ be subsets such that $B\setminus A$ is pluripolar. Then for any $u \in \mathcal{F}$-$\mathrm{PSH}_-(U)$ we have $u^*_{U\setminus A} = u^*_{U\setminus B}$.*

Proof Since $A \subset B$ we immediately find that $u_{U\setminus A} \le u_{U\setminus B}$. On the other hand, suppose that $v \in \mathcal{F}$-$\mathrm{PSH}(U)$ is such that $v \le u$ on $U\setminus B$. Since $B\setminus A$ is pluripolar, we can find by Josephson's theorem a plurisubharmonic function $\psi \in \mathrm{PSH}(U)$ such that $B\setminus A \subset \{\psi = -\infty\}$. Since U is bounded, we may assume that $\psi < 0$ on U. Now for any $a > 0$ we have $v + a\psi \in \mathcal{F}$-$\mathrm{PSH}_-(U)$ and $v + a\psi \le u$ on $U\setminus A$. Letting a tend to zero, we see by varying v that $u_{U\setminus A} = u_{U\setminus B}$ outside the pluripolar set $\{\psi = -\infty\}$. By Theorem 5.4.14, we know that $\{u^*_{U\setminus A} > u_{U\setminus A}\}$ and $\{u^*_{U\setminus B} > u_{U\setminus B}\}$ are also pluripolar. As a union of three pluripolar sets is pluripolar, we see that $u^*_{U\setminus A}$ and $u^*_{U\setminus B}$ are equal outside a pluripolar set. But both these functions are $\mathcal{F}$-continuous, and a pluripolar set has empty $\mathcal{F}$-interior, so the functions must be equal on all of U. □

5.10 Maximal $\mathcal{F}$-Plurisubharmonic and the Monge-Ampère Operator

Let $\Omega \subset \mathbb{C}^n$ be Euclidean open and let U be an $\mathcal{F}$-open subset of Ω. In view of Proposition 5.9.10 and Corollary 5.3.5, the maximalized function $u^*_{\Omega\setminus U}$ of a locally bounded plurisubharmonic function u on Ω is maximal in U and satisfies $(dd^c u^*_{\Omega\setminus U})^n|_U = 0$. On the other hand, it is well known that u is maximal if and only if it satisfies $(dd^c u)^n = 0$ on Ω, cf. e.g. [Kl, Theorem 4.4.2]. These facts lead one to ask whether one has $(dd^c u)^n = 0$ for all finite $u \in \mathcal{F}$-MPSH(U). In the next theorem we shall prove that the answer is "yes" when u is the restriction to U of a locally bounded plurisubharmonic function on Ω. In Theorem 5.10.3 below we will show that the answer is "yes" in the general case.

Theorem 5.10.1 *Let U be an $\mathcal{F}$-open subset of a Euclidean open set $\Omega \subset \mathbb{C}^n$. Let $u \in \mathrm{PSH}(\Omega)$ be locally bounded. If u is $\mathcal{F}$-maximal in U, then $(dd^c u)^n|_U = 0$.*

Proof We can cover U with sets V_i, $i \in I$, that are $\mathcal{F}$-open and satisfy $\overline{V}_i \subset U$ and $\overline{V}_i \subset W_i \subset \overline{W}_i \subset \Omega$ for some bounded Euclidean open set W_i. By subtracting a constant if necessary, we may assume that $u < 0$ on W_i, which allows us to look at the maximalized function

$$^{W_i}u^*_{W_i\setminus V_i} = (\sup\{w \in \mathrm{PSH}_-(W_i) : w \le u \text{ in } W_i\setminus V_i\})^*.$$

By Corollary 5.3.5 we have $(dd^c(^{W_i}u^*_{W_i\setminus V_i}))^n|_{V_i} = 0$. Because u is $\mathcal{F}$-maximal in U, we have $u = {}^{W_i}u^*_{W_i\setminus V_i}$ on W_i and hence $(dd^c u)^n|_{V_i} = 0$. The assertion of the theorem now follows from the quasi-Lindelöf property of the pluri-fine topology and the well known fact that the Monge-Ampère measure of a locally bounded plurisubharmonic function does not charge pluripolar sets. □

Lemma 5.10.2 *Let u be a negative $\mathcal{F}$-maximal $\mathcal{F}$-plurisubharmonic function on an $\mathcal{F}$-domain U in $\mathbb{C}^n$. On any bounded $\mathcal{F}$-open subset V of U such that $\overline{V} \subset U$ we then have $u^*_{U\setminus V} = u^*_{U\setminus \overline{V}^{\mathcal{F}}} = u$.*

Proof The equality $u^*_{U\setminus V} = u$ is an immediate consequence of the definition of $\mathcal{F}$-maximal functions. If we let W be the $\mathcal{F}$-interior of $\overline{V}^{\mathcal{F}}$, then Proposition 5.9.12 tells us that $u^*_{U\setminus \overline{V}^{\mathcal{F}}} = u^*_{U\setminus W}$. And as W is $\mathcal{F}$-open and $\overline{W}^{\mathcal{F}} = \overline{V}^{\mathcal{F}} \subset U$, we have $u^*_{U\setminus W} = u$ since u is $\mathcal{F}$-maximal. □

Theorem 5.10.3 *Let u be a finite $\mathcal{F}$-maximal $\mathcal{F}$-plurisubharmonic function on an $\mathcal{F}$-domain U in $\mathbb{C}^n$. Then we have $(dd^c u)^n = 0$.*

Proof By Theorem 5.7.17 there exists an $\mathcal{F}$-closed pluripolar set $E \subset \mathbb{C}^n$ such that u is C-strongly $\mathcal{F}$-plurisubharmonic on $U\setminus E$. Let z be an element of $U\setminus E$, then there is a bounded $\mathcal{F}$-neighborhood $V \subset \overline{V} \subset U\setminus E$ of z and a sequence (u_j) of finite continuous plurisubharmonic functions on neighborhoods of V that converges uniformly to u on V. Since u is $\mathcal{F}$-continuous, we can shrink V a bit, to ensure that u is bounded on V. By subtracting if necessary a constant from u and all u_j, we may assume that $u < 0$ on V. For each j, u_j is plurisubharmonic on some Euclidean open set O_j containing V. We may assume that the sequence (O_j) is decreasing, and that $u_{j+1} \ge u_j$ on O_{j+1} (use the uniform convergence to adjust the sequence (u_j) to become increasing on V, and then replace u_j by $\max\{u_1, \ldots, u_j\}$ on O_j). Now let m be a lower bound for u on V, and replace each u_j by $\max\{u_j, m\}$ to make these (u.s.c.) functions locally bounded. Let $\varepsilon > 0$, then we can find j_0 such that $u_j \leq u \leq u_j + \varepsilon$ on V for all $j \geq j_0$. Let O be an $\mathcal{F}$-open set such that $O \subset \overline{O} \subset V$, then on V we have for all $j \geq j_0$ (using Lemma 5.10.2)

$$u - \varepsilon \leq u_j \leq {}^{O_j}(u_j)^*_{O_j \setminus O} \leq {}^{O_j}(u_j)^*_{V \setminus O} \leq {}^{V}(u_j)^*_{V \setminus O} \leq {}^{V}u^*_{V \setminus O} = u,$$

which implies that the sequence of restrictions to V of the functions ${}^{O_j}(u_j)^*_{O_j \setminus O}$ converges uniformly to u. However, we know that $(dd^c({}^{O_j}(u_j)^*_{O_j \setminus O}))^n = 0$ on O by Corollary 5.3.5. Now note that the sequence $({}^{O_j}(u_j)^*_{O_j \setminus O})$ is increasing on V:

$${}^{O_j}(u_j)^*_{O_j \setminus O} \leq {}^{O_j}(u_j)^*_{O_{j+1} \setminus O} \leq {}^{O_{j+1}}(u_j)^*_{O_{j+1} \setminus O} \leq {}^{O_{j+1}}(u_{j+1})^*_{O_{j+1} \setminus O}.$$

Hence we can apply Corollary 5.8.13 to see that $(dd^c u)^n = 0$ on O. As we can cover all of $U \setminus E$ by such sets O, we can now see that $(dd^c u)^n = 0$, using the definition of the Monge-Ampère operator for $\mathcal{F}$-plurisubharmonic functions in Definition 5.8.6. □

Corollary 5.10.4 *Let u be a negative finite $\mathcal{F}$-plurisubharmonic function on an $\mathcal{F}$-domain U in $\mathbb{C}^n$, and V an $\mathcal{F}$-open subset of U, then we have $(dd^c u^*_{U \setminus \overline{V}^{\mathcal{F}}})^n|_V = (dd^c u^*_{U \setminus \overline{V}})^n|_V = 0$.*

Also, if U is bounded, or if V equals the $\mathcal{F}$-interior of some $\mathcal{F}$-closed set, we have $(dd^c u^*_{U \setminus V})^n|_V = 0$.

Proof This follows immediately from the above theorem and Propositions 5.9.10, 5.9.11 and 5.9.12. □

The converse of the above result is well known in the case of a Euclidean open $U \subset \mathbb{C}^n$ for locally bounded plurisubharmonic functions on U. See for example [Kl, Corollary 3.7.6]. The following theorem shows that this converse is also true when u is the restriction to an $\mathcal{F}$-open subset U of a locally bounded plurisubharmonic function on a Euclidean domain Ω containing U.

Theorem 5.10.5 *Let u be a locally bounded plurisubharmonic function on a Euclidean domain $\Omega \subset \mathbb{C}^n$. If $(dd^c u)^n = 0$ on an $\mathcal{F}$-open $U \subset \Omega$, then the restriction of u to U is $\mathcal{F}$-maximal.*

Proof Let G be a bounded $\mathcal{F}$-open set such that $G \subset \overline{G} \subseteq U$ and v an upper bounded u.s.c. on $\overline{G}^{\mathcal{F}}$ such that $v \in \mathcal{F}\text{-PSH}(G)$ and $v \leq u$ on $\partial_{\mathcal{F}} G$. Let us put

$$w = \begin{cases} \max\{u, v\} & \text{on } G \\ u & \text{on } \Omega \setminus G. \end{cases}$$

Then by Propositions 5.4.4 and 5.4.6, w is a locally bounded plurisubharmonic function on Ω. Moreover we have $(dd^c u)^n = (dd^c w)^n$ on $\Omega \setminus \overline{G}$ and $(dd^c u)^n = 0$ on

U, hence $(dd^c u)^n \leq (dd^c w)^n$. Since $u = w$ on $\Omega \backslash G$, it follows from the comparison principle [Kl, Corollary 3.7.4] that $w \leq u$, and therefore $u = w$. We then deduce that $v \leqslant u$ on G. So we have proved that u is $\mathcal{F}$-maximal. □

Now let us recall the following result of Błocki [Bl1, Theorem 2.1]:

Theorem 5.10.6 *Let Ω be a bounded Euclidean domain of $\mathbb{C}^n$ and let $u, h, v_1, \ldots, v_n \in \mathrm{PSH}(\Omega) \cap L^{\infty}_{loc}(\Omega)$ be such that $u \leq h$ and $\lim_{z \to \partial\Omega}(h(z) - u(z)) = 0$. Then*

$$\int_{\Omega} (h-u)^n dd^c v_1 \wedge \ldots \wedge dd^c v_n \leq n! \|v_1\|_{\Omega} \ldots \|v_{n-1}\|_{\Omega} \int_{\Omega} |v_n| (dd^c u)^n. \tag{5.10.1}$$

Proof Taking $h_{\varepsilon} = \max\{u, h - \varepsilon\}$ we see that the left hand of (5.10.1) with h replaced by h_{ε} converges. We can therefore assume that $h = u$ in a neighborhood of $\partial\Omega$.

Assume we have proven the theorem for smooth u and h. Now for arbitrary u and h, let u_{ε} and h_{ε} denote their smooth regularizations, so that $u_{\varepsilon} \downarrow u$ and $h_{\varepsilon} \downarrow h$ as $\varepsilon \downarrow 0$. Then $u_{\varepsilon} \leq h_{\varepsilon}$ and, if we shrink Ω a little, $u_{\varepsilon} = h_{\varepsilon}$ in a neighborhood of $\partial\Omega$. So we have

$$\int_{\Omega} (h_{\varepsilon} - u_{\varepsilon})^n dd^c v_1 \wedge \ldots \wedge dd^c v_n \leq n! \, \|v_1\|_{\Omega} \cdots \|v_{n-1}\|_{\Omega} \int_{\Omega} |v_n| (dd^c u_{\varepsilon})^n. \tag{5.10.2}$$

By Lebesgue's convergence Theorem, the left-hand side of (5.10.2) tends to

$$\int_{\Omega} (h-u)^n dd^c v_1 \wedge \ldots \wedge dd^c v_n.$$

By the convergence Theorem, cf. [Kl, Theorem 3.4.11], the measures $v_n (dd^c u_{\varepsilon})^n$ converge weakly to $v_n (dd^c u)^n$ thus, having shrunk Ω already, the right-hand side of (5.10.2) tends to

$$n! \, \|v_1\|_{\Omega} \ldots \|v_{n-1}\|_{\Omega} \int_{\Omega} |v_n| (dd^c u)^n.$$

Now it suffices to prove Theorem 5.10.6 when u and h are smooth and they coincide in a neighborhood of $\partial\Omega$. We may also assume that $\|v_j\|_{\Omega} = 1$. First, we want to show that for $p = 2, \ldots, n$ we have

$$\int_{\Omega} (h-u)^p (dd^c u)^{n-p} \wedge dd^c v_{n-p+1} \wedge \ldots \wedge dd^c v_n \tag{5.10.3}$$

$$\leq p \int_{\Omega} (h-u)^{p-1} (dd^c u)^{n-p+1} \wedge dd^c v_{n-p+2} \wedge \ldots \wedge dd^c v_n.$$

By Stokes' Theorem, the left-hand side of (5.10.3) is equal to

$$\int_\Omega v_{n-p+1}(dd^c u)^{n-p} \wedge dd^c(h-u)^p \wedge \ldots \wedge dd^c v_n,$$

while the right-hand side is greater than or equal to

$$p\int_\Omega (-v_{n-p+1})(h-u)^{p-1}(dd^c u)^{n-p+1} \wedge dd^c v_{n-p+2} \wedge \ldots \wedge dd^c v_n,$$

because $-v_{n-p+1} = |v_{n-p+1}| \leq 1$. To prove (5.10.3) it is therefore enough to prove that for $p = 2, \ldots, n$

$$-dd^c(h-u)^p \leq p(h-u)^{p-1} dd^c u, \tag{5.10.4}$$

that is, the difference between the right and the left hand of (5.10.4) is a positive $(1, 1)$-form. We have

$$\begin{aligned} dd^c(h-u)^p &= d(p(h-u)^{p-1}d^c(h-u)) \\ &= p(p-1)(h-u)^{p-2}d(h-u) \wedge d^c(h-u) \\ &+ p(h-u)^{p-1}dd^c(h-u) \\ &\geq p(p-1)(h-u)^{p-2}d(h-u) \wedge d^c(h-u) - p(h-u)^{p-1}dd^c u \end{aligned}$$

and $d(h-u) \wedge d^c(h-u) = i\partial(h-u) \wedge \bar\partial(h-u) \geq 0$, thus (5.10.4) follows and we have (5.10.3). Now by (5.10.3) and Stokes' theorem

$$\begin{aligned} \int_\Omega (h-u)^n dd^c v_1 \wedge \ldots \wedge dd^c v_n &\leq n! \int_\Omega (h-u)(dd^c u)^{n-1} \wedge dd^c v_n \\ &= n! \int_\Omega (-v_n)(dd^c u)^{n-1} \wedge dd^c(u-h) \\ &\leq n! \int_\Omega |v_n|(dd^c u)^n \end{aligned}$$

and the proof is complete. □

Corollary 5.10.7 *Let Ω be a bounded Euclidean domain of $\mathbb{C}^n$ and let $u, h \in$ $\mathrm{PSH} \cap L^\infty_{loc}(\Omega)$ be such that $u \leq h$ and $\lim_{z\to\partial\Omega}(h(z) - u(z)) = 0$. Then*

$$\|h-u\|_{L^n(\Omega)} \leq \frac{R^2}{4}\left(\int_\Omega (dd^c u)^n\right)^{\frac{1}{n}},$$

where

$$\|h-u\|_{L^n(\Omega)} = \left(\int_\Omega |h-u|^n dm\right)^{\frac{1}{n}}$$

(where m is the Lebesgue measure on Ω*) and*

$$R = \min\{r > 0 : \Omega \subset B(z_0, r) \text{ for some } z_0 \in \mathbb{C}^n\}.$$

In particular $R \leq \operatorname{diam}(\Omega)$*, and when* Ω *is a ball, then* $R = \operatorname{diam}(\Omega)/2$.

Proof Without loss of generality we may assume $\Omega \subset B(0, R)$. It is enough to apply Theorem 5.10.6 with

$$v_1(z) = \ldots = v_n(z) = |z|^2 - R^2. \qquad \square$$

Lemma 5.10.8 *Let* U *be an* $\mathcal{F}$*-open subset of* $\mathbb{C}^n$, $u \in \mathcal{F}\text{-}\mathrm{PSH}_-(U)$, *u finite, and* (O_j) *a decreasing sequence of Euclidean open sets such that* $U \subset \bigcap_j O_j$*. Let* (u_j) *be a sequence of functions such that* $u_j \in \mathrm{PSH}_-(O_j)$ *and* $u_{j+1} \leq u_j$ *on* O_{j+1} *for all j, and such that the sequence* $(u_j|_U)$ *converges uniformly to u on U. Let G be an* $\mathcal{F}$*-open such that* $\overline{G} \subset U$*, and let* (ω_j) *be a decreasing sequence of* $\mathcal{F}$*-open sets such that* $\overline{\omega_j} \subset O_j$ *for all j and* $\bigcap_j \overline{\omega_j} = \overline{G}$*. Then the sequence of the restrictions of* ${}^{O_j}(u_j)^*_{O_j\setminus\overline{\omega_j}}$ *to U is decreasing to* $u^*_{U\setminus\overline{G}}$.

Proof For any j we have ${}^{O_j}(u_j)_{O_j\setminus\overline{\omega_j}} = u_j$ on $O_j\setminus\overline{\omega_j}$. On $O_{j+1}\setminus\overline{\omega_{j+1}}$ we therefore find that ${}^{O_{j+1}}(u_{j+1})^*_{O_{j+1}\setminus\overline{\omega_{j+1}}} = u_{j+1} \leq u_j$, so $\{{}^{O_{j+1}}(u_{j+1})^*_{O_{j+1}\setminus\overline{\omega_{j+1}}} > u_j|_{O_{j+1}}\}$ is an $\mathcal{F}$-open set contained in $\overline{\omega_{j+1}} \subset \overline{\omega_j} \subset O_j$. Applying Proposition 5.4.4 allows us to glue these functions together to an $\mathcal{F}$-plurisubharmonic function h_j on O_j,

$$h_j = \begin{cases} {}^{O_{j+1}}(u_{j+1})^*_{O_{j+1}\setminus\overline{\omega_{j+1}}} & \text{on } \{{}^{O_{j+1}}(u_{j+1})^*_{O_{j+1}\setminus\overline{\omega_{j+1}}} > u_j|_{O_{j+1}}\} \\ u_j & \text{on } O_j\setminus\{{}^{O_{j+1}}(u_{j+1})^*_{O_{j+1}\setminus\overline{\omega_{j+1}}} > u_j|_{O_{j+1}}\}. \end{cases}$$

We have ${}^{O_{j+1}}(u_{j+1})^*_{O_{j+1}\setminus\overline{\omega_{j+1}}} \leq h_j$ on O_{j+1} and, as h_j is $\mathcal{F}$-plurisubharmonic and $h_j = u_j$ on $O_j\setminus\overline{\omega_j}$, we find that $h_j \leq {}^{O_j}(u_j)_{O_j\setminus\overline{\omega_j}} \leq {}^{O_j}(u_j)^*_{O_j\setminus\overline{\omega_j}}$. Hence ${}^{O_{j+1}}(u_{j+1})^*_{O_{j+1}\setminus\overline{\omega_{j+1}}} \leq {}^{O_j}(u_j)^*_{O_j\setminus\overline{\omega_j}}$ on O_{j+1}, so that the sequence $({}^{O_j}(u_j)^*_{O_j\setminus\overline{\omega_j}}|_U)$ of $\mathcal{F}$-plurisubharmonic functions on U is decreasing. Since ${}^{O_j}(u_j)^*_{O_j\setminus\overline{\omega_j}} = u_j$ on $O_j\setminus\overline{\omega_j}$, this implies that $\lim_{k\to+\infty} {}^{O_k}(u_k)^*_{O_k\setminus\overline{\omega_k}} \leq u_j$ on $U\setminus\overline{\omega_j}$ for all j. As the u_j decrease to u, and $\bigcap_j \overline{\omega_j} = \overline{G}$, we find that $\lim_{k\to+\infty} {}^{O_k}(u_k)^*_{O_k\setminus\overline{\omega_k}} \leq u$ on $U\setminus\overline{G}$. Combining this with the fact that ${}^{O_j}(u_j)^*_{O_j\setminus\overline{\omega_j}} \geq u_j \geq u$ on U for all j, we get

$$\lim_{j\to+\infty} {}^{O_j}(u_j)^*_{O_j\setminus\overline{\omega_j}} = u \text{ in } U\setminus\overline{G}.$$

As the limit of the monotonically decreasing sequence $({}^{O_j}(u_j)^*_{O_j\setminus\overline{\omega_j}})$ of $\mathcal{F}$-plurisubharmonic functions on U, all minorized by a $\mathcal{F}$-plurisubharmonic function on U, is again $\mathcal{F}$-plurisubharmonic according to Theorem 5.4.2(iv), we now find that

$$\lim_{j\to+\infty} {}^{O_j}(u_j)^*_{O_j\setminus\overline{\omega_j}} \leq u^*_{U\setminus\overline{G}} \text{ in } U.$$

Before proving the inverse inequality, note that the conclusion of the lemma only depends on $\overline{G}$ rather than G itself. Let H be the $\mathcal{F}$-interior of $\overline{G}$. Now $G \subset H \subset \overline{G}$, and hence $\overline{H} = \overline{G}$. So without loss of generality, we may replace G by H and assume that G equals the $\mathcal{F}$-interior of $\overline{G}$. Using this new assumption, the set $\overline{U \setminus \overline{G}}^{\mathcal{F}}$ has an $\mathcal{F}$-open complement, whose intersection with U is contained in $\overline{G}$, hence in G. As $\overline{G} \subset U$ and G is $\mathcal{F}$-open, it follows that $\partial_{\mathcal{F}} G \subset U \setminus G \subset \overline{U \setminus \overline{G}}^{\mathcal{F}}$.

We can now prove the inverse of the latter inequality. For all j we have, $u_{U \setminus \overline{G}} = u \le u_j$ on $U \setminus \overline{G}$, and by $\mathcal{F}$-upper semicontinuity of u_j, this implies that $u^*_{U \setminus \overline{G}} \le u_j$ on $U \setminus \overline{G}$ as well. Since both these functions are in fact $\mathcal{F}$-continuous, we can see that the inequality even holds on $\overline{U \setminus \overline{G}}^{\mathcal{F}}$, and hence on $\partial_{\mathcal{F}} G$. Then define

$$v_j = \begin{cases} \max\{u_j, u^*_{U \setminus \overline{G}}\} & \text{on } G \\ u_j & \text{on } O_j \setminus G. \end{cases}$$

By Proposition 5.4.4, v_j is $\mathcal{F}$-plurisubharmonic on O_j, and we can see that $v_j = u_j$ on $O_j \setminus \overline{\omega_j}$ since $\overline{G} \subset \overline{\omega_j}$. It then follows that $v_j \le {}^{O_j}(u_j)^*_{O_j \setminus \overline{\omega_j}}$, and so $u^*_{U \setminus \overline{G}} \le {}^{O_j}(u_j)^*_{O_j \setminus \overline{\omega_j}}$ on G. Combining this with the fact that $u^*_{U \setminus \overline{G}} \le u_j$ on $\overline{U \setminus \overline{G}}^{\mathcal{F}}$, we find that $u^*_{U \setminus \overline{G}} \le {}^{O_j}(u_j)^*_{O_j \setminus \overline{\omega_j}}$ on all of U and therefore

$$\lim_{j \to +\infty} {}^{O_j}(u_j)^*_{O_j \setminus \overline{\omega_j}} \ge u^*_{U \setminus \overline{G}} \text{ in } U. \qquad \square$$

Now we can prove a converse of Theorem 5.10.3.

Theorem 5.10.9 *Let u be a finite $\mathcal{F}$-plurisubharmonic function on an $\mathcal{F}$-domain $U \subset \mathbb{C}^n$. If $(dd^c u)^n = 0$ then there exists an $\mathcal{F}$-closed pluripolar set $F \subset U$ such that u is $\mathcal{F}$-locally $\mathcal{F}$-maximal in $U \setminus F$.*

Proof Since we are looking for an $\mathcal{F}$-local property outside a pluripolar set we may assume according to the text following Definition 5.9.1 and the quasi-Lindelöf property of the topology $\mathcal{F}$ that $u < 0$ on U. By Theorem 5.7.17 there exists an $\mathcal{F}$-closed pluripolar subset F of U such that u is C-strongly $\mathcal{F}$-plurisubharmonic on $U \setminus F$. Let $z \in U \setminus F$, then there exists an $\mathcal{F}$-open $\mathcal{F}$-neighborhood $V_z \subset \overline{V_z} \subset U$ such that $\overline{V_z}$ is compact and $u = \lim u_j$ uniformly on $\overline{V_z}$, where for every j, the function u_j is a finite continuous plurisubharmonic function on a bounded Euclidean open set O_j containing $\overline{V_z}$. By shrinking V_z, we may assume that u is bounded on $\overline{V_z}$. By shrinking the O_j, we may assume that $\overline{O_{j+1}} \subset O_j$, and using the uniform convergence on $\overline{V_z}$ we can ensure that $u_{j+1} < u_j < 0$ on V_z for all j. Replacing O_{j+1} by $\{w \in O_{j+1} : u_{j+1}(w) < u_j(w)\}$ will give us $u_{j+1} \le u_j$ on O_{j+1}. Finally, as u is bounded on $\overline{V_z}$ and the u_j are continuous and converge uniformly to u on V_z, we can see that if we shrink the O_j somewhat more, all u_j will be uniformly bounded. Let

$w \in V_z$. By Theorem 5.8.12, we find that $(dd^c u_j|_U)^n$ converges $\mathcal{F}$-locally vaguely to $(dd^c u)^n$. So we can find an $\mathcal{F}$-open W such that $w \in W \subset V_z$ and

$$\lim_{j\to+\infty} \int \varphi(dd^c u_j|_U)^n = \int \varphi(dd^c u)^n$$

for every bounded $\mathcal{F}$-continuous function φ on U with compact support $\mathrm{Supp}(\varphi) \subset W$. Now construct as in Remark 5.8.11 an $\mathcal{F}$-continuous function φ on U with compact support within W, satisfying $0 \le \varphi \le 1$ on U and $\varphi = 1$ on some $\mathcal{F}$-open $\mathcal{F}$-neighborhood G of w.

Note that $G \subset \overline{G} \subset W \subset V_z$, and let (ω_j) be a decreasing sequence of Euclidean open sets such that for any index j we have $\overline{\omega_{j+1}} \subset \omega_j \subset \overline{\omega_j} \subset O_j$ and $\bigcap_j \omega_j = \overline{G}$.

Define $h_{j,k} = {}^{O_j}(u_j)^*_{O_j\setminus\overline{\omega_k}}$ for $k \ge j$, and $h_j = {}^{O_j}(u_j)^*_{O_j\setminus\overline{G}}$. Note that as the u_j are uniformly bounded, and that $\overline{G} \subset \overline{\omega_j} \subset O_j$ for all j, we can use the maximum principle to see that all $h_{j,k}$ and all h_j are uniformly bounded by the same bounds. By applying Lemma 5.10.8 to the functions $h_{j,k}$, where we let $k \to \infty$ and keep j fixed, we see that $(h_{j,k})_k$ decreases to h_j. By applying Lemma 5.10.8 again (taking this time all ω_j in the lemma to be G), we see that (h_j) decreases to ${}^U u^*_{U\setminus G}$.

Now by Corollary 5.10.7 if $R = \min\{r > 0 : O_1 \subset B(z_0, r)$ for some $z_0 \in \mathbb{C}^n\}$, we get

$$\left\|h_{j,k+1} - u_j\right\|_{L^n(G)} \le \left\|h_{j,k+1} - u_j\right\|_{L^n(\omega_k)} \le \frac{R^2}{4}\left(\int_{\omega_k} (dd^c u_j)^n\right)^{\frac{1}{n}}$$

for all $k \ge j$. Letting $k \to +\infty$, we deduce that

$$\left\|h_j - u_j\right\|_{L^n(G)} \le \frac{R^2}{4}\left(\int_{\overline{G}} (dd^c u_j)^n\right)^{\frac{1}{n}}$$

for all j. By letting $j \to +\infty$, we obtain $\left\|{}^U u^*_{U\setminus\overline{G}} - u\right\|_{L^n(G)}$ on the left-hand side, by dominated convergence. On the right-hand side, the choice of G and the fact that $(dd^c u)^n = 0$ imply that

$$\int_{\overline{G}} (dd^c u_j)^n \le \int_W \varphi(dd^c u_j)^n \to \int_W \varphi(dd^c u)^n = 0.$$

Therefore, $u = {}^U u^*_{U\setminus\overline{G}}$ on G, so that u is $\mathcal{F}$-maximal there according to Proposition 5.9.12. Since all of $U\setminus F$ will be covered by such sets G, u is $\mathcal{F}$-locally $\mathcal{F}$-maximal in $U\setminus F$. □

Theorem 5.10.10 *Let U be an $\mathcal{F}$-domain in $\mathbb{C}^n$ and let $u \in \mathcal{F}$-$\mathrm{PSH}(U)$ be finite. Then $(dd^c u)^n = 0$ if and only if u is $\mathcal{F}$-locally $\mathcal{F}$-maximal in the complement of an $\mathcal{F}$-closed pluripolar subset of U.*

Proof The "only if part" of this theorem is just Theorem 5.10.9. For the "if part", combining Theorem 5.10.3 with the quasi-Lindelöf property of the plurifine topology shows that U can be covered by countably many $\mathcal{F}$-open sets V with $(dd^c u)^n|_V = 0$ and a pluripolar set E. Since $(dd^c u)^n(E) = 0$ (see paragraph just before Definition 5.8.6), this means that $(dd^c u)^n = 0$. □

Corollary 5.10.11 *Let $U \subset \mathbb{C}^n$ be an $\mathcal{F}$-open set, and let (u_j) be a sequence of finite $\mathcal{F}$-maximal functions in $\mathcal{F}$-PSH(U), increasing to a finite $u \in \mathcal{F}$-PSH(U). Then u is $\mathcal{F}$-locally $\mathcal{F}$-maximal outside some closed pluripolar set F.*

Proof Since all u_j are $\mathcal{F}$-maximal, Theorem 5.10.3 implies that $(dd^c u_j)^n = 0$ for all j. Applying Corollary 5.8.13, we see that $(dd^c u)^n = 0$. By Theorem 5.10.10, we can now find a closed pluripolar set F such that u is $\mathcal{F}$-locally $\mathcal{F}$-maximal in $U \setminus F$. □

It is natural to ask whether an $\mathcal{F}$-locally $\mathcal{F}$-maximal function u in an $\mathcal{F}$-open set $U \subset \mathbb{C}^n$ is $\mathcal{F}$-maximal in U. In the case of a Euclidean open set we have the following positive answer:

Proposition 5.10.12 *Let u be a locally bounded plurisubharmonic function on a Euclidean open set Ω. If u is $\mathcal{F}$-locally $\mathcal{F}$-maximal in Ω, then u is maximal.*

Proof By Theorem 5.10.10 we have $(dd^c u)^n = 0$, hence u is maximal by [Kl, Corollary 3.7.6]. □

When working with $n = 1$, we also have a positive answer if u is finite, see Proposition 5.9.7. However, this result is not valid when the function u is not finite, as seen in the following example.

Example 5.10.13 Let μ be a Borel measure on the unit disk $\mathbb{D}$ in $\mathbb{C}$, such that $\mu \neq 0$, Supp$(\mu) = K$ for some compact polar set $K \subset \mathbb{D}$, and μ has no atoms (see, for example, [Ran, p. 82], [C, Chapter IV, Theorem 1]). Let $u = -G_{\mathbb{D}}\mu$, then u is a subharmonic function on $\mathbb{D}$ and $dd^c u = 2\pi\mu$. Since $\mu(\mathbb{D} \setminus K) = 0$, the function u is harmonic on $\mathbb{D} \setminus K$, and hence maximal in $\mathbb{D} \setminus K$. By Proposition 5.9.3, u will be $\mathcal{F}$-maximal in $\mathbb{D} \setminus K$ as well. Now let $z \in K$. Since K is a polar set in $\mathbb{C}$, each of its points is finely isolated, see Corollary 1.4.19(c). So we can find a regular finely open set $V \subset \mathbb{D}$ such that $V \cap K = \{z\}$. Since μ has no atoms, we have $\mu(V) = 0$ and hence $-u$ is invariant in V according to Theorem 2.9.11, so that u is $\mathcal{F}$-maximal in V by Proposition 5.9.8. Now note that $-u = G_{\mathbb{D}}\mu = G_V\mu + \widehat{R}^{\mathbb{D}\setminus V}_{G_{\mathbb{D}}\mu}$ according to (2.6.5). However, since $\mu(V) = 0$, one has $G_V\mu \equiv 0$ because $G_V(\cdot, y) = 0$ for $y \notin V$, V being regular. So $u = -\widehat{R}^{\mathbb{D}\setminus V}_{G_{\mathbb{D}}\mu} = -\widehat{R}^{\mathbb{D}\setminus V}_{-u} = (-u)^*_{\mathbb{D}\setminus V}$ and hence u is $\mathcal{F}$-maximal in V by Proposition 5.9.10. It therefore follows that u is $\mathcal{F}$-locally $\mathcal{F}$-maximal in all of $\mathbb{D}$. However, u is not maximal in $\mathbb{D}$, since maximal functions in $\mathbb{D}$ are harmonic, and u is not harmonic as $\Delta u = -2\pi\mu \neq 0$. By Proposition 5.9.3, this means that u cannot be $\mathcal{F}$-maximal in $\mathbb{D}$ either. So we have found an $\mathcal{F}$-locally $\mathcal{F}$-maximal function which is not $\mathcal{F}$-maximal.

Remark 5.10.14 The example above also shows that the converse to Proposition 5.9.8 does not hold. The negative finely subharmonic function u on the (finely) open set $\mathbb{D} \subset \mathbb{C}$ is $\mathcal{F}$-locally $\mathcal{F}$-maximal. However, $-u$ cannot be invariant according to Theorem 2.9.11.

Note: Except Theorem 5.10.6 and its proof and corollary which are due to Błocki [Bl1], the remaining results of the Sects. 5.9 and 5.10 are due to El Kadiri and Smit [EKS].

Appendix
An Overview of Further Results in Fine Potential Theory

In classical and axiomatic setting, fine Potential Theory is the theory of finely harmonic or finely superharmonic functions in a finely open set U, which generalizes the notions of harmonic and superharmonic functions in a usual open set. This theory was initiated by Fuglede in the last sixties and the early seventies. The book of Fuglede *Finely harmonic functions* [Fu8] appeared in 1972. Since then, this theory has attracted many authors who have made so many important contributions to it by solving questions and problems arising in it and in its related topics as the theory of finely holomorphic functions in a finely open set of $\mathbb{C}$, using analytic and probabilistic methods. This section is devoted to give an overview of some substantial contributions not exposed in the present book.

A natural question raised in the very early beginning of fine Potential Theory is whether a finely [super]harmonic on a finely open set U is the restriction to U of a [super]harmonic function in an open neighborhood of U. This was solved negatively by Gardiner in [Ga3]. He proved that, more generally, if U is a non-empty finely open set in $\mathbb{R}^N$, then the following statements are equivalent:

(a) for any finely harmonic function u on U there is a non-empty finely open set $V \subseteq U$ and a harmonic function w on an open set $W \supseteq V$ such that $w = u$ on V,

(b) for any finely harmonic function u on U there is a harmonic function w on an open set W such that $W \setminus U$ is polar and $w = u$ on $U \cap W$,

(c) U is of second Baire category (with respect to the Euclidean topology),

answering so the following more precise question: Given a finely open set U without interior points, does there exist a finely harmonic function u on U such that no restriction of u to a non-empty finely open subset of U admits a harmonic extension?

The notion of fine differentiability was studied by many authors. A function f defined on a finely open set U in the classical Potential theory in $\mathbb{R}^m$ and taking values in $\mathbb{R}^k$ is said to be finely differentiable at $x_0 \in U$ if there exists a linear map $L : \mathbb{R}^m \to \mathbb{R}^k$ such that

$$\text{f} - \lim_{x \to x_0} \frac{f(x) - f(x_0) - L(x - x_0)}{|x - x_0|} = 0$$

M. El Kadiri and B. Fuglede, *Classical Fine Potential Theory*, Springer Monographs in Mathematics, https://doi.org/10.1007/978-981-96-0430-2

where $|x|$ stands for the Euclidean norm of $x \in \mathbb{R}^m$. In this case we write $d_f(x_0)$ for L, and $d_f(x_0)$ is called the fine differential of f at x_0. The question of differentiability of finely harmonic functions had been subject of investigations by many authors. Let U be a finely open set in $\mathbb{R}^N$ and h a finely harmonic on U. By combination of a result of Fuglede [Fu13] on local extension of finely superharmonic functions, the Brelot property and a result of Debiard and Gaveau [DG1], every point $x_o \in U$ has a fine neighborhood $V = V_{x_0}$ such that $V \subset U$ and $h|_V$ can be approximated uniformly on V by functions h_n, harmonic in open sets (in the usual topology) containing V. Debiard and Gaveau [DG3] proved that there exists a function denoted by $\nabla h \in L^2(V, m)$ such that $\nabla h_n \to \nabla h$ in $L^2(V, m)$, which does not depend on the sequence (h_n), where m denotes the N-dimensional Lebesgue measure on V. In the other directions, they give an example of a compact set K and a point $x_0 \in K'$ such that $|\nabla h_n(x_0)| \to \infty$ as $n \to \infty$.

The set of points of a finely open set U where a finely harmonic function on U is not differentiable was studied by Fuglede and by Davie and Oksendal. In the case of the plane $\mathbb{C}$, Fuglede [Fu20] has proved, using a theorem of Mizuta [Miz], that a finely harmonic function f on a finely open set $U \subset \mathbb{C}$ is always finely differentiable outside a set G with $C_h^*(G) = 0$, where C_h^* is the outer capacity associated to the kernel $h(z) = |z|^{-1}\log(1/|z|)$. In the opposite direction, Davie and Øksendal proved that, given any compact set E with $C_h(E) = 0$, one can find a finely open set $V \supseteq E$ and a finely harmonic function f on V such that f is not finely differentiable at any point of E. It could be conjectured that if g is the uniform limit of functions g_n harmonic in a neighborhood of a compact subset K of $\mathbb{C}$ then $(\nabla g_n(x))$ always converges outside a set of zero logarithmic capacity. Davie and Øksendal showed in [DO] that this conjecture fails. More precisely, they proved that for any compact set E with zero analytic capacity, there exists a compact set K with $E \subseteq K'$ and functions g_n harmonic in a neighborhood of K such that $g_n \to 0$ uniformly on K and $|\partial g_n/\partial \bar{z}| \to \infty$ uniformly on E.

In a probabilistic approach, Mastrangelo introduced and studied in [Ma1, Ma2] the stochastic differentiability of a real function defined on finely open set in $\mathbb{R}^N$, and proved that in some points the fine and stochastic differentials satisfies analogous properties as usual differentials on Euclidean open sets. She investigated in [Ma2] the stochastic differentiability of finely harmonic functions and the fine harmonicity of the partial derivatives. In [Ma3] she proved that, if f is finely differentiable on a finely open set U, there exists, for all real $\varepsilon > 0$, a compact set $K_\varepsilon \subset U$, such that the Lebesgue measure of $U \setminus K_\varepsilon$ is majorized by ε and on which f is differentiable, the two differentials thus obtained being equal.

More recently, the problem of fine continuous differentiability of real functions on a finely open set U of $\mathbb{R}^N$ was studied by Gardiner in [Ga2]. He established an explicit characterization of those real-valued functions on a finely open set in Euclidean space which are continuously differentiable with respect to the fine topology of the classical Potential theory. More precisely, he proved that a real function on a finely open set in $\mathbb{R}^N$ is finely continuously differentiable if and only if it is finely locally the restriction of a function of class $\mathcal{C}^k$ on the whole space $\mathbb{R}^N$. This result improves and extends previous works by Lavicka [Lav1, Lav2]. He also considered differentiability of all orders, and deduced consequences of this characterization.

In [Ly3], Lyons considered the map from the fine interior U of a compact set to the measures on the fine boundary given by Balayage of the unit point mass onto the fine boundary (the Keldyŝ measure). He showed that every point in U has a compact fine neighborhood on which the map is continuous from the initial topology on the compact set to the norm topology on measures. One consequence of this result is a Hartogs-type theorem for finely harmonic functions. The Hartogs' theorem, rational approximation theory, and other results by the author are then used to prove that the derivative of a finely holomorphic function exists everywhere and is finely holomorphic.

The stochastic characterizations of finely harmonic morphisms were obtained by Øksendal [Ok1]: First, they are exactly the functions which preserve the paths of Brownian motion $B_t; t \geq 0$. This is a fine topology version of the classical theorem of P. Lévy about conformal invariance of Brownian motion. (See McKean [McKe], Bernard, Campbell and Davie [BCD] or Csink and Øksendal [CO].) Second, the finely harmonic morphisms are exactly the functions φ such that $\varphi(B_t)$ is a conformal martingale in the sense of Getoor and Sharpe [GS], i.e. $Z_t = \varphi(B_t)$ is a martingale satisfying $\langle Z, Z \rangle = 0$. These results are applied to study the removable singularity sets and the boundary behavior of finely harmonic morphisms.

In [Ok2] Øksendal studied the behavior at certain boundaries of a finely open set $V \subset \mathbb{R}^d$ of finely harmonic functions on a finely harmonic functions with finite Dirichlet integral with respect to Gm, i.e.

$$\int_V |\nabla h(y)|^2 G(x, y) dm(y) < \infty \text{ for } x \in V, \qquad (*)$$

where m denotes the Lebesgue measure, $G(x, y)$ the Green function. Using Brownian motion and stochastic calculus, he proved that such functions h always have boundary values h^* along a.a. Brownian paths. This partially extends results by Doob, Brelot, and Godefroid, who considered ordinary harmonic functions with finite Dirichlet integral with respect to m and Green lines instead of Brownian paths. As a consequence he obtained several properties equivalent to (*), one of these being that h is the harmonic extension to V of a random "boundary" function h^* (of a certain type), i.e. $h(x) = E^x[h^*]$ for all $x \in V$. Another application is that the polar sets are removable singularity sets for finely harmonic functions satisfying (*). This is in contrast with the situation for finely harmonic functions with finite Dirichlet integral with respect to m.

In another direction, if a Green space R (in the sense of Doob [Do1, Section 4]) which admits the Green function, the problem of the behavior at the Martin boundary $\Delta(R)$ of a positive finely superharmonic function u on an unbounded finely open subset of R was considered by Masaoka in [Mas3]. He proved there that it holds that

(i) u is decomposed uniquely into the sum of a fine Potential p in U, a positive finely superharmonic function f in U, which is quasibounded finely harmonic in

$U \setminus \{z \in U : u(z) = +\infty\}$ and a positive finely superharmonic function f, in U which is singular finely harmonic in $U \setminus \{z \in U : u(z) = +\infty\}$, and that

(ii) u and f have a common fine limit at almost every point of $\Delta_1(U) = \{\zeta \in \Delta_1(R) : \widehat{R}_{k_\zeta}^{\complement U} \neq k_\zeta\}$ (where $k_\zeta = k(., \zeta)$ and K is the Martin kernel of R) with respect to ω_i (the set of minimal points of Martin boundary $\Delta(R)$ of R).

We recall here that f is called to be singular in U if a bounded positive finely harmonic function h in U dominated by f is equal to zero identically in U. The function f is called to be quasibounded in U if there exists a sequence (f_n) of bounded positive finely harmonic functions in U such the (f_n) converges increasingly to f on U.

This result is used in [Mas5] to obtain a theorem of a Littlewood type for positive finely superharmonic functions in the upper half space $\mathbf{H}$ of the d-dimensional Euclidean space ($d > 2$).

In [Fu14] Fuglede proved that for a finely continuous mapping φ from an open set U ($\subset \mathbb{R}^n$, $n \geq 2$) into $\mathbb{R}^m$ ($m \geq 2$), the followings are equivalent:

(i) φ is a harmonic morphism on U.

(ii) The components φ_j of φ ($1 \leq j \leq m$), $\varphi_i\varphi_j$ ($i \neq j$) and $\varphi_i^2 - \varphi_j^2$ are harmonic in U.

(iii) The components φ_j of φ ($1 \leq j \leq m$) are harmonic in U, and

$$\nabla\varphi_i.\nabla\varphi_j = \delta_{i,j}|\nabla\varphi_i|^2 \text{ on } U.$$

By using a probabilistic method, Masaoka extended in [Mas1] the result of Fuglede for finely harmonic morphisms in $\mathbb{R}^n$. He proved that for a finely continuous mapping φ from a finely open set U ($\subset \mathbb{R}^n$, $n \geq 2$) into $\mathbb{R}^m$ ($m \geq 2$), the followings are equivalent:

(i) φ a finely harmonic morphism on U.

(ii) The components φ_j of φ ($1 \leq j \leq m$), $\varphi_i\varphi_j$ ($i \neq j$) and $\varphi_i^2 - \varphi_j^2$ are finely harmonic in U.

(iii) The components φ_j of φ ($1 \leq j \leq m$) are finely harmonic in U, and

$$\nabla\varphi_i.\nabla\varphi_j = \delta_{i,j}|\nabla\varphi_i|^2 \text{ a.e. on } U,$$

where $\nabla\varphi_i$ is the gradient of φ_i.

The contributions of Nguyen-Xuan-Loc [Ng2, Ng3] to classical and general fine Potential theories are based upon probabilistic methods and concern among other topics the minimum principle for finely hyperharmonic and its generalizations to the general setting of Markov processes. The article [Ng1] is devoted to the study of excessive nearly Borel functions in a finely open set of the Potential theory of a Markov process and their characterization. There also are other works of Nguen-Xuan-Loc where he studied finely harmonic morphisms by probabilistic methods with applications of to the particular case of a finely open subset of $\mathbb{C}$.

A Hartogs-type theorem for separately finely [super]harmonic finely locally [lower]bound functions on a product $U_1 \times U_2$ of finely open sets U_1 and U_2 of $\mathbb{R}^m$ and $\mathbb{R}^n$ respectively, was obtained by El Kadiri in [EK3]. This result was proved by a probabilistic method. El Kadiri and Aslimani also studied the fine local uniform continuity of series and general families of finely locally uniformly bounded finely harmonic functions in [EK5, AEK]. As an application, it was proved in [AEK] that the fine Green kernel of a fine domain of a Greenian domain of $\mathbb{R}^N$ is regular in a certain sense.

The notion of finely biharmonic function on a finely open set of $\mathbb{R}^N$ (or more generally a biharmonic space in the sense of Smyrnelis satisfying the axiom D) is considered and studied in [BER, EK1] and used to characterize the functions on a compact subset K of $\mathbb{R}^N$ that are limit on K of sequences of restrictions to K of biharmonic functions in neighborhoods of K in [EK6].

The Sobolev spaces on finely open sets in classical Potential theorie and their use, probably inspired from Hilbertian technical methods of Deny, were introduced and studied by Feyel and de La Pradelle in [FL1, FL2]. In other directions, the role of fine topology in general Potential theory was highlighted in [Fe2, FL3]. In a general setting Feyel studied in [Fe1] some fine properties of strongly supermedian functions with respect to a Hunt kernel on a LCD space by purely potentialist methods. Similar studies can be found in certain works by Mokobodzki.

Kolsrud used fine topology technics in another context where he developed the notion of finely harmonic and superharmonic functions on a finely open set of a symmetric Dirichlet space in the sense of Fukushima [Fuk]. The main tools used are probabilistic and based on capacities and properties of the fine topology in this setting. This extends in particular the result in classical and axiomatic fine Potential theory to the more general Potential theory of the Dirichlet forms.

There are also many contributions of Lukeš to the study of the fine topology in the Potential theory. The Lusin-Menchoff property and the Zahorski property of fine topology in Potential theory are discussed in [Lu]. In [LZ2], Lukeš and Zajíček gave a simple proof of the Fuglede's theorem asserting that any finely continuous function in an abstract harmonic space is of the first class of Baire, with some applications of their method to the density topology.

Malý showed in his thesis in 1986 that the structure of standard H-cones (see [BBC2]) is rich enough to develop Potential theory, including Dirichlet problem, on open and finely open subsets of the underlying space. The substantial part of this thesis treats the capacity with values in the cone of all positive superharmonic functions on a finely open set and related quasi-topological notions. In this setting it is shown in [LMZ] that if U is a finely open set, then for each boundary function f that is integrable with respect to the harmonic measure, there exists a unique finely harmonic function h on U that is quasi-continuous up to the boundary extending the values of f. This function h coincides with the "Perron solution" of the considered Dirichlet problem. He also proved that if u be a quasi-l.s.c and finely l.s.c function on U such that for every $x \in U$ there is a fundamental system of fine neighborhoods V of x with the property $\varepsilon_x^{\complement V}(u) \leq u(x)$, then u is finely hyperharmonic on U.

The methods of fine topology in several topics of real analysis are also discussed in a book edited by Lukeš, J. Malý and L. Zajíček [LMZ].

Let us also mention the extension of the important properties, such as local connectivity, Baire property,... of fine topology in classical (linear) Potential theory to nonlinear Potential theory (associated with the nonlinear equation) by many authors.

Indeed, some of the works mentioned above and in the bibliography are written in a large generality including topological spaces, general capacities and families of functions, and some results thus apply also to the nonlinear theory. At the same time, many other results rely indirectly on a linear structure, e.g. through potentials, integral representation, and convex cones of superharmonic functions, which are in general not available in the nonlinear setting. The nonlinear fine potential theory started in the 1970s on unweighted $\mathbb{R}^n$, see the notes to Chapter 12 in [HKM] and Section 2.6 in [MZ]. For the fine potential theory associated with p-harmonic functions on unweighted $\mathbb{R}^n$, see [MZ] and Latvala [Lat]. The monograph [HKM] is the main source for fine potential theory on weighted $\mathbb{R}^n$ (note that Chapter 21 contains some more recent results). The study of fine potential theory on metric spaces was investigated by A. Björn, J. Björn, V. Latvala [BBL, Bj], Kinnunen–Latvala [KL], and Korte [K]. For further references to nonlinear and fine nonlinear potential theory, see the introduction to [BBL].

The notion of $\mathcal{F}$-plurisubharmonic or plurifinely plurisubharmonic functions, studied in Chap. 5, was first introduced, in analogy with the finely subharmonic functions on a fine domain of $\mathbb{R}^n$, $n \geq 1$, by El Kadiri in [EK4] where some first properties of these functions were studied. Later, El Marzguioui and Wiegerinck proved the local connectedness of the pluri-fine (topology $\mathcal{F}$) and established the pluri-fine continuity properties of the $\mathcal{F}$-plurisubharmonic functions. The most important properties of $\mathcal{F}$-plurisubharmonic functions were obtained by El Kadiri, Fuglede and Wiegerinck in [EKFW], where they also introduced and studied the notion of holomorphic functions relative to pluri-fine topology in several complex variables (the $\mathcal{F}$-holomorphic functions). The Monge-Ampère operator for finite $\mathcal{F}$-plurisubharmonic functions was defined by El Kadiri and Wiegerinck in [EKW]. If u is a finite $\mathcal{F}$-plurisubharmonic on an $\mathcal{F}$-open set $U \subset \mathbb{C}^n$, the associated Monge-Ampère measure $(dd^c u)^n$ is a positive Borel measure on U which does not charge the pluripolar subsets of U. Maximal $\mathcal{F}$-plurisubharmonic were introduced and studied by El Kadiri and Smit in [EKS], where they proved in particular that the finite $\mathcal{F}$-maximal plurisubharmonic functions are solutions of the equation $(dd^c u)^n = 0$ in the set of functions of class $\mathcal{F}\text{-}\operatorname{PSH}(U)$, and introduced the notion of local $\mathcal{F}$-maximality.

As in the Euclidian setting, it is of interest to determine the large class of plurifinely plurisubharmonic functions to which the Monge-Ampère operator can be extended. In recent works on $\mathcal{F}$-plurisubharmonic functions (cf. [CL, HCL, HTV, HVC1, HVC2, HCLL]), an attempt is made to extend the Monge-Ampère operator and some related results to not necessarily finite $\mathcal{F}$-plurisubharmonic functions on a certain class of $\mathcal{F}$-domains (so-called $\mathcal{F}$-hyperconvex domains). However, so far the only examples known of such $\mathcal{F}$-domains are trivial. It would be interesting to find non trivial examples of these domains.

References

[Ad] R.A. Adams, in Sobolev Spaces. Pure and Applied Mathematics, vol. 65 (Academic Press, New York-London, 1975)

[ACH] P. Åhag, U. Cegrell, R. Czyż, P.H. Hiep, Monge-Ampère measures on subvarieties. J. Math. Pures Appl. **423**(1), 94–105 (2015)

[Ai] H. Aikawa, Potential Analysis on non-smooth domains—Martin boundary and boundary Harnack principle, in *Complex Analysis and Potential Theory*, pp. 235–253, CRM Proc. Lecture Notes, vol. 55 (American Mathematical Society, Providence, RI, 2012)

[Al] E. Alfsen, *Compact convex sets and boundary integrals, in Ergebnisse der Mathematik und ihrer Grenzgebiete,* vol. 57 (Springer, New York, 1971)

[An] A. Ancona, Sur une conjecture concernant la capacité et l'effilement. *Théorie du Potentiel (Orsay, 1983)* 34–68. Lecture Notes in Mathematics , vol. 1096 (Springer, Berlin, 1984)

[AG] D.H. Armitage, S.J. Gardiner, *Classical Potential Theory*. Springer Monographs in Mathematics. (Springer, London Ltd., London, 2001)

[AEK] A. Aslimani, M. El Kadiri, Propriété de convergence de certaines familles de fonctions finement harmoniques et régularité du noyau de Green fin. Ann. Polon. Math. **115**(1), 43–51 (2015)

[BW] P. Baird, J.C. Wood, in *Harmonic Morphisms Between Riemannian Manifolds*. London Mathematical Society Monographs. New Series, vol. 29. (The Clarendon Press, Oxford University Press, Oxford, 2003)

[Ba1] H. Bauer, Weiterführung einer axiomatischen Potentialtheorie ohne Kern (Existenz von Potentialen). *Z. Wahrscheinlichkeitstheorie und Verw. Gebiete* 1 (1962/1963), 197–229

[Ba2] H. Bauer, Propriétés fines des fonctions hyperharmoniques en théorie axiomatique du potentiel. Ann. Inst. Fourier (Grenoble) **15**, fasc. 1, 137–154 (1965)

[Ba3] H. Bauer, *Harmonische Räume und ihre Potentialtheorie*. Lecture Notes in Mathematics, vol. 22 (Springer, Berlin, 1966)

[Be] R. Becker, *Convex cones in analysis, in Travaux en Cours*, vol. 67 (Hermann Éditeurs des Sciences et des Arts, Paris, 2006)

[BT1] E. Bedford, B.A. Taylor, The Dirichlet problem for complex Monge-Ampère equation. Invent. Math. **37**(1), 1–44 (1976)

[BT2] E. Bedford, B.A. Taylor, A new capacity for plurisubharmonic functions. Acta Math. **149**(1–2), 1–40 (1982)

[BT3] E. Bedford, B.A. Taylor, Fine topology, Šilov boundary, and $(dd^c)^n$. J. Funct. Anal. **72**(2), 225–251 (1987)

[BT4] E. Bedford, B.A. Taylor, Plurisubharmonic functions with logarithmic singularities. Ann. Inst. Fourier (Grenoble) **38**(4), 133–171 (1988)

M. El Kadiri and B. Fuglede, *Classical Fine Potential Theory*, Springer Monographs in Mathematics, https://doi.org/10.1007/978-981-96-0430-2

[BG] C. Bensouda, P.M. Gauthier, Superharmonic approximation on closed sets. J. Contemp. Math. Anal. **29**(1), 13–28 (1994)

[BER] C. Bensouda, M. El Kadiri, I. Rouchdi, Fonctions finement biharmoniques dans un espace biharmonique. Rend. Mat. Appl. **23**(7), 1, 131–161 (2003)

[Berg] C. Berg, Quelques propriétés de la topologie fine dans la théorie du potentiel et des processus standards. Bull. Sci. Math. **95**(2), 27–31 (1971)

[BCD] A. Bernard, E.A. Campbell, A.M. Davie, Brownian motion and generalized analytic and inner functions. Ann. Inst. Fourier (Grenoble) **29**(1), 207–228 (1979)

[Bet] D. Betsakos, Geometric theorems and problems for harmonic measures. Rocky Mountain J. Math. **31**(3), 773–795 (2001)

[Beu] A. Beurling, *Etudes sur un problème de majorization* (Thesis, Uppsala, 1933)

[BeB] L. Beznea, N. Boboc, *Potential theory and right processes, in Mathematics and its Applications*, vol. 572 (Kluwer Academic Publishers, Dordrecht, 2004)

[Bi] L. Bieberbach, *Lehbruch der Funktionentheorie*, vol. II (Teubner, Berlin-Leipzig, 1927)

[Bj] J. Björn, Fine continuity on metric spaces. Manuscripta Math. **125**(3), 369–381 (2008)

[BjBj] A. Björn, J. Björn, Obstacle and Dirichlet problems on arbitrary nonopen sets in metric spaces, and fine topology. Rev. Mat. Iberoam. **31**(1), 161–214 (2015)

[BBL] A. Björn, J., Björn, V. Latvala, The weak Cartan property for the p-fine topology on metric spaces. Indiana Univ. Math. J. **64**(3), 915–941 (2015)

[BH1] J. Bliedtner, W. Hansen, Simplicial cones in potential theory. Invent. Math. **29**(2), 83–110 (1975)

[BH2] J. Bliedtner, W. Hansen, Simplicial cones in potential theory, II. Approximation theorems. Invent. Math. **46**(3), 255–275 (1978)

[BH3] J. Bliedtner, W. Hansen, *Potential Theory* (Universitext (Springer, Berlin, An analytic and probabilistic approch to balayage, 1986)

[Bl1] Z. Błocki, Estimates for the complex Monge-Ampère operator. Bull. Polish. Acad **41**(2), 151–157 (1993)

[Bl2] Z. Błocki, On the definition of the Monge-Ampère operator in $\mathbb{C}^2$. Math. Ann. **328**(3), 415–423 (2004)

[Bl3] Z. Błocki, The domain of definition of the complex Monge-Ampère operator. Am. J. Math. **128**(2), 519–530 (2006)

[BB1] N. Boboc, Gh. Bucur, Natural localization and natural sheaf property in standard H-cones of functions. I. Revue Roumaine Math. Pures Appl. **30**(1), 1–21 (1985)

[BB2] N. Boboc, Gh. Bucur, Potentials and supermedian functions on finely open sets in standard H-cones. Revue Roumaine Math. Pures Appl. **31**(9), 745–775 (1986)

[BB3] N. Boboc, Gh. Bucur, Green potentials on standard H-cones. Rev. Roumaine Math. Pures Appl. **32**(4), 293–320 (1987)

[BB4] N. Boboc, Gh. Bucur, Fine potentials and supermean functions on standard H-cones. Rev. Roumaine Math. Pures Appl. **32**(10), 881–890 (1987)

[BBC1] N. Boboc, Gh. Bucur, A. Cornea, H-cones and potential theory. Ann. Inst. Fourier (Grenoble) **23**(3–4), 71–108 (1975)

[BBC2] N. Boboc, Gh. Bucur, A. Cornea, *Order and Convexity in Potential Theory: H-Cones*. Lecture notes in Mathematics, vol. 853 (Springer, Berlin, 1980)

[BCC] N. Boboc, C. Constantinescu, A. Cornea, Axiomatic theory of harmonic functions. Nonnegative superharmonic functions. Ann. Inst. Fourier **15**(1), 283–312 (1965)

[BC] N. Boboc, A. Cornea, Espaces harmoniques. Axiome D et théorème de convergence. Rev. Roumaine Math. Pures Appl. **13**, 933–947 (1968)

[Bor1] E. Borel, Sur quelques points de la théorie des fonctions, Thèse, Paris 1894. Ann. Sci. École Norm. Sup. **12**(3), 9–55 (1895)

[Bor2] E. Borel, *Leçons sur les fonctions monogènes uniformes d'une variable complexe* (Gauthier-Villars, Paris, 1917)

[Bou1] N. Bourbaki, Éléments de mathématique. Part I. Les structures fondamentales de l'analyse. Livre III. Topologie générale. Chapitres III et IV. Actualités Sci. Indust., No. 916 (Hermann & Cie, Paris, 1942)

[Bou2] N. Bourbaki, Éléments de mathématique. Intégration, chap. 5: Intégration des mesures (Hermann, Paris, 1958)
[Bou3] N. Bourbaki, Éléments de mathématique. Espaces vectoriels topologiques. Chapitres 3 à 5 (Hermann, Paris, 1966)
[Bou4] N. Bourbaki, Éléments de mathématique. Topologie générale, chapitres 1 à 2. (Hermann, Paris, 1971)
[Bou5] N. Bourbaki, Éléments de mathématique. Topologie générale. Chapitre 9. (Hermann, Paris, 1971)
[Bou6] N. Bourbaki, Éléments de mathématique. Intégration, chap. 1–4. (Hermann, Paris, 1992)
[Bou7] N. Bourbaki, Éléments de mathématique. Espaces Vectoriels Topologiques, chap. 4. La dualité dans les espaces vectoriels topologiques (Hermann, Paris, 1995)
[Br1] M. Brelot, Sur la théorie moderne du potentiel. C.R. Acad. Sci. Paris **209**, 828–830 (1939)
[Br2] M. Brelot, Points irréguliers et transformations continues en théorie du potentiel. J. Math. Pures Appl. **19**(9), 319–337 (1940)
[Br3] M. Brelot, Sur les ensembles effilés. Bull. Sci. Math. **68**(2), 12–36 (1944)
[Br4] M. Brelot, Sur le rôle du point à l'infini dans la théorie des fonctions harmoniques. Ann. Ec. Norm. Sup. **61**(4), 301–332 (1944)
[Br5] M. Brelot, Minorantes sousharmoniques extrémales et capacités. J. Math. Pures Appl. **24**(9), 1–32 (1945)
[Br6] M. Brelot, Sur l'allure des fonctions harmoniques et sousharmoniques à la frontière. Math. Nachr. **4**, 298–307 (1951)
[Br7] M. Brelot, *Eléments de la Théorie Classique du Potentiel* (Centre de Documentation Universitaire, Paris, 1959)
[Br8] M. Brelot, Axiomatiques des fonctions harmoniques et surharmoniques dans un espace localement compact. *Sém. Théorie du Potentiel*, vol. 2 (Paris, 1959), pp. 1.01–1.40
[Br9] M. Brelot, *Lectures on Potential Theory* (Tata Institute of Fundamental Research, Bombay, 1960)
[Br10] M. Brelot, Introduction axiomatique de l'effilement. Annali di Mat. Pisa **57**(4), 77–96 (1962)
[Br11] M. Brelot, Quelques propriétés et applications nouvelles de l'effilement. *Sém. Théorie du Potentiel,* vol. 6. (Paris, 1962), pp. 1.27–1.40
[Br12] M. Brelot, Aspect statistique et comparé des deux types d'effilement. An. Acad. Brasil. Ci. **37**, 1–15 (1965)
[Br13] M. Brelot, Capacity and balayage for decreasing sets. *Proceedings of the Fifth Berkeley Symposium on Mathematical Statistics and Probability (Berkley, California 1965/66), Volume II: Contributions to Probability Theory, Part 1* (University California Press, Berkeley, CA, 1967), pp. 279–293
[Br14] M. Brelot, Recherches axiomatiques sur un théorème de Choquet concernant l'effilement. Nagoya Math. J. **30**, 33–46 (1967)
[Br15] M. Brelot, La topologie fine en théorie du potentiel, in *1967 Symposium on Probability Methods in Analysis (Loutraki, 1966)*, pp. 36–47, Lecture Notes in Mathematics, vol. 31 (Springer, Berlin-New York, 1967)
[Br16] M. Brelot, Recherches sur la topologie fine et ses applications: Théorie du potentiel. Ann. Inst. Fourier (Grenoble) **17**(2), 395–423 (1967)
[Br17] M. Brelot, *On Topologies and Boundaries in Potential Theory*. Lecture Notes in Mathematics, vol. 175 (Springer, Berlin, 1971)
[BrCh1] M. Brelot, G. Choquet, Espaces et lignes de Green. Ann. Inst. Fourier (Grenoble) **3**, 199–263 (1951)
[BrCh2] M. Brelot, G. Choquet, Le théorème de convergence en théorie du potentiel. J. Madras Univ. B **27**, 277–286 (1957)
[CZ] A.P. Calderon, A. Zygmund, On the existence of certain singular integrals. Acta Math. **88**, 85–139 (1952)
[CL] H.V. Can, P.T. Lieu, The pluripolar parts of the Monge-Ampère measures of $\mathcal{F}$-plurisubharmonic functions. Ann. Polon. Math. **128**(2), 99–111 (2022)

[C] L. Carleson, Selected problems on exceptional sets, in *Van Nostrand Mathematical Studies*, vol. 13 (D. Van Nostrand Co., Inc., Princeton, N.J.-Toronto, Ont.-London, 1967)

[Ca1] H. Cartan, Capactité extéreure et suites convergentes de potentiels. C.R. Acad. Sci. Paris **214**, 944–946 (1942)

[Ca2] H. Cartan, Théorie du potentiel newtonien: énergie, capacité, suites de potentiels. Bull. Soc. Math. France **73**, 74–106 (1945)

[Ca3] H. Cartan, Théorie générale du balayage en potentiel newtonien. Ann. Univ. Grenoble. Sect. Sci. Math. Phys. (N.S.) **22**, 221–280 (1946)

[Ca4] H. Cartan, Formes différentielles. Applications élémentaires au calcul des variations et à la théorie des courbes et des surfaces (Hermann, Paris, 1967)

[Ce1] U. Cegrell, On the discontinuity of the complex Monge-Ampère operator. C. R. Acad. Sci. Paris. Série I Math. **296**(21), 869–871 (1983)

[Ce2] U. Cegrell, The general definition of the complex Monge-Ampère operator. Ann. Inst. Fourier (Grenoble) **54**(1), 159–179 (2004)

[CeW] U. Cegrell, J. Wiklund, A Monge-Ampère norm for δ-plurisubharmonic functions. Math. Scand. **97**(2), 201–216 (2005)

[CLN] S. Chern, H.I. Levine, L. Nirenberg, Intrinsic norms on a complex manifold, in *Global Analysis (Papers in honor of K. Kodaira)* (University Tokyo Press, Tokyo, 1969), pp. 119–133

[Ch1] G. Choquet, Theory of capacities. Ann. Inst. Fourier (Grenoble) **5**(1953/54), 131–295 (1955)

[Ch2] G. Choquet, Sur les fondements de la théorie fine du potentiel. *Sém. Théorie du Potentiel*, vol. 1 (1958), pp. 1.1–1.10

[Ch3] G. Choquet, Sur les points d'effilement d'un ensemble. Application à l'étude de la capacité. Ann. Int. Fourier (Grenoble) **9**, 221–280 (1959)

[Ch4] G. Choquet, *Lectures on analysis. Volume I: Integration and Topological Vector Spaces*, ed. by J. Marsden, T. Lance, S. Gelbart (W. A. Benjamin, Inc., New York-Amsterdam, 1969)

[Ch5] G. Choquet, Existence et unicité des représentations intégrales au moyen des points extrémaux dans les cônes convexes, in *Séminaire Bourbaki, Vol. 4*, Exp. No. 139, 33–47 (Societe Mathematique de France, Paris, 1995)

[Co] C. Constantinescu, Some properties of the balayage of measures on a harmonic space. Ann. Inst. Fourier (Grenoble) **17**(1), 273–293 (1967)

[CC1] C. Constantinescu, A. Cornea, Ideale Ränder Riemannscher Flächen. Ergebnisse der Mathematik und ihrer Grenzgebiete, (N.F.), Band 32. (Springer-Verlag, Berlin-Gőttingen-Heidelberg, 1963)

[CC2] C. Constantinescu, A. Cornea, Compactifications of harmonic spaces. Nagoya Math. J. **25**, 1–57 (1965)

[CC3] C. Contantinescu, A. Cornea, *Potential theory on harmonic spaces, in Die Grundlehren der mathematischen Wissenschaften, Band*, vol. 158 (Springer, New York, 1972)

[CO] L. Csink, B. Øksendal, Stochastic harmonic morphisms: Functions mapping the paths of one diffusion into the paths of another. Ann. Inst. Fourier **33**(2), 219–240 (1983)

[DO] A.M. Davie, B. Øksendal, Analytic capacity and differentiability properties of finely harmonic functions. Acta Math. **149**(1–2), 127–152 (1982)

[Da] E.B. Davies, *Spectral Theory and Differential Operators*. Cambridge Studies in Advanced Mathematics, vol. 42 (Cambridge University Press, Cambridge, 1995)

[DG1] A. Debiard, B. Gaveau, Potentiel fin et algèbres de fonctions analytiques. I. J. Funct. Anal. **16**, 289–304 (1974)

[DG2] A. Debiard, B. Gaveau, Potentiel fin et algèbres de fonctions analytiques. II. J. Funct. Anal. **17**, 296–310 (1974)

[DG3] A. Debiard, B. Gaveau, Differentiabilité des fonctions finement harmoniques. Invent. Math. **29**(2), 111–123 (1975)

[DG4] A. Debiard, B. Gaveau, Potentiel fin et algèbres de fonctions analytiques. III. J. Funct. Anal. **21**(4), 448–468 (1976)

[DG5] A. Debiard, B. Gaveau, Méthodes de contrôle optimal en analyse complexe. IV, in *Applications aux algèbres de fonctions analytiques*. Lecture Notes in Mathematics, vol. 798 (Springer, Berlin, 1980), pp. 109–140

[DM] C. Dellacherie, P.A. Meyer, *Probabilités et Potentiel, in Chapitres XII-XVI* (Hermann, Paris, 1987)

[D] J.-P. Demailly, Mesure de Monge-Ampère et caractérisation des variétés algébriques. Mém. Soc. Math. France (N.S.) No. 19 (1985)

[DeL] J. Deny, J.L. Lions, Les espaces du type Beppo Levi. Ann. Inst. Fourier (Grenoble) **21**(1945), 305–370 (1955)

[De1] J. Deny, Les potentiels d'énergie finie. Acta Math. **82**, 107–183 (1950)

[De2] J. Deny, Sur la convergence de certaines intégrales de la théorie du potentiel. Ark. Math. (Basel) **5**, 367–370 (1954)

[De3] J. Deny, Méthodes hilbertiennes en théorie du potentiel. 1970 Potential Theory (C.I.M.E., I Ciclo, Stresa, 1969) pp. 121–201. Edizioni Cremonese, Rome

[Do1] J.L. Doob, Brownian motion on Green space. Teor. Veroyatnost. i Primenen. **2**, 3–33 (1957). Theory Probab. Appl. *2*, 1–30 (1957)

[Do2] J.L. Doob, Some classical function theory theorems and their modern versions. Ann. Inst. Fourier (Grenoble) **15**(1), 113–136 (1965)

[Do3] J.L. Doob, Applications to analysis of a topological definition of smallness of a set. Bull. Am. Math. Soc. **72**, 579–600 (1966)

[Do4] J.L. Doob, *Classical potential theory and its probabilistic counterpart, in Grundlehren der mathematischen Wissenschaften*, vol. 262 (Springer, New York, 1984)

[EEW] A. Edigarian, S. El Marzguioui, J. Wiegerinck, The image of a finely holomorphic map is pluripolar. Ann. Polon. Math. **97**(2), 137–149 (2010)

[Ed] T. Edlund, Complete pluripolar curves and graphs. Ann. Polon. Math. **84**(1), 75–86 (2004)

[EJ] T. Edlund, B. Jöricke, The pluripolar hull of a graph and fine analytic continuation. Ark. Math. **44**(1), 39–60 (2006)

[EK1] M. El Kadiri, Fonctions finement biharmoniques. Rend. Accad. Naz. Sci. XL Mem. Mat. Appl. **24**(5), 43–62 (2000)

[EK2] M. El Kadiri, Sur la décomposition de Riesz et la représentation intégrale des fonctions finement surharmoniques. Positivity **4**(2), 105–114 (2000)

[EK3] M. El Kadiri, Fonctions séparément finement surharmoniques. Positivity **7**(3), 245–256 (2003)

[EK4] M. El Kadiri, Fonctions finement plurisousharmoniques et topologie plurifine. Rend. Accad. Naz. Sci. XL Mem. Mat. Appl. **27**(5), 77–88 (2003)

[EK5] M. El Kadiri, Sur les suites de fonctions finement harmoniques. Riv. Mat. Univ. Parma **2**(7), 63–68 (2003)

[EK6] M. El Kadiri, Uniform approximation of continuous functions on compact sets by biharmonic and bisuperharmonic functions in a biharmonic space. Rend. Mat. Appl. **29**(2, 7), 143–152 (2009)

[EK7] M. El Kadiri, A note on maximal plurifinely plurisubharmonic functions. Ann. Polon. Math. **120**(3), 217–225 (2017)

[EKF1] M. El Kadiri, B. Fuglede, Martin boundary of a fine domain and a Fatou-Naïm-Doob theorem for finely superharmonic functions. Potential Anal. **44**(1), 1–25 (2016)

[EKF2] M. El Kadiri, B. Fuglede, Sweeping at the Martin boundary of a fine domain. Potential Anal. **44**(2), 401–422 (2016)

[EKF3] M. El Kadiri, B. Fuglede, The Dirichlet problem at the Martin boundary of a fine domain. J. Math. Anal. Appl. **457**(1), 179–199 (2018)

[EKFW] M. El Kadiri, B. Fuglede, J. Wiegerinck, Plurisubharmonic and holomorphic functions relative to the pluri-fine topology. J. Math. Anal. Appl. **381**(2), 706–723 (2011)

[EKS] M. El Kadiri, I.M. Smit, Maximal plurifinely plurisubharmonic functions. Potential Anal. **41**(4), 1329–1345 (2014)

[EKW] M. El Kadiri, J. Wiegerinck, Plurifinely plurisubharmonic functions and the Monge-Ampère operator. Potential Anal. **41**(2), 469–485 (2014)

[EMW1] S. El Marzguioui, J. Wiegerinck, The pluri-fine topology is locally connected. Potential Anal. **25**(3), 283–288 (2006)

[EMW2] S. El Marzguioui, J. Wiegerinck, *Connectedness in the pluri-fine topology, in Functional Analysis and Complex Analysis, Istanbul 2007, 105–115, Contemporary Mathematics*, vol. 481 (American Mathematical Society, Providence, RI, 2009)

[EMW3] S. El Marzguioui, J. Wiegerinck, Continuity properties of finely plurisubharmonic functions. Indiana Univ. Math. J. **59**(5), 1793–1800 (2010)

[EW1] A. Edigarian, J. Wiegerinck, Graphs that are not complete pluripolar. Proc. Am. Math. Soc. **131**(8), 2459–2465 (2003)

[EW2] A. Edigarian, J. Wiegerinck, Determination of the pluripolar hull of graphs of certain holomorphic functions. Ann. Inst. Fourier (Grenoble) **54**(6), 2085–2104 (2004)

[Fed] H. Federer, *Geometric measure theory, in Die Grundlehren der mathematischen Wissenschaften, Band*, vol. 153 (Springer, New York, 1969)

[Fe] C. Fernströom, J.C. Polking, Bounded point evaluations and approximation in L^p by solutions of elliptic partial differential equations. J. Funct. Anal. **28**(1), 1–20 (1978)

[FL1] D. Feyel, A. de La Pradelle, Espaces de Sobolev sur les ouverts fins. C. R. Acad. Sci. Paris Sér. A-B **280**, 1125–1127 (1975)

[FL2] D. Feyel, A. de La Pradelle, Le rôle des espaces de Sobolev en topologie fine. *Séminaire de Théorie du Potentiel Paris, No. 2 (University, Paris, Paris, 1975–1976)*, pp. 43–61. Lecture Notes in Mathematics, vol. 563 (Springer, Berlin, 1976)

[FL3] D. Feyel, A. de La Pradelle, Topologies fines et compactifications associées à certains espaces de Dirichlet. Ann. Inst. Fourier (Grenoble) **27**(4), 121–146 (1977)

[Fe1] D. Feyel, Sur la théorie fine du Potentiel. Bull. Soc. Math. France **111**(1), 41–57 (1983)

[Fe2] D. Feyel, Représentation des fonctionnelles surmédianes. Z. Wahrsch. Verw. Gebiete **58**(2), 183–198 (1981)

[Fr] O. Frostman, Potentiel d'équilibre et capacité des ensembles. Comm. Sém. Math. Lund 3 (1935)

[Fu1] B. Fuglede, On the theory of potentials in locally compact spaces. Acta Math. **103**, 139–215 (1960)

[Fu2] B. Fuglede, The logarithmic potential in higher dimensions. Mat.-Fys. Medd. Danske Vid. Selsk. **33**(1), 14 pp (1960)

[Fu3] B. Fuglede, Le théorème du minimax et la théorie fine du potentiel. Ann. Inst. Fourier (Grenoble) **15**(1), 65–88 (1965)

[Fu4] B. Fuglede, Fine connectivity and finely harmonic functions. *Actes du Congrès International de Mathématiciens (Nice, 1970), Tome 2*, pp. 513–519. *Gauthier-Villars Éditeur, Paris* (1971)

[Fu5] B. Fuglede, The quasi topology associated with a countably subadditive set function. Ann. Inst. Fourier (Grenoble) **21**(1), 123–169 (1971)

[Fu6] B. Fuglede, Capacity as a sublinear functional generalizing an integral. Mat.-Fys. Medd. Danske Vid. Selsk. **38**(7), 44 pp (1971)

[Fu7] B. Fuglede, Connexion en topologie fine et balayage des mesures. Ann. Inst. Fourier (Grenoble) **21**(3), 227–244 (1971)

[Fu8] B. Fuglede, Finely harmonic functions. Lecture Notes in Mathematics, vol. 289 (Springer, Berlin, 1972)

[Fu9] B. Fuglede, Remarks on fine continuity and the base operation in potential theory. Math. Ann. **210**, 207–212 (1974)

[Fu10] B. Fuglede, Boundary minimum principles in potential theory. Math. Ann. **210**, 213–226 (1974)

[Fu11] B. Fuglede: Fonctions harmoniques et fonctions finement harmoniques. Ann. Inst. Fourier (Grenoble) **24**(4), 77–91 (1974, 1975)

[Fu12] B. Fuglede, Sur la fonction de Green pour un domaine fin. Ann. l'institut Fourier **25**(3–4), 201–206 (1975). https://doi.org/10.5802/aif.579

[Fu13] B. Fuglede, Finely harmonic mappings and finely holomorphic mappings. Ann. Acad. Sci. Fenn. Ser. I Math. **2**, 113–127 (2013). https://doi.org/10.5186/aasfm.1976.0210

[Fu14] B. Fuglede, Harmonic morphisms between riemannian manifolds. Ann. l'institut Fourier **28**(2), 107–144 (1978). https://doi.org/10.5802/aif.691

[Fu15] B. Fuglede, Harnack sets and openness of harmonic morphisms. Math. Ann. **241**(2), 181–186 (1979)

[Fu16] B. Fuglede, Invariant characterization of the fine topology in potential theory. Math. Ann. **241**(2), 187–192 (1979)

[Fu17] B. Fuglede, Asymptotic paths for subharmonic functions and polygonal connectedness of fine domains, in *Seminar on Potential Theory, Paris, No. 5*, pp. 97–116. Lecture Notes in Mathematics, vol. 814 (Springer, Berlin, 1980)

[Fu18] B. Fuglede, Sur les fonctions finement holomorphes. Ann. Inst. Fourier (Grenoble) **31**(4), 57–88 (1981)

[Fu19] B. Fuglede, Fine topology and finely holomorphic functions. *18th Scandinavian Congress of Mathematicians (Aarhus, 1980)*, pp. 22–38, Progress in Mathematics, vol. 11 (Birkhauser, Boston, MA, 1981)

[Fu20] B. Fuglede, Fonctions BLD et fonctions finement surharmoniques, in *Séminaire de Théorie du Potentiel Paris, No. 6*. ed. by F. Hirsch, G. Mokobodzki (Springer, Berlin, Heidelberg, 1982), pp.126–157. https://doi.org/10.1007/BFb0093266

[Fu21] B. Fuglede, Localization in fine potential theory and uniform approximation by subharmonic functions. J. Funct. Anal. **49**(1), 57–72 (1982). https://doi.org/10.1016/0022-1236(82)90085-4

[Fu22] B. Fuglede, Integral representation of fine potentials. Math. Ann. **262**(2), 191–214 (1983)

[Fu23] B. Fuglede: Représentation intégrale des potentiels fins. C.R. Acad. Sc. Paris Sér. I Math. **300**(5), 129–132 (1985)

[Fu24] B. Fuglede, Value distribution of harmonic and finely harmonic morphisms and applications in complex analysis. Ann. Acad. Sci. Fenn. Ser. A I **11**(1), 111–135 (1986)

[Fu25] B. Fuglede, Functions finement holomorphes de plusieurs variables - un essai. *Sem. d'analyse P. Lelong–P. Dolbeault–H. Skoda, années 1983/84,* 133–145, Lecture Notes in Mathematics, vol. 1198 (Springer, Berlin, 1986)

[Fu26] B. Fuglede, Finely holomorphic functions and finely harmonic morphisms. Aequationes Math. **34**(2–3), 167–173 (1987)

[Fu27] B. Fuglede, Finely holomorphic functions a survey. Revue Roumaine Math. Pures **33**(4), 283–295 (1988)

[Fu28] B. Fuglede, On the mean value property of finely harmonic and finely hyperharmonic functions. Aequationes Math. **39**(2–3), 198–203 (1990)

[Fu29] B. Fuglede, The Dirichlet Laplacian on finely open sets. Potential Anal. **10**(1), 91–101 (1999)

[Fu30] B. Fuglede, Continuous domain dependence of the eigenvalues of the Dirichlet Laplacian and related operators in Hilbert space. J. Funct. Anal. **167**(1), 183–200 (1999)

[Fu31] B. Fuglede, Harmonic morphisms applied to classical potential theory. Nagoya Math. J. **202**, 107–126 (2011)

[FuGW] B. Fuglede, A. Groot, J. Wiegerinck, Domains of existence for finely holomorphic functions. Potential Anal. **51**(3), 469–481 (2019)

[FuZ] B. Fuglede, N. Zorii, Green kernels associated with Riesz kernels. Ann. Acad. Sci. Fenn. Math. **43**(1), 121–145 (2018)

[Fuk] M. Fukushima, Dirichlet forms and Markov processes. *North-Holland Mathematical Library*, vol. 23 (North-Holland Publishing Co., Amsterdam, New York; Kodansha, Ltd., Tokyo, 1980)

[GL] T.W. Gamelin, T.J. Lyons, Jensen measures for $R(K)$. J. Lond. Math. Soc. **27**(2), 317–330 (1983)

[GS] T. Gamelin, N. Sibony, Subharmonicity for uniform algebras. J. Funct. Anal. **35**(1), 64–108 (1980)

[GH] S.J. Gardiner, W. Hansen, The Riesz decomposition of finely superharmonic functions. Adv. Math. **214**(1), 417–436 (2007)

[Ga1] S.J. Gardiner, Sets of harmonicity for finely harmonic functions. Potential Anal. **21**(1), 1–6 (2004)

[Ga2] S.J. Gardiner, Finely continuously differentiable functions. Math. Z. **266**(4), 851–861 (2010)

[Ga3] S.J. Gardiner, Coincidence of harmonic and finely harmonic functions. Potential Anal. **34**(1), 81–88 (2011)

[Gar] Gårding, Dirichlet's problem for linear elliptic partial differential equations. Math. Scand. **1**, 55–72 (1953)

[GaH] P.M. Gauthier, W. Hengartner, *Approximation uniforme qualitative sur des ensembles non bornés* (Presses Univ. Montréal, QC, 1982)

[GaL] P.M. Gauthier, S. Ladouceur, Uniform approximation and fine potential theory. J. Approx. Theory **72**(2), 138–140 (1993)

[Gav] B. Gaveau, Potentiel de Green fin et approximation fine dans les espaces de Sobolev. C.R. Acad. Sci. Paris Sér. A–B **281**(9), 269–271 (1975)

[GS] R.K. Getoor, M.J. Sharpe, Conformal martingales. Invent. Math. **16**, 271–308 (1972)

[Gow] K. Gowrisankaran, Extreme harmonic functions and boundary value problems. Ann. Inst. Fourier (Grenoble) **13**(2), 307–356 (1963)

[Han] W. Hansen, Valeurs propres pour l'opérateur de Schrodinger, *Séminaire de Théorie du Potentiel, Paris, No. 9,* pp. 117–134, Lectures Notes in Mathematics, vol. 1393. (Springer, Berlin, Heidelberg, 1989)

[Hau] F. Hausdorff, Uber halbstetige Funktionen und deren Verallgemeinerung. Math. Z. **5**(1), 292–309 (1919)

[Hav] V.P. Havin, Approximation in the mean by analytic functions. Dokl. Akad. Nauk SSSR **178**, 1025–1028 (1968)

[Hay] W.K. Hayman, On the characteristic of functions meromorphic in the unit disk and of their integrals. Acta Math. **112**, 181–214 (1964)

[HK] W.K. Hayman, P.B. Kennedy, *Subharmonic functions*. Volume I. London Mathematical Society Mathematical monographs, vol. 9 (Academic Press [Harcourt Brace Jovanovich, Publishers], London-New York, 1976)

[Hed1] L.I. Hedberg, Non-linear potentials and approximation in the mean by analytic functions. Math. Z. **129**, 299–319 (1972)

[Hed2] L.I. Hedberg, Approximation by harmonic functions and stability of the Dirichlet problem. Exposition. Math. **11**(3), 193–259 (1993)

[HKM] J. Heinonen, T. Kilpelainen, O. Martio, *Nonlinear Potential Theory of Degenerate Elliptic Equations* (Dover Publications Inc, Mineola, NY, 2006)

[Hel] L.L. Helms, *Introduction to potential theory, in Pure and Applied Mathematics*, vol. XXII (Wiley-Interscience, New York-London-Sydney, 1969)

[He] R.-M. Hervé, Recherches axiomatiques sur la théorie des fonctions surhahmoniques et du potentiel. Ann. Inst. Fourier (Grenoble) **12**, 415–571 (1962)

[HTV] N.X. Hong, N.V. Trao, H. Viet, Approximation of plurifinely plurisubharmonic functions. J. Math. Anal. Appl. **450**(2), 1062–1075 (2017)

[HVC1] N.X. Hong, H.V. Can, On the approximation of weakly pluri-finely plurisubharmonic functions. Indag. Math. (N.S.) **29**(5), 1310–1317 (2018)

[HVC2] N.X. Hong, H.V. Can, Weakly solutions to the complex Monge-Ampère equation on bounded plurifinely hyperconvex domains. Complex Anal. Oper. Theory **13**(4), 1713–1727 (2019)

[HCL] N.X. Hong, N.T. Lien, H.V. Can, The stability of solutions to the complex Monge-Ampère equations in bounded $\mathcal{F}$-hyperconvex domains. J. Math. Anal. Appl. **483**(1), 123606, 11 pp (2020)

[HCLL] N.X. Hong, H.V. Can, N.T. Lien, P.T. Lieu, Complex Monge-Ampère equations for plurifinely plurisubharmonic functions. Indag. Math. (N.S.) **34**(3), 588–605 (2023)

[Ho] L. Hörmander, An Introduction to Complex Analysis in Several Variables, 3rd edn. North-Holland Mathematical Library, vol. 7. (North-Holland Publishing Co., Amsterdam, 1990)

[Iv] F. Iversen, *Recherches sur les inverses des fonctions méromorphes* (Thèse, Helsingfors, 1914)

[Jan] K. Janssen, A co-fine domination principle for harmonic spaces. Math. Z. **141**, 185–191 (1975)

[J] B. Josefson, On the equivalence between locally polar and globally polar sets for plurisubharmonic functions on $\mathbb{C}^n$. Ark. Math. **16**(1), 109-115 (1978)

[KeNa] J.L. Kelley, I. Namioka, *Linear topological spaces, in The University Series in Higher Mathematic* (D. Van Nostrand Co., Inc, Princeton, N.J., 1963)

[KL] J. Kinnunen, V. Latvala, Fine regularity of superharmonic functions on metric spaces. *Future Trends in Geometric Function theory*, 157–167, RNC Workshop Jyväskylä 2003, Rep. Univ. Jyväskylä Dep. Math. Stat. 92, *University of Jyväskylä, Jyväskylä* (2003)

[Ki] C.O. Kiselman, Sur la définition de l'opérateur de Monge-Ampère complexe. *Complex analysis (Toulouse, 1983)*, 139–150. Lecture Notes in Mathematics, vol. 1094 (Springer, Berlin, 1984)

[Kl] M. Klimek, Pluripotential theoy. London Mathematical Society Monographs. New Series, vol. 6. Oxford Science Publications (The Clarendon Press, Oxford University Press, New York, 1991)

[Ko] T. Kolsrud, Bounded point evaluations and balayage. Ark. Mat. **20**(1), 137–146 (1982)

[K] R. Korte, A Caccioppoli estimate and fine continuity for superminimizers on metric spaces. Ann. Acad. Sci. Fenn. Math. **33**(2), 597–604 (2008)

[Kur] C. Kuratowski, *Topologie* (I-II. Editions Jacques Gabay, Sceaux, 1992)

[Lai] I. Laine, Covering properties of harmonic Bl-mappings, II. Ann. Acad. Sci. Fenn. Ser. A **I**, 570 (1974)

[Lan] R.S. Landkof, Foundations of modern potential theory. Die Grundlehren der mathematischen Wissenschaften, Band, vol. 180. (Springer, New York-Heidelberg, 1972)

[Lat] V. Latvala, Finely superharmonic functions of degenerate elliptic equations. Ann. Acad. Sci. Fenn. Ser. A I Math. Dissertationes 96 (1994)

[Lav1] R. Lávička, A remark on fine differentiability. Adv. Appl. Clifford Algebr. **17**(3), 549–554 (2007)

[Lav2] R. Lávička, Finely continuously differentiable functions. Expo. Math. **26**(4), 353–363 (2008)

[LM] P.D. Lax, A.N. Milgram, Parabolic equations. *Contributions to the Theory of Partial Differential Equations*. Annals of Mathematics Studies, vol. 33 (Princeton University Press, Princeton, NJ, 1954), pp. 167–190

[LeJ1] Y. Le Jan, Quasi-continuous functions associated with Hunt processes. Proc. Am. Math. Soc. **86**(1), 133–138 (1982)

[LeJ2] Y. Le Jan, Quasi-continuous functions and Hunt processes. J. Math. Soc. Jpn. **35**(1), 37–42 (1983)

[LeJ3] Y. Le Jan, Fonctions càd-lag sur les trajectoires d'un processus de Ray. *Théorie du Potentiel (Orsay, 1983)*, 412–418, Lecture Notes in Mathematics, vol. 1096 (Springer, Berlin, 1984)

[Leb] H. Lebesgue, Sur le problème de Dirichlet. Rend. Circ. Mat. Palermo **24**, 371–402 (1907)

[Lel1] P. Lelong, Définition des fonctions plurisousharmoniques. C. R. Acad. Sci. Paris **215**, 398–400 (1942)

[Lel2] P. Lelong, Les fonctions plurisousharmoniques. Ann. Sci. Ecole Norm. Sup. **62**(3), 301–338 (1945)

[Lel3] P. Lelong, Prolongement d'une fonction plurisousharmonique sur certains ensembles de capacité nulle. C. R. Acad. Sci. Paris **242**, 55–57 (1956)

[Lel4] P. Lelong, *Fonctions plurisousharmoniques et formes différentielles positives* (Gordon and Breach, Paris-London-New York, 1968)

[Lel5] P. Lelong, *Positivity in Complex Spaces and Plurisubharmonic Functions*. Queen's papers in Pure and Applied Mathematics, vol. 112. (Queen's University, Kingston, ON, 1998)

[Lev] N. Levenberg, Monge-Ampère measures associated to extremal plurisubharmonic functions in $\mathbb{C}^n$. Trans. Am. Math. Soc. **289**(1), 333–343 (1985)

[LMP] N. Levenberg, G. Martin, E.A. Poletsky, Analytic disks and pluripolar sets. Indiana Univ. Math. J. **41**(2), 515–532 (1992)

[Lu] J. Lukeš, The Lusin-Menchoff property of fine topologies. Comment. Math. Univ. Carolinae **18**(3), 515–530 (1977)

[LM1] J. Lukeš, J. Malý, Fine hyperharmonicity without axiom D. Math. Ann. **261**(3), 299–306 (1982)

[LM2] J. Lukeš, J. Malý, Thinness, Lebesgue density and fine topologies (an interplay between real analysis and potential theory). *Summer School in Potential Theory (Joensuu, 1990)*, 35–70, Joensuun Yliop. Luonnont. Julk., vol. 26 (University Joensuu, Joensuu, 1992)

[LMZ] J. Lukeš, J. Malý, L. Zajíček, *Fine Topology Methods in Real Analysis and Potential Theory*. Lecture Notes in Mathematics, vol. 1189 (Springer, Berlin, 1986)

[LZ1] J. Lukeš, L. Zajíček, The insertion of G_δ sets and fine topologies. Comment. Math. Univ. Carolinae **18**(1), 101–104 (1977)

[LZ2] J. Lukeš, L. Zajíček, When finely continuous functions are of the first class of Baire. Comment. Math. Univ. Carolinae **18**(4), 647–657 (1977)

[Ly1] T.J. Lyons, Finely holomorphic functions. J. Funct. Anal. **37**(1), 1–18 (1980)

[Ly2] T.J. Lyons, A theorem in fine potential theory and applications to finely holomorphic functions. J. Funct. Anal. **37**(1), 19–26 (1980)

[Ly3] T.J. Lyons, Cones of lower semicontinuous functions and a characterisation of finely hyperharmonic functions. Math. Ann. **261**(3), 293–297 (1982)

[Ly4] T.J. Lyons, Finely harmonic functions need not be quasi-analytic. Bull. Lond. Math. Soc. **16**(4), 413–415 (1984)

[Ly5] T.J. Lyons, An application of fine potential theory to prove a Phragmén-Lindelöf theorem. Ann. Inst. Fourier (Grenoble) **34**(2), 63–66 (1984)

[MSZ] J. Malý, D. Swanson, W.P. Ziemer, Fine behavior of functions whose gradients are in an Orlicz space. Studia Math. **190**(1), 33–71 (2009)

[MZ] J. Malý, W.P. Ziemer, *Fine regularity of solutions of elliptic partial differential equations, in Mathematical Surveys and Monographs*, vol. 51 (American Mathematical Society, Providence, RI, 1997)

[Mas1] H. Masaoka, A characterization of the finely harmonic morphism in $\mathbb{R}^n$. J. Math. Kyoto Univ. **26**(2), 223–231 (1986)

[Mas2] H. Masaoka, On the decomposition of nonnegative finely harmonic functions. J. Math. Kyoto Univ. **27**(4), 709–721 (1987)

[Mas3] H. Masaoka, On the behavior of nonnegative finely superharmonic functions at the Martin boundary. J. Math. Kyoto Univ. **30**(2), 193–205 (1990)

[Mas4] H. Masaoka, Theorems of Plessner and Riesz types for finely harmonic morphisms. J. Math. Kyoto Univ. **30**(3), 481–491 (1990)

[Mas5] H. Masaoka, A note on the boundary behavior for nonnegative finely superharmonic functions in the upper half space. Osaka J. Math. **28**(2), 461–462 (1991)

[MD] M. Mastrangelo-Dehen, D. Dehen, Étude des fonctions finement harmoniques sur des ouverts fins de C. C.R. Acad. Sci. Paris, Sér. A-B **275**, 361–364 (1972)

[Ma1] M. Mastrangelo, Différentiabilité fine, différentiabilité stochastique, différentiabilité stochastique de fonctions finement harmoniques. Ann. Inst. Fourier, (Grenoble) **28**(2), 161–186 (1978)

[Ma2] M. Mastrangelo-Dehen, Différentiabilité stochastique des fonctions finement harmoniques. Invent. Math. **51**(1), 15–27 (1979)

[Ma3] M. Mastrangelo, Différentiabilité fine et différentiabilité sur des compacts. Bull. Soc. Math. France **108**(1), 3–15 (1980)

[McKe] H.P. McKean, *Stochastic integrals, in Probability and Mathematical Statistics*, vol. 5 (Academic Press, New York-London, 1969)

[McKi] R. McKissick, A non-trival normal sup norm algebra. Bull. Am. Math. Soc. **69**, 391–395 (1963)

[Miz] T. Mizuta, Fine differentiability of Riesz potentials. Hiroshima Math. J. **8**(2), 505–514 (1978)

[Mo] G. Mokobodzki, Représentation intégrale des fonctions surharmoniques au moyen des réduites. Ann. Inst. Fourier **15**(1), 103–112 (1965)

[MS] G. Mokobodzki, N. Sibony, Cônes de fonctions continues. C.R. Acad. Sci. Paris Sér. A-B **264**, 15–18 (1967)

[Nev] R. Nevanlinna, Über eine Minimumaufgabe in der theorie der Konformen Abbildung. Nachr. Ges. Wiss. Göttingen, Math.-Phys. Kl **I.37**, 103–115 (1933)

[Ng1] N.-X. Loc, Characterization of excessive functions on finely open, nearly Borel sets. Math. Ann. **196**, 250–268 (1972)

[Ng2] N.-X. Loc, Fine boundary minimum principle and dual processes. Z. Wahrscheinlichkeitstheorie und Verw. Gebiete **27**, 233–256 (1973)

[Ng3] N.-X. Loc, Sur le principe du minimum fin. Nagoya Math. J. **54**, 135–148 (1974)

[Ng4] N.-X. Loc, Sur la théorie des fonctions finement holomorphes. (II).*Complex Analysis Joensuu 1978 (Proc. Colloq. Univ. Joensuu, Joensuu, 1978)*, pp. 289–300. Lecture Notes in Mathematics, vol. 747 (Springer, Berlin, 1979)

[Ng5] N.-X. Loc, Singularities of locally analytic processes. *Potential Theory, Copenhagen 1979 (Proc. Colloq., Copenhagen, 1979)*, pp. 267–288, Lecture Notes in Mathematics, vol. 787 (Springer, Berlin, 1980)

[Ng6] N.X. Loc, Complex stochastic analysis in several parameters and applications. *Hanoi: Publishing House for Science and Technology, Vietnam Academy of Science and Technology.*

[Ok1] B. Øksendal, Finely harmonic morphisms, Brownian path preserving functions and conformal martingales. Invent. Math. **75**(1), 179–187 (1984)

[Ok2] B. Øksendal, Finely harmonic functions with finite Dirichlet integral with respect to the Green Measure. Trans. Am. Math. Soc. **287**(2), 687–700 (1985)

[Ph] R.R. Phelps, *Lectures on Choquet's theorem*, 2nd edn. Lecture Notes in Mathematics, vol. 1757. (Springer, Berlin, 2001)

[PW] E. Poletsky, J. Wiegerinck, Graphs with multiple sheeted pluripolar hulls. Ann. Polon. Math. **88**(2), 161–171 (2006)

[PS] G. Polya, G. Szegö, Isoperimetric inequalities in Mathematical Physics. Annals of Mathematics Studies, vol. 27. (Princeton University Press, Princeton, N.J., 1951)

[Py] P. Pyrih, Finely locally injective finely harmonic morphisms. Potential Anal. **1**(4), 373–378 (1992)

[Ran] T. Ransford, Potential theory in the complex plane, in *London Mathematical Society Student Texts*, vol. 28. (Cambridge University Press, 1995)

[Re] F. Rellich, Ein Satz über mittlere Konvergenz', Nachr. Akad. Wiss. Göttingen, *Math.-Phys* Kl. (1930), pp. 30–35

[Ri] M. Riesz, L'intégrale de Riemann-Liouville et le problème de Cauchy. Acta Math. **81**, 1–223 (1949)

[Sa] A. Sadullaev, Plurisubharmonic measures and capacities on complex manifolds. Uspekhi Mat. Nauk. **36**, no. 4(220), 53–105, 247 (1981)

[Sb] N. Sibony, Quelques problèmes de prolongement de courants en analyse complexe. Duke Math. J. **52**(1), 157–197 (1985)

[Sc] L. Schwartz, Théorie des distributions. Tome II. Publications de l' Institut de Mathématiques de l'Université de Strasbourg, 10. Actualités Scientifiques et Industrielles, No. 1122. Hermann & Cie, Paris (1951)

[Si] J. Siciak, Pluripolar sets and pseudocontinuation. *Complex Analysis and Dynamical Systems II*, 385–394, Contemporary Mathematics, vol. 382, Israel Mathematical Conference Proceedings (American Mathematical Society, Providence, RI, 2005)

[T] J.C. Taylor, An elementary proof of the theorem of Fatou-Naïm-Doob. *1980 Seminar on Harmonic Analysis (Montreal, Que., 1980)*, pp. 153–163, CMS Conference Proceedings, vol. 1 (American Mathematical Society, Providence, RI, 1981)

[To1] N. Toda, Etude de fonctions méromorphes au voisinage d'un point-frontière irrégulier. Bull. Sci. Math. **89**(2), 93–102 (1965)

[To2] N. Toda, Sur l'allure des fonctions meromorphes. Nagoya Math. J. **26**, 173–181 (1966)

[U] J.L. Ullman, Orthogonal polynomials for general measures. I. *Rational approximation and interpolation (Tampa, Fla., 1983)*. Lecture Notes in Mathematics, vol. 1105 (Springer, Berlin, 1984), pp. 524–528

[VP1] Ch. de La Vallée-Poussin: Sur quelques extensions de la méthode du balayage de Poincaré et sur le problème de Dirichlet. C.R. Acad. Sci. Paris, **192**, 651–653 (1931)

[VP2] Ch. de La Vallée-Poussin, Le potentiel logarithmique. *Paris-Louvain* (1949)

[Wa] J. Wang, An approximate Taylor's theorem for $R(X)$. Math. Scand. **33**(1973), 343–358 (1974)

[Wh] H. Whitney, Analytic extensions of differentiable functions defined in closed sets. Trans. Am. Math. Soc. **36**(1), 63–89 (1934)

[W] J. Wiegerinck, Plurifine potential theory. Ann. Polon. Math. **106**, 275–292 (2012)

[Wo] E. Wolf, Bounded point evaluations and smoothness properties of functions in $R(X)$. Trans. Am. Math. Soc. **238**, 71–88 (1978)

[Z] V.P. Zaharjuta, Extremal plurisubharmonic functions, orthogonal polynomials and Bernstein-Walsh theorem for analytic functions of several variables. Ann. Polon. Math. **33**(1–2), 137–148 (1976/77)

[Zal] L. Zalcman, Analytic capacities and rational approximation. Lecture Notes in Mathematics, vol. 50. (Springer, Berlin, 1968)

[Zwo] W. Zwonek, A note on pluripolar hulls of graphs of Blaschke products. Potential Anal. **22**(2), 195–206 (2005)

Symbol Index

M. El Kadiri and B. Fuglede, *Classical Fine Potential Theory*, Springer Monographs in Mathematics, https://doi.org/10.1007/978-981-96-0430-2

Index

M. El Kadiri and B. Fuglede, *Classical Fine Potential Theory*, Springer Monographs in Mathematics, https://doi.org/10.1007/978-981-96-0430-2

The manufacturer's authorised representative in the EU is Springer
Nature Customer Service Centre GmbH, Europaplatz 3, 69115 Heidelberg,
Germany. If you have any concerns regarding our products, please
contact ProductSafety@springernature.com

Printed and bound by CPI Group (UK) Ltd, Croydon, CR0 4YY
01/10/2026
02225849-0014